Handbook of Compressible Aerodynamics

The author wishes to dedicate this book
to Marie-Catherine, Hélène and Romain

Handbook of Compressible Aerodynamics

Jean Délery

ISTE

WILEY

First published 2010 in Great Britain and the United States by ISTE Ltd and John Wiley & Sons, Inc. Adapted and updated from three volumes *Traité d'aérodynamique compressible* published 2008 in France by Hermes Science/Lavoisier © LAVOISIER 2008

ISTE Ltd
27-37 St George's Road
London SW19 4EU
UK

www.iste.co.uk

John Wiley & Sons, Inc.
111 River Street
Hoboken, NJ 07030
USA

www.wiley.com

Library of Congress Cataloging-in-Publication Data

Délery, J. (Jean), 1939-
Handbook of compressible aerodynamics / Jean Délery.
p. cm.
Includes bibliographical references and index.
ISBN 978-1-84821-141-4
1. Aerodynamics–Handbooks, manuals, etc. 2. Compressibility–Handbooks, manuals, etc. 3. Gas flow–Handbooks, manuals, etc. 4. Gases, Compressed–Handbooks, manuals, etc. 5. Expansion of gases–Handbooks, manuals, etc. I. Title.
TL574.C4D45 2010
629.132'35–dc22

2010016490

British Library Cataloguing-in-Publication Data
A CIP record for this book is available from the British Library
ISBN 978-1-84821-141-4

Printed and bound in Great Britain by CPI Antony Rowe, Chippenham and Eastbourne.

FSC www.fsc.org MIX Paper from responsible sources FSC® C013604

Table of Contents

List of Symbols

a: velocity (celerity) of sound

a_c : critical sound velocity (for Mach 1)

A: area of a stream tube or pipe cross-section

A_c : throat (critical) area

A_m : area of the cylindrical mixing chamber (supersonic ejector theory)

A_{0_p} : area of the primary flux at injection (supersonic ejector theory)

A_{0_s} : area of secondary flux at injection (supersonic ejector theory)

c: airfoil chord

C_A : axial force coefficient

C_D : drag coefficient

C_f : skin friction coefficient

C_h : heat transfer coefficient

C_L : lift coefficient

C_m : pitching moment coefficient

C_N : normal force coefficient

C_p : pressure coefficient

C_p : constant pressure specific heat

Cr : Crocco number

C_v : constant volume specific heat

C_s : specific fuel consumption

(C) : steady shock wave

D: diameter

D'_D : power transmitted by species diffusion

D_h : hydraulic diameter

D_j : diffusion velocity of species *(j)*

D: velocity (celerity) of the shock wave

$\vec{D}$: dynalpy vector

Da: Damköhler number

e: specific internal energy (per unit of mass)

e_T : specific total internal energy (per unit of mass)

E: internal energy

E_T : total internal energy

E'_M : power transmitted by electromagnetic interaction

E'_Q : power transmitted by reception (and re-emission) of thermal radiation

f: lift-over-drag ratio (*L*/*D*)

f: frequency

$\vec{f}, f_i$: body force (per unit of mass)

F : nozzle thrust

$\bar{F}$: nozzle thrust at adaptation

F_A : axial force

F_D : drag force

F_L : lift force

F_N : normal force

Fr : Froude number

g: acceleration of gravity

G: Gibbs free energy

h: specific enthalpy (per unit of mass)

h_T, h_{st} : specific stagnation (total) enthalpy

h_f : friction (recovery) enthalpy

h_w : enthalpy of the gas at the wall

H: system enthalpy

k: roughness height

Kn : Knudsen number

K_p : chemical equilibrium constant

L: reference length

Le: Lewis number

L_p : heat of reaction

m: mass

M: molar mass

M: Mach number

M^* : critical Mach number $(M^* = V / a_c)$

M: aerodynamic moment

M_j : molar mass of species *(j)*

M_s : shock Mach number

$\vec{n}$: unit vector normal to a surface element

n_j : number of moles of species *(j)*

N: characteristic relation coefficient $\left(N = -\varepsilon \dfrac{\sin\varphi \sin\alpha}{y} \right)$

$\vec{P}$: vector tension

p: pressure

p_a : ambient pressure

p_E : external pressure

Pe : Péclet number

Pr: Prandtl number

p_l : airfoil lower side pressure

p_{st} : stagnation pressure

p_u : airfoil upper side pressure

q_m : mass flow

$\vec{\dot{q}}, \dot{q}_i$: heat flux (vector)

$\dot{q}$: heat flux (scalar)

$\dot{q}_w$: wall heat flux

Q'_C : contact thermal power (conduction/convection)

Q'_E : thermal power by spontaneous radiation emission

Q_I : internal thermal power dissipated in an irreversible way

r: constant of gas state equation:

$$C_p - C_v = \frac{\Re}{M} = r$$

$$\left(r = 287 J/kg/K \text{ for air}\right)$$

Re: Reynolds number

$\Re$: universal perfect gas constant $(= 8.3145\ J/mole/K)$

s: specific entropy

S: system entropy

S: reference surface

Sc : Schmidt number

St: Stanton number

Su: Strouhal number

t: time

$\vec{t}$: unit vector tangent to a curve

T: temperature

T: drag

T_f : friction (recovery) temperature

T_{st} : stagnation temperature

T_w : wall temperature

$\overline{\overline{T}}, T_{ij}$: stress tensor

v: specific volume $\left(v = 1/\rho\right)$

u,v,w: velocity Cartesian components

u_i : velocity Cartesian components (tensorial notation)

U: velocity

$U_{\lim}$: unsteady limit velocity

$\vec{V}$: velocity vector

V: velocity (scalar)

$V_{\lim}$: steady limit velocity

V_n : velocity component normal to the shock

V_t : velocity component tangent to the shock

w: velocity difference across the shock

W' : mechanical power of external forces (contact or body forces)

W'_{int} : mechanical power of internal forces

w'_{int} : local (specific) power of the system internal forces

x_i : Cartesian coordinates ($i = 1, 2, 3$)

x,y,z: Cartesian coordinates

x_1, y_1 : intrinsic coordinates (linked to the streamline)

z: $z = \ln p$ (method of characteristics and conical flows)

Z: compressibility factor

Z: characteristic relation coefficient

$$\left(Z = \frac{\sin\alpha\cos\alpha}{\gamma}\right)$$

Z: acoustic impedance $\left(Z = \rho a\right)$

α : angle of incidence

α : Mach angle

α : degree of dissociation

γ : specific heat ratio

$\delta(\gamma)$: function in the mass flow formula: $\delta(\gamma) = \frac{\gamma}{\sqrt{\gamma - 1}}\left(\frac{\gamma + 1}{2}\right)^{-\frac{\gamma+1}{2(\gamma-1)}}$

δ_{ij} : Kronecker delta

ε : air intake mass flow coefficient

ε : $\varepsilon = 0$: two-dimensional planar flow, $\varepsilon = 1$: axisymmetric flow

Γ : circulation

η : air intake efficiency

(η) : left running characteristic

(η) : progressive wave (velocity $u + a$)

θ : wall local angle

θ : polar angle (conical flows)

Θ : conical nozzle half angle

Θ_{ij} : Lighthill stress tensor

θ_v : characteristic vibrational temperature

θ_D : characteristic dissociation temperature

θ_i : characteristic ionization temperature

κ : thermal conductivity

λ : bulk (volume) viscosity

λ : mean free (molecular) path

λ : pressure loss coefficient $(\lambda = 4C_f)$

λ : area ratio: $\lambda = \dfrac{A_m}{A_{0_p}}$ (supersonic ejector theory)

λ, μ : characteristic coordinates (Riemann invariants)

μ : molecular viscosity

μ : mass flow ratio (supersonic ejector theory)

ν : kinematic viscosity $(\nu = \mu / \rho)$

ν_j : stoichiometric coefficients

(ξ) : right running characteristic

(ξ) : regressive wave (velocity $u - a$)

π : isentropic pressure ratio $(\pi = p / p_{st})$

Π : pressure ratio across a shock wave $\left(\Pi = \dfrac{p_2}{p_1}\right)$

ρ : density

σ : shock angle with respect to the upstream velocity

$\sigma_{\max}$: shock angle for the maximum deflection angle

σ_{son} : shock angle for the sonic state behind the shock

Σ : area ratio (area–Mach number relation: $\Sigma = A / A_c$)

τ : shear stress

τ_w : wall shear stress

τ_{ij} : viscosity tensor

φ : velocity angle

φ : velocity angle with respect to the shock

Φ : thrust function: $\Phi(M,\gamma) = \pi(M,\gamma)\Sigma(M,\gamma)(1+\gamma M^2)$

Φ : velocity potential

Φ' : perturbation potential

ω : pressure number:

$$\omega(M,\gamma)=\sqrt{\frac{\gamma+1}{\gamma-1}}\tan^{-1}\sqrt{\frac{\gamma-1}{\gamma+1}\left(M^2-1\right)}-\tan^{-1}\sqrt{M^2-1}$$

$\vec{\omega}$: vorticity vector: $\vec{\omega}=\frac{1}{2}\overrightarrow{curl}\,\vec{V}$

(Ω) : unsteady shock wave

Ψ : stream function

Ψ : characteristic angle with respect to axis Ox $\left(\Psi=\varphi\pm\alpha\right)$

Acknowledgment

This treatise is based on courses given at the Ecole Nationale Supérieure de l'Aéronautique et de l'Espace (Supaéro), the University of Versailles-Saint-Quentin and Ecole Polytechnique. It has been written for PhD students and young engineers who are not yet specialized; and can thus be considered as an introduction to compressible aerodynamics, illustrated by concrete applications.

Gas dynamics teaching in France owes much to Pierre Carrière who was Director of Aerodynamics at Onera (Office National d'Etudes et de Recherches Aérospatiales) when I began my career. Having had the opportunity very early on to learn from him, he gave me – and also many others – an essential foundation for our studies in aerodynamics. This treatise draws a part of its inspiration from the courses delivered by Pierre Carrière at Supaéro and Ecole Nationale Supérieure des Techniques Avancées (ENSTA).

This treatise owes much to the numerous contacts that I have had during my professional life with colleagues from the aerospace industry in the framework of major French and European programs in which I had the opportunity to participate. However, I am especially grateful to engineers of the Onera aerodynamic departments with whom I had many exchanges and who generously gave me results to illustrate this treatise. I am greatly indebted to the theoreticians of the Numerical Simulation and Aeroacoustics Department of Onera who initiated me to the use of the simulation code *elsA* and helped me to clarify some delicate theoretical aspects. I am very grateful to engineers and scientists of the Fundamental and Experimental Aerodynamics Department of the Onera – Meudon Center for all their help. In particular, many thanks to the PhD students and assistants who I gave advice to, which helped me to develop the remaining obscure points in the course of vigorous discussions.

I would also like to express my gratitude to the Direction and Services of Onera who offered me the opportunity to work in the most favorable atmosphere and provided me with access to the incomparable means of this big institution.

Thanks also to the Director of Hermes Fluid Dynamics Collection, Jean-Luc Achard, whose valuable advice enabled me to considerably improve the original version of the manuscript.

To conclude, I would like to express my heartfelt thanks to my wife Marie-Catherine for her patience and help in the final reviewing of the French version of this treatise.

General Introduction

At a time when the dramatic progress in computer performance and numerical methods enable the resolution of the Navier–Stokes equations giving a remarkably accurate prediction of air motion (at least in the usual conditions of aerodynamics), the question may be raised of the utility of the concepts and theories developed in the present book. Is it not sufficient to present the general motion equations (established in Chapter 4) and adequate developments on numerical techniques? Indeed, there exist codes to faithfully compute complex flows, as in a separated propulsion nozzle, past a transonic airfoil, and even around a complete aircraft. Sometimes, we call this approach numerical aerodynamics to distinguish between experimental and theoretical aerodynamics. However, we can put forward some arguments to justify our approach.

The solution of the Navier–Stokes equations is still far from being a panacea. Determination of a solution, including accurately taking into account turbulent motion, is still out of reach because of the vast amount of calculation resources that would require the prediction of the flow, even past simple shapes. This ultimate weapon, known as Direct Numerical Simulation (DNS), is still at a developmental stage and only applied to basic flow cases. Thus, we have to rely on some turbulence modeling that is frequently uncertain and not always reliable in complex flows. Even with the use of a turbulence model (we speak of the RANS approach: Reynolds-Averaged Navier–Stokes), the solution of the Navier–Stokes equations remains costly in terms of computing resources and requires delicate application to the flow past complex bodies.

Thus, it may be of interest to have simple methods for the sizing of some devices or for estimating performance at the pre-design step. The one-dimensional steady flow theory gives a prediction with an accuracy of several percent of the thrust of a planar or axisymmetric nozzle, by using isentropic expansion tables and elementary calculations. The method of characteristics allows the accurate definition of the wall contour of a supersonic nozzle and requires only a fraction of a second on a PC. This advantage is noteworthy, as a great quantity of calculations must be performed before the optimized shape is obtained, due to the large number of constraints (among them, the size and weight of the nozzle which are key parameters for nozzle equipping space launcher engines). Also, it is possible to predefine, using "by hand" calculations, the geometry of a supersonic air-intake using the shock wave theory. This type of approach has been used to design the *Concorde* air-intakes whose compression efficiency is remarkable.

However, a more convincing argument can be the physical understanding of reality. Indeed, in some way, the purely numerical approach displaces the problem, sometimes on an uncertain ground, by giving a picture of the flows, which is, in fact, a virtual transposition of the real flows. By visualizing the numerical results with a plotting program or conducting an experiment in a wind tunnel, in the two circumstances one is confronted with the double problem of data evaluation and physical interpretation. Any scientific or technological progress is based on a sound physical interpretation of observations in the framework of more or less sophisticated theoretical developments which alone are accessible to true understanding by the mind. The theory of discontinuities (shock waves and slip lines) illuminates many aspects of supersonic flows, such as jet structure, drag production, or real gas effects in hypersonic flows. Mastering the propagation mechanism of perturbations and the concepts of domains of influence and dependence are the basis of the understanding of supersonic and transonic flows. In a similar way, the one-dimensional unsteady flow theory is a prerequisite to the rational analysis of wave phenomena encountered in strategic domains like aeroelasticity, aeroacoustics, the supersonic boom, and in the many circumstances (frequently damaging) where the flow is the seat of strong fluctuations.

This treatise is devoted to compressible flow aerodynamics in the context of the non-viscous fluid approximation. It constitutes an initial contact with this matter, often called gas dynamics. Each chapter gives a presentation that is simple, with a minimum of arduous mathematical developments and with the help of numerous graphical representations. The theoretical parts are most often followed by the presentation of practical application methods based on the theoretical results just established with a view to help not only the understanding of the theoretical developments but also to give immediate industrial applications. At the end of the treatise, appendices provide tables containing the main useful results and formulas deduced from the theoretical developments.

The book can be roughly divided into three parts.

Chapters 2–8 are devoted to the presentation of some basic notions in classic aerodynamics and to reviews that are indispensable for the understanding of the following parts (in particular, thermodynamics).

Chapters 9–16 present the theory of the one-dimensional steady flows and the theory of discontinuities (shock waves and slip lines). This is the heart of gas dynamics where most of the essential results are established.

Chapters 17–24 are devoted to the theory of characteristics and its applications to the study of steady two-dimensional supersonic flows and one-dimensional unsteady flows. This theory is the key to understanding pressure wave propagation phenomena in flows. Finally Chapter 25 gives some properties of transonic flows which are of interest in many domains (wings of transport aircraft, turbomachines blades, supersonic diffusers, etc.) and which are still the subject of a very active research effort.

Chapter 1

Aerodynamics: Aims, Areas of Application and Current Issues

The aim of aerodynamics is the study of the flow of air around a body, in general an air or land vehicle, this also includes an engineering structure, the rotor of a helicopter, the blades of a turbo engine or even a wind turbine, just to mention a few examples. More precisely, the intended aim is to determine the effect of the flow on the body. In fact, all fluids that flow over a material surface exert contact actions (pressure and friction) on it, which results in a force referred to as an aerodynamic force or aerodynamic resultant, obtained by integration over the whole body. This force is applied on a point called the pressure center, which generally does not coincide with the center of gravity of the body; hence an aerodynamic moment tends to make the body rotate around its center of gravity. The first goal of aerodynamics is to determine the components of force and moment. These two components play a particularly important role. The component perpendicular to the velocity of progress is the *lift*, which, for an aircraft at cruise conditions, is a force directed upwards, compensating the weight of the aircraft. It defines the lifting capacity of the plane, thus the payload (number of passengers and mass of freight). In a car, we instead look for a lift directed towards the ground, which pushes the vehicle to the road. The component of the aerodynamic resultant along to the velocity is *drag*, which opposes the progress of the vehicle. The extent of the drag determines the propulsive force necessary to accelerate or keep the vehicle moving. It is an essential piece of data, which determines the fuel consumption.

In reality, it is the vehicle that moves in an immobile atmosphere, but according to the principle of relativity of movement (based on the invariance of the equations in motion in relative movements with constant speed), it is conceptually easier to consider the air moving around the immobile vehicle. This situation is reproduced in a wind tunnel where the scale model is fixed, while the air is brought to speed by pumps or fans. This change in the point of reference, which may appear trivial, sometimes creates awkward problems for the engineers when they study land vehicles moving on a road (cars, vans, etc.) or railway tracks (high-speed train). The proximity of the ground can have substantial effect on the flow, but in a referential linked to the wind tunnel, where the vehicle is fixed, the ground has to move at the velocity of the vehicle. This sometimes requires complex devices where the scale model is mounted above a belt that reproduces this movement.

Aerodynamics also has to predict the local stresses exerted on the body, for example pressure, in order to calculate deformations of the structure. The reciprocal influence between the structure, which by deforming changes the aerodynamic field, and the flow which thus applies a changed load upon the structure is at the origin of a fluid-structure coupling. This interaction can lead to dangerous situations like flutter, a famous example of which is the destruction of the Tacoma suspension bridge in 1940 due to the effects of a storm. The study of this phenomenon, which involves almost all aspects of aerodynamics (wings of planes, blades of compressors or turbines, rotor blades of helicopters or wind turbines) is a matter of *aeroelasticity*.

The turbulent boundary layer developing over the fuselage, the possible separation on certain parts of a vehicle, the efflux of jet engines, propellers, rotors of helicopters, wind turbines, are significant noise sources. Beyond 300 km/h for trains and 120 km/h for cars, the level of aerodynamically caused noise of ground vehicles can exceed the sound of rumbling and the engine. The field of study that brings together aerodynamics and acoustics is known as *aeroacoustics*.

The transfer of heat by convection and/or radiation between the flow and the vehicle wall plays an essential role in hypersonics where the vehicle is submitted to intense heating at the time of reentry into the atmosphere. A flow can also help to cool down hot parts (for example, car radiators or heat exchangers). This mechanical fluids/thermal coupling belongs to *aerothermodynamics*.

Homing heads to guide missiles towards their target use the infrared radiation of the jets of fighter engines or the exhaust of helicopter engines. This is referred to as the *infrared signature*. Minimization of the radar signature of fighter planes imposes very specific shapes, which forces us to reconsider aerodynamics.

The impact of aerodynamics is particularly significant in the research of sustainable development and energy saving. Indeed, the reduction of external disturbances, or nuisances, affecting the environment due to the passage of the vehicle (noise, emission of pollutants) leads to the design of vehicles where the aerodynamics should be very clean (in particular, to reduce drag). From this perspective, a very active research effort is currently underway throughout the world to control flows in order to avoid their negative impact and improve the aerodynamic performance of vehicles (planes as well as automobiles).

In general, aerodynamics, seen as a science or technology of the movement of fluids, is also of interest in areas such as ventilation, air conditioning (referred to as *aeraulics*), the construction of buildings and engineering structures, the smelting of metals, a multitude of other industrial processes, etc. Aerodynamics also plays a key role in the area of propulsion and energy production (spark ignition engines for cars, jet engines for planes, turbo generators for thermal power stations, windmills, etc.) where these machines are able to get fluids moving up to supersonic velocities.

The difference between aerodynamics and hydrodynamics is sometimes a very thin line. Hydrodynamics considers the flow of an incompressible fluid – in this case water – and at low velocity, which is also the case with planes flying at low subsonic flows, dirigible balloons, sailplanes (gliders). In those two cases, the equations that govern the movement are the same, thus the solutions will be identical (for the same boundary conditions). The definition of a hydrofoil is performed in the same way as that of the profile of a wing with different constraints. The flow around the hull of a submarine is similar to the flow around the fuselage of a plane at low velocity with specific aspects (presence of a conning tower, the existence of propulsion propellers at the end of the hull). Furthermore, the *cavitation*, unknown in aerodynamics, caused by the vaporization of water, in regions of low pressure, plays an essential role in hydrodynamics.

This book concentrates on classic aerodynamics, in the sense that only the flow of monophasic fluids (the most common of the air) will be contemplated where no chemical reaction occurs and where the thermal processes do not play a crucial role. This point of view will be modified when we bring up hypersonic flows. However, we will not explore this area in detail.

In this context, it is important to distinguish between *external aerodynamics* and *internal aerodynamics*:

– *External aerodynamics* concerns the flows around the vehicle (wing, fuselage, cockpit, etc.).

– *Internal aerodynamics* covers the flow passing through the propulsion nacelles of planes, or caused by propulsion systems of missiles and space launchers.

Certain aspects fall between external and internal aerodynamics: for instance, air inlets of aircraft engines and propulsion nozzles. In fact, the functioning of these devices is strongly influenced by the external field. Similarly, the rear part of a plane or a rocket (afterbody), where the propulsive jet escapes, is the home to some adverse interactions between the internal flow leaving the nozzle and the external field along the fuselage. The reciprocal influence between the propulsion nacelle and the wing, to which the pod is attached through a pylon, modifies both the lift of the wing and the propulsive balance of the nacelle: it is the domain of the *engine integration*.

This book concentrates on the compressible aerodynamics, what we call *gas dynamics*, science adding to the equations of motion, the equation of energy, and gas thermodynamics.

Indeed, although air is a highly compressible medium, at velocities considered low in aeronautics (lower than 150 m/s or 500 km/h), the variations in density induced in the air by the moving vehicle can be considered negligible. This hypothesis, which had been well proven for aircraft up to the Second World War, enabled the development of a wide range of prediction methods still applied to the phases of flight at low velocities (even the fastest aircraft have to take off and land) and in areas such as private aircraft, wind turbines, ground vehicles, etc.

The compressibility of the air began to be a concern for aircraft manufacturers (and pilots) when the velocities of combat aircraft exceeded 700 km/h to enter into the transonic domain. Then phenomena occur (that were not well known at the time) that caused the loss of aircraft during dives where the velocity was close to the speed of sound. Of course, the compressible nature of the air played a key role when the velocities of aircraft became supersonic, which was the case for fighter planes, missiles (strategic and tactical), and space launchers from the 1960s. It was also the time when the project for a supersonic airliner was launched, which would lead to the *Concorde*. Staying in the field of aeronautics the effects of compressibility are also felt in the stages of compressors and the turbines of jet engines, as well as in propulsion nozzles.

It has long been known that the projectiles of artillery and infantry exit the barrel at supersonic velocities. The shock phenomena were known in the late nineteenth century and the governing equations were established at that time by John Macquorn Rankine (1820–1872) and Pierre Henri Hugoniot (1851–1887). The shocks accompanying a supersonic projectile were visualized by the physicist Ernst Mach (1838–1916). In the study of supersonic flows, we must also mention the contributions of Ludwig Prandtl (1875–1953), Jacob Ackeret (1898–1981), and many others. Thus, when supersonic flight took off, we had the theoretical tools enabling us to understand and predict the phenomena that we would be facing. A sensitive area was transonic region, when the aerodynamic field includes subsonic and supersonic regions, as is the case on the wings of current airliners. The mixed nature of elliptic-hyperbolic equations governing the flow then leads to difficulties and specific behaviors that are not yet fully mastered.

This book is confined to the assumption of a non-viscous fluid (also called perfect fluid) which gives an accurate description of almost every field of aerodynamics, the regions where the appearing viscosity is small for normal operating conditions. However, the assumption of a non-viscous fluid prohibits the prediction of transfer phenomena such as friction and heat flux. In addition, there are many circumstances where the viscosity of the air manifests dramatically in the formation of large separated zones, giving rise to vortex structures. The assumption of a non-viscous fluid is the zero-order approximation that allows understanding of almost all phenomena, but which is inadequate if we want to get the real issues of aerodynamics whose solutions lie in taking into account the viscosity of air.

This introductory chapter was primarily intended to show that we should not have a very limited outlook on aerodynamics. The knowledge and know-how acquired in this field can be applied to many industries where there is fluid in motion, sometimes at speeds reaching levels previously only known in aeronautics (the case of high speed trains).

Chapter 2

Aerodynamic Forces: Basic Definitions

2.1. Main aerodynamic coefficients

Before we consider the equations (or mathematical models) and the calculation methods that are used to predict the forces to which a vehicle is subjected when it moves within a fluid, it is useful to first introduce certain concepts that are repeatedly reoccurring in aerodynamics.

To illustrate the idea, let us consider a *two-dimensional* airfoil. This means that the wing is supposed to have an infinite span, so that the flow does not have a component of velocity along the dimension perpendicular to the airfoil (wing span). The velocity at *upstream infinity*, meaning a distance upstream large enough that the wing does not have a significant influence on the flow, which is denoted by $\vec{V}_\infty$.

As shown in Figure 2.1, the airfoil is set at an angle of incidence α, which means that the chord of the wing (straight line joining the leading edge and the trailing edge) makes an angle α relative to the upstream velocity vector.

From now on, we will often use the airfoil as an example. In addition, the definition of wing airfoils (or profiles) forms one of the major objectives in aerodynamics. It is obvious that the following concepts can be applied to any object, whether flying or not, that moves within a fluid. Depending on a physical process, which we will analyze further, the flow around the airfoil exerts a set of forces on it, or stresses, that can be replaced by an equivalent force $\vec{F}$, or *aerodynamic force* (also called *aerodynamic resultant*), applied to a point called the *center of pressure* (see Figure 2.2).

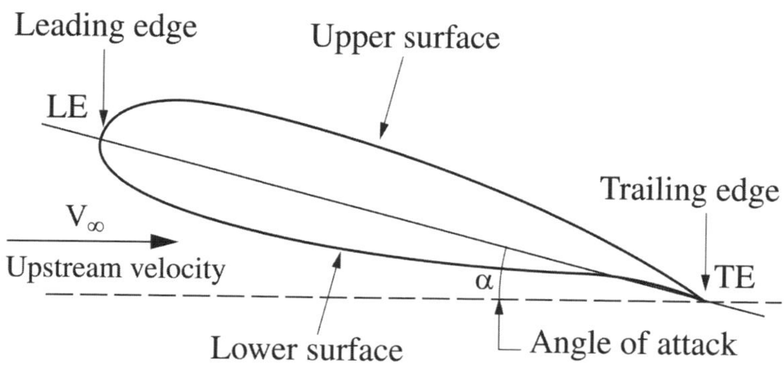

Figure 2.1. *Two-dimensional airfoil at an incidence*

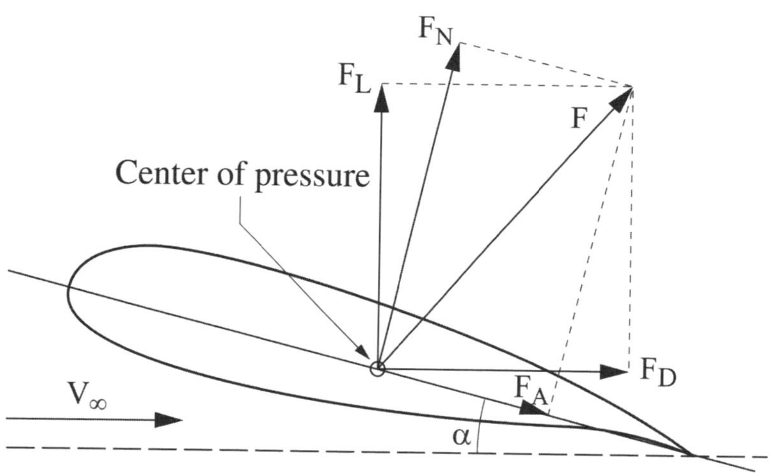

Figure 2.2. *Resultant of aerodynamic forces and decomposition*

We will break down $\vec{F}$ into two components:

– F_L, perpendicular to the velocity $\vec{V}_\infty$, which is designated as *lift*, or lift force,

– F_D, parallel to the velocity $\vec{V}_\infty$, which is called *drag* or drag force.

In certain applications, for example the aerodynamics of missiles which operate under very high and quickly variable incidences, we break down the resulting $\vec{F}$ according to the axes related to the vehicle, or:

– a *normal force* F_N, perpendicular to the chord of the airfoil or the axis of the fuselage,

– an *axial force* F_A, along the chord of the airfoil or axis of the fuselage.

Calculations of the flow around an airfoil are usually done by considering a system of axes linked to the airfoil around which the calculation grid is constructed. So it is natural to calculate the normal and axial components and then make the projection into the system tied to the upstream velocity vector.

We have the following relations between the components that have been defined as:

$$\boxed{\begin{aligned} F_z &= F_N \cos\alpha - F_A \sin\alpha \\ F_x &= F_N \sin\alpha + F_A \cos\alpha. \end{aligned}}$$

As the center of pressure does not usually coincide with the center of gravity G, the airfoil is subjected to an *aerodynamic pitching moment M.* If the moment M tends to increase the angle of incidence α, it is called a *nose-up moment* and calculated positive; otherwise, the moment is called a *nose-down moment* and calculated negative (see Figure 2.3).

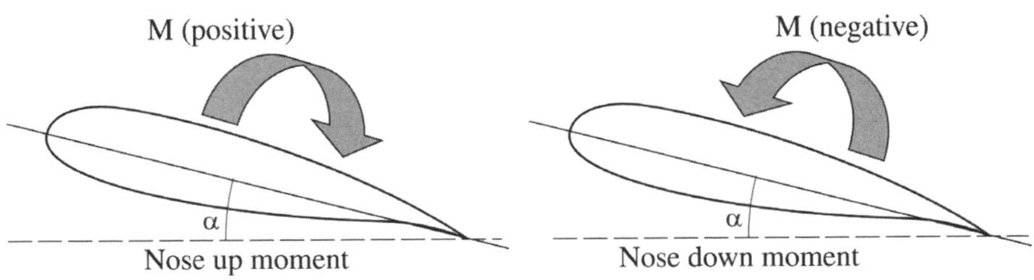

Figure 2.3. *Aerodynamic pitching moment*

Knowing that the lift of an airfoil increases with the angle of incidence α, we can see that a nose-up moment will be destabilizing since it tends to make the airfoil rotate towards an increase of α. It results in an increase of lift, therefore in a nose-up moment, and a shift toward stall. On the contrary, a nose-down moment tends to decrease the angle α, thus the lift and the moment: if during stable flight turbulences cause an increase of the angle α, the negative moment will make the aircraft return to its initial attitude.

When we define the performance of an airfoil, either by calculation or by conducting tests in a wind tunnel, we agree to express the pitching moment relative to a reference point which is often the leading edge, or a point on the chord located a quarter of the way down from the leading edge. Knowing this moment and the chosen point of its application, we can transport the moment to a true center of gravity G of the airfoil whose position depends on the distribution of the mass of the structure. It can be shown from thin airfoil theory that there is a point located on the chord at which the pitching moment is (almost) independent of the lift (that is of incidence α): this point is called *aerodynamic center.*

We assume that the intensity of the aerodynamic forces to which an airfoil is submitted (lift, drag, pitching moment) will be greater when:

– the module of velocity V_∞ is greater,

– the density of the fluid is higher,

– the size of the wing is larger.

It is then only natural to try to characterize the performances of an airfoil, and more generally those of an aircraft or a missile or any vehicle, by dimensionless coefficients, if possible independent of the quantities listed above.

If we denote with ρ_∞ the density of the fluid with the velocity V_∞, considerations of dimensional analysis show that the sought dependence must be of the form:

$$\sim V_\infty^2 \rho_\infty S.$$

Here S is a reference surface, characteristic for the size of the body (for example, the projected surface of the wing). In this way, we introduce the following aerodynamic coefficients:

– *lift coefficients* C_L defined by:

$$F_L = \frac{1}{2} \rho_\infty V_\infty^2 S \quad C_L$$

– *drag coefficient* C_D defined by:

$$F_D = \frac{1}{2} \rho_\infty V_\infty^2 S \quad C_D$$

– *normal force coefficient* C_N defined by:

$$F_N = \frac{1}{2}\rho_\infty V_\infty^2 S C_N$$

– *axial force coefficient* C_A defined by:

$$F_A = \frac{1}{2}\rho_\infty V_\infty^2 S C_A$$

– *moment coefficient* C_m defined by:

$$M = \frac{1}{2}\rho_\infty V_\infty^2 S\, L\, C_m$$

where L is a reference length that may be the chord of the wing.

The factor ½ is used to introduce the *dynamic pressure* of the relation of Bernoulli (see section 9.6.1):

$$\boxed{q_\infty = \frac{1}{2}\rho_\infty V_\infty^2.}$$

The area S and the length L are reference quantities that are in principle arbitrary. Consequently, when the aerodynamic performances of a vehicle are provided by the data of the coefficient C_L, C_D, C_m, it is important that the geometrical reference quantities are well defined. It goes without saying that the consideration of references that have no connection with the shape of the vehicle would deprive the thus calculated aerodynamic coefficients of any real meaning.

The main objective of aerodynamics will be to determine the coefficients C_L, C_D, C_m either by calculation or by experiment (there are other qualities, for example the stability of the aircraft). In other words, the aim of aerodynamics is to define the shape of the *aerodynamic object* with coefficient C_L, C_D, C_m, etc. to ensure the imposed or desired performances. For example, for an airliner it will be necessary to design a wing that allows the aircraft to carry a given payload over a given distance with minimum fuel consumption. These conditions will impose a C_L, defining the carrying capacity, a C_D, setting the propulsive force to be ensured, thus determining the fuel consumption. Additionally, the aircraft should have acceptable flight qualities that will set precise boundaries for the moment coefficient C_m.

What has just been discussed is an ultra-simplification of the real problem of designing an aircraft. Many other constraints are involved in such a design: the behavior of the structure, interior design, pricing, operating and maintenance costs, pollution, noise, etc. Moreover, the global coefficients C_L, C_D, C_m are not the only factors defining the aerodynamics of the aircraft. It is necessary to go into a much finer description of the aerodynamics of the aircraft components: wing, fuselage, fins, high-lift flaps, propulsion nacelles, etc. In addition, calculation of the structure requires knowledge of the local force (pressure distribution) to which these different elements are subjected.

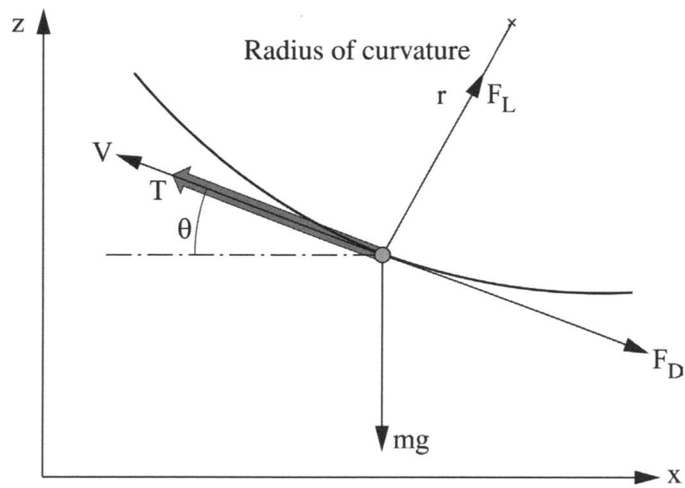

Figure 2.4. *Trajectory of an aircraft in a vertical plane*

2.2. The lift-over-drag ratio and its practical implications

To illustrate the importance of the coefficients of lift C_L and drag C_D, we write the equations governing the movement of an aircraft in a vertical plane (see Figure 2.4). We place ourselves in a system of intrinsic axes formed by the tangent to the trajectory of the aircraft and the normal to this trajectory. If V is the module of velocity, m the mass of the aircraft, T the propulsive force, g the acceleration due to gravity, and θ the slope of the trajectory, we have the following two equations of movement:

$$m\frac{dV}{dt} = T - F_D - m\ g\ \sin\theta = T - \frac{1}{2}\rho_\infty V_\infty^2\ S\ C_D - m\ g\ \sin\theta$$

$$m\ \omega^2\ r = F_L - m\ g\ \cos\theta = \frac{1}{2}\rho_\infty V_\infty^2\ S\ C_L - m\ g\ \cos\theta$$

where ω is the angular velocity:

$$\omega = \frac{d\theta}{dt} = \frac{V}{r}$$

and r is the radius of curvature of the trajectory. It is assumed that the mass of the aircraft is constant, which is not accurate because its engines consume fuel, therefore m decreases with time. We assume that this mass decrease has a negligible effect here.

In the case of straight flight at constant velocity ($\omega = 0,\ dV/dt = 0$), then:

$$T = F_D + m\ g\ \sin\theta = \frac{1}{2}\rho_\infty V_\infty^2\ S\ C_D + m\ g\ \sin\theta$$

$$F_L = \frac{1}{2}\rho_\infty V_\infty^2\ S\ C_L = m\ g\ \cos\theta\,.$$

So essentially, the thrust is used to compensate for the drag effort F_D and the lift F_L for the weight of the aircraft.

A high drag coefficient C_D will require a significant thrust, while an increased lift coefficient C_L will allow larger mass to be carried. Let us consider the case of flight without the engine, then:

$$m\frac{dV}{dt} = -F_D - m\ g\ sin\,\theta$$

$$m\ \omega^2\ r = F_L - m\ g\ \cos\theta$$

If we consider the case of a straight trajectory $(\omega = 0)$ traveled at constant velocity $(dV/dt = 0)$, then;

$$F_D = -m\ g\ sin\,\theta \qquad F_L = m\ g\ cos\,\theta$$

where:

$$cot\ \theta = -\frac{F_L}{F_D} = -\frac{C_L}{C_D}\,.$$

The slope of the trajectory (which should be negative, as the aircraft or glider loses altitude) will be much smaller in absolute value when the ratio C_L/C_D is much higher. This ratio:

$$\boxed{f = \frac{C_L}{C_D}}$$

is called the *lift-over-drag ratio*. The search for an increased lift-over-drag ratio is a major objective in aerodynamics, as this parameter characterizes the gliding qualities of a wing or an aircraft. A glider, or sailplane, during gliding can have a lift-over-drag ratio reaching 60; the lift-over-drag ratio of an airliner of the type Airbus is around 20; the *Concorde* had a lift-over-drag ratio of 7.5 at supersonic cruising flight (10 in transonic, 13 in subsonic), the *Space Shuttle*, a poor gliding craft, has a lift-over-drag ratio between 2 and 3. To illustrate the idea, when an airliner with a lift-over-drag ratio of 20 shuts off its engines at an altitude of h = 10,000 m, it will meet the ground at a distance $x = h \times \cot g\theta = h \times f = 200\text{ km}$. Lift-over-drag ratio plays an important role in aviation as it is essential for the operational efficiency and profitability of the airliner. If we consider a cruise flight at constant velocity and powered flight, the mass of the aircraft decreases due to the fuel consumption according to the relation:

$$dm = -C_s T\, dt$$

where C_s is the specific consumption (expressed according to engine manufacturers in mass of kerosene per kilo of thrust per hour: *kg/daN/h*), T the propulsive force and t the time. When velocity is constant $(V = V_\infty)$, the thrust T compensates the drag, so that:

$$T = \frac{1}{2}\rho_\infty \; V_\infty^2 \; S \; C_D \,.$$

The weight of the aircraft is offset by the lift, therefore:

$$mg = \frac{1}{2}\rho_\infty \; V_\infty^2 \; S \; C_L \,.$$

Knowing that: $dt = \dfrac{dx}{V_\infty}$, it follows from these relations:

$$\frac{dm}{m} = -\frac{g}{\frac{1}{2}\rho_\infty V_\infty^2 S C_L} \; C_s \; \frac{1}{2}\rho_\infty V_\infty^2 S C_D \; \frac{dx}{V_\infty} \,.$$

or else:

$$\frac{dm}{m} = -g \frac{C_s}{f V_\infty} dx .$$

Hence the relation, known as *Breguet's formula*, giving the flight range L for an aircraft as a function of its mass at take off m_1 and its mass at landing m_2:

$$\boxed{L = \frac{f}{g} \frac{V_\infty}{C_s} \ln\left(\frac{m_1}{m_2}\right).}$$

This famous formula shows that the distance a plane can cover, carrying a mass of kerosene $(m_1 - m_2)$, is directly proportional to its lift-over-drag ratio f. Introducing the Mach number $M_\infty = \dfrac{V_\infty}{a_\infty}$, which is the appropriate aerodynamic parameter for fast aircraft, the formula is written as:

$$L = a_\infty \frac{f}{g} \frac{M_\infty}{C_s} \ln\left(\frac{m_1}{m_2}\right).$$

This expression shows the importance of the grouping:

$$\frac{f M_\infty}{C_s}$$

that will maximize the definition of an aircraft, if the other constraints (many) are fixed. Table 2.1 below shows the values of this grouping for typical aircraft.

	Boeing 707	Airbus A340	Concorde	Fighter jet
C_S	0.95	0.65	1.21	2.5
f	16	19	7.5	3.5
M_∞	0.75	0.83	2	2
fM_∞/C_S	12.6	24.3	12.4	2.8

Table 2.1. *Lift-over-drag ratio and aerodynamic quality of typical aircraft*

We see the significant progress represented by the *Airbus A340* compared to the *Boeing 707* which was commissioned in the early 1960s. The gain was obtained by engines with lower fuel consumption, an increase in the cruising Mach number and a significant improvement of the lift-over-drag ratio. Fighter jets have very poor performance but their use is not the same.

In more "economic" terms we can write:

$$f = \frac{lift}{drag} = \frac{weight}{thrust} \approx \frac{payload}{fuel\ consumption}.$$

That is:

$$f \approx \frac{payload}{operating\ costs}.$$

Thus, the lift-over-drag ratio is also a measure of the aerodynamic efficiency of an aircraft in terms of profitability (in fact, the operating cost includes many other factors, not only consumption of kerosene).

In summary, the objective is to design an aircraft with the highest possible lift coefficient and the smallest possible drag coefficient. These objectives are often contradictory: increasing C_L, for example by flaps, also increases C_D. The concern of wing designers is to optimize the shape in order to improve the lift-over-drag ratio, given the mission of the aircraft.

The Breguet formula includes the ratio (m_1 / m_2) which must be optimized in order to increase the range *L*. The denominator m_2, equal to the mass at landing, is the sum of the payload and the mass of the aircraft structure (or dry mass). Hence, the interest to lighten the structure by using composite materials.

2.3. Curves of lift, drag, and polar

The main characteristics of an aerodynamic airfoil are generally summarized by the curves presented below.

– The *lift curve* shows the variation of the lift coefficient C_L as a function of the angle of incidence α. As shown in Figure 2.5, this variation is initially linear, then bends to pass through a maximum after which C_L decreases. The fall of the lift corresponds to the aerodynamic stall of the airfoil, caused by the separation of the boundary layer.

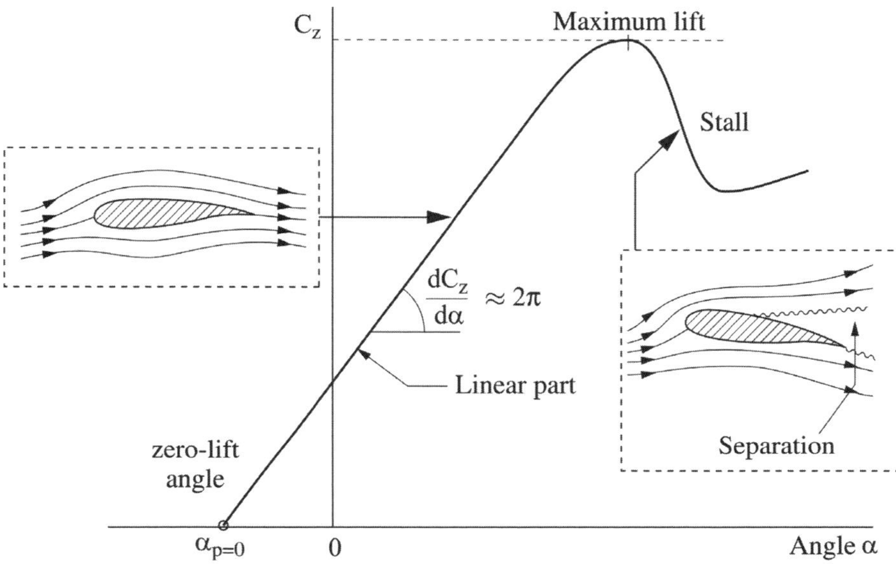

Figure 2.5. *Lift curve of an airfoil*

– The *drag curve* shows the evolution of the drag coefficient C_D as a function of the incidence α. As it is shown in Figure 2.6, it is a fairly flat curve, the drag varying little for moderate values of α. Then the curve bends, C_D beginning to grow rapidly when the angle of incidence reaches high values.

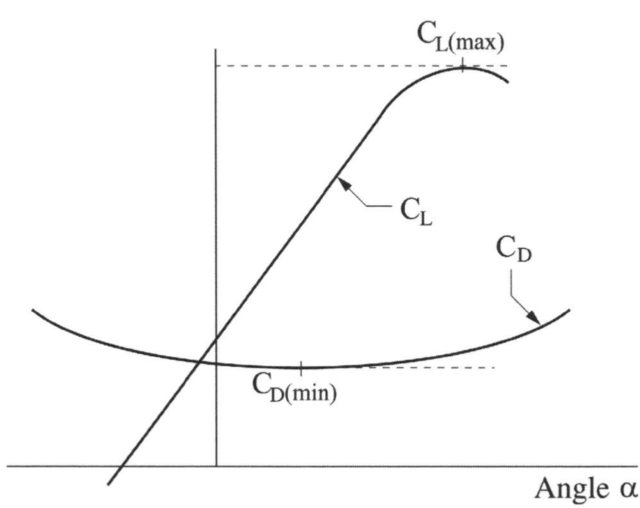

Figure 2.6. *Drag curve of an airfoil*

– The *polar curve* or polar, obtained by plotting the lift factor C_L in the ordinate and the drag coefficient C_D in the abscissa (see Figure 2.7). The polar is often graduated with values of the angle of incidence. The slope of the straight line joining the origin and a point of the polar is equal to the value of the lift-over-drag ratio for the corresponding incidence. The point of tangency of this straight line corresponds to the *maximum lift-over-drag ratio.*

– The *moment curve* shows the moment coefficient C_m as a function of the incidence angle, the moment being referred to a point, which is necessary to provide the position.

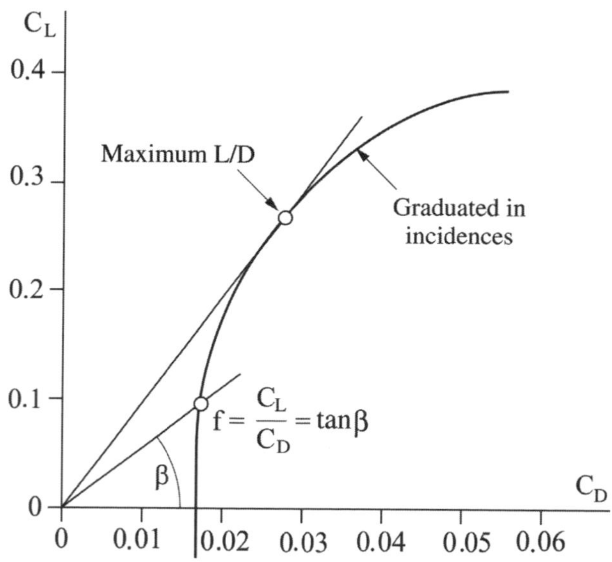

Figure 2.7. *Polar curve of an airfoil*

2.4. Other coefficients characterizing the effects of a fluid on a body

The coefficients C_L, C_D, C_m, etc. characterize global forces to which the vehicle is submitted. But it is equally important to define local effects that are of three types.

2.4.1. *Effect of pressure*

As we will see in Chapter 4, we will break down the force exerted by the fluid on the surface of the vehicle into a normal component, corresponding to the action of pressure p, and a tangential component which is the effect of friction τ. In fact, the quantities p and τ have the dimension of a force per unit of surface (they are measured in pascals). The local value of the pressure is often given in the form of a *pressure coefficient* without dimension:

$$\boxed{C_p = \frac{p - p_\infty}{q_\infty}}$$

where: $q_\infty = \frac{1}{2}\rho_\infty V_\infty^2$ is, as we know, the dynamic pressure of the fluid upstream of the velocity V_∞ and the density ρ_∞. If the gas can be regarded as calorically perfect, the velocity of sound a_∞ is given by (see section 3.8):

$$a_\infty^2 = \gamma \frac{p_\infty}{\rho_\infty}$$

and we have the equivalent expression: $q_\infty = \frac{\gamma}{2} p_\infty M_\infty^2$, where $M_\infty = V_\infty / a_\infty$ is the Mach number of the upstream flow; from which another expression of C_p derives:

$$\boxed{C_p = \frac{2}{\gamma\ M_\infty^2}\left(\frac{p}{p_\infty} - 1\right).}$$

The main benefit of using a pressure coefficient is to absorb the effects of the velocity and the density. We can thus characterize the distribution of the pressure over an airfoil using a dimensionless quantity with a universal character, depending only on the form and the incidence of the airfoil. This is true for non-viscous, incompressible flows (subsonic) providing that the *Reynolds number* (see section 6.1 for the definition of the Reynolds number) is high enough so that the viscous effects are weak or the boundary layer does not separate. When we approach the speed of sound (transonic profile), the distribution of C_p also depends on the Mach number (see Chapter 25). It is noteworthy that the dynamic pressure does not depend on the temperature of the upstream flow (therefore its stagnation temperature; see Chapter 9 for the definition of stagnation conditions), which means that the aerodynamic forces are independent of the stagnation temperature of the flow (in an initial approximation, of course, as the temperature influences the viscous effects). To be more precise, for a fixed Reynolds number, the forces do not depend on the level of the stagnation temperature. This fact is exploited by *cryogenic wind tunnels* where, by reducing the temperature, the Reynolds number is increased without increasing aerodynamic forces on the scale models (see section 6.4).

Knowledge of the pressure distribution on a vehicle is useful from many points of view. It is used to calculate the aerodynamic performance, as we will see later. It is essential to know the stresses to which the vehicle is submitted and predict local deformations of its structure in order to measure them. Furthermore, the deformations of the structure will modify the aerodynamic field, thus the forces themselves which gives rise to an *aeroelastic coupling*.

2.4.2. *The effect of skin friction*

If τ_p designates friction on the vehicle surface at a certain point, we define the *skin friction coefficient* using the relation:

$$\boxed{C_f = \frac{\tau_p}{q_\infty}.}$$

Knowledge of C_f is necessary to calculate the friction drag of the vehicle.

2.4.3. *Heat flux and heating*

In many cases the thermal situation is that of *adiabatic equilibrium*. This means that there is no heat exchange between the vehicle and the surrounding flow. Such a situation is reached for aircraft flying at velocities that can be considered moderate (Mach number lower than 2.5). The surface of the aircraft then reaches a temperature T_f called *friction temperature* or *adiabatic equilibrium* temperature (also known as *recovery temperature*).

But there are circumstances where the adiabatic equilibrium is not established, as in the case of high supersonic flight, all the more, in hypersonic. Then, the flow tends to heat the vehicle, heat being transferred from the fluid to the wall by a mechanism involving the viscosity and thermal conductivity of the fluid. Such a transfer occurs via convection and, in cases where the gas is very hot, via radiation.

The wall-flow heat transfers also occur in the nozzles of rocket engines, in thermal engines, heat exchangers, etc.

If we designate with $\dot{q}_w$ the density of the *heat flux* toward the wall (that is to say, the energy received per units of time and surface, expressed as W/m²), the heat transfer coefficient C_h is defined in a general manner by:

$$\boxed{C_h = \frac{\dot{q}_w}{\rho_e \; V_e \; \left(h_f - h_w\right)}}$$

where ρ_e and V_e are the local values of the density and velocity in the non-viscous flow adjacent to the boundary layer, h_f the friction (recovery) enthalpy, corresponding to the friction (recovery) temperature, and h_w the enthalpy of the fluid at the wall (in fact these enthalpies are specific enthalpies, i.e. enthalpies per unit of mass). When the local values ρ_e and V_e are difficult to define (ρ_e and V_e have to be taken at the boundary layer edge), we use the factor:

$$C_{h_\infty} = \frac{\dot{q}_w}{\rho_\infty \; V_\infty \; \left(h_{T_\infty} - h_w\right)}$$

where h_{T_∞} is the total enthalpy of the upstream flow (also called the stagnation enthalpy and denoted h_{st_∞}, see Chapter 9). The coefficient C_h is often called the *Stanton number* and is written *St*.

If the gas is calorically perfect (the hypothesis is acceptable for temperatures lower than 1,000 K), therefore:

$$h = C_p T$$

thus:

$$C_h = \frac{\dot{q}_w}{\rho_e \; V_e \; C_p \; \left(T_f - T_w\right)} \quad \text{and} \quad C_{h_\infty} = \frac{\dot{q}_w}{\rho_\infty \; V_\infty \; C_p \left(T_{T_\infty} - T_w\right)}.$$

The knowledge of the heat-flux distribution is essential to determine the ability of the structure to withstand aerodynamic heating and to size the thermal protection system (TPS) to be provided. These protections are provided either by a refractory or ablative material (the case of reentry vehicles into the atmosphere), or by equipping with a cooling system by circulating a fluid (the case of the nozzles of rocket engines of space launchers, for example).

2.5. Evaluation of the action exerted by a fluid on a body

2.5.1. *General relations*

Before we approach the detailed description of the movement of a fluid and the presentation of equations that govern such a movement, we will introduce some basic concepts that will enable us to explain – still in an intuitive manner – the action exerted by the fluid on a physical obstacle. To illustrate this idea we consider a two-dimensional wing airfoil, which allows us to simplify the mathematical formalism. It is clear that the concepts introduced here apply to a three-dimensional body of arbitrary shape. We place ourselves in a coordinate system *Oxy*, which originates from the leading edge of the airfoil, Ox being along the chord of the airfoil and Oy directly perpendicular to Ox (see Figure 2.9). The airfoil is placed under an angle of incidence α in a uniform flow with the velocity $\vec{V}_\infty$ at upstream infinity.

The effect of a fluid on a material surface – or *contact effect* – is represented by a vector $\vec{P}$, called the *tension vector*, the force exerted by the fluid on a surface element ds is expressed by:

$$d\vec{F} = \vec{P}\, ds$$

As shown in Figure 2.8, $\vec{P}$ is broken down into:

– a component normal to the surface ds:

$$d\vec{F}_1 = -p\, \vec{n}\, ds$$

where p is a scalar identified with the pressure and $\vec{n}$ the unit vector normal to ds oriented toward the exterior of the surface. With the convention on minus sign "–", the effect of the external fluid on a body, whose surface has a normal unit $\vec{n}$, will be given by: $-d\vec{F}_1 = p\, \vec{n}\, ds$;

– a tangential component contained in the locally tangent plane to the surface and written as:

$$d\vec{F}_2 = \tau\, \vec{t}\, ds$$

where τ is a scalar called *wall friction* or *skin friction* and $\vec{t}$ a unit vector tangential to ds.

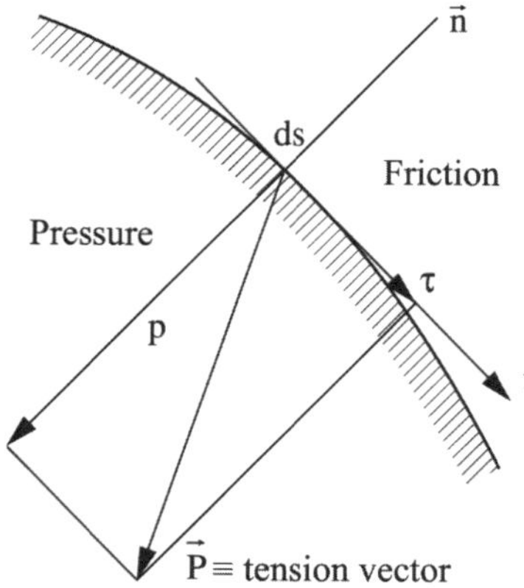

Figure 2.8. *Contact effect between a moving fluid and the surface of an obstacle*

In three-dimensional flows we divide the friction vector into two components:

$$d\vec{F}_2 = \left(\tau \ \vec{t}_1 \ + \tau_2 \ \vec{t}_2\right) \ ds$$

where $\vec{t}_1$ and $\vec{t}_2$ are two unit vectors contained in the plane tangent to the three-dimensional surface.

Thus, in two dimensions, we have locally:

$$d\vec{F} = \left(-p \ \vec{n} + \tau \ \vec{t}\right) \ ds.$$

Let us return to the airfoil and express the force exerted by the fluid onto the surface element *ds* for a wing of unit span. We designate with θ the angle between the normal vector $\vec{n}$ and the axis *Oy*, oriented according to the trigonometric sense (counter-clockwise, see Figure 2.9). If F_N and F_A are the normal (along *Oy*) and axial components (along *Ox*) of the aerodynamic force, we have for the upper surface of the airfoil:

$$dF_{N_u} = -p_u \cos\theta \ ds + \tau_u \sin\theta \ ds$$

$$dF_{A_u} = p_u \ \sin\theta \ \ ds + \tau_u \ \ \cos\theta \ \ ds.$$

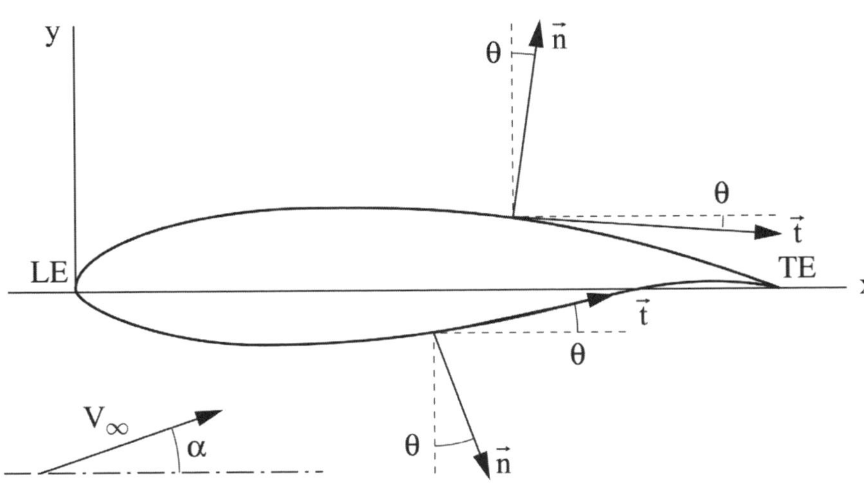

Figure 2.9. *Calculation of the aerodynamic stress over a two-dimensional airfoil*

We recall that dF_N is positive if this component is directed "upward", meaning along *Oy*; likewise dF_A is positive if the effort is along Ox.

For the lower surface:

$$dF_{N_l} = p_l \cos\theta \, ds + \tau_l \sin\theta \, ds$$

$$dF_{A_l} = -p_l \sin\theta \, ds + \tau_l \cos\theta \, ds \, .$$

The normal and axial forces exerted on the airfoil, per unit of wingspan, are obtained through integration of the previous relations from the leading edge *LE* to the trailing edge *TE*, and are:

$$F_N = \int_{LE}^{TE} \left(-p_u \cos\theta + \tau_u \sin\theta\right) ds + \int_{LE}^{TE} \left(p_l \cos\theta + \tau_l \sin\theta\right) ds$$

$$F_A = \int_{LE}^{TE} \left(p_u \sin\theta + \tau_u \cos\theta\right) ds + \int_{LE}^{TE} \left(-p_l \sin\theta + \tau_l \cos\theta\right) ds \, .$$

The lift F_L and the drag F_D are then obtained by using transposition formulas given in section 2.1.

The aerodynamic pitching moment exerted on the body depends on the point at which this moment has been taken. We remember that the moment $\vec{M}$, relative to a point *O*, of a force $\vec{F}$ applied to a point *P*, is the vector product:

$$\vec{M} = \overrightarrow{OP} \times \vec{P} \, .$$

In two-dimensional flow, the moment is a vector perpendicular to the plane containing the motion. According to this definition, the moment is positively oriented when it tends to rotate the body counter-clockwise.

This convention is contrary to that of aerodynamicists, for which a positive moment (nose-up) tends to augment the angle of incidence, meaning to rotate the airfoil clockwise. We will keep this fact in mind during the study of the following relations.

The moment M_{LE} from the leading edge of the airfoil is given by:

$$M_{LE} = \int_{LE}^{TE} \left[\left(p_u \ \cos\theta - \tau_u \ \sin\theta\right)x + \left(p_u \ \sin\theta + \tau_u \cos\theta\right)y\right] \ ds$$
$$+ \int_{LE}^{TE} \left[\left(-p_l \ \cos\theta - \tau_l \ \sin\theta\right)x + \left(-p_l \ \sin\theta + \tau_l \ \cos\theta\right)y\right] \ ds.$$

Here, ds, represents the curvilinear abscissa along the contour of the airfoil, M_{LE} being a moment per wing span unit. This is positive if the moment is nose-up (striving to increase the angle of incidence) and negative if the moment is nose-down (striving to decrease the angle of incidence).

The point in the above formulas is to demonstrate that the *physical origin* of the aerodynamic efforts (lift, drag, and moment) is in the contact effects between the body and the moving fluid. As we have seen, such an effect can be broken down into a normal force at the wall, result of the pressure, and a tangential force, created by friction. We should never lose sight of the physical cause of the aerodynamic efforts. This comment concerns both the *external aerodynamics*, where we focus on lift and drag, and the *internal aerodynamics*, where the major concern is the propulsion (more precisely the thrust-drag balance of the propulsion nacelle, see section 5.2). To evaluate these quantities, we will have to use general theorems applying the concept of momentum and dynalpy. These are certainly very powerful mathematical "stratagems", but they should not mask the physical reality.

2.5.2. *Lift and drag: drag decomposition*

The previous relations give the expressions of the normal and axial components of the aerodynamic resultant in a system of axes linked to the airfoil. Lift and drag are actually defined in a system linked to the upstream velocity vector $\vec{V}_\infty$. By using the formulas of axis change given in section 2.1 and taking into account that α is the angle of incidence, we have the following expressions where the terms introducing pressure on one hand and friction on the other have been regrouped.

Lift:

$$F_L = \left[\int_{LE}^{TE} -p_u \cos\theta \ ds + \int_{LE}^{TE} p_l \cos\theta \ ds \right] \cos\alpha - \left[\int_{LE}^{TE} p_u \sin\theta \ ds - \int_{LE}^{TE} p_l \sin\theta \ ds \right] \sin\alpha$$

$$+ \left[\int_{LE}^{TE} \tau_u \sin\theta \ ds + \int_{LE}^{TE} \tau_l \sin\theta \ ds \right] \cos\alpha - \left[\int_{LE}^{TE} \tau_u \cos\theta \ ds + \int_{LE}^{TE} \tau_l \cos\theta \ ds \right] \sin\alpha.$$

Drag:

$$F_D = \left[\int_{LE}^{TE} -p_u \cos\theta \, ds + \int_{LE}^{TE} p_l \cos\theta \, ds \right] \sin\alpha + \left[\int_{LE}^{TE} p_u \sin\theta \, ds - \int_{LE}^{TE} p_l \sin\theta \, ds \right] \cos\alpha$$

$$+ \left[\int_{LE}^{TE} \tau_u \sin\theta \ ds + \int_{LE}^{TE} \tau_l \sin\theta \ ds \right] \sin\alpha + \left[\int_{LE}^{TE} \tau_u \cos\theta \ ds + \int_{LE}^{TE} \tau_l \cos\theta \ ds \right] \cos\alpha.$$

During cruise flight, when the angle of incidence does not exceed a few degrees, lift and drag are very close to normal and axial components.

The objective of the above expressions is to highlight the roles played by the pressure and the friction in the formation of the lift and drag. We see that it is the pressure that contributes most to the lift (the angle α is generally low and for a thin airfoil θ is small at most points, except in a small region at the leading edge). We also see that the drag can contribute to the lift, in a weak proportion and for the given reasons.

Pressure and friction contribute to the drag in variable proportions, depending on the circumstances. By regrouping the terms as we did, we thus distinguish:

– *pressure drag:*

$$F_{D_{(p)}} = \left[\int_{LE}^{TE} - p_u \cos\theta \, ds + \int_{LE}^{TE} p_l \cos\theta \, ds\right] \sin\alpha + \left[\int_{LE}^{TE} p_u \sin\theta \, ds - \int_{LE}^{TE} p_l \sin\theta \, ds\right] \cos\alpha$$

– *friction drag:*

$$F_{D_{(f)}} = \left[\int_{LE}^{TE} \tau_u \sin\theta \, ds + \int_{LE}^{TE} \tau_l \sin\theta \, ds\right] \sin\alpha + \left[\int_{LE}^{TE} \tau_u \cos\theta \, ds + \int_{BA}^{BF} \tau_l \cos\theta \, ds\right] \cos\alpha \, .$$

For a thin airfoil, with a small angle of incidence, the friction drag is predominant. In contrast, for a blunt projectile (with a rounded nose), the drag is mainly produced by the high pressure exerted on the nose and the depression prevailing at the base; the contribution of the friction is only a few percent. This is also the case for automobiles. In the approximation framework of the non-viscous fluid, the only contact effect is the pressure. Then, in the absence of shock wave, the projection along the upstream velocity of the pressure integral over the entire body must be zero: there is compensation between the regions with over pressure and those with depression (in relation to the upstream pressure p_∞). The drag of the vehicle is zero: this is the *paradox of d'Alembert* which states that in non-viscous fluids the drag of a body is zero (see Chapter 5).

The above break down, or decomposition, has a great practical importance because it allows us to work on a vehicle shape to best optimize lift and/or drag by identifying the regions that contribute to those terms. They also reflect the physical origins of the lift and drag.

2.5.3. *Distribution of pressure on an airfoil and lift*

Except for the cases of extreme operations for aircraft, we are often able to neglect the friction term, retaining only the pressure contribution (but in this case we will not have access to the drag). This hypothesis assumes that the tension vector $\vec{P}$ is carried by the normal to the considered surface element: it is the hypothesis of the perfect fluid, or better *non-viscous*, where the viscosity is neglected.

If the angle of incidence α is small – which is the case for an airliner at cruising flight whose incidence is in few degrees – we can write:

$$F_L = F_N \cos\alpha - F_A \sin\alpha \cong F_N .$$

In addition, if the fluid is non-viscous:

$$F_N = \int_{LE}^{TE} -p_u \cos\theta \, ds + \int_{LE}^{TE} p_l \cos\theta \, ds .$$

Therefore, if the airfoil is thin, which implies $\theta \cong 0$ (except in the region of the leading edge):

$$\boxed{F_L \cong F_N = \int_{LE}^{TE} (p_l - p_u) \, ds.}$$

We see that the lift results mainly from the difference of pressure between the lower and upper surfaces of the airfoil. When designing an airfoil, we therefore will endeavor to reduce the pressures on the upper side as much as possible by ensuring that the airfoil causes an acceleration of the flow which leads to a decrease of the pressure. At the same time, we will limit the lower surface acceleration in such a way as to not have too low levels of p_{st} (here we have used the fact that the pressure diminishes when the velocity increases). Figure 2.10 shows the distribution of upper and lower surface pressure of a typical airfoil in the form of the pressure coefficient C_p. We have observed the habit of airfoil designers consisting of plotting $-C_p$ as a function of the abscissa x along the chord (x is often normalized by the length of the chord c). A stagnation point (local velocity zero) exists at the front of the airfoil, a little below the leading edge. If the trailing edge of the airfoil is sharp, another stagnation point is present at the trailing edge. The coefficient C_p is maximum at these two points, its value corresponding to what is called the *stagnation pressure* of the flow. On the upper side the flow is strongly accelerated by the leading edge contour which leads to a very fast decrease of C_p ($-C_p$ passes through a more or less pronounced peak). As the velocity again becomes zero at the trailing edge, it is necessary for the flow to slow down, C_p increases after passing through a minimum (maximum for $-C_p$).

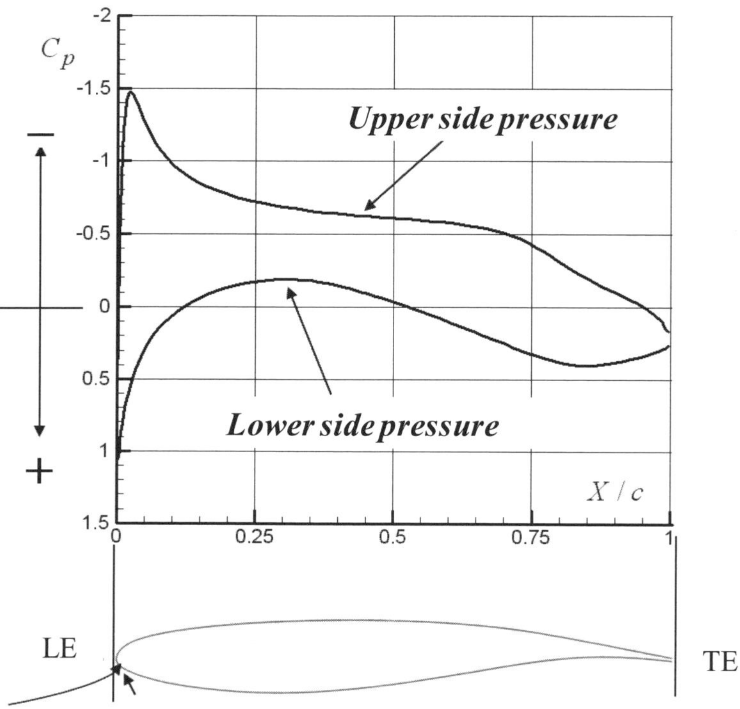

Figure 2.10. *Distribution of pressure coefficient on an airfoil (airfoil Onera OAT15A at Mach 0.4 and an angle of incidence 4°)*

The amplitude of the evolution of C_p is smaller $(C_p)_{lower}$ at the lower surface, always greater than $(C_p)_{upper}$. We can see that the lift is represented by the area contained between the distributions of upper and lower surface pressure, the coefficient C_L being given by:

$$C_L = \int_{LE}^{TE} \left(C_{p_l} - C_{p_u}\right) d\left(\frac{x}{c}\right).$$

Here C_L is defined by taking the product chord $c \times$ wing span b as a reference surface.

An airfoil design with high lift aims is used to increase this area. However, we have constraints while searching for high lift, such as:

– not to increase the drag of the airfoil at the same time (keeping a good lift-over-drag ratio),

– succeed in an actually feasible wing, not very thin (fuel reserves are in the wings of an aircraft; also they pass the commands actuating the flaps),

– ensure structural resistance (the wing is subjected to a considerable bending moment at the junction on the level of the fuselage),

– avoid *separation of the boundary layer* which causes loss of lift (stalling of the airfoil) and a sudden increase of the drag, while causing more or less strong fluctuations (buffeting).

The last point leads us to anticipate the following but the phenomenon of separation plays an important role in aerodynamics and it is crucial to be aware of it for this discipline.

2.5.4. *Aerodynamic stall of an airfoil*

In classic aerodynamics the fluid (in this case air, but the following can be applied to other fluids such as water in hydrodynamics) can be considered as non-viscous almost everywhere in the space around the body (these concepts will be outlined in Chapter 7). In other words, calculations done neglecting viscosity – which greatly simplifies the problem – give results closely representing reality. In fact, at the risk of reaching physical inconsistencies, the viscosity cannot be completely neglected. However, under "normal" situations this viscosity makes itself felt in small areas: they are the *boundary layers* which develop on the fuselage of the vehicle, the *wake* escaping from the vehicle and the *shock waves*.

As these viscous regions are very small (compared to the size of the vehicle), it can be admitted that they have little influence on the flow. This is true as long as no separation happens. However, if a boundary layer is submitted to an adverse pressure gradient, i.e. in a field where the pressure increases, its thickness δ can increase dramatically when the pressure gradient exceeds a critical value. The viscous regions then occupy a notable region and influence the entire aerodynamic field. Under these conditions the hypothesis of the completely non-viscous fluid gives very wrong results.

For an airfoil the increase of the angle of incidence α results in a much higher acceleration at the leading edge; from which follows at the upper side a much stronger depression, thus increasing the lift, which is a well-known effect. A much stronger depression means a subsequently much more important compression, since the stagnation pressure must be found at the trailing edge. Therefore, the boundary layer developing on the upper surface of the airfoil will be submitted to a much more intense adverse pressure gradient and, if α continues to grow, a situation will be reached where the boundary layer separates.

This phenomenon can disrupt the flow around the airfoil, which results in a profound alteration of the pressure distribution. In a non-separated situation, the boundary layer is subjected to the pressure field imposed by the non-viscous field, but only to a certain limit. If this field becomes too *adverse* (compresses too much), then the boundary layer separates and will impose the pressure field and at least its presence should be taken into account in the establishment of the flow.

These phenomena are illustrated in Figure 2.11. First we have portrayed the airfoil with a small angle of incidence α on which the boundary layer remains attached. An overly large increase of α causes the separation of the upper side boundary layer. This results in the formation of a more or less extended separated zone that affects the pressure distribution over the airfoil, certainly on the upper side. Hence drop of lift and increased drag: we say that the *airfoil stalls*.

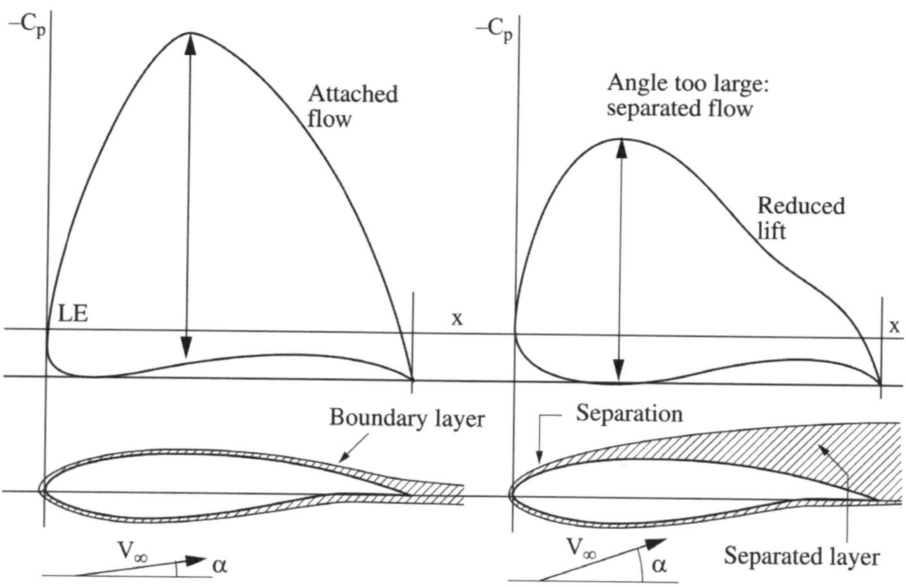

a – attached boundary layer　　*b – separated boundary layer*

Figure 2.11. *Aerodynamic stalling of an airfoil*

2.5.5. *Return to the aerodynamic moment: center of pressure*

We have seen that the efforts of the flow around the airfoil are the result of the integration of the pressure and friction stresses acting on the vehicle surface. These effects are equivalent to a unique force $\vec{F}$ applied to a point called the center of pressure. In terms of flight mechanics the trajectory of the aircraft and its attitude are determined by the transport of the force $\vec{F}$ at the center of gravity *G*. Unless the center of pressure coincides with the center of gravity (exceptional circumstances), the vehicle is subjected to a moment relative to its center of gravity. In practice and by considering the case of the airfoil, the force $\vec{F}$ can be placed at the leading edge *LE* and then we have to provide its moment M_{LE} relative to *LE*, this moment being evaluated in section 2.5.1. In fact, the application point of $\vec{F}$ can be chosen at any other point *C* of the chord of the airfoil. It is necessary to give the value of the moment M_c at this point which should be such that the moment of $\vec{F}$ (here replaced by its normal component F_N) has the value M_{LE}, relative to the leading edge. If x_c is the abscissa of *C* relative to the leading edge, then:

$$M_{LE} = M_c - x_c\, F_N\,.$$

The symbol "–" is introduced as, by convention, a positive force F_N pointing upward creates a negatively counted pitching moment.

The center of pressure is the point *CP* of the chord when the moment of *F*, relative to *CP*, is zero. This point has a position x_{cp} such that:

$$M_{LE} = -x_{cp}\, F_N\,.$$

where:

$$x_{cp} = -\frac{M_{LE}}{F_N}\,.$$

The center of pressure is not always convenient to use because when the normal force tends toward zero, $x_{cp} \to \infty$. F_N is often placed at a quarter of the chord, we then have:

$$M_{LE} = M_{c/4} - \frac{c}{4} F_N\,.$$

In summary:

$$M_{LE} = M_{c/4} - \frac{c}{4} F_N = -x_{cp} \ F_N$$

Figure 2.12 summarizes the different moments used:

$M_{LE} = x_{CP} F_N$ is the moment relative to the leading edge,

$M_{CG} = M_{LE} - x_{CG} F_N$ is the moment relative to the center of gravity,

$M_{CP} = M_{LE} - x_{CP} F_N = 0 \rightarrow x_{CP} = -\frac{M_{LE}}{F_N}$ defines the center of pressure,

$M_{CR} = M_{LE} - x_{CR} F_N$ is the moment relative to the point of forces reduction, that is, the point relative to which the moment is measured in the wind tunnel by dynamometric devices (see below).

Also, we provide the moment relative to the *aerodynamic center* which is the point of the chord for which the moment relative to this point is independent of the angle (as long as it is not very large).

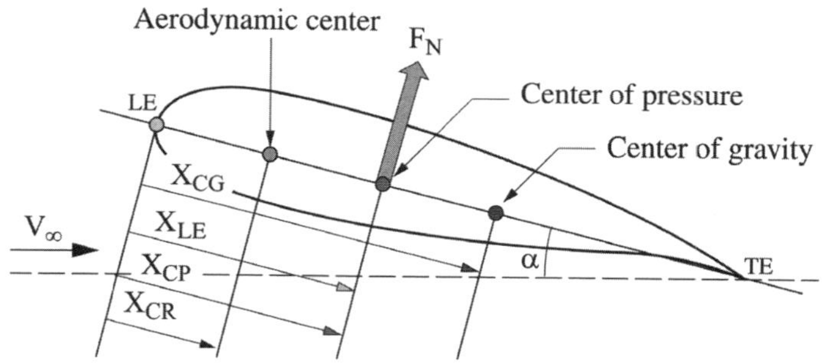

Figure 2.12. *Different systems for the reduction of the aerodynamic resultant on an airfoil*

2.6. The complete aircraft case: aerodynamic torsor

It is not our intention here to dwell on the expression of the aerodynamic forces in the case of an aircraft, which is to say the case of a three-dimensional body. This would lead us too far, particularly considering stability issues. However, it is important to give some definitions about the decompositions used for the aerodynamic forces. The contact action resulting from the flow around the aircraft can be reduced to one force, or aerodynamic force, $\vec{F}$ applied to a point C called *center of pressure*. The problem presented to an aerodynamicist (among many others!) will be to know $\vec{F}$ and C. The movement of the aircraft, the problem belonging to the flight mechanic teams, will be resolved by moving $\vec{F}$ to the center of gravity G of the aircraft. The effect of the aerodynamic actions will thus be reduced to a force $\vec{F}$ applied in G and a moment $\vec{M}$ applied the same in G (see Figure 2.13). This provides the elements needed to calculate the trajectory of the aircraft (movement of the center of gravity) and movements around G.

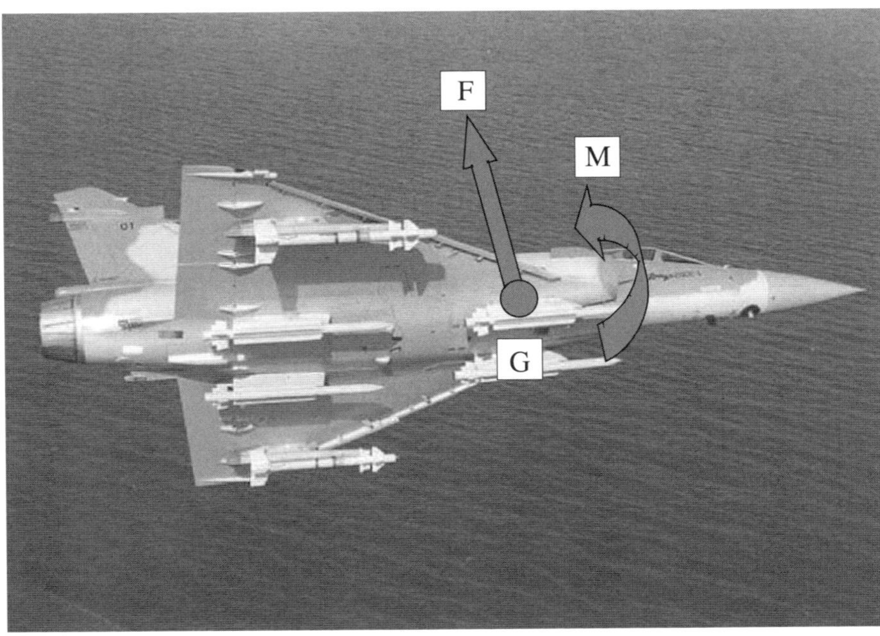

Figure 2.13. *The aerodynamic torsor applied to the center of gravity of an aircraft (image Dassault Aviation)*

The kinematic data can thus be summarized:

– a force applied in G,

– a moment resulting from moving $\vec{F}$ from C to G, being $\vec{M} = \overrightarrow{CG} \times \vec{F}$.

The set $\{\vec{F}, \vec{M}\}$ forms what is known as the *aerodynamic torsor*, the application point is generally different from the center of gravity of the aircraft since this center is not known to aerodynamicist.

The components of the aerodynamic torsor are measured in the wind tunnel with devices called *wind tunnel balances*. The stresses measured by the balance (the aerodynamic force and its moment) are given relative to the point called *center of reduction* (of the forces). The position of this point must of course be provided to proceed to moving the moment to any other point.

Generally, the vectors are divided into their three components of the three-dimensional space, which means that the aerodynamic torsor has six components (three for the force, three for the moment). This break down can be done along different systems of coordinates. Thus we distinguish:

– the aerodynamic system linked to the upstream velocity, called wind axes;

– the aircraft system linked to the aircraft or a scale model, called body axes;

– the wind tunnel system linked to the wind tunnel in which tests are run;

– the balance system linked to the balance used to measure the aerodynamic forces.

Here we will consider only the aerodynamic system which is the one we have been using in the preceding paragraphs. The axes of this system are defined in the following way (see Figure 2.14):

– $\vec{X}_a$ longitudinal axis carried by the velocity $\vec{V}_\infty$ and forward facing;

– $\vec{Z}_a$ vertical axis located in the plane of symmetry, normal at $\vec{X}_a$ and pointing toward the lower surface;

– $\vec{Y}_a$ transverse axis, normal to $\vec{X}_a$ and to $\vec{Z}_a$ oriented towards the right side of the aircraft (sometimes called pilot right-hand direction).

In this system the force $\vec{F}$ is decomposed as follows:

– projection of $\vec{F}$ along $\vec{X}_a \rightarrow$ drag F_D

– projection of $\vec{F}$ along $\vec{Z}_a \rightarrow$ lift F_L

– projection of $\vec{F}$ along $\vec{Y}_a \rightarrow$ side force F_Y

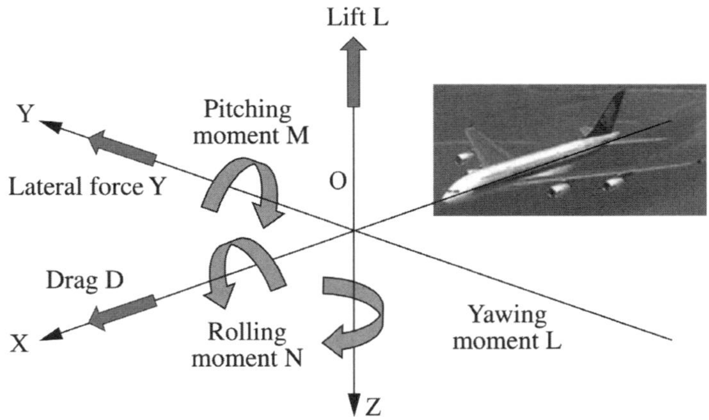

Figure 2.14. *Aerodynamic axes and components of the aerodynamic torsor*

The moment $\vec{M}$ is broken down in an identical way:

– projection of $\vec{M}$ along $\vec{X}_a \rightarrow$ rolling moment M_R

– projection of $\vec{M}$ along $\vec{Z}_a \rightarrow$ yawing moment M_Y

– projection of $\vec{M}$ along $\vec{Y}_a \rightarrow$ pitching moment M_P.

If the components of the aerodynamic torsor are known in a certain system of coordinates T_1, its components in any other system T_2 are deduced from the transformation matrix $[R]$ by relations of the type:

$$\begin{bmatrix} X' \\ Y' \\ Z' \end{bmatrix}_{T_2} = [R] \begin{bmatrix} X \\ Y \\ Z \end{bmatrix}_{T_1}.$$

Trimming the aircraft is ensured when the moment of aerodynamic forces with respect to the center of gravity G is zero. Thus, trimming pitch is obtained by the tailplane which has a pitching moment relative to G that compensates for the pitching moment of the wing. If the center of gravity is located aft of the wing, as in the case of Figure 2.15, the tail plane must actually have a positive lift. If G were located forward of the wing, then the tail plane must have a negative lift (having a downward pointing lift). The centering or position of center of gravity, is very important for piloting airplanes. This centering varies with the size of the load and its repartition in the plane.

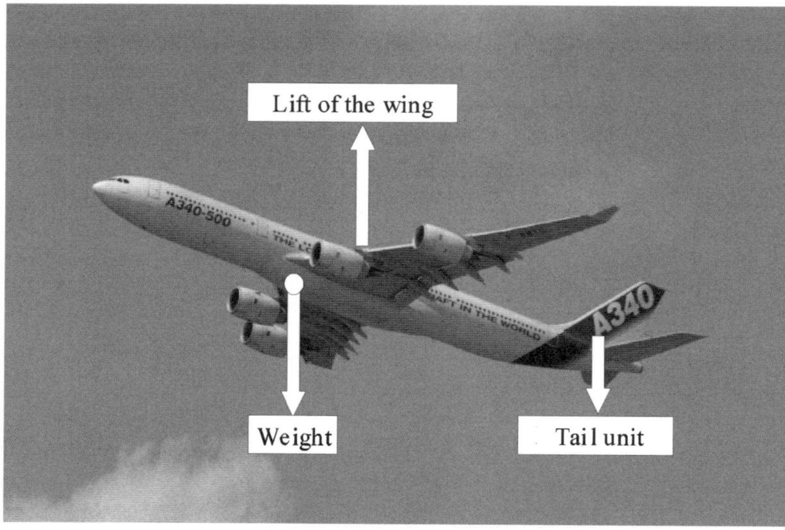

(a)

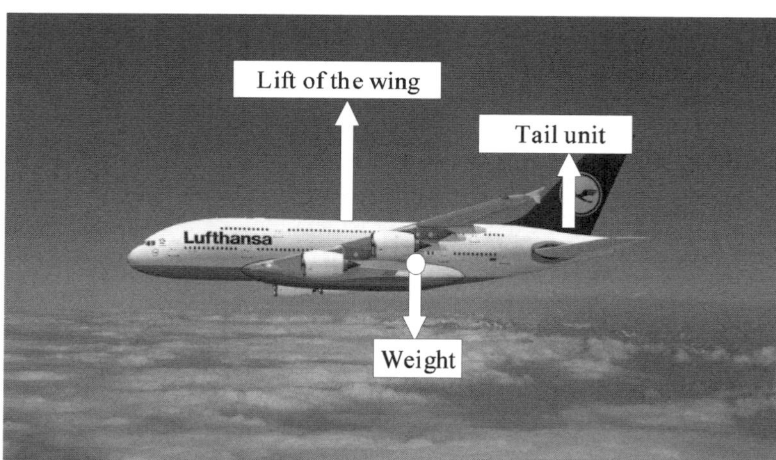

(b)

Figure 2.15. *Trimming of the aircraft pitching moment. (a) Forward centering: the center of gravity is forward of the center of pressure. (b) Aft centering: the center of gravity is behind the center of pressure*

On the *Concorde* it was necessary to transfer fuel from one tank to another to compensate for the changes in the centering resulting from the displacement of the aerodynamic resultant during the transition from subsonic to supersonic.

A high stability is a security feature, but at the expense of maneuverability of the aircraft. Indeed, rapid maneuvers require an almost immediate response of the aircraft to a command on its controls, implying a narrow stability margin. Great maneuverability is sought for fighter planes, which are naturally unstable. The ongoing action on the controls that is needed to keep control of the aircraft is ensured by automation since the pilot could not cope with working steering, given his other tasks.

To conclude this section, we show in Table 2.2 some values for the coefficients of lift and drag of typical aircraft.

	Cruising			**Take-off**		
	C_L	C_D	**Lift-over-drag ratio**	C_L	C_D	**Lift-over-drag ratio**
Subsonic flight	0.50	0.0270	18.5	1.5	0.13	11.5
Supersonic flight	0.12	0.012	10	0.40	0.045	8.9

Table 2.2. *Transport aircraft coefficients*

At cruise conditions, a modern airliner has a lift coefficient C_L of about 0.5 and a lift-over-drag ratio close to 20. At take-off, we look for as high a lift as possible (C_L can reach 2 for certain configurations) rather than the lift-over-drag ratio which is no longer an essential parameter. Table 2.3 provides the same information for a fighter jet in cruise flight. During combat phases the data can change significantly. Table 2.4 shows the very instructive case of a sailplane.

	Interception		
	C_L	C_D	**Lift-over-drag ratio**
Delta wing fighter plane	0.14	0.020	7

Table 2.3. *Aerodynamic coefficients for a fighter plane*

	Transient			**Spiral**		
	C_L	**C_D**	**Lift-over-drag ratio**	**C_L**	**C_D**	**Lift-over-drag ratio**
sailplane	0.3	0.01	30	1.0	0.066	15.2

Table 2.4. *Aerodynamic coefficients of a sailplane*

During the flight phase called transient, the glider, after leaving one ascent, looks for another: the pilot has to adopt a configuration allowing him to go as far as possible with the given starting altitude. Thus he places himself at the maximum lift-over-drag ratio (equal to 30 in the shown example). When he found an ascent he has to climb as high and as fast as possible: therefore he flies with maximum lift at the expense of the lift-over-drag ratio which is here divided by nearly 2.

	Front surface *S* (m^2)	**C_D**	**SxC_D (m^2)**
Touring motorcycle	0.7	0.90	0.63
Sedan	1.8	0.30	0.54
Minivan	2.6	0.50	1.30
Semi-trailer	9	0.90	8.10
Formula 1	1.6	0.90	1.44
Touring cyclist	0.5	1.00	0.50

Table 2.5. *Drag of some ground vehicles*

To conclude, Table 2.5 shows the drag coefficient C_D of some ground vehicles. For these we use the product $S\,C_D$ of the drag coefficient and the front surface of the vehicle. Note that this product is almost three times higher for a minivan (called one-body car) than for a conventional sedan (or three-body car), which is not in the interest of fuel economy! The difference in resistance to advancement is not so great, a good part of the resistance against advancement coming from contact effects between the tyres and the road.

Chapter 3

Review of Thermodynamics

3.1. Macroscopic systems, basic state variables, and absolute temperature

The systems studied in thermodynamics are macroscopic material systems containing a large number of material particles. They are defined in space and time and can be described by a finite number of functions of space and time: these are the state variables of the system that can be dependent. Their set at a given moment constitutes the macroscopic state of the system. The state of a system evolves over time under the influence of other systems forming the exterior, this influence manifests in various forms: temperature differences, electromagnetic potential, acceleration, deformation, etc. The interaction can also happen through microscopic exchanges of matter (diffusion) without an apparent macroscopic transport or differences in the temperature or the potential. It is agreed that a measurable physical quantity, called *energy*, can be associated with all these forms of action. This quantity is transported and exchanged between the systems like matter.

The macroscopic state of a system is formed by the following properties:

– the system occupies a certain volume (ϑ) in space at each moment t, (ϑ) being measurable and always positive,

– it has a measurable and positive mass m,

– a positive density ρ can be associated to the mass m and the volume (ϑ) (see the concept of continuous medium in section 4.1),

– a position and a velocity can be associated with each element of the system, together with an impulse,

– at the macroscopic level, it can be said that the matter always has a positive thermodynamic temperature *T*,

– in addition to the mechanical state variables and the temperature, the system is characterized by its physico-chemical state (chemical composition, physical phases, energy states of molecules and atoms) and its electromechanical state.

The chemical state can be characterized at equilibrium by the masses of the chemical components (in view of possible chemical reactions). More generally, the system could be in chemical non-equilibrium; its description should then be supplemented by parameters such as the degrees of the reaction progress. This may also be true for the physical state.

The electromagnetic state can be defined by two local variables, which are:

– the electric charge and the intensity of the electric current when these quantities can be observed at the macroscopic level,

– the electric and magnetic polarization in the case of charges related to the microscopic level (for example, the two atoms of a dielectric molecule) or circuits at the atomic scale (magnets).

In classic aerodynamics, corresponding to the most common flight conditions, the gas (in this case air) is considered to be chemically inert and not an electric conductor. However, as we shall see later, the air can heat up at hypersonic to extreme temperatures by the passage of shock waves. This heating triggers physico-chemical or electric processes that deeply affect its thermodynamics. It is similar to ionizing air to create plasma; this is in order to exercise a controlling action on the flow in order to minimize the drag or to avoid separation.

3.2. Energy exchanges

We consider a system at a given time *t* and energy flows or amount of energy per unit of time, thus power.

The energetic interaction or energy exchanges are classified according to the following aspects:

– appearance of a macroscopic motion (involving the choice of a frame of reference), this aspect is related to a so-called mechanical interaction,

– need for physical contact, from which actions such as stress forces in a liquid follow (see section 4.4),

– actions at a distance, for example gravity,

– modification of the physico-chemical state if the medium is reactive,

– modification of the electromagnetic state (otherwise it would be neutral systems).

Six modes of energy interactions are distinguished at the physical level of basic mechanisms, each of which being governed by specific rules:

1. Interaction by mechanical power W', caused by external forces: it obeys the law of dynamics and can be exerted at contact or over distance.

2. Interaction by thermal power at contact $Q_C^{'}$: it appears when there are temperature differences (thermal conduction and convection).

3. Interaction by diffusion between two media when there are differences in the concentration of certain components. The exchanged power $D_D^{'}$ is called *heat transfer*, although it does not necessarily involve temperature differences.

4. Spontaneous thermal radiation by the matter with power $Q_E^{'}$: it is an action over distance with electromagnetic nature of microscopic origin and disordered character. This emission is always a loss of energy.

5. Reception of thermal radiation $E_Q^{'}$ by the matter (and its re-emission).

6. Electromagnetic interaction (of non-thermal origin) defined at the microscopic level with the power noted as $E_M^{'}$.

The physical characteristic of contact actions (the tension within a moving fluid, see section 4.4) is that they work over a very short distance. Thus, at the macroscopic level, their local resultants are proportional to the contact surface, while remote actions, such as gravity, are proportional to the volume. An important property is derived from this characteristic: in all areas where there is continuity in the macroscopic state of the matter, the sum of the reciprocal contact interactions is zero. Thus, in a continuous medium:

– the contact forces are equal and opposite to both sides of the surface,

– the fluxes of contact heat are equal and opposite.

The first point is no longer true if there is no continuity in the matter, such as both sides of a shock wave that travels at a different velocity to that of the flow. In Chapter 13, we will see that the contact action, which is the pressure, will attain different values on both sides of the shock wave. In contrast, for a discontinuity of the contact surface, which moves at the velocity of the fluid, there is contact action continuity, therefore pressure.

In Chapter 4, we will see that in fluid mechanics the power W' corresponds to the work of contact forces within the fluid (friction, pressure) and the work of gravity, the latter as a general rule is ignored in aerodynamics. The term $Q_C^{'}$ is the classic heat flux through conduction or convection. The radiation flux $Q_E^{'}$ has to be taken into account in hypersonic flows, where the gas brought to high temperatures can radiate strongly. The electromagnetic interactions, origins of the term $E_M^{'}$, occur when the air is ionized, which can also happen in hypersonics or in circumstances where a plasma is created. In classic aerodynamics the term $D_D^{'}$ occurs more rarely.

3.3. Postulates of thermodynamics

3.3.1. *Postulate of the total energy or first law of thermodynamics*

The energy actions on a given system are measured at every moment by the values of the applied forces. From these actions results a modification of the state characterizing the system (the set of the state variables). Experience has shown that there are two general laws linking the energy actions to the change of the system. These laws involve two scalar quantities: the total energy E_T and the entropy S which, by hypothesis, depends only on state variables.

The change in the total energy E_T is governed by an equation which is a universal law, known as the *first law of thermodynamics*, which we will write in the following way:

$$\boxed{\frac{dE_T}{dt} = \Sigma\left(interaction \;\; powers \;\; with \;\; the \;\; exterior\right)}$$

or:

$$\boxed{\frac{dE_T}{dt} = W^{'} + Q_C^{'} + D_D^{'} + Q_E^{'} + E_Q^{'} + E_M^{'}.}$$

If the system is isolated, this relation expresses the conservation of energy $(E_T = \text{constant})$.

Introducing the total energy per unit of mass or the total specific energy $e_T(M,t)$, which is a function of the point M, of the volume (ϑ) and time, we get:

$$E_T(t) = \iiint_\vartheta e_T\left(M,t\right)\rho\, d\vartheta .$$

The rules of total derivation regarding time show that we can write:

$$\frac{dE_T}{dt} = \iiint_{(\vartheta)} \frac{\partial(\rho e_T)}{\partial t} d\vartheta + \iint_{(S)} \rho e_T \vec{V}_T . \vec{n}\, dS$$

where $\vec{V}_T$ is the velocity of the matter, $\vec{n}$ the unit normal to the surface element dS pointing outside, (S) the surface limiting the volume (ϑ). The velocity $\vec{V}_T$ can be decomposed into a velocity $\vec{V}$ of the macroscopic transport of matter, that is to say its center of gravity (barycentric velocity), and a diffusion velocity $\vec{V}_j$ corresponding to the average diffusion velocity of each species *j*, if the medium is composed of several species; from which follows:

$$\frac{dE_T}{dt} = \iiint_{(\vartheta)} \frac{\partial(\rho e_T)}{\partial t} d\vartheta + \iint_{(S)} \rho e_T \vec{V} . \vec{n}\, dS + \sum_j \left(\iint_{(S)} \rho_j e_{T_j} \vec{V}_j . \vec{n}\, dS \right).$$

If the medium is homogenous, the diffusion flux is zero, from which follows the relation we will be using in classic aerodynamics:

$$\frac{dE_T}{dt} = \iiint_{(\vartheta)} \frac{\partial(\rho e_T)}{\partial t} d\vartheta + \iint_{(S)} \rho e_T \vec{V} . \vec{n}\, dS \,.$$

3.3.2. *Internal energy*

If C designates the kinetic energy of the system, the theorem of the kinetic energy can be expressed by the relation:

$$\frac{dC}{dt} = W' + W'_{\text{int}} \,.$$

This relation brings into play the power of the interaction forces with the exterior W' (power associated with the work of the external forces acting on the system) and the power of the internal forces of the system W'_{int}. We define the *internal energy* of the system as:

$$E = E_T - C \,.$$

Given the first law, we obtain:

$$\frac{dE}{dt} + W'_{\text{int}} = Q'_C + D'_D + Q'_E + E'_Q + E'_M \,.$$

If the medium is homogenous:

$$C = \iiint_{(\vartheta)} \rho \vec{V}.\vec{V}\, d\vartheta$$

and:

$$E = E_T - \iiint_{(\vartheta)} \rho \vec{V}.\vec{V}\, d\vartheta .$$

Thus, by introducing the specific internal energy e, the local expressions:

$$e_T = e + \frac{V^2}{2} \quad \rightarrow \quad e = e_T - \frac{V^2}{2} .$$

The internal energy of a gas is stored in different forms associated with the degrees of liberty of the molecules. We thus distinguish the translation energy, resulting from their overall movement, the rotation energy due to the possibility of the molecule to rotate around itself, the vibration energy, resulting from the relative movement – or vibration – of the atoms.

Thus, we have:

$$e = e_T + e_R + e_V + e_E$$

the relation showing the translation, rotation, vibration, and electric energies of the molecule. In the normal state the exchanges between these different types of energy will offset each other, leading to a state of energy balance or thermodynamic equilibrium. In which case we show that the gas is divariant, which means that its thermodynamic properties can be deduced from the knowledge of two other properties. This result remains true if chemical reactions occur between the components of the gas, provided that these reactions are at equilibrium.

3.3.3. *Enthalpy*

The quantity, called *enthalpy* of a system with uniform pressure p, is:

$$H = E + p\vartheta$$

where ϑ is the volume occupied by the system, or if we consider specific local quantities and by denoting with v the volume occupied by the unit mass of the fluid (we get $v = \frac{1}{\rho}$):

$$h = e + pv = e + \frac{p}{\rho}.$$

We shall see that this quantity is naturally introduced when we apply the first law to moving fluids, particularly when the introduction and extraction works are introduced, which are associated with the effect of the pressure on the fluid passing through the device in question.

3.3.4. *Postulate on entropy and the second law of thermodynamics*

This law is stated by the following additional hypothesis:

– there is a locally defined absolute temperature *T*,

– there is an intrinsic quantity *S*, called *entropy*, such that the thermal exchanges are tied to the variation of *S* by the local relation:

$$\boxed{T\frac{dS}{dt} \geq Q_C^{'} + Q_E^{'}.} \qquad [3.1]$$

Let us recall that $Q_C^{'}$ is the thermal power resulting from the temperature differences (conduction and convection flux) and $Q_E^{'}$ the thermal power provided by radiation.

The relation above is a universal law that is accepted like Newtons laws.

Relation [3.1] can also be written as:

$$\boxed{T\frac{dS}{dt} = Q_C^{'} + Q_E^{'} + Q_I^{'}}$$

where the term $Q_I^{'}$, regarded as intrinsic to the system, corresponds to the internal heat caused by irreversible dissipation of energy.

Relation [3.1] is of a local nature, with the temperature *T* being defined locally. This no longer makes sense for a system with a non-uniform temperature, therefore, the statement has to be completed by the following additional hypothesis.

The entropy of a system is the sum of the entropies of its macroscopic elements: it is an extensive quantity. For a system with non-uniform temperature, the entropy does not only vary due to heat exchange with the exterior and the dissipation of energy into heat, but also because of heat transfer within the system. In Chapter 13 we will provide an example of the shock wave which produces entropy, although the phenomenon is adiabatic (no heat input from the outside).

Entropy is also defined by unit mass or specific entropy s. The quantities e, h, and s are *state functions* that depend only on the thermodynamic state of the fluid. In more specific terms, this means that if a thermodynamic system, in this case consisting of a unit mass of gas, undergoes a transformation from a state (*1*) to a state (*2*), the finite variations:

$$e_2 - e_1 \quad , \quad h_2 - h_1 \quad , \quad s_2 - s_1$$

depend only on states (*1*) and (*2*). They are independent of how the system has moved from state (*1*) to state (*2*).

In what follows, we will have to distinguish:

1. the *adiabatic process* during which there is no heat exchange,

2. the *reversible process* during which no dissipative phenomenon manifests, thus corresponding to $Q_I^{'} = 0$ (absence of viscosity, thermal conductivity or mass transfer),

3. the *isentropic process*, both adiabatic and reversible.

If a system change occurs between (*1*) and (*2*) without heat exchange, we must have:

$$\boxed{s_2 - s_1 \geq 0}$$

or in a differential form:

$$\boxed{ds \geq 0.}$$

If the process is also reversible:

$$ds = 0 .$$

The *second law of thermodynamics* postulates that, during any transformation, the entropy of an isolated system can only increase or, at best, remain constant (the case of isentropic changes). It leads us to reject as physically unacceptable solutions that are mathematically eligible, leading to a decrease in entropy. The second law imposes a direction of certain evolutions that could be made reversible while satisfying the fundamental equations of mechanics. As we will discuss in Chapter 7, Euler equations have this property of reversibility. The normal direction of change of things is restored in the Navier–Stokes equations through the terms of transfer (friction and heat flux). The second law is in fact imposed at the level of comportment laws by stating that the viscosity and thermal conductivity are exclusively positive quantities.

3.4. Expression of the entropy

Considering a reversible change, such as:

$$T\frac{dS}{dt} = Q_C^{'} + Q_E^{'}.$$

If we consider a change where the heat exchanges happen only via the terms of conduction/convection $Q_C^{'}$, and radiation $Q_E^{'}$, then according to the first law:

$$\frac{dE}{dt} + W_{\text{int}}^{'} = Q_C^{'} + Q_E^{'}.$$

By introducing enthalpy, the expression becomes:

$$\frac{dH}{dt} - \frac{d(p\vartheta)}{dt} + W_{\text{int}}^{'} = \frac{dH}{dt} - p\frac{d\vartheta}{dt} - \vartheta\frac{dp}{dt} + W_{\text{int}}^{'} = T\frac{dS}{dt}.$$

We write this relation in a local form by introducing specific quantities:

$$\frac{dh}{dt} - \frac{d(pv)}{dt} + w_{\text{int}}^{'} = \frac{dh}{dt} - p\frac{dv}{dt} - v\frac{dp}{dt} + w_{\text{int}}^{'} = T\frac{ds}{dt}$$

where v is the specific volume and $w_{\text{int}}^{'}$ the power of the internal forces in the volume v. We show that in a reversible change the power $w_{\text{int}}^{'}$ is given by:

$$w_{\text{int}}^{'} = p\frac{dv}{dt}$$

from which follows the relation:

$$\frac{dh}{dt} - \frac{d(pv)}{dt} + w_{\text{int}}^{'} = \frac{dh}{dt} - p\frac{dv}{dt} - v\frac{dp}{dt} + p\frac{dv}{dt} = T\frac{dS}{dt}$$

or:

$$T\frac{ds}{dt}=\frac{dh}{dt}-v\frac{dp}{dt}$$

with the density:

$$\boxed{T\frac{ds}{dt}=\frac{dh}{dt}-\frac{1}{\rho}\frac{dp}{dt}.}$$

3.5. Equation of state and the perfect gas model

Application of the general laws expressed in what precedes to a particular system needs to introduce relations between the quantities that have been defined. Such relations are called constitutive equations and their structure is based on three principles:

– the law of determinism, which states that behavior can be influenced by the present or past state, future states playing no role,

– the law of invariance according to which the behavior laws, which are intrinsic to the system, must be expressed in a way independent of the frame of reference chosen to identify the spatiotemporal change of the system.

We refer the reader to the literature for a thorough study of the constitutive equations in thermodynamics. Here we shall merely state the properties of the perfect gas which is a very good approximation of reality for the vast majority of current aeronautical applications.

A gas consists of particles (molecules, atoms, ions, and electrons) moving in a more or less random way. If these particles are close to each other, they exert a mutual action due to a field effect: they are the intermolecular forces. However, if the gas is sparse (diluted), the particles are quite far apart so that these mutual forces have no influence on their movement. We then say that the gas behaves like an *ideal gas* or a *perfect gas*.

By definition, a perfect gas is characterized by the following behavior:

– the state variables are its volume ϑ (or the specific volume v) and the temperature T,

– the intensive variable associated with the volume is the pressure p,

– it is a homogenous medium,

– for the molar mass *M* the product $p\vartheta$ only depends on the temperature and is independent of the nature of the gas.

We have the relation called *equation of state*:

$$\boxed{p\vartheta = n\Re T}$$

where $\Re$ is the universal constant of the perfect gas and *n* the number of moles of gas occupying the volume ϑ. By introducing the density ρ, the equation of state is written as:

$$\boxed{\frac{p}{\rho} = \frac{\Re}{M} T}$$

with:

$$\boxed{\Re = 8.3145 \text{ J/mole/K.}}$$

Often the equation of state is written in the following form:

$$\frac{p}{\rho} = r\,T \qquad \text{by stating:} \qquad r = \frac{\Re}{M}.$$

In this equation, *r* is a constant dependent on the gas in question: it is no longer universal. In the case of the air, $M = 0.029$ kg/mole, so that:

$$r = 287\ J / kg / K\ .$$

For any, except divariant ideal gas:

$$e = e(T,p) + e_0(T_0, p_0)$$

$$h = h(T,p) + h_0(T_0, p_0)$$

if we take the temperature and pressure as variables describing the gas. The quantities $e_0(T_0, p_0)$ and $h_0(T_0, p_0)$ are the energy and enthalpy for the reference state (p_0, T_0) relative to which the values of $e(T,p)$ and $h(T,p)$ for the state (p,T) have been evaluated. Indeed, the laws of thermodynamics are expressed in the form of balance relations between two states. Therefore, the state functions such as $e(T,p)$, $h(T,p)$, as well as the entropy $s(T,p)$ are defined by a constant which is the reference state. In all applications, it is considered that changes between the two states (*1*) and (*2*) (or total derivatives of the thermodynamic functions) are made, so that the reference state is eliminated. In practice, the state functions are often expressed by writing that the constant is zero.

The gas will be said to be *thermally perfect* when *e* and *h* are only functions of the temperature *T*:

$$e = e(T) + e_0(T_0)$$

$$h = h(T) + h_0(T_0).$$

If *de* and *dh* are the differentials of *e* and *h*, we will write:

$$de = C_v\, dT$$
$$dh = C_p\, dT$$

where:

– C_v is the specific heat at *constant volume*,

– C_p is the specific heat at *constant pressure.*

In general, C_v and C_p are functions of the temperature *T*.

Thus for a thermally perfect gas:

$$e(T) = \int_{T_0}^{T} C_v(T)\,dT + e_0(T_0)$$

$$h(T) = \int_{T_0}^{T} C_p(T)\,dT + h_0(T_0).$$

The specific entropy is expressed from the relation:

$$ds = \frac{1}{T}dh - \frac{1}{T\rho}dp = C_p\frac{dT}{T} - r\frac{dp}{p}.$$

Hence:

$$s(T,p) = \int_{T_0}^{T} C_p\frac{dT}{T} - rLog\frac{p}{p_0} + s_0(T_0,p_0)$$

we obtain the following equivalent form:

$$s(T,\rho) = \int_{T_0}^{T} C_v \frac{dT}{T} - rLog\frac{\rho}{\rho_0} + s_0(T_0,\rho_0).$$

In these relations, according to the case, $s_0(T_0,p_0)$ or $s_0(T_0,\rho_0)$, is the entropy for (T_0,p_0) or (T_0,ρ_0).

Note that the specific entropy of a perfect gas depends on both temperature and pressure (or density). The integral depending only on the temperature is sometimes called *entropy of temperature.*

If the temperature T varies within moderate limits (to illustrate the idea, from ambient to 1,000 K), we can consider C_p and C_v to be constants: the gas is then said to be *calorically perfect*. Under these conditions:

$$e = C_v\, T + e_0$$

$$h = C_p\, T + h_0$$

$$s = C_p\, Log\frac{T}{T_0} - r\, Log\frac{p}{p_0} + s_0$$

and in an analogous manner:

$$s = C_v\, Log\frac{T}{T_0} - r\, Log\frac{\rho}{\rho_0} + s_0.$$

For a perfect gas, C_p and C_v are related by the Mayer relation:

$$\boxed{C_p - C_v = \frac{\Re}{M} = r.}$$

If we introduce the ratio of specific heats:

$$\gamma \equiv \frac{C_p}{C_v}$$

the above relations give:

$$1-\frac{C_v}{C_p}=\frac{r}{C_p} \quad \text{or} \quad 1-\frac{1}{\gamma}=\frac{r}{C_p}.$$

Hence the two following relations, which can be useful:

$$C_p=\frac{\gamma\ r}{\gamma-1}$$

$$C_v=\frac{r}{\gamma-1}.$$

For a calorically perfect gas, γ is a constant sometimes called an *isentropic exponent* (in the case of air $\gamma=1.4$).

We can deduce the following expressions for the internal energy and enthalpy:

$$\boxed{\begin{aligned} e&=\frac{r}{\gamma-1}T+e_0\\ h&=\frac{\gamma\ r}{\gamma-1}T+h_0.\end{aligned}}$$

3.6. Isentropic relations for a calorically perfect gas

For a calorically perfect gas the variation between two states (*1*) and (*2*) is such that:

$$s_2-s_1=C_p\ \ Log\frac{T_2}{T_1}-r\ \ Log\frac{p_2}{p_1}.$$

If the change between (*1*) and (*2*) is isentropic, by definition $s_2=s_1$, and thus:

$$C_p\ \ Log\frac{T_2}{T_1}=r\ \ Log\frac{p_2}{p_1}$$

or:

$$Log\frac{p_2}{p_1}=\frac{C_p}{r}Log\frac{T_2}{T_1}=\frac{C_p}{C_p-C_v}Log\frac{T_2}{T_1}=\frac{\gamma}{\gamma-1}Log\frac{T_2}{T_1}$$

where:

$$\frac{p_2}{p_1} = \left(\frac{T_2}{T_1}\right)^{\frac{\gamma}{\gamma-1}}.$$

In a similar manner, we have:

$$C_v \; Log\frac{T_2}{T_1} = r \; Log\frac{\rho_2}{\rho_1}$$

where:

$$Log\frac{\rho_2}{\rho_1} = \frac{C_v}{r} Log\frac{T_2}{T_1} = \frac{C_v}{C_p - C_v} Log\frac{T_2}{T_1} = \frac{1}{\gamma-1} Log\frac{T_2}{T_1}.$$

Thus:

$$\frac{\rho_2}{\rho_1} = \left(\frac{T_2}{T_1}\right)^{\frac{1}{\gamma-1}}$$

from this the fundamental isentropic relations follow:

$$\boxed{\frac{p_2}{p_1} = \left(\frac{\rho_2}{\rho_1}\right)^{\gamma} = \left(\frac{T_2}{T_1}\right)^{\frac{\gamma}{\gamma-1}}.}$$

Isentropic flows are not uncommon in aerodynamics. This applies to most of the flow field surrounding an airfoil at subsonic velocity. We will see in Chapter 4 that the viscosity is one source of entropy. However, under normal flight conditions (that is in the absence of a massive separation of the boundary layer), the viscous effects are only significant in small regions: the boundary layers and wakes. Thus, the flow is essentially isentropic. Even in supersonic flows, which can contain shock waves that are sources of entropy, there are situations where the flow is nearly isentropic; for example in propulsion nozzles. Often, the flow can also be considered isentropic in the regions between two rectilinear shock waves. This issue will be discussed in more detail in the coming chapters.

We will repeatedly use the notion of entropy that occurs via the laws of thermodynamics, in the so-called dissipative processes involving fluid viscosity. Indeed, as we shall see, the growth (or generation) of the entropy in a flow is caused by viscous terms either at boundary layers or shock waves. However, the loss of stagnation pressure (crucial for air intake, see Chapter 16) and drag are directly related to entropy production. There is also entropy generation during the heat supply by combustion or heat transfers.

3.7. Comments on the thermodynamics of studied gases

3.7.1. *Introductory remarks*

In aerodynamics often circumstances are encountered where the air, subjected to extreme temperature conditions (and sometimes pressure), ceases to behave like a perfect gas. This is the case of hypersonic flows where the air is heated to extreme temperatures by shock waves forming in front of the vehicle or re-entry body (see Chapter 13). Then we talk about a real gas, although in principle real gases are gases where intermolecular forces are felt, and we can no longer apply the perfect gas equation based on the assumption of a diluted gas. We must use more complex equations of which the best known is the *Van der Waals equation*:

$$p = \frac{rT}{v-b} - \frac{a}{v^2}.$$

The so-called *Viriel* effects are observed as soon as the gas pressure becomes greater than (about) 20 bars, at a low enough temperature. In such circumstances, we still use a state law similar to the Mariotte–Gay–Lussac law (Boyle–Mariotte law) which is written as:

$$pV = Z n \mathfrak{R} T$$

where n represents the number of moles contained in the volume V and Z is the compressibility factor expressed as:

$$Z = 1 + B\left(\frac{n}{V}\right) + C\left(\frac{n}{V}\right)^2 + D\left(\frac{n}{V}\right)^3 \ldots$$

The coefficients B, C, D, called virial coefficients, are functions of the temperature. This expression shows that for high dilutions (few moles per volume V), the virial coefficients tend to become negligible compared to the unit.

In the behavior of a gas at high temperature, which is of particular interest here, the first phenomena observed are excitations of the vibration and rotation of molecules which makes the gas no longer calorically perfect. Then the gas adopts the behavior of a thermally perfect gas whose specific heats are temperature dependent. If the temperature increases, the gas molecules dissociate and its properties become the functions of both temperature and pressure. The gas ceases to behave like a perfect gas.

3.7.2. *Variation of the specific heats of a thermally perfect gas*

Classic statistic mechanics enables us to express the specific heat at constant pressure of a gas according to the number of degrees of freedom q of the appropriate molecular model for the gas in question.

Thus we have for *one mole of gas*:

$$C_p = \frac{q+2}{2}\Re \quad \text{and} \quad C_v = \frac{q}{2}\Re .$$

A monatomic gas (this is the case of helium or argon), for which the model is a sphere, has only three degrees of freedom (translational motion), thus:

$$C_p = \frac{5}{2}\Re \quad \text{and} \quad C_v = \frac{3}{2}\Re \quad \text{from which follows: } \gamma = \frac{C_p}{C_v} = \frac{5}{3} = 1.6667.$$

Experiments confirm this expression throughout the temperature range between condensation and ionization. These gases are called thermally perfect before ionization occurs.

The molecular model for the simplest diatomic gas (this is the case for oxygen or nitrogen) is a system of two spheres located at a fixed distance. There are then 5 degrees of freedom: 3 translations and 2 rotations, thus:

$$C_p = \frac{7}{2}\Re \quad \text{and} \quad C_v = \frac{5}{2}\Re \quad \text{of which follows: } \gamma = \frac{C_p}{C_v} = \frac{7}{5} = 1.4 .$$

This expression is suitable for moderate temperatures. When the temperature increases, the bond between two atoms ceases to be rigid and two additional degrees of freedom appear, called vibration, so that C_p tends to $\frac{9}{2}\Re$.

To understand the change of C_p, we have to involve quantum mechanics, which allows us to determine the onset and development of the degrees of vibration freedom. We obtain the following expression:

$$\frac{C_p}{\Re} = \frac{\gamma_{perf}}{\gamma_{perf} - 1} + \left[\frac{\theta_v / 2T}{\sinh \ (\theta_v / 2T)} \right]^2$$

where γ_{perf} denotes the ratio of specific heats in the case of a calorically perfect gas and θ_v a temperature characteristic of the vibration of the considered molecule (vibrational temperature). This equation shows that if $T << \theta_v$, then $\frac{C_p}{\Re} \rightarrow \frac{\gamma_{perf}}{\gamma_{perf} - 1} = \frac{7}{2}$ for a diatomic gas $\left(\gamma_{perf} = \frac{7}{5} = 1.4 \right)$. As $\Re = \left(C_p \right)_{perf} - \left(C_v \right)_{perf} = \left(C_p \right)_{perf} \frac{\gamma_{perf} - 1}{\gamma_{perf}}$, it follows that $C_p \rightarrow \left(C_p \right)_{perf}$. If $T >> \theta_v$, then $\frac{C_p}{\Re} \rightarrow \frac{2\gamma_{perf} - 1}{\gamma_{perf} - 1} = \frac{9}{2}$ for a diatomic gas.

After simple calculations we arrive at the equivalent expression:

$$\frac{C_p}{\Re} = \frac{\gamma_{perf}}{\gamma_{perf} - 1} + \left(\frac{\theta_v}{T} \right)^2 \times \exp\left(\frac{\theta_v}{T} \right) \times \left(\exp\left(\frac{\theta_v}{T} \right) - 1 \right)^{-2}.$$

We deduce the relationship giving the specific heat at constant pressure for a thermally perfect gas:

$$C_p = \left(C_p \right)_{perfect} \left\{ 1 + \frac{\gamma_{perf} - 1}{\gamma_{perf}} \left[\left(\frac{\theta_v}{T} \right)^2 \times \exp\left(\frac{\theta_v}{T} \right) \times \left(\exp\left(\frac{\theta_v}{T} \right) - 1 \right)^{-2} \right] \right\}.$$

The specific heat at constant volume is given by:

$$C_v = \left(C_v \right)_{perfect} \left\{ 1 + \left(\gamma_{perf} - 1 \right) \left[\left(\frac{\theta_v}{T} \right)^2 \times \exp\left(\frac{\theta_v}{T} \right) \times \left(\exp\left(\frac{\theta_v}{T} \right) - 1 \right)^{-2} \right] \right\}$$

from which follows the ratio of specific heats:

$$\gamma = 1 + \frac{\gamma_{perf} - 1}{1 + \left(\gamma_{perf} - 1\right)\left[\left(\frac{\theta_v}{T}\right)^2 \times \exp\left(\frac{\theta_v}{T}\right) \times \left(\exp\left(\frac{\theta_v}{T}\right) - 1\right)^{-2}\right]}.$$

This relation shows that if $T << \theta_v$, then $\gamma \to \gamma_{perf} = 1.4$ for a diatomic gas. If $T >> \theta_v$, then $\gamma_{perf} \to \frac{2\gamma_{perf} - 1}{\gamma_{perf}} = 1.2857$ always for a diatomic gas.

The table below gives the values of θ_v for the diatomic gases most often encountered in aerodynamics:

	O_2	N_2	NO	H_2
θ_v (K)	2230	3340	2690	6100

Table 3.1. *Vibrational characteristic temperature of usual gases*

For air, which is a mixture of gases, we can use the value: $\theta_v = 3{,}060$ K. Figure 3.1 shows the change of γ of air due to the vibration excitation. We observe that γ quickly becomes smaller than 1.4.

A triatomic gas contains in its simplest form 6 degrees of freedom and $C_p = 4\Re$. The appearance of vibrational degrees then obeys a law similar to that of diatomic gases.

The next step of the increase of C_p with temperature, for polyatomic gases, occurs when dissociation phenomena appear, which we will now examine.

3.7.3. *Dissociation phenomena*

3.7.3.1. *Mixture of perfect gases and the Dalton law*

The dissociated air, which no longer behave like a perfect gas, is considered to be a perfect gas mixture in pressure and temperature equilibrium. Dalton's law states that the pressure of the mixture is equal to the sum of the partial pressures of each component or species, the partial pressure being the pressure that a component would have were it to occupy alone the volume containing the mixture.

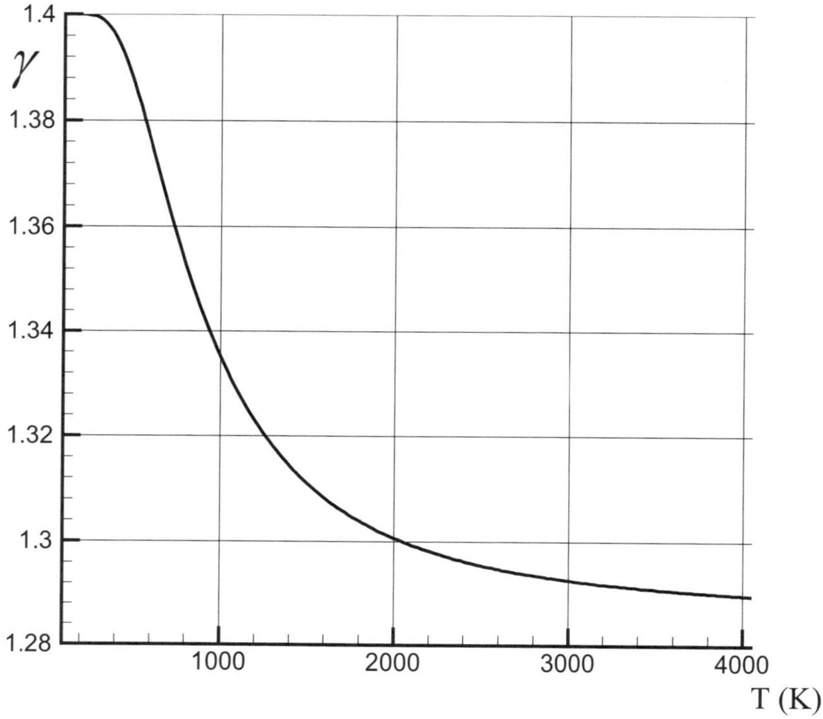

Figure 3.1. *Effect of the vibration excitation on γ of air*

To demonstrate this law, we use Gibbs' concept of free energy *F*, which is defined as:

$$F = E - TS$$

from which, knowing that $TdS = dE + pdV$ follows:

$$dF = dE - TdS - SdT = -SdT - pdV \ .$$

It follows that the pressure is:

$$p = -\left(\frac{\partial F}{\partial V}\right)_T .$$

Let us consider a mixture (*J*) of perfect gases with the respective masses m_j, occupying the volume *V*, so that the specific volume of each component is $v_j = \frac{V}{m_j}$ and its specific free energy (per mass unit) $f_j = f_j\left(T, v_j\right)$.

The partial pressure of each component is then:

$$p_j = -\left(\frac{\partial f_j}{\partial v_j}\right)_T .$$

The free energy of the mixture is the sum of those of its components, thus:

$$F = \sum_{j=1}^{j=J} m_j f_j\left(T, v_j\right)$$

and its pressure is:

$$\boxed{p = -\left(\frac{\partial F}{\partial V}\right)_T = -\sum_{j=1}^{j=J} m_j \left(\frac{\partial f_j}{\partial V}\right)_T = -\sum_{j=1}^{j=J}\left(\frac{\partial f_j}{\partial v_j}\right)_T = \sum_{j=1}^{j=J} p_j .}$$

Let us write now the equations of state for the various species in volume *V*. We obtain, by designating with n_j the number of moles of the species *j*:

$$p_1 V = n_1 \Re T$$

$$-------$$

$$p_j V = n_j \Re T$$

$$-------$$

Adding the above equation by taking account Dalton's law, we get:

$$V \sum_{j=1}^{j=J} p_j = pV = \left(\sum_{j=1}^{j=J} n_j\right) \Re T$$

and dividing the equation of the rank j by this result:

$$\boxed{\frac{p_j}{p} = \frac{n_j}{\sum_{j=1}^{j=J} n_j}.}$$

Hence the theorem: the ratio of the partial pressure of a species to the total pressure of the mixture is equal to the relative molar concentration of this species.

3.7.3.2. *General equations of the chemical equilibrium*

The study of chemical equilibria is of great importance for problems in aerodynamics at high velocities, especially in hypersonics. We then find ourselves in the presence of gaseous media in equilibrium at high temperatures and moderate pressures so that these media can be treated as perfect gas mixtures (high dilution rates). Let us consider a mixture of perfect gases in equilibrium of pressure p and temperature T, as well as chemical equilibrium, and, to simplify, assume that there is one chemical equation of the form:

$$\sum \nu_i (I) \leftrightarrow \sum \nu_i' (I'). \qquad [3.2]$$

Here ν_i are the stoichiometric coefficients of the reaction, (I) are the reactive species, and (I') resulting species. For example, in the reaction:

$$O_2 + 2H_2 \quad \leftrightarrow \quad 2H_2O$$

(I) designates oxygen and hydrogen, (I') water, and the stoichiometric coefficients are:

$$\nu_{O_2} = 1, \quad \nu_{H_2} = 2, \quad \nu'_{H_2O} = 2.$$

We assume by designating (J) any reactive species or result:

– m_j mass of the species (J)

– p_j partial pressure of (J)

– M_j molar mass of (J)

– $r_j = \dfrac{\Re}{M_j}$.

The chemical reaction [3.2] requires that in infinitesimal forward $\vec{D}$ or reverse $\overleftarrow{R}$ reaction, we have:

$$\frac{dm_1}{M_1\nu_1} = \cdots = \frac{dm_i}{M_i\nu_i} = \cdots = -\frac{dm_i^{'}}{M_i^{'}\nu_i^{'}} = d\lambda\,. \qquad [3.3]$$

The parameter λ characterizes the degree of progress of the reaction. Equations [3.3] correspond to the binding conditions of the system. According to the general theory of the thermodynamic equilibrium, equilibrium is achieved when the free enthalpy (Gibbs free energy) $G = H - TS$ of the system is minimal, given the binding conditions:

$$\frac{\partial G}{\partial \lambda} = 0\,.$$

If there were n possible chemical reactions of the type [3.2], we would have n distinct λ parameters $(\lambda_1, \lambda_2, \ldots, \lambda_n)$ and n equations of the type [3.3]. Let us return to the case of a single equation and write the free enthalpy of the mixture by introducing the specific free enthalpies:

$$G = \sum_j m_j g_j\,.$$

By definition, we have for the unit mass of the species: $g_j = h_j - Ts_j$, according to the results of section 3.5

$$g_j = \int_0^T C_{p_j} dT - T\int_0^T C_{p_j}\frac{dT}{T} dT + \frac{\Re}{M_j} T \;\ln\; p_j$$

we can write

$$g_j = \overline{g}_j\left(T\right) + \frac{\Re}{M_j} T \ln p_j\,.$$

By designating as $\overline{g}_j\left(T\right)$, the expression $\overline{g}_j = \int_0^T C_{p_j} dT - T \;\; \overline{s}_j\left(T\right)$.

We obtain for g:

$$g = \sum_j m_j \overline{g}_j\left(T\right) + \sum_j \frac{\Re T\, m_j}{M_j} \ln\, p_j\,.$$

The condition of the equilibrium is written by taking into account [3.3] and separating reactant species from the resulting species:

$$\frac{\partial G}{\partial \lambda}=\sum\left[M_i\nu_i\bar{g}_i-M_i'\nu_i'\bar{g}_i'\right]+\Re T\sum\left[\nu_i\ln p_i-\nu_i'\ln p_i'\right]$$

$$+\Re T\sum\left[\frac{m_i}{M_i p_i}\frac{\partial p_i}{\partial \lambda}+\frac{m_i'}{M_i' p_i'}\frac{\partial p_i'}{\partial \lambda}\right]=0.$$

Or, according to the equation of state:

$$\frac{m_i'\Re T}{M_i' p_i'}=\frac{m_i\Re T}{M_i p_i}=V$$

therefore the last term is written (with the index j, the sum being done on the reactant and resulting species):

$$V\sum\frac{\partial p_j}{\partial \lambda}=V\frac{\partial}{\partial \lambda}\left(\sum p_j\right)$$

and, according to the Dalton equation $\sum p_i = p = cst$, this term is zero. Ultimately, the equilibrium equation takes the form of Guldberg and Waage where $\prod$ denotes a product:

$$\boxed{\frac{\prod_{i'}\left(p_i'\right)^{\nu_i'}}{\prod_{i}\left(p_i\right)^{\nu_i}}=K_p\left(T\right)} \qquad [3.4]$$

where the constant of the equilibrium is defined by:

$$\ln K_p\left(T\right)=\frac{1}{\Re T}\sum\left[M_i\nu_i\bar{g}_i-M_i'\nu_i'\bar{g}_i'\right]=-\frac{\Delta\tilde{G}}{\Re T} \qquad [3.5]$$

denoting $\Delta\tilde{G}$ the change of the free enthalpy (Gibbs free energy) of a system that is assumed to be in stoichiometric proportions according to equation [3.2] and allowing this reaction completely from left to right (so-called forward reaction $\vec{D}$).

COMMENT.– We can write: $\Delta\tilde{G} = \Delta\tilde{H} - T\Delta\tilde{S}$ the two terms on the right-hand side characterize the changes in enthalpy and entropy of temperature respectively defined by the stoichiometric complete reaction $\vec{D}$. Let us then put:

$$\frac{d}{dT}\left(\frac{\Delta\tilde{G}}{\Re T}\right) = \frac{d}{dT}\left(\frac{\Delta\tilde{H}}{\Re T}\right) - \frac{d}{dT}\left(\frac{\Delta\tilde{S}}{\Re}\right) = -\frac{\Delta\tilde{H}}{\Re T^2} + \frac{1}{\Re T}\left[\frac{d\Delta\tilde{H}}{dT} - T\frac{d\Delta\tilde{S}}{dT}\right]$$

Going back to definitions, we verify that the second term is zero because:

$$\Delta\tilde{H} = \int_0^T \tilde{C}_p dT \ , \Delta\tilde{S} = \int_0^T \tilde{C}_p \frac{dT}{T}$$

from which follows the van't Hoff equation:

$$\boxed{\frac{d\left(\ln\ K_p\right)}{dT} = \frac{\Delta\tilde{H}}{\Re T^2} = \frac{-L_p}{\Re T^2}} \qquad [3.6]$$

because $-\Delta\tilde{H}$ is precisely the heat of the reaction at constant pressure as understood by chemists:

$$\Delta\tilde{H} = \sum M_i^{'} \nu_i^{'} h_i^{'} - \sum M_i \nu_i h_i = -L_p \ .$$

This reaction heat depends on the temperature as indicated by the law of Kirschoff (application of the first law of thermodynamics):

$$L_p\left(T\right) = -\Delta\tilde{H} = -\Delta\tilde{H}\left(0\right) + \int_0^T \left(M_i \nu_i C_{p_i} - M_i^{'} \nu_i^{'} C_{p_i}^{'}\right)\ dT \qquad [3.7]$$

where $L_p\left(0\right) = -\Delta\tilde{H}\left(0\right)$ is the heat of reaction at $T = 0$ (heat difference of formation of reactant and resulting species). The integral represents the difference in heat of reactant and resulting species due to the temperature rise from $T = 0$ to T. This second term is often less important than the first. Equation [3.6] enables us to calculate $\ln K_p$ once L_p is known as a function of temperature, the constant of integration being provided by a particular experiment. The kinetic theory allows the direct calculation of $\ln K_p$ by [3.5] from quantum mechanics data.

Application to the study of the dissociation of a diatomic gas

We consider the chemical equation:

$$A_2 \leftrightarrow 2A$$

where A is a perfect diatomic gas with the molar dissociation heat $\Delta\tilde{H} = L_D \; (L_D > 0)$. In principle (Kirschoff's law), L_D depends on the temperature and can be written as:

$$L_D = L_D^{(0)} + \varepsilon(T)$$

where $\varepsilon(T)$ is a correction term, usually with a very small value. We put:

$$\theta_D = \frac{L_D^{(0)}}{\Re}$$

which is a characteristic dissociation heat. Equation [3.6] of van't Hoff then integrates as:

$$\ln \; K_p = \int_0^T \frac{\varepsilon(T)}{\Re T^2} dT - \frac{\theta_D}{T}.$$

We can simplify and obtain the general results:

– either by neglecting $\varepsilon(T)$,

– or by taking the first term of its expansion as a function of *T*, which assumes that C_{p_A} and $C_{p_{A_2}}$ are constant. This more accurate model is due to Lighthill.

In the first case we have:

$$K_p = Cst \times \exp\left(-\frac{\theta_D}{T}\right).$$

In the second (Lighthill gas) we obtain:

$$K_p = Cst \times T \exp\left(-\frac{\theta_D}{T}\right).$$

We will only use the second case which is very useful in aerodynamics. Denoting the degree of dissociation as α, that is to say the mass of the element A present in the unit of mass of the mixture, the partial pressures p_A and p_{A_2} are given by:

$$p_A = \alpha \frac{\Re}{M_A} T \rho$$

$$p_{A_2} = (1-\alpha) \frac{\Re}{M_{A_2}} T \rho$$

where ρ denotes the density of the mixture. As the molar masses are $2M_A = M_{A_2}$ and as $p = p_A + p_{A_2}$ (Dalton's law), we obtain by adding the above relations:

$$p = (1+\alpha) \frac{\Re}{M_{A_2}} T \rho$$

from which follows:

$$\frac{p_A}{p} = \frac{2\alpha}{1+\alpha} \quad \frac{p_{A_2}}{p} = \frac{1-\alpha}{1+\alpha}.$$

The Guldberg and Waage equation [3.4] then gives:

$$\frac{p_A^2}{p_{A_2}} = \frac{4\alpha^2}{1-\alpha^2} p = K_p$$

and is for the Lighthill gas:

$$\boxed{\frac{\alpha^2}{1-\alpha^2} = Cst \times \frac{\exp\left(-\frac{\theta_D}{T}\right)}{\frac{p}{T}}}.$$

If the equation of state is taken into account, this relation can be written as:

$$\frac{\alpha^2}{1-\alpha^2} = \frac{\rho_D}{\rho} \times \exp\left(-\frac{\theta_D}{T}\right).$$

We can also write the equation of state as:

$$\frac{p}{p_D} = (1+\alpha)\frac{T}{T_D}\frac{\rho}{\rho_D} \qquad [3.8]$$

where p_D and ρ_D refer to the pressure and density characteristics of the gas under study:

$$p_D = Cst \times \theta_D \text{ , } \rho_D = Cst \times \frac{M_{A_2}}{\Re}.$$

We see that the dissociation rate is a function of the temperature and density $\alpha\left(\frac{\rho}{\rho_D}, \frac{T}{\theta_D}\right)$, or equivalently of the temperature and pressure $\alpha\left(\frac{p}{p_D}, \frac{T}{\theta_D}\right)$. We calculate the specific enthalpy using:

$$h = \alpha h_A + (1-\alpha) h_{A_2}.$$

If we normalize the specific enthalpies by $e_D^0 = \frac{L_D^{(0)}}{M_{A_2}}$ and if we neglect $\varepsilon(T)$ compared to L_D^0 (then: $h_A - h_{A_2} = e_D^0$), it leads to:

$$\frac{h}{e_D^0} = \frac{h_{A_2}}{e_D^0} + \alpha\left(\frac{p}{p_D}, \frac{T}{\theta_D}\right).$$

Thus, the reduced enthalpy is a linear function of α and depends on the temperature and pressure at the same time.

The entropy of the mixture is obtained in the form of:

$$s = \alpha s_A + (1-\alpha) s_{A_2}$$

with the specific entropies of the two elements:

$$s_A = \overline{s}_A(T) - \frac{2\Re}{M_{A_2}} \ln p_A$$

$$s_{A_2} = \overline{s}_{A_2}(T) - \frac{\Re}{M_{A_2}} \ln p_{A_2}.$$

We can complete the calculation of s by expressing p_A and p_{A_2} as functions of p and α in such a way that s appears as a function of $\frac{T}{\theta_D}$ and $\alpha\left(\frac{p}{p_D},\frac{T}{\theta_D}\right)$.

COMMENTS.–

1. A diatomic gas, although composed of two perfect gases A and A_2, no longer behaves like a perfect gas during dissociation. Equation [3.8] directly shows this:

$$\frac{p}{\rho T} = Cst \times \left[1+\alpha\left(\frac{p}{p_D},\frac{T}{\theta_D}\right)\right].$$

Under these circumstances, the enthalpy depends on the temperature and the pressure.

2. The evolution of $\alpha\left(\frac{p}{p_D},\frac{T}{\theta_D}\right)$ shows that at a low pressure, the dissociation takes place over a very narrow temperature interval: for example, when $p = 10^3\,Pa\left(\approx 10^{-2}\,atm\right)$, the dissociation begins almost at $0.035\theta_D$ and is completed at $0.05\theta_D$. Table 3.2 shows the values of θ_D for various diatomic gases.

	O_2	N_2	NO	H_2
$\theta_D\,(\mathrm{K})$	59 000	113 300	75 500	52 400

Table 3.2. *Characteristic dissociation temperature of some diatomic gases*

Figure 3.2 shows the evolution of the dissociation degree of oxygen as a function of temperature at different pressure levels. For example, oxygen under a pressure of $10\,\mathrm{Pa}$ ceases to be a perfect gas at a temperature of $\approx 2{,}000\,\mathrm{K}$ and is completely dissociated above $\approx 3{,}000\,\mathrm{K}$. Under the same pressure, nitrogen begins to dissociate at $4{,}000\,\mathrm{K}$ and is completely dissociated at $5{,}600\,\mathrm{K}$.

We see that at low pressures (corresponding to the flight of hypersonic vehicles and re-entry bodies at high altitude); the complete dissociation of oxygen molecules occurs in a very narrow temperature range. This behavior can lead to numerical difficulties in solving the equations of fluid mechanics coupled with those of chemistry, the initiation of the chemical equations taking place over a very short time of integration.

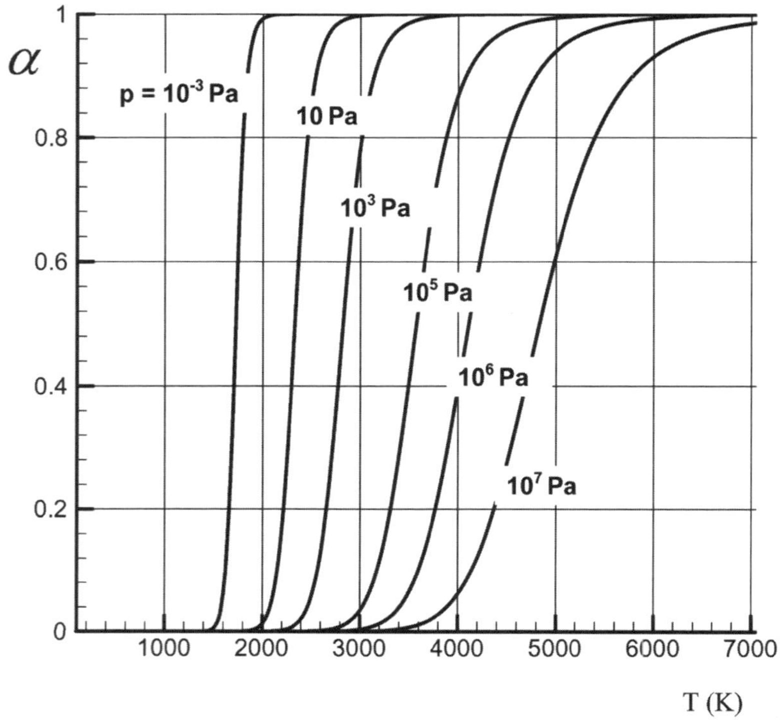

Figure 3.2. *Dissociation rate of oxygen for different pressure levels*

If we consider air, oxygen dissociates first, nitrogen remaining neutral. Then nitrogen dissociates, oxygen being entirely in an atomic state. However, this scheme is approximate, because the table above indicates that the dissociation of NO occurs between 2,500 K and 3,500 K, which complicates the phenomenon.

Application to the ionization of a monatomic gas

We consider the ionization reaction of an atom:

$$A \quad \leftrightarrow \quad A^{+} + e$$

with α being the degree of ionization: the mass of A^{+}, present in the unit of the mass of the mixture, m being the corresponding mass of the electrons e $(m << \alpha)$.

The number of electrons is equal to the number of ionized atoms, therefore, if M_e is the molar electronic mass:

$$\frac{m}{M_e} = \frac{\alpha}{M_A}$$

knowing that $M_A \cong M_{A^+}$. On the other hand, the non-ionized mass is $1-\alpha-m \cong 1-\alpha$ and, the partial pressures following the conditions are:

$$p_e V = \frac{m}{M_e}\Re T = \frac{\alpha}{M_A}\Re T$$

$$p_{A^+} V = \frac{\alpha}{M_A}\Re T$$

$$p_A V = \frac{1-\alpha}{M_A}\Re T$$

$$pV = \frac{1+\alpha}{M_A}\Re T .$$

The law of mass action is written as:

$$\frac{\alpha^2}{1-\alpha^2} = \frac{K_p(T)}{p}$$

where $K_p(T)$ is obtained by integrating the law of van't Hoff (the heat of ionization L_i is assumed to be constant):

$$K_p = Cst \times \exp\left(-\frac{\theta_i}{T}\right) \quad \text{with} \quad \theta_i = \frac{L_i}{\Re}$$

where θ_i represents a characteristic ionization temperature of the element A. The constant involved in the expression K_p is calculated from statistical mechanics, and the expression of the equilibrium equation is written in short (Saha's equation):

$$\frac{\alpha^2}{1-\alpha^2} = \frac{(2\pi m)^{\frac{3}{2}}}{ph^3}(kT)^{\frac{5}{2}} \exp\left(-\frac{\theta_i}{T}\right)$$

where:

– m : electron mass $= 9\times10^{-31}\,\text{kg}$

– h: Planck's constant $= 6.626\times10^{-34}\,\text{J}\times\text{s}$

– k: Boltzmann's constant $= 1.380\times10^{-23}\,\text{J/K}$.

Table 3.3 gives the values of θ_i for common gases.

	O	N	H	He	A	C
$\theta_i\,(\text{K})$	158 000	168 800	157 800	285 400	182 900	130 800

Table 3.3. *Characteristic ionization temperature of some typical gases*

We observe that these values are significantly higher than the corresponding values of θ_D, the ionization occurring at much higher temperatures than the dissociation. This is shown in Figure 3.3, which gives the ionization rate of oxygen for different pressure levels.

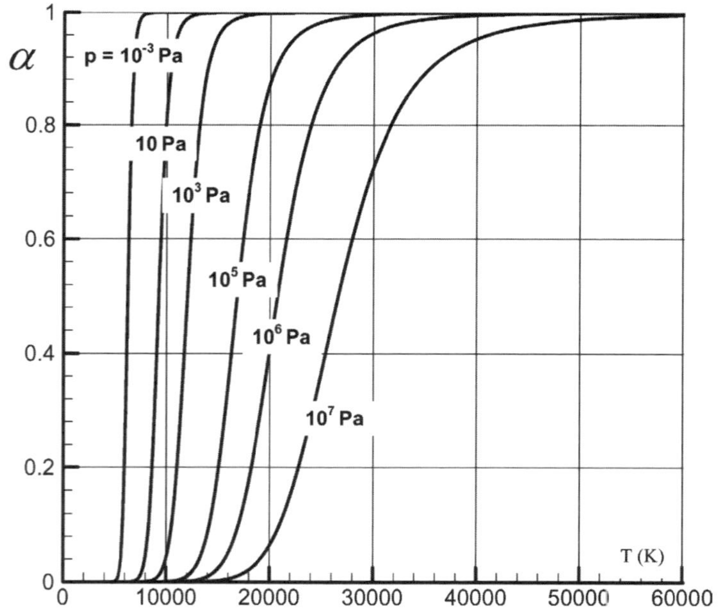

Figure 3.3. *Ionization rate of oxygen for different pressure levels*

3.7.3.3. *Composition of a mixture of reactive perfect gases at equilibrium*

We concentrate on the case of air for which the four essential equilibrium equations are the following:

$$r=1 \quad : \quad O_2 \leftrightarrow O+O$$
$$r=2 \quad : \quad N_2 \leftrightarrow N+N$$
$$r=3 \quad : \quad NO \leftrightarrow N+O$$
$$r=4 \quad : \quad NO \leftrightarrow NO^+ + e.$$

The four equations above, far from exhausting all possible combinations, are sufficient for up to 8,000 K to discuss the equilibrium. We consider n_j to be the chemical variables, which is defined as the number of moles of the species (*J*) in the unit of mass of the mixture which contains a total number of species equal to *J*. The equation of state shows that for the unit of mass we have:

$$p_j = n_j \Re T \rho .$$

Each of the four equations gives us an equilibrium equation of the form:

$$\prod_{j,r} \left(n_j \Re T \rho\right)^{\nu_{J,r}} = K_{p,r}$$

the positive ν_J are counted positive for the resulting species and the second index *r* indicating that it is one of the reactions. As $\Re T \rho$ is constant, we also get:

$$\prod_{j,r} \left(n_j\right)^{\nu_{J,r}} = \frac{1}{\rho^{\sum \nu_{J,r}}} K_{c,r}$$

where we introduced a new equilibrium constant:

$$K_{c,r} = K_{p,r} \left(\Re T\right)^{-\sum \nu_{J,r}} .$$

For example, it appears in the case of the equation $r=1$ for which $\sum \nu_{J,1} = 2-1=1$:

$$K_{c,1} = K_{p,1} \left(\Re T\right)^{-1}$$

and:

$$\frac{n_O^2}{n_{O_2}} = \frac{1}{\rho} K_{c,1} .$$

Then likewise:

$$\frac{n_N^2}{n_{N_2}} = \frac{1}{\rho} K_{c,2}$$

$$\frac{n_N n_O}{n_{NO}} = \frac{1}{\rho} K_{c,3} \qquad [3.9]$$

$$\frac{n_e n_{NO^+}}{n_{NO}} = \frac{1}{\rho} K_{c,4} .$$

Added to these four equilibrium equations are the three equations of the conservation of elementary species and electric charges, which gives by designating with $2\left(n_{O_2}\right)^0$ and $2\left(n_{N_2}\right)^0$ the initial situation in atomic mass of oxygen and nitrogen:

$$2n_{O_2} + n_O + n_{NO} + n_{NO^+} = 2\left(n_{O_2}\right)^0$$

$$2n_{N_2} + n_N + n_{NO} + n_{NO^+} = 2\left(n_{N_2}\right)^0 \qquad [3.10]$$

$$n_{NO} + n_e = 0 .$$

The last equation expresses that the initial electrical neutrality is preserved.

The seven equations [3.9] and [3.10] determine the seven unknown quantities $n_{O_2}, n_O, n_{NO}, n_N, n_{NO}, n_{NO^+}, n_e$ thus the problem is determined. For a given ordered pair $[T, \rho]$, the K_p being known, thus also the K_c, we solve the system of the seven equations. We then calculate the pressure through the equation of state:

$$p = \left(\sum n_j\right) \Re T \rho$$

then the enthalpy and specific entropy of the mixture can be given by:

$$h = \sum n_j h_j .$$

$$s = \sum n_j \bar{s}_j(T) - \Re \sum n_j Log\, p_j$$

or from the state equation: $p_j = n_j \Re T \rho$.

$$s = \sum n_j \bar{s}_j(T) - \Re\left[\sum n_j Log\, n_j - \sum n_j Log \sum n_j\right] - \Re \sum n_j Log\, p\,.$$

The overall dissociation rate of the gas is represented by the *compressibility factor*:

$$\boxed{Z = \frac{\sum(n_j)}{\sum(n_j)^0}}$$

which is equal to the number of moles in the mixture under equilibrium conditions, divided by the number of moles of the initial mixture before the chemical reactions. The factor Z is a function of the temperature T and the pressure p.

The reactive gases, comprising a mixture of perfect gases occupying the volume V, we have according to Dalton's law:

$$pV = \left(\sum n_j\right)\Re T\,.$$

Furthermore, the gaseous mixture before the reaction behaves like a perfect gas, hence:

$$pV = \left(\sum n_j^0\right)\Re T\,.$$

For the unit of mass, by designating by M_j^0 the molar masses of the species, then:

$$pv = \frac{p}{\rho} = \frac{\left(\sum n_j^0\right)}{\left(\sum M_j^0 n_j^0\right)}\Re T\,.$$

From which follows the equation of state:

$$pv = \frac{p}{\rho} = rT \quad \rightarrow \quad p = r\rho T$$

Taking into account the reaction mixture we get:

$$pv = \frac{\left(\sum n_j\right)}{\left(\sum n_j M_j\right)} \Re T = \frac{\left(\sum n_j\right)}{\left(\sum n_j^0\right)} \frac{\left(\sum n_j^0\right)}{\left(\sum n_j M_j\right)} \Re T .$$

The mass being conserved:

$$\sum n_j^0 M_j^0 = \sum n_j M_j$$

so that the equation of state for the mixture of reactive gases can be written as:

$$\boxed{pv = \frac{p}{\rho} = Z(T,p) r T \quad \rightarrow \quad p = Z(T,p) r \rho T}$$

where r is the constant relative to the unit of mass of the gas mixture *before the chemical reactions*.

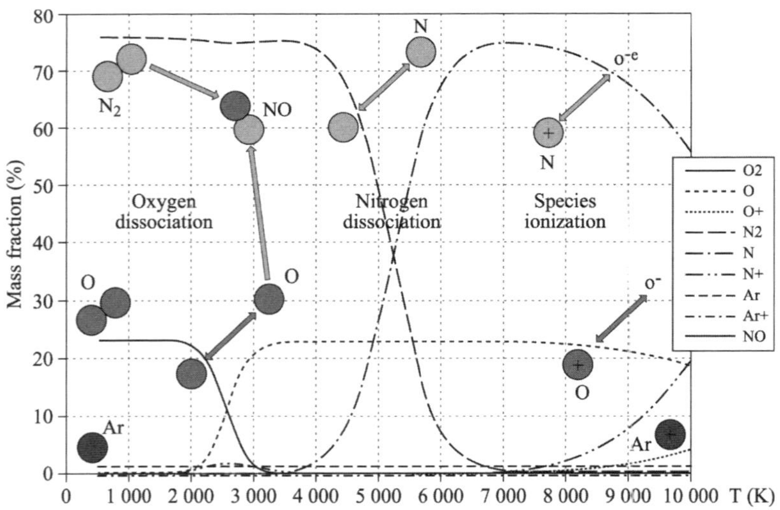

Figure 3.4. *Dissociation of air at high temperature (document Von Karman Institute, Rhode-Saint-Genèse, Belgium)*

Figure 3.4 shows the distribution of mass concentrations of the above species as a function of the temperature. It is found that molecular oxygen begins to dissociate at 2,000 K and it is entirely in atomic form beyond 3,000 K. The dissociation of nitrogen begins at 3,800 K and ends at 6,500 K. NO is formed in small quantities at about 2,000 K. Beyond 7,500 K, the species are partially ionized, the electrons leaving the atoms are becoming free electrons. The air becomes an electric conductor, called *plasma*. Such a plasma has very specific properties which attempts have been made to use to improve the aerodynamics of aircrafts and missiles, the major difficulty being to create a plasma under normal flight conditions. A plasma, like any other conducting body, tends to be impervious to electromagnetic waves. This phenomenon is the origin of the *blackout* during atmospheric re-entry when the vehicle, still at high speed, is surrounded by ionized air. This causes a period of several minutes during which radio communication between the vehicle and earth is interrupted.

3.7.4. *Mollier diagram*

In the study of flows, it is often useful to use as characteristics of the thermodynamic state, the fluid's (specific) enthalpy h and (specific) entropy s. In many circumstances, we can consider the entropy s as a constant (called isentropic flow), thus it is convenient to represent the properties of the fluid on a $[s,h]$ diagram where the entropy is the abscissa and the enthalpy ordinate.

As we have seen above, for real gases at high temperature and in equilibrium, the enthalpy depends on both temperature T and pressure p. The resolution of the equilibrium having been performed for a sufficient number of the values (T,p), we are able to draw the three main families of curves on this diagram:

– the isobars $p = Cst$

– the iso-densities $\rho = Cst$

– the isotherms $T = Cst$.

We often trace the *iso-Z* curves in preference to the iso-densities. From such a diagram we can read at each point (h,s) the values of p, T, Z and deduce that:

$$\rho = \frac{p}{Z\,r\,T}.$$

Figure 3.5 shows a Mollier diagram of air. It can be shown that at point P the isobaric slope always is greater than that of the isotherm. The slope of the iso-density curve is between that of the isobar and the isotherm. We should mention that Mollier diagram is very useful for discussing some thermodynamic properties of gas at equilibrium. One of the interesting points of the Mollier diagram is that an isentropic evolution is represented by a line parallel to the axis of the enthalpies. In other words, when a gas changes in an isentropic way, its representative point moves on a straight line s = constant. The cycles of engines are often shown in a Mollier diagram, which gives a vivid picture of the change of the fluid during its passage through the various components of the engine (air intake, compressor, combustion chamber, turbine and nozzle). The deviation from the vertical s = constant of the representative point of the system during an evolution is a measure of the entropy production in any part of the engine, thus of its efficiency.

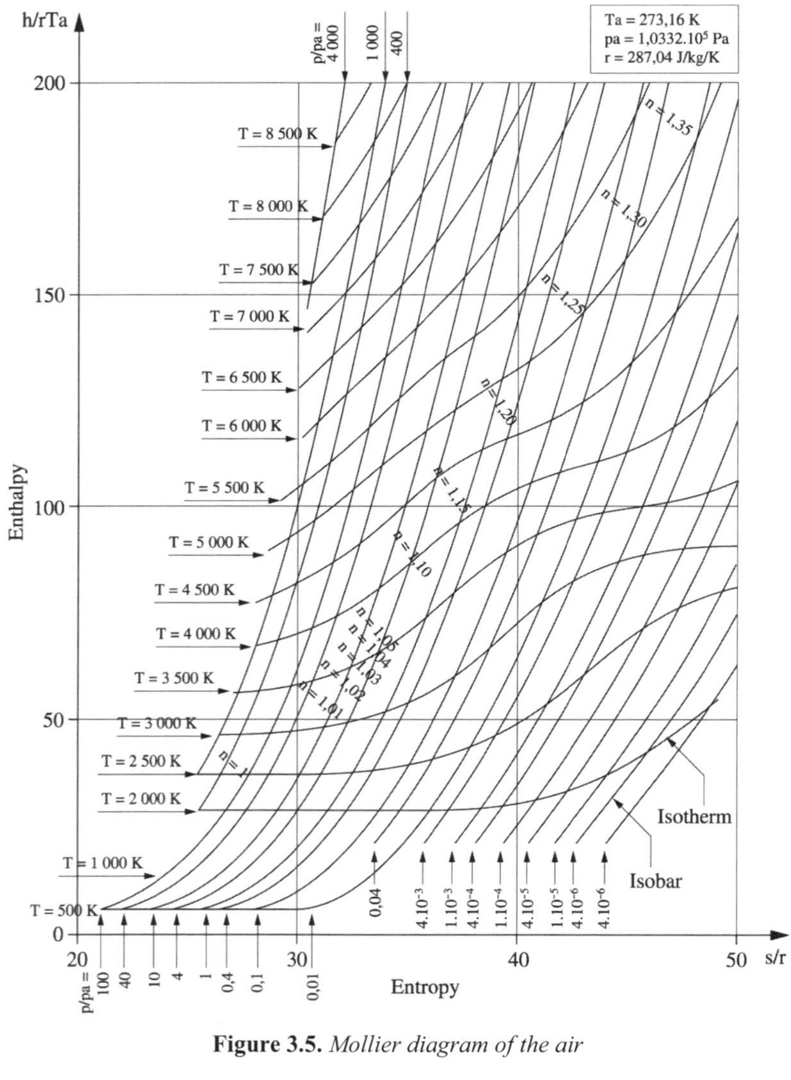

Figure 3.5. *Mollier diagram of the air*

In aerodynamics, the graphic form of the Mollier diagram is hardly used any longer, except in teaching. In computer applications, the codes use a digitized form where different curves are represented by analytical expressions. Sometimes the composition of the mixture is calculated as a function of (p,T), then the thermodynamic functions are determined by theoretical relations (partition functions). The input of the program is sometimes different from the enthalpy and entropy; for example, this can be the internal energy or the pressure. It is no longer a Mollier diagram *per se* but a tool that gives the thermodynamic properties of a gas knowing two other properties, which ultimately is the same.

3.7.5. *Differential parameters of a dissociated gas in equilibrium*

The following are the most important thermodynamic parameters:

$$C_p = \left(\frac{\partial h}{\partial T}\right)_p, \qquad l = \left(\frac{\partial h}{\partial p}\right)_T$$

$$\bar{\gamma} = \left(\frac{\partial \ln p}{\partial \ln \rho}\right)_s, \qquad \chi = \left(\frac{\partial \ln p}{\partial \ln \rho}\right)_h$$

$$D_p = \left(\frac{\partial \ln Z}{\partial \ln p}\right)_T, \qquad D_T = \left(\frac{\partial \ln Z}{\partial \ln T}\right)_p.$$

We consider, for example, a diatomic gas, investigated in section 3.7.3.2, represented by the simplified equation $\left[\varepsilon(T)=0\right]$:

$$\frac{4\alpha^2}{1-\alpha^2} = Cst\frac{\exp\left(-\frac{\theta_D}{T}\right)}{p}. \qquad [3.11]$$

In the present case $Z = 1+\alpha$ The partial differentiation of the above relations with respect to $\ln T$ and $\ln p$ gives:

$$D_T = \frac{\alpha(1-\alpha)}{2}\frac{\theta_D}{p}$$

$$D_p = -\frac{\alpha(1-\alpha)}{2}.$$

We then find:

$$h(T,p)=\alpha h_A+(1-\alpha)h_{A_2}$$

where h_A and h_{A_2} are the specific enthalpies of the species A and A_2 in the state (T,p); thus we get:

$$C_p=\alpha C_{p_A}+(1-\alpha)C_{p_{A_2}}+\left[h_A-h_{A_2}\right]\frac{\partial\alpha}{\partial T}$$

where C_{p_i} represents the specific heat of each species.

We find from [3.11] that:

$$\frac{\partial\alpha}{\partial T}=\frac{\alpha\left(1-\alpha^2\right)}{2}\frac{\theta_D}{T^2}$$

and: $h_A-h_{A_2}=\Re\theta_D$ according to the definition of θ_D (see equation [3.7]).

From which follows:

$$C_p=\alpha C_{p_A}+(1-\alpha)C_{p_{A_2}}+\frac{\alpha\left(1-\alpha^2\right)}{2}\Re\left(\frac{\theta_D}{T}\right)^2.$$

According to the equation of state: $p=Z\,r\rho T$

$$\frac{dp}{p}=\frac{dZ}{Z}+\frac{d\rho}{\rho}+\frac{dT}{T}$$

or:

$$\frac{d\rho}{\rho}=\left(1-D_p\right)\frac{dp}{p}-\left(1+D_T\right)\frac{dT}{T}. \qquad [3.12]$$

On the other hand, as $dh=Tds+\dfrac{dp}{\rho}$, by taking into account the equation of state, it follows:

$$ds = \frac{1}{T}\left(\frac{\partial h}{\partial p}dp + \frac{\partial h}{\partial T}dT\right) - \frac{dp}{\rho T} = \frac{1}{T}\left(\frac{\partial h}{\partial p} - \frac{1}{\rho}\right)dp + \frac{1}{T}\frac{\partial h}{\partial T}dT =$$
$$\left(\frac{1}{T}\frac{\partial h}{\partial p} - \frac{Z\,r}{p}\right)dp + \frac{1}{T}\frac{\partial h}{\partial T}dT.$$

Then, expressing ds as an exact total differential, we get:

$$\frac{\partial}{\partial T}\left(\frac{1}{T}\frac{\partial h}{\partial p} - \frac{Z\,r}{p}\right) = \frac{\partial}{\partial p}\left(\frac{1}{T}\frac{\partial h}{\partial T}\right)$$

from which:

$$\frac{1}{T}\frac{\partial h}{\partial p} = -\frac{D_T}{\rho T}\,.$$

On the other hand, $\frac{\partial h}{\partial T} = C_p$, leading to:

$$ds = C_p\frac{dT}{T} - \frac{1 + D_T}{\rho T}dp \pm$$
$$dh = C_p dT - \frac{D_T}{\rho}dp.$$

Making $ds = 0$ and taking into account [3.12] to remove $\frac{dT}{T}$, we obtain after simple calculations:

$$\boxed{\overline{\gamma} = \left(\frac{\partial \ln p}{\partial \ln \rho}\right)_s = \frac{C_p}{C_p\left(1 - D_p\right) - \frac{p}{\rho T}\left(1 + D_T\right)^2}.}$$

Then making $dh = 0$:

$$\chi = \frac{C_p}{C_p\left(1 - D_p\right) - \frac{p}{\rho T}D_T\left(1 + D_T\right)}.$$

The speed of sound is defined by:

$$a^2 = \left(\frac{\partial p}{\partial \rho}\right)_s$$

and according to the definition of $\bar{\gamma}$ we find that:

$$a^2 = \bar{\gamma}\frac{p}{\rho}.$$

Thus, in general, $\bar{\gamma}$ is a function of (p, ρ). Only in the case of a perfect gas does $\bar{\gamma}$ equal $\dfrac{C_p}{C_v}$, thus it is either constant, or a function of the temperature (see section 3.7.2).

Chapter 4

Fundamental Equations of Fluid Mechanics

4.1. Continuous regime and molecular regime

A gas such as air consists of molecules moving in a random way. At the atomic level, the gas is a discontinuous medium. The mean distance separating the molecules is characterized by the (molecular) mean free path λ, which in general we know to calculate from thermodynamic quantities. In the case where a body with the characteristic dimension L (diameter of the cylinder, chord of the airfoil, length of the vehicle) is placed in a flow where the mean free path is λ, the following two extreme situations can be met.

– The mean free path is very small compared to the size of the body $\lambda/L << 1$: then the molecules are so close to each other that their action on the body is seen as a continuous action, the impact of each molecule cannot be distinguished. The fluid is considered as a continuous medium of which local properties can be defined by going to the limit $\delta\upsilon \to 0$ of a small volume $\delta\upsilon$. If the volume $\delta\upsilon$ contains a mass δm:

$$\frac{\delta m}{\delta\vartheta} \to \rho$$

when $\delta\upsilon \to 0$, which defines the density ρ. The fluid motion is amenable to the Navier–Stokes equations, which are the theoretical basis of conventional aerodynamics.

– The mean free path is of the same order as the size of the body $\lambda/L = O(1)$ then every impact of a molecule is felt isolated by the body, therefore, the definition of local quantities from a passage to the limit $\delta\upsilon \to 0$ no longer makes sense. We need to find another way of modeling the flow, for example, by simulating fluid particles colliding with each other and with the walls. The motions of particles are then calculated by basic mechanics equations. As it is not possible to take into account all molecules of a gas, a statistical approach is used, known as the Monte Carlo method (here we refer to *DSMC*: *direct simulation by Monte Carlo*).

The aforementioned distinction is quantified by introducing a dimensionless number called the Knudsen number which is defined by:

$$Kn = \frac{\lambda}{L}.$$

If $Kn << 1$, the flow is said to take place as a *continuous flow*. This case corresponds to the vast majority of applications encountered in fluid mechanics.

If $Kn = O(1)$, the flow is said to take place as a *molecular flow*. Such a flow is found at very high altitudes (greater than 100 km) during the re-entry of a hypersonic vehicle, for example, the Space Shuttle. In the intermediate situation, where the mean free path is not very small compared to *L*, while much lower than *L*, we talk about a *transitional regime*. The flow is then the seat of rarefaction effects. If they are small, the flow can be calculated with the methods applicable in continuous flow. It then is necessary to modify the boundary conditions applied to the wall by tolerating a certain slip in velocity (the no slip condition $\vec{V} = 0$ relative to the wall no longer applies, see section 4.9).

The free mean path is given by the formula:

$$\lambda = \frac{KT}{2\pi p D^2}$$

where K is the Boltzmann constant ($K = \Re/N$ where $\Re$ is the perfect gas constant and N is the Avogadro number), p is the pressure, and D the diameter of the molecule (in the considered gas model). As a guide, the following table gives the mean free path (in meters) versus altitude for a standard atmosphere.

Z (km)	λ (m)
20	10^{-6}
70	10^{-3}
110	1
150	10

Table 4.1. *Mean free path versus altitude for a standard atmosphere*

At “aeronautical” altitudes, i.e. roughly less than 20 km, the mean free path is less than 1 μm, i.e. much smaller than the size of the vehicle, which is of the order of several meters. The Navier–Stokes equations can be safely used. At about 70 km, λ becomes of the order of mm, which is, compared to the size of the re-entry body, not very small but is still small enough to consider the air as a continuous medium, except perhaps in some regions where the flow strongly expands. At these altitudes, the DSMC approach is applicable, allowing useful crosschecks with the resolution of the Navier–Stokes equations. Beyond 100 km, the mean free path becomes comparable to the length of the vehicle, and the assumption of the air as continuous medium becomes false.

According to the kinetic theory of gases, molecular viscosity is given by the relationship:

$$\mu = \rho\lambda\sqrt{\frac{2}{\gamma\pi}}\,a$$

where a is the velocity of sound and γ is the ratio of specific heats. The Reynolds number, calculated using the length L, is expressed as (see section 6.1 for the definition of the Reynolds number):

$$\mathrm{Re} = \frac{\rho VL}{\mu} = \frac{\rho VL}{\rho\lambda a}\sqrt{\frac{\gamma\pi}{2}} = \sqrt{\frac{\gamma\pi}{2}}\,\frac{VL}{a\lambda}.$$

For a given gas, the square root is a constant with the value 1.48 for the air, hence, the Knudsen number expression:

$$Kn = 1.48\frac{M}{\mathrm{Re}}$$

where M is the Mach number. The Knudsen number is of the form:

$$Kn \approx \frac{M}{\mathrm{Re}}.$$

4.2. The continuous regime and Navier–Stokes equations

In very general terms, the motion of a fluid, which can be regarded as a continuous medium, is governed by Navier–Stokes equations, which can be found in treatises on fluid mechanics. They will be presented here without going into greater detail. These equations are obtained as follows.

1. First, it uses basic physical laws, e.g.:

– the conservation of mass, according to which there is neither creation nor destruction of mass;

– *Newton's law*, or the fundamental law of mechanics, which is expressed by:

$$force = mass \times acceleration$$

– the conservation of energy, according to which energy can only pass from one form into another, a condition known as the first law of thermodynamics (see Chapter 3).

2. A model is adopted to describe the properties of the fluid, which means that the fluid obeys certain constitutive laws, allowing us to: evaluate its deformation under the influence of internal forces, or *tensions*, determine how the heat is propagating and to express its thermodynamic variables.

There are different levels of description for the behavior of a fluid in motion.

1. First, we can consider a *macroscopic level* by considering a volume (ϑ) containing a finite mass of fluid m. At this level, this is possible to adopt two perspectives:

– either the volume (ϑ) is fixed in space and time or it is permanently traversed by the fluid, the mass m of the fluid varying with time (except in the case of a stationary flow);

– or the volume (ϑ) follows the fluid in its motion in such a way that it contains always the same particles (atoms, molecules).

2. It is also possible to consider a small elementary volume $d\vartheta$, in the sense of an infinitesimal calculation, but large enough to contain a very large number of particles, if the fluid in $d\vartheta$ is treated as a continuous medium. Also, in this case the volume $d\vartheta$ can either be fixed or swept along by the fluid.

3. Finally, at the *microscopic level* we consider the fluid as consisting of particles, the physical laws being applied to atoms and molecules. This approach is excellent at the fundamental level, because it enables a subtle description of the constitutive laws of the medium. This has to be adopted if the fluid can no longer be treated as a continuous medium, which happens, as we have seen, when the average distance between molecules, measured by the mean molecular free path, is comparable to the size of the body. In what follows, we will exclude this case, the fluid being treated as a continuous medium.

4. There is a more fundamental approach to describe the motion of a fluid by describing the fluid with a distribution function $f(r,c,t)$, depending on the position r and velocity c of the molecules, as well as the time t. The evolution of the distribution function is governed by the Boltzmann equation, the macroscopic quantities such that density, velocity, and energy can be defined at any moment by the distribution function. If the fluid is "diluted", that is, if the number of inter-molecular collisions is small and if these collisions are mostly binary (only involving two molecules), then we can find an expression for the distribution function, and it is possible to demonstrate that the gas satisfies Navier–Stokes equations. Joining the two approaches is a way to establish the Navier–Stokes equations from a description of the macroscopic behavior of the fluid.

Customarily, we call the three previously mentioned equations as Navier–Stokes equations. In reality, only the equation of the motion made up by the laws of behavior expressing internal tensions, should be attributed to Navier and then Stokes (*Law of the Equilibrium and Motion of Solid Elastic Bodies*, 1821 by Henri Navier). The other two equations are the universal application of the law of mass conservation and the first law of thermodynamics. This observation does not detract from the merits of Henri Navier (1785–1836) and Georges Gabriel Stokes (1819–1903).

4.3. Law of mass conservation and continuity equation

We consider a control volume (ϑ), *finite* and *fixed* in time and space (see Figure 4.1). In general, the mass of fluid contained in (ϑ) will vary, as opposed to cases where we consider a volume containing a fixed mass that moves with the fluid. The law of mass conservation states that there is no creation or loss of matter, we should express that any change of mass contained in (ϑ) is compensated for by a contribution or removal of fluid mass across the surface (S) delimiting (ϑ).

If dS is an element of the surface, the mass flow through dS is given by:

$$\rho \vec{V} \cdot \vec{n} \, dS$$

where according to the common convention, ρ is the density of the fluid, $\vec{V}$ is the velocity vector on (S), and $\vec{n}$ is the normal unit vector of the element dS, directed to the exterior of (ϑ). Knowing that ρ is a mass per unit of volume (dimension $[M][L]^{-3}$), the mass flow is a mass per unit of time (dimension $[M][T]^{-1}$).

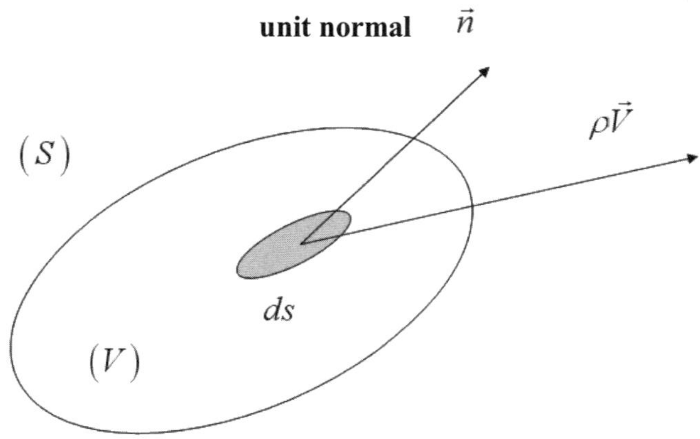

Figure 4.1. *Control volume for the mass conservation*

As $\vec{n}$ is directed outward, the expression $\rho\vec{V}.\vec{n}\,dS$ denotes a mass flow going out of the control volume if the velocity $\vec{V}$ is also directed from the interior to the exterior of (S); conversely, if $\vec{V}$ is directed to the interior of (ϑ), the mass flow is ingoing. The total mass flow crossing the surface (S) is given by:

$$\iint_{(S)} \rho\,\vec{V}.\vec{n}\,dS\,.$$

Moreover, the mass contained in (ϑ) is:

$$\iiint_{(\vartheta)} \rho\,d\vartheta$$

and its variation per unit of time is expressed by:

$$\frac{\partial}{\partial t}\iiint_{(\vartheta)} \rho\,d\vartheta\,.$$

Remembering that a decrease in mass contained in (ϑ) corresponds to a total mass flow going out through (S), expressed by a positive number, we get:

$$\frac{\partial}{\partial t}\iiint_{(\vartheta)} \rho\,d\vartheta = -\iint_{(S)} \rho\,\vec{V}.\vec{n}\,dS\,.$$

From which the *continuity equation*, which is applied to the finite volume (ϑ) follows:

$$\boxed{\frac{\partial}{\partial t}\iiint_{(\vartheta)} \rho \, d\vartheta + \iint_{(S)} \rho \, \vec{V}.\vec{n} \, dS = 0.}$$

This integral form is useful in certain applications because its use does not imply knowledge of local properties of the fluid inside the control volume. However, most of the time a local form of the continuity equation is used. Because the volume (ϑ) is fixed in time:

$$\frac{\partial}{\partial t}\iiint_{(\vartheta)} \rho \, d\vartheta = \iiint_{(\vartheta)} \frac{\partial \rho}{\partial t} d\vartheta .$$

Furthermore, the divergence theorem allows us to write:

$$\iint_{(S)} \rho \, \vec{V}.\vec{n} \, dS = \iiint_{(\vartheta)} \nabla.\left(\rho \vec{V}\right) d\vartheta$$

where ∇ designated the operator nabla: $\nabla.\vec{f} = div\left(\vec{f}\right)$. It is therefore:

$$\iiint_{(\vartheta)} \left[\frac{\partial \rho}{\partial t} + \nabla.\left(\rho \vec{V}\right)\right] d\vartheta = 0 .$$

Since this relationship must be satisfied *irrespective of the volume* (ϑ) – no assumptions were made in the choice of this volume – it follows that:

$$\boxed{\frac{\partial \rho}{\partial t} + \nabla.\left(\rho \vec{V}\right) = 0}$$

which constitutes the local continuity equation, written at one point of the fluid.

Using tensor notation and considering an *orthonormal system of coordinates*, the continuity equation is written as:

$$\frac{\partial \rho}{\partial t} + \frac{\partial}{\partial x_i}\left(\rho u_i\right) = 0 .$$

The summation convention relatively to the repeated indices is applied, which means that:

$$\frac{\partial}{\partial x_i}(\rho u_i) = \frac{\partial}{\partial x_1}(\rho u_1) + \frac{\partial}{\partial x_2}(\rho u_2) + \frac{\partial}{\partial x_3}(\rho u_3)$$

for a three-dimensional space with the coordinates (x_1, x_2, x_3) (see the mathematical revision in Appendix 1). In what follows, we will often use this tensor notation, which has the advantage of being compact without concealing the physical meaning of terms involved in equations. It also identifies quite well the forms used to discretize these equations by numerical methods. However, the equations are written in a system of orthonormal axes, which is a restriction in that the equations are usually solved in curvilinear coordinate systems, fitted to the geometry of the body. Then it is necessary to come back to the equations written in intrinsic form (see section 4.8) and express the mathematical operators in the chosen system of axes. In our opinion, the intrinsic form, although elegant, tends to conceal the physical laws and principles underlying the equations.

In many applications, the flow can be regarded as stationary, which means that the properties of a fixed point in space with the coordinates (x, y, and z) do not depend on time t. Thus, follows:

$$\iiint_{(\vartheta)} \nabla.\left(\rho\vec{V}\right) d\vartheta = 0$$

for which follow the local equations:

$$\boxed{\nabla.\left(\rho\vec{V}\right) = 0} \quad \text{or} \quad \boxed{\frac{\partial}{\partial x_i}(\rho u_i) = 0.}$$

In the case of an *incompressible* flow, we again find the well-known relation:

$$\nabla.\vec{V} = 0 \qquad \text{or} \qquad \frac{\partial u_i}{\partial x_i} = 0$$

which shows the velocity field is solenoidal, according to a sometimes-used denomination.

COMMENT.– In some applications involving different phases, such as a gas and a solid or a gas and a liquid, the continuity equation applied to the *gas only* may include a nonzero right-hand side member (source term) expressing the fact that the gas is created or destroyed, for example, as a result of a chemical reaction with the other phase or a vaporization. This occurs in a combustion chamber of and engine or in a rocket engine. It is obvious that the total mass of the system, including the two phases is preserved.

4.4. Equation of motion

The equation of motion, sometimes called momentum equation, is an expression of Newton's law:

$$\vec{F} = m\,\vec{\gamma}$$

where $\vec{F}$ is the force exerted on a body with the mass *m*, undergoing an acceleration $\vec{\gamma}$. In the case of a system with *variable mass*, the law takes the more general form of:

$$\vec{F} = \frac{d}{dt}\left(m\vec{V}\right)$$

where $m\vec{V}$ is the impulse or momentum of the system.

To apply Newton's law to our problem, we must assess on one hand the forces to which the system is submitted and on the other hand the time variation of its momentum. Consider the mass *m* of the fluid contained in the fixed volume (ϑ) (see Figure 4.2).

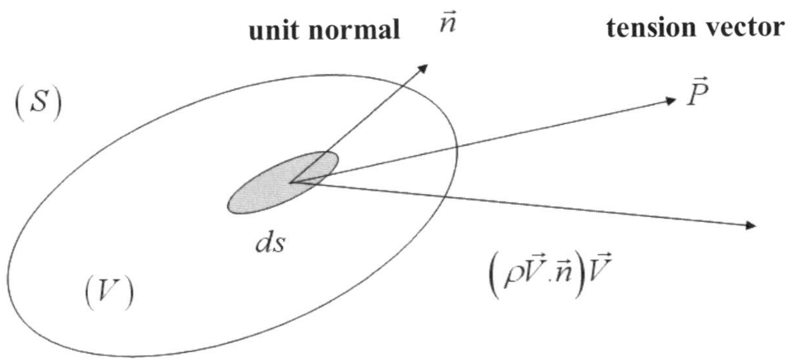

Figure 4.2. *Control volume for the application of the equation of motion*

The forces acting on this system are as follows:

– the *body forces* (or at distance), which are exerted within the fluid itself in the interior of (ϑ), are gravity and electromagnetic forces, in the case of an electric conductor;

– the *contact forces*, exerted on the system by the exterior fluid surrounding the volume (ϑ). These forces act by contact action on the surface (S) delimiting (ϑ).

If $\vec{f}$ denotes a body force exerted on the *unit of mass of fluid*, the first term of forces applied to the system is written as:

$$\iiint_{(\vartheta)} \rho \, \vec{f} \, d\vartheta$$

since $\vec{f}$ acts on all the fluid contained in (ϑ).

At some point on the surface (*S*), the fluid *within* (ϑ) exerts on (*S*) a surface force (per surface unit) called *tension*, represented by the vector $\vec{P}$ (see section 2.5.1). The overall action of these contact forces is expressed by the surface integral:

$$\iint_{(S)} \vec{P} \, dS \, .$$

Let us evaluate the rate of change of the momentum of the system. This variation has the following two causes:

– on the one hand, the modification of the velocity field within the volume (ϑ), which gives the term:

$$\frac{\partial}{\partial t} \iiint_{(\vartheta)} \rho \, \vec{V} \, d\vartheta$$

– on the other hand, a variation in the mass of the fluid contained in (ϑ), resulting from the mass flows entering and leaving across (*S*). The mass flow across a surface element dS is expressed by:

$$\rho \vec{V} . \vec{n} \, dS .$$

It corresponds to a flow momentum:

$$\left(\rho \vec{V}.\vec{n}\right)\vec{V}\, dS$$

so that the total flux of the momentum across (*S*) is:

$$\iint_{(S)} \left(\rho\vec{V}.\vec{n}\right)\vec{V}\, dS\,.$$

Thus, Newton's law leads to the expression:

$$\boxed{\frac{\partial}{\partial t}\iiint_{(\vartheta)} \rho\,\vec{V}\, d\vartheta + \iint_{(S)} \left(\rho\vec{V}.\vec{n}\right)\vec{V}\, dS = -\iint_{(S)} \vec{P}\, dS + \iiint_{(\vartheta)} \rho\,\vec{f}\, d\vartheta}$$

which is a first integral form of the equation of motion. It is necessary to put a minus sign before the integral of contact actions, as $\vec{P}$ is the tension exerted by the fluid contained in (ϑ) on the exterior fluid of (ϑ), while in the application of the fundamental equation, they are forces involved outside of the system and acting on the fluid contained in (ϑ).

Taking into account the equation of continuity, it is possible to write:

$$\boxed{\iiint_{(\vartheta)} \rho\frac{d\vec{V}}{dt} d\vartheta = -\iint_{(S)} \vec{P}\, dS + \iiint_{(\vartheta)} \rho\,\vec{f}\, d\vartheta.}$$

COMMENT.– The transition from one expression to another is easily demonstrated using the tensor notation. Under the divergence theorem, we can write:

$$\frac{\partial}{\partial t}\iiint_{(\vartheta)} \rho u_i\, d\vartheta + \iint_{(S)} u_i\left(\rho u_j n_j\right)dS = \frac{\partial}{\partial t}\iiint_{(\vartheta)} \rho u_i\, d\vartheta + \iiint_{(\vartheta)} \frac{\partial}{\partial x_j}\left(\rho u_i u_j\right) d\vartheta\,.$$

Thus:

$$\frac{\partial}{\partial t}\iiint_{(\vartheta)} \rho u_i\, d\vartheta + \iiint_{(\vartheta)} \frac{\partial}{\partial x_j}\left(\rho u_i u_j\right) d\vartheta = \iiint_{(\vartheta)} \left[\frac{\partial}{\partial t}\left(\rho u_i\right) + \frac{\partial}{\partial x_j}\left(\rho u_i u_j\right)\right] d\vartheta\,.$$

Since the volume (ϑ) is fixed, the term in brackets can be written:

$$u_i\frac{\partial \rho}{\partial t} + \rho\frac{\partial u_i}{\partial t} + \rho u_j\frac{\partial u_i}{\partial x_j} + u_i\frac{\partial\left(\rho u_j\right)}{\partial x_j} = u_i\left[\frac{\partial \rho}{\partial t} + \frac{\partial\left(\rho u_j\right)}{\partial x_j}\right] + \rho\left(\frac{\partial u_i}{\partial t} + u_j\frac{\partial u_i}{\partial x_j}\right).$$

Given the equation of continuity, it remains:

$$\rho\left(\frac{\partial u_i}{\partial t}+u_j\frac{\partial u_i}{\partial x_j}\right)=\rho\frac{du_i}{dt}.$$

Thus, we have:

$$\frac{\partial}{\partial t}\iiint_{(\vartheta)}\rho\vec{V}\,d\vartheta+\iint_{(S)}\left(\rho\vec{V}.\vec{n}\right)\vec{V}\,ds=\iiint_{(\vartheta)}\rho\frac{d\vec{V}}{dt}.$$

In continuous fluid mechanics, we introduce the *stress tensor* $\overline{\overline{T}}$ by which the tension $\vec{P}$ is expressed in the form of:

$$\vec{P}=-\overline{\overline{T}}.\vec{n}$$

where $\vec{n}$ is the normal unit vector of the surface element on which $\vec{P}$ is exerted. Here, the symbol (.) represents the contracted product of the tensor $\overline{\overline{T}}$ and the vector $\vec{n}$, i.e. the result is a vector (see Appendix 1). Thus we have:

$$\iint_{(S)}\vec{P}\,dS=-\iint_{(S)}\overline{\overline{T}}.\vec{n}\,dS$$

or by applying the divergence theorem:

$$\iint_{(S)}\vec{P}\,dS=-\iiint_{(\vartheta)}div\left(\overline{\overline{T}}\right)d\vartheta=-\iiint_{(\vartheta)}\nabla.\overline{\overline{T}}\,d\vartheta.$$

It is recalled that the divergence of a tensor is a vector (see Appendix 1). Finally, the equation of motion reads:

$$\boxed{\iiint_{(\vartheta)}\left[\rho\frac{d\vec{V}}{dt}-\nabla.\overline{\overline{T}}-\rho\vec{f}\right]=0.}$$

The above equation has to be true regardless of the volume (ϑ), from which the local equation follows:

$$\boxed{\rho\frac{d\vec{V}}{dt}=\nabla.\overline{\overline{T}}+\rho\vec{f}.}$$

The equation of motion is a vectorial equation that leads to three scalar equations projected on the coordinate axes (x, y, and z) of the three-dimensional space.

According to the intended use, it is possible to give different forms for this equation. For example, from the expression of the total (material) derivative with respect to time of the velocity vector (see Appendix 1) we can write:

$$\frac{d\vec{V}}{dt} = \frac{\partial \vec{V}}{\partial t} + \frac{1}{2}\overrightarrow{grad}\left(V^2\right) + \overrightarrow{curl}\left(\vec{V}\right) \times \vec{V}$$

where ($\times$) denotes the vector product. We can write:

$$\rho\left[\frac{\partial \vec{V}}{\partial t} + \frac{1}{2}\overrightarrow{grad}\left(V^2\right) + \overrightarrow{curl}\left(\vec{V}\right) \times \vec{V}\right] = \nabla.\overline{\overline{T}} + \rho\,\vec{f}.$$

If the tensorial notation is used:

$$\rho\frac{du_i}{dt} = \frac{\partial T_{ij}}{\partial x_j} + \rho f_i$$

or, by taking into account the continuity equation:

$$\frac{\partial\left(\rho u_i\right)}{\partial t} + \frac{\partial}{\partial x_j}\left(\rho u_i u_j\right) = \frac{\partial T_{ij}}{\partial x_j} + \rho f_i.$$

COMMENT.– The repetition of the index in the term $\frac{\partial T_{ij}}{\partial x_j}$ denotes a divergence lowering of the tensor $\frac{\partial T_{ij}}{\partial x_j}$ at the rank of a vector. A double repetition lowers the rank of the tensor to a scalar (divergence of a vector).

4.5. Expression of the stress tensor

At this stage, and by assuming that we know how to calculate the body forces $\vec{f}$, the system of equations resulting from the application of fundamental laws of mechanics is *open*: it can only be solved if we have a constitutive law to express the stress tensor:

$$\overline{\overline{T}} = T_{ij} .$$

To find an expression for this tensor belonging to the fundamental fluid mechanics, we shall recall here only the results relevant to aerodynamics.

The tensor T is *symmetric*:

$$T_{ij} = T_{ji} .$$

If we put into evidence the hydrostatic pressure p, T_{ij} can be written in the form:

$$T_{ij} = -p\,\delta_{ij} + \tau_{ij}$$

with $\delta_{ij} = 0$ if $i \neq j$ and $\delta_{ij} = 1$ if $i = j$ (*Kronecker* symbol).

The tensor τ_{ij} is called the *viscosity tensor* (or viscous stress tensor). It is also symmetric. The term $p\delta_{ij}$ is a spherical tensor of which all the terms are zero, except those of the main diagonal, which are equal to p (see Appendix 1).

All fluids we will consider here have a *Newtonian* behavior. This means that there is a linear dependence between the viscosity tensor and the *tensor of deformation rate* or *strain tensor*.

$$\frac{\partial u_j}{\partial x_k}$$

being

$$\tau_{ij} = \in_{ijkl} \frac{\partial u_k}{\partial x_l}$$

where $\in_{ijkl}$ is a tensor of the fourth order, characteristic for the considered fluid. If the medium does not allow favored directions (called isotropic medium), it shows that the tensor $\in_{ijkl}$ is reduced to two coefficients, λ and μ, called *Lamé coefficients*. Thus, we can write:

$$\tau_{ij} = \lambda \frac{\partial u_k}{\partial x_k} \delta_{ij} + \mu \left(\frac{\partial u_i}{\partial x_j} + \frac{\partial u_j}{\partial x_i} \right).$$

This relation is often put into the form of:

$$\tau_{ij} = \left(\lambda + \frac{2}{3} \mu \right) \delta_{ij} \frac{\partial u_k}{\partial x_k} + \mu \left[\left(\frac{\partial u_i}{\partial x_j} + \frac{\partial u_j}{\partial x_i} \right) - \frac{2}{3} \delta_{ij} \frac{\partial u_k}{\partial x_k} \right].$$

The first term of this expression is called *bulk viscosity* or *volume viscosity*. Its coefficient is zero for monatomic gases. The bulk viscosity is equally zero for incompressible fluids, since $\frac{\partial u_k}{\partial x_k} \equiv div \vec{V} = 0$. The Stokes hypothesis allows that this is a general property, thus, we state: $3\lambda + 2\mu = 0$. For common fluids, and thus, for air, the viscosity tensor is given by:

$$\tau_{ij} = \mu \left[\left(\frac{\partial u_i}{\partial x_j} + \frac{\partial u_j}{\partial x_i} \right) - \frac{2}{3} \delta_{ij} \frac{\partial u_k}{\partial x_k} \right].$$

Determination of the stress tensor requires knowledge of the *molecular viscosity* μ (also called laminar viscosity, or short, viscosity). This is a property of a fluid, which is defined at the molecular and atomic level. Its determination is based either on theories on the constitution of the microscopic medium (kinetic theory for gases) or on measurements. In the case of air in a wide range of temperatures (from 100 K to 3,000 K), μ can be calculated by using the *Sutherland formula*

$$\mu = 1.454\ 10^{-6} \frac{T^{3/2}}{T + 110.4}.$$

The absolute temperature T is expressed in Kelvin, μ (dimension $N \times s / m^2$) is given in MKS (SI system).

Figure 4.3 shows, according to the formula of Sutherland, the evolution of the viscosity of air with temperature. We observe that for air (and gases in general) μ increases with the temperature, which is not true for liquids whose viscosity decreases with increasing temperature (starting a cold car in winter is sometimes difficult!). Cooling air down lowers the viscosity, which is used in cryogenic wind tunnels to simulate high Reynolds numbers (see section 6.4).

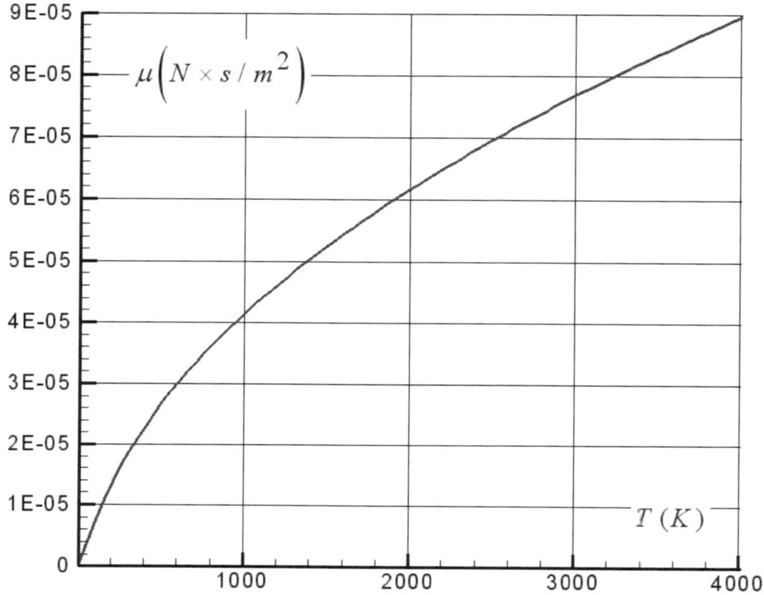

Figure 4.3. *Sutherland law for the viscosity of air*

4.6. Energy equation

In an incompressible flow (or considered as such), the only unknowns are the pressure p and velocity vector $\vec{V}$. We are dealing with four unknowns (p and the three components of the velocity vector $\vec{V}$ for which we have four equations: the continuity equation and the three scalar equations resulting from the motion equation (plus a constitutive law expressing the stress tensor). The system is "closed", the flow can be determined through appropriately defined initial and boundary conditions .

In a compressible flow, the density ρ is one of the unknowns. To calculate it, another equation, the energy equation, should be considered. This should be made up by the additional constitutive laws: the fluid equation of state and the law of thermal conduction.

The energy equation follows from applying the *first law of thermodynamics*, which is stated in section 3.3.1.

Let us recall that what we call the specific *total energy* (that is, for the unit of mass) is the sum:

$$\boxed{e_T = e + \frac{V^2}{2}}$$

of the internal and the kinetic energies of the moving fluid. In the case of a heavy fluid, or a mass of gas of great thickness (atmosphere), the specific potential energy should be included: $g \times z$, where g is the acceleration of gravity and z the altitude. This energy is neglected in classic aerodynamics applications.

In common use, the specific adjective is often left out, that is, per unit of mass, by calling e the fluid energy. This is a source of error in the application of the first law, which considers "true" energies for a system with the mass m. Thus, the internal energy of the system is $E_i = me$, and its kinetic energy is $E_c = 1/2\, mV^2$. To avoid confusion, the specific energy, enthalpy, and entropy of the fluid should be designated with lower case letters: e, h and s.

An elementary volume $d\vartheta$ of the system contains a total energy:

$$\rho\left(e + \frac{V^2}{2}\right) d\vartheta$$

and the finite volume (ϑ) the total energy:

$$E_T = \iiint_{(\vartheta)} \rho\left(e + \frac{V^2}{2}\right) d\vartheta .$$

The first law of thermodynamics states that the total variation of the energy of the system E_T with respect to time $\frac{dE_T}{dt}$ results from the energy exchanged per unit of time (or powers) from outside (see Chapter 3).

The total energy E_T will vary due to internal thermodynamic processes and changes in the velocity of the fluid contained within (ϑ), hence the term:

$$\frac{\partial}{\partial t}\left[\iiint_{(\vartheta)} \rho\left(e + \frac{V^2}{2}\right) d\vartheta\right] = \iiint_{(\vartheta)} \frac{\partial}{\partial t}\left[\rho\left(e + \frac{V^2}{2}\right)\right] d\vartheta .$$

We recall that because the volume (ϑ) is fixed, it is possible to pass the differentiation operator with respect to time within the integral. However, since (ϑ) is fixed, the surface (S) limiting this is crossed by a mass flow carrying a certain amount of energy. If dS is an element of (S), the energy flux across dS is given by:

$$\left(e+\frac{V^2}{2}\right)\rho\,\vec{V}.\vec{n}\,dS\,.$$

The flux represents here an energy per unit of time and unit of surface, that is, a power per unit of surface, which is expressed in W/m^2. For the surface (S), this is:

$$\iint_{(S)}\left(e+\frac{V^2}{2}\right)\rho\,\vec{V}.\vec{n}\,dS\,.$$

Thus, the total derivative with respect to time of the energy E_T contained in the volume (ϑ) is:

$$\frac{dE_T}{dt}=\iiint_{(\vartheta)}\frac{\partial}{\partial t}\left[\rho\left(e+\frac{V^2}{2}\right)\right]d\vartheta+\iint_{(S)}\rho\left(e+\frac{V^2}{2}\right)\vec{V}.\vec{n}\,dS\,.$$

According to the first law of thermodynamics, $\dfrac{dE_T}{dt}$ must be equal to the sum of the powers of the external forces applied to the system and the thermal powers exchanged by the system with the exterior. If the powers associated with the diffusion of species (heat transfer) and electromagnetic radiation can be neglected, the first law is expressed by the relationship (see section 3.3.1):

$$\boxed{\frac{dE_T}{dt}=W'+Q_C'+E_Q'}$$

where W' is the power of the external forces, Q_C' is the thermal power exchanged by thermal conduction, and E_Q' is the power exchanged by radiation.

Power of external forces: First, this is the work of contact forces, or tensions, that are exerted on the surface (*S*). Given the direction of the tension vector $\vec{P}$ the work of the force $\vec{P}$ applied to a surface element in motion with the velocity $\vec{V}$ during a unit of time, is expressed by (power per unit of surface):

$$-\vec{P}.\vec{V}\,.$$

Thus, for the entire surface (S):

$$W_1^{'} = -\iint_{(S)} \vec{P}.\vec{V}\, dS\,.$$

The power of body forces:

$$W_2^{'} = \iiint_{(\vartheta)} \rho\, \vec{f}.\vec{V}\, d\vartheta\,.$$

If $\vec{q}$ denotes the heat flux vector on the surface (*S*), then the heat exchanged by the system through (*S*) is (given the orientation of $\vec{n}$):

$$Q_C^{'} = -\iint_{(S)} \vec{q}.\vec{n}\, dS\,.$$

The heat flux $\vec{q}$ is a power per unit of surface (this is measured in W/m^2), then the product $\vec{q}.\vec{n}\, dS$ is a power: the homogenity is well ensured.

Ultimately, the energy equation is written as:

$$\iiint_{(\vartheta)} \frac{\partial}{\partial t}\left[\rho\left(e+\frac{V^2}{2}\right)\right] d\vartheta + \iint_{(S)} \rho\left(e+\frac{V^2}{2}\right)\vec{V}.\vec{n}\, dS =$$
$$-\iint_{(S)} \vec{P}.\vec{V}\, dS + \iiint_{(\vartheta)} \rho\, \vec{f}.\vec{V}\, d\vartheta - \iint_{(S)} \vec{q}.\vec{n}\, dS + E_Q^{'}.$$

This equation implies powers, this is a scalar equation.

The radiation flux, represented by $E_Q^{'}$, only is important in special situations: in the case of a re-entry body at hypersonic speed, where the extremely heated air around the vehicle radiates or highly heated gas propellants exiting the nozzle of a missile or rocket. In what follows, we will only retain the conduction flux.

To go further in explaining the various terms of the energy equation, it is advantageous to introduce the tensor notation. Thus, we will write:

$$\iiint_{(\vartheta)} \frac{\partial}{\partial t}\left[\rho\left(e+\frac{V^2}{2}\right)\right] d\vartheta = \iint_{(S)}\left[-\rho\, u_i\left(e+\frac{V^2}{2}\right)+u_j T_{ji}-\dot{q}_i\right] n_i\, dS + \iiint_{(\vartheta)} \rho u_i f_i\, d\vartheta .$$

By applying the divergence theorem to the surface integral, it becomes:

$$\iiint_{(\vartheta)} \frac{\partial}{\partial t}\left[\rho\left(e+\frac{V^2}{2}\right)\right] d\vartheta = \iiint_{(\vartheta)}\left\{-\frac{\partial}{\partial x_i}\left[\rho u_i\left(e+\frac{V^2}{2}\right)-u_j T_{ji}+\dot{q}_i\right]+\rho f_i u_i\right\} d\vartheta .$$

The volume (ϑ) being arbitrary, we deduce the local equation:

$$\boxed{\frac{\partial}{\partial t}\left[\rho\left(e+\frac{V^2}{2}\right)\right]+\frac{\partial}{\partial x_i}\left[\rho u_i\left(e+\frac{V^2}{2}\right)\right]=\rho\, f_i u_i+\frac{\partial}{\partial x_i}\left(u_j T_{ji}-q_i\right).}$$

COMMENT.– The term $u_j T_{ji}$ is the contracted product of the vector u_j by the tensor T_{ji}, the repetition index indicating that the result is a vector. The application of the operator $\frac{\partial}{\partial x_i}$ leads to a new repetition index, thus the end result is a scalar in accordance to the other terms of the equation.

Taking into account the above given expression for the stress tensor, we obtain:

$$\frac{\partial}{\partial t}\left[\rho\left(e+\frac{V^2}{2}\right)\right]+\frac{\partial}{\partial x_i}\left[\rho\, u_i\left(e+\frac{V^2}{2}\right)\right]=\rho\, f_i u_i-\frac{\partial\left(p u_i\right)}{\partial x_i}+\frac{\partial}{\partial x_i}\left(u_j \tau_{ji}-\dot{q}_i\right).$$

In many applications, it is convenient to introduce the *specific total enthalpy,* also called *specific stagnation enthalpy* since it corresponds to the enthalpy of the fluid if decelerated to zero velocity in an adiabatic process (see section 9.5.2.1):

$$\boxed{h_T \equiv h_{st} = e+\frac{V^2}{2}+\frac{p}{\rho}.}$$

Therefore, we will have:

$$\rho\, h_{st} = \rho \left(e + \frac{V^2}{2} \right) + p\,.$$

Simple calculations lead to the following form for the energy equation:

$$\frac{\partial}{\partial t}\left(\rho\, h_{st}\right) + \frac{\partial}{\partial x_i}\left(\rho\, u_i\, h_{st}\right) = \frac{\partial p}{\partial t} + \frac{\partial}{\partial x_i}\left(u_j\, \tau_{ji} - \dot{q}_i\right) + \rho\, f_i\, u_i\,.$$

If we develop the first member, by taking into account the continuity equation, it becomes:

$$\begin{aligned}\frac{\partial}{\partial t}\left(\rho\, h_{st}\right) + \frac{\partial}{\partial x_i}\left(\rho\, u_i\, h_{st}\right) &= h_{st}\left[\frac{\partial \rho}{\partial t} + \frac{\partial}{\partial x_i}\left(\rho\, u_i\right)\right] + \rho\left[\frac{\partial h_{st}}{\partial t} + \frac{\partial}{\partial x_i}\left(u_i\, h_T\right)\right] \\ &= \rho\left[\frac{\partial h_{st}}{\partial t} + \frac{\partial}{\partial x_i}\left(u_i\, h_{st}\right)\right] = \rho \frac{dh_{st}}{dt}.\end{aligned}$$

Finally, the energy equation can also be written as:

$$\boxed{\frac{dh_{st}}{dt} = \frac{1}{\rho}\frac{\partial p}{\partial t} + \frac{1}{\rho}\frac{\partial}{\partial x_i}\left(u_j\, \tau_{ji} - \dot{q}_i\right) + f_i\, u_i.}$$

This equation highlights the factors that cause the total enthalpy of a fluid to vary, namely:

– the unsteady pressure variations,

– the viscosity terms,

– the thermal transfers,

– the body forces.

In situations, found frequently in practice, where such actions are non-existent (or of very low amplitude), the stagnation enthalpy of the fluid is conserved.

COMMENT.– If the system is traversed by a transmission shaft (the case of engines and rotating machinery), then the power (algebraically) transmitted by the shaft should be taken into account the energy balance.

4.7. Entropy equation and expression of the heat flux

We can give to the energy equation a very instructive form by introducing the *specific entropy* s. We place ourselves into the rather general case of a flow with chemical kinetics, where species k are present with the concentrations ξ_k (or molar masses). The rate of reaction for the species k can be measured by:

$$\alpha_k \frac{d\xi_k}{dt}$$

where α_i is a function given by the thermochemistry, which we do not need to explain here. Thermodynamics allows us to establish the relationship (see section 3.4):

$$\frac{dh_{st}}{dt} = T\frac{ds}{dt} + \frac{1}{\rho}\frac{dp}{dt} + \vec{V}.\frac{d\vec{V}}{dt} + \alpha_i \frac{d\xi_i}{dt}.$$

According to the equation of motion, we have:

$$\rho\frac{d\vec{V}}{dt} = \nabla.\overline{\overline{T}} + \rho\,\vec{f}$$

or, in tensor notation:

$$\rho\frac{du_i}{dt} = \frac{\partial T_{ij}}{\partial x_j} + \rho\, f_i\,.$$

As:

$$\rho\frac{du_i}{dt} = \frac{\partial T_{ij}}{\partial x_j} + \rho\, f_i$$

or by expressing the strain tensor:

$$\vec{V}.\frac{d\vec{V}}{dt} \equiv u_i\frac{du_i}{dt} = \frac{u_i}{\rho}\frac{\partial \tau_{ij}}{\partial x_j} + f_i\, u_i\,.$$

Thus, the relation:

$$\vec{V}.\frac{d\vec{V}}{dt} = -\frac{u_i}{\rho}\frac{\partial p}{\partial x_i} + \frac{u_i}{\rho}\frac{\partial \tau_{ij}}{\partial x_j} + f_i\, u_i$$

leads to:

$$T\frac{ds}{dt} = \frac{1}{\rho}\frac{\partial p}{\partial t} + \frac{1}{\rho}\frac{\partial}{\partial x_i}\left(u_j\tau_{ij} - \dot{q}_i\right) + f_i u_i - \frac{1}{\rho}\frac{dp}{dt} + \frac{u_i}{\rho}\frac{\partial p}{\partial x_i} - \frac{u_i}{\rho}\frac{\partial \tau_{ji}}{\partial x_j} - f_i u_i - \alpha_k\frac{d\xi_k}{dt}$$

from where:

$$T\frac{ds}{dt} = \frac{1}{\rho}\left[\frac{\partial}{\partial x_i}\left(u_j\tau_{ij}\right) - u_i\frac{\partial \tau_{ij}}{\partial x_j} - \frac{\partial \dot{q}_i}{\partial x_i}\right] - \alpha_k\frac{d\xi_k}{dt}$$

and finally:

$$\boxed{T\frac{ds}{dt} = \frac{1}{\rho}\left(\tau_{ji}\frac{\partial u_j}{\partial x_i} - \frac{\partial \dot{q}_i}{\partial x_i}\right) - \alpha_k\frac{d\xi_k}{dt}.}$$

This equation highlights the *factors of entropy production* along the streamlines of a flow, these being:

– the viscosity ($\tau_{ji}\dfrac{\partial u_j}{\partial x_i}$),

– the heat transfers (involving $\dot{q}_i$),

– the chemical non-equilibrium phenomena (involving $\dfrac{d\xi_k}{dt}$).

The coefficients μ and κ (thermal conductivity, see in the following) are positive to satisfy the second law of thermodynamics, it is shown that $\tau_{ji}\dfrac{\partial u_j}{\partial x_i}$ is also positive.

Therefore, the viscosity always contributes to generate entropy. In contrast, the heat transfer term can be negative, the entropy decreasing locally. However the change in entropy between an upstream and a downstream equilibrium state must be positive or zero (see the shock structure theory, section 13.6).We shall see in section 5.3 that the drag of a body is directly related to the increase in entropy that the presence of the body produces within the flow. Therefore, there is a close link between the above equation and the physical mechanisms behind the drag.

To be completely defined, the energy equation requires, due to thermal conduction, an expression for the *heat flux*. If the medium is assumed to be isotropic, $\vec{\dot{q}}$ is expressed by the Fourier law:

$$\vec{\dot{q}} = -\kappa\, \overrightarrow{grad}\, T \quad \text{or} \quad \dot{q}_i = -\kappa \frac{\partial T}{\partial x_i}$$

where T is the absolute temperature and κ is the coefficient of *thermal conductivity*. Like the viscosity μ, conductivity κ is a physical property of the medium provided, either by theory or experiments.

It is common to express κ by introducing the *Prandtl number*, which is defined by (see section 6.1):

$$\boxed{\Pr = \frac{\mu C_p}{\kappa}}$$

where C_p is the specific heat at constant pressure. For many fluids, excluding extreme temperature conditions, the Prandtl number can be considered constant. In the case of air $\Pr = 0.75$. Thus, knowledge of the viscosity leads to knowledge of the thermal conductivity.

Ultimately, the energy equation, written with the total enthalpy, takes the form:

$$\boxed{\frac{dh_{st}}{dt} = \frac{1}{\rho}\frac{\partial p}{\partial t} + \frac{1}{\rho}\frac{\partial}{\partial x_i}\left(u_j\, \tau_{ji} + \kappa \frac{\partial T}{\partial x_i}\right) + f_i\, u_i.}$$

At this level, the quantities specifying the physical properties of the fluid are the transfer coefficients μ and κ which we assume to be calculable by appropriate formulas and the thermodynamic variables:

– pressure p;

– density ρ;

– temperature T;

– specific internal energy e;

– specific enthalpy h;

– specific entropy s.

The above variables are not independent. They are connected by thermodynamic relations, which can be either analytical formulas or tables or diagrams (see Chapter 3). If we consider the case of a gas in *thermodynamic and chemical equilibrium* (if this gas is the seat of thermochemical processes, this implies that they are in equilibrium), then the gas is said to be *divariant*. Under these conditions, all thermodynamical properties can be expressed by just two of them. At this level, we will say no more about it.

In the vast majority of aerodynamic applications, air can be regarded as a perfect gas obeying the equation of state (see Chapter 3):

$$\frac{p}{\rho} = \frac{\Re}{M} T$$

where:

– $\Re$ is the universal perfect gas constant: $\Re = 8.3145$ J/mole/K;

– M is the molar mass of the considered gas. For air: $M = 0.029$ kg/mole

In the particular case of air, the equation of state is often written in the form of:

$$\frac{p}{\rho} = r\,T \text{ with: } r = \frac{\Re}{M} = 287 \text{ J/kg/K.}$$

4.8. Summary of basic equations

The fundamental equations are written here in an intrinsic form, independent of the coordinate system. This way of writing involves operators whose expression, when they are made explicit, depends on the coordinate system chosen according to the nature of the problem to be solved. Thus, one is led to adopt curvilinear coordinates in order to best represent the solution near the body (wing, fuselage or complete aircraft):

– *continuity equation:*

$$\frac{\partial \rho}{\partial t} + \nabla.\left(\rho \vec{V}\right) = 0;$$

– *equation of motion:*

$$\frac{\partial\left(\rho\vec{V}\right)}{\partial t}+\nabla.\left[\left(\rho\vec{V}\right)\otimes\vec{V}\right]-\nabla.\overline{\overline{T}}=0\,;$$

–*energy equation:*

$$\frac{\partial\left(\rho\, e_T\right)}{\partial t}+\nabla.\left(\rho\, e_T\,\vec{V}\right)+\left(\nabla.\vec{q}-\nabla.\vec{V}.\overline{\overline{T}}\right)=0\,.$$

The symbol $\overline{\overline{T}}$ represents the stress tensor, $\nabla.\overline{\overline{T}}$ is the divergence of this tensor, which gives a vector, the symbol $\otimes$ represents the tensor product and ∇ is the nabla operator (see Appendix 2).

The strain tensor is given using:

$$\overline{\overline{T}}=-\left(p+\frac{2}{3}\mu\,\nabla.\vec{V}\right)\overline{\overline{I}}+\mu\left[\nabla\vec{V}+\left(\nabla\vec{V}\right)^T\right].$$

The symbol $\nabla\vec{V}$ represents the gradient tensor of $\vec{V}$ which has in tensor notation the components $\dfrac{\partial u_i}{\partial u_j}$ (see Appendix 1).

The heat flux is expressed through:

$$\vec{q}=-\frac{\mu\, C_p}{\mathrm{Pr}}\nabla T\,.$$

4.9. Boundary conditions

In most applications for external aerodynamics, we impose that at an infinite distance from the body the fluid is at rest, which means that long-distance perturbations created by the body disappear. If we consider a vehicle moving with a constant velocity, and placing ourselves in a frame of reference linked to the vehicle, we will write that at infinity the fluid has a velocity $\vec{V}_\infty$, its pressure p_∞ and temperature T_∞ having imposed values (these are conditions at the altitude of the flight in question).

In internal aerodynamics, displaying upstream and downstream conditions is more difficult because in these applications we are dealing with flows in "pipes" where the conditions in the entry and exit sections are sometimes difficult to express.

Here, we will say no more on the conditions to be satisfied at a great distance from the body, or at the entry and exit of the region to calculate. This is a delicate issue requiring careful consideration of each discussed case, the conditions to be imposed depending in particular on the subsonics or supersonics nature of the flow. Moreover, in practical applications, the equations of motion are usually solved by numerical methods. Then we have to define a calculation domain of finite dimensions encompassing the body (for obvious reasons, this is not possible to perform a calculation in an infinite domain, which would include an infinite number of mesh points). Consequently, conditions at infinity should be replaced by conditions expressed on the boundary located at a finite distance. These conditions, whose expression is often subtle, must be such that the solution obtained in the finite domain is identical to that which would be obtained for boundaries at the infinity, or at least not too different.

Let us consider the conditions to be satisfied on the body. In the vast majority of applications, in external and internal aerodynamics, the fluid is considered to be a continuous medium amenable to the equations given in the preceding sections. Thus, the physical observation imposes a *condition of adherence* or *no-slip condition* on the surface of a body. This consists to write at any point of the body, that the velocity of the fluid relative to the body is zero. If we place ourselves in a frame of reference linked to the body – which we do most of the time – we write that the velocity vector is zero on the surface of the body:

$$\vec{V}_p = 0\text{, where } u = v = w = 0 \text{ or in tensor mode } \left(u_i\right)_w = 0\,.$$

We must also write a condition for the energy equation, which should rather be expressed by involving the temperature, a quantity easier to measure than the energy or enthalpy. Two cases have to be distinguished:

– *Imposed wall temperature*: $T = T_w$, where T_w is given. This condition corresponds to a wall which is maintained at a constant temperature by internal heating or cooling. This is the case, for example, for the nozzle of the *Vulcain* engine of Ariane 5, which is cooled by circulating liquid hydrogen in order to maintain a wall temperature of about 1,000 K. This situation also occurs in *blow-down wind tunnels*, where the test time is very short, so that the model does not have time to warm up significantly.

– *Adiabatic wall*: this condition shows that the heat transfer is zero at the surface of the body. Such a situation, which corresponds to the adiabatic equilibrium between fluid and vehicle, is written:

$$\left(\frac{\partial T}{\partial n}\right)_w = 0\,.$$

It shows the derivative of the temperature along the local normal to the wall is zero, which is the condition of zero-heat transfer according to the Fourier law. The adiabatic equilibrium is achieved when a fluid flows for a long time over a calorically insulated wall, or when thermal equilibrium eventually establishes between the flow and structure of the vehicle (aircraft in cruise flight, car, etc.). The wall then takes on a temperature called adiabatic equilibrium or friction or recovery temperature, designated by T_f.

The problem, has five main unknowns $(u, v, w, T$ and p or $\rho)$, a fifth condition should be imposed on the wall. The most natural condition, dictated by physics, is to write:

$$\left(\frac{\partial p}{\partial n}\right)_w = 0\,.$$

The usual boundary conditions are recalled in Figure 4.4.

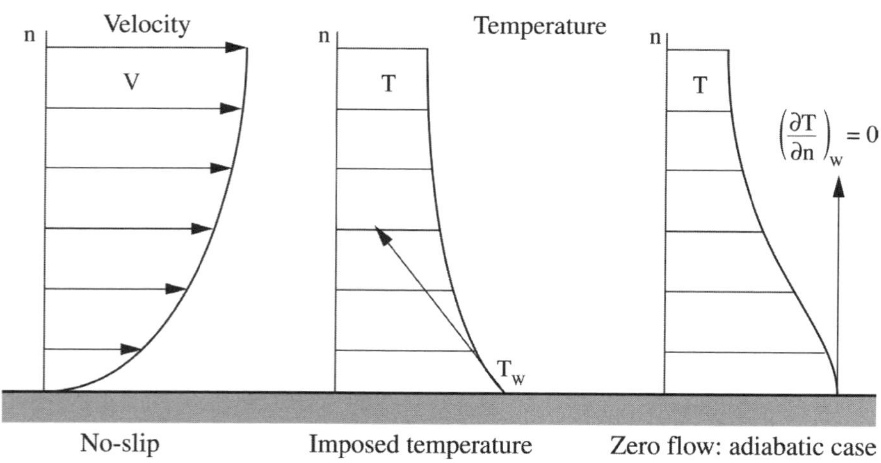

Figure 4.4. *Boundary conditions at a wall*

4.10. Acoustic analogy and Lighthill's equation

It is instructive to mention here a remarkable form of the Navier–Stokes equations, established by Lighthill in the early 1950s, which form the basis of aeroacoustics, a discipline proposing to identify sources of noise in aerodynamics and predict conditions of noise propagation. The reduction of noise generated by aircraft engines, but also wings, landing gear, has indeed become a major issue in the control and reduction of noise pollution caused by air traffic. This also concerns car and railway manufacturers searching for greater comfort, especially at high speeds where the flow noise reaches a level comparable to that of rolling over the road or rail.

We recall the equation of continuity (see section 4.3):

$$\frac{\partial \rho}{\partial t}+\frac{\partial}{\partial x_i}(\rho u_i)=0$$

and the equation of motion (see section 4.4):

$$\frac{\partial(\rho u_i)}{\partial t}+\frac{\partial}{\partial x_j}(\rho u_i u_j)=\frac{\partial T_{ij}}{\partial x_j}+\rho f_i$$

where the stress tensor is expressed in the form of:

$$T_{ij}=-p\,\delta_{ij}+\tau_{ij}$$

with (in the case of a Newtonian fluid):

$$\tau_{ij}=\mu\left[\left(\frac{\partial u_i}{\partial x_j}+\frac{\partial u_j}{\partial x_i}\right)-\frac{2}{3}\delta_{ij}\frac{\partial u_k}{\partial x_k}\right].$$

We apply the operator $\dfrac{\partial}{\partial t}$ to the equation of continuity:

$$\frac{\partial^2 \rho}{\partial t^2}+\frac{\partial^2}{\partial x_i \partial t}(\rho u_i)=0$$

then the operator $\dfrac{\partial}{\partial x_i}$ to the equation of motion:

$$\frac{\partial^2 (\rho u_i)}{\partial t \partial x_i} + \frac{\partial^2}{\partial x_j \partial x_i}(\rho u_i u_j) = \frac{\partial^2 T_{ij}}{\partial x_j \partial x_i}.$$

Subtracting these two equations, we get:

$$\frac{\partial^2 \rho}{\partial t^2} + \frac{\partial^2}{\partial x_i \partial t}(\rho u_i) - \frac{\partial^2}{\partial t \partial x_i}(\rho u_i) - \frac{\partial^2}{\partial x_j \partial x_i}(\rho u_i u_j) = -\frac{\partial^2 T_{ij}}{\partial x_j \partial x_i}$$

or by making T_{ij} explicit:

$$\frac{\partial^2 \rho}{\partial t^2} - \frac{\partial^2}{\partial x_j \partial x_i}(\rho u_i u_j) = \frac{\partial^2}{\partial x_j \partial x_i}(p\delta_{ij} - \tau_{ij}).$$

Now we subtract the expression:

$$a_0^2 \Delta\rho \equiv a_0^2 \frac{\partial^2 (\rho\delta_{ij})}{\partial x_i \partial x_j}$$

from the two members of this equation to introduce a propagation operator (a_0 is the velocity of sound in the ambient medium at rest and Δ is the Laplacian operator, see Appendix 1), which gives:

$$\boxed{\frac{\partial^2 \rho}{\partial t^2} - a_0^2 \Delta\rho = \frac{\partial^2 \Theta_{ij}}{\partial x_j \partial x_i}.}$$

We have written:

$$\Theta_{ij} = \rho u_i u_j + \left(p - a_0^2 \rho\right)\delta_{ij} - \tau_{ij}.$$

This term is called the *Lighthill turbulent stress tensor.*

Lighthill's equation is exact: it includes no approximation and is derived directly from the Navier–Stokes equations. This equation is the basis of modern aeroacoustics because it governs the propagation of pressure fluctuations (via those of the density) in an environment and highlights the source of these changes, which is the Lighthill tensor precisely. Indeed the terms on the right-hand side are in general non–zero in a limited volume, for example in the exhaust jet of an engine and its immediate vicinity. When we move away from the jet, the velocities in an atmosphere at rest are zero, and the terms involving viscosity disappear. Moreover, the medium being isentropic (except in the exhaust jet), it follows that the right hand side of the equation equals zero. We will say no more about this equation, referring the reader to the works of acoustics.

Chapter 5

First Applications of the Conservation Equations

5.1. Theorem of the dynalpy

In what follows, we restrict ourselves to flows where body forces are neglected (neither a heavy fluid nor an electrical conductor). If the motion is steady and (*S*) is a closed surface enclosing a finite volume ($\mathcal{V}$) that *does not contain a body*, then we write the equation of motion of section 4.4 as:

$$\iint_{(S)} \left(\rho\vec{V}.\ \vec{n}\right)\ \vec{V}\ \ dS = -\iint_{(S)}\ \vec{P}\ \ dS$$

where $\vec{P}$ is the tension vector representing the contact forces exerted by the fluid contained in (*S*) on the external medium (hence the minus sign to the right-hand side). This relation can also be written as:

$$\iint_{(S)} \left[\vec{P} + \left(\rho\,\vec{V}.\vec{n}\right)\vec{V}\right] dS = 0$$

The *dynalpy vector* is given by:

$$\boxed{\vec{D} = \vec{P} + \left(\rho\,\vec{V}.\vec{n}\right)\vec{V}.}$$

The relation above shows the following result:

$$\iint_{(S)} \vec{D}\ \ dS = 0$$

which is read as:

> In a steady flow, and in the absence of body forces, the flux of the dynalpy vector through a closed surface not containing any body is zero.

Let us consider an impermeable body (no mass exchange between the body and the external medium) placed in a flow and apply the above result to the fluid contained in the volume ($\mathcal{V}$) defined by the surface (S) composed of two parts (see Figure 5.1):

– (S_1), which is located at a certain distance from the body,

– (S_2), which coincides with the surface of the body.

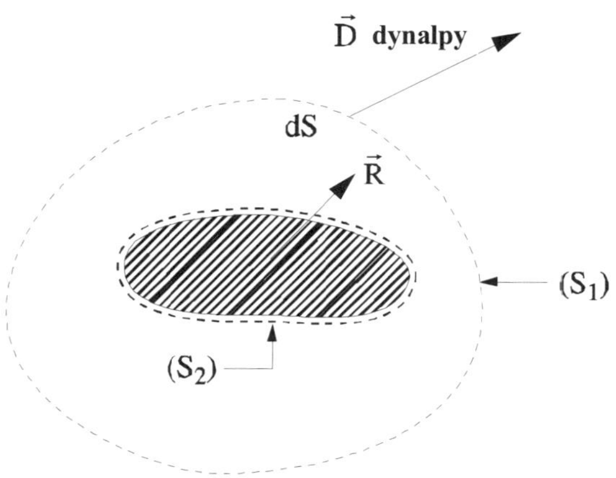

Figure 5.1. *Application of the dynalpy theorem to a surface surrounding a material body*

The general theorem allows us to write:

$$\iint_{(S)} \vec{D}\, dS = \iint_{(S_1)\cap(S_2)} \vec{D}\, dS = \iint_{(S_1)} \vec{D}\, dS + \iint_{(S_2)} \vec{D}\, dS = 0 .$$

The surface of the body is impermeable (we exclude the case where an injection or fluid aspiration is performed through either (S_2) or the case discussed in section 5.2, where there is captation or ejection of fluid), the flow momentum on (S_2)

is zero. Indeed, over (S_2), the no-slip condition requires $\vec{V}=0$. It is noteworthy that in the approximation of the non-viscous fluid, the velocity is tangent to the surface (slip condition) and in this case, $\vec{V}.\vec{n}=0$, which means that the momentum flux term is also zero. Thus, on the surface of the body:

$$\iint_{(S_2)}\left[\vec{P}+\left(\rho\,\vec{V}.\vec{n}\right)\vec{V}\right]dS=\iint_{(S_2)}\vec{P}\,dS\,.$$

also:

$$\iint_{(S_2)}\vec{P}\;\;dS=\vec{R}$$

is a vector representing the resultant of the aerodynamic forces exerted on the body by the fluid in motion. Thus:

$$\iint_{(S_1)}\vec{D}\;\;dS+\iint_{(S_2)}\vec{P}\;\;dS=0$$

and hence:

$$\vec{R}=-\iint_{(S_1)}\vec{D}\,dS\,.$$

We can state the *dynalpy theorem*:

> In steady flow, without body forces, the resultant of the aerodynamic forces applied to a body is equal to the opposite flux of the dynalpy vector through a surface surrounding the body.

5.2. Action of a flow through an engine nacelle: thrust and captation drag

The following result is very important for propulsion applications where the efforts have to be calculated that are caused by the fluid passing through or being ejected by the propelling device. Figure 5.2 shows a sketch of a propulsion nacelle including

– an air intake capturing the upstream air,

– a power unit (engine), secured to the nacelle,

– an air outlet or *nozzle* through which gases are ejected out of the engine.

In supersonic aircraft, such as the *Concorde* or fighter jets, the power unit is extended by an afterburner or post combustion chamber. A conduit positioned downstream of the turbine where kerosene is injected. The combustion that then occurs in incompletely burnt air contributes to the propulsion by increasing the velocity of the ejected gases. On the *Concorde*, the afterburner was only lit for takeoff.

A continuous fluid flow passes through the nacelle from the air intake (*I*) to the exit nozzle (*E*). We propose to evaluate the effort that this flow exerts on the nacelle. For this, we must single out the leading edge *LE* of the air intake which is the line where the upstream flow streamlines split up into those entering the nacelle and those passing outside. These streamlines constitute a stream surface which delimits the flow entering the nacelle.

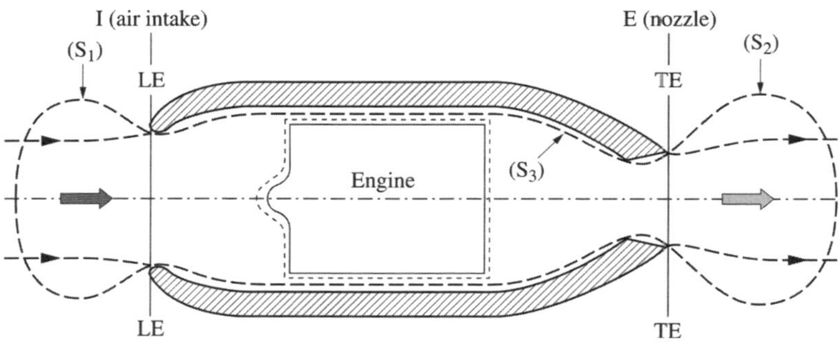

Figure 5.2. *Forces exerted by the fluid passing through a nacelle*

Similarly, we must distinguish the streamlines issuing from the nozzle which together form a surface that defines the fluid exiting the nacelle. We denote with *TE* the line on which this surface is based. To apply the dynalpy theorem, we consider a surface (*S*) consisting of the following elements:

– (S_1) is in the upstream fluid and is based on the leading edge *LE*,

– (S_2) is in the downstream fluid and is based on the trailing edge *TE*,

– (S_3) coincides with the internal physical surface of the device, including the walls of the engine, the internal wall of the nacelle, and the struts of the engine mounts.

By construction, the surfaces (S_1), (S_2) and (S_3) define a volume containing no material parts. Applying the general dynalpy theorem, we can write:

$$\iint_{(S_1)\cap(S_2)\cap(S_3)} \vec{D} dS = 0$$

or:

$$\iint_{(S_1)} \vec{D}\ dS + \iint_{(S_2)} \vec{D}\ dS + \iint_{(S_3)} \vec{D}\ dS = 0.$$

For the same reasons as above, since the momentum flux through (S_3) is zero, there remains:

$$\iint_{(S_3)} \vec{P}\ dS$$

which is precisely the propulsion force, i.e. the integral of the contact actions that the fluid exerts on the material parts of the nacelle, including the engine. We recall that these contact actions result from both pressure and friction, and hence, the propulsion term encompasses these two terms. Therefore, we have:

$$\iint_{(S_1)} \left[\vec{P} + \left(\rho\,\vec{V}.\vec{n}\right)\vec{V}\right] dS + \iint_{(S_2)} \left[\vec{P} + \left(\rho\,\vec{V}.\vec{n}\right)\vec{V}\right] dS = -\iint_{(S_3)} \vec{P}\ dS.$$

By writing:

$$\vec{R} = \iint_{(S_3)} \vec{P}\ dS$$

we get:

$$\boxed{\vec{R} = -\left[\iint_{(S_2)} \vec{D}\ dS + \iint_{(S_1)} \vec{D}\ dS\right].}$$

This formula gives the force exerted on the propulsion unit by the air flow passing through it. In fact, what we call propulsive force or *thrust* P, is the projection of $\vec{R}$ in the direction of the upstream velocity $\vec{V}_\infty$. If $\vec{x}$ denotes the unit vector oriented along the direction of $\vec{V}_\infty$, then the thrust P is given by:

$$P = \vec{R}\,.\,\vec{x}.$$

The unit vector $\vec{n}$ is usually directed along the exterior normal to the considered surface, and as in (S_2) velocities point outward, it follows that:

$$-\iint_{(S_2)} \vec{D}\ dS$$

is a vector *directed in the upstream direction*, which corresponds to a thrust. The scalar product:

$$-\left(\iint_{(S_2)} \vec{D}\ dS\right) . \ \vec{x}$$

will be a negative number. But, as usual, we state that an effective thrust force is positive and we write:

$$F = \left(\iint_{(S_2)} \vec{D}\ dS\right) . \ \vec{x}.$$

This term is called the *nozzle thrust*. This thrust is equal to the component along the direction of $\vec{x}$ of the dynalpy flux exiting through the control surface (S_2) based on the lips of the nozzle exit. This is usually positive. Not to be forgotten is that the thrust vector is opposite to $\vec{x}$ and is thus headed upstream.

For similar reasons, we find that:

$$-\iint_{(S_1)} \vec{D}\ dS$$

is a downstream-directed vector corresponding to a drag force. This drag T is also considered as a positive scalar, so:

$$T = -\left(\iint_{(S_1)} \vec{D} dS\right) . \ \vec{x}.$$

In this case, the term is called *captation drag*. This is equal to the component along the direction of the upstream velocity $\vec{V}_\infty$ of the dynalpy flux entering through the control surface (S_1), which is based on the lip of the air-intake, which is a positive scalar.

Finally, we have as the *internal thrust* of the propulsion nacelle:

$$\boxed{P = F - T.}$$

Thus, the effective thrust P is the difference between the two terms:

– F: called nozzle thrust,

– T: called captation drag.

COMMENT.– The relation obtained for the force $\vec{R}$ is a vector, and the propulsion effect is a vector that may have a component perpendicular to the axis of the nacelle. Propulsion can, therefore, have both a thrust as well as a side force, thus creating a moment that has to be countered by using control surfaces. This force can also be used to increase the maneuverability of combat aircraft by fitting them with movable nozzles or by deflecting the jet by fluidic action. This is the case of thrust vectoring.

COMMENT.– To get the effectively obtained thrust, we must also take into account the action of the fluid flowing over the external part of the nacelle. This question is discussed in section 10.5.

5.3. Drag of a vehicle in uniform flow: origin of drag and the Oswatitsch relation

We consider a vehicle moving at a constant velocity $\vec{V}_\infty$ in an atmosphere with a pressure p_∞. The fluid exerts on the vehicle a force $\vec{R}$ whose component along the velocity $\vec{V}_\infty$ is the drag T. Thus:

$$T = \vec{R} \cdot \frac{\vec{V}_\infty}{\left|\vec{V}_\infty\right|}$$

If (S) is a surface surrounding the vehicle, we have under the result established above:

$$\vec{R} = -\iint_{(S)} \vec{D} \ dS$$

where:

$$\vec{D} = \vec{P} + \left(\rho \vec{V} \cdot \vec{n}\right) \vec{V} \ dS$$

is the dynalpy vector. To evaluate T, we consider the surface consisting of the following elements (see Figure 5.3):

– (S_1) is part of a plane (P_1) perpendicular to $\vec{V}_\infty$ limited by a closed curve (C_1). The plane (P_1) is sufficiently upstream of the vehicle for the flow on (S_1) to be considered as uniform (velocity $\vec{V}_\infty$, pressure p_∞, and temperature T_∞).

– (S_3) is a stream tube based on the contour (C_1).

– (S_2) is the surface located on a downstream plane (P_2), which is perpendicular to $\vec{V}_\infty$, defined by the trace (C_2) of the surface (S_3) in the plane (P_2).

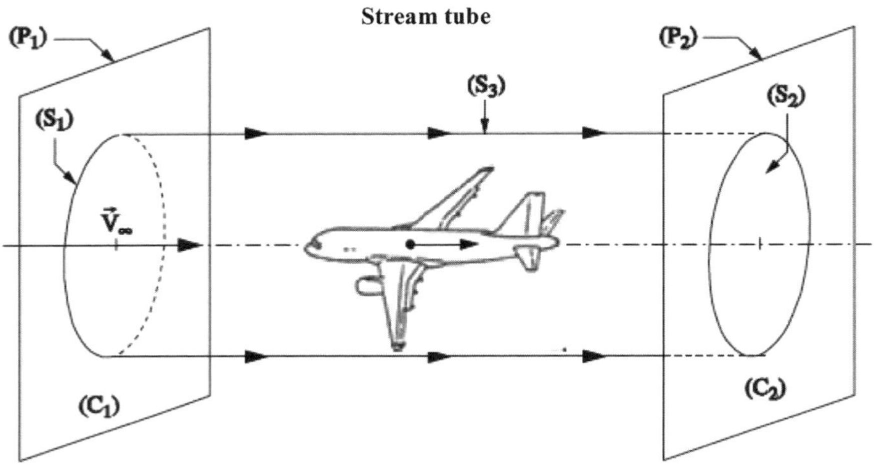

Figure 5.3. *Drag of a body in a uniform flow*

Let us evaluate the flux of the dynalpy vector through (S_1) + (S_2). If these surfaces are chosen far from the body, we can admit that the tension reduces to the pressure, and the action of viscosity is now negligible because of the tendency toward uniformity of the velocity field. Additionally, if (S_2) is chosen sufficiently far downstream of the body, pressure is again uniform and equal to p_∞. A plane like (S_2) satisfying these conditions is called a *Trefftz plane*. Therefore, we get:

$$\iint_{(S_1)\cap(S_2)} \vec{D}\, dS = \iint_{(S_1)\cap(S_2)} p_\infty\, \vec{n}\, dS + \iint_{(S_1)} \left(\rho_\infty \vec{V}_\infty . \vec{n}\right) \vec{V}_\infty\, dS + \iint_{(S_2)} \left(\rho \vec{V} . \vec{n}\right) \vec{V}\, dS$$

If the surface (S_3) is a stream tube, the mass flow is the same through (S_1) and (S_2), and therefore:

$$\iint_{(S_1)} \left(\rho_\infty \vec{V}_\infty . \vec{n}\right) dS = -\iint_{(S_2)} \left(\rho \vec{V} . \vec{n}\right) dS$$

Since $\vec{V}_\infty$ is a constant vector, it follows that:

$$\iint_{(S_1)} \left(\rho_\infty \vec{V}_\infty . \vec{n}\right) \vec{V}_\infty\, dS = -\iint_{(S_2)} \left(\rho \vec{V} . \vec{n}\right) \vec{V}_\infty\, dS$$

Consequently:

$$\iint_{(S_1)\cap(S_2)} \vec{D}\, dS = \iint_{(S_1)\cap(S_2)} p_\infty\, \vec{n}\, dS + \iint_{(S_2)} \left(\rho\vec{V}.\vec{n}\right)\left(\vec{V}-\vec{V}_\infty\right) dS.$$

As (S_3) is a stream surface, no mass flow crosses it [$\rho\vec{V}.\vec{n}=0$ all over (S_3)]. Additionally, if (S_3) is chosen sufficiently far from the body, the viscous terms are negligible there so that the tension is reduced to the pressure term, hence:

$$\iint_{(S_3)} \vec{D}\, dS = \iint_{(S_3)} p\, \vec{n}\, dS.$$

Now we enlarge the curve (C_1) so that (S_3) is in a region located far away from the body. We thus can assume that the pressure on (S_3) is equal to p_∞, with the pressure field again becoming uniform rather rapidly away from the vehicle (at least in subsonic flows). Ultimately, we will have:

$$\vec{R} = -\left[\iint_{(S_1)\cap(S_2)\cap(S_3)} p_\infty \vec{n}\, dS + \iint_{(S_2)} \left(\rho\vec{V}.\vec{n}\right)\left(\vec{V}-\vec{V}_\infty\right) dS\right].$$

As the integral of a uniform pressure over a closed surface is zero, there remains:

$$\vec{R} = -\iint_{(S_2)} \left(\rho\vec{V}.\vec{n}\right)\left(\vec{V}-\vec{V}_\infty\right) dS.$$

The drag T is expressed as:

$$\boxed{T = -\frac{\vec{V}_\infty}{\left|\vec{V}_\infty\right|}\left[\iint_{(S_2)} \left(\rho\vec{V}.\vec{n}\right)\left(\vec{V}-\vec{V}_\infty\right) dS\right].}$$

To give the above integral a thermodynamic interpretation, we now place ourselves in the framework of small perturbations by stating that:

$$\left[\frac{V-V_\infty}{V_\infty}\right] \leq 1$$

where the letters that are not over barred by an arrow designate the vector modules. We set:

$$\delta\vec{V} = \vec{V} - \vec{V}_\infty$$

By introducing the stagnation enthalpy:

$$h_{st} = h + \frac{V^2}{2}$$

we have:

$$\delta h_{st} = \delta h + \vec{V} . \delta \vec{V} .$$

Moreover, a classic thermodynamic relation (see section 3.4) gives:

$$\delta h = T \ \delta s + \frac{\delta p}{\rho}$$

The assumed small differences are calculated between the downstream surface (S_2) and the upstream surface (S_1). By linearizing, we can write to the first order:

$$\delta h_{st} - \delta h = \vec{V}_{\infty} . \overrightarrow{\delta V} = \vec{V}_{\infty} . \left(\vec{V} - \vec{V}_{\infty} \right)$$

and:

$$\delta h = T_{\infty} \delta s \qquad \text{since} \quad \delta p = 0 \ \text{(by hypothesis).}$$

If we write:

$$dq_m = \left(\rho \vec{V} . \vec{n} \right) dS$$

to designate the mass flow across the surface element *dS*, we obtain the following expression for the drag of the body:

$$\boxed{T = \frac{1}{V_{\infty}} \iint_{(S_2)} \left(T_{\infty} \delta s - \delta h_{st} \right) dq_m .}$$

This relation, established by Oswatitsch, is very interesting because it explains the source of drag and thrust of a vehicle.

If the flow is *adiabatic*, then $\delta h_{st} = 0$ and:

$$T = \frac{1}{V_\infty} \iint_{(S_2)} \left(T_\infty \delta s\right) dq_m .$$

The origin of the drag is the production of entropy in fluid due to the presence of the vehicle. Hence in a steady, iso-energetic ($\delta h_{st} = 0$) and isentropic flow, the drag of the vehicle is zero: we remember the *paradox of d'Alembert*.

Thus, we can state the following proposition, known as *Oswatitsch's theorem*:

> In a steady flow where the pressure is again uniform at great distance, the aerodynamic drag of a vehicle is equal to the flux across a surrounding surface of the entropy rise which it has generated.

The calculation of this flux can be limited to portions of the surface (S) on which the entropy generated is non-zero. For example, for a plane flying in a still atmosphere, the flux of entropy is zero everywhere except in the wake downstream of the aircraft. We can then limit the evaluation of the entropy production to a surface placed downstream of the aircraft (see below). The theorem applies also to supersonics, but, it is necessary to consider a surface far away from the aircraft as shock waves (entropy generators) propagate over very long distances before being dissipated by the viscosity and the pressure again becomes uniform. Once generated, the entropy is not destroyed; it is permanent and always convected by the flow! These results reflect the fact that all motion is transformed into heat by the action of viscosity.

We recall the energy equation written using the entropy (see section 4.7):

$$T\frac{ds}{dt} = \frac{1}{\rho}\left(\tau_{ji}\frac{\partial u_j}{\partial x_i} - \frac{\partial q_i}{\partial x_i}\right)$$

where we have not kept the terms associated with a possible chemical relaxation. We see that the factors of entropy generation are the viscosity and heat transfer. If we omit the second factor, which is important for very high velocities, the only source of drag is the *viscosity of the fluid*. In a non-viscous fluid, there is no drag.

In classic aerodynamics, there are two sources of entropy, namely:

– the boundary layers where the viscous terms become important because of the rapid changes of the velocity near a surface,

– the shock waves which, although calculable from the equations of non-viscous fluids, are viscous phenomena (see section 13.6). Shock waves are formed almost inevitably when the flow contains supersonic regions.

We use to call:

– viscous drag as the drag associated with the generation of entropy in the boundary layers,

– wave drag as the drag resulting from the entropy generation by shock waves.

The name viscous drag in this sense is, in fact, misleading since any drag is caused by the viscosity (if the flow is adiabatic), remembering that the shocks are viscous phenomena.

We return to the expression:

$$T = \frac{1}{V_\infty} \iint_{(S_2)} \left(T_\infty \, \delta s - \delta h_{st} \right) dq_m$$

which shows that the production of stagnation enthalpy δh_{st} can cause a change of sign of the drag T, i.e. producing a propulsive effect. This is being done by power plants where the injection of enthalpy is performed by the combustion of a suitable fuel. The transfer of this energy to the fluid is made via a more or less complex device: propeller, compressor, air inlet, and nozzle. Certain drag reduction systems work by production or injection of a plasma at the nose of a body: this is still an injection of stagnation enthalpy into the fluid which can only reduce drag (the production of a plasma on an aircraft in flight is another matter!). However, as any heat input is accompanied by the generation of entropy, it is important that the resulting production term δs does not compensate δh_{st}, which is the source of the propulsion effect. A major problem in the design of air-breathing hypersonic vehicles, i.e. using oxygen from atmospheric air to propel themselves, is to create an effective propulsive force T, the generation of entropy across the shocks of the air intake being very high (see Chapter 16). The captation drag can then exceed the thrust of the nozzle.

If the combustion provides an amount of heat $\delta Q = \delta h_{st}$, then, according to the second law of thermodynamics, an increase in entropy corresponds to it:

$$\delta s \leq \frac{\delta Q}{T}$$

where T is the combustion temperature (temperature at which the heat supply is made).

Therefore, we will have:

$$T = \frac{1}{V_\infty} \iint_{(S_2)} \left(T_\infty - T\right) \delta s \, dq_m .$$

There can only be a propulsion effect if: $T > T_\infty$.

5.4. The Betz method for determining drag

The drag of a properly designed airfoil is a small force at normal flight angles of incidence. As a result, the prediction by calculation is difficult because it requires accurate modeling of viscous effects, difficult even with the most advanced theoretical methods. We must therefore use wind tunnel experiments to safely determine the drag of both air and land vehicles. As we have seen in Chapter 2, the aerodynamic force is the result of the integration of contact actions between the fluid and the wall of the vehicle. We can, theoretically, deduce the drag (and lift) by integrating local pressure and skin friction values. In reality, the method is not applicable as it would require the measurement of these quantities in a large number of points (thousands for an aircraft model). In addition, the methods of measuring skin friction still lack precision. For this reason, we usually directly measure the drag force by mounting the model of the airfoil on an *aerodynamic balance*.

However, the use of such a balance may be impossible for different reasons. Then, an indirect method, based on the relation demonstrated above, is used:

$$T = \frac{1}{V_\infty} \iint_{(S_2)} \left(\rho \vec{V} . \vec{n}\right) \left(\vec{V} - \vec{V}_\infty\right) . \vec{V}_\infty \, dS.$$

The surface (S_2) must be located on a plane far enough downstream of the airfoil so that the pressure there becomes again constant and equal to p_∞ (in practice, a distance of 3 chord lengths is enough for an airfoil). By denoting with u the velocity component parallel to $\vec{V}_\infty$ in this plane, we can write by considering a two-dimensional configuration (for example, an airfoil):

$$T = -\frac{1}{V_\infty} \iint_{(S_2)} \rho \, u \left(u - V_\infty\right) V_\infty \, dS$$

where V_∞ represents the module of $\vec{V}_\infty$ and T is a scalar quantity.

Introducing the drag coefficient C_D, defined by:

$$C_D = \frac{T}{\frac{1}{2}\rho_\infty V_\infty^2 A}$$

where A is a reference surface, that is most often the projected surface of the airfoil, leading to:

$$C_D = 2\iint_{(S_2)} \frac{\rho u}{\rho_\infty V_\infty}\left(1 - \frac{u}{V_\infty}\right)\frac{dS}{A}.$$

This formula shows that the drag of a vehicle is derived from the knowledge of the velocity field downstream of it in a plane where the pressure is again equal to the upstream pressure p_∞ (Trefftz plane). It is possible to measure the local velocity in a fluid by various procedures, multi-hole pressure probes, LDA (*Laser Doppler Anemometry*), PIV (*Particle Image Velocimetry*). The drag coefficient is deduced by a survey of the flow carried out in an appropriate plane chosen downstream of the vehicle.

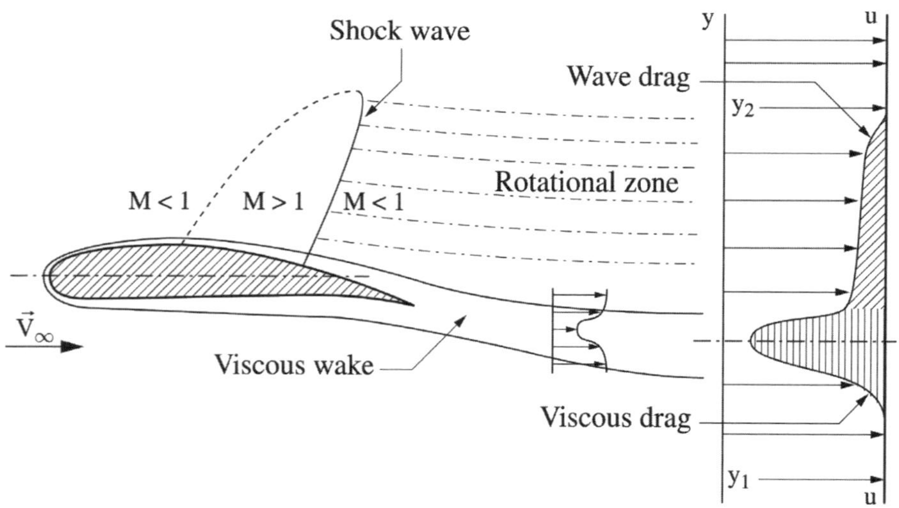

Figure 5.4. *Betz method to determine the drag of an airfoil*

The problem is the choice of the surface (S_2) to be explored and its extent. In fact, outside of the wake produced by the boundary layers and shock waves, u is almost equal to V_∞ everywhere; thus, the term under the integration sign vanishes outside the wake. In the case of a two-dimensional airfoil it is enough to probe the flow on a single line, normal to $\vec{V}_\infty$, since the flow is invariant in the spanwise direction. Then:

$$C_D = \frac{2}{c}\int_{y_1}^{y_2} \frac{\rho u}{\rho_\infty V_\infty}\left(1 - \frac{u}{V_\infty}\right) dy$$

where y_1 and y_2 are two such ordinates so that $u \approx V_\infty$ if $y \le y_1$ and $y \ge y_2$. Here, c denotes the chord of the airfoil. Figure 5.4 shows schematically a u/V_∞ velocity profile downstream of a transonic airfoil. We can see two velocity gaps or deficits: one associated with the wake created by the boundary layer, which is the source of *viscous drag*, and the other due to the loss of stagnation pressure through the shock wave, which corresponds to the *wave drag*.

It is not always possible to perform the survey in the Trefftz plane that is far enough downstream of the model that the pressure in the plane (P_2) is everywhere equal to p_∞. If the pressure is again equal to p_∞ on the lateral surface (S_3), and the non-uniformity confined to a region within (S_2), by considering the relation:

$$\vec{R} = -\left[\iint_{(S_1)\cap(S_2)\cap(S_3)} p\,\vec{n}\,dS + \iint_{(S_2)}\left(\rho\vec{V}.\vec{n}\right)\left(\vec{V} - \vec{V}_\infty\right)dS\right]$$

we can write:

$$\vec{R} = -\left[\iint_{(S_1)\cap(S_3)} p_\infty \vec{n}\,dS + \iint_{(S_2)} p\,\vec{n}\,dS + \iint_{(S_2)}\left(\rho\vec{V}.\vec{n}\right)\left(\vec{V} - \vec{V}_\infty\right)dS\right]$$

or:

$$\vec{R} = -\left[\iint_{(S_1)\cap(S_2)\cap(S_3)} p_\infty \vec{n}\,dS + \iint_{(S_2)} p\,\vec{n}\,dS - \iint_{(S_2)} p_\infty\,\vec{n}\,dS + \iint_{(S_2)}\left(\rho\vec{V}.\vec{n}\right)\left(\vec{V} - \vec{V}_\infty\right)dS\right].$$

After simple calculations:

$$C_D = \frac{2}{A}\iint_{(S_2)}\left[\frac{1}{\gamma M_\infty^2}\left(1 - \frac{p}{p_\infty}\right) + \frac{\rho u}{\rho_\infty V_\infty}\left(1 - \frac{u}{V_\infty}\right)\right] dS$$

knowing that:

$$q_\infty = \frac{1}{2}\rho_\infty V_\infty^2 = \frac{\gamma}{2} p_\infty M_\infty^2.$$

The method for determining the drag of the vehicle is the same as in the previous case. We proceed with a measure of the velocity distribution in a plane downstream of the model. But, given the non-uniformity of pressure, it will also be necessary to determine the static pressure field.

The generalized Betz method is now used to "extract" the various components of the drag of an aircraft, as it allows, by careful examination of the velocity fields and pressure readings downstream of a model, to distinguish the different contributions to the drag, namely:

– the viscous drag, generated by the boundary layers,

– the wave drag, generated by the shock waves,

– the so-called induced drag, a result of the lift arising from vortex detachment at the wing tips.

Although the origin of the drag lies in the viscosity, this break down allows the engineer to improve the design of a wing by distinguishing the phenomena origin of the drag on which we hope to act. The distinguishing characteristics are: weakening shocks in transonics, reducing induced drag by the installation of winglets at the wing tips, and laminar flow control to reduce the drag due to boundary layers by maintaining laminar flow. This extraction method is applied to measured fields, as well as to those obtained by calculation from the resolution of Navier–Stokes equations.

Figure 5.5 shows in the form of isovalue contours, the distribution of the stagnation pressure measured downstream of the wing of an aircraft model tested in a wind tunnel (the stagnation pressure, which is the pressure resulting from an isentropic local retardation of the flow, to zero velocity, is directly related to entropy, see section 9.5.2.1).

The first case corresponding to subsonic flight highlights the viscous wake formed by the boundary layers and containing viscous drag and the vortex at the wing tip, which contains the induced drag. The second case corresponding to transonic speed shows the wake produced by the shock wave that contains the wave drag.

The method of determining the aerodynamic forces, mainly the drag, from an integration of local forces exerted by the flow on the surface of the vehicle, is sometimes called near-field method (which sometimes involves the use of a balance). The extraction of the drag by applying the theorems of conservation and integration of the field variables (pressure and velocity) at some distance from the vehicle is called the far-field method, although the integration is often on surfaces near the body.

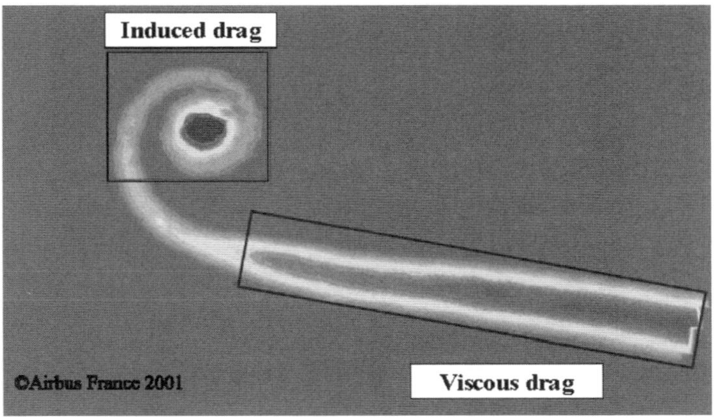

a – Subsonic flight Mach number: $M_0=0.70$ *–* $C_z=0.45$

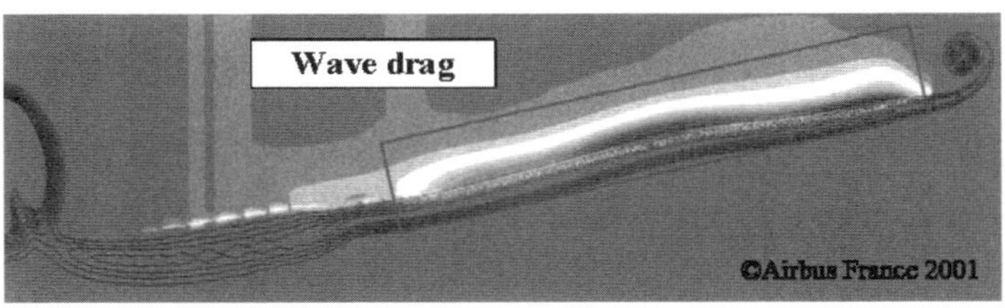

b – Transonic flight Mach number: $M_0=0.85$ *–* $C_z=0.55$

Figure 5.5. *Extraction of the drag by measuring the stagnation pressure losses in the wake of wing models*

Chapter 6

Dimensionless Equations: Similarity Parameters

6.1. Normalization of equations: Reynolds, Froude, and Prandtl numbers

We consider the following basic equations, written in tensorial form:

– *continuity equation:*

$$\frac{\partial \rho}{\partial t}+\frac{\partial}{\partial x_i}(\rho\, u_i)=0$$

– *equation of motion:*

$$\frac{\partial}{\partial t}(\rho\, u_i)+\frac{\partial}{\partial x_j}\left(\rho\, u_i\, u_j\right)=-\frac{\partial p}{\partial x_i}+\frac{\partial}{\partial x_j}\left(\tau_{ij}\right)+\rho f_i\,.$$

The last term of this equation represents the body forces acting within the fluid. In what follows, we consider this to be gravity whose expression is:

$$\rho f_i=\rho g e_i$$

where g is the gravity acceleration and e_i the unit vector indicating the direction of its action, generally the local vertical.

– *energy equation:*

$$\frac{\partial}{\partial t}\left[\rho\left(e+\frac{V^2}{2}\right)\right]+\frac{\partial}{\partial x_i}\left[\rho\, u_i\left(e+\frac{V^2}{2}\right)\right]=\frac{\partial}{\partial x_i}\left(u_j\, \tau_{ij}-\dot{q}_i\right)-\frac{\partial}{\partial x_i}\left(p\, u_i\right)+\rho g e_i u_i\,.$$

We assume that the gas is calorically perfect and the stress tensor τ_{ij} can be written as:

$$\tau_{ij}=\mu\left[\left(\frac{\partial u_i}{\partial x_j}+\frac{\partial u_j}{\partial x_i}\right)-\frac{2}{3}\delta_{ij}\frac{\partial u_k}{\partial x_k}\right].$$

The purpose here is to give these equations dimensionless form by introducing non-dimensional quantities. This *normalization* operation is intended, in particular, to evaluate the relative importance of terms occurring in these equations that can lead to simpler forms if certain terms are very small, and therefore negligible, compared to other terms. Also, normalization will allow us to define *similarity rules* that enable us to transpose the obtained results from a certain configuration to another configuration (for example, to transpose the results obtained from a scale model to the real aircraft). For this normalization process to be meaningful, we must use for each variable (dependent or independent) a *scale* that is characteristic for the studied configuration:

– for the distances x_i a length L, i.e. the chord of the airfoil, the diameter of the missile, or the length of the car,

– for the velocity components u_i the constant upstream velocity V_∞,

– for the density ρ the upstream value ρ_∞,

– for the time t, a time interval t_∞ for the non-defined moment,

– for the viscosity μ, the value in the upstream flow μ_∞.

We introduce the non-dimensional variables:

$$\overline{x}_i=\frac{x_i}{L}\quad,\quad \overline{u}_i=\frac{u_i}{V_\infty}\quad,\quad \overline{\rho}=\frac{\rho}{\rho_\infty}\quad,\quad \overline{t}=\frac{t}{t_\infty}$$

which leads to the following equations:

– *continuity equation:*

$$\frac{\rho_\infty}{t_\infty}\frac{\partial\overline{\rho}}{\partial t}+\frac{\rho_\infty V_\infty}{L}\frac{\partial}{\partial\overline{x}_i}\left(\overline{\rho}\,\overline{u}_i\right)=0$$

– *equation of motion:*

$$\frac{\rho_\infty V_\infty}{t_\infty}\frac{\partial}{\partial \overline{t}}\left(\overline{\rho}\,\overline{u}_i\right)+\frac{\rho_\infty V_\infty^2}{L}\frac{\partial}{\partial \overline{x}_j}\left(\overline{\rho}\,\overline{u}_i\,\overline{u}_j\right)=-\frac{\partial p}{\partial x_i}+\frac{\mu_\infty V_\infty}{L^2}\frac{\partial}{\partial \overline{x}_j}\left(\overline{\tau}_{ij}\right)+\rho_\infty\overline{\rho}ge_i$$

– *energy equation:*

$$\frac{\rho_\infty}{t_\infty}\frac{\partial}{\partial \overline{t}}\left(\overline{\rho}\,e_T\right)+\frac{\rho_\infty V_\infty}{L}\frac{\partial}{\partial \overline{x}_i}\left(\overline{\rho}\,\overline{u}_i\,e_T\right)=\frac{V_\infty^2\mu_\infty}{L^2}\frac{\partial}{\partial \overline{x}_i}\left(\overline{u}_j\,\overline{\tau}_{ij}\right)-\frac{1}{L}\frac{\partial \dot{q}_i}{\partial \overline{x}_i}-\frac{V_\infty}{L}\frac{\partial}{\partial \overline{x}_i}\left(p\,\overline{u}_i\right)+\rho_\infty V_\infty\frac{L}{t_\infty{}^2}\overline{g}\,\overline{\rho}\overline{u}_i e_i$$

where:

$$\overline{\tau}_{ij}=\frac{\mu}{\mu_\infty}\left[\left(\frac{\partial \overline{u}_i}{\partial \overline{x}_j}+\frac{\partial \overline{u}_j}{\partial \overline{x}_i}\right)-\frac{2}{3}\delta_{ij}\frac{\partial \overline{u}_k}{\partial \overline{x}_k}\right]$$

or:

$$\overline{\tau}_{ij}=\overline{\mu}\left[\left(\frac{\partial \overline{u}_i}{\partial \overline{x}_j}+\frac{\partial \overline{u}_j}{\partial \overline{x}_i}\right)-\frac{2}{3}\delta_{ij}\frac{\partial \overline{u}_k}{\partial \overline{x}_k}\right].$$

The continuity equation invites us to take for the timescale t_∞ the ratio L/V_∞. Hence, this equation is then reduced to:

$$\frac{\partial \overline{\rho}}{\partial \overline{t}}+\frac{\partial}{\partial \overline{x}_i}\left(\overline{\rho}\,\overline{u}_i\right)=0$$

a form that does not involve any similarity parameters. Thus if we adopt the above scales, any solution of the reduced continuity equation transposes unconditionally to another case.

We multiply the motion equation by $\dfrac{L}{\rho_\infty V_\infty^2}$, which gives:

$$\frac{\partial}{\partial \overline{t}}\left(\overline{\rho}\,\overline{u}_i\right)+\frac{\partial}{\partial \overline{x}_j}\left(\overline{\rho}\,\overline{u}_i\,\overline{u}_j\right)=-\frac{L}{\rho_\infty V_\infty^2}\frac{\partial p}{\partial x_i}+\frac{\mu_\infty}{\rho_\infty V_\infty L}\frac{\partial}{\partial \overline{x}_j}\left(\overline{\tau}_{ij}\right)+\frac{Lg}{V_\infty^2}\overline{\rho}e_i$$

As $\rho_\infty V_\infty^2$ has the dimension of a pressure (this is twice the dynamic pressure of the upstream flow), it is natural to take this quantity as the pressure scale, from which follows:

$$\overline{p} = \frac{p}{\rho_\infty V_\infty^2}$$

and the non-dimensional equation:

$$\frac{\partial}{\partial \overline{t}}(\overline{\rho}\,\overline{u}_i) + \frac{\partial}{\partial \overline{x}_j}(\overline{\rho}\,\overline{u}_i\,\overline{u}_j) = -\frac{\partial \overline{p}}{\partial \overline{x}_i} + \frac{\mu_\infty}{\rho_\infty V_\infty L}\frac{\partial}{\partial \overline{x}_j}(\overline{\tau}_{ij}) + \frac{Lg}{V_\infty^2}e_i\,.$$

We consider the energy equation written in the form of:

$$\frac{\rho_\infty V_\infty}{L}\frac{\partial}{\partial \overline{t}}(\overline{\rho}\,e_T) + \frac{\rho_\infty V_\infty}{L}\frac{\partial}{\partial \overline{x}_i}(\overline{\rho}\,\overline{u}_i\,e_T) = \frac{V_\infty^2 \mu_\infty}{L^2}\frac{\partial}{\partial \overline{x}_i}(\overline{u}_j\,\tau_{ij}) - \frac{1}{L}\frac{\partial \dot{q}_i}{\partial \overline{x}_i} - \frac{\rho_\infty V_\infty^3}{L}\frac{\partial}{\partial \overline{x}_i}(\overline{p}\,\overline{u}_i) + \frac{\rho_\infty V_\infty^3}{L}\overline{g}\,\overline{\rho}\,\overline{u}e_i\,.$$

If we multiply this equation by $\dfrac{L}{\rho_\infty V_\infty^3}$, then e_T will be divided by V_∞^2 which has the dimension of a specific energy. Thus, we adopt the dimensionless variable:

$$\overline{e_T} = \frac{e_T}{V_\infty^2} = \frac{e}{V_\infty^2} + \frac{1}{2}\frac{V^2}{V_\infty^2}.$$

From this follows the form of the energy equation:

$$\frac{\partial}{\partial \overline{t}}(\overline{\rho}\,\overline{e}_T) + \frac{\partial}{\partial \overline{x}_i}(\overline{\rho}\,\overline{u}_i\,\overline{e}_T) = \frac{\mu_\infty}{\rho_\infty V_\infty L}\frac{\partial}{\partial \overline{x}_i}(\overline{u}_j\overline{\tau}_{ij}) - \frac{1}{\rho_\infty V_\infty^3}\frac{\partial \dot{q}_i}{\partial \overline{x}_i} - \frac{\partial}{\partial \overline{x}_i}(\overline{p}\,\overline{u}_i) + \frac{Lg}{V_\infty^2}\overline{\rho}\overline{u}_i e_i\,.$$

According to the Fourier law the heat flux is written as:

$$\dot{q}_i = -\kappa\frac{\partial T}{\partial x_i}\,.$$

But for a calorically perfect gas:

$$T = \frac{e}{C_v} = \frac{C_p}{C_v}V_\infty^2\frac{\overline{e}}{C_p} = \gamma\frac{V_\infty^2}{C_p}\overline{e}\,.$$

We thus get:

$$\dot{q}_i = -\kappa \frac{\partial T}{\partial x_i} = -\frac{\gamma}{C_p} \frac{V_\infty^2}{L} \kappa \frac{\partial \overline{e}}{\partial \overline{x}_i}$$

and:

$$\frac{1}{\rho_\infty V_\infty^3} \frac{\partial \dot{q}_i}{\partial \overline{x}_i} = -\frac{\gamma}{\rho_\infty V_\infty L} \frac{\kappa}{C_p} \frac{\partial^2 \overline{e}}{\partial \overline{x}_i \partial \overline{x}_i} .$$

This expression is slightly transformed by writing:

$$-\frac{\gamma}{\rho_\infty V_\infty L} \frac{\kappa}{C_p} \frac{\partial^2 \overline{e}}{\partial \overline{x}_i \partial \overline{x}_i} = -\gamma \frac{\mu_\infty}{\rho_\infty V_\infty L} \frac{\mu}{\mu_\infty} \frac{\kappa}{\mu C_p} \frac{\partial^2 \overline{e}}{\partial \overline{x}_i \partial \overline{x}_i} .$$

At this level, we introduce the following dimensionless groupings:

$$\mathrm{Re} = \frac{\rho_\infty V_\infty L}{\mu_\infty}$$ called the Reynold number

$$Fr = \frac{V_\infty}{\sqrt{Lg}}$$ called the Froude number

$$\mathrm{Pr} = \frac{\mu C_p}{\kappa}$$ called the Prandtl number

The square of the Froude number is a measure of the ratio between the terms of inertia (or acceleration) and the gravitational forces within the fluid. A large Froude number implies that the action of gravity is negligible in the first approximation. As water is a heavy fluid, the Froude number plays an important role in hydrodynamics especially in issues of air-water interfaces involving a light fluid, air, and heavy fluid, water. As air is much lighter, most of the time we neglect the gravity forces in problems involving air, except when modeling the atmosphere. In aerodynamics, the action of gravity is ignored.

Hence, the new expression of the equations where only dimensionless variables appear:

– *continuity equation:*

$$\frac{\partial \overline{\rho}}{\partial \overline{t}} + \frac{\partial}{\partial \overline{x}_i} (\overline{\rho}\, \overline{u}_i) = 0$$

– *equation of motion:*

$$\frac{\partial}{\partial t}(\bar{\rho}\,\bar{u}_i)+\frac{\partial}{\partial x_j}\left(\bar{\rho}\,\bar{u}_i\,\bar{u}_j\right)=-\frac{\partial \bar{p}}{\partial x_i}+\frac{1}{\mathrm{Re}}\frac{\partial}{\partial x_j}\left(\bar{\tau}_{ij}\right)+\frac{1}{Fr^2}\bar{\rho}\,e_i$$

– *energy equation:*

$$\frac{\partial}{\partial t}(\bar{\rho}\,\bar{e}_T)+\frac{\partial}{\partial x_i}(\bar{\rho}\,\bar{u}_i\,\bar{e}_T)=\frac{1}{\mathrm{Re}}\frac{\partial}{\partial x_i}\left(\bar{u}_j\bar{\tau}_{ij}\right)+\frac{\bar{\mu}}{\mathrm{Re}\,\mathrm{Pr}}\gamma\frac{\partial^2\bar{e}}{\partial x_i \partial x_i}-\frac{\partial}{\partial x_i}(\bar{p}\,\bar{u}_i)+\frac{1}{Fr^2}\overline{\rho u_i e_i}.$$

We denote by the *Péclet number* with Pe and the grouping:

$$\boxed{Pe = \mathrm{Re}\times\mathrm{Pr} = \frac{\rho_\infty V_\infty L C_p}{\kappa}.}$$

It is a measure of the relative importance of dissipation and conduction terms in the energy exchange in a fluid.

We express the ratio $\dfrac{\mu}{\mu_\infty}$ through the Sutherland law, which we recall:

$$\mu = 1.454\times 10^{-6}\frac{T^{3/2}}{T+110.4}$$

thus follows:

$$\bar{\mu}=\left[\frac{T}{T_\infty}\right]^{3/2}\frac{1+\dfrac{110.4}{T_\infty}}{\dfrac{T}{T_\infty}+110.4}.$$

The temperature T is calculated from

$$T=\frac{e}{C_v}=\frac{1}{C_v}\left[e_T-\frac{V^2}{2}\right]=\frac{V_\infty^2}{C_v}\left[\bar{e}_T-\frac{\bar{V}^2}{2}\right].$$

It follows that:

$$\frac{T}{T_\infty}=\frac{V_\infty^2}{C_v T_\infty}\left[\bar{e}_T-\frac{\bar{V}^2}{2}\right].$$

But:

$$a_\infty^2 = \gamma\left(C_p - C_v\right)T_\infty = \gamma(\gamma-1)\,C_v T_\infty$$

therefore:

$$\frac{T}{T_\infty} = \gamma(\gamma-1)\frac{V_\infty^2}{a_\infty^2}\left[\overline{e}_T - \frac{\overline{V}^2}{2}\right] = \gamma(\gamma-1)\,M_\infty^2\left[\overline{e}_T - \frac{\overline{V}^2}{2}\right].$$

That way:

$$\overline{\mu} = \frac{\mu}{\mu_\infty} = f\left(M_\infty, \gamma, T_\infty, \overline{e}\right)\cdot$$

Moreover the equation of state of the gas is written:

$$\frac{p}{\rho} = r\,T = \left(C_p - C_v\right)T = (\gamma-1)\,C_v\,T = (\gamma-1)\,e$$

and is with dimensionless variables:

$$\frac{\rho_\infty V_\infty^2}{\rho_\infty}\frac{\overline{p}}{\overline{\rho}} = (\gamma-1)\,V_\infty^2\,\overline{e}$$

or:

$$\boxed{\frac{\overline{p}}{\overline{\rho}} = (\gamma-1)\,\overline{e}.}$$

The equation of state remains unchanged in the present change of variables.

If we introduce the dimensionless temperature $\overline{T} = T / T_\infty$, we see that:

$$\overline{e} = \frac{1}{\gamma(\gamma-1)M_\infty^2}\overline{T}\,.$$

6.2. Other notable numbers

6.2.1. *Temporal similarity and the Strouhal number*

We have seen that the time scale suitable to obtain the dimensionless form of the continuity equations is:

$$t_\infty = \frac{L}{V_\infty}.$$

This scale, or characteristic time, is interpreted as the time taken by a fluid particle moving at the speed V_∞ to travel a distance equal to the length scale L which is usually the chord of the airfoil (or any other characteristic length typical of the size of the vehicle).

In other words, if the aerodynamic phenomenon changes with a characteristic duration that is long compared to the scale t_∞, then we can consider the derivatives compared to time as negligible and admit that this change is a sequence of stationary states: we say that we are dealing with a pseudo-stationary flow. For example, for an aircraft at final approach, flying at estimated speed of 100 m/s and if we take for L a chord length of 3 m, we get $t_\infty = 0,03s$. This time is very short compared to the speed of change of the aircraft, including its configuration changes due to flap deflections. Under these conditions, we can treat the descent as a series of stationary states, with the airplane barely moving in view of the flow around it.

This is no longer true for a combat aircraft in rapid manoeuvre or a helicopter rotor, propeller, turbomachine, etc. Then we must solve the full unsteady equations.

In some circumstances the airflow is affected by self-induced unsteadiness, with a periodic character (the case of buffeting, flutter, buzz of air intakes).

The frequency f of these instabilities can also take a dimensionless form by introducing the grouping:

$$\boxed{Su = \frac{fL}{V_\infty}}$$

called the *Strouhal number*. This is a way to normalize the period $t = 1/f$ by the scale t_∞.

For a given phenomenon occurring in defined circumstances, the Strouhal number is often (almost) constant. This implies that if unsteady tests are conducted on a model N times smaller than the real device, the observed frequencies will be N times larger. If we want to simulate the effects of a forced oscillation, the oscillation frequency of the model must be N times higher.

6.2.2. *Reactive flows and the Damköhler number*

In hypersonic aerodynamics, when crossing shock waves, intense heating initiates dissociation-recombination reactions of its components O_2 and N_2 (see section 3.7 and Chapter 13). These reactions are not instantaneous, but take place with a certain reaction velocity α_j, which represents the quantity (mass or molar) of the species (*j*) created or destroyed during the time interval *dt*. Thus, we write by denoting with ξ_j the molar concentration of the species (*j*):

$$\alpha_j = \left(\frac{d\xi_j}{dt} \right).$$

If the reactive fluid has a velocity V_∞ the concentration of the species (*j*) will have changed over the length of travel L:

$$\Delta\xi_j = \alpha_j \Delta t = \alpha_j L / V_\infty .$$

If we define a rate of formation for the species (*j*):

$$K = \frac{1}{\xi_j} \left(\frac{d\xi_j}{dt} \right)$$

we obtain:

$$\frac{\Delta\xi_j}{\xi_j} = \frac{KL}{V_\infty} .$$

We introduce the grouping:

$$\boxed{Da = \frac{KL}{V_\infty}}$$

called *Damköhler's number*: it represents the ratio between the chemical and aerodynamic velocities. If this number is high, this means that over the path of L a large quantity of species have been formed (in the algebraic sense), or in more precise terms, that the chemical reactions involving these species have had time to proceed. If the chemical velocities are much larger than the aerodynamic velocity, the chemical equilibrium may have time to settle each time. On the other hand, if this number is small, it implies that over the path of L, the delay or *residence time* Δt was too short for the chemical reaction to progress significantly. The gas composition does not vary; it is said that this composition is frozen.

In summary:

– if $Da >> 1$: *equilibrium hypothesis*, concentrations are calculated from the Guldberg and Waage laws of mass action (see section 3.7.3.2),

– if $Da << 1$: *frozen hypothesis*, the gas composition is considered invariant and it is possible to use the model of a calorically perfect gas with a constant γ.

In intermediary situations, we have to simultaneously solve the equations of motion and chemical non-equilibrium, involving the laws of chemical kinetics. We will say nothing more on this issue that concerns extreme situations in aerodynamics.

6.2.3. *Multi-species flows and the Lewis number*

In some applications, the flow involves several gases that mix with each other to possibly react. This occurs at hypersonic velocities where the temperature rise due to shock waves that cause the dissociation of components of the air. Such a circumstance occurs also between a jet and a surrounding flow whose composition is different (exhaust jet engine in the atmosphere). Therefore, it is necessary to know the composition of the mixture at any point of the field, i.e. the relative proportion of each of its component species (j).

Consider a mixture of J species, each moving in the flow with a velocity v_i^j. We call ρ_j the density of the species (j) and ρ the total density and define the mass concentration of the species (j) by:

$$\alpha_j = \frac{\rho_j}{\rho}.$$

It is clear that:

$$\sum_{j=1}^{J} \rho_j = \rho \quad \text{and} \quad \sum_{j=1}^{J} \alpha_j = 1.$$

The barycentric velocity u_i, i.e. the entrainment velocity of the total of all species is defined by the relation:

$$\sum_{j=1}^{J} v_i^j \rho_j = u_i \sum_{j=1}^{J} \rho_j = \rho u_i.$$

Hence by introducing the concentration, it becomes:

$$u_i = \sum_{j=1}^{J} \alpha_j v_i^j$$

we define the *relative diffusion velocity* of the species (j) by:

$$V_i^j = v_i^j - u_i .$$

The overall flow obeys the fundamental equations for mass, motion and energy, equation for the mass being:

$$\frac{\partial \rho}{\partial t} + \frac{\partial}{\partial x_i}(\rho u_i) = 0 .$$

The density of the species (j) satisfies a continuity equation similar to the one written for the total density but with an additional term $\dot{w}_j$ representing creation (or destruction) of the considered species by the chemical reactions ($\dot{w}_j$ has the dimension of a mass per unit of time and unit of volume). Thus the continuity equation for the species (j) is written:

$$\frac{\partial \rho_j}{\partial t} + \frac{\partial}{\partial x_i}\left(\rho_j v_i^j\right) = \dot{w}_j.$$

The global mass conservation implies that:

$$\sum_{j=1}^{J} \dot{w}_j = 0 .$$

Since:

$$V_i^j = v_i^j - u_i$$

thus results:

$$\frac{\partial \rho_j}{\partial t} + \frac{\partial}{\partial x_i}\left(\rho_j u_i\right) = -\frac{\partial}{\partial x_i}\left(\rho_j V_i^j\right) + \dot{w}_j .$$

Let us rewrite the continuity equation for the species (*j*) with the mass concentration:

$$\frac{\partial\left(\rho \alpha_j\right)}{\partial t} + \frac{\partial}{\partial x_i}\left(\rho \alpha_j u_i\right) = -\frac{\partial}{\partial x_i}\left(\rho \alpha_j V_i^j\right) + \dot{w}_j .$$

The relative migration velocity V_i^j, or *diffusion velocity*, of species (*j*) is expressed by using a Fick's law-type relation:

$$V_i^j = -D_j \frac{1}{\alpha_j} \frac{\partial}{\partial x_i}\left(\alpha_j\right).$$

The minus sign expresses the fact that the diffusion takes place from regions with the highest concentration to regions with the lowest concentration. Therefore, the continuity equation is written as:

$$\frac{\partial\left(\rho \alpha_j\right)}{\partial t} + \frac{\partial}{\partial x_i}\left(\rho \alpha_j u_i\right) = +\frac{\partial}{\partial x_i}\left[\rho D_j \left(\frac{\partial \alpha_j}{\partial x_i}\right)\right] + \dot{w}_j .$$

This equation is often written by introducing the *Lewis number*, defined by:

$$\boxed{Le = \frac{\kappa}{\rho D_j C_p}.}$$

According to the Fourier law for thermal conduction:

$$\dot{q}_i = -\kappa \frac{\partial T}{\partial x_i} = -\frac{\kappa}{C_p} \frac{\partial h}{\partial x_i}$$

the Lewis number represents the ratio between the diffusion of energy through the transmission by molecules agitation (thermal conductivity) and molecular diffusion causing mass transport. Knowing that the Prandtl number is defined by:

$$\Pr = \frac{\mu C_p}{\kappa}$$

it follows that:

$$\rho D_j = \frac{\mu}{Le \Pr}$$

therefore the equation:

$$\frac{\partial(\rho\alpha_j)}{\partial t} + \frac{\partial}{\partial x_i}(\rho\alpha_j u_i) = \frac{\partial}{\partial x_i}\left[\frac{\mu}{Le \Pr}\left(\frac{\partial \alpha_j}{\partial x_i}\right)\right] + \dot{w}_j.$$

The product $Le \times \Pr$ is equal to the *Schmidt number* which is defined by:

$$Sc = Le \Pr = \frac{\kappa}{\rho D_j C_p} \frac{\mu C_p}{\kappa} = \frac{\mu}{\rho D_j}.$$

By introducing the kinetic viscosity $\nu = \dfrac{\mu}{\rho}$:

$$\boxed{Sc = \frac{\nu}{D_j}.}$$

The equation of motion is not altered by the fact that the gas is multispecies as it applies to the total mass contained in the control volume, thus:

$$\frac{\partial(\rho u_i)}{\partial t} + \frac{\partial}{\partial x_j}(\rho u_i u_j) = -\frac{\partial p}{\partial x_i} + \frac{\partial \tau_{ij}}{\partial x_j}.$$

The total stagnation enthalpy of the gas mixture h_T is expressed as:

$$h_{st} = h + \frac{V^2}{2} = \sum_{j=1}^{J} \alpha_j \left(h_j - h_j^0\right) + \frac{V^2}{2}$$

where h_j^0 is the enthalpy released by the formation of the elements at 0 K. By convention, h_j^0 often is set equal to zero for molecules and to minus the dissociation heat for atoms (for details, we refer to treaties of thermochemistry).

Under these conditions, the energy equation, written with the enthalpy, takes the form:

$$\frac{\partial}{\partial t}(\rho\ h_{st})+\frac{\partial}{\partial x_i}(\rho\ u_i\ h_{st})=\frac{\partial p}{\partial t}+\frac{\partial}{\partial x_i}\left(\frac{\mu}{\Pr}\frac{\partial h_{st}}{\partial x_i}\right)+\frac{\partial}{\partial x_i}\left[\mu\left(1-\frac{1}{\Pr}\right)\frac{\partial}{\partial x_i}\left(\frac{V^2}{2}\right)\right]$$
$$+\frac{\partial}{\partial x_i}\left[\rho D_j(1-Le)\sum_{j=1}^{J}\left(h_j-h_j^0\right)\frac{\partial \alpha_j}{\partial x_i}\right]$$

$$\frac{\partial}{\partial t}(\rho\, h_T)+\frac{\partial}{\partial x_i}(\rho\, u_i\ h_T)=\frac{\partial p}{\partial t}+\frac{\partial}{\partial x_i}\left(\frac{\mu}{\Pr}\frac{\partial h_T}{\partial x_i}\right)+\frac{\partial}{\partial x_i}\left[\mu\left(1-\frac{1}{\Pr}\right)\frac{\partial}{\partial x_i}\left(\frac{V^2}{2}\right)\right]$$
$$+\frac{\partial}{\partial x_i}\left[\rho D_j(1-Le)\sum_{j=1}^{J}\left(h_j-h_j^0\right)\frac{\partial \alpha_j}{\partial x_i}\right].$$

or:

$$\frac{\partial}{\partial t}(\rho\ h_{st})+\frac{\partial}{\partial x_i}(\rho\, u_i\ h_{st})=\frac{\partial p}{\partial t}+\frac{\partial}{\partial x_i}\left(\frac{\mu}{\Pr}\frac{\partial h_{st}}{\partial x_i}\right)+\frac{\partial}{\partial x_i}\left[\mu\left(1-\frac{1}{\Pr}\right)\frac{\partial}{\partial x_i}\left(\frac{V^2}{2}\right)\right]$$
$$+\frac{\partial}{\partial x_i}\left[\mu\left(\frac{1}{\Pr Le}-\frac{1}{\Pr}\right)\sum_{j=1}^{J}\left(h_j-h_j^0\right)\frac{\partial \alpha_j}{\partial x_i}\right].$$

We will say no more about these issues which are addressed in the specialized literature.

6.3. Similarity conditions and the importance of the Reynolds number

If we consider two different aerodynamic flows obeying the same law of viscosity, and if the physical quantities representing these flows are normalized as described in section 6.1, then these flows are governed by identical equations provided there is an equality of:

– geometries (similarity),

– Reynolds numbers Re,

– Prandtl numbers Pr,

– Mach numbers M_∞,

– ratios of specific heats γ,

– levels of the temperature T_∞.

The equality of Mach numbers is necessary for there to be an identity between the phenomena of compressibility since these phenomena (including, in particular, the shock waves) are governed by the Mach number. If the above conditions are met, then the equations have identical solutions in the space of normalized variables, if in addition the boundary conditions are identical.

The carried out change of variables does not change the direction of the velocity vector (the u_i components are divided by the same constant): therefore, the streamlines and shapes of the bodies are not changed. In other words, if both bodies are geometrically similar in a manner ensuring identical boundary conditions on the wall, then the solutions with the dimensionless (or normalized) variables are identical. The above findings warrant testing of a model in a wind tunnel. The model must have the same shape as the real object and have the same angle of attack with respect to the incoming flow. But, it may be smaller than the real object (wing, aircraft, missiles, etc.). If the above parameters are identical in the wind tunnel and in reality, then the results obtained from the model are transferable to the vehicle by simple upscaling. At low velocities, where the density and viscosity can be regarded as constant, the energy equation becomes useless (except for heat transfer problems). So, if we consider two geometrically similar bodies, the flow field expressed in dimensionless variables will be the same around these two bodies *provided that the Reynolds numbers are identical.* This conclusion remains valid at high velocities but additionally equal Mach numbers have to be achieved.

A key number in aerodynamics is thus the Reynolds number:

$$\boxed{Re_\infty = \frac{\rho_\infty V_\infty L}{\mu_\infty}.}$$

The equations of the preceding section show that the inverse of this number is a factor of the terms involving the viscosity and thermal conductivity of the fluid. In other words, *with the adopted normalization*, Re_∞ characterizes the relative importance of the viscous terms in the equations governing the movement: it is a measure of the ratio between the viscous terms and the terms of inertia and pressure. The lower the Reynolds number, the more the viscous effects become important.

Conversely, if the Reynolds number is large, the viscous effects tend to become secondary. The limit $Re_\infty \to \infty$ corresponds to the case of a perfect fluid in the non-viscous sense. The flow is then governed by the following equations, called the *Euler equations*:

– *continuity equation:*

$$\frac{\partial \bar{\rho}}{\partial \bar{t}} + \frac{\partial}{\partial \bar{x}_i}(\bar{\rho}\,\bar{u}_i) = 0$$

– *equation of motion:*

$$\frac{\partial}{\partial \bar{t}}(\bar{\rho}\,\bar{u}_i) + \frac{\partial}{\partial \bar{x}_j}\left(\bar{\rho}\,\bar{u}_i\,\bar{u}_j\right) = -\frac{\partial \bar{p}}{\partial \bar{x}_i}$$

– *energy equation:*

$$\frac{\partial}{\partial \bar{t}}(\bar{\rho}\,\bar{e}_T) + \frac{\partial}{\partial \bar{x}_i}(\bar{\rho}\,\bar{u}_i \bar{e}_T) = -\frac{\partial}{\partial \bar{x}_i}(\bar{p}\,\bar{u}_i)\,.$$

Deleting the transfer terms (not only viscosity but also thermal conduction) lowers the order of the differential system which is of the first order, whereas the Navier–Stokes equations are of second order. Consequently, we cannot satisfy an equal number of boundary conditions. The Euler equations are discussed in greater detail in the next Chapter.

Given the largely non-viscous character of the field around an aircraft, we can ask ourselves why so much importance is attached to the Reynolds number and why we are not just content with the Euler model of the non-viscous fluid, thus avoiding expensive development of wind tunnels to achieve the similarity of Reynolds number. In fact, the Euler flow around a body is influenced by the effect of the boundary layer, which has major repercussions for supercritical airfoils in transonics where any variation of the thickness of the boundary layer causes a large displacement of the shock wave (see section 25.10). We know that the development of the boundary layer depends on the Reynolds number. Furthermore, we saw that the drag has its origin in the viscosity acting in the boundary layers (and shock waves for transonic and supersonic flows). The drag is not predictable by non-viscous fluid models. Finally, the phenomena at the approach of separation, appearance of buffeting, nuisances of all kinds, etc. are conditioned by the viscosity, and thus the Reynolds number. The Reynolds number plays a crucial role in aerodynamics as it allows us to ensure the similarity between actual flight and testing of models in wind tunnel. However, we should be aware of a more complex physical reality that cannot be represented by taking into account a unique number.

Thus, when the flow separates, even at a high Reynolds number, viscous areas take important dimensions of the order of the size of the airfoil in the case of a wing. In that case, the Euler equations are wrong, although the Reynolds number is high. Thus, a high Reynolds number does not necessarily mean viscous regions with a small extent. Moreover, the Reynolds number is high in a fluid such as air because the viscosity of air is low. However, the fact that the viscous terms are negligible, is not a unique consequence of the smallness of μ, indeed, the terms involving the viscosity are in the form of:

$$\frac{\partial}{\partial x_j}(\tau_{ij})$$

where:

$$\tau_{ij} = \mu\left[\left(\frac{\partial u_i}{\partial x_j}+\frac{\partial u_j}{\partial x_i}\right)-\frac{2}{3}\delta_{ij}\frac{\partial u_k}{\partial x_k}\right].$$

The viscosity tensor τ_{ij} is the product of μ by a tensor whose components are derivatives of the velocity field (the strain tensor). Thus, τ_{ij} becomes significant not because the viscosity is high (it still is small in gases), but because the velocity field varies rapidly.

In a more precise manner, the viscous effects are negligible when the magnitude of the term:

$$\frac{\partial}{\partial x_i}\left(\tau_{ij}\right)$$

is much less than that of the other terms of the Navier–Stokes equations that constitute the so-called Euler part of the equations.We can no longer neglect the viscous effects when the magnitude of the term:

$$\frac{\partial}{\partial x_i}\left(\tau_{ij}\right)$$

becomes comparable to the magnitude of the Euler term, as occurs in regions with a high velocity gradient. If regions appear in a flow where the velocity varies rapidly, they result in an “awakening” of the viscous terms that are “dormant” elsewhere in the field where the velocity is changing more gradually. This awakening takes place in boundary layers, wakes, vortex sheets, and shock waves, i.e. around the discontinuities of the Euler type solution (see Chapter 13).

The process of making the equations dimensionless, by dividing the physical quantities by appropriate scales, is essential to conduct a discussion of the magnitudes of the terms of these equations. Such a discussion obviously is only meaningful if the different terms are numbers that can be compared with each other. We then can judge whether certain terms may be neglected before others. The choice of scales (length, time, velocity, density, etc.), in principle is arbitrary. However, we arranged to take quantities of reference that make dimensionless variables vary within a "reasonable" interval of the order 1. Thus, dividing the lengths x_i by the chord of the airfoil (for a wing problem) ensures that the normalized variable $\overline{x}_i$ does not take values much larger than 1 in the area influenced by the airfoil. The same is true for the components u_i if they are divided by V_∞, the density if it is divided by ρ_∞, the pressure if it is related to the dynamic pressure of the flow, etc. (see section 6.1). The time scale L / V_∞ is the time taken by a fluid particle moving with a velocity V_∞ to travel a distance equal to the chord of the airfoil: it is a characteristic time of the problem.

The adoption of normalized variables in this way is also beneficial for numerical applications, since it avoids the very large or very small values that physical variables expressed in metric units can adopt.

6.4. Simulation of the Reynolds number in wind tunnels

The equality of the Reynolds numbers between the aircraft and an experimental model is the condition to satisfy to be able to reproduce the viscous effects in the wind tunnel, and therefore to measure the values of realistic aerodynamic coefficients. However, for obvious reasons, the models tested have dimensions much less than that of the aircraft. Thus, for the *Airbus A380* with a wingspan of 80 m, it is not possible to test models with wingspans greater than 4 m, even in the transonic wind tunnel S1 at the Onera Modane Centre whose test section has a diameter of 8 m, with the risk of seeing the flow disturbed by the wind tunnel walls (blockage effect). In a very favorable case, taking into account the existing test facilities, there is a difference of 20 between reality and the test, resulting in a Reynolds number 20 times smaller if the conditions of velocity, pressure, and temperature are met. Considering the expression of the Reynolds number:

$$Re_\infty = \frac{\rho_\infty V_\infty L}{\mu_\infty}$$

when the reference length L is imposed by the maximum size of the model, which are the parameters on which we can act to increase the Reynolds number?

– It is not possible to increase the speed for at least two reasons: first, because the power expended to put a fluid in motion is proportional to the cube of the speed to achieve, then because in high subsonics (domain of airliners), any increase in speed will cause the premature appearance of transonic and supersonic phenomena.

– It remains to increase the density, which under the equation of state:

$$\rho_\infty = \frac{p_\infty}{rT_\infty}$$

can be achieved by increasing the stagnation (reservoir) pressure p_{st_∞} of the wind tunnel and/or decreasing the stagnation temperature T_{st_∞} The local quantities are proportional to the stagnation conditions corresponding to temperature and pressure conditions in the fluid before its acceleration, these conditions being measured in the settling chamber of the wind tunnel (see section 9.5.2).

Increasing the stagnation pressure is achieved in *pressurized wind tunnels*, where the stagnation pressure can reach $4\times10^5\,Pa$ in the wind tunnel F1 at the Onera Fauga-Mauzac Centre. This allows a 4-fold increase in the Reynolds number. In this case a limitation is the capability of the wind tunnel to resist effort due to a high pressure. In addition, the aerodynamic forces being proportional to the flow dynamic pressure (see section 2.1), the pressurization increases the risk of deformation or even breaking of the models. In transonics, the model angle of incidence should be known with accuracy down to a tenth of a degree, which excludes any perceptible distortion of the model and its support.

The density can be increased by lowering the temperature, which is achieved in the so-called *cryogenic* wind tunnels where the system is fed with cold nitrogen, close to the vaporization temperature of around 100 K (nitrogen has a specific heat ratio γ equal to that of air). Thus, the density relative to ambient conditions $\left(T_{st_\infty} \approx 300K\right)$. is multiplied by about 3. The cooling of the flow also has the effect of decreasing its viscosity μ_∞ (see section 4.4). Thus, we win on both counts. The dynamic pressure is expressed as:

$$q_\infty = \frac{1}{2}\rho_\infty V_\infty^2$$

which can be written for a calorically perfect gas as:

$$q_\infty = \frac{\gamma}{2} p_\infty M_\infty^2$$

Since q_∞ does not depend on the temperature level of the gas, cryogenization does not change the aerodynamic forces on the models. This is another advantage of the process.

We may also note that the experimental Mach number M_∞ being fixed (similarity conditions to be respected in compressible flow) and the flow velocity being given by:

$$V_\infty = a_\infty M_\infty$$

where:

$$a_\infty = \sqrt{\gamma r T_\infty}$$

cryogenization diminishes as $V_\infty = a_\infty M_\infty$ and so does the power required to operate the facility. However, the decrease of V_∞ has a negative impact on the Reynolds number, but as dependence is in $\sqrt{T_\infty}$ this negative effect is largely compensated for by the increase of ρ_∞ and the decrease of μ_∞ The diagram in Figure 6.1 summarizes these results. There are two major wind tunnels, both cryogenic and pressurized: the National Transonic Facility (NTF) in the United States and the European Transonic Wind Tunnel (ETW) based in Cologne. This is a facility operated in cooperation with France, Germany, Great Britain, and The Netherlands whose test section has an area of $2.4 \text{ m} \times 2 \text{ m}$.

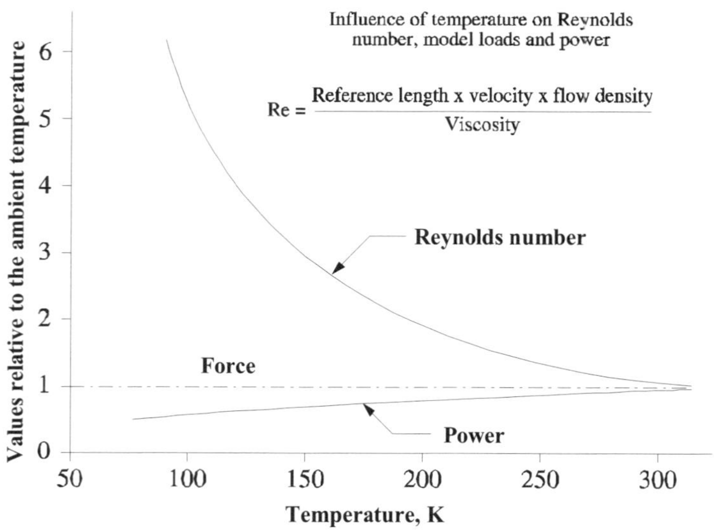

Figure 6.1. *Effect of cryogenization in wind tunnels (origin: ETW wind tunnel in Cologne)*

Figure 6.2 shows the range covered by the ETW wind tunnel in terms of simulation parameters [Mach number, Reynolds number]. By combining pressurization and cryogenization, ETW multiplies the experimental Reynolds number by 10, compared to conventional wind tunnels.

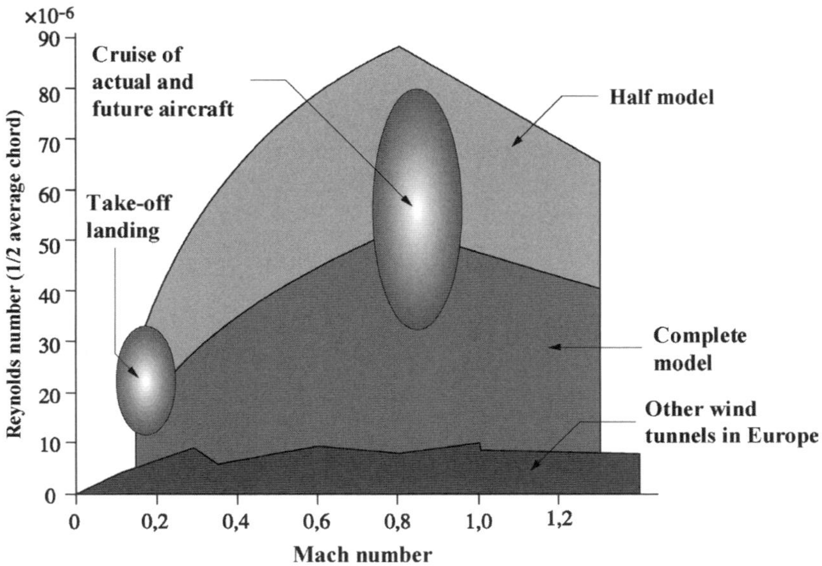

Figure 6.2. *Flight envelope simulated in the ETW wind tunnel in Cologne (origin: ETW wind tunnel in Cologne)*

6.5. Conclusion

There are many other numbers serving as parameters of similarity in aerodynamics, and more generally in fluid mechanics. Overall, these numbers express the relation between two scales or velocities of competing phenomena. Thus, the Reynolds number expresses the ratio between scales characterizing the inertial and viscous terms of the Navier–Stokes equations. If the Reynolds number is large, this means that the viscous terms (they have as a factor the inverse of the Reynolds number $1/\mathrm{Re}$) are negligible compared to the inertia and pressure terms, leading to the Euler equations. We have seen that this leads to a loss of physics, because the no-slip condition could not be satisfied with the Euler equations. So there will be regions where the viscosity cannot be neglected, especially in boundary layers. We can make similar interpretations for other numbers. As is known, the Mach number is the ratio between fluid velocity V_∞ and the speed of sound a_∞ Now, a_∞ is a measure of internal energy e of the gas (for a calorically perfect gas:

$a^2 = \gamma r T = \gamma(\gamma - 1)e$), which reflects the molecular excitation motion. Thus, saying that the Mach number is small means that the gas velocity is small compared to the velocity of molecular motion.

The table in Appendix 2 summarizes most of the numbers encountered in aerodynamics. Some appear only in very particular problems far away from the concerns of the "classic" aerodynamicist. This is the case of the Peclet number, used primarily in thermics, and the Froude number, involved in hydrodynamics in the problems of air–water interface. The Stanton number is a dimensionless coefficient used to correlate the wall heat flux, similar to the friction coefficient, for example. We included it into this table because of its designation. The Crocco number (see section 9.5) is interesting because it includes the effect γ and remains finite, unlike the Mach number. However, it is rarely used.

Chapter 7

Classification of Flows

7.1. Viscous and non-viscous flows: Euler equations

Real fluids are viscous and heat conductors. These properties, known as transport properties, are the result of movement of molecules, which causes a transfer of momentum and heat within the fluid. However, experience shows that in many situations, these transport effects have little influence on the fluid motion. It is said that we then deal with a non-viscous fluid whose motion is governed by equations where the terms involving viscosity and thermal conductivity have been deleted. They are the Euler equations that are written (for an orthonormal coordinate system):

– *continuity equation:*

$$\frac{\partial \rho}{\partial t} + \frac{\partial}{\partial x_i}(\rho\, u_i) = 0\,,$$

– *equation of motion:*

$$\frac{\partial}{\partial t}(\rho\, u_i) + \frac{\partial}{\partial x_j}(\rho\, u_i\, u_j) = -\frac{\partial p}{\partial x_i}\,,$$

– *energy equation:*

$$\frac{\partial}{\partial t}(\rho\, e_T) + \frac{\partial}{\partial x_i}(\rho\, u_i e_T) = -\frac{\partial}{\partial x_i}(p\, u_i)\,.$$

These equations play an important role in aerodynamics as they represent the zero-order approximation of the Navier-Stokes equations corresponding to the limit of the infinite Reynolds number (see section 6.1). In fact, in most classic applications of aerodynamics, the Reynolds numbers are very high, Re_∞ being of the order of tens of millions for an airliner (the reference length is the chord of the wing). The transport term of standard equations (see section 6.1), whose factor is the inverse of the Reynolds number, then are an order of magnitude lower than the other terms of the equations, known as *Euler terms*. Thus, the viscosity can be considered ineffective in most areas of the aerodynamics flow field. Under these conditions, the Euler equations give, *almost everywhere*, an excellent description of the flow, except in some regions that are small compared to the body size (they are the boundary layers, wakes, and shock waves). Because of the omission of the transport terms, the Euler equations do not contain any parameters of similarity whose presence is related to these terms, so no length and velocity scales. In the absence of physico-chemical phenomena, the only conditions to be maintained, for example, between a real vehicle and a model, are similarities in shape, equal numbers of Mach and γ, and of course, identical boundary conditions.

The Euler equations involve partial derivatives of the first order. The terms containing second-order derivatives of the Navier-Stokes equations have been omitted. This leads to a decrease in the number of boundary conditions that can be satisfied: the no-slip condition on a wall can no longer be met. This is replaced by a less stringent slip condition, consisting of writing that the velocity vector is locally tangent to the surface, which is if $\vec{n}$ denotes the normal unit vector to the surface:

$$\vec{V}.\vec{n} = 0 .$$

Additionally, there will be no temperature conditions (see Figure 7.1).

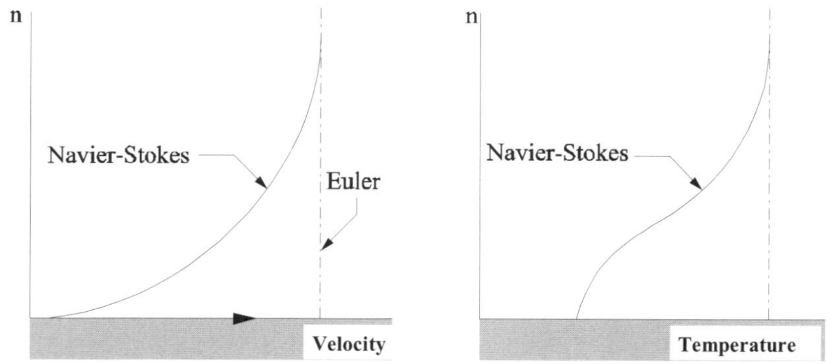

Figure 7.1. *Boundary conditions for the Euler equations*

Use of the Euler equations leads to a substantial simplification of the problem, both mathematically and numerically. However, if we want the no-slip condition to be taken into account and obtain a physically correct description of the flow, we cannot neglect the viscosity at all without causing inconsistencies. It must exist in regions with a small thickness δ compared to the characteristic dimension L of the body, the regions being the boundary layers, wakes, and shock waves.

Figure 7.2 shows the flow around a transonic airfoil for the angles of incidence of 0° to 9°, calculated by a numerical solution of the Navier-Stokes equations. In the first case, the viscous effects are confined to the upper and lower surfaces in thin boundary layers and a thin wake; therefore, we can consider that the field is quasi-totally non-viscous and we get a good prediction for the lift of the airfoil by using the Euler equations (but the prediction of viscous drag is impossible). However, at an angle of incidence of 9°, the boundary layer separates from the upper surface at about 25% of the chord, which greatly enhances the viscous zone, which then extends into a very thick wake. Now, the viscous field occupies a large area and we must expect a poor prediction of the lift by the Euler equations: we have to resort to the use of the Navier-Stokes equations.

It is noteworthy that in both cases, the Reynolds number characteristic for the flow, calculated with the upstream and the airfoil chord conditions, is the same. A high Reynolds number is not a guaranty of an almost negligible effect of the viscosity: when an extended separation occurs, the Reynolds number loses its original meaning.

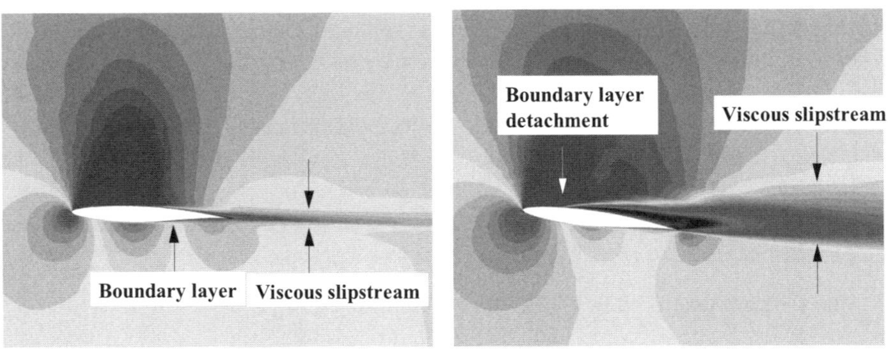

a – zero incidence: limited effects

b – angle of incidence of 9°: important effects

Figure 7.2. *Viscous effects on a transonic airfoil (calculated with the Onera code elsA)*

Reversibility of the Euler equations. It is easy to verify that the Euler equations are invariant under the change of sign of the time, the dimensions of space remaining unchanged. This has the effect of changing the sign of derivatives compared to time, thus, that of velocity. Therefore, the equations of continuity, motion, and energy are invariant.

To illustrate the idea, we consider a transonic flow around a stationary airfoil, which is the solution of the Euler equations, the flow flowing "normally" from the leading edge to the trailing edge, and we assume that a shock wave forms on the upper side (see Chapter 25). If we change the direction of the velocity field, the flow (which is then in effect flowing from the trailing edge to the leading edge), is still the solution of the Euler equations. But reversing the sign of the velocity causes the shock to become an expansion shock, the Mach number on its upstream face being subsonic and supersonic on the downstream, such a circumstance being ruled out by the second law of thermodynamics (see Chapter 13). The Euler equations must be supplemented by the explicit statement of the second law. This reversibility disappears with the Navier-Stokes equations, the so-called diffusion terms containing second derivatives of the velocity, which change sign with the reversal of time. Additionally, the Navier-Stokes equations contain the second principle, which is applied in the expression of the constitutive laws for viscosity and thermal conduction. It is not longer outside of the equations but contained within them.

Non-uniqueness of the solution. For given boundary conditions, there is no single solution to the Euler equations; they can allow an infinite number of solutions. The well-known example is that of the airfoil for which, in non-viscous fluid, a solution is obtained for any given value of the circulation. This is the *external condition*, known as the Kutta-Joukowsky condition, which sorts by imposing the solution closest to the physical reality, that is, a flow with a stagnation point at the trailing edge. We consider another simple case, that of the rearward facing step within a two-dimensional supersonic flow with the pressure p_0 and Mach number M_0 (see Figure 7.3). If we give a value p_1 of the pressure just downstream of the step such that $p_1 < p_0$, it is possible to construct a solution of the Euler equations made of (see chapters relative to supersonic flows):

– the uniform upstream flow (0) with Mach number M_0,

– a centered supersonic expansion caused by step-shoulder D and ensuring the expansion p_0 to p_1,

– the uniform flow (1) with Mach number M_1 at pressure p_1,

– an isobaric boundary (f) (slip line) at the pressure p_1 separating the flow (1) from the *dead water* (stagnant region) in contact with the base, this region being without velocity, and the pressure p_1,

– a downstream flow (2) with pressure p_2 and Mach number M_2 flowing along the downstream wall and separated from the flow (1) by a shock wave (*C*) starting off from the point *R* where (*f*) impacts with the wall.

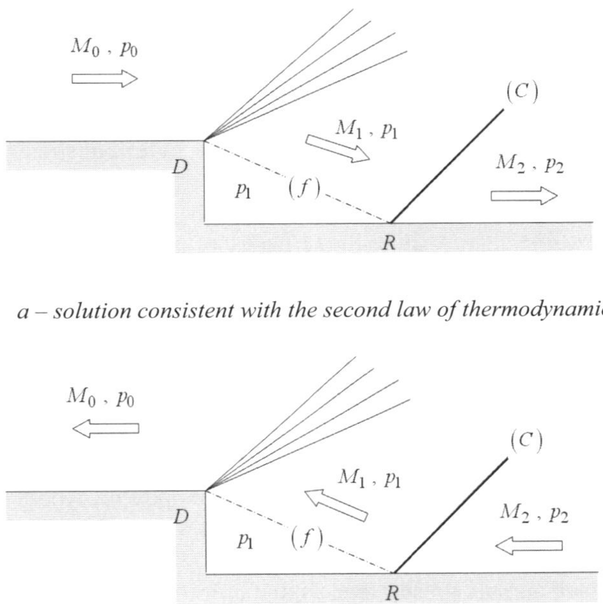

a – solution consistent with the second law of thermodynamics

b – reversed solution, contrary to the second law of thermodynamics

Figure 7.3. *Downstream reattachment, a rearward facing step: example of a non-unique and reversible solution*

This "patchwork" is made up of flows satisfying the Euler equations, since a uniform state is the solution of these equations. In addition, the boundary conditions (slip on the wall) are satisfied and the flow contains discontinuities compatible with the Euler equations. It is clear that such a solution will be obtained for any value of the pressure p_1, within limits we will not specify here (see section 14.2).

The problem of the supersonic rearward facing step has an infinite number of solutions in the context of Euler equations. In this case, as well as in that of the airfoil, it is the introduction of viscosity that will provide a unique solution, as observed in reality. We note that in the rearward facing step problem, we can change the direction of flow starting from an upstream state (2) flowing in the direction of step. However, in this case, the oblique shock (*C*) starting from *R* will be crossed in the way leading to a reduction of the entropy.

It is noteworthy that these behaviors are usually not observed when the Euler equations are solved numerically. Indeed, on one hand, the discretization of the equations introduces an apparent numerical viscosity. On the other hand, the addition of artificial viscosity to stabilize the calculation in regions with high gradient leads us to the conclusion that the solved equations are actually "viscous". An investigation of this problem would lead us too far; the reader is referred to treatises on numerical analysis.

Other properties. The system consisting of the Euler equations is a system of partial differential equations called pseudo-linear (or quasi-linear) in the sense that it does not involve products of derivatives. This remarkable property allows mathematical developments leading to the theory of characteristics (see Chapters 17 and 21). The full Euler equations form a *hyperbolic system in time*, which means that we can integrate from a given flow field at an initial time t_0 by advancing in time without taking into account the conditions at a later time: the present depends on the past and not on the future. This property is used in the numerical integration methods called *time marching*.

In steady flow (suppression of the time derivatives) the Euler equations form, according to the local value of Mach number, an elliptic or hyperbolic differential system. In the first case (Mach number less than 1), a modification of the boundary conditions (shape of the obstacle, for example) is felt throughout the space. In the second case (Mach number greater than 1), the influence of such changes is felt only in a finite domain. The elliptical nature of the Euler equations in subsonics makes their numerical solution difficult because of the continuing influence of boundary conditions throughout the calculation domain. One method is to integrate the unsteady equations, which have the property of being hyperbolic in time. Starting from an arbitrary initial field, we advance the solution in time until it reaches a steady state, if it exists for the imposed boundary conditions, that is, until the time derivatives become zero. It is clear that convergence is obtained more rapidly if the initial field is not too far from the steady solution.

Finally, the Euler equations admit solutions with discontinuities of the first order, so-called weak solutions: they are the shock waves and slip lines. These properties are studied in detail in the forthcoming chapters.

7.2. Compressible and incompressible flows

Some fluids, liquids in this case, are incompressible (but not completely), so that their density ρ can be taken as constant with a good degree of accuracy. Then the assumption $\rho = Cste$ is justified, which greatly simplifies the equations. In particular, except for thermal problems, there is no need to consider the energy equation: $\rho = \rho_\infty$ is a given quantity that no longer has to be connected to the pressure and temperature through an equation of state. Although the air, like all gases, is compressible, experience shows that ρ can be regarded as constant in flows whose velocity is small or moderate.

In a more precise manner, the change of ρ in a flow, or compressibility effect, can be characterized by the value of the local Mach number M:

$$M = \frac{V}{a}.$$

Anticipating the results to come (see section 9.5), we can show that the variation $\delta\rho$ of ρ is related to the variation δM of the Mach number by:

$$\frac{\delta\rho}{\rho} = -M^2 \frac{\delta M}{M}.$$

If the Mach number is not too high:

$$\frac{\delta M}{M} \approx \frac{\delta V}{V}.$$

Thus:

$$\frac{\delta\rho}{\rho} \approx -M^2 \frac{\delta V}{V}.$$

Accordingly, in a flow at low Mach number, and where velocity variations are not too large, which is connected to the hypothesis on the Mach number, the density varies very little and can be assumed constant maintaining a good degree of accuracy. In practice, the air flows, where $M < 0{,}3\ (V \approx 100\ \text{m/s})$, are classified as incompressible. This hypothesis can be extended to flight Mach numbers close to 0.5. However, we must be careful because around the studied body, for example a wing profile, there are regions of strong acceleration where the Mach number reaches locally much higher values, close to one, and where due to this, the compressibility effects are felt. For example, when the Mach number goes from 0.5 to 0.8, ρ varies by 20%, which is no longer negligible.

The assumption of total incompressibility is the basis of considerable theoretical developments in the field of fluid mechanics. It leads to the Laplace equation for the non-viscous and non-rotational flows, which paved the way for very successful mathematical solutions that are still the basis of predictive tools such as the method of singularities (also called panel method). The Laplace equation also underlies the thin airfoils theory and the lifting line of the pre-computer era, which has contributed so much to the aerodynamics of the wing with a finite span. However, with the development of airliners whose cruising Mach number lies in the high subsonics, not speaking of supersonic aircraft and hypersonic vehicles, the assumption of incompressibility is less and less usable, which tends to relegate to marginal situations in aviation the incompressible approaches. The use of increasingly massive numerical methods increased such marginalization by allowing a direct approach of real compressible problems, often by solving the complete Euler equations when taking into account that the viscosity is not essential.

Thus, paradoxically, the assumption of incompressibility adds a difficulty since a constant ρ corresponds to an infinite velocity for the propagation of pressure disturbances, hardly compatible with the methods of solving a system of hyperbolic equations by time marching (we remember that the speed of sound is given by $a^2 = \left(\dfrac{\partial p}{\partial \rho}\right)_s$, i.e. $a \to \infty$ if $\rho = constant$). Therefore, we sometimes used methods known as artificial compressibility methods, consisting of reintroducing a variable density via a circuitous process.

The effects of rarefaction (see section 4.1) viscosity, compressibility, occur gradually: there is no barrier between molecular and continuous regimes, between viscous and non-viscous fluids, between compressible and incompressible flows. There is a gradual transition from one situation to another, the distinctions corresponding to certain types of modeling or approximating of equations. However, the difference between subsonic and supersonic flows is associated with distinct behaviors within the framework of the same level of modeling or approximation.

7.3. Subsonic and supersonic flows

One, simplistic, way to understand this difference is to examine the propagation of a disturbance of small intensity (acoustic wave), by considering first an atmosphere at rest (these behaviors are analyzed more rigorously in Chapter 21). Such a disturbance propagates in the gas with a velocity that is, by definition, the speed of sound a_0. Emitted at an initial time at a point P, the disturbance attains a circle (in a two-dimensional space) with the radius $a_0 \Delta t$ at a later time $t = t_0 + \Delta t$ (see Figure 7.4a). Thus, the disturbance will spread to infinity if the fluid is non-dissipative (non-viscous), or to a great distance in the case of a real viscous fluid, where, little by little, the motion energy is dissipated into heat.

We now consider the propagation in a fluid moving with constant velocity $\vec{V}_\infty$ whose module V_∞ is less than $a_0 (|\vec{V}_\infty| < a_0)$, thus in a subsonic flow. The absolute velocity of the disturbance propagating in a direction identified by the unit vector $\vec{n}$ is $a_0 + \vec{V}_\infty . \vec{n}$. The locus of points reached by the disturbance at time Δt is a circle with the radius $a_0 \Delta t$; whose center is at a distance $\vec{V}_\infty \Delta t$ from the source point P (see Figure 7.4b). The disturbance propagates with the greatest speed $(a_0 + V_\infty)$ in the direction of $\vec{V}_\infty$ and with the minimum speed $(a_0 - V_\infty)$ in the opposite direction when it moves against the stream. Again in this case, the disturbance will eventually concern the whole plane (the whole three-dimensional space), perceived equally at upstream as well as downstream infinity.

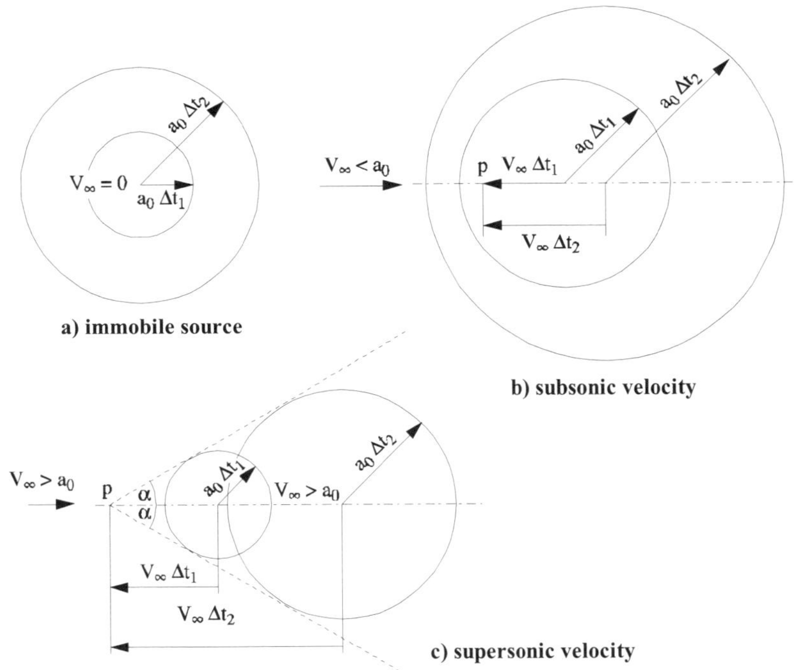

Figure 7.4. *Subsonic and supersonic flows*

In the third situation, $|\vec{V}_\infty|$ is greater than a_0, the flow is supersonic. Then the locus of points reached by the disturbance at a time Δt is a circle with the radius $a_0 \Delta t$ moved toward the right, according to the convention adopted in Figure 7.4c, and completely located to the right of the source point P. We see that the disturbance is only perceived downstream of P: it does not propagate against the stream. We verify that the points reached by the disturbance at times $n\Delta t$ are on circles which are enveloped by two lines issuing from *P*, making an angle α with respect to the velocity $\vec{V}_\infty$ with:

$$\sin\alpha = \frac{a_0 \; n\Delta t}{V_\infty \; n\Delta t} = \frac{a_0}{V_\infty}.$$

Knowing that $\frac{V_\infty}{a_0} = M$ is the Mach number, we see that:

$$\alpha = \sin^{-1}\frac{1}{M}.$$

The angle α is called the Mach angle (see Chapter 17). Thus, the disturbances emitted from a fixed point *P* will be perceived inside the sector of angle 2α, or cone angle in a three-dimensional space. In particular, what happens in *P* will not affect the upstream. The sonic transition state corresponding to:

$$\frac{V_\infty}{a_0} = M = 1$$

causes a dramatic change in the behavior of the flow. This behavior is controlled by the non-viscous part of the Navier-Stokes equations (Euler terms), where the *stationary* nature changes when the Mach number *M* passes through the unit value. These equations are:

– of an *elliptic type* if $M < 1$,

– of a *hyperbolic type* if $M > 1$.

The result is a profound change in the nature of the solutions.

Figure 7.5 plots the waves emitted by a source according to its travel velocity. It is observed that for a travel velocity equal to the speed of sound, there is an accumulation of waves at the source (see Figure 7.5c). This critical regime occurs whenever a source of disturbances moves at the velocity of propagation of the disturbances it emits, whether pressure waves as in this case, or deformation or displacement of the medium, such as waves on a pond.

The interpretation of subsonic and supersonic behavior is based on the assumption of an acoustic propagation of pressure waves. In fact, in classic acoustics pressure disturbances have a very low intensity (a few Pascals), i.e. they do not affect the properties of the medium in which they propagate so that their velocity is constant and equal to a_0. For intense disturbance, the wave modifies the properties of the medium so that the sound for successive waves has different propagation velocities. We enter the field of nonlinear acoustics: shock waves are formed by this mechanism (section 21.6). The plots in Figure 7.5 correspond to the propagation of a gentle "music"; airplanes cause much stronger disturbances and are accompanied by shock waves at supersonic velocities.

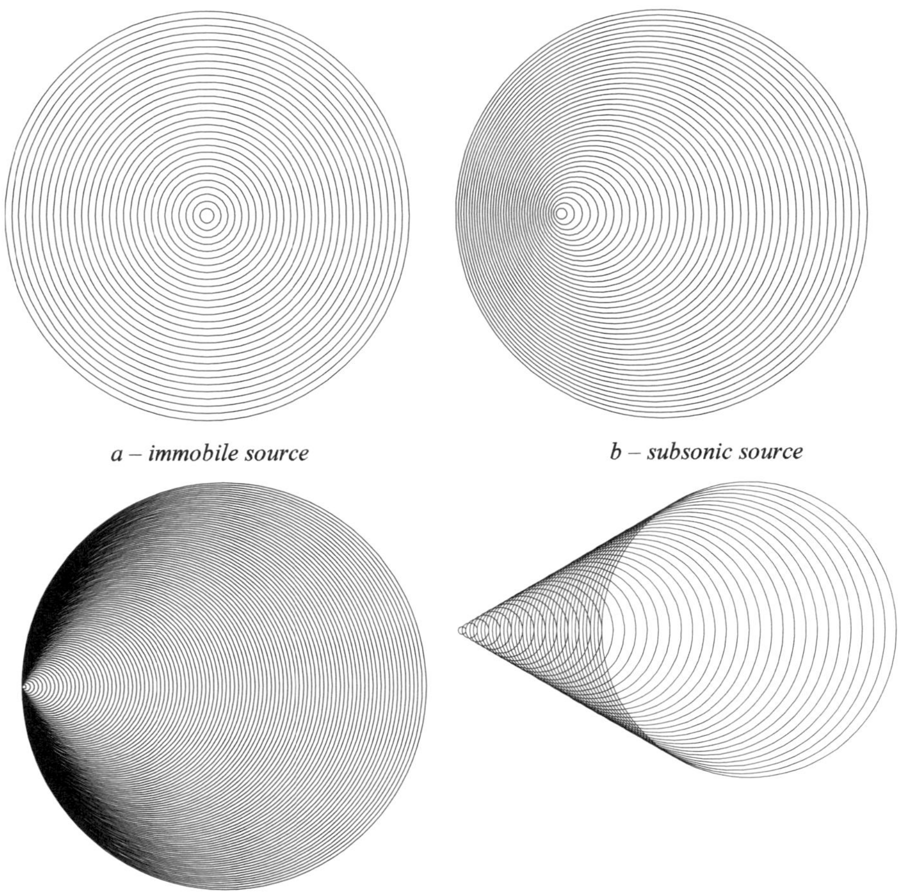

a – immobile source *b – subsonic source*

c – source at the velocity of sound *d – source at supersonic velocity*

Figure 7.5. *Propagation of acoustic waves according to the velocity of the source*

7.4. Physical and mathematical approximations

Concluding this chapter, Figure 7.6 summarizes the levels of approximation in the physical description of the motion of fluids. At the top is the Boltzmann equation, which is placed at the level of particles constituting the fluid, whose behavior is described by a distribution function. Depending on the density of particles, we can consider two less complex approaches: one is the case of rarefied fluids in molecular regime. Then a particle collides "occasionally" with the vehicle, which will be dealt with by direct simulation of the kind of Monte Carlo (*DSMC*). The other is applied to the fluid dense enough to be felt as a continuous medium to which the Navier-Stokes equations are applied.

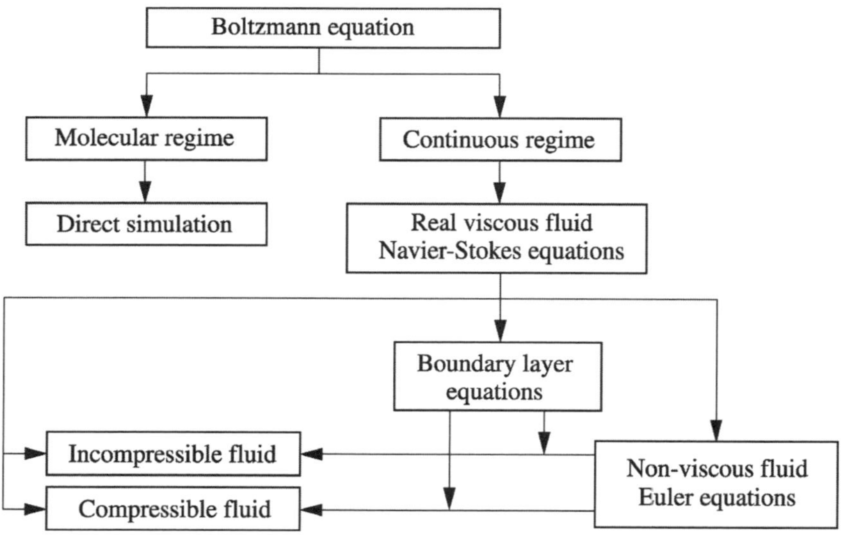

Figure 7.6. *The levels of physical approximation*

At this stage, three ways open as follows:

– the global Navier-Stokes approach, but with sublevels in the case of turbulent flows,

– the non-viscous approach, leading to considerable simplification of the Euler equations, but with the limitations we have seen,

– then an intermediate way where viscous effects are confined to thin layers, the rest of the flow being amenable to the Euler equations: they are the Prandtl equations of the boundary layer.

With each step down the stair toward simplification, we lose a part of the physical reality contained in the original description, limiting the fidelity and accuracy of the prediction. Such simplifications also lead to inconsistencies or unusual behavior that will have to be circumvented by more or less efficient tricks. Thus, with the Euler equations, there is no longer uniqueness of the solution: the no-slip condition on a surface is lost, making it impossible to calculate the drag, and there may be discontinuity surfaces (shock wave, wake sheets, and slip lines). These phenomena pose problems for behavior at the limit which, in fact, are artificial, as in reality the changes are always continuous, at least if the hypothesis of a continuous medium applies. Under the assumption of incompressibility, the velocity of the propagation of pressure disturbances is infinite: then the difficulties are overcome by *ad hoc* techniques.

Levels of mathematical complexity are associated with each level of physical simplification (see Figure 7.7). Starting from the general case of the three-dimensional space and time, we are led to distinguish the non-steady and steady cases for which the flow is assumed to depend only on the space variables. It is a common situation, for example, in the case of an aircraft in cruising flight or moving so slow, compared to the characteristic aerodynamic time, that at each moment, the flow can be considered steady. This is called the pseudo-steady approach.

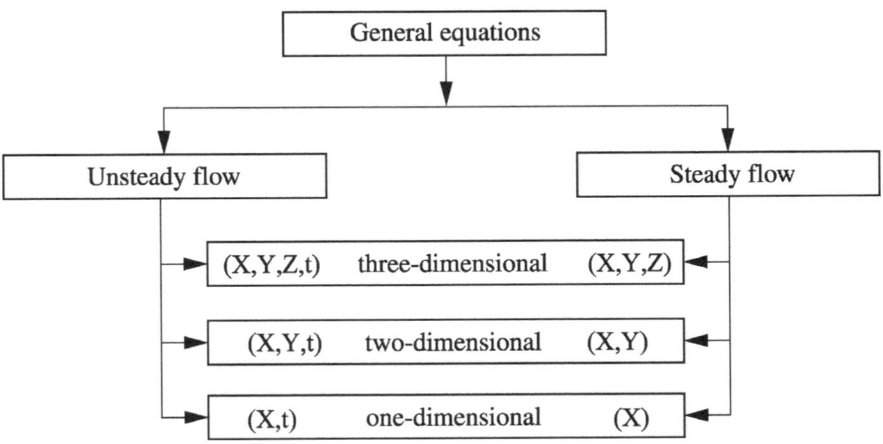

Figure 7.7. *The levels of mathematical approximation*

Nevertheless, only the non-steady approach allows us to understand certain behaviors, all flows having a starting and an establishment stage: it is the transitory (or transient) regime (see the starting process of a nozzle, section 24.5). The three-dimensional space can be reduced to two-dimensional to simplify the solution of equations; although in reality the situations where the flow can be considered two-dimensional are rare (the cases of axisymmetric nozzles of rocket engines and of airfoils).

However, the two dimensions are a prerequisite for understanding complex phenomena, where, despite the assistance provided by modern visualization graphics, their conceptualization and perception in a three-dimensional space are difficult.

The case of a one-dimensional space seems to be the ultimate avatar: nevertheless, the one-dimensional assumption allows us to establish general fundamental results and to assimilate the phenomena of wave propagation.

Chapter 8

Fundamental Concepts of Fluid Mechanics

8.1. Trajectories, streamlines and streaklines

The understanding of the organization of a flow cannot be based solely on knowledge of its local properties, velocity, pressure, density, temperature, etc. at each point in space (x, y, z) and every time t. We often need a more "dynamic" representation by using the motion of fluid particles, or at least the direction of the displacement. Such a visualization of the flow is a valuable aid in the design of vehicles or systems using fluids. It plays a vital role in the study of separated three-dimensional flows whose understanding is based on the topological analysis of the velocity and skin friction fields. To assist this representation, we use the concepts of trajectory and streamlines, and to a lesser extent, streak lines.

8.1.1. *Trajectories in a moving fluid*

We consider an *unsteady* flow in the three-dimensional space whose velocity field $\vec{V}(x, y, z, t)$ is known at any point (x, y, z) and at any time t. Given a fluid element (E_1) passing through the point P_0 at time t_0 (see Figure 8.1a), at a later time $t = t_0 + \Delta t$, that same element can be found at a point P_1. The path followed by (E_1) on its displacement from P_0 to P_1 is called a *trajectory*.

We consider another fluid element (E_2) passing through the same point P_0 at a later time t_1. At time $t = t_0 + \Delta t$, (E_2) has no reason to be in P_1 if the velocity field depends on time. It will have reached a position P_2.

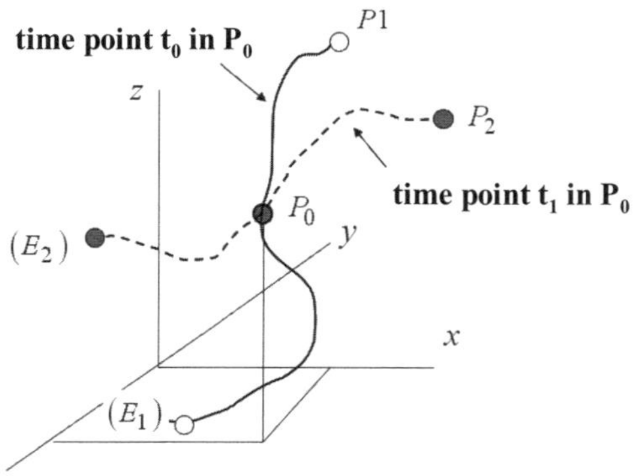

a – Eulerian description: the particles passing through P_0 at times t_1 and t_2 are not the same

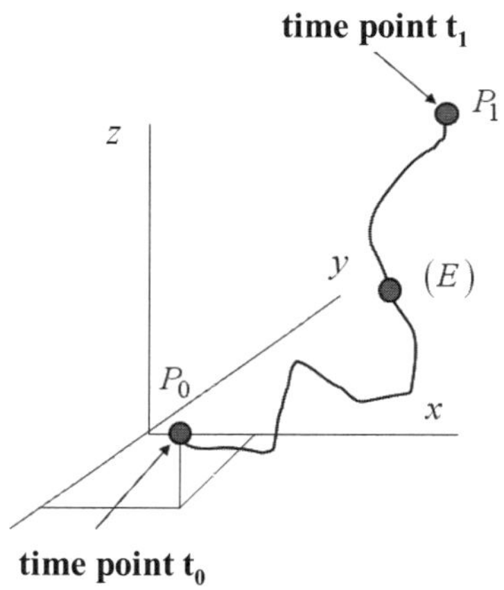

b – Lagrangian description: we follow the same particle in its movement

Figure 8.1. *Trajectory in unsteady flow*

The trajectories are the curves followed by fluid elements during their movement under the effect of the velocity field. We can give a mathematical definition of trajectories by observing that this concept corresponds to a description of motion using *Lagrange variables*, where we follow a fluid element in its movement that passes through a position (x_0, y_0, z_0) at time t_0 (see Figure 8.1b). The description of motion is then obtained by determining the coordinates (x, y, z) of the same element at any later time *t*, giving:

$$x = f(x_0, y_0, z_0, t)$$

$$y = g(x_0, y_0, z_0, t) \qquad [8.1]$$

$$z = h(x_0, y_0, z_0, t)$$

The independent variables (x_0, y_0, z_0) are called Lagrange variables. The velocity field $\vec{V}$ is given by:

$$\vec{V}\begin{cases} \dfrac{\partial f}{\partial t}(x_0, y_0, z_0, t) \equiv u\,(x, y, z, t) \\ \dfrac{\partial g}{\partial t}(x_0, y_0, z_0, t) \equiv v\,(x, y, z, t) \\ \dfrac{\partial h}{\partial t}(x_0, y_0, z_0, t) \equiv w\,(x, y, z, t) \end{cases} \qquad [8.2]$$

The trajectories are defined by equations [8.1] by taking as conditions at a time t_0 the values (x_0, y_0, z_0), satisfying:

$$\begin{aligned} x_0 &= f\,(x_0, y_0, z_0, t_0) \\ y_0 &= g\,(x_0, y_0, z_0, t_0) \\ z_0 &= h\,(x_0, y_0, z_0, t_0) \end{aligned}$$

The trajectory is also the solution of the differential system:

$$\boxed{\frac{dx}{u(x, y, z, t)} = \frac{dy}{v(x, y, z, t)} = \frac{dz}{w(x, y, z, t)} = dt} \qquad [8.3]$$

with as initial conditions the position (x_0, y_0, z_0) at time t_0. The solution of the system [8.3] allows us to determine the displacement of the fluid element, which was in (x_0, y_0, z_0) at $t = t_0$ as a function of time. We can also describe the movement by considering the point P_0, fixed in space. Then the fluid particles at different times will not be the same, such a description is called *Eulerian* (see Figure 8.1a). In this case, we describe the movement from components of the velocity, called *Euler variables*, which will be determined as a function of time and space coordinates. Therefore, the solution appears as a field of velocity vectors. This is the description that is most often used in fluid mechanics of continuous and monophasic media.

We can construct the trajectory of a fluid element by injecting into the flow a tracer, made of a solid particles, sufficiently small to be carried by the fluid with the velocity of the flow. The tracer, made visible by illumination, is filmed under different angles. The frame by frame analysis of films allows us to reconstruct the position of the particle tracer in the three-dimensional space and to trace the curve made of the points it has occupied for the different instants of the recorded images. This is its trajectory precisely.

8.1.2. *Streamlines of a moving fluid*

By definition, a streamline is a curve, which, in each of its points, is tangential to the velocity vector at this point. To use the vocabulary of electromagnetism, it is a line of force of the velocity field. A streamline *is defined at a time T*, where the velocity field is "frozen" (see Figure 8.2). If *ds* is an elementary displacement on a streamline (components *dx*, *dy*, and *dz*), this line is a curve, which is the solution of the differential system:

$$\boxed{\frac{dx}{u(x,y,z,T)} = \frac{dy}{v(x,y,z,T)} = \frac{dz}{w(x,y,z,T)} = \frac{ds}{V(x,y,z,T)}} \qquad [8.4]$$

where (u,v,w) are the components of the velocity in the Cartesian system (x,y,z) and V is the velocity module, given as:

$$V = \sqrt{u^2 + v^2 + w^2}\ .$$

The differential system [8.4] is integrated from an initial point with the coordinates (x_0, y_0, z_0) and by progressing in the velocity field. Thus, by choosing a set of N departure points, it is possible to calculate N streamlines of the velocity field, frozen at the time T. In general, the streamlines vary as a function of time: a set of streamlines corresponds to each time T, defining the topology of the field at the time T.

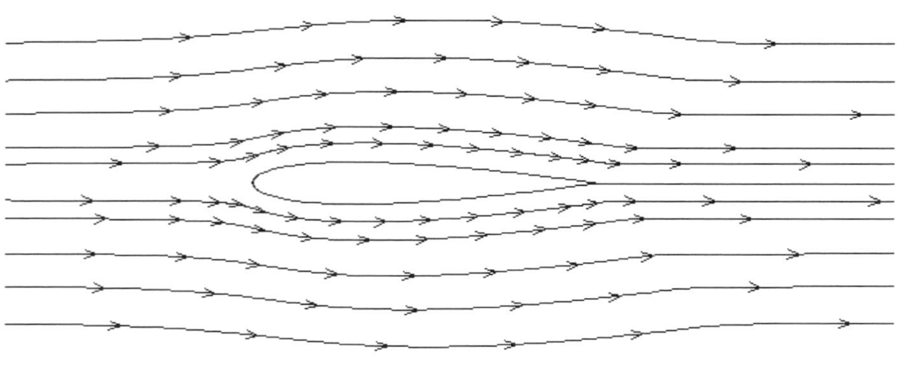

Figure 8.2. *Streamlines of the velocity field around an airfoil*

The resolution of [8.4] allows us to determine the streamline passing through a point of coordinates (x_0, y_0, z_0). In principle, it is possible to get this line in the form of a function: $f(x,y,z) = 0$. In fact, an analytical representation of streamlines is rare, the vast majority of solutions to problems of fluid mechanics being obtained numerically or by probing the flows. In reality, to construct the streamlines of a known velocity field, we use step-by-step integration methods. With the field $\vec{V}(x, y, z, T)$ frozen at time *T*, we choose a point of origin P_0 of the streamline to draw. The displacement on the streamline is defined by the variable *s*, we determine the point P_1 close to P_0 with:

$$\overrightarrow{P_0P_1} = \vec{V}_{P_0}\, ds$$

where $\vec{V}_{P_0} \equiv \vec{V}\left(x_{P_0}, y_{P_0}, z_{P_0}, T\right)$ is the velocity in P_0. Having determined P_1, it is possible to calculate the velocity $\vec{V}_{P_1}$ in this point (the velocity field is known), which allows us to improve the precision by starting from P_0 with a velocity equal to the average of $\vec{V}_{P_0}$ and $\vec{V}_{P_1}$ from which follows a new point P_1:

$$\overrightarrow{P_0P_1} = \frac{1}{2}\left(\vec{V}_{P_0} + \vec{V}_{P_1}\right) ds\,.$$

The accuracy of the construction can be improved by iterating until the position P_1 no longer moves within a given tolerance. This construction is difficult in the regions of the field, where the velocity vector varies rapidly in direction, as in vortex flows. Consequently, we then have to adjust the step *ds* in such a way as to be sure that the so constructed curve follows the streamline. It may also be necessary to interpolate additional points if the points where the velocity is known are too far apart to allow a "continuation" of the streamline.

In *unsteady* flow, trajectories and streamlines are distinct curves associated with different concepts:

– In the tracing of a trajectory, we follow the movement of a fluid element over time, which allows a cinematographic view of the flow.

– A streamline gives information on the velocity field at a given time: we take a snapshot of the movement.

8.1.3. *Streak lines of a moving fluid*

We call streak line (or emission line) the locus of points occupied at time *t* by all fluid particles which passed through a given point *P* since the time t_0 prior to *t*. We visualize a streak line by photographing with a short exposure time a dye introduced into a water flow from an orifice placed on a model. At the shooting time, the image freezes the dye that had been emitted since an earlier time. We get the same result by taking a snapshot of smoke particles emitted punctually into an airflow.

Figure 8.3 is a photograph showing the traces left by a dye emitted from orifices placed at the ends of the blades of a helicopter rotor model, tested in a water tunnel. As the rotor is rotating, the flow is unsteady. The picture shows at time t of the shooting the traces of the dye that were issued from each blade tip since a time t_0, which was the start of the rotation of the rotor. Therefore, we clearly see the location of the particles emitted at a point P (each blade tip) for a time prior to the one looked at. The dye displays streak lines.

Figure 8.3. *Streak lines from the blade tips of a helicopter rotor (image by Henri Werlé, document Onera)*

8.1.4. *The case of steady flow: stream surface and stream tube*

A flow is said to be steady when its properties at a point (x, y, z) in space are independent of time. It is easy to convince us that in such a flow, all trajectories passing through the same point 1 are merged. These curves are, in fact, defined by the system [8.3], which says that the solution curves are tangential to the local velocity vector in each of their points. Therefore, if we consider a path through point 1, as the velocity field $\vec{V} = \vec{V}(x, y, z)$ is independent of time, the trajectory will be invariant with time.

Under these conditions, the streamlines taken at time T will also be invariant and coincide with the trajectories. It is the same for streak line. Hence, the important result:

In a steady flow, trajectories, streamlines and streak lines coincide.

Figure 8.4 shows a visualization of the flow around an airfoil done in a water tunnel. A white dye is emitted upstream of the profile from orifices arranged on a rake (not visible in the picture). The picture therefore shows emission lines of the tracer. As the flow is steady, these lines are streamlines of the flow around an airfoil.

The streamlines of a steady flow are the solution curves of a differential system:

$$\frac{dx}{u(x, y, z)} = \frac{dy}{v(x, y, z)} = \frac{dz}{w(x, y, z)} \qquad [8.5]$$

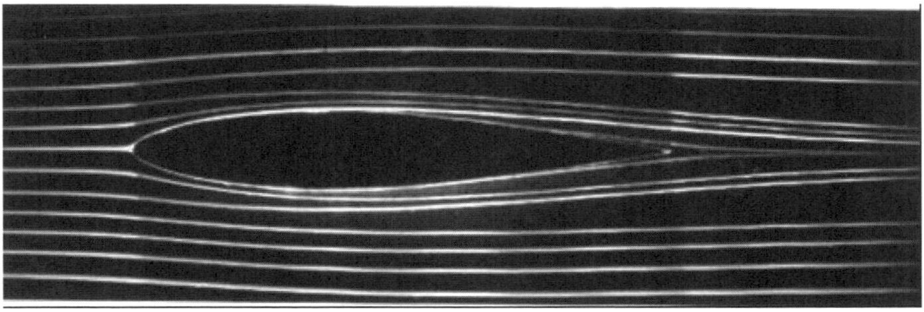

Figure 8.4. *Visualization of the flow around an airfoil (image by Henri Werlé, document Onera)*

which shows the vector $\vec{V}$ is collinear with the trajectory element *ds*, which can also be expressed as:

$$\vec{V} \times \overrightarrow{ds} = 0$$

where the symbol "$\times$" denotes the vector product. The system [8.5] can be written as follows:

$$\begin{cases} wdy - vdz = 0 \\ udz - wdx = 0 \\ vdx - udy = 0 \end{cases}$$

If the velocity field is being known, the integration of the above system gives the trajectory from a point P_0 that has been fixed. The solution is rarely obtained in analytical form and has to be made as a point after point construction of the streamline (see above). In case of a two-dimensional flow, the differential equation defining a streamline is simply:

$$vdx - udy = 0$$

or:

$$\frac{dy}{dx} = \frac{v}{u}$$

which shows that at every point the streamline is tangential to the velocity vector.

We also use the concept of the *stream surface* that is defined by considering a curve (C) of the three-dimensional space (see Figure 8.5). Through each point of (C) a streamline passes, which together form a surface called the stream surface. If (C) is a closed curve, then the stream surface is called a *stream tube*. We call sometimes stream tube the fluid domain limited by a stream surface, which closes on itself: in one case, it is the container, in the other, it is the content. The distinctive feature of a stream surface is that it is not traversed by any fluid flow, since on such a surface $\vec{V}.\vec{n} = 0$, if $\vec{n}$ is the normal unit vector to the surface. It follows that in steady flow, the mass flow through any section of the stream tube is constant.

Streamlines and stream surfaces play a fundamental role in the description of three-dimensional flows, where the perception of the field is sometimes difficult.

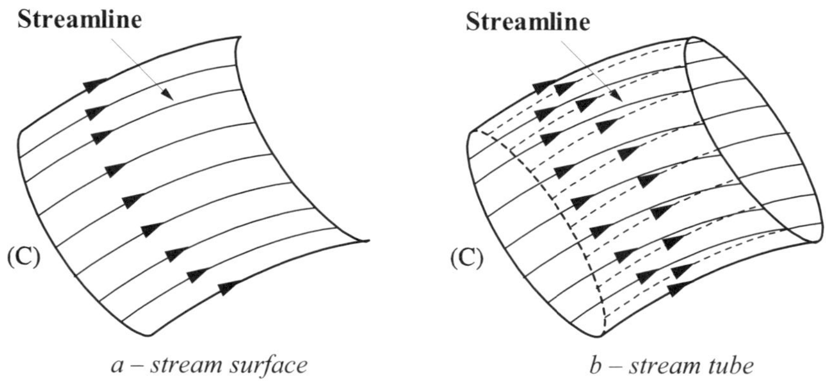

Figure 8.5. *Stream surface and stream tube*

8.2. Vorticity, circulation, stream function and velocity potential

8.2.1. *Vorticity and Crocco's theorem*

The vorticity vector $\vec{\omega}$, or more simply vorticity, whose use is almost constant in fluid mechanics, is a mathematically well-defined concept but whose physical interpretation always leaves some dissatisfaction. Interpreting the vorticity as a local rotation of a fluid element, itself a rather vague concept, is often difficult to perceive: some vortices, where the very fast flow rotates, are irrotational! Here, we adopt a purely formal introduction to the vorticity, the reader is referred to books of fluid mechanics for a more rigorous definition and a thorough interpretation of the concept of rotational flow. In three-dimensional flows, separation is at the origin of vortex structures where the vorticity, produced in the *boundary layers* under the action of viscosity, tends to concentrate.

We consider the intrinsic acceleration expression (see Appendix 1):

$$\frac{d\vec{V}}{dt} = \frac{\partial \vec{V}}{\partial t} + \overrightarrow{grad}\,\frac{V^2}{2} + \overrightarrow{curl}\,\vec{V} \times \vec{V}\,.$$

The above relation involves the curl of the velocity vector, which is written in the following way in a system of orthogonal axes of base vectors $(\vec{i}, \vec{j}, \vec{k})$:

$$\overrightarrow{curl}\,\vec{V} \equiv \vec{\omega} = \left(\frac{\partial w}{\partial y} - \frac{\partial v}{\partial z}\right)\vec{i} + \left(\frac{\partial u}{\partial z} - \frac{\partial w}{\partial x}\right)\vec{j} + \left(\frac{\partial v}{\partial x} - \frac{\partial u}{\partial y}\right)\vec{k}$$

We can also write in a symbolic way through the nabla operator:

$$\vec{\omega} = \nabla \times \vec{V}$$

The dimension of the vorticity vector, is the inverse of time: it is a frequency. In a two-dimensional flow, due to invariance of u and v according to the transverse coordinate z and the fact that $w = 0$ (by definition), the vorticity has a single component:

$$\omega_z = \left(\frac{\partial v}{\partial x} - \frac{\partial u}{\partial y} \right) \vec{k}$$

which is perpendicular to the plane containing the flow.

A flow is called *irrotational* if the vorticity, i.e. the curl of the velocity vector is zero. This property can either be local and only verified in certain regions of the flow, or global, that is, true everywhere in the field. The vorticity term is far from being purely formal, since the fact that the flow to be irrotational can greatly facilitate the establishment of a solution to the equations of motion. But what is remarkable is that irrotationality is a very common property, characteristic for a large number of flows of practical interest. Furthermore, we will see that vorticity, entropy, and viscosity are closely related quantities.

We consider a flow without body forces and introduce the specific total or stagnation enthalpy:

$$h_{st} = h + \frac{V^2}{2}.$$

We have for the gradient of this quantity:

$$\overrightarrow{grad}\ h_{st} = \overrightarrow{grad}\ h + \overrightarrow{grad}\ \frac{V^2}{2}.$$

According to thermodynamics (see section 3.4):

$$\overrightarrow{grad}\ h = T\ \overrightarrow{grad}\ s + \frac{1}{\rho}\ \overrightarrow{grad}\ p$$

where s is the specific entropy.

If the fluid is *non-viscous,* the equation of motion is reduced to:

$$\frac{d\vec{V}}{dt} = -\frac{1}{\rho}\,\overrightarrow{grad}\; p\,.$$

By combining the above relations and the expression $\dfrac{d\vec{V}}{dt}$, we obtain the *Crocco equation or theorem*:

$$\boxed{\overrightarrow{grad}\; h_{st} = T\,\overrightarrow{grad}\; s + \vec{V} \times \overrightarrow{curl}\; \vec{V} = T\,\overrightarrow{grad}\; s + \vec{V} \times \vec{\omega}}$$

which establishes a relation between the vorticity and entropy changes in a flow, where viscosity plays a negligible role and no body forces exist.

Assuming that a constant stagnation enthalpy is not a very restrictive assumption, it is indeed verified in a non-viscous and non-heat-conducting steady flow, in the absence of body forces. Then, the energy equation shows that h_{st} is constant on a streamline (see section 4.6). In addition, if the flow is uniform upstream, which corresponds to the displacement of a vehicle in an atmosphere at rest, the stagnation enthalpy is constant everywhere. Such a flow is called isoenergetic. Thus, the Crocco equation is reduced to:

$$\boxed{T\,\overrightarrow{grad}\; s + \vec{V} \times \overrightarrow{curl}\; \vec{V} = 0}$$

with the relation showing that if the entropy is constant in the flow:

– either the vorticity vector is aligned with the velocity vector,

– or the vorticity vector is zero.

The first conclusion concerns very special flows. In contrast, the second result concerns a large number of situations. Indeed, as noted in section 4.7, the energy equation written with the entropy s shows that the only terms, likely to change s, are those involving viscosity and thermal conductivity (excluding flows with chemical relaxation). In a classic aerodynamic flow, these effects occur in regions with dimensions most often small compared to the size of the vehicle: boundary layers, wakes, and shock waves. If we consider a flow without shock waves, the entropy is almost constant in the entire field. The Crocco relation then establishes that it is irrotational almost everywhere.

This conclusion must be modified if the flow contains a curved shock wave. The shock wave theory (see Chapter 13) shows that the entropy production across a shock depends on the local slope of the shock with respect to the upstream velocity vector. Thus, downstream of a curved shock, the streamlines have different levels of entropy. Behind the shock the flow can again be considered non-viscous, but the term $\overrightarrow{grad}\ s$ is not zero and therefore the flow is rotational.

This situation occurs on the upper surface of an airfoil on which a curved shock wave forms in high subsonic (see Figure 8.6). In this case, the region where the vorticity is non-zero has large dimensions, a fact which must be taken into account in the calculation. It should be noted that downstream of a straight shock, the field has constant entropy levels, thus it is irrotational.

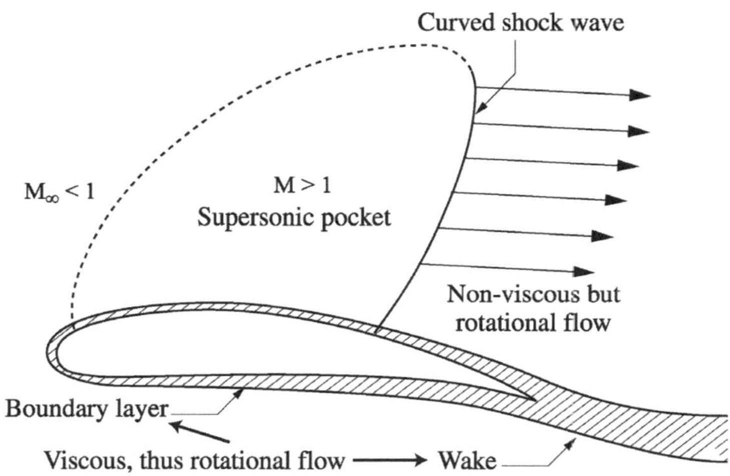

Figure 8.6. *Regions with non-zero vorticity around a transonic airfoil*

A rotational region exists also downstream of the detached curved shock, which forms in front of a blunt ogive. In hypersonic flows, where shocks are very intense, this region is the seat of a strong gradient of entropy: it is called *entropy layer.*

Crocco's theorem also shows the connection between a stagnation enthalpy gradient and vorticity. Such a situation is found downstream of a propeller, a wind turbine, or any rotating device causing a non-uniform distribution of the stagnation enthalpy within the flow. This causes to take a few precautions before deciding a flow is irrotational when its entropy is constant.

8.2.2. *Circulation: Kelvin, Lagrange and Helmholtz theorems*

We consider any closed curve (*C*) within a flow. At any point of (*C*), there is a velocity vector $\vec{V}$ (see Figure 8.7). The circulation, denoted by Γ, is defined by:

$$\boxed{\Gamma = -\oint_{(C)} \vec{V}.\vec{t}\ dl}$$

where $\vec{t}$ is the unit vector tangent to (*C*) and *l* the curvilinear abscissa on (*C*). The circulation plays an important role in determining the lift of an airfoil.

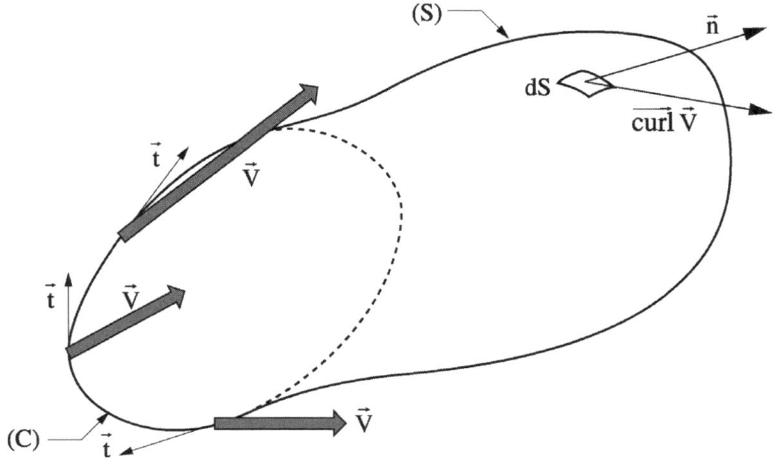

Figure 8.7. *Circulation over a closed curve (C)*

Given any open surface (*S*) based on the contour (*C*), at each point *P* of (*S*) there is a velocity $\vec{V}$, and according to the Stokes' theorem, we can write:

$$\Gamma = -\oint_{(C)} \vec{V}.\vec{t}\ dl = -\iint_{(S)} \overrightarrow{curl}\ \vec{V}.\vec{n}\ dS$$

where $\vec{n}$ is the unit vector normal to the surface element *dS*.

We can state:

The circulation around a closed curve (*C*) is equal to the flux of vorticity through any open surface (*S*), based on (*C*).

leading to:

The circulation is zero in a flow that is irrotational within the domain delimited by the contour (*C*).

We consider that the curve (*C*) moves by being carried along by the fluid; each of its points moving with the velocity $\vec{V}$. Let us evaluate the resulting variation of Γ over time. It is obtained from the particle (material) derivative:

$$\frac{d\Gamma}{dt} = -\oint_{(C)} \frac{d\vec{V}}{dt}.\vec{t}\ dl - \oint_{(C)} \vec{V}.\frac{d}{dt}\left(\vec{t}\ dl\right).$$

However, as the segment $\vec{t}\ dl$ follows the fluid in its movement, it moves with the velocity $\vec{V}$ in each of its points, thus:

$$\frac{d}{dt}\left(\vec{t}\ dl\right) = d\vec{V}.$$

Simply this term is the change in velocity when moving around the contour (*C*). Thus:

$$\oint_{(C)} \vec{V}.\frac{d}{dt}\left(\vec{t} dl\right) = \oint_{(C)} \vec{V}.d\vec{V} = \oint_{(C)} d\left(\frac{V^2}{2}\right).$$

If the scalar function $\frac{V^2}{2}$ is uniform, its integral over the closed contour is zero. Thus:

$$\frac{d\Gamma}{dt} = -\oint_{(C)} \frac{d\vec{V}}{dt}.\vec{t}\ dl$$

Knowing that for a *non-viscous* fluid and in the absence of body forces:

$$\frac{d\vec{V}}{dt} = -\frac{1}{\rho}\overrightarrow{grad}\ p = -\overrightarrow{grad}\ h + T\ \overrightarrow{grad}\ s$$

it follows:

$$\frac{d\Gamma}{dt} = \oint_{(C)} \left(\overrightarrow{grad}\ h - T\ \overrightarrow{grad}\ s\right)\vec{t}\ dl\ .$$

But:

$\overrightarrow{grad}\ h.\vec{t}\ dl = dh$ (variation of *h* for a displacement over (*C*)

$\overrightarrow{grad}\ s.\vec{t}\ dl = ds$ (variation of *s* for a displacement over (*C*)

and as *h* is a uniform scalar function, it remains:

$$\frac{d\Gamma}{dt} = -\oint_{(C)} T\ ds.$$

Thus, in general, the circulation on a closed curve (*C*) does not remain constant over time, even for a non-viscous fluid. The velocity of change of Γ is related to variations of entropy and temperature distributions on (*C*). The derivative $d\Gamma / dt$ disappears only if the fluid is *isothermal* or if the flow is *isentropic*.

If these conditions are satisfied, then:

$$\boxed{\frac{d\Gamma}{dt} = 0}$$

which is the Kelvin theorem; or, which is the same:

In a non-viscous and isentropic fluid, and in the absence of body forces, the circulation along a closed curve, consisting of the same fluid particles, is conserved.

The above theorem is sometimes established in the framework of the hypothesis of a *barotropic* fluid and then reads: for a non-viscous barotropic fluid, and in the absence of body forces, the circulation along a closed curve consisting of the same fluid particles is conserved. We call a fluid whose pressure p depends only on the density ρ barotropic. Thus, the isobars are also iso-density curves. We know that for a divariant fluid, all thermodynamic properties are deduced from two of them (see Chapter 3). Thus:

$$p = p(\rho, s)$$

So, if the entropy is constant throughout the flow there is a direct relation between p and ρ. For a calorically ideal gas, we saw that:

$$\frac{p}{\rho^{\gamma}} = constant$$

In an isentropic flow, the fluid is barotropic. Thus, both statements are equivalent.

Considering a contour (C_0) enclosing at the moment t_0 a surface (D_0) belonging to a region (D) where the vorticity is everywhere zero. The circulation Γ_0 around (C_0) is also zero. At a later time t, the contour (C_0) is now (C) with an associated circulation Γ and according to Kelvin's theorem:

$$\frac{d\Gamma}{dt} = 0.$$

Therefore, $\Gamma = \Gamma_0 = 0$. It follows that (C) encloses a surface through which the vorticity is zero. This property is true for any curve of (D) as the flow is irrotational everywhere. Hence, the theorem of Lagrange:

In a non-viscous, isentropic fluid and without body forces, an irrotational flow at a given time remains irrotational at a later time.

This result is crucial because it demonstrates that if a flow upstream of a vehicle is irrotational everywhere, corresponding to flying in an atmosphere at rest, then the flow around the vehicle will remain irrotational (subject to the made assumption), which implies the absence of boundary layers and shock waves. We could study this flow from the theory of irrotational flow that leads to simple and powerful methods.

Application of the Kelvin theorem

We recall that Stokes' theorem shows that the circulation Γ along a curve (*C*), on which a surface (*S*) is based is equal to the vorticity flux through this surface (see above). Kelvin's theorem on the conservation of the circulation does not imply that the vorticity must be zero for any surface (*S*), based on (*C*) but:

– that (*C*) is located in a region where the flow is non-viscous and isentropic,

– that the vorticity flux through (*S*) is zero.

If we follow the contour (*C*) carried along by the fluid, it is possible that vorticity is created in the region delimited by (*C*). The condition for the application of Kelvin's theorem is that the region where the vorticity is created does not reach (*C*).

We find ourselves in such a situation during the flight of an aircraft in the atmosphere, far from the ground. The wings of the aircraft emit from their tips very strong vortices (we should rather say vortex structures). These vortices are a direct consequence of the wing lift and concentrate the entropy, or the vorticity produced in the boundary layers. Thus, the aircraft is followed by two large vortices forming what is called a *vortex wake*. In fact, other vortices are created by the aircraft, but generally, they are less strong than the vortices at the wing tips. Here, we give a simplified picture of reality.

Imagine a closed contour (*C*) of vast dimensions originally located upstream of the aircraft, where the flow is everywhere irrotational (see Figure 8.8), and consider the air flowing around a fixed aircraft. The circulation Γ on this contour is zero. The contour (*C*) is carried along by the flow and passes around the aircraft and then continues its way downstream. If this contour has been chosen large enough, it never meets the wakes, that is, regions where the vorticity is not zero. According to Kelvin's theorem, the circulation on (*C*) is always zero and the vorticity flux across the surface based on (*C*) downstream of the plane is zero. This means that the vorticity produced by the boundary layer has a zero flux through any plane intersecting the trajectory downstream of the aircraft. In other words, regions with positive vorticity are compensated by the regions with negative vorticity. This implies the existence of two vortices rotating in opposite directions and balancing each other.

For an object less symmetric and less refined than an aircraft, for example, a car (but then with attention to the boundary layer of the road), the wake can be composed of a large number of completely different vortices: the rule is that the vorticity flux for this set of vortices is zero. To put it in simple terms, the vortices rotating in one direction compensate those turning in the other direction.

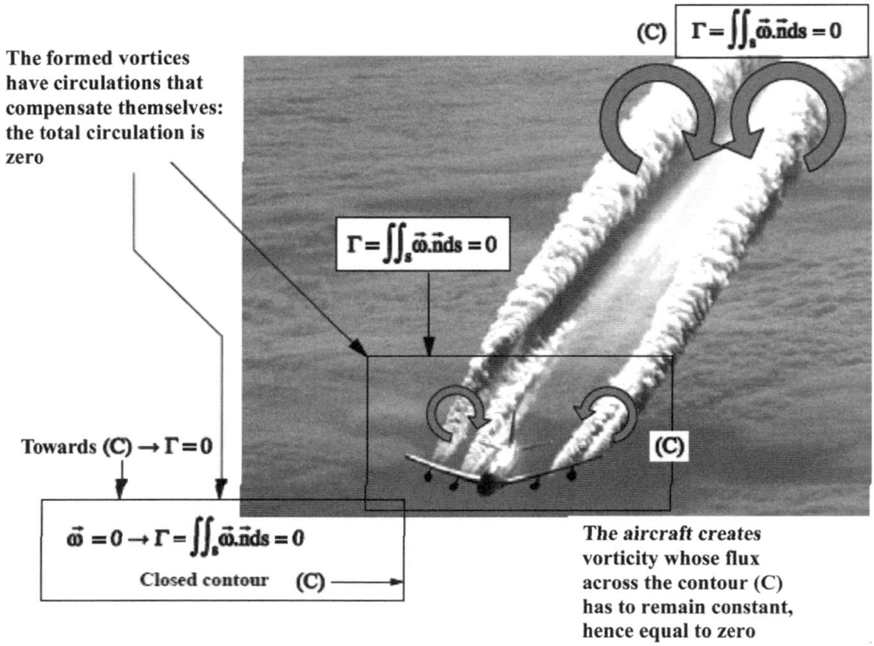

Figure 8.8. *Vortex wake of an aircraft and the Kelvin theorem*

Vortex lines and vortex surfaces: Helmholtz theorem

A vorticity field can be associated with any vector field, the vorticity possibly being zero everywhere or by domains. Following the streamlines definition, it is possible to define vortex lines as lines tangent in each of their point to the local vorticity vector. Then by considering a curve (*C*) we can introduce vortex surfaces and if the curve (*C*) is closed, vortex tubes.

If $\vec{\omega}$ is the vorticity vector and $\vec{n}$ the normal unit vector to the vortex surface (Σ), then by definition:

$$\vec{\omega}\,.\,\vec{n}\;=0.$$

We consider the vortex surface $(\mathrm{T}) \equiv (\mathrm{T}_0)$ at the time t_0 on which the closed contour (C_0) circumscribing a part (S_0) of the surface is traced (see Figure 8.9). We have, according to Stokes' theorem:

$$\Gamma[(C_0),t_0] = -\oint_{(C_0)} \vec{V}.\vec{t}\,dl = -\iint_{(S_0)} \vec{\omega}.\vec{n}\,dS = 0\,.$$

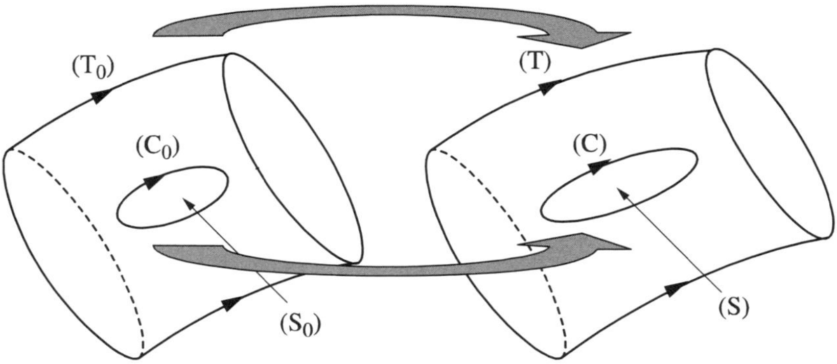

Figure 8.9. *Demonstration of the Helmholtz theorem*

In time, fluid particles that made up (T_0) move, carried away by the fluid, and form at the time t a surface (T). The curve (C_0) becomes a curve (*C*) plotted on (T) and defining a surface (*S*) part of (T). According to Kelvin's theorem:

$$\Gamma[(C),t] = \Gamma[(C_0),t_0] = 0$$

Thus:

$$\iint_{(S)} \vec{\omega}.\vec{n}\, dS = 0$$

This relation is true for any curve (C_0) plotted on (T_0), i.e. for any curve (*C*) plotted on (T). Thus, at each point (T):

$$\vec{\omega}.\vec{n} = 0$$

The surface (T) is still a vortex surface. Hence, the Helmholtz theorem:

> For a non-viscous fluid in isentropic flow and in absence of body forces, the vortex surfaces and vortex tubes are fluid surfaces following the fluid in its motion.

We consider a vortex tube based on a curve (C) and a surface (Σ) delimited by (C), as drawn in Figure 8.10. The circulation along (C) is given by:

$$\Gamma(C) = -\iint_{(\Sigma)} \vec{\omega}.\vec{n}dS \ .$$

Given another curve (C') embracing the vortex tube and a surface (Σ') defined by (C'). To calculate the circulation over (C), we can consider the vorticity flux through the surface based on (C), in particular, the surface consisting of (Σ') and of the part of the vortex tube (T) included between (C) and (C'). It follows that:

$$\Gamma(C) = -\left[\iint_{(\Sigma')} \vec{\omega}.\vec{n}dS + \iint_{(T)} \vec{\omega}.\vec{n}dS\right] = -\iint_{(\Sigma')} \vec{\omega}.\vec{n}dS$$

because:

$$\iint_{(T)} \vec{\omega}.\vec{n}\, dS = 0$$

But:

$$\Gamma(C') = -\iint_{(\Sigma')} \vec{\omega}.\vec{n}dS$$

therefore:

$$\Gamma(C) = \Gamma(C') \ .$$

The value of the circulation on the curve (C), which remains constant, is called the *vortex tube intensity*.

We consider the case of a vortex tube bounded by a surface, outside of which the vorticity is zero, or gradually tends toward zero. The vortex is immersed in an otherwise irrotational flow, as is in the case of vortices emitted by an aircraft wing. If we calculate the circulation on a curve (C) surrounding the vortex at a distance sufficient to be in the irrotational region, the contribution to the vorticity flux of the irrotational regions is of course zero. The intensity of a vortex can be calculated on any contour surrounding the vortex and placed far enough that it embraces all the vorticity attached to the vortex. In practice, we take simple contours on which the calculation of the flow motion is facilitated.

Returning to the calculation made above for introducing the intensity of the vortex, we imagine that beyond the contour (C') the vortex tube is interrupted, the flow now being irrotational. We can calculate the circulation $\Gamma(C')$ by evaluating the flow through a surface (Σ'') based on (C') but located on the side where the vorticity is zero everywhere. According to Stokes' theorem:

$$\Gamma(C') = 0$$

in contradiction with the previous result which has shown that $\Gamma(C) = \Gamma(C') \neq 0$. Hence, the theorem:

A vortex tube can neither begin nor end in a fluid: either it closes on itself, or it rests on a solid surface, or a free surface.

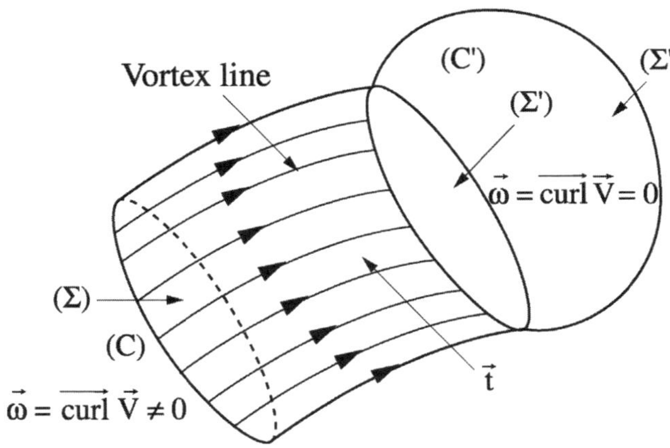

Figure 8.10. *Vortex intensity and consequence of the Helmholtz theorem*

In the first case, we speak of a *toroidal vortex (ring vortex)*, and in the second a *tornado-like vortex*. The vortex called the horseshoe vortex, whose two branches escape downstream, is actually a vortex ring closing up at a great distance. A spectacular case of vortex structure obeying this rule is the atmospheric tornado, which always reaches to the ground and ends in the upper layers of the atmosphere where air gets scarce almost to vacuum conditions.

8.2.3. *Stream function*

The concept of the stream function only applies to two-dimensional flows, planar or axisymmetric. If the flow is *steady*, it satisfies the following form of the continuity equation (see section 4.3):

$$\nabla.\left(\rho\,\vec{V}\right)=0$$

We introduce a scalar function $\psi(x,y)$ such that:

$$\boxed{\begin{aligned}\frac{\partial\psi}{\partial y}&=\rho\ u\\ \frac{\partial\psi}{\partial x}&=-\rho\ v.\end{aligned}}$$

Under the rule of a cross derivation:

$$\frac{\partial^2\psi}{\partial y\ \partial x}=\frac{\partial^2\psi}{\partial x\ \partial y}.$$

Therefore:

$$\frac{\partial^2\psi}{\partial y\,\partial x}-\frac{\partial^2\psi}{\partial x\,\partial y}=\frac{\partial(\rho u)}{\partial x}+\frac{\partial(\rho v)}{\partial y}=0$$

The introduction of the function ψ, called stream function, allows us to satisfy the continuity equation.

The differential of the function ψ is written:

$$d\psi=\frac{\partial\psi}{\partial x}\,dx+\frac{\partial\psi}{\partial y}\,dy=\rho\,u\,dx-\rho\,v\,dy=\rho\,(u\,dx-v\,dy)$$

Therefore, $d\psi$ is zero on curves such that:

$$\frac{dy}{dx}=\frac{u}{v}.$$

As we know, these curves are the streamlines of the steady flow. Thus, the curves $\psi=\text{constant}$ coincide with the streamlines. We show without difficulty that the mass flow passing between two streamlines (l_1) and (l_2) is equal to the difference of the values ψ_1 and ψ_2 of the stream function on (l_1) and (l_2), respectively.

In a given flow, the stream function ψ is defined within an arbitrary constant, most often we assign the value $\psi = 0$ to a streamline that is chosen for reasons of convenience: this is the case, for example, for the body surface. The definition of ψ within an arbitrary constant is due to the fact that ψ intervenes only by its derivatives and only the difference between two values of ψ has a physical meaning. It is same for thermodynamic functions such as enthalpy, internal energy, or entropy. The concept of a stream function, which only exists for steady two-dimensional flows (planar or axisymmetric) is rather general since it results from the continuity equation. It applies to both compressible as well as incompressible fluids, which can be viscous or non-viscous.

COMMENT.– It is possible to extend the stream function concept to three-dimensions by introducing a *potential vector*. In incompressible flow, we define the potential vector $\vec{A}$ by:

$$\vec{V} = \overrightarrow{curl}\,\vec{A}\,.$$

Considering the general property due to which the divergence of a curl vector is zero (see Appendix 1):

$$div\left(\overrightarrow{curl}\right) = 0\,.$$

We see that the introduction of $\vec{A}$ allows us to satisfy the incompressible continuity equation:

$$div\ \ \vec{V} = 0.$$

We show that $\vec{A}$ is defined within the gradient of an arbitrary scalar function *f*:

$$\vec{A}' = \vec{A} + \overrightarrow{grad}\ \ (f).$$

Indeed:

$$\overrightarrow{curl}\,\vec{A}' = \overrightarrow{curl}\vec{A} + \overrightarrow{curl}\left[\overrightarrow{grad}\left(f\right)\right] = \overrightarrow{curl}\vec{A} = \vec{V}$$

since the curl of a gradient is zero. We find the similar property according to which a stream function is defined within and arbitrary constant. To our knowledge, the potential vector generalizing the two-dimensional stream function has been little used in fluid mechanics. This is not the case in electromagnetism where the concept is common.

8.2.4. *Potential function for the velocity or velocity potential*

We consider the framework of an *irrotational flow*, which can be compressible, three-dimensional, and unsteady.

If $\vec{V}$ is the velocity field of such a flow, it satisfies:

$$\overrightarrow{curl}\ \vec{V}=0\,.$$

If $\Phi(x,y,z)$ is a scalar function, we have the classic result (see the appendix 1):

$$\overrightarrow{curl}\ \left(\overrightarrow{grad}\ \Phi\right)=0\,.$$

By comparing the two previous relations, we can state:

$$\boxed{\vec{V}=\overrightarrow{grad}\ \Phi\ \left(x,y,z\right)}$$

In other words, for an irrotational flow, there is a scalar function $\Phi(x,y,z)$ such that the velocity is given by:

$$\vec{V}=\overrightarrow{grad}\ \Phi\ \left(x,y,z\right).$$

According to the definition of the gradient of a function, we have in Cartesian coordinates:

$$u=\frac{\partial\Phi}{\partial x}\quad,\quad v=\frac{\partial\Phi}{\partial y}\quad,\quad w=\frac{\partial\Phi}{\partial z}$$

The existence of a potential function can greatly simplify the solution of a problem of fluid mechanics since it allows us to replace an unknown vector $\vec{V}$ by an unknown scalar Φ. Thus, instead of three scalar equations for the components (u,v,w) of $\vec{V}$, we have only one scalar equation for Φ, hence, the importance of *irrotational flows*. In particular, the potential function is used for the study of transonic flows (see Chapters 19 and 25).

In a two-dimensional flow, the lines on which Φ = constant are called *equipotential lines*, or simply equipotentials. As at a point *P*:

$$\vec{V} = \overrightarrow{grad}\ \Phi$$

and as $\vec{V}$ is tangential to the streamline at *P*, it follows that at *P*, *the streamline and the equipotential* are orthogonal, which follows from the property that the vector $\overrightarrow{grad}\ f$ is perpendicular to the curve $f = constant$ (see Appendix 1). Thus, the streamlines and equipotentials of a two-dimensional flow form two sets of orthogonal curves (see Figure 8.11).

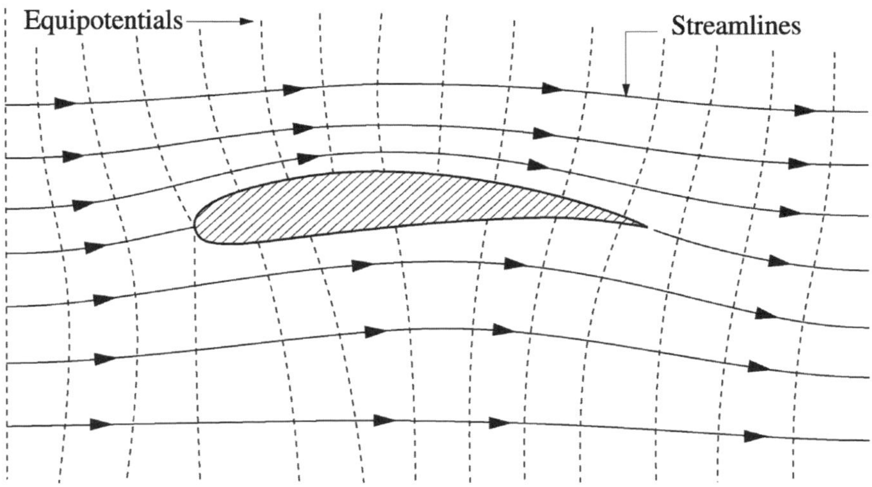

Figure 8.11. *Streamlines and equipotentials around an airfoil*

The existence of a *stream function* is an interesting property because it can also replace a vector equation by a scalar equation. However, the fact that the stream function is defined only for two-dimensional steady flows, greatly limits the field of application.

The condition of irrotationality can severely limit the use of the potential function as its use is excluded in the viscous regions and in all areas where the vorticity is present, such as downstream of a curved shock wave. There is a certain duality between the advantages and disadvantages of the potential function and stream function. The current trend is to directly solve the Euler equations to be free of restrictions implied by the concepts of the stream function and velocity potential.

Chapter 9

One-Dimensional, Non-Viscous and Adiabatic Steady Flows

9.1. Definition and basic hypotheses

We call a flow depending only on one space variable one-dimensional for example, x. Therefore, we concern ourselves with the flow in a stream tube aligned along Ox and whose section $A(x)$ at a point of the abscissa x is normal to Ox. The axis Ox is often straight but may also be slightly curved (see Figure 9.1).

The basic hypothesis is to assume that the flow properties are constant in any section $A(x)$. Such a situation may correspond to a very slender stream tube, isolated in a flow. In practice, the one-dimensional approximation applies to a flow in a duct whose area $A(x)$ is slowly varying along an axis Ox.

The one-dimensional approximation is often made in internal aerodynamics when dealing with flows in ducts with a slowly varying section, air inlet, nozzle, diffuser, ejector, etc. Its advantage is that it leads to very simple calculations, which is very convenient for applications where we do not look for great accuracy, for example at the pre-project level. Additionally, one-dimensional theory allows us to establish general results of great importance at the fundamental level. However, we have to be wary of using this approximation in situations where the two-dimensional character of the flow is essential. Assuming a one-dimensional flow can then lead to inconsistent or paradoxical results.

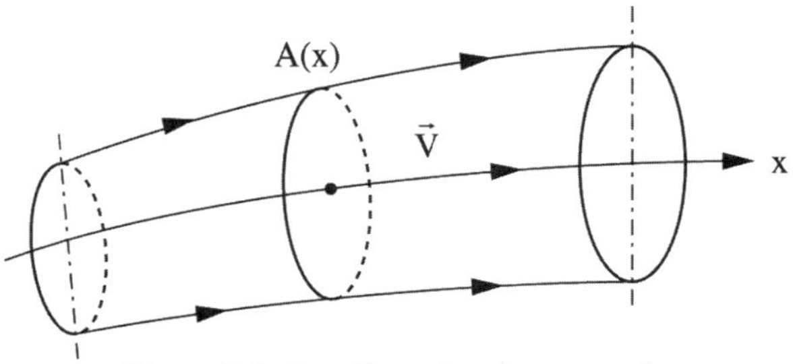

Figure 9.1. *One-dimensional stream tube*

Given p, ρ, V, T, etc. the pressure, the density, the velocity, the temperature, etc. in the section $A(x)$, these quantities are constant by hypothesis. Thus:

$$p = p(x)$$

$$\rho = \rho(x)$$

$$\vec{V} = \vec{V}(x)$$

$$T = T(x)$$

etc.

The hypothesis of a one-dimensional flow implies that the velocity $\vec{V}$ is constant over the section $A(x)$ and aligned with the x-axis, that is, normal to $A(x)$. In what follows, the velocity, therefore, will be treated as a scalar V. The following assumptions are made:

– the flow is permanent (of steady),

– the fluid is non-viscous and non-heat conducting,

– the fluid is non-reactive (no chemical reactions).

From the perspective of thermodynamics, the gas is considered divariant, which means: either thermochemical equilibrium or the frozen assumption (see Chapter 3). This assumption is not essential; in hypersonics, we calculate flows that are in non-equilibrium by assuming that they are one-dimensional. But then the thermodynamics is more complex.

The description of the movement arises from three fundamental principles:

– conservation of mass or continuity,

– equation of motion or fundamental law of mechanics,

– energy or first law of thermodynamics.

These equations must be supplemented by appropriate thermodynamic relations.

9.2. The fundamental equations

The fundamental equations for a one-dimensional flow can be derived directly from the general equations established in Chapter 4. However, to give this part a certain autonomy, we will derive them by simple reasoning.

9.2.1. *Mass conservation: continuity equation*

The mass flux or mass flow (mass per unit of time in kg/s) through a surface (S) is given by the following equation if the flow is steady:

$$q_m = \iint_{(S)} \rho \vec{V}.\vec{n}\, dS$$

where $\vec{n}$ is the unit vector normal to (S). In the case where the flow is one-dimensional, its properties are constant on (S) and if (S) is on a plane surface of area A, then the mass flow through this section A is given by:

$$\boxed{q_m = \rho VA = constant.}$$

In the absence of sources, or sinks, of mass, the mass flow is conservative in steady flow, i.e. it is an invariant of the flow. A logarithmic differentiation of the above expression gives:

$$\boxed{\frac{d\rho}{\rho} + \frac{dV}{V} + \frac{dA}{A} = 0}$$

which is a differential form of the continuity equation in steady conditions.

9.2.2. *Equation of motion or Euler equation*

In the second order, the mass of the fluid comprised at time t between the abscissa x and $x + dx$ contained in the volume $MLL'M'$ (see Figure 9.2), is given by:

$$\rho \, A \, dx.$$

We apply the fundamental law of mechanics to this mass: $\vec{f} = m\vec{\gamma}$.

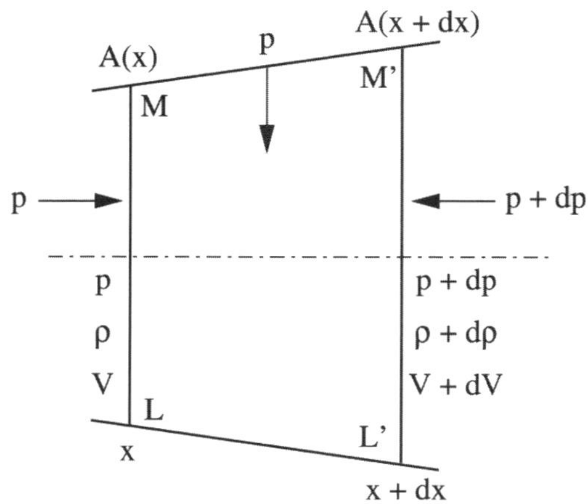

Figure 9.2. *Establishing the Euler equation*

In this case, the equation has the scalar form:

$$dm \, \gamma \equiv dm \frac{dV}{dt} = \rho A dx \frac{dV}{dt}$$

where *dm* is the mass of the fluid contained in a slice with the width *dx*.

The acceleration is expressed as:

$$\frac{dV}{dt} = \frac{\partial V}{\partial t} + \frac{\partial V}{\partial x}\frac{dx}{dt}$$

hence for a steady flow:

$$\frac{dV}{dt} = \frac{dV}{dx}\frac{dx}{dt} = V\frac{dV}{dx}.$$

Let us evaluate the resultant of the forces f along the direction x. These forces have as expressions:

– on the input side $A(x)$: pA ,

– on the output side: $-\left[pA + d(pA)\right]$.

In the absence of viscosity, the tension reduces to the pressure p. The resulting force is normal to the wall of the stream tube, and has the component along Ox:

$$p \ dA.$$

Ultimately:

$$\rho\, A\, dx\, V\frac{dV}{dx} = p\,A - \left[p\,A + d\left(p\,A\right)\right] + p\ dA = -A\ dp$$

from where:

$$\boxed{V\,dV + \frac{dp}{\rho} = 0}$$

which is called the *Euler equation*.

In the above relation, the area A of a stream tube has disappeared. Thus, the Euler equation is a local relation valid on any streamline, even in a flow (non-viscous) that is not one-dimensional. Under the Euler equation, if $dV > 0$ then $dp < 0$, the pressure decreases in a flow whose velocity increases, and vice versa, of course. Linking the variations of velocity and pressure in a fluid is a first important consequence of this equation. In an incompressible flow, as density is constant, we set $\rho = \rho_0$ and the Euler equation integrates to give:

$$p + \rho_0 \frac{V^2}{2} = constant\,.$$

The constant can be equaled to the fluid pressure in a region where the velocity is zero or very low. We state:

$$\boxed{p_{st} = p + \rho_0 \frac{V^2}{2}}$$

where p_{st} is the *total pressure* or *stagnation pressure* of the flow. It is the pressure in a tank where the fluid is stored at rest before expansion due to the velocity V. The above equation is the well known Bernoulli relation.

9.2.3. *Energy equation*

We consider the mass of fluid contained in the volume $MLL'M'$, used to establish the Euler equation. The flow is steady between the time t and time $t + \delta t$ *the total energy contained in this volume has not changed.* During δt the volume has been traversed by the mass δm, so that it entered through the side LM, the total energy being:

$$\delta m\left(e + \frac{V^2}{2}\right)$$

and it left through the side $L'M'$, the total energy being:

$$-\delta m\left[e + de + \frac{V^2}{2} + d\left(\frac{V^2}{2}\right)\right].$$

Ultimately, the amount of total energy received by convective transfer is:

$$\Delta E_{conv} = \delta m\left[de + d\left(\frac{V^2}{2}\right)\right].$$

The energy equation follows from applying the first law of thermodynamics:

$$\Delta E_T = \Delta W + \Delta Q$$

which shows that change in the total energy ΔE_T of the system is equal to the sum of the energy received in the form of work of external forces applied to the system ΔW and the heat ΔQ (see Chapter 3).

In this case, where the flow is permanent and where there is no input of energy by chemical reactions, the variation in total energy ΔE_T of the system results from the balance of convective flux:

$$\Delta E_T = \Delta E_{conv} = \delta m\left[de + d\left(\frac{V^2}{2}\right)\right].$$

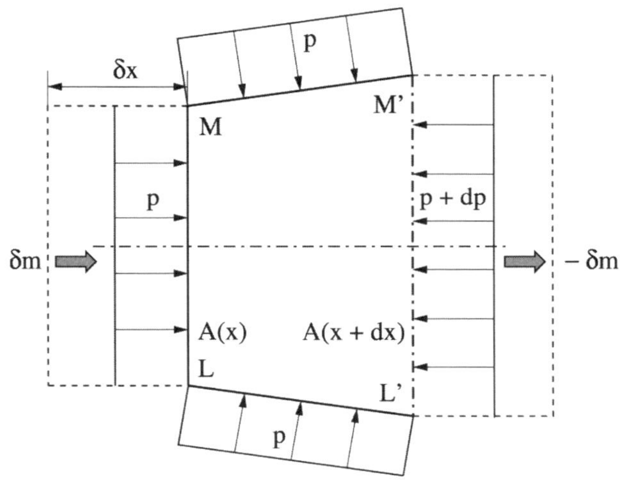

Figure 9.3. *Establishing the energy equation*

We evaluate the work of external forces acting on the volume $MLL'M'$ (see Figure 9.3):

– On the lateral side $(LL' \; or \; MM')$, the force reduces to the influence of the pressure, which is normal to the surface element on which it applies (non-viscous fluid hypothesis). In this case, such a force does not contribute any work to the system because it is normal to the velocity.

– On the input side A acts the force pA from which follows the work:

$$pA\delta x.$$

The motion δx is associated with the mass δm, introduced into the system by:

$$\delta m = \rho A \delta x .$$

from which follows the *introduction work*:

$$\frac{p}{\rho}\delta m.$$

– On the output side a pressure force acts that leads to the *extraction work*:

$$-(p+dp)(A+dA)\delta x = -pA\delta x - pdA\delta x - Adp\delta x - dpdA\delta x$$

hence to second order:

$$-pA\delta x - d(pA)\delta x = -\left(\frac{p}{\rho}\right)\delta m - d\left(\frac{p}{\rho}\right)\delta m$$

or

$$-\delta m\left[\frac{p}{\rho} + d\left(\frac{p}{\rho}\right)\right].$$

Ultimately, the work received by the fluid contained in $LL'MM'$ is given by:

$$-\delta m\, d\left(\frac{p}{\rho}\right).$$

The fluid is assumed to be a non-conductor of heat, and in the absence of radiation, there is no energy input by heat exchange, so that:

$$\boxed{\delta m\left[de + d\left(\frac{V^2}{2}\right)\right] = -\delta m\, d\left(\frac{p}{\rho}\right)}$$

or

$$\delta m\left[de + d\left(\frac{V^2}{2}\right)\right] + \delta m\, d\left(\frac{p}{\rho}\right) = 0$$

and after division by δm:

$$\boxed{d\left(e + \frac{V^2}{2} + \frac{p}{\rho}\right) = 0.}$$

We recall that the quantity:

$$e + \frac{p}{\rho} = h$$

is the enthalpy (specific), which is a state function of the fluid, hence, the new form of the energy equation:

$$d\left(h + \frac{V^2}{2}\right) = 0\,.$$

This expression shows that in a steady and adiabatic flow the total – or stagnation – enthalpy:

$$\boxed{h_{st} = h + \frac{V^2}{2}}$$

is constant. This is the second invariant of the flow, more restrictive than the mass flow since it implies an adiabatic flow. Like the Euler equation, the equation of energy is a local relation, valid on any streamline of the flow, for the assumptions made at the beginning of this chapter.

The invariance of the stagnation enthalpy in an adiabatic evolution, which is often the case, reflects the fact that the sum of kinetic and thermodynamic energies of the fluid remains constant. Any increase in velocity comes at the expense of internal energy and vice versa. Thus, the nozzle of a wind tunnel, or rocket engine, is an energy converter transforming the enthalpy into kinetic energy. To achieve high velocities, the case of hypersonic flow, we must start from a high stagnation enthalpy state. Conversely, the slowing of a stream at high velocity will result in a significant increase in its thermodynamic energy, or true enthalpy, therefore its temperature. The relation for the stagnation enthalpy makes no assumption about the thermodynamics of the fluid: it applies when real gas effects are present.

If the gas is calorically perfect, then:

$$C_p T_{st} = C_p T + \frac{V^2}{2}$$

from which follows:

$$T_{st} = T + \frac{V^2}{2C_p}$$

The temperature associated with the stagnation enthalpy is called *total temperature* or *stagnation temperature*. The term “reservoir temperature” is also in use. The stagnation temperature is measured by placing in the flow a probe whose sensing element is a thermocouple junction. This junction is often seen as a small spherical ball, with a diameter of a few tenths of a millimeter, formed by the welding of the two metals, associated in the thermocouple effect. The flow, with the upstream velocity V relatively to the probe, is stopped by the ball, which thus measures the temperature T_{st} of the flow. In fact, for accurate measurements, care must be taken and the probe is regularly calibrated.

Before the re-entry of a spacecraft into the atmosphere, the velocity achieved is around 10,000 m/s. We calculate the stagnation enthalpy of air reaching the vehicle knowing that the static temperature T_∞ at the considered altitude is close to 250 K (the exact value is irrelevant as we shall see):

$$h_{st} = C_p T_\infty + \frac{V_\infty^2}{2} = 1{,}000 \times 250 + 0.5 \times (10{,}000)^2 = 0.25 \times 10^6 + 5 \times 10^7 \quad \text{(in J/kg)}.$$

The stagnation enthalpy of the air is almost entirely in the form of kinetic energy. The quantity h_{st} is invariant, even when crossing the shock forming ahead of the vehicle (see Chapter 13). At the nose of the vehicle, which is a stagnation point where the velocity relative to the vehicle vanishes (see Figure 9.4), there is a conversion of kinetic energy into real or static enthalpy h so that:

$$h = h_{st} \approx 5 \times 10^7 \ \text{J/kg}$$

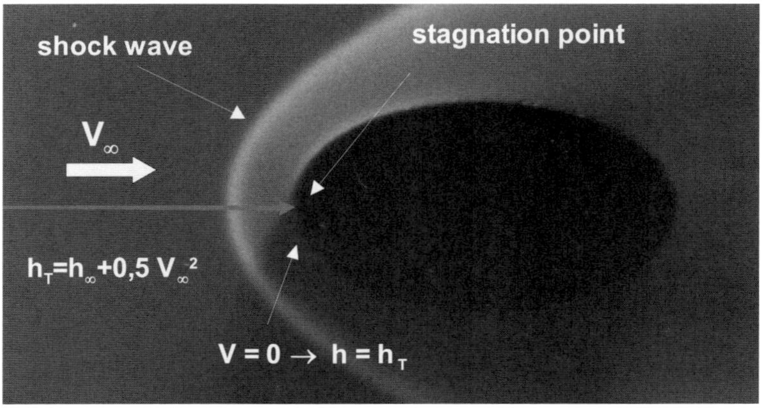

Figure 9.4. *Stagnation point at the nose of a hypersonic vehicle*

If we assume that air behaves like a calorically perfect gas with $C_p =$ $1{,}000$ J/kg/K, a temperature at the nose results:

$$T = \frac{h}{C_p} \approx \frac{5 \times 10^7}{1{,}000} \approx 50{,}000 \text{ K}.$$

This simple calculation shows that a body entering the atmosphere is subjected to extremely high temperatures on some of its parts (nose, leading edges of wings, and flaps). In reality, the assumption of a calorically perfect gas under these conditions is grossly false. Once the temperature exceeds 2,000 K, thermodynamics of air is complicated by the real gas effects (see section 3.7). From a certain threshold temperature, the molecules of oxygen and then nitrogen are dissociated, which has the effect of "pumping" energy with the (fortunate), result of a very significant lowering of the temperature at the stagnation point that is pretty close to 8,000 K. This helps implement thermal protection, as no material could withstand 50,000 K. Without the real gas effects, there would have never been a space adventure.

As already noted, the equations of motion and energy no longer contain the section $A(x)$. They have a local character and can be applied to the limit $A(x) \to 0$, which means that they are valid on any streamline of flow subject to the general assumptions made above.

We see that h_{st} represents the value of the enthalpy at a point where the flow velocity is zero. If in an overall steady and adiabatic flow there is a uniform region then all the streamlines coming from this region have the same stagnation enthalpy level. Particularly, if the flow is from a region at rest where velocities are almost zero (a tank), the stagnation enthalpy is equal to the enthalpy of the fluid in this region: hence, the term *reservoir* enthalpy, sometimes used.

We know that the equations governing fluid motion are insufficient to calculate its properties. In the case of a non-viscous fluid, there are four unknown quantities, p, ρ, V, and e (or h), for three equations. The system must be supplemented by a relation of thermodynamic origin, which, in the case of a divariant gas will be, for example, in the form of:

$$h = h(p, \rho) .$$

At this stage we have not made very restrictive assumptions regarding the thermodynamics of the gas. Divariant behavior is indeed quite general if we exclude the case where non-equilibrium phenomena happen, which occurs in extreme circumstances, in particular, hypersonics. The relation $h = h(p, \rho)$ can either be explicit as in the case of a calorically perfect gas, or provided in the form of tables or diagrams if real gas effects occur (see Chapter 3).

9.2.4. *Isentropy of the flow*

For a divariant fluid, we have the general thermodynamic relation:

$$dh = Tds + \frac{dp}{\rho}$$

therefore:

$$Tds = dh - \frac{dp}{\rho}.$$

But, according to the energy and Euler equations:

$$dh = -VdV = \frac{dp}{\rho}$$

from which follows:

$$\boxed{Tds = 0 \quad hence \quad ds = 0.}$$

The flow of a non-viscous and adiabatic fluid is also isentropic. This local property is true on each streamline. If the flow is from a region (or reservoir) where the entropy is everywhere the same, the flow remains isentropic as a whole during a subsequent development, unless it crosses a shock wave. We sometimes say that it is homentropic.

9.3. The Hugoniot theorem

The results above allow us to deduce the consequences of great practical importance in the study of compressible flows. Before going any further, we specify new assumptions under which the following results are valid:

– the flow is steady,

– it is continuous, implying the absence of shock waves or discontinuity surfaces,

– the fluid is non-viscous and not a heat conductor,

– the gas is divariant from a thermodynamic point of view, and

– absence of chemical reactions.

These assumptions may seem very restrictive but, in fact, they are verified by many flows under conditions of conventional aerodynamics. They must be revised in hypersonics or even in flows with combustion.

The one-dimensional theory that we have seen leads to the relations:

$$\frac{d\rho}{\rho}+\frac{dV}{V}+\frac{dA}{A}=0$$

$$V\,dV+\frac{dp}{\rho}=0\,.$$

The second relation is also written as:

$$VdV+\frac{dp}{d\rho}\frac{d\rho}{\rho}=0$$

which gives, by taking into account the first:

$$VdV+\frac{dp}{d\rho}\left(-\frac{dV}{V}-\frac{dA}{A}\right)=0$$

or:

$$\frac{dV}{V}\left[1-\frac{V^2}{\left(dp/d\rho\right)}\right]+\frac{dA}{A}=0\,.$$

If the flow is isentropic:

$$a^2=\left(\frac{\partial p}{\partial\rho}\right)_s=\frac{dp}{d\rho}$$

where a is the velocity of sound. Finally, we reach the following relation known as *Hugoniot theorem*:

$$\boxed{\frac{dV}{V}\left(1-\frac{V^2}{a^2}\right)+\frac{dA}{A}=0.}$$

The ratio V/a is called Mach number:

$$\boxed{M=\frac{V}{a}}$$

– a value $M < 1$ characterizes a subsonic flow,

– a value $M > 1$ characterizes a supersonic flow, and

– the value $M = 1$ corresponds to the sonic state.

The Hugoniot relation can also be written as:

$$\boxed{\frac{dV}{V}\left(1-M^2\right)+\frac{dA}{A}=0.}$$

This fundamental equation establishes a relation between the variations in velocity dV and the variations dA of the area of the stream tube. Two cases are distinguished (see Figure 9.5):

– *subsonic flow* $M < 1$, corresponding to $V < a$.

We then see that dA and dV are of opposite sign, the factor $(1-M^2)$ before dV/V being positive, therefore:

In a subsonic flow the velocity increases when the area of the cross-section diminishes (and vice versa).

– *supersonic flow* $M > 1$, corresponding to $V > a$.

Then, dA and dV have the same sign, the factor $(1-M^2)$ before dV/V is negative, therefore:

In a supersonic flow velocity increases when the area of the cross-section increases (and vice versa).

– *the limit case of a sonic flow $M = 1$, corresponding to $V = a$.*

We can write:

$$\frac{dA}{dV} = -\frac{A}{V}\left(1 - M^2\right) = 0 .$$

If $dV \neq 0$, then $dA = 0$, whatever the value of dV:

For a sonic flow the area of the cross-section is stationary.

As the acceleration of the flow from a subsonic speed implies a reduction of the area $A\left(dA < 0\right)$, it follows that the condition $dA = 0$ of the sonic state corresponds to a minimum of the cross section. This minimum section is called *a throat*. In the case where V is different from *a*, the situation $dA = 0$ leads to $dV = 0$, the velocity passes through an extremum, minimum, or maximum as appropriate, when the section of the stream tube is stationary (see Figure 9.5).

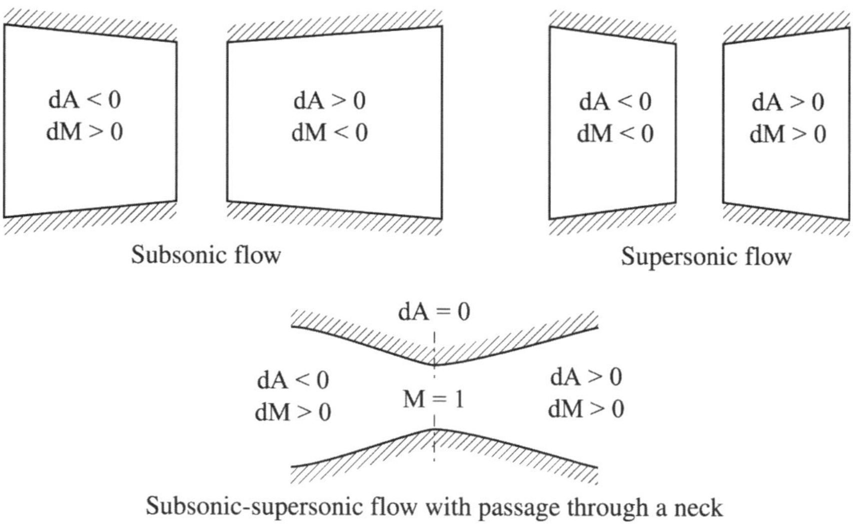

Figure 9.5. *Hugoniot theorem*

If we want to accelerate the flow beyond the velocity of sound, we must use a nozzle with a convergent to increase the speed in subsonics, followed by a divergent for further acceleration to supersonics, with, between the two parts, a throat where the section is stationary. Of course, it is necessary to have a sufficiently low pressure downstream of the nozzle to keep the flow supersonic. A nozzle of this kind is sometimes called a *de Laval nozzle*. We shall see in Chapter 19 how to calculate the shape of the nozzle to achieve a uniform supersonic flow in the outlet section of the nozzle. Figure 9.6 shows a convergent-divergent nozzle equipping a supersonic wind tunnel.

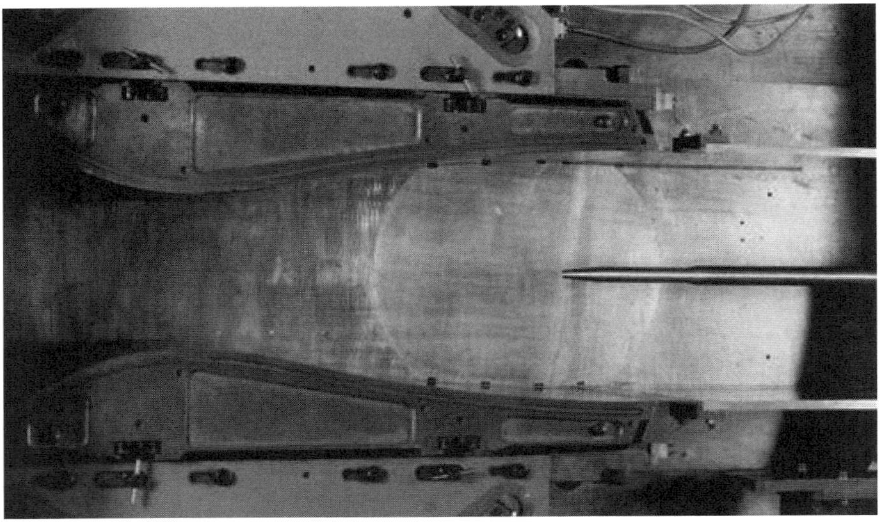

Figure 9.6. *Convergent-divergent nozzle equipping the supersonic wind tunnel S5 of the Onera Meudon Center*

9.4. Consequences of the Hugoniot theorem: sonic throat and starting a supersonic nozzle

The behavior of the solution curve of the Hugoniot equation is shown in Figure 9.7, which shows the evolution of the Mach number M in a convergent-divergent nozzle with the section $A(x)$. The integration starts from an initial section A_0 where the flow has a subsonic Mach number M_0. In the convergent section, the Mach number increases and the solution curve meets the geometric throat where $\frac{dA}{dx} = 0$ when M is still less than one. Then, in the divergent, the Mach number decreases according to the Hugoniot theorem. There is, in the plane of the solutions

$[A, M]$, a singular point of the saddle point type, where, at the same time, $\frac{dA}{dx} = 0$ and $M = 1$. All but two of the solution curves avoid the saddle point for which we have at the same time $\frac{dA}{dx} = 0$ and $M = 1$.

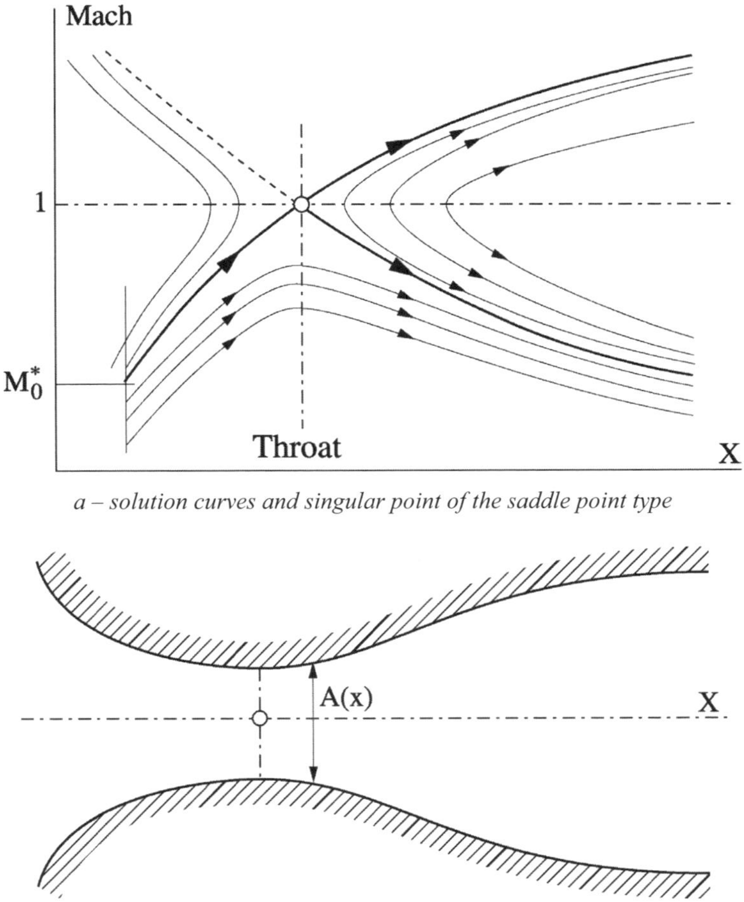

a – solution curves and singular point of the saddle point type

b – evolution of the nozzle section

Figure 9.7. *Flow solution in a supersonic convergent–divergent nozzle*

In our case, there exists a value M_0^* of the Mach number in the initial section such that the corresponding solution curve passes through the saddle point. Beyond that, as $\frac{dA}{dx}$ and $\left(M^2-1\right)$ change signs at the same time, the solution continues with the rise of the Mach in supersonics. If the initial value is greater than M_0^*, then the solutions are not physical. Reaching the saddle point, the solution may follow the second solution curve passing through the saddle point so that the Mach number decreases to subsonics after having passed through the value one. There is only one initial value that allows the passage of the solution curve through the saddle point and the achievement of a supersonic flow. This is an eigenvalue problem or a two point boundary value problem where one condition has to be fulfilled by the solution curve during being solved. The condition of passage through the saddle point (throat in the physical plane) "blocks" the solution, that is insures its unicity compatible with a subsonic-supersonic flow.

We consider the problem of the starting of a supersonic nozzle by adopting the same representation, where the pressure replaces the Mach number (see Figure 9.8). The pressure p_0 in the inlet section being fixed, the pressure p_e in the outlet section is gradually lowered. If p_e is somewhat lower than p_0, then the flow in the nozzle remains subsonic, the velocity passing through a maximum at the throat. It is a mode of operation of the Venturi type (see section 9.6.3). We note that we can maintain such a subsonic flow with a downstream level equal to the entry level, the pressure recovery occurring with an efficiency of one due to the neglect of viscosity. In reality, the viscous effects present in the boundary layers are such that complete recovery is not possible.

By continuing to lower the pressure p_e, we reached a level p_e^* such that the pressure distribution goes through the throat, the downstream flow becoming supersonic. If the pressure is low enough, then the Mach number is supersonic to the output and the pressure distribution in the nozzle no longer depends on the downstream pressure. If the downstream pressure is greater than the pressure of the supersonic solution, a shock wave must occur in the nozzle so as to make the pressure distribution jump to a subsonic solution. The shock will gradually occur more and more downstream as the pressure p_e decreases, until the complete starting of the nozzle (here, we anticipate on the shock wave theory presented in Chapter 13).

Understanding of the starting process of a supersonic nozzle actually requires a description of unsteady phenomena showing the establishment of the flow and its time evolution. Such a description will be given in section 24.5, once the necessary knowledge has been acquired.

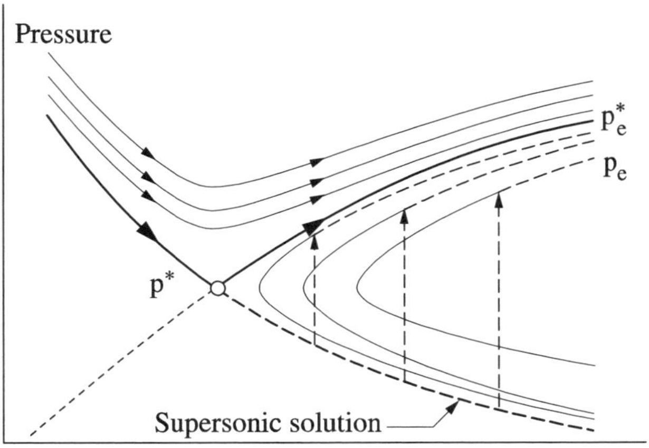

a – pressure in the nozzle

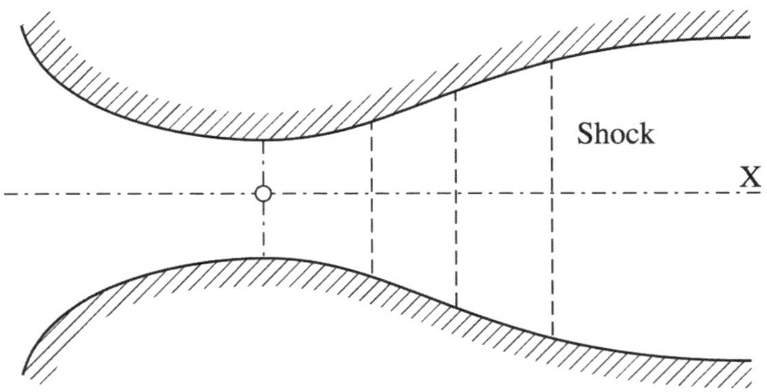

b – displacement of the shock in the nozzle

Figure 9.8. *Starting a supersonic nozzle*

We see that the establishment and maintenance of a supersonic flow with the Mach number M_0 are not solely due to the presence of a throat and a suitable area ratio. We must also achieve a downstream pressure low enough so that the so-called starting shock, that makes the flow become subsonic flow is placed downstream of the useful part of the nozzle. For that, a downstream pressure should be maintained lower than the pressure p_1 right downstream of the shock occurring for the Mach number M_0 (or at the limit equal to p_1).

In wind tunnels, this condition is achieved with pumps creating a vacuum downstream of the nozzle, or by creating a sufficiently high upstream stagnation pressure p_{st_0}, for example, by discharging compressed air, stored at high pressure in a reservoir.

To start a Mach 2 nozzle, a ratio of $\frac{p_{st_0}}{p_1} = 1.739$ is enough, which is obtained without difficulty with pumps if the stagnation pressure is the atmosphere. At Mach 10, it is necessary that $\frac{p_{st_0}}{p_1} = 364$ which is obtained by combining a source of high-pressure compressed air upstream and a vacuum sphere downstream. The flow with the Mach number M_0 is maintained as long as the pressure in the tank with compressed air is high enough and the pressure in the vacuum sphere sufficiently low. A wind tunnel operating on this principle is known as *intermittent wind tunnel* or *blow-down wind tunnel.* Test durations varying from a few seconds to several minutes can be obtained, depending on the Mach number.

9.5. Isentropic expansion or compression of a flow

9.5.1. *Limit velocity concept*

It is convenient to represent the thermodynamic properties of a divariant gas in a Mollier diagram in which the variables characterizing the state of the system are the enthalpy h and entropy s (see section 3.7.4). Usually, s is given in the abscissa and h in the ordinate (see section 3.7.4). Through each point of the plane $[s-h]$ passes an isobaric line $(p = const)$, an isothermal line $(T = constant)$, etc. Knowing the entropy and enthalpy of a gas, we can locate its representative point (or image) in the plane $[s-h]$ and thus derive all its other thermodynamic properties.

We imagine that a fluid expands, thus, accelerates, from a state of rest where the velocity is zero or nearly zero. This occurs when one relaxes a gas stored under high pressure in a reservoir to feed the nozzle of a wind tunnel or even a rocket engine where the gases in the combustion chamber have low velocities compared to the ejection velocities at the nozzle exit. An enthalpy h_{st} and an entropy s_{st} correspond to a state of rest or stagnation state from which follows a representative (or image point) point (E_{st}) in the Mollier diagram (see Figure 9.9). From h_{st} and s_{st}, we can determine other stagnation quantities of the gas: pressure p_{st}, temperature T_{st}, density ρ_{st}, etc. In practice, it is the pressure p_{st} and temperature T_{st} that are known because these quantities are directly measurable. The enthalpy and entropy can only be determined indirectly. However, consideration of the Mollier diagram facilitates the following discussion.

According to the Euler equation the setting in motion of the gas $(dV > 0)$ causes a decrease in pressure from which follows a decrease of temperature T and density ρ in an adiabatic process. This results in a decrease in enthalpy. These behaviors are classic results of thermodynamics, which need not be demonstrated here. The representative point (E) of the flow in the Mollier diagram moves in the direction of h decreasing along an isentropic line of equation $s = s_{st}$. The trajectory of (E) is thus a particular curve in the Mollier diagram.

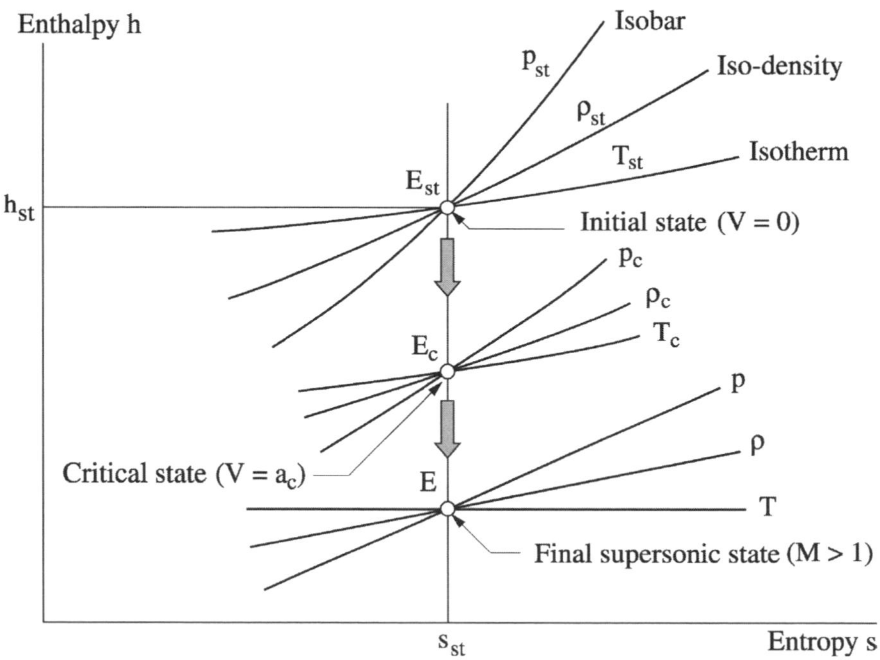

Figure 9.9. *Image of the isentropic expansion of a gas in the Mollier diagram*

The expansion of the flow may continue until a state (E) where the thermodynamic quantities attain the values h, p, ρ, T and the velocity the value V. The energy equation written with the enthalpy:

$$h_{st} = h + \frac{V^2}{2}$$

shows us that the velocity V of a fluid is given by:

$$\boxed{V = \sqrt{2\left(h_{st} - h\right)}.}$$

This is a very general relation subject to the assumptions made above, which allows us to calculate the velocity of a flow. This relation does not make very restrictive assumptions regarding the thermodynamics of the gas and can be used in extreme conditions (hypersonic flows, and rocket nozzles).

If we imagine an expansion to the vacuum, then p, ρ, T and h tend toward zero. The above relation shows that the gas velocity tends to a finite *limit velocity* $V_{\lim}$ such that:

$$\boxed{V_{\lim} = \sqrt{2\ h_{st}}.}$$

The limit velocity is used to characterize the level of the maximum velocity that we can hope to achieve by expanding a gas close to the vacuum. It is a measure of the level of its stagnation enthalpy. During an adiabatic expansion, a gas quickly reaches a velocity close to the limit velocity. Thus, for air $(\gamma = 1.4)$ at Mach 10, $V/V_{\lim} \approx 0.976$. A further acceleration to Mach 20 will only increase the velocity by 3%.

Physically, the expansion of the gas cannot be continued close to the vacuum while remaining under the assumptions made to establish the above results. During the adiabatic expansion, the temperature will decrease until liquefaction of the gas, which will change its thermodynamic behavior. Moreover, at very low pressures, the concept of continuous medium is no longer valid, the interatomic or intermolecular distances becoming of the same order as the characteristic dimensions of the system, for example, the diameter of the nozzle, from which follow rarefaction effects (see Chapter 4). However, experience shows that velocity values very close to the limit velocity can be reached before the basic assumptions from which we started are in default. The above formula provides information on the maximum velocity, which we can hope to achieve by starting from a state where the gas has a certain level of enthalpy or energy. Thus, we can see that achieving high velocities, typical for the re-entry of a spacecraft (about 10 km/s), involves very large stagnation enthalpies. This is one of the difficulties encountered when simulating hypersonic flows in ground facilities.

For a gas assumed to be calorically perfect, the velocity of sound is given by:

$$a = \sqrt{\left(\frac{\partial p}{\partial \rho}\right)_s} = \sqrt{\gamma\, r\, T}\ .$$

If the temperature T tends toward zero, the velocity of sound a also tends toward zero. Consequently, in an expansion close to the vacuum, *the Mach number becomes infinite*. Thus, during an expansion from the rest, the Mach number M varies from zero to infinity, passing through the value $M_c = 1$ when the velocity is equal to the sonic value a_c, reached at the throat.

As the mass flow $q_m = \rho V A$ is invariant, if A passes through a minimum for the critical state (where $V = a_c$), then ρV is maximum for this state. Finally, at the limits $V = 0$ and $V = V_{\lim}$ (since ρ tends to zero), the section A tends toward infinity, as we can be convinced easily. These results are summarized in Figure 9.10.

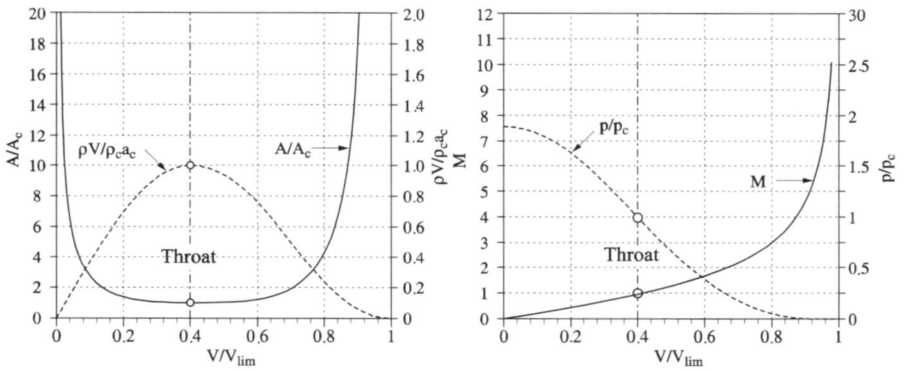

Figure 9.10. *Expansion in a convergent-divergent supersonic nozzle*

We note that the evolution of the area ratio $\dfrac{A}{A_c}$ in function of $\dfrac{V}{V_{\lim}}$ is very flat near the critical velocity, the passage through the throat being very progressive. This phenomenon is causing difficulties encountered in transonic flow simulation since in this velocity range the slightest variation in the section causes a rapid change in velocity which makes the flow passing in supersonics. We actually pass from a subsonic system with elliptic solutions to a supersonic system with hyperbolic solutions with radical changes in the behavior of the aerodynamic field (see Chapter 7). Thus, in a transonic wind tunnel where we wish to study flows at a subsonic Mach number close to 1, the reduction of the test section area, due to the presence of the model, can make the flow around it become supersonic. This fact (called blockage effect) leads us to design a test section, where the flow is "unchoked" by equipping the test section with perforated or slotted walls allowing a partial escape of the flow, hence, a decrease of its velocity. It also makes use of adaptive walls presenting a local deformation compensating the obstruction of the model.

9.5.2. *Isentropic relations and the area-Mach number relation*

9.5.2.1. *Local stagnation conditions*

As we have seen earlier, during an isentropic change from the stagnation state to the supersonic velocity V, there is a passage through a critical state where the velocity is equal to the local velocity of sound. The pressure, temperature, and density then pass through the said critical values p_c, T_c, ρ_c. In general, if we consider a flow whose properties are known (pressure, temperature, density, and velocity), it is possible to attach to this flow a state resulting from an *isentropic stagnation of the fluid*, i.e. an isentropic retardation of the flow to zero velocity. To do this, it is enough to imagine the change of a fluid where its entropy is conserved. This change may be real or fictitious. The flow might not have changed isentropically since the reservoir, if it has traversed shock waves, enables us to define local stagnation conditions by calculating the flow evolution up to $V = 0$. Thus:

– p_{st} = local stagnation pressure,

– ρ_{st} = local stagnation density,

– T_{st} = local stagnation temperature, and

– a_{st} = local stagnation sound velocity,

These concepts are important and in permanent use in the aerodynamics of compressible flow.

9.5.2.2. *Application to a calorically perfect gas*

In the case of a calorically perfect gas, it is possible to establish explicit relations giving the changing properties of a fluid during an isentropic expansion or compression. We recall that the state equation of a perfect gas is written:

$$\frac{p}{\rho} = \frac{\Re}{M} T$$

where $\Re$ is the perfect gas universal constant and here M is the molar mass of the gas in question. We write more often, by stating $r = \Re / M$:

$$\frac{p}{\rho} = rT .$$

The gas is called calorically perfect if its specific heats at constant pressure and constant volume are constant:

$$C_p = \text{constant}, \qquad C_v = \text{constant}$$

Consequently, the ratio of specific heats γ is also constant.

The thermodynamic properties of the gas can thus be calculated by the following explicit relations (see Chapter 3):

– Mayer relation: $r = C_p - C_v$,

– enthalpy: , $h = C_p T$

– internal energy: $e = C_v T$,

– entropy: $s = C_p \, lnT - r \, ln \, p$ or $s = C_v \, ln\left(p\rho^{-\gamma}\right)$

– velocity of sound: $a^2 = \gamma \dfrac{p}{\rho} = \gamma \, rT = (\gamma - 1)h$.

Under the perfect gas hypothesis, the energy equation is written as:

$$h + \frac{V^2}{2} = h_{st} = constant \;\; or \;\; C_p \;\; T + \frac{V^2}{2} = C_p \;\; T_{st} = constant$$

which gives:

$$V^2 + 2C_p T = constant\,.$$

Knowing that:

$$2C_p T = \frac{2C_p}{\gamma \;\; r} a^2 = \frac{2C_p}{\gamma\left(C_p - C_v\right)} a^2 = \frac{2}{\gamma - 1} a^2$$

we deduce the *equation of Saint Venant:*

$$\boxed{V^2 + \frac{2}{\gamma - 1} a^2 = constant.}$$

If another state of the fluid is known during an adiabatic change (no heat exchange), the constant of the preceding equation can be calculated. We can, for example, take the stagnation state for which $V = 0$ and $a = a_{st}$, which gives:

$$\boxed{V^2 + \frac{2}{\gamma - 1} a^2 = \frac{2}{\gamma - 1} a_{st}^2}$$

from which follows the values of the limit velocity:

$$V_{\lim} = \sqrt{\frac{2}{\gamma - 1}} \, a_{st}.$$

In considering the critical states for which $V = a_c$, it follows that:

$$V^2 + \frac{2}{\gamma - 1} a^2 = a_c^2 + \frac{2}{\gamma - 1} a_c^2$$

giving:

$$\boxed{V^2 + \frac{2}{\gamma - 1} a^2 = \frac{\gamma + 1}{\gamma - 1} a_c^2.}$$

We now will look for relations giving the ratios $T / T_{st}, p / p_{st}, \rho / \rho_{st}$ with Mach number M as parameter (and γ).

The relation:

$$V^2 + \frac{2}{\gamma - 1} a^2 = \frac{2}{\gamma - 1} a_{st}^2$$

gives:

$$\frac{V^2}{a^2} + \frac{2}{\gamma - 1} = \frac{2}{\gamma - 1} \frac{a_{st}^2}{a^2}$$

But:

$$\frac{a_{st}^2}{a^2} = \frac{T_{st}}{T}$$

from which follows, knowing that $M = V / a$:

$$\boxed{\frac{T}{T_{st}} \equiv \theta(M,\gamma) = \left(1 + \frac{\gamma - 1}{2} M^2\right)^{-1}.}$$

If the flow is isentropic:

$$p\rho^{-\gamma} = p_{st}\rho_{st}^{-\gamma}.$$

Since the gas is a perfect gas:

$$\frac{p}{p_{st}} = \frac{\rho}{\rho_{st}} \frac{T}{T_{st}}$$

then:

$$\frac{p}{p_{st}} = \left(\frac{\rho}{\rho_{st}}\right)^{\gamma} = \left(\frac{p}{p_{st}}\right)^{\gamma} \left(\frac{T}{T_{st}}\right)^{-\gamma}$$

which gives:

$$\frac{p}{p_{st}} = \left(\frac{T}{T_{st}}\right)^{-\frac{\gamma}{\gamma-1}}$$

and therefore:

$$\boxed{\frac{p}{p_{st}} \equiv \pi(M,\gamma) = \left(1 + \frac{\gamma - 1}{2} M^2\right)^{-\frac{\gamma}{\gamma-1}}.}$$

Thus:

$$\boxed{\frac{\rho}{\rho_{st}} \equiv \Re(M,\gamma) = \left(1 + \frac{\gamma - 1}{2} M^2\right)^{-\frac{1}{\gamma-1}}.}$$

9.5.2.3. *The area-Mach number relation or law of sections*

The conservation of the mass flow allows us to write:

$$q_m = \rho V A = \rho_c V_c A_c = \rho_c a_c A_c$$

where the mass flow is expressed in the local section A and in the critical section A_c. We get:

$$\frac{A}{A_c} = \frac{\rho_c}{\rho}\frac{a_c}{V}$$

or by introducing the stagnation conditions:

$$\frac{A}{A_c} = \frac{\rho_c}{\rho_{st}}\frac{\rho_{st}}{\rho}\frac{a_c}{a_{st}}\frac{a_{st}}{a}\frac{a}{V}.$$

The relations established above allow us to write:

$$\frac{\rho_c}{\rho_{st}} = \left(1+\frac{\gamma-1}{2}\right)^{-\frac{1}{\gamma-1}} = \left(\frac{\gamma+1}{2}\right)^{-\frac{1}{\gamma-1}}$$

$$\frac{a_c}{a_{st}} = \left(\frac{T_c}{T_i}\right)^{\frac{1}{2}} = \left(\frac{\gamma+1}{2}\right)^{-\frac{1}{2}}$$

$$\frac{a_{st}}{a} = \left(1+\frac{\gamma-1}{2}M^2\right)^{\frac{1}{2}}$$

$$\frac{\rho_{st}}{\rho} = \left(1+\frac{\gamma-1}{2}M^2\right)^{\frac{1}{\gamma-1}}$$

which gives:

$$\frac{A}{A_c} = \left(\frac{\gamma+1}{2}\right)^{-\frac{1}{\gamma-1}}\left(\frac{\gamma+1}{2}\right)^{-\frac{1}{2}}\left(1+\frac{\gamma-1}{2}M^2\right)^{\frac{1}{\gamma-1}}\left(1+\frac{\gamma-1}{2}M^2\right)^{\frac{1}{2}}M^{-1}.$$

Which ultimately is:

$$\boxed{\frac{A}{A_c} = \left(\frac{2}{\gamma+1}\right)^{\frac{\gamma+1}{2(\gamma-1)}}\frac{1}{M}\left(1+\frac{\gamma-1}{2}M^2\right)^{\frac{\gamma+1}{2(\gamma-1)}}}.$$

This formula known as the area-Mach number relation shows that for a flow with a choked throat (i.e. a flow where the Mach number is equal to 1 in the critical section A_c), the local Mach number depends only on the ratio A/A_c. This is the case of supersonic nozzles of rocket engines whose exit Mach number is determined by the section ratio $\sigma = A_E / A_c$ where A_E is the area of the exit section.

9.5.2.4. *Expressions for the velocity*

It is often useful to know the expressions V/a_{st} and V/a_c as a function of the Mach number. By writing:

$$\frac{V}{a_{st}} = \frac{V}{a}\frac{a}{a_{st}}$$

and taking the above expressions into account, it becomes:

$$\boxed{\frac{V}{a_{st}} = M\left(1+\frac{\gamma-1}{2}M^2\right)^{-\frac{1}{2}}.}$$

Similarly, we obtain:

$$\boxed{\frac{V}{a_c} = M\left(\frac{\gamma+1}{2}\right)^{\frac{1}{2}}\left(1+\frac{\gamma-1}{2}M^2\right)^{-\frac{1}{2}}.}$$

The expressions that have been established are permanently used in aerodynamics of compressible flows. For this reason, the various ratios T/T_{st}, p/p_{st}, ρ/ρ_{st} are tabulated as a function of the Mach number *M*, most often for $\gamma = 1.4$, which corresponds to air under normal conditions.

We sometimes use as a parameter the ratio of the local velocity to the limit velocity, $V/V_{\lim}$. For a calorically perfect gas:

$$\frac{V}{V_{\lim}} = \sqrt{\frac{C_pT_{st} - C_pT}{C_pT_{st}}} = \sqrt{1-\frac{T}{T_{st}}} = \sqrt{1-\frac{1}{1+\frac{\gamma-1}{2}M^2}}$$

which gives:

$$\boxed{\frac{V}{V_{\lim}} \equiv Cr = \sqrt{\frac{\frac{\gamma-1}{2}M^2}{1+\frac{\gamma-1}{2}M^2}}.}$$

The parameter Cr is called *Crocco number*. It varies in the finite interval [0,1] when the Mach number varies from 0 to infinity. It also allows us to include the effect of γ, which may be useful for some applications.

COMMENTS.–

1. As we have seen in this chapter, the stagnation enthalpy h_{st} of a steady adiabatic flow remains constant even if the flow undergoes irreversible changes that modify its entropy. The property $h_{st} = cst$ uniquely results of the first law of thermodynamics. Consequently, for a calorically perfect gas, the stagnation temperature is constant.

2. For a calorically perfect gas, the specific entropy is given by the following equation applied to the local stagnation state:

$$s_{st} = C \, ln \, T_{st} - r \, ln \, p_{st} = \bar{s}(T_{st}) - r \, ln \, p_{st}$$

where the term $\bar{s}\,(T_{st})$ is called the entropy of temperature. If we consider a flow of the kind we are considering here, where irreversibility can occur, the local entropy will be different on either side of the irreversibility. According to the second law of thermodynamics, entropy s_2 downstream of irreversibility is greater than the entropy s_1 upstream. As T_{st} remains unchanged, we get:

$$s_2 - s_1 = -r\ln \; p_{st_2} + r\ln p_{st_1} = r\ln\frac{p_{st_1}}{p_{st_2}} > 0\,.$$

It follows that $p_{st_1} / p_{st_2} > 1$, hence $p_{st_2} < p_{st_1}$. Thus, in an adiabatic flow, any irreversibility leads to a decrease in local stagnation pressure. This result has very important practical consequences in the field of supersonic air intakes and propulsion in general (see Chapter 16).

3. The dynamic pressure, normalized by the stagnation pressure of the flow, is expressed in the case of a calorically perfect gas:

$$q_v = \frac{\rho V^2}{2 p_{st}} = \frac{1}{2}\frac{\rho}{\rho_{st}}\frac{V^2}{a_{st}^2}\frac{\rho_{st}a_{st}^2}{p_{st}}\rho_{st}a_{st}^2 = \frac{1}{2}\frac{\rho}{\rho_{st}}\frac{V^2}{a_{st}^2}\frac{p_{st}}{rT_{st}}\frac{\gamma r T_{st}}{p_{st}} = \frac{\gamma}{2}\frac{\rho}{\rho_{st}}\frac{V^2}{a_{st}^2}$$

or by using the expressions given above:

$$q_v = \frac{\rho V^2}{2 p_{st}} = \frac{1}{2}\gamma \; M^2 \left[1 + \frac{\gamma - 1}{2} M^2\right]^{-\frac{\gamma}{\gamma - 1}}.$$

By taking the logarithmic derivative we obtain:

$$\frac{dq_v}{q_v} = 2\frac{dM}{M} - \frac{\gamma}{\gamma - 1}\frac{(\gamma - 1)\, MdM}{1 + \frac{\gamma - 1}{2} M^2} = \frac{2 - M^2}{M\left[1 + \frac{\gamma - 1}{2} M^2\right]}.$$

This relation shows that the dynamic pressure q_v is maximum for a Mach number such that:

$$\boxed{M = \sqrt{2}}$$

regardless of the value of γ.

9.5.2.5. *One application of the area-Mach number relation: the sonic flow meter*

Frequently in internal aerodynamics, a mass flow has to be measured, which is the case of the mass flow feeding the propulsion nozzles of a model aircraft or the flow entering an air intake model. This is also the case or the mass flow used in fluidic active flow control devices. Accurate knowledge of the mass flow is essential not only for assessing the performance of the device (thrust of the nozzle, mass flow coefficient of the air intake), but also to design the experimental setup. Where possible, the mass flow can be measured with great accuracy by installing in the supply or extraction lines a sonic flow meter consisting of a diaphragm with a circular hole with the diameter D_c through which passes the flow whose mass flow has to be measured. This diaphragm is mounted within a casing as shown in Figure 9.11 to ensure its interchangeability. It consists of a plate in which an aperture is machined in the shape of a convergent-divergent nozzle with most often a circular contour. If we have a sufficient pressure difference, it is possible to achieve sonic flow in the aperture with the diameter D_c.

In steady flow, and under the one-dimensional theory, the mass flow (*kg/s*) is expressed by the relation:

$$q_m = \rho V A$$

which is convenient to write in the critical section of area A_c where the Mach number is equal to 1 and the velocity equal to the speed of sound a_c, which gives:

$$q_m = \rho_c \; a_c \; A_c \, .$$

Thus, introducing the stagnation conditions:

$$q_m = \frac{\rho_c}{\rho_{st}} \frac{a_c}{a_{st}} \rho_{st} \, a_{st} \, A_c \, .$$

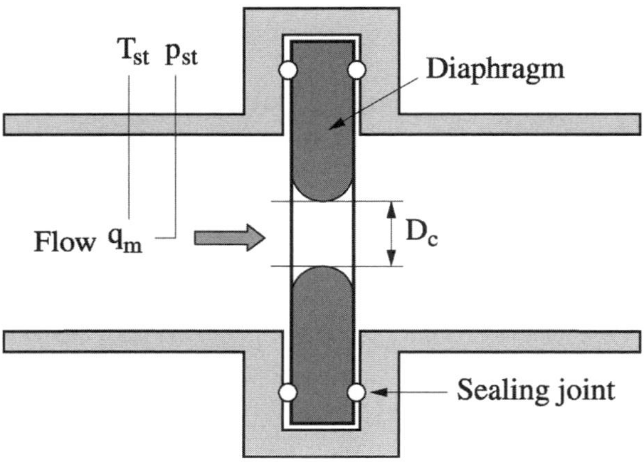

Figure 9.11. *Flow meter with sonic throat*

By expressing the ratios using the isentropic expansion relations above, where we set $M = 1$, and by introducing the stagnation pressure and temperature, which are directly measurable quantities, it follows:

$$q_m = \left(\frac{\gamma+1}{2}\right)^{-\frac{1}{\gamma-1}} \left(\frac{\gamma+1}{2}\right)^{-\frac{1}{2}} \frac{p_{st}}{r\,T_{st}} \sqrt{\gamma r T_{st}} \, A_c \, .$$

where:

$$\delta(\gamma) = \frac{\gamma}{\sqrt{\gamma-1}} \left(\frac{\gamma+1}{2}\right)^{-\frac{\gamma+1}{2(\gamma-1)}} .$$

Hence, the formula of an "ideal" mass flow meter:

$$\boxed{\overline{q}_m = \frac{\delta(\gamma)}{\sqrt{C_p\,T_{st}}}\,p_{st}\,A_c}$$

in the case of air: $\delta(\gamma) = 1.281$.

Knowing the throat area $A_c = \dfrac{\pi}{4} D_c^2$, the mass flow passing through the flow meter is determined by measuring the stagnation pressure p_{st} and temperature T_{st} upstream of the diaphragm by the means of judiciously placed probes. As the flow is not strictly One-dimensional in the region of the throat (see Chapter 19) and as there are boundary layers on the walls, we introduce a correction coefficient C_D; thus, the actual mass flow is given by:

$$q_m = C_D\,\overline{q}_m = C_D\,\frac{\delta(\gamma)}{\sqrt{C_p\,T_{st}}}\,p_{st}\,A_c\,.$$

If the throat of the diaphragm is carefully tooled and the radius of the contour large enough to ensure smooth change of the section, the coefficient C_D is very close to 1, often greater than 0.99.

The interchangeable diaphragm enables adjustment of the diameter of the sonic aperture to the value of the mass flow to be measured in order to remain within a range of upstream stagnation pressures, compatible with the means to produce available compressed air. If the upstream pressure becomes too large, a diaphragm with a larger hole diameter is adopted. Thus the interchangeability allows for varying the mass flow for fixed upstream conditions.

As we have emphasized, the use of a sonic flow meter requires having a supply of compressed air at high pressure or a high pumping capacity, if a captation capability is measured as in the case of an air intake (see Chapter 16). In addition, the diaphragm causes a significant drop in the stagnation pressure resulting from the establishment of a supersonic flow. Therefore, it is often necessary to measure the mass flow, or volume flow when in incompressible, by using a device called a Venturi tube or *Venturi*, whose principle is explained in the following.

9.6. Incompressible case: the Bernoulli relation and application

9.6.1. *The Bernoulli relation*

Although the air is a highly compressible fluid, changes in its density in a flow whose velocity varies from 0 to 150 *m/s* do not exceed several percent (see section 7.2). Therefore, we can solve many problems of aerodynamics assuming a purely incompressible fluid, which leads to very interesting simplifications. This section is devoted to the application of the one-dimensional theory to flows that can be regarded as purely incompressible. The theory of incompressible flows represents a "block" itself which underwent tremendous development and led to major results. Here we will only infer results by specialization of some of the above relations where the density can be regarded as constant.

This relation, which is probably the most famous in fluid mechanics, relates the pressure and velocity of an incompressible non-viscous and *steady* flow. By applying the equation of motion, we established in section 9.2.2.1 the Euler equation:

$$V\,dV + \frac{dp}{\rho} = 0\,.$$

If the density ρ is constant and such that $\rho = \rho_0$, then the Euler equation integrates in the form:

$$\boxed{p + \frac{1}{2}\rho_0\, V^2 = constant}$$

this equation, verified on a streamline, is called the *Bernoulli relation*.

We sometimes state:

$$p + \frac{1}{2}\rho_0\, V^2 = p_T = p_{st}$$

where p_T is the total pressure, also called the *stagnation pressure* and noted p_{st}. This pressure is measured by placing a tube, connected to a pressure transducer, facing the flow. The velocity in the tube is decreased to zero between the inlet section and the transducer, which then gives a measure of the quantity p_{st}. Such a probe is called a *Pitot tube* in tribute to Henri Pitot the inventor of the device in 1724.

In incompressible flows, stagnation pressure p_{st} and total pressure p_{st} are equivalent concepts and terms that are used indiscriminately. In a supersonic flow, we have to be more rigorous in the choice of words; the pressure measured by a Pitot tube is, in fact, the stagnation pressure behind the normal shock that forms in front of the Pitot tube (see Chapter 13). This pressure is different from the stagnation pressure p_{st} of the flow. We must therefore clarify what is meant by the term stagnation pressure. It is the pressure given by the Pitot tube, or the *isentropic* stagnation pressure? We avoid any ambiguity by speaking of the pitot pressure and stagnation pressure.

If viscous effects are of importance, as in boundary layers and wakes, the Pitot tube measures a local stagnation pressure, the information from the probe varying from one point to another of the aerodynamic field. In subsonics, this stagnation pressure is the one that would be obtained by imagining an isentropic retardation of the fluid down to zero velocity. This pressure p_{st} is different from the stagnation pressure p_{st_0} of the flow in the non-viscous part of the field, outside the boundary layers and wakes: usually $p_{st} < p_{st_0}$. There is an intermediate situation where the fluid is non-viscous but rotational, the entropy being different from one streamline to another, while being constant along the same streamline. Such a flow exists downstream of a incurved shock wave, where the stagnation pressure varies when we change the streamline (see Chapter 13).

Thus, in general, for an incompressible flow, the constant of the Bernoulli equation, or stagnation pressure, varies from one point to another of the field or from one streamline to another. However, in an isentropic *irrotational* flow (see the Crocco relation in Chapter 8), the constant is the same for all streamlines, so

$$p + \frac{1}{2}\rho_0 V^2 = \text{constant in the whole flow.}$$

9.6.2. *Measurement of the velocity: the Pitot probe*

According to Bernoulli relation, the velocity module V of the fluid is deduced from the difference between stagnation pressure and static pressure by the formula:

$$\boxed{V = \sqrt{\frac{2(p_{st} - p)}{\rho_0}}.}$$

As the density ρ is here a constant of known magnitude, the measurement of p and p_{st} allows access to the velocity V. The classic Pitot tube measures the stagnation pressure p_{st}, the static pressure p is given, for example, by a static pressure orifice, placed on one wall of the channel as shown in Figure 9.12. In the case of a ground vehicle, the pressure p may be the atmospheric pressure provided that the tip of the Pitot tube is placed far enough in front of the vehicle so that the pressure level is not affected by the flow around the vehicle.

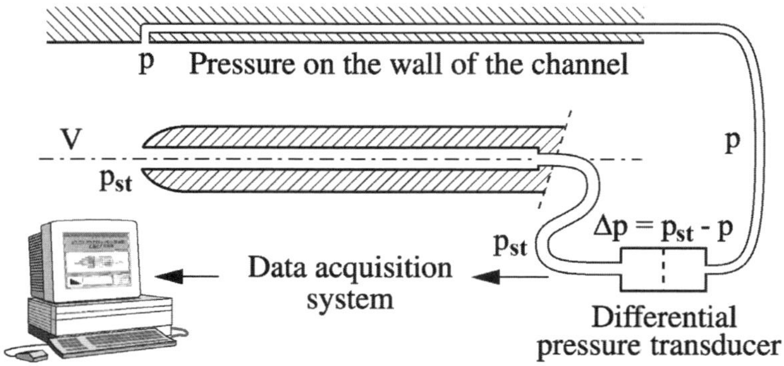

Figure 9.12. *Pitot tube for the measurement of the velocity in a channel*

In the case of an aircraft in flight, the static pressure is unknown so that it must be measured to apply the Bernoulli relation. However, measuring the static pressure p within the fluid is difficult because it is an intrinsic property of the fluid that is defined for an observer, which follows the fluid as it moves at the velocity V. We determine the static pressure on a body by placing an orifice (or pressure tap) at the surface of this body, the orifice being connected to a sensor through a tube. It is possible to measure the pressure p in a flow by placing the aperture on a cylinder whose axis is aligned with the local velocity vector. If the orifice is far enough downstream from the cylinder front end (at least 10 times its diameter), the pressure at the orifice is practically that of the flow upstream of the cylinder.

The probe called a *Prandtl antenna* associates the measurement of the stagnation pressure with a measurement of the static pressure through the device shown in Figure 9.13, p_{st} is given by the probe nose and p by the orifice located downstream. The two pressures are in communication with two transducers, or connected to a differential transducer which directly measures the difference $\Delta p = (p_{st} - p)$. Knowing Δp, the velocity is determined from the formula above. In general, to make the probe less sensitive to angle of incidence (which anyway must be small), several static orifices are placed around the cylindrical part. These orifices are all in communication with the same transducer which then measures an average pressure taking into account a possible incidence effect.

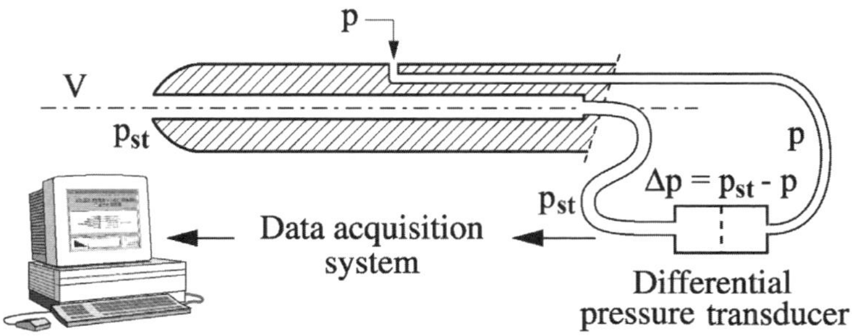

Figure 9.13. *Prandtl antenna for measuring the velocity in a fluid*

9.6.3. *The Venturi*

We consider a duct axisymmetric or symmetric with respect to the axis *X'X*, shown in Figure 9.14. It includes an entry with the area A_1 followed by a gradual decrease to a minimum section or throat with the area A_2. Then, the section increases again continuously until the exit area A_s. Such a device is called a *Venturi*. The flow in the Venturi is assumed to be one-dimensional, implying that its properties are constant in sections normal to the axis *X'X*. If the flow is steady, the mass flow q_m is the same in all sections and we have:

$$\rho_0 \ A_1V_1 = \rho_0 \ A_2V_2.$$

In incompressible, we often work with the volume flow:

$$q_v = AV$$

which is also invariant in a steady flow (expressed in m^3/s). Furthermore, the relation of Bernoulli gives:

$$p_1 + \frac{\rho_0}{2}V_1^2 = p_2 + \frac{\rho_0}{2}V_2^2.$$

Hence, after elementary calculations, the following expression for the velocity V_1:

$$\boxed{V_1 = \sqrt{\frac{2(p_2 - p_1)}{\rho_0 \left[1-\left(\frac{A_1}{A_2}\right)^2\right]}}.}$$

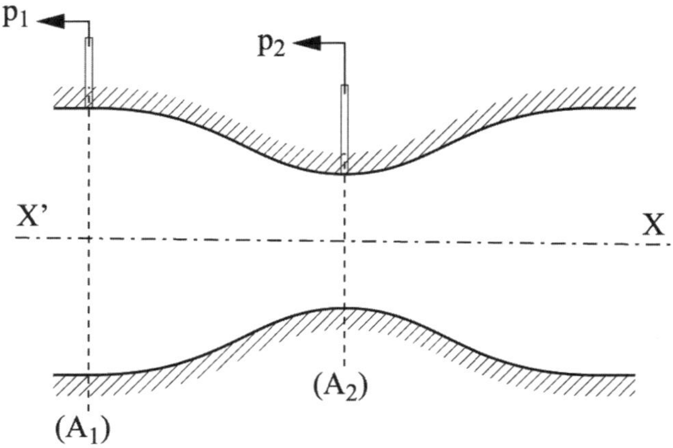

Figure 9.14. *Venturi tube*

Knowing the geometry of the duct, i.e. sections A_1 and A_2, as well as the density ρ_0, the measurement of the pressure difference $\Delta p = p_2 - p_1$ enables us to determine the velocity at the entrance to the Venturi. The device can measure the velocity of the flow, aspirated for example by a pipe to fuel an engine or by a cooling system, the volume flow being derived from the knowledge of V_1. From the value of the volume flow q_v indicated by the Venturi we can also regulate this flow. Venturis are commonly used in industry.

Chapter 10

Application of the One-Dimensional Theory to the Calculation of Supersonic Nozzles

10.1. General definition of a supersonic nozzle

A supersonic propulsion nozzle with a convergent–divergent shape ensures the expansion of propellants from a subsonic state, in the combustion chamber, turbine outlet, or even after the postcombustion chamber, to a supersonic state in the exit plane (*E*). In most cases, the nozzles are axisymmetric. Such a supersonic nozzle is shown in Figure 10.1. It must include a throat (*C*) where the Mach number is equal to 1. The nozzles of this type also equip supersonic wind tunnels.

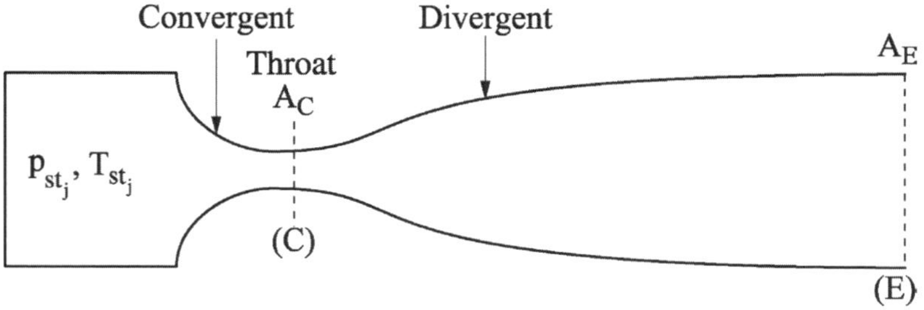

Figure 10.1. *Layout of a supersonic nozzle*

The nozzle is fed under the initial (stagnation) conditions:

– pressure: p_{st_j},

– temperature: T_{st_j}.

In these notations, we have introduced the index j to denote a propulsion jet. We denote with the index E the conditions in the exit plane whose section has an area A_E:

– pressure: p_E,

– density: ρ_E,

– temperature: T_E,

– gas velocities: V_E,

– Mach number: M_E.

The area of the throat is denoted by: A_C.

Two quantities are of particular importance for the nozzles:

– the mass flow: q_m,

– the thrust: F.

These quantities can be calculated using the theory of one-dimensional flows. The results thus obtained are approximate, the flow in a real nozzle being not strictly one-dimensional. A precise calculation of the performance of a nozzle must use methods taking into account the two-dimensional character of the actual flow (see Chapter 19). However, in this application, the one-dimensional theory provides a good approximation often sufficient at the preliminary design stage.

Two types of nozzles are more commonly used:

– Conical nozzles where the divergent is made of a simple truncated cone. We denote by Θ the *half-angle* of the cone (see Figure 10.2a).

– Contoured (or ideal) nozzles whose wall has a contour calculated so that the flow in the exit plane of the nozzle is uniform (see Figure 10.2b). The method used to determine the contour of such a nozzle is given in Chapter 19.

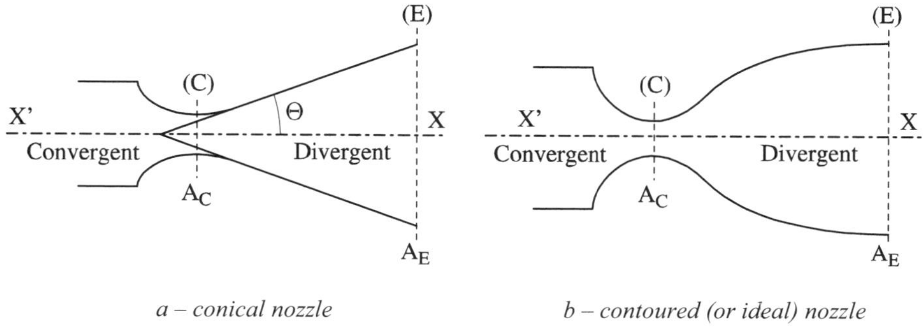

Figure 10.2. *Two types of propulsion nozzles*

10.2. Calculation of the flow in the nozzle

The one-dimensional assumption means that the properties of the flow are constant across a section A normal to the axis $X'X$ of the nozzle. We further assume that the fluid is non-viscous and the flow permanent. A nozzle ensures an *isentropic expansion* of the flow from the initial conditions (p_{st_j}, T_{st_j}) to the conditions in the exit plane $(p_E, T_E, \rho_E \ldots)$. The assumption of constant entropy implies:

– the absence of chemical reactions, which means that gases from the combustion chamber are chemically inert. This condition is not always satisfied in the nozzles of rocket engines where chemical reactions can proceed in the nozzle;

– there is no exchange of heat, especially at the walls. This condition is not met for rocket engines where the nozzle must be cooled to withstand gas of the jet whose stagnation temperature exceeds 3,000 K. However, at the scale of the flow, these transfers bring only weak energies into play; we thus can neglect them;

– the viscous effects are absent, implying the absence of shock waves, as is in the case of well-designed nozzles. However, along the wall, there is a boundary layer where viscous dissipation occurs. This effect can be regarded as of second-order, the boundary layers in nozzles usually being very thin.

The expansion can be represented in a Mollier diagram [entropy s, enthalpy h] (see Figure 10.3). Starting from the initial state located on the isobar p_{st_j} and the isotherm T_{st_j}, the representative point of the expansion in the nozzle moves along the vertical line: $s = s_{st_j} = constant$. This leads to the condition (E) corresponding to the generally imposed nozzle exit conditions.

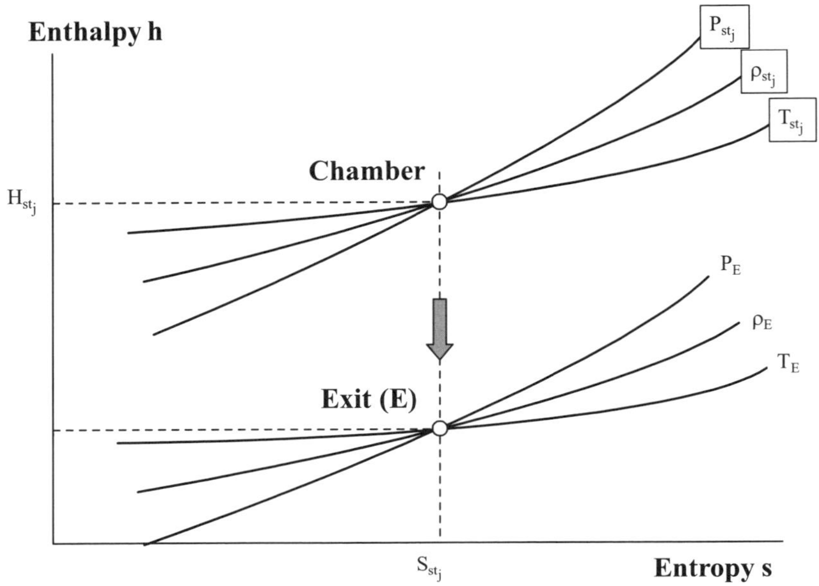

Figure 10.3. *Image of the isentropic expansion in a Mollier diagram*

As we have seen, in the context of the one-dimensional hypothesis, the flow is governed by the following equations:

– *continuity equation:*

$$q_m = \rho VA = constant$$

or in differential form:

$$\frac{d\rho}{\rho} + \frac{dV}{V} + \frac{dA}{A} = 0,$$

– *equation of motion (Euler equation):*

$$V \ dV + \frac{dp}{p} = 0,$$

– *energy equation:*

$$h + \frac{V^2}{2} = h_{st_j} = constant.$$

These equations where the unknowns are ρ, p, V, and h must be supplemented by a thermodynamic relation.

The law of section A (x) being known by construction, we can calculate the flow in the nozzle step by step, using, in the general case, a Mollier diagram to define the gas thermodynamics. This calculation will be performed starting from the throat where the nozzle is choked (the local Mach number is *1*). We can thus determine the flow in the nozzle down to the exit plane (E).

If the gas can be regarded as a calorically perfect gas, the usual assumption at the predesign stage, then there are analytic relations for the different ratios:

$$\frac{p}{p_{st_j}} \equiv \pi(M,\gamma) = \left(1 + \frac{\gamma-1}{2} M^2\right)^{-\frac{\gamma}{\gamma-1}}$$

$$\frac{\rho}{\rho_{st_j}} \equiv \Re(M,\gamma) = \left(1 + \frac{\gamma-1}{2} M^2\right)^{-\frac{1}{\gamma-1}}$$

$$\frac{T}{T_{st_j}} \equiv \theta(M,\gamma) = \left(1 + \frac{\gamma-1}{2} M^2\right)^{-1}$$

supplemented by the area–Mach number relation:

$$\frac{A}{A_c} \equiv \Sigma(M,\gamma) = \frac{1}{M}\left(\frac{2}{\gamma+1} + \frac{\gamma-1}{\gamma+1} M^2\right)^{\frac{\gamma+1}{2(\gamma-1)}}.$$

The area–Mach number relation shows that the Mach number in any section of the nozzle depends only on the area of this section, or more precisely on the ratio between the area of the considered section and the area of the throat. This ratio $\frac{A}{A_c}$ is sometimes written as σ. This is the law of the sections $A(x)$ which completely defines the flow in the nozzle (of course with the stagnation conditions). Generally, the gas velocity results from the energy equation:

$$h_{st_j} = h + \frac{V^2}{2}$$

from which follows:

$$V = \sqrt{2(h_{st_j} - h)}.$$

The above relation does not assume that the flow is isentropic and the gas perfect. It only assumes the absence of energy exchange during the evolution (adiabatic flow). We recall that the so-called *limit velocity* is the velocity reached by the gas when the expansion is continued close to a vacuum (see section 9.5.1). Therefore, as $h \to 0$:

$$V_{\text{lim}} = \sqrt{2h_{st_j}}.$$

In the case of a calorically perfect gas, the velocity will be calculated by:

$$\frac{V}{a_{st_j}} = \frac{V}{a}\frac{a}{a_{st_j}} = \frac{V}{a}\sqrt{\frac{T}{T_{st_j}}}$$

which gives:

$$\frac{V}{a_{st_j}} = \frac{M}{\sqrt{1+\frac{\gamma-1}{2}M^2}}.$$

The limit velocity corresponds to $M \to \infty$ and therefore is given by:

$$V_{\text{lim}} = \sqrt{\frac{2}{\gamma-1}}\, a_{st_j}.$$

10.3. Calculation of the nozzle mass flow

In steady state, and under the one-dimensional theory, the mass flow (kg/s) is expressed by the equation:

$$q_m = \rho V A.$$

As has been demonstrated in section 9.5.2.5, q_m can be calculated by the relation:

$$q_m = \overline{q}_m = \frac{\delta(\gamma)}{\sqrt{C_p \ T_{st_j}}} p_{st_j} \ A_c$$

where:

$$\delta(\gamma) = \frac{\gamma}{\sqrt{\gamma - 1}} \left(\frac{\gamma + 1}{2} \right)^{-\frac{\gamma+1}{2(\gamma-1)}}.$$

Table 10.1 gives the values of the function $\delta(\gamma)$ for values of γ usual in propulsion problems:

γ	1.15	1.20	1.25	1.30	1.35	1.40
$\delta(\gamma)$	1.7683	1.5886	1.4714	1.3890	1.3279	1.2810

Table 10.1. *Values of the function* $\delta(\gamma)$ *of the mass flow formula*

The above formula gives the mass flow in the case of a one-dimensional flow of a non-viscous and calorically perfect fluid. In real nozzles, these conditions are not fulfilled, the flow being actually two-dimensional (generally axisymmetric), there are boundary layers, and real gas effects can be present. We should then calculate the actual mass flow using the relation:

$$q_m = C_d \, \overline{q}_m$$

where C_d is the *mass flow coefficient* (also called *discharge coefficient*) representing the mass flow deviation from the ideal case. This ratio, determined from experiments or calculations, is usually close to 1 (higher than 0.95 for good nozzles).

10.4. Calculation of the nozzle thrust

If (S_2) and (S_1) are two surfaces based on the lips of the nozzle and the air intake and if $\vec{D}$ denotes the dynalpy vector $\vec{D} = p\vec{n} + (\rho\vec{V}.\vec{n})\,\vec{V}$, the overall propulsion force of a nacelle is given by (see section 5.2):

$$\vec{R} = -\left[\iint_{(S_2)} \vec{D} \; dS + \iint_{(S_1)} \vec{D} \; dS \right].$$

The thrust F is obtained by projecting $\vec{R}$ on the upstream velocity $\vec{V}_\infty$, or in this case, the axis of the nozzle, supposed to be axisymmetric. If $\vec{x}$ designates the unit vector in this direction, then:

$$F = -\left[\iint_{(S_2)} \vec{D}\ dS + \iint_{(S_1)} \vec{D}\ dS \right] \cdot \vec{x}.$$

The first integral is conventionally called nozzle thrust, the second captation drag. In the case of a single nozzle operating in a vacuum, there is no captation drag, since the nozzle is fed by the combustion chamber.

Therefore:

$$F = -\left[\iint_{(S_2)} \vec{D}\ dS \right] \cdot \vec{x}.$$

We call *vacuum thrust* or *intrinsic thrust* of a nozzle the axial component of the dynalpy flux leaving the nozzle:

$$F = \left(\iint_{(S_E)} \vec{D}\, dS \right) \cdot \vec{x}_E$$

where $\vec{x}_E$ is the unit vector carried by the axis, oriented in the downstream–upstream direction (while $\vec{x}$ is in the opposite direction). F is by convention a positive scalar, the thrust force being opposed to the upstream velocity vector $\vec{V}_\infty$ of the vehicle. Thus, the thrust vector is in the opposite direction of $\vec{x}$ if by convention $\vec{x}$ is oriented along the velocity vector $\vec{V}_\infty$. To evaluate the dynalpy flux, it is wise to adopt a surface, if possible of simple form, on which the flow is easy to calculate. In this case, we choose for (S_E) the section of the nozzle exit in which, by construction, the flow is uniform, and hence:

$$\left(\iint_{(S_E)} \vec{D}\, dS \right) \cdot \vec{x}_E = \iint_{(S_E)} \left[\vec{P} + \left(\rho \vec{V} . \vec{n} \right) \vec{V} \right] dS = \left(p_E + \rho_E V_E^2 \right) A_E$$

since under the assumption of a non-viscous fluid, the tension vector $\vec{P}$ is reduced to the pressure term $p\,\vec{n}$ (see Chapters 2 and 4).

Hence, the following expression of the vacuum thrust:

$$\boxed{F = \left(p_E + \rho_E V_E^2 \right) A_E}$$

or:

$$\boxed{F = p_E (1+\gamma_E M_E^2)\ A_E.}$$

If the gas is calorically perfect, we write:

$$F = p_{st_j}\ A_c \frac{p_E}{p_{st_j}} \left(1+\gamma_E M_E^2\right) \frac{A_E}{A_c}$$

or:

$$F = \frac{p_E}{p_{st_j}} \frac{A_E}{A_c} \left(1+\gamma_E M_E^2\right)\ p_{st_j}\ A_c.$$

We can introduce the function:

$$\boxed{\Phi(M,\gamma) = \pi(M,\gamma)\ \Sigma(M,\gamma)\ (1+\gamma M^2)}$$

known as the *thrust function*. This function tends towards infinity if $M \to 0$, passes through a minimum for $M = 1$, reincreases in supersonic to tend toward a finite asymptotic value when $M \to \infty$. This asymptotic value is a function of γ. Figure 10.4 shows the variation of $\Phi(M,\gamma)$ with the indication of the limit value when the Mach number tends to infinity.

The vacuum thrust will be expressed by the relationship:

$$\boxed{F = \Phi\left(M_E,\gamma_E\right) p_{st_j} A_c.}$$

This formula shows that the thrust of a rocket engine is proportional to the area of the throat of the nozzle, i.e. its size (obviously a nozzle twice as big will push twice as much), and the stagnation pressure of the ejected gas, i.e. the pressure in the combustion chamber. The size of the nozzle is quickly defined by the dimensions of the launcher and constructive data, such as weight, mechanical constraints, etc. We can therefore improve the thrust by increasing the pressure in the chamber, which may call into question the definition of the rocket motor itself. The vacuum thrust can also be increased by acting on the Mach number M_E at nozzle exit. Knowing that (in supersonic) the thrust function $\Phi\left(M_E,\gamma_E\right)$ is an increasing function of the Mach number M_E, it seems sufficient to increase M_E. However, an increase in M_E implies an increase of the exit section A_E, from which follow weight and size problems for the nozzle. In addition, the increase in M_E causes the decrease in exit pressure p_E, from which follow other constraints arising from the non-adaptation of the nozzle (see section 10.5).

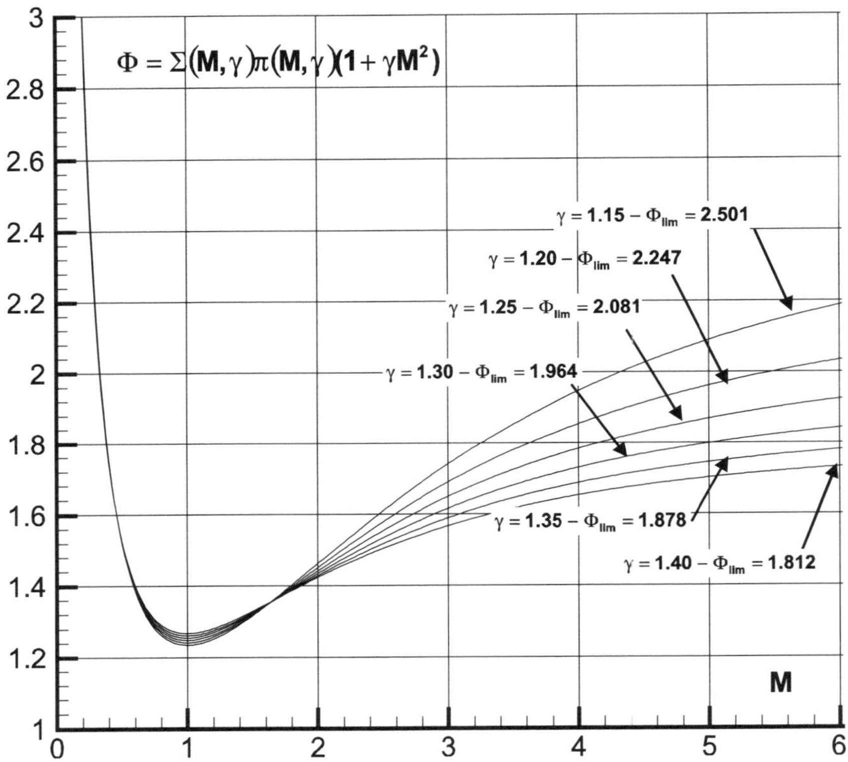

Figure 10.4. *The thrust function*

In the case of a conical nozzle, it is not quite accurate to assume that the flow is uniform in the plane of the exit section. Calculations by the method of characteristics show that it is more accurate to consider the flow leaving the nozzle to be uniform on the portion of a sphere based on the lip of the nozzle exit. This amounts to an assumption of a *source flow*, assuming that this is as if the flow at the exit originates from the point Ω, which is at the apex of the cone with the half-angle Θ (see Figure 10.5). This approximation is even more justified when the nozzle is longer. In contrast, in the throat region, it is more accurate to consider a flow uniform in planes normal to the axis.

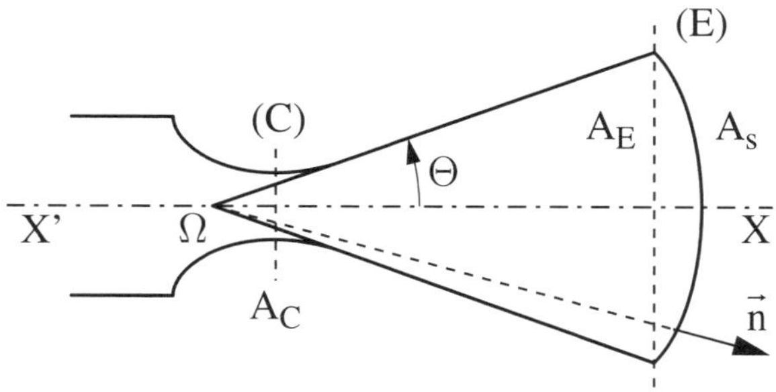

Figure 10.5. *Calculation of a conical nozzle*

If A_S designates the area of the spherical part, the Mach number M_S of the flow over this part will be calculated from:

$$\Sigma\left(M_s, \gamma_s\right) = \frac{A_S}{A_c} \rightarrow M_s.$$

Here, it is advantageous to calculate the dynalpy flux through the spherical part, since on it the flow is uniform; hence, for the thrust *F*, obtained by projecting the dynalpy flux on the axis X'X:

$$F = \left(\iint_{(S_E)} \vec{D}\, dS \right) \cdot \vec{x} = \iint_{(S_E)} \left(p_s + \rho_s\, V_s^2 \right) \vec{n} \cdot \vec{x}\, dS$$

or:

$$F = \left(p_s + \rho_s \;\; V_s^2 \right) \;\; \iint_{(S_s)} \;\; \vec{n} \cdot \vec{x} \;\; dS.$$

It is easy to convince ourselves that the integral is equal to the projected frontal area A_E. Therefore:

$$\iint_{(S_s)} \;\; \vec{n} \cdot \vec{x} \;\; dS = A_E$$

so that:

$$F = \left(p_s + \rho_s\, V_s^2 \right) A_E$$

or:

$$F = \left(p_s + \rho_s \; V_s^2 \right) \; A_s \; \frac{A_E}{A_S}.$$

The geometry shows that:

$$\frac{A_E}{A_S} = \frac{1 + \cos \Theta}{2}.$$

Thus:

$$\boxed{F = \frac{1 + \cos \Theta}{2} \left(p_s + \rho_s \; V_s^2 \right) \; A_S}$$

or by introducing the thrust function:

$$\boxed{F = \frac{1 + \cos \Theta}{2} \Phi \; \left(M_s, \gamma_s \right) \; p_{st_j} \; A_S.}$$

The factor $(1 + \cos \Theta)/2$ is called the *conicity factor*. This is less than 1 and represents the loss of thrust resulting from the use of a conical instead of a contoured nozzle. This loss arises from the fact that, for the conical nozzle, the momentum on the spherical part has a component perpendicular to the axis *X'X* not contributing to the thrust. Table 10.2 gives the values of the conicity factor as a function of the half-angle of the conical nozzle. For an angle $\Theta = 20°$, the thrust loss is 3%, which is not negligible. For a fixed exit Mach number, a small opening angle minimizes the loss of thrust but leads to very long nozzles, resulting in a loss of thrust by friction due to the boundary layer and the increased size of the nozzle can be prohibitive.

$\Theta(°)$	0	5	10	15	20	25	30
$\frac{1 + \cos \Theta}{2}$	1	0.9981	0.9924	0.9829	0.9698	0.9532	0.9330

Table 10.2. *Conicity factor or loss of thrust of a conical nozzle*

10.5. Conventional thrust at the pressure p_a and adaptation of a nozzle

We consider the propulsion nacelle, represented in Figure 10.6, with the entry (*I*) and exit (*E*), placed in an atmosphere with uniform pressure p_a. The projection of the force resulting from the uniform external pressure p_a along the axis of the nacelle is equal to that pressure multiplied by the difference in the areas $(A_E - A_I)$.

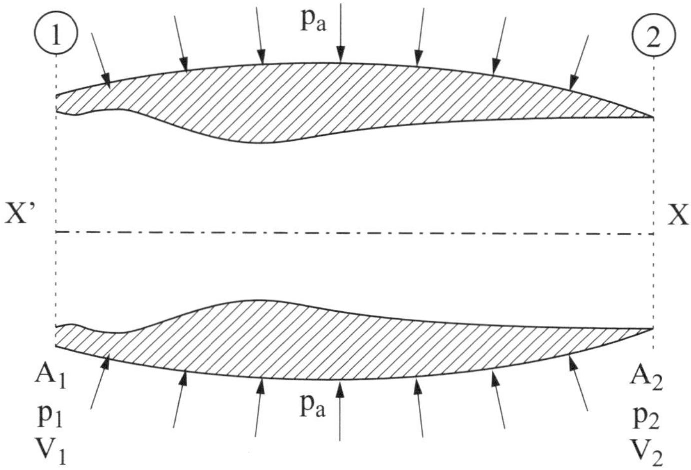

Figure 10.6. *Conventional thrust at the pressure* p_a

The axial force (or thrust) of the nacelle in the atmosphere with the pressure p_a will thus be given by the following expression:

$$F_I^E = \left(p_E + \rho_E\, V_E^2\right) A_E - \left(p_I + \rho_I\, V_I^2\right) A_I - p_a\left(A_E - A_I\right)$$

or even:

$$F_I^E = \left[\rho_E\ V_E^2\ A_E + \left(p_E - p_a\right)\ A_E\right] - \left[\rho_I\ V_I^2\ A_I + \left(p_I - p_a\right)\ A_I\right].$$

We call *conventional thrust* at the pressure p_a or *gross nozzle thrust* the quantity:

$$\boxed{\bar{F} = \left[\rho_E\ V_E^2 + \left(p_E - p_a\right)\right]\ A_E.}$$

If we note that:

$$\rho_E\, V_E\ A_E = q_m$$

we can also write:

$$\boxed{\overline{F} = q_m \; V_E + (p_E - p_a) A_E.}$$

The nozzle, with a fixed throat section, and fed under given stagnation conditions, operates in an atmosphere with the pressure p_a. This raises the following question: what should be the value of the pressure p_E in the exit plane so that this nozzle provides the maximum thrust $\overline{F}$? Often, it is this condition of optimum thrust that defines the nozzle, i.e. the pressure that has to be achieved in its exit plane.

If the mass flow q_m is fixed, we get:

$$\frac{d\overline{F}}{dp_E} = q_m \frac{dV_E}{dp_E} + \left(p_E - p_a\right) \frac{dA_E}{dp_E} + A_E.$$

But according to the Euler equation:

$$V_E \, dV_E + \frac{dp_E}{\rho_E} = 0$$

from which follows:

$$\frac{dV_E}{dp_E} = -\frac{1}{\rho_E \; V_E} = -\frac{A_E}{q_m}.$$

Moreover, according to the equation of continuity:

$$\frac{dA_E}{A_E} + \frac{d\rho_E}{\rho_E} + \frac{dV_E}{V_E} = 0.$$

Hence:

$$\frac{dA_E}{A_E} = -\frac{d\rho_E}{\rho_E} - \frac{dV_E}{V_E} = -\frac{1}{\gamma}\frac{dp_E}{p_E} + \frac{1}{\rho_E \; V_E^2} dp_E.$$

In this relation, we take into account the isentropicity of the flow. Thus:

$$\frac{dA_E}{A_E} = -\frac{1}{\gamma}\frac{dp_E}{p_E} + \frac{p_E}{\rho_E \; V_E^2}\frac{dp_E}{p_E}.$$

But as: $\rho_E V_E^2 = \gamma\, p_E\, M_E^2$:

$$\frac{dA_E}{A_E} = \left(-\frac{1}{\gamma} + \frac{1}{\gamma\, M_E^2} \right) \frac{dp_E}{p_E} = \frac{1 - M_E^2}{\gamma\, M_E^2} \frac{dp_E}{p_E}$$

and we finally obtain:

$$\frac{d\overline{F}}{dp_E} = -A_E + \left(p_E - p_a \right) \frac{1 - M_E^2}{\gamma\, M_E^2} \frac{A_E}{p_E} + A_E$$

$$\frac{d\overline{F}}{dp_E} = \frac{\left(1 - M_E^2\right)}{\gamma} \frac{A_E}{M_E^2\, p_E} \left(p_E - p_a \right) = -\frac{\left(M_E^2 - 1\right)}{\gamma\, M_E^2} \frac{A_E}{p_E} \left(p_E - p_a \right).$$

We find that $\overline{F}$ passes through an extremum, which is a maximum when:

$$\boxed{p_E = p_a.}$$

This condition is called the *nozzle adaptation condition*, which then provides the maximum thrust for operation at ambient pressure p_a. By adapting:

$$\boxed{\overline{F} = q_m\, V_E.}$$

The Mach number M_E being greater than 1, if $p_E < p_a$, $d\overline{F} / dp_E$ is positive, and for $p_E > p_a$, $d\overline{F} / dp_E$ is negative; therefore, $\overline{F}$ passes well through a maximum for $p_E = p_a$ (see Figure 10.7).

The condition of adaptation can be understood in a physical manner by examining the distribution of pressure inside the nozzle. We will not forget that this is the difference $(p - p_a)$ between the pressure inside the nozzle and the external pressure which is the source of the axial force (see Figure 10.8). We remember also that the flow in the nozzle expands continuously. We denote by (a) the section where $p_E = p_a$. If we extend the divergent beyond (a), the pressures on the wall downstream of (a) will be lower than p_a.

Consequently, the axial force:

$$-\left(p - p_a \right) dA$$

will be positive, i.e. in the upstream–downstream direction, which corresponds to a drag. Thus, the divergent part of the nozzle downstream of (a) has a negative thrust.

Conversely, if we cut the nozzle upstream of (a), we will lose a portion of the divergent where:

$$-\left(p-p_a\right)dA$$

is negative, which then could produce an effective thrust. The maximum thrust is obtained when the exit section is such that:

$$p_E = p_a.$$

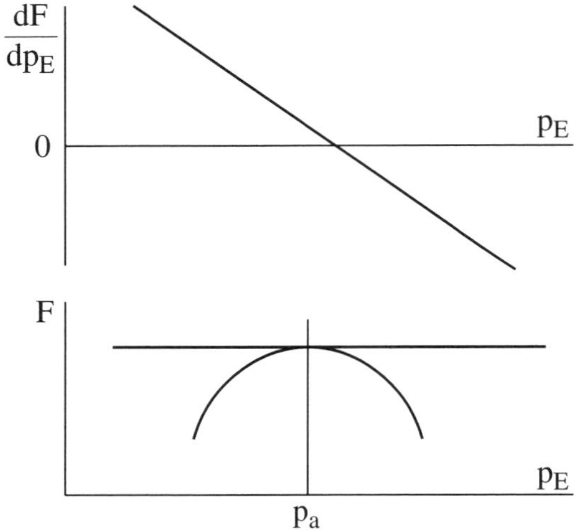

Figure 10.7. *Extremum of the thrust at adaptation condition*

In the case where the pressure p_E is greater than the external pressure p_a, the nozzle is said to be *under-expanded*, which means it does not expand enough to reach the condition:

$$p_E = p_a.$$

In the opposite case, where p_E is lower than p_a, the nozzle is called *over-expanded*, which means that it expands too much compared to the condition of adaptation.

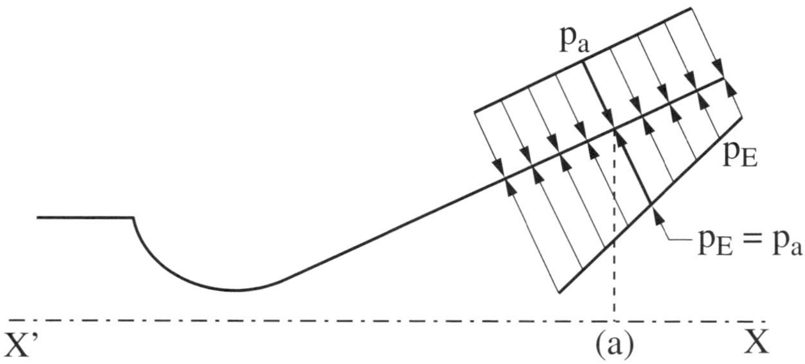

Figure 10.8. *Pressure on the wall close to the nozzle exit*

The operation of nozzles under non-adaptation conditions is a concern for propulsion aerodynamics due to important thrust losses that may result. Thus, the nozzle of an aircraft that is adapted to cruising altitude will be strongly over-expanded at the ground during takeoff and landing, resulting in a significant loss of thrust. This loss is minimized by limiting the expansion in the nozzle by changing its geometry. The nozzles of supersonic aircraft are equipped with variable geometry divergents with which we can adjust the angle of divergence to limit the supersonic expansion at low altitude.

The problem of adaptation is crucial for space launchers. Indeed, for obvious technological constraints, it is not possible to adapt a space rocket nozzle for all flight altitudes, which would require varying the exit Mach number (and therefore the exit pressure) by a variable geometry. In general, the rocket nozzles are adapted to ground level or low altitude, where we need the maximum thrust. They then operate in a state of under-expansion, which is less detrimental than over-expansion.

Figure 10.9 summarizes the circumstances of operation of a contoured nozzle when the viscosity effects are neglected. If the pressure p_E at the exit plane is equal to the external pressure p_a (the case of an adapted nozzle), then the flow leaves the nozzle in the form of a cylindrical supersonic jet separated from the atmosphere by a slip line, or jet line (see Figure 10.9a). If the pressure p_E is greater than p_a (under-expanded nozzle), the flow undergoes at the lip of the nozzle exit a centered expansion and a deflection causing a pluming of the supersonic jet (see Figure 10.9b). Conversely, if p_E is lower than p_a (over-expanded nozzle), a shock wave and a deflection of the jet line corresponding to a contraction of the jet form at the exit lip (see Figure 10.9c) is seen. These phenomena will be interpreted in Chapters 13 and 14.

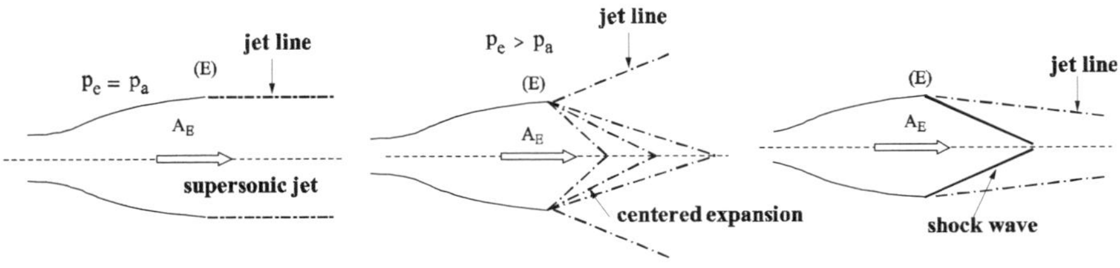

a – adapted nozzle *b – under-expanded nozzle* *c – over-expanded nozzle*

Figure 10.9. *Operating conditions of a propulsion nozzle*

The over-expansion of launcher or missile nozzles is usually avoided because the compression resulting from the adjustment of the jet with the external pressure p_a can cause the separation of the boundary layer that develops on the wall of the nozzle. The flow in the nozzle is then much affected by this phenomenon, which is also a source of disturbances, strong fluctuations, in particular.

10.6. Concluding remarks

We assumed that the flow in the nozzle is isentropic. This is not true if shock waves are formed inside the nozzle because of a poor design of the nozzle shape. However, it is not possible to avoid the production of entropy in the boundary layers developing on the nozzle wall. This leads to a loss of thrust of a few percent, although low but not negligible, for an accurate assessment of the performance.

Contoured, or ideal nozzles, are those that produce maximum thrust (all things being equal) because the outflow is parallel to the axis of the nozzle. Therefore, there is no loss of thrust capacity due to a divergence of the flow, as in the conical nozzle. The reality is actually more nuanced because obtaining the output flow perfectly uniform and parallel to the axis requires a shape with a great length (see Chapter 19 for the method of calculation of the contour of an ideal nozzle). This leads to:

– a prohibitive size and weight of the nozzle for equipping the engine of a space launcher or strategic missile,

– a significant thickening of the boundary layer at the nozzle wall causing a loss of thrust in measure to compensate and beyond the gain from a conical nozzle.

The nozzles of real launchers actually result from a compromise reached by truncating (shortening) the nozzle while seeking an exit flow as axially as possible. Thus, we talk of a *bellshaped nozzle*. The true ideal nozzles are reserved for supersonic wind tunnels where the quality and uniformity of the flow prevail and where considerations of weight and size are secondary.

Chapter 11

One-Dimensional Flows with Friction and Heat Transfer

11.1. Friction force and heat transfer on a wall

We study a one-dimensional flow in a pipe with the section $A(x)$, where a friction force is exerted on the wall at the same time as heat is exchanged with the fluid. A supply can be made by burning kerosene or any other mode of exchange: wall heating, radiation, electromagnetic effect, and mixing of hot gases. If this exchange is negative, the wall extracts heat from the fluid and cools it down (cooling problems). The case of the friction effect without heat supply corresponds to flows in pipes transporting fluids, such as gas or oil pipelines.

To address these problems, one-dimensional models have been proposed where the properties of the flow are assumed to be constant in any section $A(x)$. They combine a great simplicity of implementation with the advantage of providing sufficient accuracy for many industrial applications. In reality, the flow is two-dimensional, or three-dimensional, a boundary layer with the thickness δ increasing with distance x forming on the pipe wall. The skin friction τ_w is a consequence of the viscosity of the fluid creating a force, or contact action exerted by the fluid on the wall (see Chapter 4). The effect of the wall on the fluid is, of course, equal but with an opposite sign. The skin friction τ_w is expressed within the approximation framework of the boundary layer:

$$\tau_w = \mu_w \left(\frac{\partial u}{\partial y} \right)_w$$

where μ_w is the value of the molecular viscosity at the wall, y the normal distance to the wall, and $u(y)$ the streamwise velocity profile. If ρ_e, V_e denote the conditions at the edge of the boundary layer, we define the skin friction coefficient (see section 2.4):

$$C_f = \frac{\tau_w}{1/2\,\rho_e V_e^2}.$$

If the wall temperature T_w is different from the flow temperature T_e, a heat transfer occurs through an exchange mechanism or conduction, described by Fourier's law involving the thermal conductivity κ (see section 4.6). The expression of heat flux $\dot{q}_w$ received by the wall (within the framework of the boundary layer approximation) is then:

$$\dot{q}_w = -\kappa_w \left(\frac{\partial T}{\partial y}\right)_w$$

where $T(y)$ is the temperature profile in the boundary layer. If the derivative $\left(\frac{\partial T}{\partial y}\right)_w$ is negative, the wall gives heat to the flow; if $\left(\frac{\partial T}{\partial y}\right)_w$ is positive, the wall receives heat, thus cooling the flow. In the case where the adiabatic equilibrium is established between the wall and the fluid, $\dot{q}_w = 0$, thus $\left(\frac{\partial T}{\partial y}\right)_w = 0$ (see Figure 11.1). The wall then adopts an adiabatic equilibrium temperature also called *friction temperature* T_f (also called *recovery temperature*).

The heat transfer is represented by a transfer coefficient C_h defined by:

$$C_h = \frac{\dot{q}_w}{\rho_e V_e C_p \left(T_f - T_w\right)}.$$

In this expression, T_w and T_f denote, respectively, the temperature of the fluid at the wall and the friction temperature.

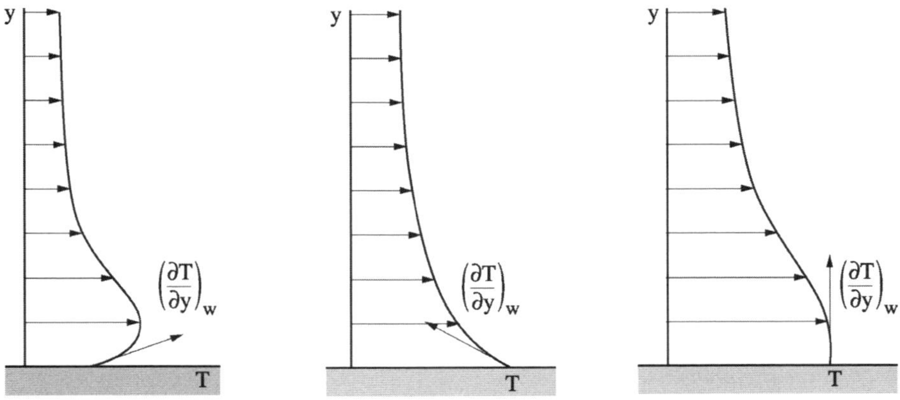

a – $(\partial T/\partial y)_w > 0$ *the fluid heats the wall* *b –* $(\partial T/\partial y)_w < 0$ *the wall heats the fluid* *c -* $(\partial T/\partial y)_w = 0$ *thermal equilibrium*

Figure 11.1. *Thermal boundary conditions between a wall and a flow*

The development of a boundary layer is controlled (in part) by the Reynolds number defined in section 6.1. For problems of pipes, we use a Reynolds number calculated with the average velocity and density in the section $A(x)$, taking as a reference length the diameter D of the tube. If we denote by $\bar{\rho}$ the density, $\bar{V}$ the velocity, and $\bar{\mu}$ the viscosity of the average uniform flow through the section $A(x)$, the Reynolds number is calculated by:

$$\mathrm{Re} = \frac{\bar{\rho}\bar{V}D}{\bar{\mu}}.$$

At the entrance of the pipe, the Reynolds number is low and a laminar boundary layer establishes. In the case of industrial pipes made of more or less rough materials, the laminar-turbulent transition is rapid and the turbulent state is established after a course of a few diameters. For this reason, problems of flow in a pipe are almost always treated for the turbulent regime. With the course in the pipeline, the boundary layer thickens and a situation is reached where the boundary layers meet. We are then in a state of joined boundary layers, also called *established turbulent regime*. In this case, the only characteristic length is the diameter of the pipe or, more generally, to treat the case of pipes with non-circular section, the hydraulic diameter defined by:

$$\boxed{D_h = \frac{4A}{L_h}}$$

where A is the area of the section and L_h the wetted perimeter of this section. In the case of a circular section, $A = \pi D^2 / 4$ and $L_h = \pi D$, leading to: $D_h = D$.

Laws of correlations, originally due to Nikuradse, are available giving the *friction coefficient*:

$$\boxed{\lambda = 4C_f}$$

as a function of the Reynolds number:

$$\mathrm{Re} = \frac{\bar{\rho}\bar{V} D_h}{\bar{\mu}}$$

and the ratio D_h / k involving the height k of the roughness. For the established turbulent regime, the coefficient λ depends slightly on the Reynolds number and can be considered constant along the pipeline even over long lengths. However, it depends strongly on the friction effect of the roughness.

With large Reynolds number, we can use the Nikuradse formula:

$$\frac{1}{\sqrt{\lambda}} = 2.035 \log_{10}\left(\frac{D_h}{k}\right) + 1.067$$

The following general relation, taking into account the Reynolds number, was proposed by Nackab:

$$\boxed{\lambda = \left\{ 2\log_{10} \left[0.27\frac{k}{D_h} + \frac{2.51}{\mathrm{Re}\left(0.4\,\mathrm{Re}^{-0.3} + 0.0053\right)^{\frac{1}{2}}} \right] \right\}^{-2}}.$$

Figure 11.2 illustrates this relation where we verify that at large Reynolds numbers, the friction coefficient depends only on the roughness of the pipe.

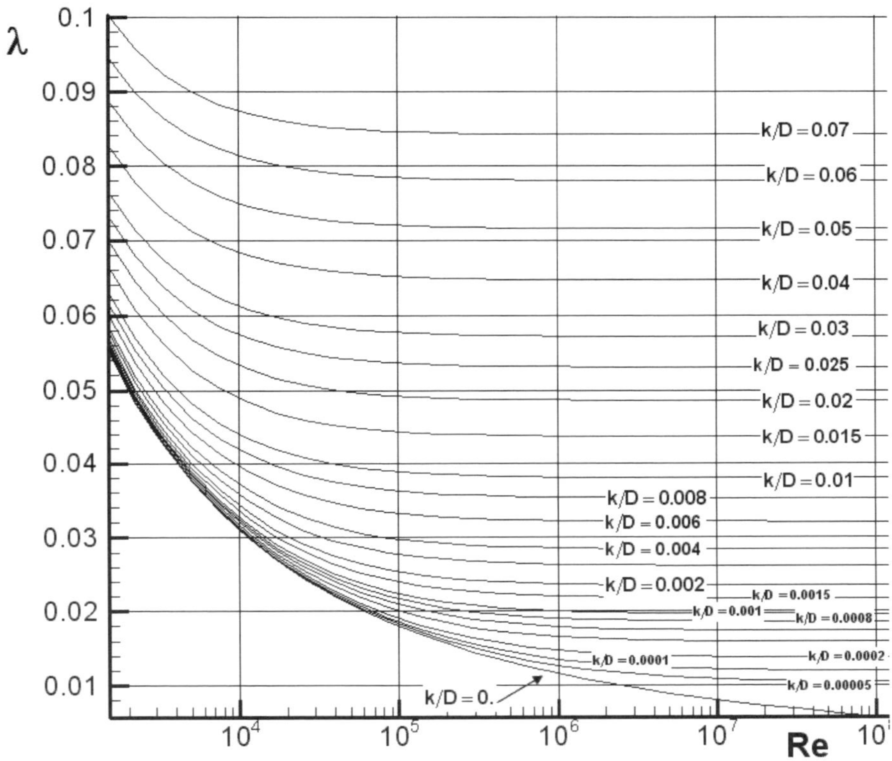

Figure 11.2. *Friction coefficient according to Nackab*

Often, the heat transfer coefficient is deduced from the friction coefficient by using the Reynolds analogy:

$$C_h = \frac{1}{2} C_f = \frac{\lambda}{8}.$$

11.2. One-dimensional theory and the generalized Hugoniot relation

We consider a one-dimensional flow in a pipe section $A(x)$ with a friction effect on the wall and heat input; this input can be positive if it is effective or negative in the case of a cooling of the fluid. Referring to Figure 9.1, in the case of a viscous fluid where friction is exerted on the wall, we have to consider an elementary force

dF representing the friction force exerted on the element of the tube with the thickness dx. The equation of motion in section 9.2.2 becomes:

$$\rho A dx V \frac{dV}{dx} = pA - \left[pA + d\left(pA \right) \right] + p dA - dF = -A dp - dF.$$

The friction effect, which is a retardation action exerted by the wall on the fluid, must be expressed by involving the sign - before the force dF which is positive by convention. Hence, introducing the friction force per unit area of the section A:

$$df = \frac{dF}{A}$$

$$VdV + \frac{dp}{\rho} + \frac{df}{\rho} = 0.$$

The friction effect alters neither the mass flow, which remains constant, nor the stagnation enthalpy h_{st} of the fluid since it is a property intrinsic to the fluid. Another way to express the invariance of the stagnation enthalpy with respect to the friction is to realize that at the wall the velocity being zero (the no-slip condition of the real fluid), there is no work of the friction force. In contrast, the heat input changes the level of the stagnation enthalpy of the fluid. This results in the system:

$$\frac{d\rho}{\rho} + \frac{dV}{V} + \frac{dA}{A} = 0 \qquad [11.1]$$

$$VdV + \frac{dp}{\rho} + \frac{df}{\rho} = 0 \qquad [11.2]$$

$$dh_{st} = dh + VdV = d\dot{q}. \qquad [11.3]$$

The term $d\dot{q}$ represents the energy (heat) fed to the fluid on the pipe element dx per unit of mass of fluid. Assuming a divariant fluid, we can write [11.3] in the following form, which introduces the specific entropy s:

$$\frac{dp}{\rho} + Tds + VdV = d\dot{q}.$$

Thus, by combining [11.3] and [11.2], it follows:

$$Tds = \frac{df}{\rho} + d\dot{q}$$

from which:

$$ds = \frac{df}{T\rho} + \frac{d\dot{q}}{T}.$$

The flow is no longer isentropic since there is entropy production through the friction effect and heat transfer. We find again the result shown by the energy equation written with the entropy: entropy production by viscous effects and heat conduction (see section 4.7). For a divariant fluid, we have:

$$dp = \left(\frac{\partial p}{\partial \rho}\right)_s d\rho + \left(\frac{\partial p}{\partial s}\right)_\rho ds.$$

In the case of a calorically perfect gas:

$$ds = C_p \frac{dT}{T} - r\frac{dp}{p} = C_p\left(\frac{dp}{p} - \frac{d\rho}{\rho}\right) - r\frac{dp}{p} = C_v \frac{dp}{p} - C_p \frac{d\rho}{\rho}$$

from which follows:

$$dp = \frac{p}{C_v} ds + \frac{C_p}{C_v}\frac{p}{\rho} d\rho$$

and:

$$\left(\frac{\partial p}{\partial s}\right)_\rho = \frac{p}{C_v}.$$

For the majority of fluids $\left(\frac{\partial p}{\partial s}\right)_\rho$ being positive, we can state:

$$\left(\frac{\partial p}{\partial s}\right)_\rho = K^2 = \frac{p}{C_v}$$

which gives:

$$dp = a^2 d\rho + K^2 ds.$$

By combining the equations above, we obtain:

$$VdV + \frac{dp}{\rho} + \frac{df}{\rho} = VdV + a^2\frac{d\rho}{\rho} + K^2\frac{ds}{\rho} + \frac{df}{\rho} = VdV + a^2\frac{d\rho}{\rho} + \frac{1}{\rho}\left(1+\frac{K^2}{\rho T}\right)df + \frac{K^2}{\rho}\frac{d\dot{q}}{T} = 0.$$

Taking into account the continuity equation, we obtain the generalized Hugoniot relation for a flow with friction and heat transfer:

$$\boxed{\frac{dV}{V}\left(1-M^2\right)+\frac{dA}{A}=\frac{1}{\rho a^2}\left[\left(1+\frac{K^2}{\rho T}\right)df+\frac{K^2}{T}d\dot{q}\right].}$$

For a calorically perfect gas:

$$\frac{K^2}{\rho T} = \frac{p}{C_v}\frac{r}{p} = \frac{C_p - C_v}{C_v} = (\gamma - 1)$$

from which follows:

$$\boxed{\frac{dV}{V}\left(1-M^2\right)+\frac{dA}{A}=\frac{\gamma}{\rho a^2}df+\frac{\gamma-1}{a^2}d\dot{q}.}$$

11.3. Flow with friction without heat input: stagnation pressure drop in a pipe

The right-hand side of the above expressions being positive, we see that the effect of friction is analogous to a decrease in section A: friction contributes to the increase in the flow velocity in the pipe. We consider the case of a pipe with a constant section without heat transfer, then:

$$\frac{dV}{V}\left(1-M^2\right)=\left(1+\frac{K^2}{\rho T}\right)\frac{df}{\rho a^2}.$$

If the flow in the inlet section of the pipe is subsonic, its velocity will increase under the friction effect. Physically, the phenomenon is interpreted by the fact that the wall friction tends to thicken the boundary layers. Thus, to maintain a constant mass flow, it is necessary for the average velocity of flow to increase. It can happen that the flow thus becomes sonic ($M = 1$). As seen in section 9.4, for this solution to be regular, it is necessary that simultaneously $df = 0$. This condition is verified only at one end of the pipe where the wall disappears. We summarize the following facts:

– An initially subsonic flow accelerates in a cylindrical pipe. If the Mach number reached at the end of the pipe is still subsonic, it is thus necessary that the pressure is equal to the imposed external pressure (for example, atmospheric pressure). This condition defines the velocity at the entrance of the pipe, thus, the mass flow. Any variation of the final pressure results in a variation of the mass flow.

– Under the same conditions, if the Mach number becomes sonic, it is necessary for this state to be reached at the open end of the pipe. In other words, the sonic state can only be reached at the open end of a pipe. The mass flow is then independent of the outside pressure; it only depends on the initial conditions and the section of the pipe.

We assume the pressure p_1 fixed at the entrance of the pipe and the entry velocity subsonic. Starting from a pressure at the end of the pipe $p_2 < p_1$, we gradually decrease p_2. Then, the velocity in the pipeline will increase at the same time as the mass flow and it may happen that the exit velocity is sonic. From this state on, any reduction in external downstream pressure p_2 will not affect the flow in the pipeline, in particular, as the mass flow is now constant. We say there is sonic choking, a situation that may become critical in the process of emptying a tank.

The initial conditions at the entry and exit of the pipe are such that (see section 9.5.2.1):

$$s_2 - s_1 = C_p \ln \frac{T_{st_2}}{T_{st_1}} - r \ln \frac{p_{st_2}}{p_{st_1}} = \int_{x_1}^{x_2} ds = \int_{x_1}^{x_2} \frac{df}{T\rho}$$

where x_1 and x_2 are the abscissas of the upstream and downstream ends of the pipe.

The flow being adiabatic $T_{st_2} = T_{st_1}$:

$$-r \ln \frac{p_{st_2}}{p_{st_1}} = \int_{x_1}^{x_2} ds = \int_{x_1}^{x_2} \frac{df}{T\rho}.$$

It follows:

$$\frac{p_{st_2}}{p_{st_1}} = \exp\left[-\frac{1}{r}\int_{x_1}^{x_2}\frac{df}{T\rho}\right].$$

The integral being positive, this ratio is less than one. There is a decrease in the stagnation pressure along the pipeline. This decrease is called *total pressure loss* or *total pressure drop* (the total pressure is here identical to the stagnation pressure). If the stagnation pressure at the exit is not sufficiently greater than the static pressure, the velocity will be insufficient to maintain a certain flow through the pipeline. Hence, the need for pumps to periodically raise the level of pressure along a gas or oil pipeline.

The friction force dF exerted on the pipe element of the length dx is given by:

$$dF = \tau_w \, l \, dx$$

where τ_w is the wall friction (force per unit area) and l the circumference of the pipe. Thus, ldx is the surface on which the friction τ_w is exerted. If the pipe has a circular section with the diameter D: $l = \pi D$. In this case:

$$df = \frac{dF}{A} = \tau_w \frac{ldx}{A} = \tau_w \frac{\pi D}{\pi D^2/4} dx = \frac{4}{2}\rho V^2 C_f \frac{dx}{D}$$

where we introduced the skin friction coefficient C_f. Stating that $\lambda = 4C_f$, we write:

$$df = \frac{dF}{A} = \frac{\lambda}{2}\rho V^2 \frac{dx}{D}$$

from which follows, knowing that $\rho V^2 = \gamma \, p \, M^2$:

$$\boxed{\frac{p_{st_2}}{p_{st_1}} = \exp\left[-\frac{\gamma}{2}\int_{x_1}^{x_2} M^2 \lambda \frac{dx}{D}\right].}$$

In an established turbulent regime, the friction coefficient is nearly constant so that we can write:

$$\frac{p_{st_2}}{p_{st_1}} \cong \exp\left[-\frac{\gamma\lambda}{2}\int_{x_1}^{x_2} M^2 \frac{dx}{D}\right].$$

This relation shows that the stagnation pressure loss is much higher, the longer the pipe, and it increases very rapidly with the flow Mach number (roughly as the square of the velocity for usual transport velocities of fluids).

The calculation of the flow with friction in a cylindrical pipe can be developed further in the case of a calorically perfect gas. The Hugoniot relation is then written:

$$\frac{dV}{V}\left(1-M^2\right)=\frac{\gamma}{\rho a^2}df. \qquad [11.4]$$

We have the following relation expressing the velocity as a function of the Mach number and speed of sound for the stagnation state (see section 9.5.2.4):

$$\frac{V}{a_{st}}=M\left(1+\frac{\gamma-1}{2}M^2\right)^{-\frac{1}{2}}$$

from which follows:

$$\frac{dV}{V}=\frac{1}{\left(1+\frac{\gamma-1}{2}M^2\right)}\frac{dM}{M}.$$

We found another way to express df:

$$df=\frac{dF}{A}=\frac{\lambda}{2}\rho V^2\frac{dx}{D}.$$

Thus, considering the right-hand side of equation [11.4], we have:

$$\frac{\gamma}{\rho a^2}df=\frac{\gamma}{\rho a^2}\frac{\lambda}{2}\rho V^2\frac{dx}{D}=\frac{\lambda}{2}\gamma M^2\frac{dx}{D}.$$

Taking these results into account, equation [11.4] is written:

$$\frac{\left(1-M^2\right)}{\gamma M^2\left(1+\frac{\gamma-1}{2}M^2\right)}\frac{dM}{M}=\frac{\lambda}{2}\frac{dx}{D}.$$

This equation is integrated over the length L of the pipe between the abscissas x_1 and x_2 to give:

$$\int_{x_1}^{x_2} \frac{\lambda}{2} \frac{dx}{D} = \int_{M_1}^{M_2} \frac{(1-M^2)}{\gamma M^2 \left(1+\frac{\gamma-1}{2}M^2\right)} \frac{dM}{M} \qquad [11.5]$$

where M_1 and M_2 are the Mach numbers at the abscissas x_1 and x_2. The integral of the right-hand side is analytically integrated by:

$$\boxed{g(M,\gamma) = \int^{M} \frac{(1-M^2)}{\gamma M^2 \left(1+\frac{\gamma-1}{2}M^2\right)} \frac{dM}{M} = -\frac{1}{2\gamma M^2} + \frac{\gamma+1}{4\gamma} \ln\left[\frac{\left(1+\frac{\gamma-1}{2}M^2\right)}{M^2}\right].}$$

This function has the following properties:

– $g(M) \to \infty$ when $M \to 0$,

– $g(M) \to$ finite limit: $g(\infty) = \frac{\gamma+1}{4\gamma} \ln\left(\frac{\gamma-1}{2}\right)$ when $M \to \infty$, and

– passes through a minimum $g(1)$ when $M = 1$.

The minimum for $M = 1$ takes the value:

$$g(1) = -\frac{1}{2\gamma} + \frac{\gamma+1}{4\gamma} \ln\left(\frac{\gamma+1}{2}\right).$$

Figure 11.3 shows the graph of the function $g(1) - g(M)$ for $\gamma = 1.4$.

The relation [11.5] is thus written:

$$\int_{x_1}^{x_2} \frac{\lambda}{2} \frac{dx}{D} = g(M_2) - g(M_1).$$

As we already noted, the coefficient λ varies little for a pipe with a given diameter, especially if the wall is very rough (which is often the case for industrial pipes). We can write, by substituting λ with an estimated average value $\overline{\lambda}$:

$$\frac{\overline{\lambda}}{2} \int_{x_1}^{x_2} \frac{dx}{D} = g(M_2) - g(M_1)$$

from which follows:

$$\boxed{\frac{L}{D}=\frac{x_2}{D}-\frac{x_1}{D}\cong\frac{2}{\overline{\lambda}}\left[g(M_2)-g(M_1)\right]}$$

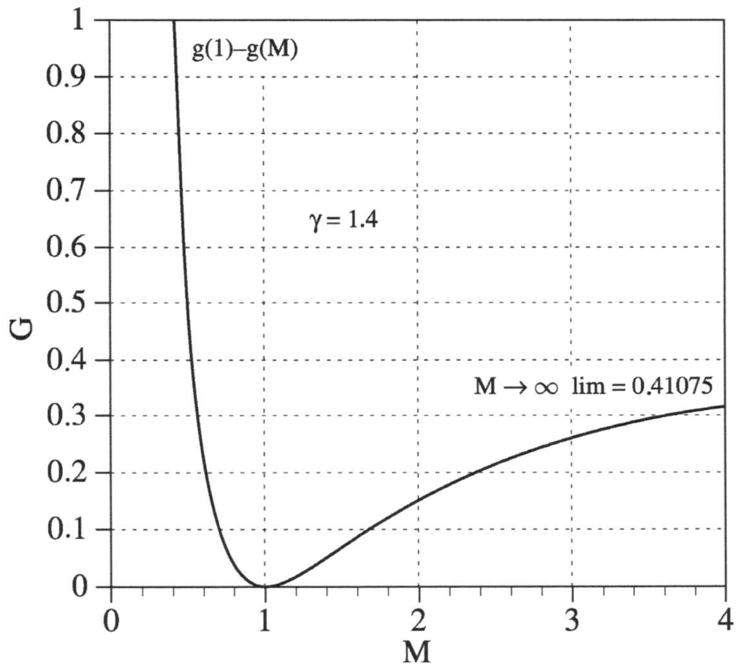

Figure 11.3. *Graph of the function* $g(1)-g(M)$ *for* $\gamma=1.4$

This relation allows us to calculate the Mach number M_2 after a length L of a pipe for a Mach number M_1 given at the entrance by:

$$g(M_2)=g(M_1)-\frac{\overline{\lambda}}{2}\frac{L}{D}.$$

According to the shape of the function $g(M)$:

– if $M_1<1$, as $g(M_2)<g(M_1)$, then $M_2>M_1$;

– if $M_1>1$, as $g(M_2)<g(M_1)$, then $M_2<M_1$.

In the subsonic state, the Mach number M_2 will increase even faster when the friction coefficient λ is much higher. In contrast, in the supersonic case, the flow slows down. For a low subsonic value of M_1 (low velocity), we find a subsonic Mach number M_2 at the end of the length L. If we gradually increase the mass flow, that is, M_1, then $g(M_1)$ decreases, and therefore, also $g(M_2)$. Thus, a situation can be attained where $g(M_2)=g(1)$. The sonic state $M_2=1$ has been reached at the end of the pipe which presents the mass flow from increasing: the pipe is choked.

11.4. Flow with heat transfer without friction: thermal choking

In this case:

$$\boxed{\frac{dV}{V}\left(1-M^2\right)+\frac{dA}{A}=\frac{1}{\rho a^2}\frac{K^2}{T}d\dot{q}.}$$

For a calorically perfect gas, we have:

$$\boxed{\frac{dV}{V}\left(1-M^2\right)+\frac{dA}{A}=\frac{\gamma-1}{a^2}d\dot{q}.}$$

If there is an effective heat input (the fluid is heated), then the right-hand side is positive. Thus, the effect is analogous to a decrease of the section and consequently:

– if the velocity is initially subsonic, the flow is accelerated,

– if the velocity is initially supersonic, the flow is decelerated.

For a pipe with a constant section:

$$\frac{dV}{V}\left(1-M^2\right)=\frac{1}{\rho a^2}\frac{K^2}{T}d\dot{q}.$$

If the velocity V at the entrance of the pipe is subsonic ($M<1$), then $dV>0$ and the flow accelerates. If the Mach number equals 1 and if $d\dot{q}\neq 0$, then a steady solution of the type studied here cannot exist. The flow becomes unsteady: we say that there is *thermal choking*. Conversely, a supersonic flow to which we add heat can be slowed down to unstarting due to the thermal choking effect. These phenomena are particularly important for the operation of combustion chambers.

In the case with simultaneous friction and heat supply, the Hugoniot equation for a pipe with constant cross section is written:

$$\frac{dV}{V}\left(1-M^2\right)=\frac{1}{\rho a^2}\left[\left(1+\frac{K^2}{\rho T}\right)df+\frac{K^2}{T}d\dot{q}\right].$$

We see that cooling the wall $\left(d\dot{q}<0\right)$ tends to counteract the effect of friction. In this way, we can limit the acceleration of the flow and delay, or avoid choking the pipe under the effect of viscosity. Physically, this effect is due to a decrease in the displacement effect of the boundary layer due to the cooling of the wall. This contraction is a consequence of the increase in density when approaching a wall colder than the flow. The loss of mass flow in the boundary layers is therefore lower than in the adiabatic case.

11.5. Calculating method for a one-dimensional flow with friction and heat transfer

In general, the flow in a pipe with constant section must be calculated step by step by integrating the following equations. The generalized Hugoniot relation gives for a calorically perfect gas:

$$\frac{dV}{V}\left(1-M^2\right)=\frac{\gamma}{\rho a^2}df+\frac{\gamma-1}{a^2}d\dot{q}.$$

But it has been established that:

$$\frac{df}{\rho}=\frac{\lambda}{2}V^2\frac{dx}{D_h}=V^2dX \quad \text{with: } dX=\frac{\lambda}{2}\frac{dx}{D_h}.$$

On the other hand, $a^2=\gamma rT=C_p\left(\gamma-1\right)T$

from which follows the relation:

$$\boxed{\frac{dV}{V}=\frac{\gamma M^2}{\left(1-M^2\right)}dX+\frac{1}{\left(1-M^2\right)}\frac{d\dot{q}}{C_pT}.} \qquad [11.6]$$

We have found:

$$dp=\frac{p}{C_v}ds+a^2d\rho$$

from which follows:

$$\frac{dp}{p} = \frac{ds}{C_v} + \frac{a^2}{rT}\frac{d\rho}{\rho} = \frac{ds}{C_v} + \gamma\frac{d\rho}{\rho}.$$

By introducing the expression of the entropy change ds:

$$ds = \frac{df}{T\rho} + \frac{d\dot{q}}{T} = V^2 dX + \frac{d\dot{q}}{T}$$

as well as the expression $\frac{d\rho}{\rho}$, we obtain:

$$\frac{dp}{p} = \frac{V^2}{C_v T}dX + \frac{d\dot{q}}{C_v T} - \gamma\left[\frac{\gamma M^2}{\left(1-M^2\right)}dX + \frac{1}{\left(1-M^2\right)}\frac{d\dot{q}}{C_p T}\right].$$

By regrouping the terms of the friction and heat transfer:

$$\frac{dp}{p} = \left[\frac{V^2}{C_v T} - \gamma\frac{\gamma M^2}{\left(1-M^2\right)}\right]dX + \frac{1}{C_p T}\left[\frac{C_p}{C_v} - \frac{\gamma}{\left(1-M^2\right)}\right]d\dot{q}$$

from which follows:

$$\boxed{\frac{dp}{p} = -\gamma M^2\frac{1+(\gamma-1)M^2}{\left(1-M^2\right)}dX - \frac{\gamma M^2}{\left(1-M^2\right)}\frac{d\dot{q}}{C_p T}.} \qquad [11.7]$$

The relation for the temperature change can be deduced from the equation of state in the form of:

$$\frac{dT}{T} = \frac{dp}{p} - \frac{d\rho}{\rho}.$$

This gives subsequens calculations that we will not detail here:

$$\boxed{\frac{dT}{T} = -\gamma\frac{(\gamma-1)M^4}{\left(1-M^2\right)}dX + \frac{1-\gamma M^2}{\left(1-M^2\right)}\frac{d\dot{q}}{C_p T}.} \qquad [11.8]$$

The relation for the Mach number is deduced from:

$$\frac{dM}{M} = \frac{dV}{V} - \frac{da}{a} = \frac{dV}{V} - \frac{1}{2}\frac{dT}{T}$$

which leads to the relation:

$$\boxed{\frac{dM}{M} = \gamma M^2 \frac{1 + \frac{(\gamma - 1)}{2} M^2}{\left(1 - M^2\right)} dX + \frac{1 + \gamma M^2}{2\left(1 - M^2\right)} \frac{d\dot{q}}{C_p T}.} \qquad [11.9]$$

The flow in the pipe is calculated step-by-step by numerically integrating equations [11.6]–[11.9] for the given laws $\lambda(x)$ and $\dot{q}(x)$. We verify that the system is singular if $M = 1$. Then the derivatives become infinite, unless if at the same time $dX = 0$ and $d\dot{q} = 0$, that is, at the ends of the pipe.

Chapter 12

Application of the One-Dimensional Theory to the Calculation of Supersonic Ejectors

12.1. Introduction

This chapter is about the application of the theory of one-dimensional steady flows to the design of what is called an ejector, and more precisely, a supersonic ejector. We will see that by simple relations, it is possible to predict the aerodynamic performance of such a device, without the need for detailed modeling of flows that requires resolution of the Navier-Stokes equations. The principle is to replace the actual flows in the device by equivalent average uniform flows in each section to which we apply the relations of mass conservation, momentum, and energy. This model, which enables a remarkably precise prediction of the characteristics of the ejector, is still used to optimize ejectors.

An ejector is a device with one or more driver jets at a high velocity driving an adjacent fluid by transferring some of their energy to it. The aim is to increase the momentum (or stagnation pressure) of the driven fluid. Entrainment is provided, either by turbulent mixing along the boundary of the different flows, and which can extend to all involved flows, or by a suction effect resulting from a significant expansion of the primary flow which imposes a very low pressure at the confluence of the two streams. These two effects are usually combined.

There are many practical applications of ejectors, particularly in the aerospace area: extractors of various types, devices to increase the thrust of a propulsive unit, or the lift of an aircraft wing, setting in motion the air stream in a wind tunnel, etc. The advantage of such systems is their simplicity (no moving parts) and low costs. They are mostly characterized by insensitivity to environmental effects such as temperature, vibration, chemical aggressivity of the driven fluid, etc. The major drawback is a low thermodynamic efficiency and the loud noise they generate. We will concentrate on the supersonic ejectors on which are based most of the applications in aeronautics.

12.2. Technological and aerodynamic characteristics

As shown in Figure 12.1, an ejector in its simplest design is arranged as follows: The driver (or primary) flow (E_p), with a mass flow q_{m_p}, stagnation pressure p_{st_p}, and stagnation temperature T_{st_p}, is provided by a nozzle called an injector nozzle (or more simply an injector). This flow drives the secondary flow (E_s), with a mass flow q_{m_s}, stagnation pressure p_{st_s}, and stagnation temperature T_{st_s}, in a duct of suitable shape called the *mixing chamber*. It is in this part of the device where the essential part of the energy exchange between the flows takes place. When the ejector operates as an extractor, a diffuser extends the mixing chamber in such a way as to recover a part of the kinetic energy of the two flows into pressure. We can thus recompress fluids down to the ambient pressure p_a. In propulsion, the propulsive flow is accelerated in a propulsive nozzle like in two-flux nozzles.

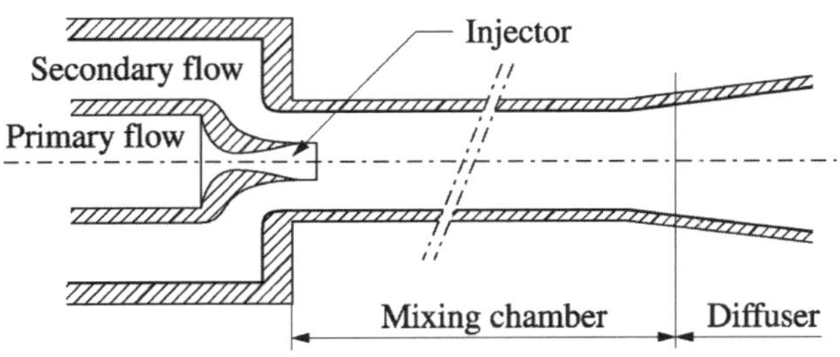

Figure 12.1. *Layout of a supersonic ejector*

The performance of an ejector with a fixed geometry is expressed as a function of the following aerodynamic parameters:

– mass flow coefficient: $\mu = \dfrac{q_{m_s}}{q_{m_p}}$,

– stagnation pressure ratio: $\overline{\pi} = \dfrac{p_{st_s}}{p_{st_p}}$,

– stagnation temperature ratio: $\theta = \dfrac{T_{st_s}}{T_{st_p}}$, and

– ambient pressure at the outlet: $\dfrac{p_a}{p_{st_p}}$.

For a fixed value θ, the relation $F\left(\mu, \dfrac{p_{st_s}}{p_{st_p}}, \dfrac{p_a}{p_{st_p}}\right)$ enables us to represent the characteristic surface of the ejector, which defines the different modes of operation. We note that there are at least two domains of operation. In the first domain, the performances of the ejector are independent of the ambient pressure at the outlet, hence, the name of *supersonic operating regime* in the sense that operation will depend solely on conditions at the entry of the ejector. In contrast, in the second domain, the pressure at the outlet of the ejector intervenes on the ejector performance. Then we talk about a *mixed regime*, where the operation depends at the same time on the upstream and downstream conditions.

12.3. Physical analysis of the supersonic ejector

This analysis is based on the schlieren visualizations in Figure 12.2, which relate to experiments on a two-flux aviation nozzle, where the wall of the mixing chamber diverges to create thrust (aviation nozzles are most often axisymmetric; the present experiments were made on a two-dimensional arrangement to allow flow visualizations). However, the phenomena are qualitatively the same in a cylindrical ejector. These visualizations show the change of the flow inside the two-flux nozzle depending on the mass flow coefficient $\mu = \dfrac{q_{m_s}}{q_{m_p}}$.

If the conditions of the primary flow are fixed, we start from a situation where the mass flow coefficient μ is zero or very low, the value of the ratio $\dfrac{p_{st_s}}{p_{st_p}}$ being accordingly low. Then the primary jet strongly expands at the nozzle exit A and re-attaches in R on the wall of the mixing chamber (see Figure 12.2a). In this case, the pressure, which is established in the secondary outlet section, practically coincides with the stagnation pressure p_{st_s}. The supersonic jet is bounded by an isobaric boundary (a slip line, see section 13.4) at the pressure p_{st_s}, along which a turbulent mixing layer develops. The secondary flow is then only driven by this turbulent mixing effect along the boundary between the two flows. The reattachment pressure at the jet impact on the wall of the mixing chamber being well over p_{st_s}, a part of the flow crossing the secondary entrance section A_{0_s} is not affected by the mixing effect. Able to overcome the compression in R, this flow forms a recirculation bubble (or vortex) that bypasses the driven flow.

We will not further analyze this operating condition, whose calculation involves more complex methods than those developed here. This condition is also found in two-flux aviation nozzles when the secondary mass flow is very small. If this mass flow is zero, a zone with very low pressure develops downstream of A. Then the system can be used in test stands for rocket engines, the jet of the engine helping to create a low pressure simulating a high altitude.

When μ increases, there is a transition from a system where the primary flow reattaches on the wall of the nozzle to a system where the two flows are contiguous but distinct within the nozzle. This corresponds to an operation similar to that of the supersonic ejector system.

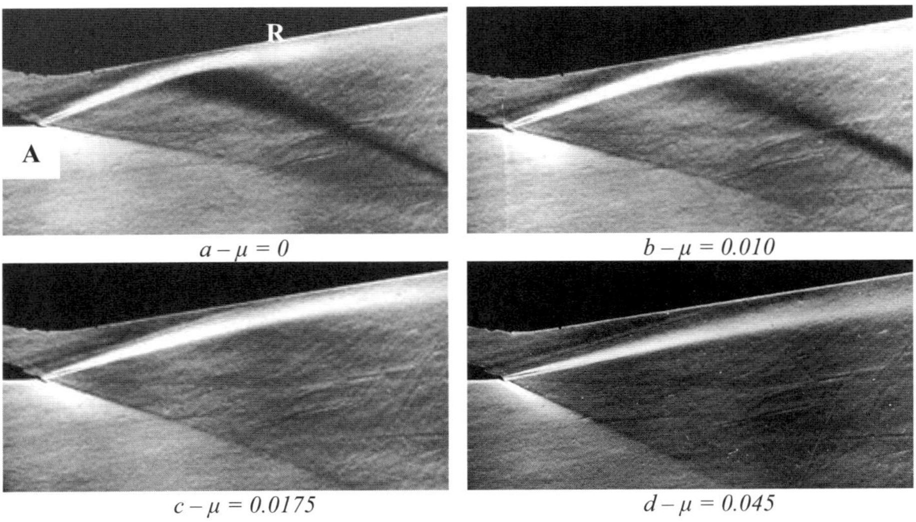

Figure 12.2. *Change of the flow in a two-flux nozzle with a transition from the reattachment regime to that of the type of a supersonic ejector (document Onera)*

When we increase the secondary pressure p_{st_s}, the mass flow q_{m_s} increases, the vortex contracts, and all the flow having passed through A_{0_s} is finally driven downstream, the part submitted to the entrainment effect losing more and more relative importance until it disappears. Henceforth, there is reattachment no longer of the primary flow on the wall of the mixing chamber and the two streams are distinct inside the ejector, separated by a slip line along which a mixing zone develops (see Figure 12.2d). Due to the acceleration imparted to the secondary flow by the entrainment effect and by the depression created through the expansion of the primary flow, the Mach number of the secondary flow increases and a sonic state appears in a section A_{c_s} (see Figure 12.3). Under such conditions, the supersonic secondary flow downstream of the throat becomes independent of the downstream pressure, hence, the name of supersonic regime.

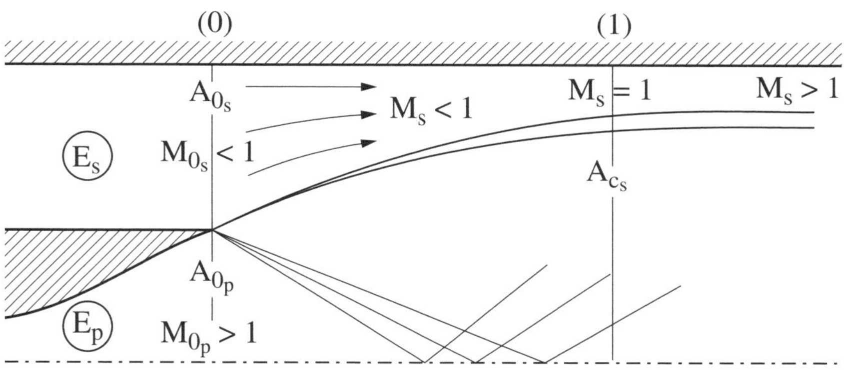

Figure 12.3. *Ejector in supersonic regime*

If the pressure p_{st_s} continues to grow, the pressure in the section A_{0_s} increases, which reduces the expansion in A of the primary jet and increases the critical section A_{c_s}, whose value approaches A_{0_s}. These two sections eventually merge and the secondary flow becomes independent of the primary flow as it is entirely determined by the pressure p_{st_s}, the temperature T_{st_s}, and the geometrical section A_{0_s}. Such a regime is called *saturated supersonic regime* (see Figure 12.4).

Downstream of the initial expansion region of the non-saturated system, the primary jet has the appearance of an under-expanded jet crossed by a series of expansion and shock waves (see section 20.8). The reflections of these waves on the border separating the two flows contribute to the increase in the turbulence mixing, thus promoting the formation of a single mixed flow. The more homogenous it is, the longer the mixing chamber is. If the mean flow (E_m) that results is supersonic, the flow retains an overall supersonic behavior as long as the exhaust pressure is low enough.

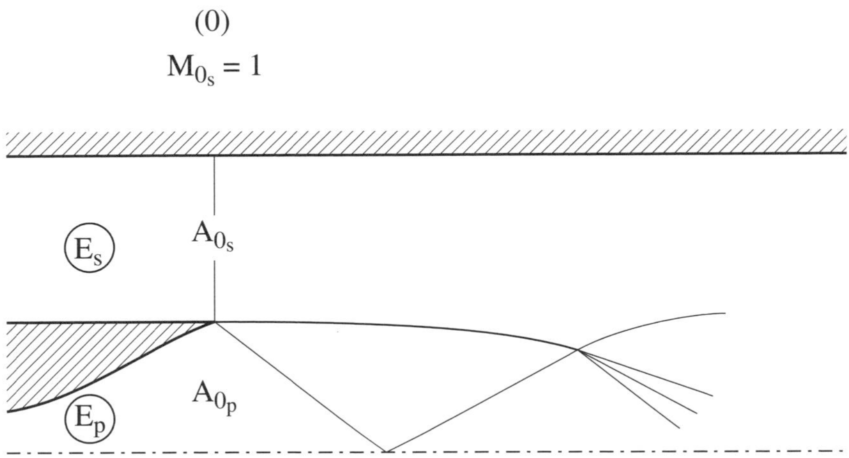

Figure 12.4. *Ejector in a saturated supersonic regime*

Let us assume that the mixing chamber discharges directly into a downstream reservoir with the pressure p_a. If the pressure p_m of the mixed flow is greater than p_a, the jet easily penetrates the downstream atmosphere. If p_m stays fixed and p_a increases and becomes greater than p_m, the one-dimensional theory, applied to the mean flow (E_m), indicates that the mode of operation will be changed as soon as p_a becomes greater than the pressure downstream of a steady normal shock forming in (E_m). In reality, the unstarting of the supersonic system occurs gradually as a result of acoustic waves that propagate back in the marginal subsonic zone, causing the recompression of the primary jet. As long as this effect does not reach the sonic section of the secondary flow, the system maintains a global supersonic behavior. The complete unstarting occurs abruptly for some values $(p_a)_{\lim}$ of the downstream pressure. This boundary can be moved if the mixing chamber is extended by an appropriate diffuser (see further down).

12.4. Physical analysis of the mixed regime

The new system, which establishes after unstarting, is called a *mixed regime* and the resulting mixed flow has a subsonic behavior with respect to the downstream pressure, whereas the primary flow is still supersonic at the exit of the injector (see Figure 12.5). The experimental analysis of the new flow shows that the secondary flow is now entirely subsonic, while the primary flow is recompressed through a complicated structure of shock waves. They cause a considerable thickening of the mixing zone, which quickly occupies an important part of the mixing chamber. If this one is long enough, the resulting strong mixing leads to a nearly homogenous flow of both the primary and secondary flows.

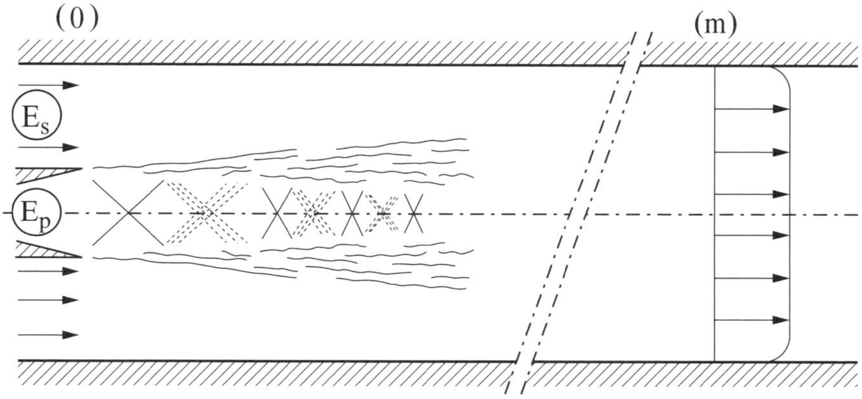

Figure 12.5. *Ejector in mixed regime*

***Comments on the application of a diffuser*:** from a practical point of view, it is necessary to design the ejector in order to evacuate a secondary flow that is as close as possible to the highest possible pressure. The application of a diffuser extending the mixing chamber is essential in this respect. If the mixed flow is subsonic (mixed regime), or low supersonic, a simple divergent with a moderate aperture is enough. However, if the flow is largely supersonic, it may be advantageous to recompress it very gradually in a convergent-divergent diffuser (see Figure 12.6).

According to the Hugoniot theorem (see section 9.3), the mixed supersonic flow slows quasi-isentropically in the convergent if this one has a relatively small slope (no shock waves formation). Then in the minimum section $A_{\min}$, it can reach a supersonic Mach number that is expedient to choose close to 1. The flow is re-accelerated in the divergent (always Hugoniot) and finally passes through a shock wave (*C*) or a system of shocks equivalent to a single shock. After the shock (*C*), it slows down to subsonic before discharging at the pressure p_a. The critical level $\overline{p}_a$ of p_a corresponds theoretically to the formation of a shock in the section $A_{\min}$.

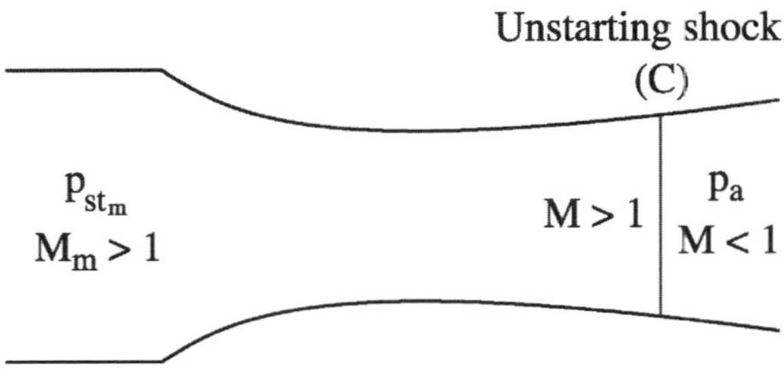

Figure 12.6. *Supersonic diffuser*

During the starting phase of the supersonic diffuser, an assumed normal shock wave (*C*) forms in the mixing chamber with the section A_m, the Mach number on the upstream side of (*C*) being M_m. The mass flow conservation for flows upstream and downstream of (*C*) results in the relation (see section 9.5.2):

$$q_m = \frac{\delta(\gamma_m)}{\sqrt{C_{p_m} T_{st_m}}}\left(p_{st_m} A_{c_m}\right)_{upstream} = \frac{\delta(\gamma_m)}{\sqrt{C_{p_m} T_{st_m}}}\left(p_{st_m} A_{c_m}\right)_{downstream}.$$

The stagnation temperature does not change when passing through the shock (see section 13.3) and if the gas properties are assumed unchanged, we then deduce:

$$\left(p_{st_m} A_{c_m}\right)_{upstream} = \left(p_{st_m} A_{c_m}\right)_{downstream}.$$

The shock relations give (see section 13.5.2):

$$\frac{\left(p_{st_m}\right)_{donwstream}}{\left(p_{st_m}\right)_{upstream}} = \Pi\left(M_m, \gamma_m\right)$$

where $\Pi\left(M_m, \gamma_m\right)$ represents the stagnation pressure loss across a normal shock with $\Pi\left(M_m, \gamma_m\right) \leq 1$. Thus:

$$\left(A_{c_m}\right)_{downstream} \equiv A_{c_m}^* = \frac{\left(A_{c_m}\right)_{upstream}}{\Pi\left(M_m, \gamma_m\right)}.$$

The device can only be started if the mass flow through the minimum section $A_{\min}$ of the diffuser is compatible with the above relation, which requires:

$$A_{\min} > A^*_{c_m}.$$

12.5. Calculation of an ejector with a cylindrical mixing chamber

12.5.1. *General calculation conditions*

Here, we neglect losses through friction resulting from the presence of a boundary layer along the wall of the mixing chamber. These losses are generally low; however, they have to be taken into account if the velocities of the secondary flow are high and if the mixing chamber is very long. The estimation of such losses would lead us too far here.

We assume that the studied mixing process is relative to calorically perfect gases that may have different properties. To facilitate direct reading of the theoretical analysis which follows, we recall some results of the one-dimensional theory presented in Chapter 9.

Isentropic expansion relations for a steady calorically perfect gas:

$$\frac{p}{p_{st}} \equiv \pi(M,\gamma) = \left(1+\frac{\gamma-1}{2}M^2\right)^{-\frac{\gamma}{\gamma-1}}$$

$$\frac{T}{T_{st}} = \left(1+\frac{\gamma-1}{2}M^2\right)^{-1}$$

$$\frac{A}{A_c} \equiv \Sigma(M,\gamma) = \left(\frac{2}{\gamma+1}\right)^{\frac{\gamma+1}{2(\gamma-1)}} \frac{1}{M}\left(1+\frac{\gamma-1}{2}M^2\right)^{\frac{\gamma+1}{2(\gamma-1)}}.$$

Mass flow formula:

$$q_m = \frac{\delta(\gamma)}{\sqrt{h_{st}}} p_{st} A_c = \frac{\delta(\gamma)}{\sqrt{C_p T_{st}}} p_{st} A_c$$

with:

$$\delta(\gamma) = \frac{\gamma}{\sqrt{\gamma - 1}} \left(\frac{\gamma + 1}{2} \right)^{-\frac{(\gamma+1)}{2(\gamma-1)}}.$$

Definition of the axial dynalpy flux in the section A:

$$D = A\left(p + \rho V^2\right) \quad \text{or:} \quad D = \frac{p}{p_{st}} \frac{A}{A_c} \left(1 + \gamma M^2\right) p_{st} A_c$$

and by introducing the thrust function (see section 10.4):

$$\Phi(M,\gamma) = \pi(M,\gamma)\Sigma(M,\gamma)\left(1 + \gamma M^2\right)$$

$$D = p_{st} A_c \Phi(M,\gamma).$$

12.5.2. *The case of the supersonic regime*

In the case of the supersonic regime, the calculation of the performances from known geometric and aerodynamic data (section of the mixing chamber, exit section of the ejector, and Mach number at the exit of the ejector) is independent of the counter pressure and reduces to the determination of the curve:

$$\frac{q_{m_s}}{q_{m_p}} = f\left(\frac{p_{st_s}}{p_{st_p}}\right) \quad \text{or:} \quad \mu = f\left(\bar{\pi}\right).$$

Here, we will not consider the calculation of the characteristic of the ejector for very low secondary mass flows, where the primary jet reattaches on the wall of the mixing chamber. Once the secondary mass flow is substantial, the mixing layer has a small thickness compared to the transverse dimensions of the two flows; thereby, we can neglect the viscous effects in a first approximation. We will make the following assumptions:

– The mixing chamber is cylindrical.

– The two flows are formed by calorically perfect gases.

– In the critical section (i.e. the section where the secondary flow is sonic), the structure of each of the flows is assumed to be nearly uniform.

Thus, it is possible to consider a one-dimensional and isentropic evolution of the flows between the sections (0) and (1), the local continuity of the pressures on the slip surface separating the two flows (see section 13.4) not necessarily entailing equality between the two mean pressures $\overline{p}_p$ and $\overline{p}_s$. We actually consider mean uniform flows in each section, which is only an approximation of reality.

The skin friction on the walls of the mixing chamber and the heat transfer between the flows, on one hand, and through the walls, on the other hand, are negligible.

More generally, the minimum section A_{1_s} of the secondary flow is not *a priori* supposed to be sonic. The following conditions are given:

– for the primary flow (E_p): $M_{0_p}, p_{st_p}, q_{m_p}, T_{st_p}, C_{p_p}, \gamma_p, A_{0_p}$;

– for the secondary flow (E_s): $M_{0_s}, p_{st_s}, q_{m_s}, T_{st_s}, C_{p_s}, \gamma_s, A_{0_s} = A_m - A_{0_p}$; and

– geometry of the mixing chamber: $\lambda = \dfrac{A_m}{A_{0_p}} \rightarrow \dfrac{A_{0_s}}{A_{0_p}} = \lambda - 1.$

The calculation of the ejector characteristic is done by applying the conservation theorems.

Conservation of the mass flow: we can express the different mass flows with the relations:

$$q_{m_p} = \frac{\delta(\gamma_p)}{\sqrt{C_{p_p} T_{st_p}}} p_{st_p} A_{c_p}, \qquad q_{m_s} = \frac{\delta(\gamma_s)}{\sqrt{C_{p_s} T_{st_s}}} p_{st_s} A_{c_s}, \qquad q_{m_p} + q_{m_s} = \frac{\delta(\gamma_m)}{\sqrt{C_{p_m} T_{st_m}}} p_{st_m} A_{c_m}.$$

We state:

$$\mu = \frac{q_{m_s}}{q_{m_p}}, \tau = \frac{\delta^2(\gamma_p)}{\delta^2(\gamma_s)} \frac{h_{st_s}}{h_{st_p}} = \frac{\delta^2(\gamma_p)}{\delta^2(\gamma_s)} \frac{C_{p_s} T_{st_s}}{C_{p_p} T_{st_p}}, \qquad \tau_m = \frac{\delta^2(\gamma_p)}{\delta^2(\gamma_m)} \frac{h_{st_m}}{h_{st_p}} = \frac{\delta^2(\gamma_p)}{\delta^2(\gamma_m)} \frac{C_{p_m} T_{st_m}}{C_{p_p} T_{st_p}}$$

and use the relations:

$$A_{c_p} = \frac{A_{0_p}}{\Sigma\left(M_{0_p}, \gamma_p\right)}, \quad A_{c_s} = \frac{A_{0_s}}{\Sigma\left(M_{0_s}, \gamma_s\right)} = \frac{A_{0_p}(\lambda - 1)}{\Sigma\left(M_{0_s}, \gamma_s\right)},$$

$$A_{c_m} = \frac{A_m}{\Sigma(M_m, \gamma_m)} = \frac{\lambda A_{0_p}}{\Sigma(M_m, \gamma_m)}$$

from which follow the expressions:

$$\frac{p_{st_s}}{p_{st_p}} = \frac{q_{m_s}}{A_{c_s}} \sqrt{C_{p_s} T_{st_s}} \frac{A_{c_p}}{q_{m_p}} \frac{1}{\sqrt{C_{p_p} T_{st_p}}} \frac{\delta(\gamma_p)}{\delta(\gamma_s)}$$

or:

$$\boxed{\frac{p_{st_s}}{p_{st_p}} = \frac{\mu\sqrt{\tau}}{\lambda - 1} \frac{\Sigma(M_{0_s}, \gamma_s)}{\Sigma(M_{0_p}, \gamma_p)}} \qquad [12.1]$$

and:

$$\boxed{\frac{p_{st_m}}{p_{st_p}} = \frac{1 + \mu}{\lambda} \sqrt{\tau_m} \frac{\Sigma(M_m, \gamma_m)}{\Sigma(M_{0_p}, \gamma_p)}.} \qquad [12.1m]$$

The values of C_{p_m} and γ_m for a mixing of different fluids can be determined from a thermodynamic model suitable for the present gases.

Geometric constraint: constant total section, which is expressed by:

$$A_{1_p} + A_{1_s} = A_m = \lambda A_{0_p}.$$

The laws of isentropic expansion between the sections (0) and (1) involve in each flow the conservation of the critical section A_c. Hence:

$$\frac{A_{1_p}}{\Sigma(M_{1_p}, \gamma_p)} = \frac{A_{0_p}}{\Sigma(M_{0_p}, \gamma_p)} \quad \text{and} \quad \frac{A_{1_s}}{\Sigma(M_{1_s}, \gamma_s)} = \frac{A_{0_s}}{\Sigma(M_{0_s}, \gamma_s)} = \frac{(\lambda - 1) A_{0_p}}{\Sigma(M_{0_s}, \gamma_s)}$$

from which follows:

$$\boxed{\frac{\Sigma\left(M_{1_p},\gamma_p\right)}{\Sigma\left(M_{0_p},\gamma_p\right)}=\lambda-(\lambda-1)\frac{\Sigma\left(M_{1_s},\gamma_s\right)}{\Sigma\left(M_{0_s},\gamma_s\right)}.} \qquad [12.2]$$

In the particular case, $M_{1_s}=1$ (sonic choking in the secondary flow), $\Sigma\left(M_{1_s},\gamma_s\right)=1$, and:

$$\boxed{\frac{\Sigma\left(M_{1_p},\gamma_p\right)}{\Sigma\left(M_{0_p},\gamma_p\right)}=\lambda-\frac{(\lambda-1)}{\Sigma\left(M_{0_s},\gamma_s\right)}.} \qquad [12.2c]$$

Equation of motion: by design, the pressure force on the walls of the cylindrical mixing chamber has a zero axial resultant. In addition, we neglect the friction exerted on the tube wall. Thus, the dynalpy flux keeps the same value in all sections. Therefore, we have between the entry section (0) of the mixing chamber and any intermediate section (1):

$$\left[p_s\left(1+\gamma_s M_s^2\right)A_s+p_p\left(1+\gamma_p M_p^2\right)A_p\right]_0^1=0.$$

This relation shows the variation of the dynalpy flux between (0) and (1) is zero. We have a similar relation between the sections (0) and (m), leading to:

$$\begin{aligned}p_m\left(1+\gamma_m M_m^2\right)A_m&=p_{p_0}\left(1+\gamma_p M_{p_0}^2\right)A_{p_0}+p_{s_0}\left(1+\gamma_s M_{s_0}^2\right)A_{s_0}=\\&p_{p_1}\left(1+\gamma_p M_{p_1}^2\right)A_{p_1}+p_{s_1}\left(1+\gamma_s M_{s_1}^2\right)A_{s_1}.\end{aligned}$$

By introducing the above-defined thrust function:

$$\begin{aligned}p_{st_m}A_{c_m}\Phi\left(M_m,\gamma_m\right)&=p_{st_p}A_{c_p}\Phi\left(M_{p_0},\gamma_p\right)+p_{st_s}A_{c_s}\Phi\left(M_{s_0},\gamma_s\right)\\&=p_{st_p}A_{c_p}\Phi\left(M_{p_1},\gamma_p\right)+p_{st_s}A_{c_s}\Phi\left(M_{s_1},\gamma_s\right).\end{aligned}$$

By using the mass flow formula:

$$q_m=\frac{\delta(\gamma)}{\sqrt{C_p T_{st}}}p_{st}A_c$$

the previous relation can be written:

$$\left(q_{m_p}+q_{m_s}\right)\Phi\left(M_m,\gamma_m\right)\frac{\sqrt{C_{p_m}T_{st_m}}}{\delta\left(\gamma_m\right)}=q_{m_p}\Phi\left(M_{0p},\gamma_p\right)\frac{\sqrt{C_{p_p}T_{st_p}}}{\delta\left(\gamma_p\right)}+q_{m_s}\Phi\left(M_{0s},\gamma_s\right)\frac{\sqrt{C_{p_s}T_{st_s}}}{\delta\left(\gamma_s\right)}$$

$$=q_{m_p}\Phi\left(M_{1p},\gamma_p\right)\frac{\sqrt{C_{p_p}T_{st_p}}}{\delta\left(\gamma_p\right)}+q_{m_s}\Phi\left(M_{1s},\gamma_s\right)\frac{\sqrt{C_{p_s}T_{st_s}}}{\delta\left(\gamma_s\right)}.$$

After division by $q_{m_p}\sqrt{C_{p_p}T_{st_p}}/\delta\left(\gamma_p\right)$:

$$(1+\mu)\Phi\left(M_m,\gamma_m\right)\sqrt{\tau_m}=\Phi\left(M_{0p},\gamma_p\right)+\mu\Phi\left(M_{0s},\gamma_s\right)\sqrt{\tau}=\Phi\left(M_{1p},\gamma_p\right)+\mu\Phi\left(M_{1s},\gamma_s\right)\sqrt{\tau}.$$

We deduce the relations:

$$\boxed{\mu\sqrt{\tau}=\frac{\Phi\left(M_{1_p},\gamma_p\right)-\Phi\left(M_{0_p},\gamma_p\right)}{\Phi\left(M_{0_s},\gamma_s\right)-\Phi\left(M_{1_s},\gamma_s\right)}}\qquad [12.3]$$

$$\boxed{\Phi\left(M_m,\gamma_m\right)=\frac{\Phi\left(M_{0_p},\gamma_p\right)+\mu\sqrt{\tau}\Phi\left(M_{0_s},\gamma_s\right)}{(1+\mu)\sqrt{\tau_m}}.}\qquad [12.3m]$$

Energy equation: if the walls are assumed to be adiabatic (no heat exchange), the fluids being non-reactive, then the flux of stagnation enthalpy is conserved, leading to:

$$\left(q_{m_p}+q_{m_s}\right)h_{st_m}=q_{m_p}h_{st_p}+q_{m_s}h_{st_s}.$$

For calorically perfect gases, we have:

$$\left(q_{m_p}+q_{m_s}\right)C_{p_m}T_{st_m}=q_{m_p}C_{p_p}T_{st_p}+q_{m_s}C_{p_s}T_{st_s}$$

a relation which is also written as:

$$\boxed{(1+\mu)\frac{\delta^2\left(\gamma_m\right)}{\delta^2\left(\gamma_p\right)}\tau_m=1+\mu\frac{\delta^2\left(\gamma_s\right)}{\delta^2\left(\gamma_p\right)}\tau.}\qquad [12.4]$$

If the two flows have the same specific stagnation enthalpy ($h_{st_s} = h_{st_p}$), then:

$$h_{st_m} = h_{st_p} = h_{st_s}.$$

12.5.3. *Algorithm for the calculation of the characteristic of the supersonic regime*

We begin with a set of data specifying the geometry of the mixing chamber, the thermodynamic, and thermal conditions, as well as the Mach number at the exit of the injector nozzle, i.e.:

$$\lambda, T_{st_p}, T_{st_s}, C_{p_p}, C_{p_s}, \gamma_p, \gamma_s, M_{0_p} > 1.$$

First, we calculate the *saturated supersonic regime* which is such that $A_{c_s} = A_{0_s}$ with the result $M_{0_s} = 1$ and $\Sigma(M_{0_s}, \gamma_s) = 1$. Therefore, relation [12.1] gives:

$$\boxed{\mu = \frac{\lambda - 1}{\sqrt{\tau}} \Sigma\left(M_{0_p}, \gamma_p\right) \frac{p_{st_s}}{p_{st_p}}.}$$

For this regime, the characteristic $\mu = f\left(\bar{\pi}\right)$ is rectilinear. Then, for each value of μ, the relation [12.4] gives τ_m, then [12.3m] gives $\Phi(M_m, \gamma_m)$ from which follows M_m. The relation [12.1m] allows us to determine $\frac{p_m}{p_{st_p}}$, which gives the maximum exhaust pressure of the system. If the pressure p_a exceeds p_m, a shock wave is formed at the exit that might unstart the system.

The calculation of the ejector for the unsaturated supersonic regime can be done by taking as calculation parameter the Mach number M_{0_s} in the secondary entry section. In this regime, there is sonic choking at the section denoted by (1) so that $M_{1_s} = 1$ and $\Sigma\left(M_{1_s}, \gamma_s\right) = 1$. The calculation chain is as follows, the number referring to the relations established above (see Table 12.1):

$M_{0_s} \leq 1$	
$\downarrow$	
[12.2c]	
$\downarrow$	
$\Sigma\left(M_{1_p}, \gamma_p\right) \rightarrow M_{1_p}$	
$\downarrow$	
[12.3]	
$\downarrow$	
$\mu\sqrt{\tau} \quad \rightarrow$	[12.4]
$\downarrow$	$\downarrow$
[12.1]	τ_m
$\downarrow$	$\downarrow$
$\overline{\pi} = \dfrac{p_{st_s}}{p_{st_p}}$	[12.3m]
	$\downarrow$
	$\Phi_m\left(M_m, \gamma_m\right) \rightarrow M_m > 1 \rightarrow \Sigma\left(M_m, \gamma_m\right)$
	$\downarrow$
	[12.1m]
	$\downarrow$
	$\dfrac{p_{st_m}}{p_{st_p}}, \; A_{c_m} = \dfrac{\lambda A_{0_p}}{\Sigma\left(M_m, \gamma_m\right)}$

Table 12.1. *Calculation chain for the supersonic regime*

The second part of the algorithm allows us to calculate the maximum recoverable pressure after mixing of the two streams and to verify, according to the exit configuration (presence or absence of a supersonic diffuser), if the calculated regime is compatible with the ambient pressure. Incidentally, this gives the critical value $A^*_{c_m}$, which is the minimum value of the throat section $A_{\min}$ of a convergent diffuser extending the mixing chamber to help the exhaust (see section 12.4).

If p_a becomes greater than the limit unstarting value, or if $A^*_{c_m}$ becomes higher than $A_{\min}$, there is either transition to a mixed regime for which the extracted secondary flow is dependent on the downstream conditions (in fact on the exit pressure p_a), or geometric choking ($A_{\min} = A^*_{c_m}$) in the case of a convergent-divergent diffuser.

12.5.4. *Algorithm for the calculation of the characteristic of the mixed regime*

The primary flow is assumed to be supersonic at the exit of the injector, but the secondary and mixed flows are subsonic, that is:

$$M_{0_p} > 1, \quad M_{0_s}, M_m < 1.$$

This system is controlled by the ratio $\dfrac{p_a}{p_{st_p}}$ (in fact by the downstream pressure p_a). The calculation of this system requires taking into account the operation of a diffuser placed at the end of the mixing chamber to achieve a pressure recovery. For a given external pressure p_a and a Mach number at the end of mixing $M_m < 1$, the diffuser associates a stagnation pressure p_{st_m} with a recompression curve $\dfrac{p_{st_m}}{p_a} = g\left(M_m, \gamma_m\right)$. The operation of the diffuser is defined by a characteristic curve (D) giving $\dfrac{p_{st_m}}{p_a}$ as a function of M_m (see Figure 12.7). When M_m is largely subsonic, the pressure recovery is maximum. When M_m increases, this ratio decreases because of losses due to friction in the diffuser (boundary layers) which increase with the Mach number. Below a certain value of p_a, the Mach number M_m becomes greater than 1 and the flow expands at supersonic velocity in the diffuser whose section grows. When the pressure reached in this expansion becomes less than p_a, a shock wave occurs in the divergent leading to a rapid drop of the ratio $\dfrac{p_{st_m}}{p_a}$ (see loss of stagnation pressure through a normal shock in section 13.5.2.3).

At the same time, the Mach number M_m at the end of the diffuser will have been fixed by a throat, preventing the pressure p_a from having an effect on M_m.

We start from a set of data specifying the geometry of the mixing chamber, the thermodynamic and thermal conditions, the Mach number at the exit of the injector nozzle, i.e.:

$$\lambda \, , T_{st_p} \, , T_{st_s} \, , C_{p_p} \, , \gamma_p \, , C_{p_s} \, , \gamma_s \, , \lambda_s \, , M_{0_p} > 1$$

as well as: p_{st_p}, p_{st_s}.

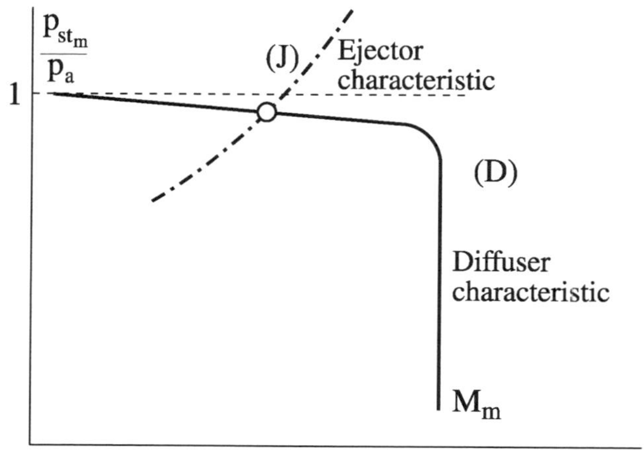

Figure 12.7. *Operating point of the ejector in a mixed regime with a downstream diffuser*

The algorithm for a given value of M_{0_s} is given in Table 12.2.

$M_{0_s}, \bar{\pi} \rightarrow$	[12.1]$\rightarrow$	$\mu\sqrt{\tau} \rightarrow$	$\mu \rightarrow$	[12.4]$\rightarrow$	$\tau_m \rightarrow$	[12.3m]$\rightarrow$	$\Phi(M_m, \gamma_m)$
	$\rightarrow \quad \Phi(M_m, \gamma_m) \quad \rightarrow$			$M_m < 1 \rightarrow$	[12.1m]$\rightarrow$	$p_{st_m} \rightarrow$	$\dfrac{p_a}{p_{st_m}}$

Table 12.2. *Calculation chain for the mixed regime*

This calculation provides a point on the curve $f\left(M_m, \dfrac{p_a}{p_{st_p}}\right)$ which we then can construct step-by-step by varying M_{0_s} [this curve (J) corresponds to a value of the ratio $\bar{\pi} = \dfrac{p_{st_s}}{p_{st_p}}$]. The intersection of (J) with the characteristic (D) of the diffuser gives a solution consistent with the data (see Figure 12.7). We note that the section (1) does not appear in the calculation. Repeating the calculation for other values of the ratio $\bar{\pi} = \dfrac{p_{st_s}}{p_{st_p}}$, we define a characteristic that depends on the exit pressure p_a. Thus, there will be a characteristic curve for each value of the ratio $\dfrac{p_a}{p_{st_p}}$.

The calculation chain is simplified if we impose the pressure p_m at the outlet of the mixing chamber. Then, the relation [12.1m] can be written:

$$\boxed{\frac{p_m}{p_{st_p}} = \frac{p_m}{p_{st_m}}\frac{p_{st_m}}{p_{st_p}} = \frac{1+\mu}{\lambda}\sqrt{\tau_m}\,\frac{\pi(M_m,\gamma_m)\Sigma(M_m,\gamma_m)}{\Sigma\left(M_{0_p},\gamma_p\right)}.} \qquad [12.1m]$$

For any value of μ, this relation associates the product of the function $\pi(M_m,\gamma_m)\Sigma(M_m,\gamma_m)$, which is a function of the Mach number whose inversion gives the Mach number M_m. Then the relation [12.3m] gives $\Phi_{0_s}(M_{0_s},\gamma_s)$, leading to M_{0_s}. Finally, [12.1] gives $\bar{\pi} = \dfrac{p_{st_s}}{p_{st_p}}$, leading to the calculation chain in Table 12.3.

$\boxed{\mu \rightarrow}$	$[12.1m]_{bis} \rightarrow$	$\pi(M_m,\gamma_m)\Sigma(M_m,\gamma_m) \rightarrow$	$M_m \rightarrow$	$[12.3m] \rightarrow$	$\Phi\left(M_{0_s},\gamma_s\right) \rightarrow$
	$\rightarrow \Phi\left(M_{0_s},\gamma_s\right) \rightarrow$		$M_{0_s} \rightarrow$	$[12.1] \rightarrow$	$\bar{\pi} = \dfrac{p_{st_s}}{p_{st_p}}$

Table 12.3. *Calculation chain for the mixed regime: variant*

12.6. Application examples

Figure 12.8 shows the operating characteristics of an air-air ejector so that $\lambda = 4$ for supersonic and saturated supersonic regimes. The curves are calculated for the values of the Mach number at the ejector exit M_{0_p} varying from 1 to 3. With a low coefficient μ, the regime is supersonic. When the secondary mass flow increases, the Mach number M_{0_s} increases in the secondary entry section until it reaches the sonic value. We then pass to the saturated supersonic regime. The transition from one regime to another is reflected in a change of the slope of the characteristic. We note that the supersonic characteristic is almost straight. The characteristic of the saturated regime is a straight line passing through the origin.

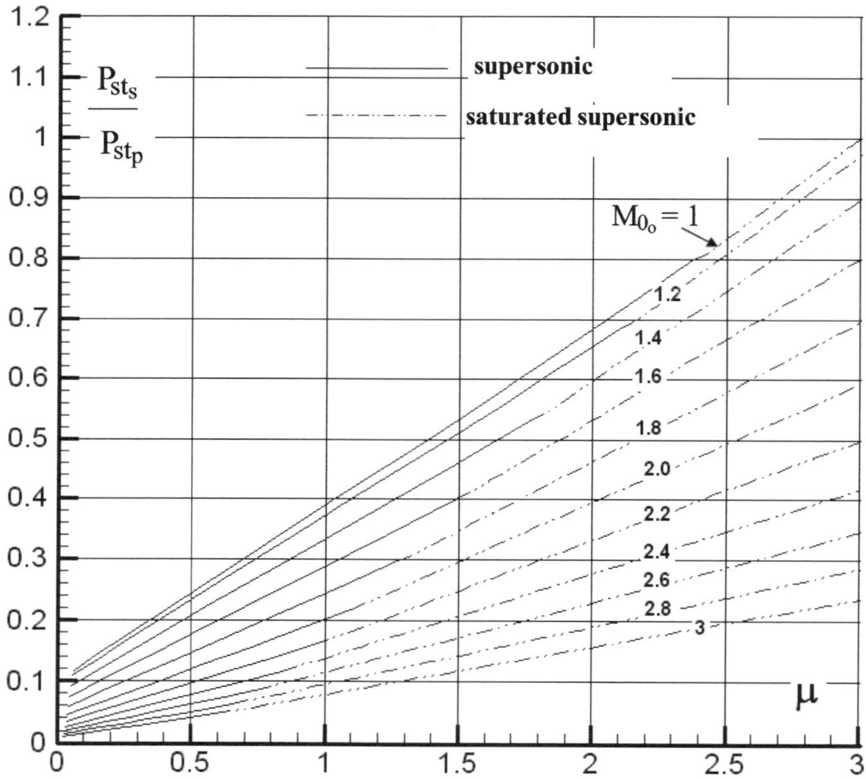

Figure 12.8. *Ejectors in supersonic and saturated supersonic regimes. Effect of the Mach number at the exit of the injector (λ = 4)*

The second application (see Figure 12.9) corresponds to a mixed regime for an air-air ejector such that $M_{0_p} = 3$ and $\lambda = 4$. The calculation is done for a mixing chamber without a downstream diffuser, by imposing the pressure p_m at the end of the mixing chamber.

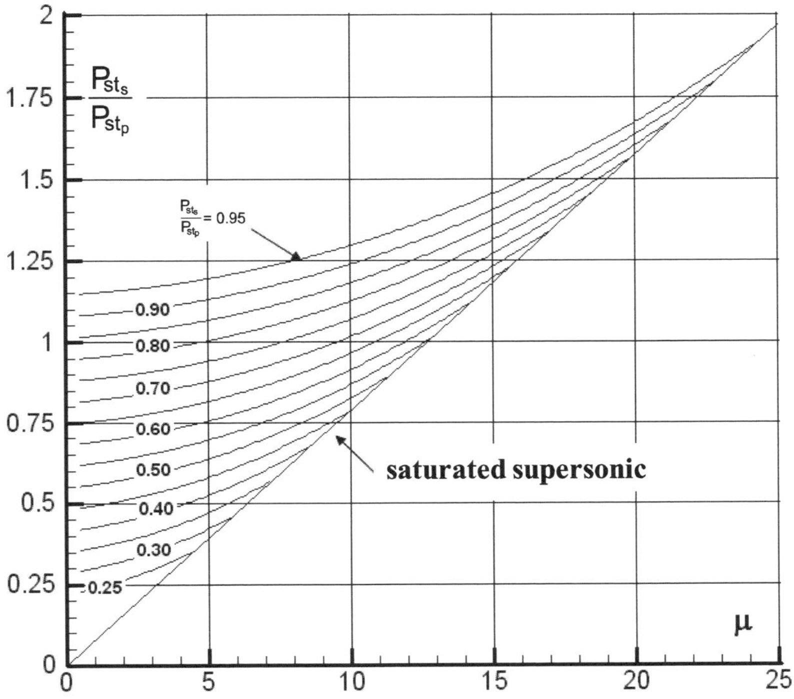

Figure 12.9. *Ejector in mixed regime. Effect of the pressure at the exit of the mixing chamber (conditions: $\lambda = 4$, $M_{0_p} = 3$)*

This time, there is a characteristic curve for each value of the ratio $\frac{p_m}{p_{st_p}}$. For a fixed value $\frac{p_m}{p_{st_p}}$, the flow coefficient μ increases with the pressure ratio $\frac{p_{st_s}}{p_{st_p}}$, the Mach number M_{0_s} increasing correspondingly to attain the value *1*. Thus, there is a transition to the saturated supersonic regime, the characteristics tending to the characteristic corresponding to this regime.

Chapter 13

Discontinuity Surfaces: Shock Wave and Slip Line

13.1. Existence of shock waves

For reasons we will not analyze here, a supersonic flow can be the seat of shock waves. This phenomenon results from the hyperbolic nature of equations governing the supersonic flow of non-viscous fluids. In this case, the disturbances propagating at the speed of sound cannot propagate against the flow and, thus, are not perceived upstream of the body that has induced them. Thus, the supersonic flow arriving ahead of the body is not "warned" of the presence of the obstacle which inflicts a velocity reduction. It follows that the adjustment of the velocity, and other properties of the flow, will have to happen abruptly through a quasi-discontinuity. The requirement of shock waves can be introduced in a less intuitive way by the wave theory described in Chapter 21 dealing with wave propagation. Here, we accept the existence of shock waves as a fact or evidence, characteristic for supersonic flow and proven by experience. Thus, we seek the relations which have to exist between the properties of the fluid downstream and upstream of the shock wave, understanding that these properties, or more precisely their variations, must satisfy the fundamental equations of fluid mechanics.

The visualizations in Figure 13.1 show some configurations where shock waves are formed in aerodynamic flows. In the case of Figure 13.1a, a sudden change of the direction of a wall causes the formation of an oblique shock wave through which the flow undergoes a direction discontinuity equal to the angle of the ramp. Such shocks occur at the ramps of a supersonic air intake or a control surface during supersonic flight.

A normal or straight shock wave occurs when a counterpressure forces a supersonic flow to change to subsonic. As an illustration, Figure 13.1b shows an interferogram of a normal shock wave formed in a transonic channel. The interference fringes here are lines of equal density value. They visualize the expansion in the upstream part of the image where the flow is almost one-dimensional (see Chapter 9). The interferogram shows the shock wave and boundary layers on the upper and lower walls of the channel whose outer boundary is indicated by the breaking of the fringes. The apparent thickness of the shock originates in its interaction with the boundary layers on the lateral windows of the test section. In reality, its thickness is less than one micrometer for the conditions of this test (see section 13.6). The normal shock waves are found on transonic airfoils (see Chapter 25), in interblade channels of turbomachine cascades, in supersonic diffusers.

A shock wave is formed when a supersonic flow encounters a pressure change, such as at the exit of an overexpanded nozzle, i.e. operating under conditions such that the pressure in the exit plane is lower than the external pressure (see Figure 13.1c). The discontinuity of the pressure generates a shock wave that equalizes the pressures in the nozzle exit. The issuing jet is confined by a boundary on which the pressure is equal to the external pressure: a slip line as we will see below. Here, this is the change in pressure which induces a deflection of the flow.

A shock wave is also formed in front of a supersonic vehicle, the shock detaching from the blunt vehicle's nose, as shown in Figure 13.1d for a planetary probe model in a flow at Mach 10.

If we assume that such a discontinuity surface exists, the local fluid properties vary with infinite derivatives. This is in contradiction with the assumption of negligible viscous terms that is the basis of the Euler equations (see section 7.1). At the level of the discontinuity, there is actually an "awakening" of viscous effects which, in fact, make a shock wave a region of *finite thickness* in which the fluid properties vary continuously. We will demonstrate in section 13.6 that the thickness is of the order of several mean free molecular paths that is extremely thin under usual aerodynamic conditions (a few micrometers or less). Thus, at the scale of the vehicle, the shock wave can be considered as a perfect discontinuity surface. During the atmospheric re-entry at very high altitudes, greater than 100 km, the mean free molecular path is large because of the low value of density and the shocks have a thickness that is no longer negligible compared to the vehicle dimensions. The assumption of continuous medium then falls into default and the calculation of shocks requires other methods that we will not present here.

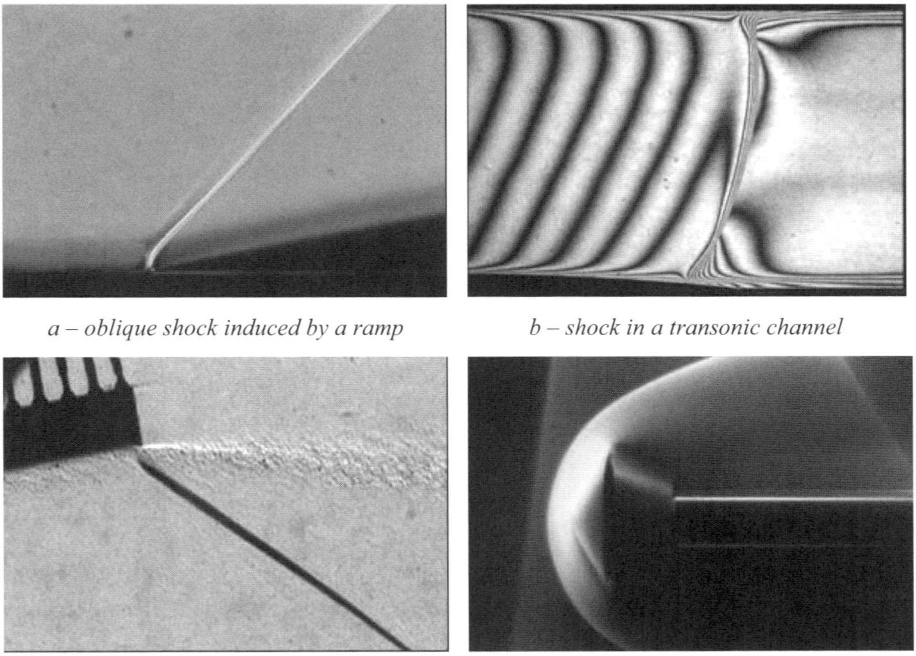

a – oblique shock induced by a ramp *b – shock in a transonic channel*

c – adaptation shock at the exit of a nozzle *d – shock in front of a planetary probe*

Figure 13.1. *Examples of the formation of shock waves in supersonic flows (documents Onera)*

We will show that the quantities characterizing the mechanical and thermodynamic states of a gas before and after the shock must be linked by compatibility relations. These relations arise from three general principles:

– mass conservation,

– fundamental equation of mechanics or equation of motion,

– energy conservation.

Here, we will assume that the flow is non-viscous and a non-conductor of heat, but bear in mind that the viscosity inevitably plays a strong role at the level of the shock. It is its intervention that is causing the jump in entropy occurring at the crossing of a shock wave, and hence the drag produced by the shocks, called *wave drag* or better *shock drag* (see Chapter 2). Within the present theory, we consider a state before and after shock, without attempting to describe the shock phenomenon itself. This can be done by using the theory of shock structures presented in section 13.6.

We will see that the conservation equations also govern a second type of discontinuity separating two flows of different nature but with parallel velocities on both sides of the discontinuity. This is the *slip-line or -surface* on which the flows "slip" on each other. Such surfaces are frequently encountered in aerodynamics: they separate the jet emerging from a propulsion nozzle from the outer flow, the flows having traversed the upper and lower surfaces of an airfoil, etc. They are formed at the intersection of shock waves (see Chapter 15). The surface of a body, in non-viscous fluid, can also be regarded as a slip line separating the wall from the flow slipping along it. Hereon, we will discuss the shock waves and slip lines in a unified way.

13.2. Discontinuity relations: Rankine–Hugoniot equations

We recall the general relations of conservation (see Chapter 4) established by restricting *to the steady case* and by considering the fluid as non-viscous. The fixed surface (S) defines a fluid volume (ϑ) containing a certain mass of fluid. We have:

– *continuity equation:*

$$\iint_{(S)} \rho \ \vec{V}.\vec{n}dS = 0.$$

This relation shows that the mass contained in the fixed volume (ϑ) is constant, the incoming fluxes being compensated by the outgoing fluxes.

– *equation of motion:*

According to our assumptions, this equation is written as:

$$\iint_{(S)} \left(\rho\vec{V}.\vec{n}\right)\vec{V}dS = -\iint_{(S)} p\vec{n}dS$$

or:

$$\iint_{(S)} \left[\left(\rho\vec{V}.\vec{n}\right) \ \vec{V} + p\vec{n}\right] \ dS = 0.$$

– *energy equation:*

This equation, expressing the first law of thermodynamics, is written under the present assumption that:

$$\iint_{(S)} \rho\vec{V}.\vec{n}\left(e + \frac{V^2}{2}\right)dS = -\iint_{(S)} p\vec{V}.\vec{n}dS$$

or by introducing the specific enthalpy:

$$\iint_{(S)} \rho\vec{V}.\vec{n}\left(h+\frac{V^2}{2}\right)dS = 0.$$

The conservation equations are applied considering the volume (ϑ) defined by the surface (S) and crossed by the discontinuity surface (Σ).

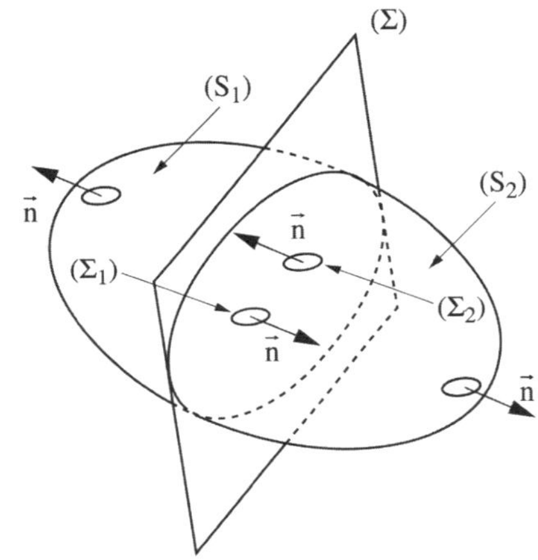

Figure 13.2. *Control volume for the discontinuity relations*

As shown in Figure 13.2, the volume (ϑ) is thus divided into two volumes:

– the volume (ϑ_1) defined by the surface (S_1) and the portion of (Σ) intercepted by (S), which will be designated by (Σ_1),

– the volume (ϑ_2) defined by the surface (S_2) and the portion of (Σ) intercepted by (S) designated by (Σ_2).

We have the following conservation equations:

For the total volume (ϑ):

$$\iint_{(S_1)\cap(S_2)} \rho\vec{V}.\vec{n}dS = 0$$

$$\iint_{(S_1)\cap(S_2)}\left[\left(\rho\vec{V}.\vec{n}\right)\vec{V} + p\vec{n}\right]dS = 0$$

$$\iint_{(S_1)\cap(S_2)} \rho\vec{V}.\vec{n}\left(h+\frac{V^2}{2}\right)dS = 0.$$

For the volume (ϑ_1):

$$\iint_{(S_1)\cap(\Sigma_1)} \rho\vec{V}.\vec{n}dS = 0$$

$$\iint_{(S_1)\cap(\Sigma_1)} \left[\left(\rho\vec{V}.\vec{n}\right)\vec{V} + p\vec{n}\right]dS = 0$$

$$\iint_{(S_1)\cap(\Sigma_1)} \rho\vec{V}.\vec{n}\left(h + \frac{V^2}{2}\right)dS = 0.$$

For the volume (ϑ_2):

$$\iint_{(S_2)\cap(\Sigma_2)} \rho\vec{V}.\vec{n}dS = 0$$

$$\iint_{(S_2)\cap(\Sigma_2)} \left[\left(\rho\vec{V}.\vec{n}\right)\vec{V} + p\vec{n}\right]dS = 0$$

$$\iint_{(S_2)\cap(\Sigma_2)} \rho\vec{V}.\vec{n}\left(h + \frac{V^2}{2}\right)dS = 0$$

as by construction:

$$(S) = (S_1)\cap(\Sigma_1)\cap(S_2)\cap(\Sigma_2)$$

it follows that:

$$\iint_{(S_1)\cap(\Sigma_1)} \rho\vec{V}.\vec{n}dS + \iint_{(S_2)\cap(\Sigma_2)} \rho\vec{V}.\vec{n}dS = 0$$

$$\iint_{(S_1)\cap(\Sigma_1)} \left[\left(\rho\vec{V}.\vec{n}\right)\vec{V} + p\vec{n}\right]dS + \iint_{(S_2)\cap(\Sigma_2)} \left[\left(\rho\vec{V}.\vec{n}\right)\vec{V} + p\vec{n}\right]dS = 0$$

$$\iint_{(S_1)\cap(\Sigma_1)} \rho\vec{V}.\vec{n}\left(h + \frac{V^2}{2}\right)dS + \iint_{(S_2)\cap(\Sigma_2)} \rho\vec{V}.\vec{n}\left(h + \frac{V^2}{2}\right)dS = 0.$$

We consider the first relation of the mass conservation which can be written as:

$$\iint_{(S_1)} \rho\vec{V}.\vec{n}dS + \iint_{(S_2)} \rho\vec{V}.\vec{n}dS + \iint_{(\Sigma_1)} \rho\vec{V}.\vec{n}dS + \iint_{(\Sigma_2)} \rho\vec{V}.\vec{n}dS = 0.$$

But as:

$$\iint_{(S_1)\cap(S_2)} \rho\vec{V}.\vec{n}dS = 0$$

it follows that:

$$\iint_{(\Sigma_1)} \rho\vec{V}.\vec{n}dS + \iint_{(\Sigma_2)} \rho\vec{V}.\vec{n}dS = 0$$

or even:

$$\iint_{(\Sigma_1)} \rho\vec{V}.\vec{n}dS = -\iint_{(\Sigma_2)} \rho\vec{V}.\vec{n}dS.$$

When writing the conservation laws, the unit normal vectors are directed outward from the considered surface, the vectors normal to (Σ_1) and (Σ_2) are in the opposite direction. By adopting the same direction then:

$$\iint_{(\Sigma_1)} \rho\vec{V}.\vec{n}dS = \iint_{(\Sigma_2)} \rho\vec{V}.\vec{n}dS.$$

The same calculation can be made for dynalpy and stagnation enthalpy fluxes, leading to the relations:

$$\iint_{(\Sigma_1)}\left[\left(\rho\vec{V}.\vec{n}\right)\vec{V} + p\vec{n}\right]dS = \iint_{(\Sigma_2)}\left[\left(\rho\vec{V}.\vec{n}\right)\vec{V} + p\vec{n}\right]dS$$

$$\iint_{(\Sigma_1)}\rho\vec{V}.\vec{n}\left(h+\frac{V^2}{2}\right)dS = \iint_{(\Sigma_2)}\rho\vec{V}.\vec{n}\left(h+\frac{V^2}{2}\right)dS.$$

These relations having to be verified whenever $(\Sigma_1)=(\Sigma_2)=(\Sigma)$, it follows that if (Σ) is a discontinuity surface within a non-viscous flow, then the conservation equations impose through (Σ):

$$\boxed{\begin{aligned}&\rho\vec{V}.\vec{n} = constant\\ &p\vec{n}+\left(\rho\vec{V}.\vec{n}\right)\vec{V} = constant\\ &\rho\ \vec{V}.\vec{n}\left(h+\frac{V^2}{2}\right) = constant.\end{aligned}}$$

These equations are called *Rankine–Hugoniot equations*. We sometimes give them the following form by noting $[f]$ the discontinuity or jump of the quantity f through (Σ), for example, $([f] = f_2 - f_1)$. Knowing that if a is constant across (Σ), we have the relation:

$$[a\, f] = a[f]$$

and:

$$[f + g] = [f] + [g].$$

The Rankine–Hugoniot equations can be written as:

$$\left[\rho \vec{V}.\vec{n}\right] = 0$$

$$[p]\vec{n} + \rho \vec{V}.\vec{n}\left[\vec{V}\right] = 0$$

$$\rho \vec{V}.\vec{n} \left[h + \frac{V^2}{2}\right] = 0.$$

13.3. Shock wave type solution

We concern ourselves with solutions of Rankine–Hugoniot equations such that there is *mass flow across the discontinuity*:

$$\rho \vec{V}.\vec{n} \neq 0.$$

Here, we restrict ourselves to the case of two-dimensional flows. We denote with indices 1 and 2 the conditions on both sides of the discontinuity. Thus:

$$V_{t_1} \quad and \quad V_{t_2}$$

are the velocity components contained in the plane of the discontinuity (tangential components, see Figure 13.3) and:

$$V_{n_1} \quad and \quad V_{n_2}$$

the components of velocity along the normal to the discontinuity (normal components).

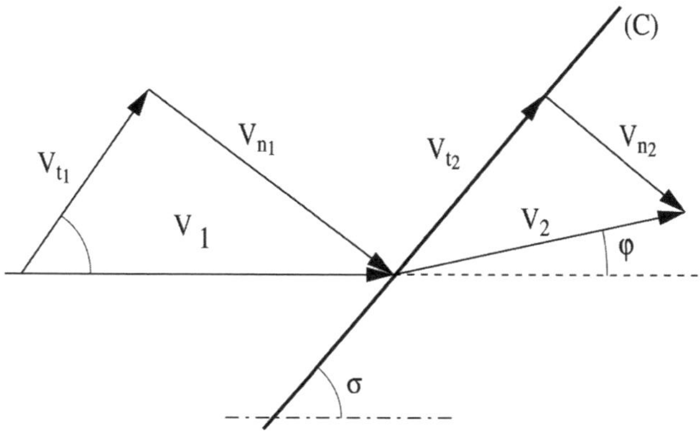

Figure 13.3. *Oblique planar shock wave*

The continuity equation leads to the relation:

$$\boxed{\rho_1 \; V_{n_1} = \rho_2 \; V_{n_2}.}$$

The projection of the (vectorial) motion equation along the discontinuity (Σ) gives the pressure term $p\vec{n}$, which is normal to the considered surface, having a zero projection on the surface:

$$(\rho_1 \; V_{n_1}) \; V_{t_1} = (\rho_2 \; V_{n_2}) \; V_{t_2}.$$

Hence, given the continuity equation

$$\boxed{V_{t_1} = V_{t_2}.}$$

The tangential component of the velocity is conserved across the shock wave.

The projection of the momentum equation along the normal to (Σ) gives:

$$\boxed{p_1 + \rho_1 \; V_{n_1}^2 = p_2 + \rho_2 \; V_{n_2}^2.}$$

The energy conservation equation gives, after division by $\rho_1 \; V_{n_1} = \rho_2 \; V_{n_2}$:

$$\boxed{h_1 + \frac{V_1^2}{2} = h_2 + \frac{V_2^2}{2}.}$$

We may have to use the form with the specific internal energy:

$$e_1 + \frac{p_1}{\rho_1} + \frac{V_1^2}{2} = e_2 + \frac{p_2}{\rho_2} + \frac{V_2^2}{2}.$$

By denoting:

$$h_{st} = h + \frac{V^2}{2}$$

which is the specific stagnation enthalpy of the flow, we see that:

$$\boxed{h_{st_1} = h_{st_2}.}$$

Hence, the essential property:

The specific stagnation enthalpy is conserved across a shock wave. The stagnation temperature is conserved if the gas is calorically perfect.

We note that we have:

$$V_1^2 - V_2^2 = \left(V_{t_1}^2 + V_{n_1}^2\right) - \left(V_{t_2}^2 + V_{n_2}^2\right) = V_{n_1}^2 - V_{n_2}^2$$

because $V_{t_1} = V_{t_2}$. Hence, another form of the energy equation:

$$\boxed{h_1 + \frac{V_{n_1}^2}{2} = h_2 + \frac{V_{n_2}^2}{2}.}$$

13.4. Slip line type solution

We rewrite the Rankine–Hugoniot equations by keeping the term of the mass flux through the control volume which intervenes when we write the balances, not to forget that the balance of mass, momentum, and energy applies to a certain *mass of fluid*, or in a steady state, to a mass per time unit, i.e. a mass flow.

– for the continuity equation:

$$\rho_1 V_{n_1} = \rho_2 V_{n_2} \qquad [13.1]$$

– for the projection of the equation of motion along the discontinuity surface (Σ):

$$\left(\rho_1 V_{n_1}\right) V_{t_1} = \left(\rho_2 V_{n_2}\right) V_{t_2} \qquad [13.2]$$

– for the projection along the normal to (Σ):

$$p_1 + \left(\rho_1 V_{n_1}\right) V_{n_1} = p_2 + \left(\rho_2 V_{n_2}\right) V_{n_2} \qquad [13.3]$$

– for the energy:

$$\rho_1 \ V_{n_1}\left(h_1 + \frac{V_1^2}{2}\right) = \rho_2 \ V_{n_2}\left(h_2 + \frac{V_2^2}{2}\right). \qquad [13.4]$$

A solution to the equation $\rho\vec{V}.\vec{n} = constant$ is:

$$\rho\vec{V}.\vec{n} = 0$$

or:

$$V_{n_1} = V_{n_2} = 0$$

which indicates that there is no mass flow through the surface (Σ). It follows that the flows (1) and (2) on both sides of (Σ) are parallel to (Σ), and thus between them. The relation [13.2] being satisfied by $V_{n_1} = V_{n_2} = 0$, it follows that the components V_{t_1} and V_{t_2} are arbitrary: they can have independent values on each side of (Σ) (The velocity is identified with its tangential component since the normal components to (Σ) are zero).

The normal motion equation [13.3] reduces to:

$$\boxed{p_1 = p_2.}$$

Hence, the fundamental conclusion: the pressure is continuous across (Σ).

According to the energy equation [13.4], we see that the stagnation enthalpy:

$$h_{st} = h + \frac{V^2}{2}$$

may have different values that are independent of each other and on each side of (Σ). Since equation [13.4] is satisfied by the condition $V_{n_1} = V_{n_2} = 0$, thus it may be whatever h_{st}. The surface (Σ) separates two flows with independent properties: density, velocity, temperature, stagnation enthalpy, composition, or nature. These two flows flow over each other with the condition that they have the *same pressure* (see Figure 13.4).

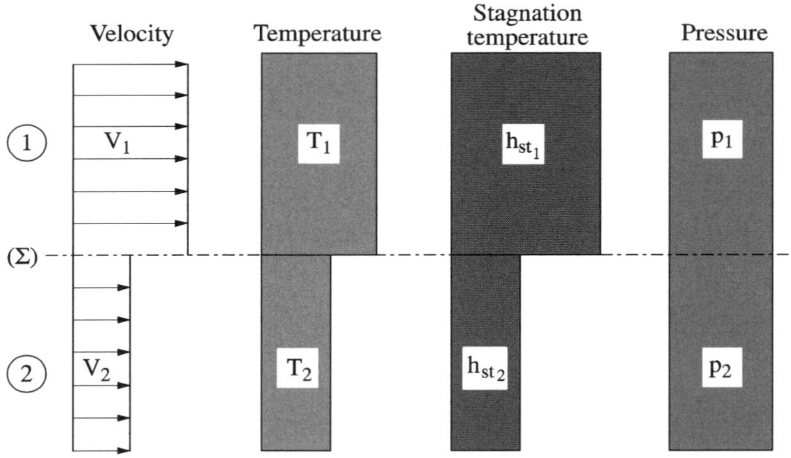

Figure 13.4. *Slip line separating two flows 1 and 2*

A slip line is a discontinuity surface, governed by the Rankine–Hugoniot equations, separating two flows that may have different properties and may be composed of different gases. The two conditions to be satisfied are:

– the parallelism of the velocities $\vec{V}_1$ and $\vec{V}_2$,

– the equality of the pressures $p_1 = p_2$.

$V_2 \neq V_1$	$\varphi_1 = \varphi_2$	$T_1 \neq T_2$	$\rho_2 \neq \rho_1$	$p_2 = p_1$	$T_{st_2} \neq T_{st_1}$	$p_{st_2} \neq p_{st_1}$	$\gamma_2 \neq \gamma_1$

The condition $\varphi_1 = \varphi_2$ indicates that flows 1 and 2 have the same direction.

We note that no conditions are imposed on the Mach numbers, and a slip line can exist between two subsonic flows.

Discontinuities such as slip lines occur in many circumstances; it is the case of:

– the surface separating the propulsion jet leaving the nozzle of a jet engine or rocket engine from the outer flow, which may be the atmosphere at rest in the case of fixed-point conditions. Then, the velocity, temperature, the stagnation pressure, and the stagnation temperature are much higher for the jet than for the outer flow. Similarly, the propulsion jet consists of a gas with different properties than the outer atmosphere (composition and thermodynamics). In the case where the jet exits into an atmosphere at rest, the flow adjacent to the jet has the velocity zero, the pressure in the jet at the contact with the atmosphere being equal to the atmospheric pressure p_a,

– a wake downstream of the trailing edge of an airfoil separating the upper and lower flows which have different velocities,

– the discontinuity beginning at the intersection of shock waves with different intensities and directions (see Chapter 15).

In reality, the discontinuity (Σ) cannot be maintained because the velocity leap between $\vec{V}_1$ and $\vec{V}_2$ implies an infinite gradient along the normal to (Σ) which has the effect of awakening the viscosity terms that are ignored in the Euler equations. Along (Σ), a viscous region, called a *mixing layer*, is formed across which the flow properties vary continuously between the states (1) and (2), as shown in Figure 13.5. The situation at the contact of a wall is similar, the slip condition imposed on the non-viscous fluid leading to a jump of the velocity $\vec{V}_1$ in the flow to $\vec{V}_2 = 0$ at the wall. In addition, the pressure in the flow is equal to the pressure at the wall. In fact, the discontinuity of velocity is not maintained and a continuous transition region forms where the viscosity is at work: this is the *boundary layer*.

In an unsteady flow, surfaces analogous to (Σ) are referred to as contact surfaces (see Chapter 22). For example, they separate two gases flowing inside a pipe. The slip line concept is generalized for the three-dimensional as the slipstream layer. Then, the velocity component in the plane (Σ) is a vector that undergoes a discontinuity in intensity and direction across (Σ), the other conditions being the same.

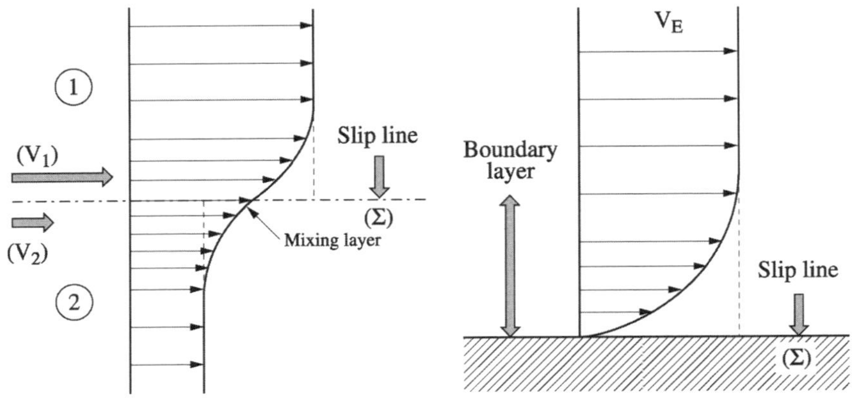

Figure 13.5. *Manifestation of the viscosity along a slip line: mixing layer and boundary layer*

Figure 13.6 shows schlieren photography of an axisymmetric jet emerging from a supersonic nozzle placed in a plenum chamber with constant pressure. The very short exposure time (less than 1 microsecond) allows us to freeze the flow and, in particular, the vortices. This jet is separated from the air at rest in the wind tunnel plenum chamber by a slip surface along which develops a turbulent mixing layer whose vortices are frozen by the short-duration photography.

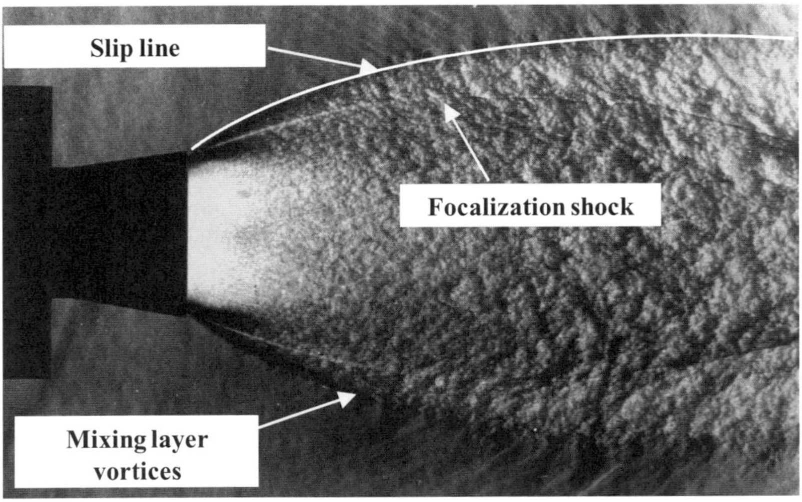

Figure 13.6. *An example of a slip line: axisymmetric jet emerging from a supersonic nozzle (document Onera)*

13.5. Normal shock wave case

13.5.1. *General relations*

Returning to the discontinuity of the shock wave type, we consider the form taken by the previous equations when the shock wave is normal to the upstream velocity, the angle of the shock thus has the value $\sigma = \frac{\pi}{2}$ (see Figure 13.3). This is a particularly important case because it is the situation of one-dimensional flows, or the one found in front of the blunt nose of a vehicle in supersonic flight. This case also corresponds to the stagnation pressure probe which exploits the relations of the normal shock to determine the local Mach number within a supersonic flow (see section 13.5.2.3). We then will return to the oblique shock wave and thus take advantage of the results obtained for the normal shock.

If $\sigma = \frac{\pi}{2}$, $V_{t_1} = 0$: then, according to the relation $V_{t_1} = V_{t_2}$, it follows that $V_{t_2} = 0$. A normal shock wave does not rotate the velocity vector, the downstream velocity component V_{n_2} remaining normal to the shock wave. Thus, we can consider that the normal component V_n is identified with the velocity itself, leading to the shock equations where the index *n* is omitted:

$$\rho_1 V_1 = \rho_2 V_2$$

$$p_1 + \rho_1 V_1^2 = p_2 + \rho_2 V_2^2$$

$$h_1 + \frac{V_1^2}{2} = h_2 + \frac{V_2^2}{2}.$$

If the upstream flow is known, the equations of the normal shock contain four unknowns: ρ_2, V_2, p_2, h_2. Therefore, we have to supplement them by a thermodynamic relation which, for a divariant gas, is of the form:

$$h = h(p, \rho).$$

The problem is then "closed" and we can calculate the properties of the flow downstream of the shock. If thermodynamics does not translate into explicit relations, the equations of the shock must be solved by an iterative method. To perform this calculation, it is useful to state:

$$\varepsilon = \frac{\rho_1}{\rho_2} = \frac{V_2}{V_1}.$$

It follows that:

$$p_2 = p_1 + \rho_1 V_1^2 \left(1-\varepsilon\right) \qquad [13.5]$$

$$h_2 = h_1 + \frac{V_1^2}{2} \left(1-\varepsilon^2\right). \qquad [13.6]$$

The recommended iterative procedure, which converges very quickly, is as follows:

– we begin a *priori* by setting $\varepsilon < 1$,

– equations [13.5] and [13.6] give p_2 and h_2,

– the thermodynamics allows the calculation of ρ_2, leading to a new estimate of ε and the cycle is repeated until convergence.

13.5.2. *Case of the calorically perfect gas*

13.5.2.1. *The Prandtl relation*

We have the shock relations:

$$\rho_1 V_1 = \rho_2 V_2$$

$$p_1 + \rho_1 V_1^2 = p_2 + \rho_2 V_2^2$$

$$C_p T_1 + \frac{V_1^2}{2} = C_p T_2 + \frac{V_2^2}{2} = C_p T_{st}$$

with (equation of state):

$$p_1 = r\rho_1 T_1 \quad and \quad p_2 = r\rho_2 T_2.$$

Taking the equation of state into account, the motion equation reads:

$$\rho_1\left(rT_1 + V_1^2\right) = \rho_2\left(rT_2 + V_2^2\right)$$

or, by taking into account the continuity equation:

$$V_2\left(rT_1 + V_1^2\right) = V_1\left(rT_2 + V_2^2\right).$$

By expressing the temperature with the energy equation, we obtain:

$$V_2\left[r\left(T_{st}-\frac{V_1^2}{2C_p}\right)+V_1^2\right]=V_1\left[r\left(T_{st}-\frac{V_2^2}{2C_p}\right)+V_2^2\right]$$

from which follows by expanding:

$$V_2 r\left(T_{st}-\frac{V_1^2}{2C_p}\right)+V_2V_1^2=V_1 r\left(T_{st}-\frac{V_2^2}{2C_p}\right)+V_1V_2^2$$

given:

$$rT_{st}\left(V_2-V_1\right)=\frac{r}{2C_p}\left(V_1^2V_2-V_2^2V_1\right)+V_1V_2^2-V_2V_1^2$$

$$rT_{st}\left(V_2-V_1\right)=\frac{r}{2C_p}V_1V_2\left(V_1-V_2\right)+V_1V_2\left(V_2-V_1\right)$$

which gives:

$$rT_{st}=V_1V_2\left(-\frac{r}{2C_p}+1\right)=V_1V_2\frac{\gamma+1}{2\gamma}$$

and:

$$V_1V_2=\frac{2\gamma}{\gamma+1}rT_{st}.$$

By expressing the speed of sound a_{st} from the relation: $a_{st}^2=\gamma\, r\, T_{st}$, it follows that:

$$V_1V_2=\frac{2}{\gamma+1}\ a_{st}^2.$$

The energy equation written for the critical state gives:

$$a_c^2+\frac{2}{\gamma-1}a_c^2=\frac{2}{\gamma-1}a_{st}^2$$

leading to:

$$a_c^2 = \frac{2}{\gamma+1} a_{st}^2.$$

We finally obtain the following relation, known as the *Prandtl relation*:

$$\boxed{V_1 V_2 = a_c^2}$$

which restates the fundamental result:

The product of the velocities on both sides of a normal shock wave is equal to the square of the critical speed.

The Prandtl relation can also be written:

$$\frac{V_1}{a_c} \frac{V_2}{a_c} = 1.$$

The dimensionless velocity V / a_c is linked to the Mach number M through the relation (see section 9.5.2.4):

$$\frac{V}{a_c} = M \sqrt{\frac{(\gamma+1)/2}{\left[1+\frac{(\gamma-1)}{2} M^2\right]}}.$$

The ratio of the velocity to the speed of sound for the critical conditions is called the *critical Mach number* and is denoted by M^*. The Prandtl relation can also be written as:

$$\boxed{M_1^* \ M_2^* = 1.}$$

It is shown that if $M = 1$, then $V = a_c$ and we can show that if $V / a_c < 1$, then $M < 1$ and if $V / a_c > 1$, then $M > 1$. Therefore:

The Mach numbers on both sides of the normal shock wave are subsonic on one side and supersonic on the other side.

13.5.2.2. *Relations between the upstream and downstream states*

Ratio of the pressures p_2 / p_1

According to the motion equation:

$$\frac{p_2}{p_1} = 1 + \frac{\rho_1 V_1^2 - \rho_2 V_2^2}{p_1}$$

or taking into account the continuity equation:

$$\frac{p_2}{p_1} = 1 + \frac{\rho_1 V_1}{p_1}(V_1 - V_2) = 1 + \frac{\rho_1}{p_1}\left(V_1^2 - V_1 V_2\right).$$

Knowing that:

$$\frac{p_1}{\rho_1} = \frac{a_1^2}{\gamma}$$

and expressing the product $V_1 \, V_2$ with the Prandtl relations, we obtain:

$$\frac{p_2}{p_1} = 1 + \frac{\gamma}{a_1^2}\left(V_1^2 - 2\frac{\gamma - 1}{\gamma + 1} C_p T_{st}\right)$$

$$\frac{p_2}{p_1} = 1 + \gamma\left(\frac{V_1^2}{a_1^2} - 2\frac{\gamma - 1}{\gamma + 1}\frac{C_p}{a_1^2} T_{st}\right)$$

$$\frac{p_2}{p_1} = 1 + \gamma\left(\frac{V_1^2}{a_1^2} - 2\frac{\gamma - 1}{\gamma + 1}\frac{C_p}{\gamma \;\, r}\frac{T_{st}}{T_1}\right).$$

As (see section 9.5.2.2):

$$\frac{T_{st}}{T_1} = 1 + \frac{\gamma - 1}{2} M_1^2$$

it follows after some simple calculations that:

$$\boxed{\frac{p_2}{p_1} = 1 + \frac{2\gamma}{\gamma + 1}\left(M_1^2 - 1\right).}$$

Ratio of the densities $\varepsilon = \rho_1 / \rho_2$

The above relations enable us to write:

$$\frac{\rho_1}{\rho_2} = \frac{V_2}{V_1} = \frac{V_1 V_2}{V_1^2} = \frac{1}{V_1^2} 2 \frac{\gamma - 1}{\gamma + 1} C_p T_{st}$$

or:

$$\frac{\rho_1}{\rho_2} = \frac{a_1^2}{V_1^2} 2 \frac{\gamma - 1}{\gamma + 1} \frac{C_p}{\gamma r} \frac{T_{st}}{T_1}$$

leading to, after simple manipulations:

$$\boxed{\frac{\rho_1}{\rho_2} = \frac{\gamma - 1}{\gamma + 1} \left[1 + \frac{2}{(\gamma - 1) \; M_1^2} \right].}$$

For the ratio ρ_2 / ρ_1:

$$\boxed{\frac{\rho_2}{\rho_1} = \frac{(\gamma + 1) M_1^2}{2 + (\gamma - 1) \; M_1^2}.}$$

Ratio of the temperatures T_2 / T_1

Using the gas equation of state:

$$\frac{T_2}{T_1} = \frac{p_2}{p_1} \frac{\rho_1}{\rho_2}$$

we immediately obtain:

$$\boxed{\frac{T_2}{T_1} = \left[1 + \frac{2\gamma}{\gamma + 1} \left(M_1^2 - 1 \right) \right] \left[\frac{2 + (\gamma - 1) M_1^2}{(\gamma + 1) M_1^2} \right].}$$

Mach number after the shock M_2

We can deduce the following expression for the Mach number after the shock from the Prandtl relation:

$$\boxed{M_2^2 = \frac{1+\dfrac{\gamma-1}{2}M_1^2}{\gamma\ M_1^2 - \dfrac{\gamma-1}{2}}}.$$

Entropy jump across the shock

According to the thermodynamic relation, we have:

$$s_2 - s_1 = C_p \ln\frac{T_2}{T_1} - r\ln\frac{p_2}{p_1}$$

and using the expressions obtained for the ratios of the temperatures and pressures, we obtain:

$$s_2 - s_1 = C_p \ln\left\{\left[1+\frac{2\gamma}{\gamma+1}\left(M_1^2-1\right)\right]\left[\frac{2+(\gamma-1)M_1^2}{(\gamma+1)M_1^2}\right]\right\} - r\ln\left[1+\frac{2\gamma}{\gamma+1}\left(M_1^2-1\right)\right].$$

COMMENTS.– According to the second law of thermodynamics, the jump $\Delta s = s_2 - s_1$ must be positive or zero. However, the shock equations are symmetric and indifferent to the upstream value of the Mach number, which can be either subsonic or supersonic. The expression for the entropy jump allows us to show that if M_1 is lower than 1, then $\Delta s < 0$, which is contrary to the second principle. Conversely, we can verify that if $M_1 > 1$, then $\Delta s > 0$. The subsonic shock waves are rejected as physically unacceptable. Consequently, according to the Prandtl relation, the downstream Mach number M_2 is always subsonic. In a phenomenon such as a shock wave, the conservation equations are symmetric and provide both subsonic → supersonic as well as supersonic → subsonic shocks, this property being general for the Euler equations (see Chapter 7). It is the second law of thermodynamics which imposes the direction of the change.

The energy equation written with the entropy (see section 4.7):

$$T\frac{ds}{dt} = \frac{1}{\rho}\left(\tau_{ij}\frac{\partial u_j}{\partial x_i} - \frac{\partial q_i}{\partial x_i}\right)$$

shows that entropy is produced by the viscous terms and the thermal conductivity. For slowly changing flow, these terms are usually negligible; hence the assumption of a non-viscous fluid leading to the Euler equations. Across a shock wave of finite but very small thickness, the spatial derivatives $\partial / \partial x_i$ become very large. This results in a local "activation" of the terms implying viscosity and thermal conductivity. It is the action of these terms that is responsible for the rise of entropy across a shock wave (see the analysis of the shock structure in section 13.6).

Thus, although this concept is introduced in the framework of non-viscous fluids, *the shock wave is a viscous phenomenon*. As shown in section 5.3, the aerodynamic drag of a body is directly related to the entropy production by the body in motion. We often agree to distinguish the wave drag, due to entropy production by shock waves, and the viscous drag, generated by the viscous effects manifested in the form of boundary layers. On a fundamental level, this distinction does not arise; in all cases, the drag results from the generation of entropy by viscosity.

13.5.2.3. *Stagnation pressure loss across a shock wave*

The formula for the entropy jump across a normal shock can be written in the more compact form:

$$\frac{s_2 - s_1}{C_v} = \ln\left\{\left[1 + \frac{2\gamma}{\gamma+1}\left(M_1^2 - 1\right)\right]\left[1 - \frac{2}{\gamma+1}\left(1 - \frac{1}{M_1^2}\right)\right]^\gamma\right\}.$$

If we introduce the upstream and downstream stagnation states (st_1) and (st_2), the changes before and after the shock being isentropic, we have:

$$s_{st_1} = s_1 = C_p \ln T_{st_1} - r \ln p_{st_1}$$

$$s_{st_2} = s_2 = C_p \ln T_{st_2} - r \ln p_{st_2}.$$

Knowing that the stagnation temperature is conserved across the shock wave:

$$s_2 - s_1 = s_{st_2} - s_{st_1} = r \ln\left(\frac{p_{st_1}}{p_{st_2}}\right)$$

and:

$$\frac{p_{st_2}}{p_{st_1}} = \exp\left(-\frac{s_2 - s_1}{r}\right).$$

As the jump $(s_2 - s_1)$ is positive, it follows that the ratio p_{st_2} / p_{st_1} is smaller than 1. The stagnation pressure of the flow *decreases* across a shock wave; hence the concept of losses through shocks. Taking into account the above given expression $(s_2 - s_1)/C_v$, we get:

$$\frac{p_{st_2}}{p_{st_1}} \equiv \Pi(M_1,\gamma) = \left[1 + \frac{2\gamma}{\gamma+1}\left(M_1^2 - 1\right)\right]^{-\frac{1}{\gamma-1}} \left[1 - \frac{2}{\gamma+1}\left(1 - \frac{1}{M_1^2}\right)\right]^{-\frac{\gamma}{\gamma-1}}.$$

This relation is used to determine the local Mach number in a supersonic flow using a stagnation pressure probe, also called Pitot tube (see section 9.6.2 for the application to a subsonic flow). Such a probe consists of a tube placed facing the flow and aligned with its velocity. The hole of the probe is connected to a pressure transducer via a tube. Since this tube is obturated by the transducer, the flow arriving head on the tube is stagnated so that a normal shock wave forms in front of the probe. The shock will then incline when propagating in the supersonic flow to become conical, the probe being axisymmetric (see Figure 13.7). The transducer measures the stagnation pressure p_{st_2} downstream of the normal shock, the stagnation to the transducer taking place in subsonics in an isentropic way. Moreover, if the upstream stagnation pressure p_{st_1} is measured by a second transducer, placed in the settling chamber of the wind tunnel, the ratio p_{st_2} / p_{st_1} is known. The inversion of the above formula gives the Mach number M_1 of the flow upstream of the shock caused by the probe.

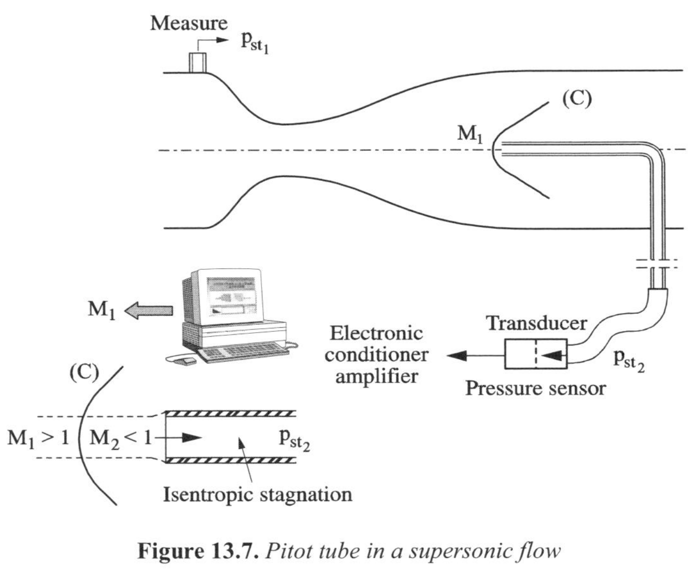

Figure 13.7. *Pitot tube in a supersonic flow*

The photograph in Figure 13.8a shows the front end of a Pitot probe with a round sensing part used in supersonic flows. Figure 13.8b shows a flattened Pitot probe used to probe boundary layers. The thickness of the nozzle is close to 0.3 mm so as to approach the wall as close as possible.

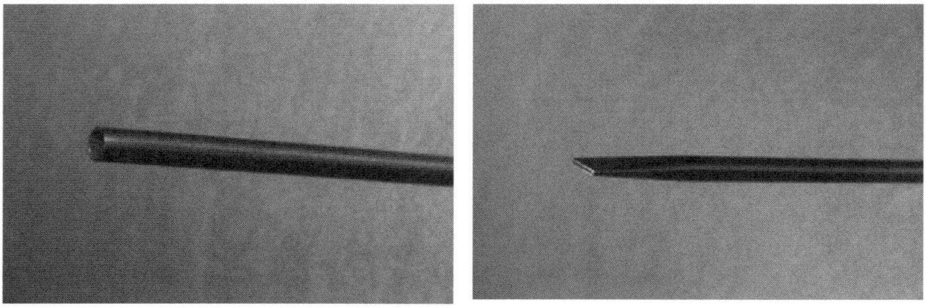

a – probe with a cylindrical end *b – probe with a flattened end for boundary layers*

Figure 13.8. *Pitot probes used in supersonic wind tunnels (documents Onera)*

Such a probe is ineffective if the upstream stagnation pressure p_{st_1} is unknown. This is what happens in non-isentropic flows (wakes or boundary layers) or in flows having been the seat of shock waves. The local stagnation pressure p_{st_1} is then different to the "reservoir" pressure of the fluid feeding the wind tunnel. It is also the situation for an aircraft in flight when the upstream stagnation pressure of the flow is unknown since it depends on the flight Mach number which is exactly what we seek to determine. Under these conditions, we have to use in conjunction with the Pitot tube a static pressure probe to measure the pressure p_1 (such a probe will not be described here). As we know, the ratio:

$$\frac{p_{st_2}}{p_1} = \frac{p_{st_2}}{p_{st_1}} \frac{p_{st_1}}{p_1}$$

is a function of the Mach number M_1 which has the expression:

$$\boxed{\frac{p_{st_2}}{p_1} = \left[1 + \frac{2\gamma}{\gamma+1}\left(M_1^2 - 1\right)\right]^{-\frac{1}{\gamma-1}} \left[1 - \frac{2}{\gamma+1}\left(1 - \frac{1}{M_1^2}\right)\right]^{-\frac{\gamma}{\gamma-1}} \left(1 + \frac{\gamma-1}{2} M_1^2\right)^{\frac{\gamma}{\gamma-1}}.}$$

This formula, called the *Rayleigh formula*, enables us to determine the Mach number M_1 from the ratio p_{st_2} / p_1. On board of an aircraft, measurements of p_{st_2} and p_1 are often performed by a single probe similar to the Prandtl antenna described in section 9.6.2 (see Figure 13.9). A static pressure port is placed on the cylindrical body of the probe whose frontal port measures p_{st_2}. If the static port is far enough downstream, the pressure it measures is very close to p_1. Such probes are usually calibrated in a wind tunnel, as a high accuracy for determining the Mach number in flight is required.

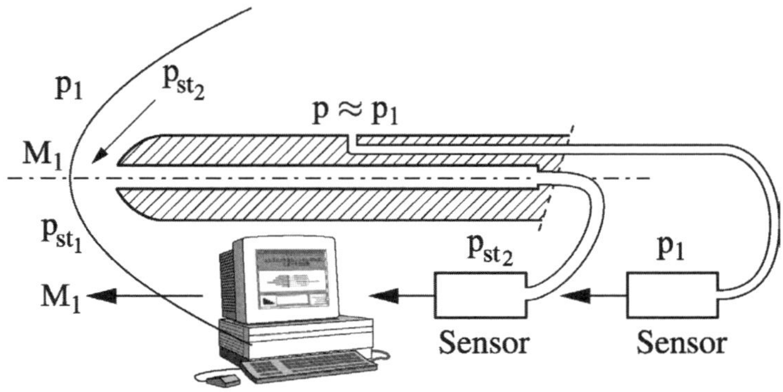

Figure 13.9. *Probe to determine the Mach number in supersonic flight*

13.5.2.4. *Some instructive passages at the limit*

The upstream Mach number tends to 1: the vanishingly weak shock

The previous expressions show that the ratios:

$$\frac{p_2}{p_1},\ \frac{\rho_2}{\rho_1},\ \frac{T_2}{T_1},\ \frac{p_{st_2}}{p_{st_1}}$$

tend toward 1 when the Mach number M_1 tends toward 1. For a Mach number equal, or very close, to 1, the intensity of the shock becomes vanishingly small. When M_1 increases from 1, the shock intensity grows gradually and the ratios above move away from 1.

The upstream Mach number tends to infinity: the infinitely strong shock

As we can see, when the Mach number M_1 tends toward infinity, the ratios:

$$\frac{p_2}{p_1} \text{ and } \frac{T_2}{T_1}$$

tend toward infinity. On the other hand, the ratio:

$$\frac{p_{st_2}}{p_{st_1}}$$

tends toward zero: the stagnation pressure losses across the shock become very large.

The density jump ρ_2 / ρ_1 tends toward a finite nonzero value, thus:

$$\frac{\rho_2}{\rho_1} \rightarrow \frac{\gamma+1}{\gamma-1} \quad for \quad M_1 \rightarrow \infty.$$

In the case of the air behaving like a calorically perfect gas $(\gamma = 1.4)$, this limit is 6.

Similarly, the downstream Mach number M_2 tends toward a finite limit:

$$M_2 \rightarrow \sqrt{\frac{\gamma-1}{2\gamma}} \quad for \quad M_1 \rightarrow \infty.$$

This limit is equal to 0.378 for $\gamma = 1.4$.

The limit values for the infinitely strong shock are interesting indications that have to be used with caution. In an air flow with a temperature close to the ambient temperature and with a high Mach number, for which $\gamma \cong 1.4$, we find that for $M_1 = 10$, the ratio ρ_2 / ρ_1 is 5.71, a value already close to the limit equal to 6. This limit is sometimes called the *hypersonic limit* because it gives a fairly good approximation of the value of the density downstream of the detached shock forming in front of a hypersonic vehicle, provided that we can neglect the *real gas* effects (see section 13.5.3). There are approximate theories that exploit this fact by considering the flow downstream of the shock as incompressible since ρ there varies little around the limit value.

In fact, during re-entry, at a point of the trajectory where the Mach number is equal to 20 and the temperature 280 K, the normal shock theory applied for $\gamma = 1.4$ leads to a temperature of $T_2 \cong 22{,}000\,K$ after the shock. Such a high value is inconsistent with the assumption of a calorically perfect gas. In reality, the thermodynamic behavior of gas changes, so that the assumption $\gamma = constant$ is grossly wrong. The so-called real gas effects occur that render the above formulas no longer applicable. To calculate the properties downstream of the shock, we have to resort to the general method using a Mollier diagram, if the gas is at equilibrium downstream of the shock. We then find a much lower temperature T_2, which is fortunate for spacecraft! The ratio ρ_2 / ρ_1 may, in turn, exceed the value 6.

13.5.2.5. *Results to retain*

The table below summarizes the behavior of the main properties of a flow across a normal shock wave.

$M_1 > 1$	$M_2 < 1$	$p_2 > p_1$	$T_2 > T_1$	$\rho_2 > \rho_1$	$p_{st_2} < p_{st_1}$	$T_{st_2} = T_{st_1}$

13.5.3. *Normal shock wave with real gas effects and chemical kinetics*

When a gas undergoes a very rapid change making its pressure and temperature vary strongly, it can lead to dissociation of the gas components. In some circumstances, the change of conditions can be so fast that equilibrium reactions have no time to establish. Thus, the law of mass action ceases to be applicable for calculating the degrees of dissociation; we have to take into account the laws of chemical kinetics describing the reaction progress as a function of time.

Such non-equilibrium phenomena occur when a shock wave suddenly raises the pressure and temperature of the gas. Thus, during a re-entry at Mach 20 at an altitude where the temperature is 250 K, under the assumption of a calorically perfect gas (with $\gamma = 1.4$), a shock wave forms in front of the spacecraft increasing the temperature to 19,700 K. This quasi-instantaneous heating of the gas triggers various reactions, including those of the dissociation of air components (see section 3.7.1), until a state corresponding to the thermochemical equilibrium is reached. Thus, downstream of the shock, exists a so-called *relaxation region* characterized by non-equilibrium processes. Non-equilibrium phenomena also occur in the rocket engine nozzles where the combustion gases, at high temperature and high pressure in the combustion chamber (approximately 3,500 K and 115×10^5 Pa for the *Vulcain 2* engine of the European launcher *Ariane 5*), expand rapidly into the nozzle reaching conditions with much lower pressure. During the expansion in the nozzle, the species formed in the combustion chamber are still reactive so that chemical reactions continue to occur in the flow but in variable conditions. The behavior of such reactive flows is characterized by the Damköhler number (see section 6.2.2) which is the ratio between the velocity of the chemical reactions progress and the velocity of the flow. If this number is high, then we can assume equilibrium and apply the laws of mass action (Guldberg and Waage) to calculate the gas composition. However, if this number is small, we can assume freezing by admitting that the composition of the gas does not vary (frozen gas assumption). Between these two cases, we have to calculate the flow by solving the chemical kinetics equations in tandem to that of the flow.

The laws of chemical kinetics are still poorly understood because of experimental difficulties encountered when characterizing the phenomenon: duration of the order of microseconds, very high temperatures and pressures, and complex composition of the gas medium. The shock tube (see Chapter 23) is often used to analyze these processes. Here, we only give a qualitative indication of these phenomena as a more detailed study would lead us too far.

We consider a uniform supersonic flow of a diatomic gas A_2 likely to undergo the dissociation reaction:

$$A_2 \leftrightarrow A + A$$

by assuming that the other degrees of freedom of the gas (translation, rotation, and vibration) are instantaneously in equilibrium.

A normal shock wave (C) occurs in this flow where it does not introduce, as we know, any deviation of the velocity (see section 13.5.1). If the upstream Mach number M_1 is high, a sharp rise in temperature happens that will excite the dissociation reaction. However, as the reaction rate is finite, this occurs gradually. The parameters of the flow downstream of the shock will therefore change until the chemical equilibrium is reached. Given that q is the mass proportion of the constituent A and $(1-q)$ that of the constituent A_2. Then, upstream of the shock, the gas being in equilibrium, q is known. Usually, the gas is not dissociated by the temperature and low-pressure conditions of the atmosphere. Thus $q = 0$. This parameter starts to vary in the crossing of the shock, but undergoes no discontinuity as the reaction velocity, according to the relaxation equations, is finite. The calculation of the initial data on the downstream side of the shock is the result of the classic equations applied to frozen gas, i.e. considering that its properties and composition do not have the time to change across the shock which is usually excessively thin.

Now let us consider far enough downstream so that equilibrium is reached. The general shock equations apply between the upstream state and the equilibrium downstream state, these equations are budget equations between upstream and downstream states that ignore what happens in between. Thus, we can solve the Rankine–Hugoniot equations, as shown in section 13.5.1, by providing the thermodynamics of the dissociated gas at equilibrium which we determine through the resolution of the equations of the mass action law or even a Mollier diagram, for example, the one of the air used for aerospace applications.

But if we look at the detail of the phenomenon, especially the thickness of the non-equilibrium zone, a step by step calculation is necessary.

The one-dimensional flow (no transverse pressure gradient can occur) is defined by the quantities $V(x), p(x), T(x), \rho(x), q(x)$, where x denotes the coordinate along the streamline. The conservation equations between the initial section (1) and the downstream section (x), are written as:

– *continuity equation*: $\rho V = \rho_1 V_1 = \dot{m}$

– *equation of motion*: $p + \rho V^2 = p_1 + \rho_1 V_1^2 = d$

– *energy equation*: $h + \dfrac{V^2}{2} = h_1 + \dfrac{V_1^2}{2} = h_{st}$.

The physicochemical equations defining the evolution of gas are the following:

– *equation of state*: $p = Z(q) r\rho T$

– *thermodynamics*: $h = h(\rho, T, q)$

– *chemical kinetics*: $\dfrac{Dq}{Dt} = V\dfrac{dq}{dx} = F(\rho, T, q)$.

The function $Z(q)$ is the compressibility factor defined in section 3.7.3.3.

The initial known value of q is q_1. Injected in the thermodynamic relation, it gives the enthalpy h_1, the gas being considered frozen. The final value of q is q_2 corresponding to $\dfrac{Dq}{Dt} = 0$, i.e. to equilibrium. This last equation corresponds to the equilibrium equation of Guldberg and Waage. This value of q inserted in the thermodynamic relation gives the function h relative to the equilibrium function which is calculable from the thermodynamic relations, tables, or even a Mollier diagram.

The conditions on the downstream side of the shock are calculated by considering the gas as frozen; in practice, by considering the gas with a value γ fixed to its upstream value, and therefore we deduce $V_2'(0), p_2'(0), T_2'(0), \rho_2'(0), q_2'(0) = q_1$. The equilibrium quantities are determined as above by considering the thermodynamics of a gas in equilibrium, and therefore $V_2(\infty)$, $p_2(\infty), T_2(\infty), \rho_2(\infty), q_2(\infty) = 0$, which means the equilibrium is theoretically reached after an infinite time.

To integrate the above system between these two states step by step, we can proceed as follows:

1. Express p, ρ, T as a function of V and the invariants m, d, h_{st} using:

$$p = d - \dot{m}V$$

$$\rho = \frac{\dot{m}}{V}$$

$$T = \frac{d - \dot{m}V}{r\ Z(q)} \times \frac{V}{\dot{m}}.$$

2. The equation:

$$h(\rho, T, q) = h_{st} - \frac{V^2}{2}$$

where replacing ρ and T as a function of the velocity V provides an implicit relation $V(q)$ calculable in numerical form.

3. Thus, we can then evaluate the second side of the chemical equation:

$$\frac{dq}{dx} = \frac{F(\rho, T, q)}{V} = G(q)$$

as a function of q. The problem is then reduced to the integral:

$$x = \int_{q_1}^{q} \frac{dq}{G(q)}$$

which provides the function $q(x)$ between q_1 and q_2.

The impact of real gas effects on conditions downstream of a normal shock is illustrated by the following results, considering the equilibrium conditions reached after the relaxation region. They are relative to air and were obtained by solving the Rankine–Hugoniot equations with thermodynamics given by a Mollier diagram provided in numerical form. The upstream conditions are: pressure $p_1 = 10^5$ Pa, temperature $T_1 = 300$ K, and $\gamma = 1.4$.

The calculation is done for an upstream velocity varying in the range of $500 \text{ m/s} < V_1 < 9{,}000 \text{ m/s}$, given a Mach number M_1 ranging from 1.43 to 25.8.

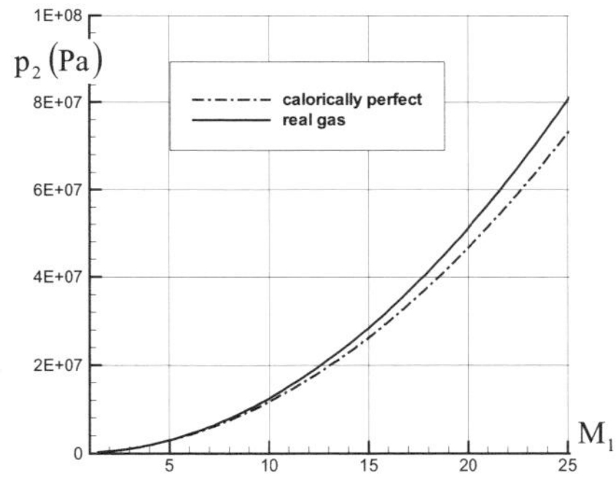

a – pressure downstream of the shock

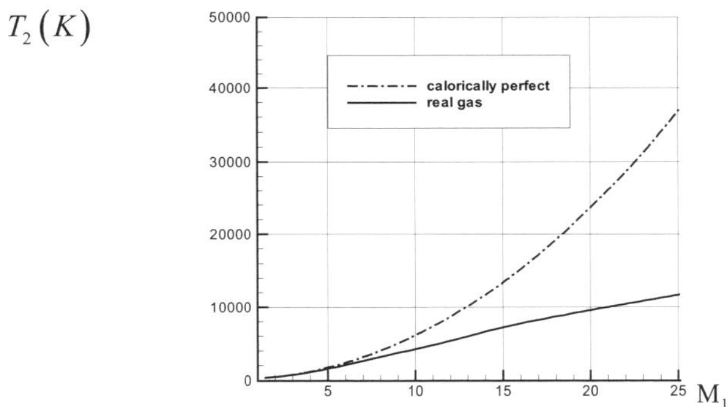

b – temperature downstream of the shock

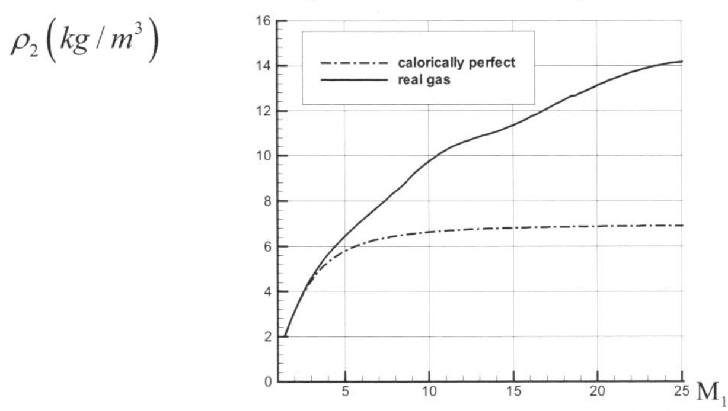

c – density downstream of the shock

Figure 13.10. *Real gas effects on the quantities downstream of a normal shock wave ($p_1=10^5$ Pa, $T_1=300$ K, $\gamma_1=1.4$)*

Figure 13.10a shows the change of the pressure p_2 downstream of the shock, compared to the case of a calorically perfect gas. We note that the real gas effects have little impact on this quantity. However, considering the temperature (see

Figure 13.10b), they lead to considerable differences: the temperature reached at Mach 25 is around $11,000\text{ K}$ instead of $36,700\text{ K}$ in the case of a calorically perfect gas. The difference is explained by the energy absorbed by the dissociation of molecules, the recombination reactions releasing energy becoming less and less likely with increasing temperature (see Figure 3.1). We note that this circumstance is fortunate, because without real gas effects, the space adventure would probably have been impossible, and no thermal protection could have withstood temperatures in the case of the calorically perfect gas. As shown in Figure 13.10c, the same difference is reflected in the density after the shock which can reach values much greater than the asymptotic value $\dfrac{\rho_2}{\rho_1} = \dfrac{\gamma+1}{\gamma-1}$, equal to 6 for $\gamma = 1.4$ (see section 13.5.2.4).

13.6. Shock structure: the role of viscosity

In the previous theory, we postulated the existence of a discontinuity of the shock wave type and established relations binding properties of the flow on both sides of the discontinuity given the fundamental equations for mass, motion, and energy. These shock relations, established by neglecting the viscosity and thermal conductivity, are reversible; the flow approaching the shock can be subsonic as well as supersonic. It is the introduction of the second law of thermodynamics which allows choosing the solution in agreement with the reality that leads to increased entropy. The following analysis highlights the mechanism of entropy production in a shock wave.

We consider an upstream state (1) and a downstream state (2), both uniform and connected by the following relations in steady flow:

– *continuity equation:*

$$\rho_1 u_1 = \rho_2 u_2$$

– *equation of motion:*

$$p_1 + \rho_1 u_1^2 = p_2 + \rho_2 u_2^2$$

– *energy equation:*

$$h_{st} = h_1 + \frac{u_1^2}{2} = h_2 + \frac{u_2^2}{2}.$$

To give a little more generality to the following results, we consider an unsteady but one-dimensional flow, where the shock propagates with a constant velocity D relative to the upstream state (1). We return to the steady case by writing the conservation equations in a reference frame moving with the shock, which gives by denoting w the velocity change through the shock (see Chapter 22 for more details):

$$w = u_2 - u_1.$$

– *continuity equation:*

$$\rho_1 D = \rho_2 (D - w)$$

– *equation of motion:*

$$p_2 - p_1 = \rho_1 D w$$

– *energy equation:*

$$h_1 + \frac{D^2}{2} = h_2 + \frac{1}{2}(D - w)^2.$$

The problem of the *shock structure* is to seek a solution which asymptotically connects the states (1) and (2). We consider that the fluid is viscous by considering the framework of the one-dimensional and unsteady flow theory. The x axis is oriented left to right and we assume that the shock propagates in the positive x direction. As shown in Figure 13.11, it meets flow (1), the uniform state (1) is reached when $x \to +\infty$ and the uniform state (2) when $x \to -\infty$.

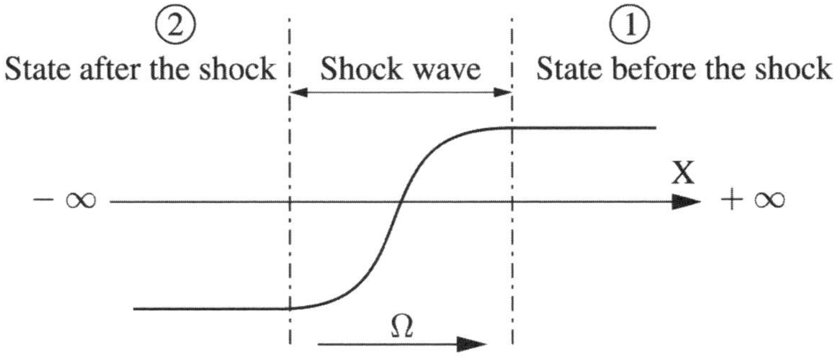

Figure 13.11. *Orientation of the x axis for the study of the shock structure*

The Navier–Stokes equations, given in Chapter 4, are written in this case as:

– *continuity equation:*

$$\frac{\partial \rho}{\partial t}+\frac{\partial}{\partial x}(\rho u)=0$$

– *equation of motion:*

$$\frac{\partial}{\partial t}(\rho u)+\frac{\partial}{\partial x}\left(\rho u^2\right)=-\frac{\partial p}{\partial x}+\frac{\partial \tau}{\partial x}$$

– *energy equation:*

$$\frac{\partial}{\partial t}(\rho e_T)+\frac{\partial}{\partial x}(\rho u e_T)=\frac{\partial}{\partial x}(u\tau-\dot{q})-\frac{\partial}{\partial x}(pu).$$

We recall that e_T is the total energy:

$$e_T=e+\frac{u^2}{2}.$$

For a flow depending only on one space variable x, the viscosity term is written as:

$$\tau=\frac{4}{3}\mu\frac{\partial u}{\partial x}$$

and according to the Fourier law:

$$\dot{q}=-\kappa\frac{\partial T}{\partial x}.$$

We obtain the equations:

– *continuity equation:*

$$\frac{\partial \rho}{\partial t}+\frac{\partial}{\partial x}(\rho u)=0$$

– *equation of motion:*

$$\frac{\partial}{\partial t}(\rho u)+\frac{\partial}{\partial x}\left(\rho u^2+p-\frac{4}{3}\mu\frac{\partial u}{\partial x}\right)=0$$

– *energy equation:*

$$\frac{\partial}{\partial t}(\rho e_T)+\frac{\partial}{\partial x}\left(\rho u e_T + pu - \frac{4}{3}\mu u\frac{\partial u}{\partial x} - \kappa\frac{\partial T}{\partial x}\right)=0.$$

The above equations are written *in conservative form*, since they express the conservation of quantities between parentheses in a steady flow. If the flow can be considered non-viscous, we find the normal shock equations. Let us write the equations in a frame of a reference system following the shock wave in its motion. If Ω denotes the absolute velocity of the wave in the "fixed" frame, we consider the moving frame X so that:

$$X = x - \Omega t.$$

We will have:

$$\frac{\partial}{\partial x}=\frac{\partial}{\partial X},\quad \frac{\partial}{\partial t}=\frac{\partial}{\partial X}\frac{\partial X}{\partial t}=-\Omega\frac{\partial}{\partial t}$$

from which follow the new forms of the equations:

$$-\Omega\frac{\partial \rho}{\partial X}+\frac{\partial}{\partial X}(\rho u)=0$$

$$-\Omega\frac{\partial}{\partial X}(\rho u)+\frac{\partial}{\partial X}\left(\rho u^2 + p - \frac{4}{3}\mu\frac{\partial u}{\partial X}\right)=0$$

$$-\Omega\frac{\partial}{\partial X}(\rho e_T)+\frac{\partial}{\partial X}\left(\rho u e_T + pu - \frac{4}{3}\mu u\frac{\partial u}{\partial X} - \kappa\frac{\partial T}{\partial X}\right)=0.$$

As the velocity of the shock Ω is assumed constant, the above system can be integrated for the first time to give the equations:

$$-\Omega\rho+\rho u = A \qquad [13.7]$$

$$-\Omega\rho u+\rho u^2 + p - \frac{4}{3}\mu\frac{\partial u}{\partial X}=B \qquad [13.8]$$

$$-\Omega\rho e_T+\rho u e_T + pu - \frac{4}{3}\mu u\frac{\partial u}{\partial X} - \kappa\frac{\partial T}{\partial X}=C \qquad [13.9]$$

where A, B, and C are constants. It is easy to verify that the previous equations are identical to the equations of a normal unsteady shock if the fluid is non-viscous (see Chapter 22). The solution to the system [13.7]–[13.9] must tend toward the state (1) when $X \to +\infty$ and the state (2) when $X \to -\infty$. The continuity equation meets this requirement since it is identical to the continuity equation for the shock. For equations [13.8] and [13.9], it must be that:

$$u,\, p,\, \rho,\, e_T,\, \mu,\, \kappa \;\to\; u_1,\, p_1,\, \rho_1,\, e_{T_1},\, \mu_1,\, \kappa_1 \qquad \textit{when} \qquad X \to +\infty$$

$$u,\, p,\, \rho,\, e_T,\, \mu,\, \kappa \;\to\; u_2,\, p_2,\, \rho_2,\, e_{T_2},\, \mu_2,\, \kappa_2 \qquad \textit{when} \qquad X \to -\infty.$$

We calculate the change of entropy in the shock structure. The energy equation written with the specific entropy s generally gives (see section 4.7):

$$T\frac{ds}{dt} = \frac{1}{\rho}\left(\tau_{ji}\frac{\partial u_j}{\partial x_i} - \frac{\partial \dot{q}_i}{\partial x_i}\right).$$

In the one-dimensional assumption, it follows that:

$$\rho\frac{ds}{dt} = \frac{1}{T}\left[\frac{4}{3}\mu\left(\frac{\partial u}{\partial X}\right)^2 + \frac{\partial}{\partial X}\left(\kappa\frac{\partial T}{\partial X}\right)\right].$$

Also:

$$\rho\frac{ds}{dt} = \rho\left(\frac{\partial s}{\partial t} + u\frac{\partial s}{\partial X}\right) = -\rho\Omega\frac{\partial s}{\partial X} + \rho u\frac{\partial s}{\partial X} = -\frac{d}{dX}\left[\rho(\Omega - u)s\right]$$

given the continuity equation and the fact that the problem is stationary in the frame X. We can write:

$$\frac{1}{T}\frac{\partial}{\partial X}\left(\kappa\frac{\partial T}{\partial X}\right) = \frac{\partial}{\partial X}\left[\frac{\kappa}{T}\left(\frac{\partial T}{\partial X}\right)\right] + \frac{1}{T^2}\kappa\left(\frac{\partial T}{\partial X}\right)^2$$

from which the equation for the entropy follows:

$$\frac{d}{dX}\left[\rho(\Omega - u)s + \frac{\kappa}{T}\left(\frac{dT}{dX}\right)\right] = -\frac{1}{T^2}\left[\frac{4}{3}\mu T\left(\frac{du}{dX}\right)^2 + \kappa\left(\frac{dT}{dX}\right)^2\right].$$

The variation of s across the shock is obtained by integrating the above equation between the upstream state (1), corresponding to $X \to +\infty$, and the downstream state (2), for which $X \to -\infty$. As the solution must tend toward uniform states, the temperature derivative is cancelled out at the limit, which gives:

$$\boxed{\left[\rho(\Omega-u)s\right]_1^2 = \int_{-\infty}^{+\infty} \frac{1}{T^2}\left[\frac{4}{3}\mu T\left(\frac{du}{dX}\right)^2 + \kappa\left(\frac{dT}{dX}\right)^2\right] dX.}$$

This expression gives the entropy change caused by the shock wave: it is the result of the action of dissipative terms, namely, viscosity and thermal conduction. Furthermore, we see that:

$$\Delta s = s_2 - s_1$$

is always positive, the term under the sign $\int$ being strictly positive. This term also gives the sign of the change in specific entropy. It seems that the second law of thermodynamics is "bypassed" in this analysis. In fact it is used on the level of the constitutive laws for viscosity and conductivity where we impose that μ and κ are positive (momentum is transferred from high speed to lower speed regions and heat must flow from warmer to colder regions).

We state:

$$V = \Omega - u$$

which represents the fluid velocity relative to the shock. The system is rewritten as:

$$\rho V = Q$$

$$\rho v^2 + p + \frac{4}{3}\mu\frac{dV}{dX} = P$$

$$\rho v\left(h + \frac{V^2}{2}\right) + \frac{4}{3}\mu v\frac{dV}{dX} + \kappa\frac{dT}{dX} = E.$$

The last equation has been written with enthalpy h. For a calorically perfect gas:

$$h = C_p T = \frac{\gamma}{\gamma - 1} rT$$

leading to the following forms of the equations:

$$\frac{4}{3}\mu\frac{dV}{dX} = P - Q\left(V + \frac{\gamma-1}{\gamma}\frac{h}{v}\right) \quad [13.10]$$

$$\frac{\kappa}{C_p}\frac{dh}{dX} + \frac{4}{3}\mu V\frac{dv}{dX} = E - Q\left(h + \frac{1}{2}V^2\right). \quad [13.11]$$

Introducing the Prandtl number defined in section 6.2:

$$P_r = \frac{\mu C_p}{\kappa}$$

it follows that:

$$\frac{\mu}{\Pr}\frac{dh}{dX} + \frac{4}{3}\mu v\frac{dV}{dX} = E - Q\left(h + \frac{V^2}{2}\right). \quad [13.12]$$

Assuming that $P_r = 3/4 = 0.75$, which is a good approximation in the case of air, equation [13.12] becomes:

$$\frac{4}{3}\mu\frac{d}{dX}\left(h + \frac{V^2}{2}\right) = E - Q\left(h + \frac{V^2}{2}\right). \quad [13.13]$$

The stagnation enthalpy:

$$h_{st} = h + \frac{V^2}{2}$$

has to tend toward equal values on both sides of the shock:

$$h_{st} = h_1 + \frac{V_1^2}{2} = h_2 + \frac{V_2^2}{2}.$$

However, the general solution of [13.13] being of exponential form, the only remaining solution finite is:

$$h_{st} = h + \frac{V_2^2}{2} = \frac{E}{Q}.$$

The stagnation enthalpy does not vary, even across the shock wave, according to the hypothesis on the Prandtl number.

The above equations can be solved analytically using a method that we will not detail here. Thus, we find that the change of the velocity V in the shock layer is given by:

$$\frac{3}{4}\frac{Q}{\mu}\frac{\gamma+1}{2\gamma}X = \frac{V_2}{V_1 - V_2}\ln\left(V - V_2\right) - \frac{V_1}{V_1 - V_2}\ln\left(V_1 - V\right).$$

We can define the thickness of the shock by evaluating the abscissas X_1 and X_2 where V reaches the limit values V_1 and V_2 within a tolerance ε. Knowing that $Q = \rho_1 V_1$, this gives the thickness of the shock:

$$\delta X \propto \frac{4}{3}\frac{\mu}{\rho_1}\frac{2\gamma}{\gamma+1}\frac{1}{\left(V_1 - V_2\right)}.$$

We see that, the smaller the velocity difference $w = V_2 - V_1$ or the more viscous the fluid is, the wider spread out the shock will be. In contrast, the denser the fluid is, the thinner is the thickness. If λ is the mean molecular free path, we have (see section 4.1):

$$\frac{\mu}{\rho} = a\sqrt{\frac{2}{\gamma\pi}}\lambda$$

where a is the speed of sound; thus, for a given velocity difference:

$$\delta X \propto \lambda.$$

Under the conditions of classic aerodynamics, the shock thickness is less than one micrometer, and is thus much smaller than the characteristic dimension of the vehicle. Therefore, we can then consider the shock wave as a surface of discontinuity.

Specifically, for $\gamma = 1.4$:

$$\delta X \approx \frac{4}{3}a_1\sqrt{\frac{2}{\gamma\pi}}\frac{2\gamma}{\gamma+1}\frac{1}{\left(V_1 - V_2\right)}\lambda \approx 1.05\frac{a_1}{V_1 - V_2}\lambda.$$

For an intense shock: $V_1 - V_2 \approx V_1$, leading to:

$$\boxed{\delta X \sim \frac{\lambda}{M_1}}$$

with M_1 the Mach number: $M_1 = \dfrac{V_1}{a_1}$.

The mean free path at 20 km altitude is close to 10^{-6} m, and if the Mach number of the vehicle is 10, the thickness of the shock will be around 0.1 μm, thus zero for a vehicle with several meters length. In contrast, at 150 km, the mean free path is 10 m. For the same Mach number, we see that the thickness of the shock ($\delta X \approx 1$ m) is of the order of the vehicle size.

Figure 13.12 shows the result of a numerical integration of the shock structure equations for an upstream Mach number $M_1 = 2$ in the case of air. The abscissa, in values of the mean molecular free path λ, shows that for the conditions of this calculation, the thickness of the shock is about 7λ. We observe that the entropy reaches a maximum before tending toward a downstream level higher than the upstream level. Considering the case of steady shock, for the entropy equation is written as:

$$\rho u \frac{ds}{dx} = \frac{1}{T^2}\left[\frac{4}{3}\mu T\left(\frac{du}{dx}\right)^2 + \kappa\left(\frac{dT}{dx}\right)^2\right] + \frac{d}{dx}\left[\frac{\kappa}{T}\left(\frac{dT}{dx}\right)\right].$$

Referring to Figure 13.12, we find that at the beginning of the shock structure, the temperature increases in such a way that its first and second derivatives are positive. Around mid-shock, the temperature curve has an inflection before passing through a maximum and then decreasing until the final constant level. Thus, the last term of the expression of the derivative $\dfrac{ds}{dx}$ becomes negative leading to a decrease of s. The heat exchange occurring within the shock structure leads to a change of the local behavior of the entropy. Ultimately, the entropy ends up with a final level higher than the initial level. The jumps given by the Rankine–Hugoniot equations show a correct asymptotic behavior for the solution.

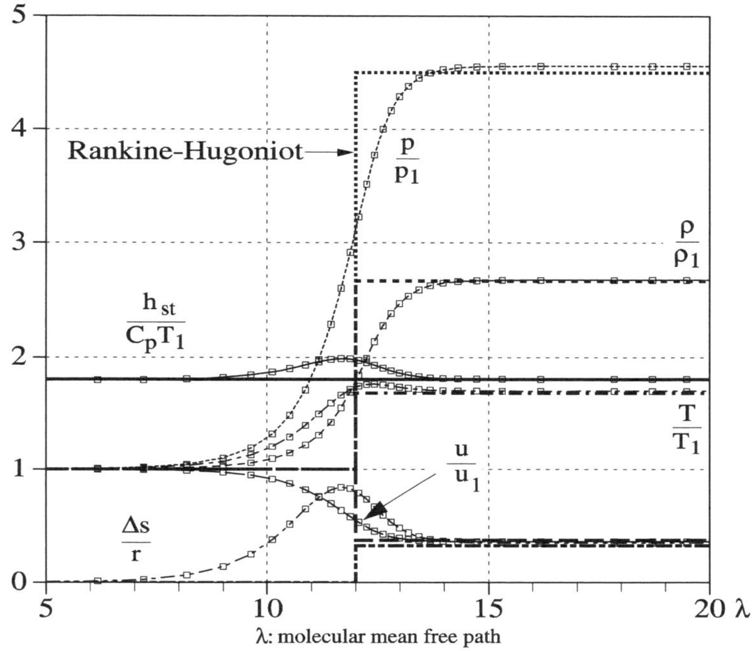

Figure 13.12. *Shock structure in air at Mach 2 and comparison with the Rankine–Hugoniot solution*

It can be shown that the ratio of λ to the characteristic length L of the vehicle is such that (see section 4.1):

$$\frac{\lambda}{L} \propto \frac{M_1}{\mathrm{Re}}$$

where Re is the Reynolds number: $\mathrm{Re} = \dfrac{\rho_1 V_1 L}{\mu_1}$. Therefore, the thickness of the shock varies as $\dfrac{1}{\mathrm{Re}}$, while the thickness $\dfrac{\delta}{L}$ of the boundary layer varies as $\dfrac{1}{\sqrt{\mathrm{Re}}}$, this last result being established within the framework of the boundary layer theory. Thus, the shock-like discontinuities, which are normal to the velocity, are thinner in the ratio $\frac{1}{\sqrt{\text{Re}}}$ than the slip line-like discontinuities affecting the tangential component of velocity.

The previous theory, based on the Navier–Stokes equations, leads to the result that the thickness of a shock wave is a few mean free molecular paths. However, the Navier–Stokes equations are based on the assumption of a continuous medium and consider objects whose size is large in comparison to λ (see Chapter 4). Thus, the above result has been obtained from theoretically inapplicable equations. It must be considered qualitative, giving only an order of magnitude of a shock thickness and showing how entropy is produced by a shock wave.

Chapter 14

Oblique Shock Wave and Shock Polar

14.1. General solution

Based on the results obtained for the normal shock, we will reconsider the problem of oblique shock (see Figure 13.3). We recall the equations given in the previous chapter:

$$\rho_1 V_{n_1} = \rho_2 V_{n_2}$$

$$V_{t_1} = V_{t_2}$$

$$p_1 + \rho_1 V_{n_1}^2 = p_2 + \rho_2 V_{n_2}^2$$

$$h_1 + \frac{V_{n_1}^2}{2} = h_2 + \frac{V_{n_2}^2}{2}.$$

We note that in these equations (except the second), the only intervention is by the normal components V_{n_1} and V_{n_2} of the upstream and downstream velocities. Thus, written with normal velocities, the oblique shock equations are identical to the equations of the normal shock. It is thus possible to use relations established for the normal shock under the condition that we consider the flow arriving normally at the shock wave. In particular, the upstream Mach number here is the *normal Mach number*, defined from the component V_{n_1} :

$$M_{n_1} = \frac{V_{n_1}}{a_1} = \frac{V_1 \sin\sigma}{a_1}$$

$$\boxed{M_{n_1} = M_1 \sin\sigma.}$$

In a similar way, the downstream Mach number M_2 is replaced by the normal Mach number, so that:

$$\boxed{M_{n_2} = M_2 \sin(\sigma - \varphi).}$$

We remember that σ is the shock angle with respect to the upstream velocity and φ the deflection of the velocity vector across the shock. Without repeating them all, the formulas giving the variations of various quantities become:

$$\frac{p_2}{p_1} = 1 + \frac{2\gamma}{1+\gamma}\left(M_1^2 \sin^2\sigma - 1\right)$$

$$\frac{\rho_2}{\rho_1} = \frac{(\gamma+1) M_1^2 \sin^2\sigma}{(\gamma-1) M_1^2 \sin^2\sigma + 2}.$$

The relation between the normal Mach numbers:

$$M_{n_2}^2 = \frac{1 + \frac{\gamma-1}{2} M_{n_1}^2}{\gamma M_{n_1}^2 - \left(\frac{\gamma-1}{2}\right)}$$

which gives:

$$M_2^2 \sin^2(\sigma - \varphi) = \frac{1 + \frac{\gamma-1}{2} M_1^2 \sin^2\sigma}{\gamma \; M_1^2 \sin^2\sigma - \left(\frac{\gamma-1}{2}\right)}$$

where the downstream normal Mach number M_{n_2} must be less than 1. On the other hand, the Mach number M_2 of the flow downstream of an oblique shock is *mostly supersonic*. It is only subsonic in a limited area corresponding roughly to strong shock-like solutions (see below).

Referring to Figure 13.3, which gives the decomposition of velocities, we obtain from elementary geometric considerations, since $V_{t_2} = V_{t_1}$:

$$\frac{\tan\ (\sigma-\varphi)}{\tan\sigma} = \frac{V_{t_1}}{V_{n_1}}\frac{V_{n_2}}{V_{t_2}} = \frac{V_{n_2}}{V_{n_1}}.$$

According to the continuity equation:

$$\frac{V_{n_2}}{V_{n_1}} = \frac{\rho_1}{\rho_2}$$

from which follows the relation:

$$\boxed{\frac{\tan\ (\sigma-\varphi)}{\tan\sigma} = \left(\frac{\gamma-1}{\gamma+1}\right)\left[1+\frac{2}{(\gamma-1)M_1^2\sin^2\sigma}\right].}$$

The above equation allows calculation of the shock angle σ for a given deflection φ. Knowing the upstream flow, as well as σ and φ, all properties of the downstream flow can thus be determined.

In practice, the problem arises most often in the following way. We know a uniform supersonic flow with the Mach number M_1 (as well as all its properties) flowing along a straight wall. At a point *A*, this wall has a discontinuity with the slope $\Delta\varphi = \varphi$ (see Figure 14.1). The problem: calculating the flow that is established along the wall downstream of *A*. We show in Chapter 17 that a solution made of a progressive compression wave type leads to a flow unacceptable for physical reasons. We seek here a solution in the form of a uniform flow (2) with the Mach number M_2 parallel to the downstream wall, thus satisfying the slip condition, and separated from the upstream flow (1) by an oblique planar shock wave.

Knowing M_1 and φ, the first step of the calculation is to solve the equation linking σ and φ. By stating $x = \sin\sigma$, it is easy to show that the equation for the shock angle is of the third degree in $\sin^2\sigma$ and may have three real solutions. The strict problem of fluid mechanics has three solutions. The angle σ having been calculated, the oblique shock relations then enables us to determine all the properties of the flow (2): the solution is fully defined.

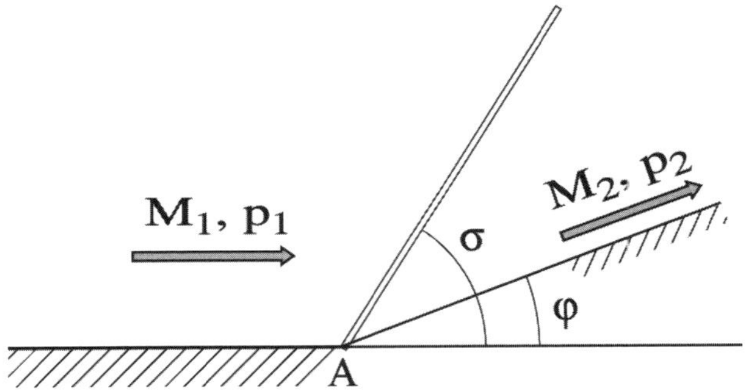

Figure 14.1. *Attached shock wave created by a wedge*

14.2. Shock polar

14.2.1. *Polar in the plane (pressure – deflection angle)*

It is instructive to discuss the nature of solutions to oblique shock equations using a *hodographic representation* where the flow is characterized by the dependent variables which can be the velocity components, or the modulus and the direction of the velocity, or any other equivalent couple of quantities. Assuming the stagnation conditions are known (they usually are data of the problem), two variables suffice to completely define the flow. To clarify, if the pressure p is known, we can calculate the ratio p/p_{st}, which, thus, according to the relations concerning isentropic evolutions (see section 9.5.2), leads to the Mach number M, the ratios ρ/ρ_{st}, which leads to ρ, and T/T_{st}, which leads to T, and thus to the modulus of the velocity V. In addition to the pressure p, we also have to know the direction φ of the velocity vector. This is true as long as the flow is isentropic (giving stagnation conditions is equivalent to giving the entropy). Across a shock wave, there is a change of entropy which leads to a new stagnation pressure. Downstream of the shock, and until the following shock, the hodographic representation with two variables becomes again valid.

Studying the properties of the oblique shock, we use the pressure p and the direction of the velocity φ as hodographic variables, the upstream flow (1) has the hodographic coordinates p_1 and φ_1. Thereafter, we take the direction φ_1 as a reference by stating $\varphi_1 = 0$, which is not a restriction. The flow (1) of the physical plane will therefore have as *image* in the hodographic plane the point (1) of coordinates $[p_1, \varphi_1]$ or $[p_1, 0]$.

In the plane $[p, \varphi]$, the states (2) linked to (1) by the oblique shock equations have as an image the points (2) of coordinates $[p_2, \varphi_2, = \varphi]$. At a deflection φ, we correspond to three image points, as the oblique shock equations can have three real roots. When we change the value of φ, the image points move on a curve $[\Gamma_1]$ called *shock polar*. Such a polar is drawn in Figure 14.2. It has the shape of a closed curve, symmetric with respect to the axis $\varphi = 0$, two opposite values of φ leading to shocks of equal intensity, one moving "up" the other "down". The particular conditions of the problem, for example, the sense of the deflection imparted by the wall, get imposed on either shock.

To a value of φ thus corresponds three points of the polar (Γ_1): 1, 2, and 3 images of the three possible solutions. We can show that the solution 3, for which the pressure is lower than the pressure before the shock p_1 (expansion shock), involves a decrease of entropy. This solution is rejected because it violates the second law of thermodynamics.

Both alternatives 1 and 2 are physically eligible:

– solution 1 is associated with the smallest pressure jump: because of this it is called a *weak shock* solution,

– solution 2 corresponds to the highest pressure jump: which leads to it being called a *strong shock* solution.

Two solutions are associated with the value $\varphi = 0$ (no deflection):

– One coincides with the origin point which is the image of the infinitely small shock, reducing it to a Mach wave (see Chapter 17). By making φ tends toward 0, the equation for the shock angle takes the limit form:

$$1 = \left(\frac{\gamma - 1}{\gamma + 1}\right)\left[1 + \frac{2}{(\gamma - 1) M_1^2 \sin^2 \sigma}\right]$$

which, after simple calculations, leads to:

$$M_1^2 \sin^2 \sigma = 1 \rightarrow \sigma = \sin^{-1} \frac{1}{M_1}.$$

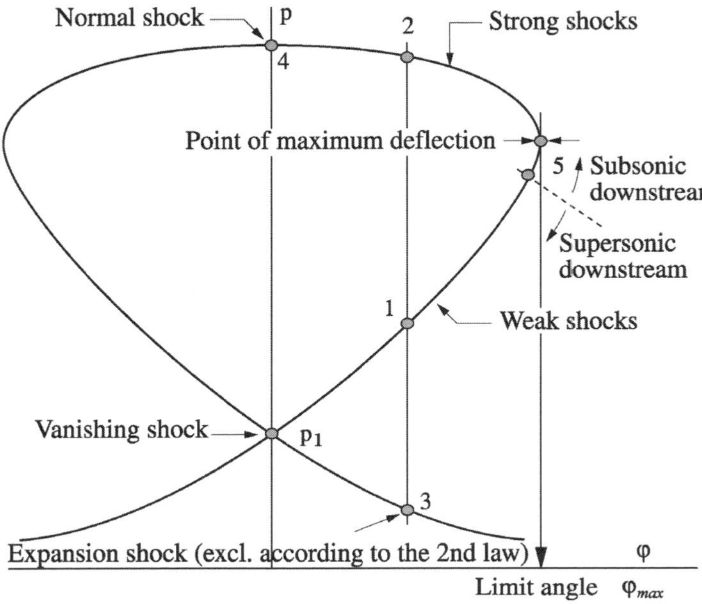

Figure 14.2. *Shock polar in the plane* $[p, \varphi]$

The shock angle tends toward the Mach angle α_1, which is the angle of propagation of a wave, i.e. a disturbance whose intensity is very small and gradually vanishing.

– The other, point 4, is the normal shock causing the most important pressure jump, the other limit for $\varphi \to 0$ being $\sigma = \dfrac{\pi}{2}$.

The shock polar highlights the following property: there is a *maximum deflection* $\varphi_{\max}$ beyond which the equations of oblique shocks have no real (admissible) solutions. This does not mean that the non-viscous fluid problem does not have a solution. It means that if $\varphi > \varphi_{\max}$, the above problem no longer allows an oblique planar shock wave, attached in *A*, as a solution. As shown in Figure 14.3, if $\varphi > \varphi_{\max}$, the shock wave (C_1) is formed upstream of *A* and becomes curved: we say that the shock is *detached*. First, normal to the wall, the shock bends gradually to become almost straight at a distance of *A*, depending on the value of φ and the upstream Mach number. The flow downstream of (C_1) is not uniform any longer; in particular, a subsonic pocket is formed near the wall. The calculation of this solution is more complex than in the previous case. Then, we have to use numerical techniques for solving Euler equations. However, the shock relations apply locally at each point of the curved shock.

The shock angle $\sigma_{\max}$ corresponding to the deflection $\varphi_{\max}$ is given by:

$$\sin^2 \sigma_{\max} = \frac{1}{4\ \gamma\ M_1^2}\left\{(\gamma+1)M_1^2 - 4 + \sqrt{(\gamma+1)[(\gamma+1)M_1^4 + 8(\gamma-1)M_1^2 + 16]}\right\}.$$

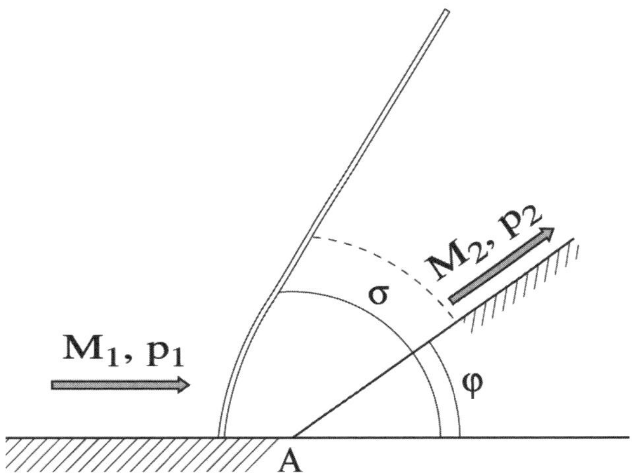

Figure 14.3. *Detached shock wave created by a wedge*

The part of the polar between the origin 0 and point 5, the image of the maximum deflection, corresponds to weak oblique shocks, the part between 5 and the normal shock 4, to strong oblique shocks. When we follow the polar from the point of origin 0 to point 4, which is the image of the normal shock, the Mach number of the downstream flow changes from supersonic to subsonic. On the polar, there is a point where the downstream state of the shock is just sonic. It corresponds to a shock angle σ_{son}, given by:

$$\sin^2 \sigma_{son} = \frac{1}{4\ \gamma\ M_1^2}\left\{(\gamma+1)M_1^2 - (3-\gamma) + \sqrt{(\gamma+1)\left[(\gamma+1)M_1^4 - 2(3-\gamma)M_1^2 + (\gamma+9)\right]}\right\}.$$

The representative points of the maximum deflection and sonic downstream state are very close to each other on the polar.

For the given γ, the maximum deflection φ_{max} depends only on the upstream Mach number M_1 and increases with M_1. The curves in Figure 14.4 show the evolution of φ_{max} and the corresponding shock angle σ_{max} as a function of the Mach number M_1 for $\gamma = 1.4$. We note that φ_{max} varies rapidly between $M_1 = 1$ (then $\varphi_{max} = 0$) and $M_1 \approx 6$. Beyond that, the increase becomes more gradual, φ_{max} tending toward a finite limit when $M_1 \to \infty$. The variation of σ_{max} is similar: first, σ_{max} decreases rapidly from the value $\sigma_{max} = \frac{\pi}{2}$, corresponding to the shock with zero intensity for $M_1 = 1$ (then, σ is equal to the Mach angle with the value $\frac{\pi}{2}$), passes through a minimum, and then, for $M_1 \to \infty$, tends to a limit such that:

$$\sin^2 \sigma_{max} = \frac{\gamma + 1}{2\gamma}$$

which gives: $\left(\sigma_{max}\right)_{lim} = 67.792° \quad for \ \gamma = 1.4.$

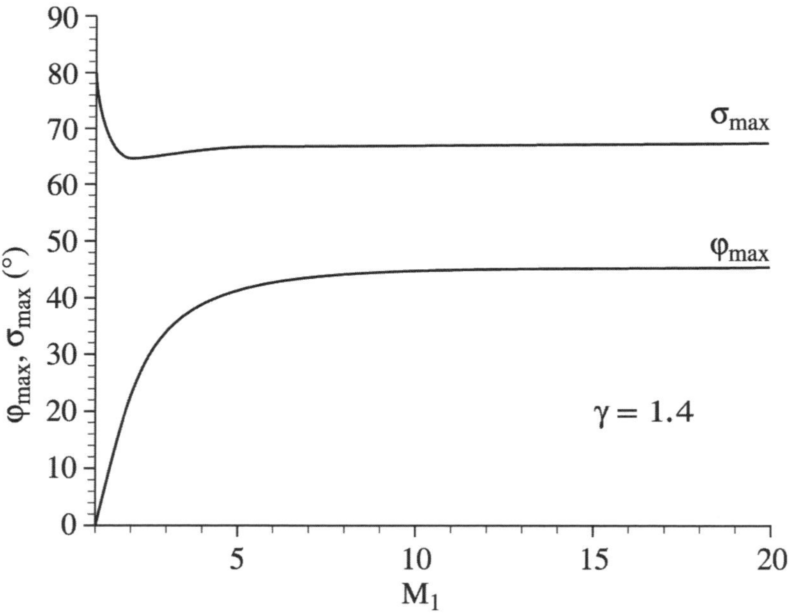

Figure 14.4. *Limit deflection angle and corresponding shock angle*

We deduce the limit value of φ_{max} from the limit form of the equation for the shock angle when $M_1 \to \infty$:

$$\tan(\sigma-\varphi)=\left(\frac{\gamma-1}{\gamma+1}\right)\tan\sigma$$

which gives: $\left(\varphi_{max}\right)_{lim} = 45.585° \ \ for \ \ \gamma = 1.4.$

There is no explicit relation for φ_{max} ; an approximate value can be obtained from the relation:

$$\varphi_{max} \cong \frac{4}{3\sqrt{3}(\gamma+1)}\frac{\left(M_1^2-1\right)^{3/2}}{M_1^2}.$$

The values of φ_{max} given by this relation are plotted in Figure 14.5. We note that the approximation is only good for $M_1 < 2$; beyond, it becomes grossly wrong.

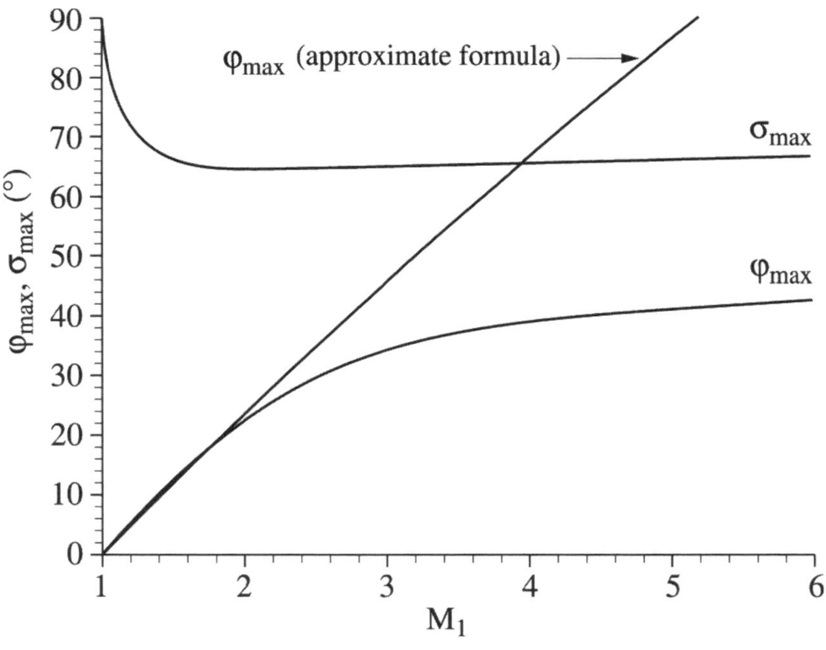

Figure 14.5. *Limit deflection angle and corresponding shock angle (Reduced upstream Mach number range)*

We also encounter the cases of the shock attached or detached to the nose of an axisymmetric supersonic vehicle. If the nose, or ogive, is conical, a *conical flow* forms around it which has the property of being invariant over straight lines coming from the apex of the cone. The shock is also conical and is calculated using the method described in section 13.7. If the nose is sharp enough, the shock is attached to the tip as shown in Figure 14.6a. If the angle of the nose exceeds the limit deflection angle, then the shock is detached and has a curved part, as shown in Figure 14.6b. The detachment also occurs if, for a fixed ogive angle, the upstream Mach number decreases.

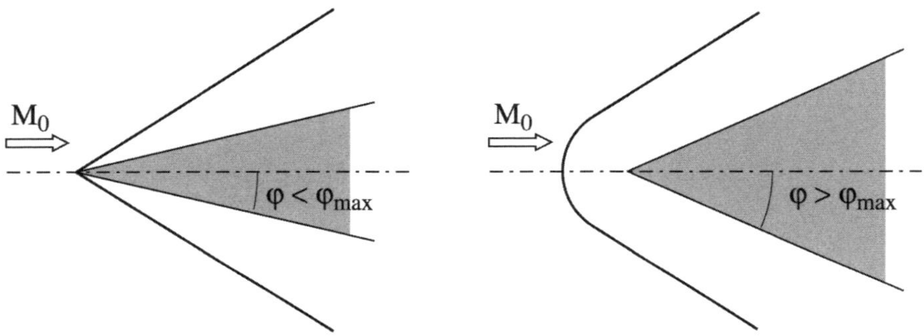

a – attached shock *b – detached shock*

Figure 14.6. *Shock wave ahead of a sharp obstacle*

Figure 14.7 shows a shadowgraph visualization of the flow around a projectile in flight in a ballistic range. Because of the drag, the velocity of the projectile decreases during the trajectory. For the image a, the Mach number of the flight is around 6. Then, the shock is attached, its angle being close to that of the conical tip. For the image b, the Mach number is 2.05 and the shock is about to detach. For the image c, the Mach number is close to 1.24 and the shock is clearly detached. Here, we should consider the axisymmetry effect on the shape of the shock, the relations established above staying valid but locally.

The hodographic representation of a flow in the plane $[p,\varphi]$ is of great interest for discussing the problems of reflection and intersection of shock waves, also called shock–shock interference. Indeed, these situations lead to the formation of flows with different properties and separated by *slip lines*. As noted in section 13.4, such contact discontinuities (Σ), separating two flows (E) and (E'), are also governed by the Rankine–Hugoniot equations. The compatibility of (E) and (E') results in equal directions (the velocity vectors are parallel on each side of (Σ)) and pressures. These conditions imply the coincidence of the image points of (E) and (E') in the hodographic plane $[p,\varphi]$.

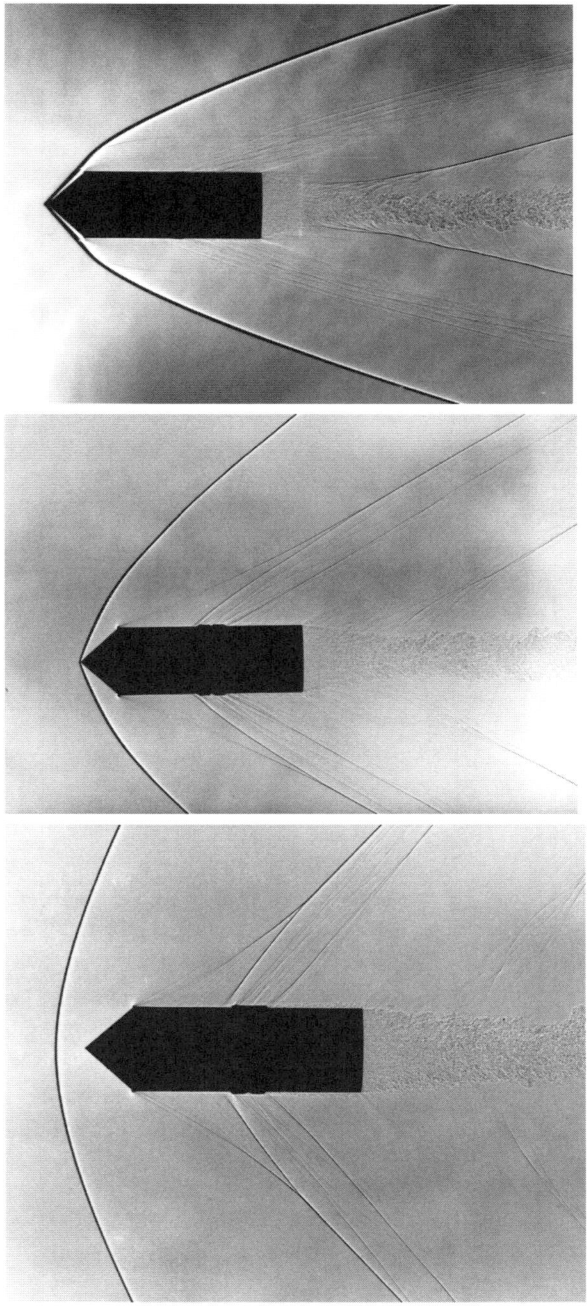

$M_0 \approx 0$

$M_0 = 2.05$

$M_0 = 1.24$

Figure 14.7. *Shadowgraph visualization of a projectile in flight in a ballistic range (document ISL – Franco-German Institute of Saint-Louis)*

Figure 14.8 shows a set of polars $[p,\varphi]$ for upstream Mach numbers M_1 varying in the range 2–10. We note that the maximum pressure jump corresponding to the normal shock increases rapidly and regularly with M_1, p_2/p_1 being equal to 114.4 for $M_1 = 10$ (if $\gamma = 1.4$). The maximum deflection angle φ_{max} is higher the larger the Mach number M_1 is, the detachment of the shock occurring as soon as $\varphi_{max} = 22.974°$ for $M_1 = 2$. Then, φ_{max} first increases rapidly with the Mach number, the growth becoming very gradual beyond $M_1 = 5$ (see Figure 14.4).

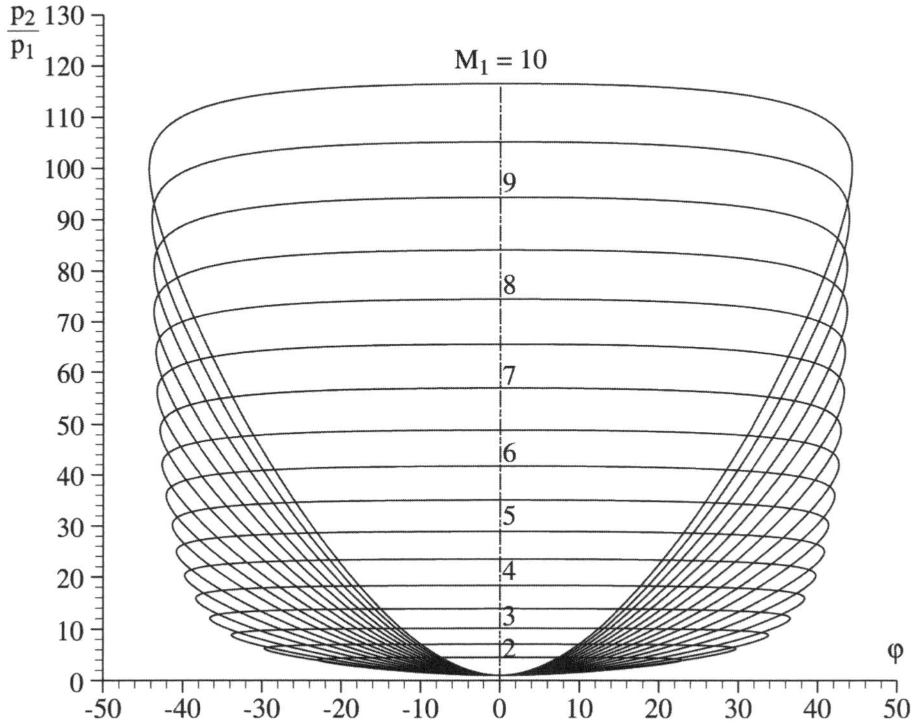

Figure 14.8. *Shock polars in the plane* $[p,\varphi]$ *for variable upstream Mach number (case of air:* $\gamma = 1.4$ *)*

14.2.2. *Polar in the plane (deflection angle – shock angle)*

We also use the shock polar in the plane (deflection angle – shock angle) constructed by plotting the shock angle σ as a function of φ for given upstream Mach numbers (with fixed γ). Such a polar $[\sigma,\varphi]$ is also a closed curve whose one-half is plotted (see Figure 14.9). This polar highlights weak and strong shock solutions, as well as the limit deflection φ_{max} which corresponds to a shock angle σ_{max}. We also note in Figure 14.9 that the point for the sonic downstream state is very close to the limit deflection. For $\varphi = 0$, the two solution points correspond, one to the Mach angle which is the direction of propagation of an infinitesimal disturbance, the other to the normal shock.

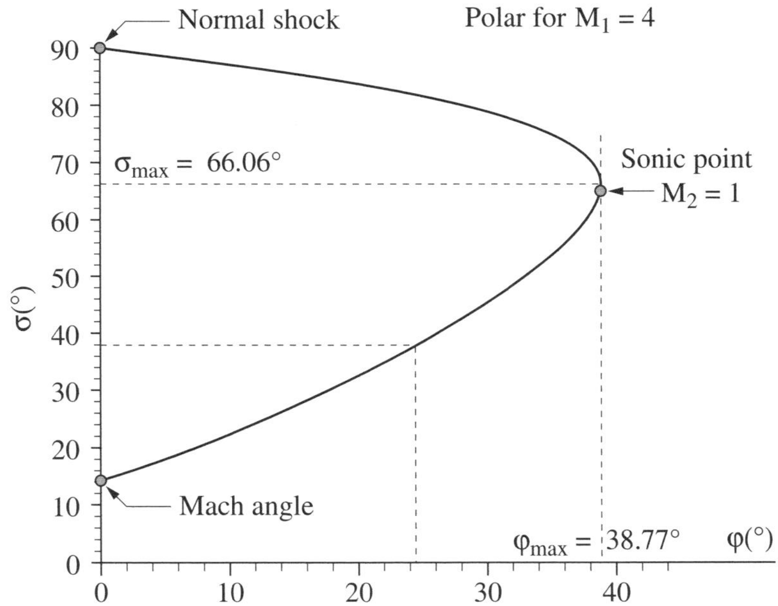

Figure 14.9. *Shock polars in the plane* $[\varphi, \sigma]$ *for an upstream Mach number* $M_1 = 4$ *(case of air:* $\gamma = 1.4$*)*

In Figure 14.10 is plotted a set of polars $[\varphi, \sigma]$ for M_1 taking values ranging from 1.05 to infinity. On this graph are also shown the locations of the sonic and maximum deflection points. We verify that these two points are close to each other, the sonic downstream state being found before the maximum deflection. Often, a graph of this type is provided along with shock tables giving the jump of quantities across a normal shock. For given values of the Mach number M_1 and the deflection φ, it indicates the corresponding shock angle σ. We deduce the normal Mach number $M_{n_1} = M_1 \sin\sigma$ from it, thereby entering into the normal shock tables that will give all the conditions after the shock.

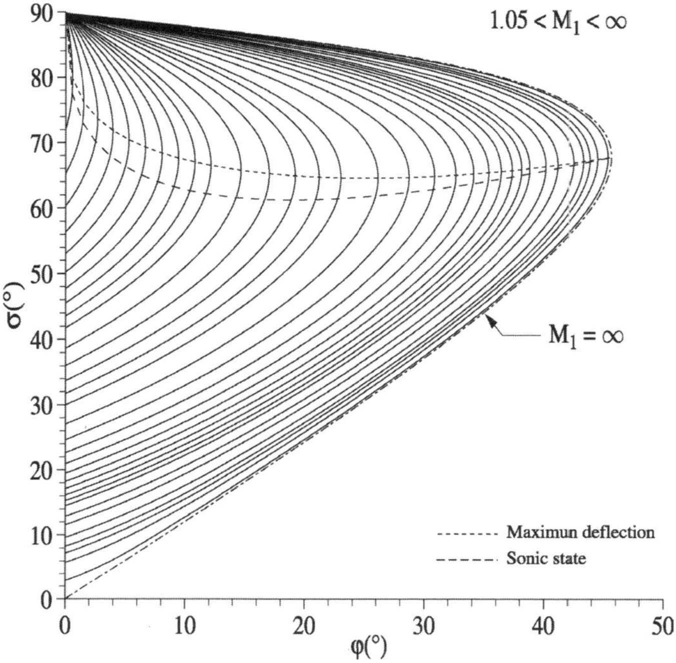

Figure 14.10. *Shock polars in the plane $[\varphi,\sigma]$ for different values of the upstream Mach number (case of air: $\gamma = 1.4$)*

When the Mach number tends to infinity, the equation of the oblique shock:

$$\frac{\tan(\sigma-\varphi)}{\tan\sigma}=\left(\frac{\gamma-1}{\gamma+1}\right)\left[1+\frac{2}{(\gamma-1)M_1^2\sin^2\sigma}\right]$$

takes the limit form:

$$\frac{tg(\sigma-\varphi)}{tg\sigma}=\left(\frac{\gamma-1}{\gamma+1}\right)$$

which defines the limit polar of the plane $[\varphi,\sigma]$. When $\varphi\to 0$, this relation becomes:

$$\tan\sigma=\left(\frac{\gamma-1}{\gamma+1}\right)\tan\sigma$$

which is satisfied for $\tan\sigma = 0$ and $\tan \rightarrow \infty$. Hence, either $\sigma = 0$, which is the limit value of the Mach angle when the Mach number tends toward infinity, or $\sigma = \frac{\pi}{2}$ corresponding to the normal shock.

14.3. Other properties of the shock polar (pressure – deflection angle)

We evaluate the slope $(dp/d\varphi)_0$ at the origin of the shock and for that we state:

$$\varepsilon = \frac{\rho_1}{\rho_2} = \frac{\tan(\sigma - \varphi)}{\tan\sigma} \qquad [14.1]$$

where we introduce the parameter ε that will measure the intensity of the shock. The normal momentum equation is written as:

$$p_2 = p_1 + \rho_1 V_1^2 \sin^2\sigma - \rho_2 V_2^2 \sin^2(\sigma - \varphi) = p_1 + \rho_1 V_1^2 \sin^2\sigma \left[1 - \frac{\rho_2 V_2^2 \sin^2(\sigma - \varphi)}{\rho_1 V_1^2 \sin^2\sigma}\right]$$

which leads to:

$$p_2 = p_1 + \rho_1 V_1^2 \sin^2\sigma (1 - \varepsilon).$$

The energy equation takes the form:

$$h_2 = h_1 + \frac{1}{2} V_1^2 \sin^2\sigma \left[1 - \frac{V_2^2 \sin^2(\sigma - \varphi)}{V_1^2 \sin^2\sigma}\right].$$

Hence:

$$h_2 = h_1 + \frac{1}{2} V_1^2 \sin^2\sigma (1 - \varepsilon^2).$$

Assuming:

$$p_2 = p_1 + \rho_1 V_1^2 \sin^2\sigma (1 - \varepsilon)$$

for the limit of the vanishing shock: $\varphi \to 0$, $\sigma \to \alpha_1$:

$$p_2 = p_1 + \rho_1 V_1^2 \sin^2 \sigma (1-\varepsilon) \to dp = -\rho_1 V_1^2 \sin^2 \alpha_1 \, d\varepsilon$$

which leads to:

$$\frac{dp}{d\varepsilon} = -\rho_1 V_1^2 \sin^2 \alpha_1. \qquad [14.2]$$

According to the energy equation: $dh = -V_1^2 \sin^2 \alpha_1 \varepsilon d\varepsilon$

$$\to \quad \frac{dh}{d\varepsilon} = -V_1^2 \sin^2 \alpha_1 \ \ \varepsilon.$$

Differentiating [14.1] in the limit conditions, we obtain:

$$\frac{-d\varphi}{\cos^2 \alpha_1} \frac{\cos \alpha_1}{\sin \alpha_1} = d\varepsilon$$

which leads to: $\dfrac{d\varphi}{d\varepsilon} = -\cos \alpha_1 \ \ \sin \alpha_1.$

Taking into account [14.2], it becomes:

$$\frac{dp}{d\varphi} = \frac{\rho_1 V_1^2 \sin^2 \alpha_1}{\sin \alpha_1 \cos \alpha_1} = \rho_1 V_1^2 \tan \alpha_1.$$

By replacing $\rho_1 V_1^2$ with $\rho_1 a_1^2 M_1^2 = \dfrac{\rho_1 a_1^2}{\sin^2 \alpha_1}$:

$$\frac{dp}{d\varphi} = \rho_1 V_1^2 \tan \alpha_1 = \frac{\rho_1 a_1^2}{\sin \alpha_1 \cos \alpha_1}.$$

The extrema of this derivative as a function of α_1 are obtained when:

$$\frac{d}{d\alpha_1}\left(\frac{dp}{d\varphi}\right) = 0 \ \ with \ \ \frac{d}{d\alpha_1}\left(\frac{dp}{d\varphi}\right) = -\rho_1 a_1^2 \frac{\cos^2 \alpha_1 - \sin^2 \alpha_1}{(\sin \alpha_1 \cos \alpha_1)^2}.$$

We verify that this derivative is zero when $\alpha_1 = \dfrac{\pi}{4}$.

As $\sin\alpha_1 = \dfrac{1}{\sqrt{2}}$, it follows that: $\boxed{M_1 = \sqrt{2}.}$

We can also verify that at the origin of the shock polar:

$$\frac{dp}{d\varphi} \to \infty \text{ if } \alpha_1 = \frac{\pi}{2} \; (M_1 = 1) \quad \text{and} \quad \alpha_1 = 0 \; (M_1 \to \infty.)$$

The slope $(dp/d\varphi)_{(0)}$ of the shock polars at the origin passes through a minimum $(dp/d\varphi)_{\min}$ for the upstream Mach number $M_1 = \sqrt{2} = 1.414$, with $(dp/d\varphi)_0 > (dp/d\varphi)_{\min}$ for $M_1 > \sqrt{2}$ and for $M_1 < \sqrt{2}$. Therefore, when $M_1 > \sqrt{2}$, the successive polars are each above the previous when the Mach number increases. This order changes when $M_1 < \sqrt{2}$, the polars for lower Mach numbers is above the previous one (see Figure 14.11).

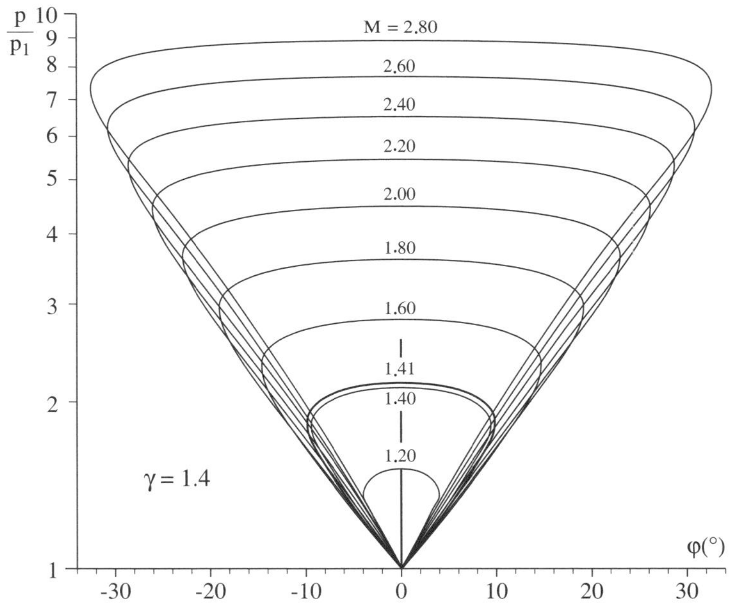

Figure 14.11. *Relative positions of shock polars in the vicinity of the polar of minimum slope at the origin*

Knowing that $\gamma p_1 = \rho_1 a_1^2$:

$$\frac{1}{p}\frac{dp}{d\varphi} = \frac{d(\ln p)}{d\varphi} = \frac{\gamma}{\sin\alpha_1 \cos\alpha_1}.$$

If the shock polar is represented in the semi-logarithmic coordinates $[\ln p, \varphi]$, then, for $M_1 = \sqrt{2}$, the slope of the polar at the origin is such that:

$$\frac{1}{p}\frac{dp}{d\varphi} = \frac{d(\ln p)}{d\varphi} = 2\gamma$$

which is the minimum value.

It is also possible to represent an isentropic evolution (expansion or compression) by an isentropic polar in the plane $[\varphi, p]$. For a two-dimensional supersonic flow, such a curve is represented by the equation (see Chapter 17):

$$\omega(M,\gamma) \pm \varphi = constant$$

where $\omega(M,\gamma)$ is the Busemann number:

$$\omega(M,\gamma) = \sqrt{\frac{\gamma+1}{\gamma-1}}\ \tan^{-1}\sqrt{\frac{\gamma-1}{\gamma+1}(M^2-1)} - \tan^{-1}\sqrt{M^2-1}.$$

We can show that the shock polar and isentropic polar have a third-order contact at their origin, thus they remain very close to one another until relatively high deflections. For moderate Mach numbers, the solutions of the weak shock type can be regarded as practically isentropic. The isentropic polar is below the shock polar for upstream Mach numbers close to 2.34. It passes above for higher Mach numbers (see Figure 14.12).

By using the thermodynamic relation:

$$Tds = dh - \frac{dp}{\rho}$$

we can write:

$$T\frac{ds}{d\varepsilon} = \frac{dh}{d\varepsilon} - \frac{1}{\rho}\frac{dp}{d\varepsilon}$$

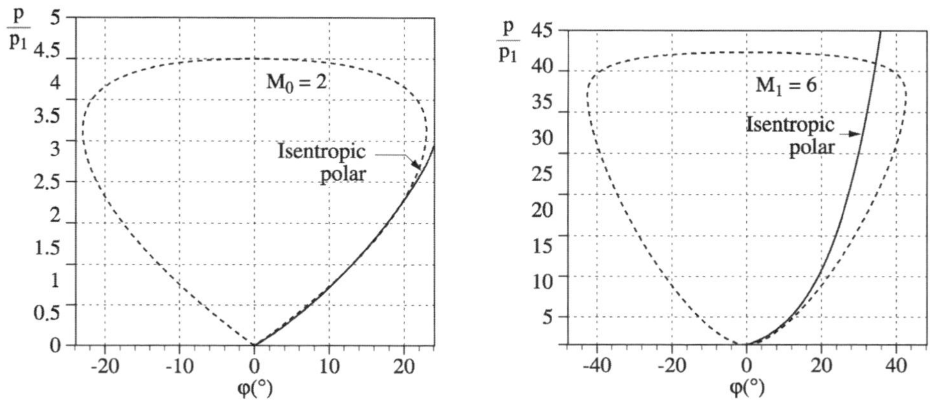

a – the isentropic polar is below the shock polar

b – the isentropic polar is above the shock polar

Figure 14.12. *Relative position of the shock polar and the isentropic polar*

within the present case (see above):

$$\frac{dp}{d\varepsilon} = -\rho_1 V_1^2 \sin^2 \alpha_1.$$

In addition:

$$h_2 = h_1 + \frac{1}{2} V_1^2 \sin^2 \sigma (1 - \varepsilon^2)$$

from which follows under the limit condition $\varepsilon \to 1$:

$$\frac{dh}{d\varepsilon} = -V_1^2 \sin^2 \sigma.$$

Thus:

$$T \frac{ds}{d\varepsilon} = \frac{dh}{d\varepsilon} - \frac{1}{\rho} \frac{dp}{d\varepsilon} = -V_1^2 \sin^2 \alpha_1 + V_1^2 \sin^2 \alpha_1 = 0.$$

At the considered order of magnitude, within the limit $\varepsilon \to 1$, the shock is an isentropic process. This explains the proximity of the polars near their origin.

14.4. The oblique planar shock wave reflection problem

We consider an oblique incident planar shock wave (C_1) propagating in a uniform flow (1) with the velocity $\vec{V}_1$ where it causes a negative deflection $\Delta\varphi = \varphi_2$. It impacts at a point *I* a rectilinear wall parallel to $\vec{V}_1$. The problem is to determine the flow downstream of *I*. As the direction φ_2 of the flow (2), downstream of (C_1) is not compatible with the direction $\varphi = 0$ of the wall, a shock wave (C_2) inducing a deflection $\Delta\varphi = -\varphi_2$ has to originate from *I* in such a way as to turn the flow (2) in a flow (3) such that $\varphi_3 = 0$. The resulting flow is represented in the physical plane in Figure 14.13. Figure 14.14 shows the situation in the hodographic plane, (Γ_1) being the polar associated with the state (1) and (Γ_2) the one associated with the state (2). The image of the flow (3) is at the intersection of (Γ_2) with the axis $\varphi = 0$.

A reflection of that type, shown in Figure 14.13, is called a *regular reflection*.

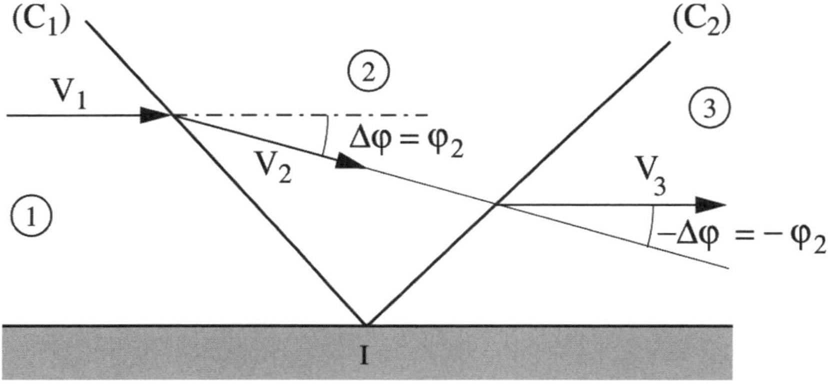

Figure 14.13. *Regular reflection of an oblique shock wave and image in the physical plane*

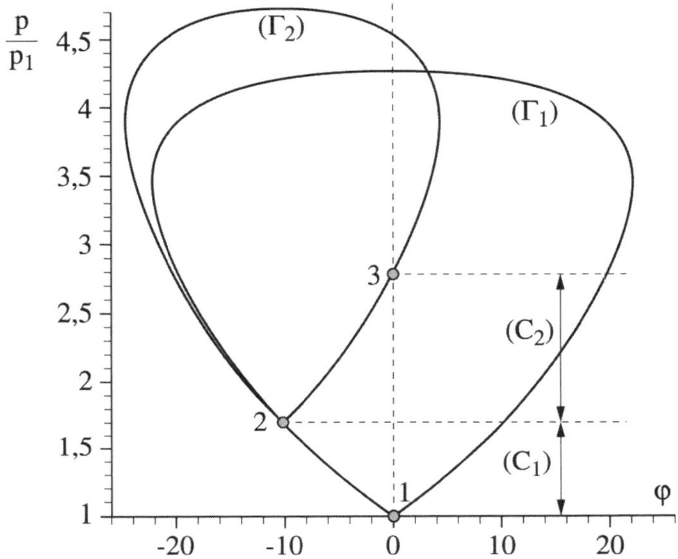

Figure 14.14. *Regular reflection of an oblique shock wave for $M_1 = 2$ and $\varphi = -10°$ and image in the plane of polars*

The hodographic representation shows us that if the deflection φ_2 increases, the intensity of the incident shock (C_1) increases. So that a situation can be reached when the polar (Γ_2) no longer intersects the axis $\varphi = 0$ (see Figure 14.15). From now on, the polar (Γ_2) intersects (Γ_1) [representing the incident shock (C_1)] at a point 3 which thus is the image of the flow resulting from the propagation of the "reflected" shock (C_2) through the flow (2). But, here, the reflection point of the shock (C_1) can no longer be at the wall because, as seen on the polars, the angle after (C_2) is not zero but negative. The construction that has been made does not satisfy the slip boundary condition on the wall. The solution takes the form drawn in Figure 14.16 to which corresponds the hodographic image of Figure 14.17. Within the flow exists a triple point *T*, which is a pseudoreflection point of (C_1), where four states coexist:

– the upstream flow (1),

– the flow (2), separated from (1) by the shock wave (C_1), which has the image point 2 in the plane $[p, \varphi]$,

– the flow (3), separated from (2) by the shock wave (C_2), whose polar (Γ_2) intersects the polar (Γ_1) at the point 3, image of (3). The direction of the velocity in (3) is different from zero, thus incompatible with the one on the wall,

– the flow (4), separated from (1) by the shock wave (C_3), whose image 4 is located on the polar (Γ_1).

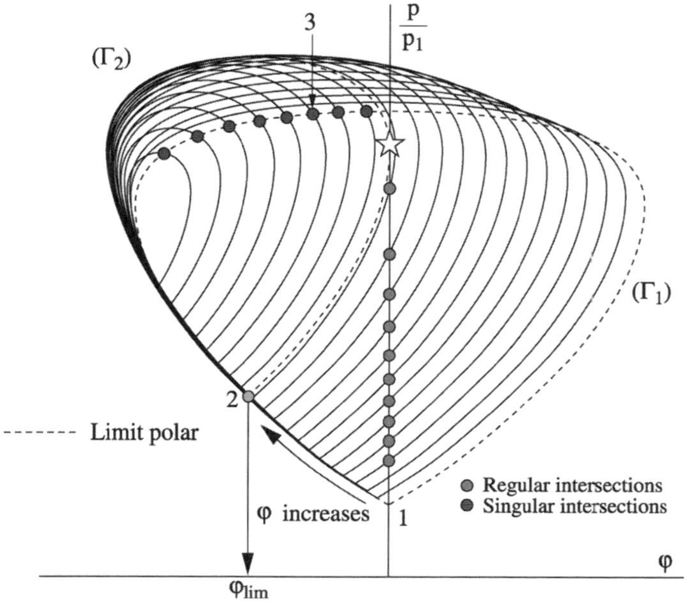

Figure 14.15. *Transition from regular reflection to singular reflection*

Therefore, we build a solution where two different flows, (3) and (4), coexist downstream of *T*. We know that such a circumstance is possible if (3) and (4) are separated by a slip line-like discontinuity, the conditions to be met being: same direction and equal pressures. For (3) and (4) to be compatible, it is necessary that their images are coincident, implying that the images 3 and 4 are at the intersection of the polars (Γ_1) and (Γ_2). As the flows (3) and (4) result from the crossing of different shocks, their entropies are different, and hence also their velocities, densities, etc. They have to be separated by the slip line (Σ). The emergence of a triple point with the formation of a nearly-normal shock close to the wall is called a *Mach phenomenon*. We note that 4 is found on the strong shocks branch on (Γ_1), the shock (C_3) being almost normal in *T*. In fact, it is slightly curved between *T* and the wall it meets in *H* where it has to be normal, since in *H* the deflection is zero. Between *T* and *H*, the shock (C_3) has as image the part of the polar (Γ_1) comprised between 4 and 5, the image of the flow being downstream of *H*.

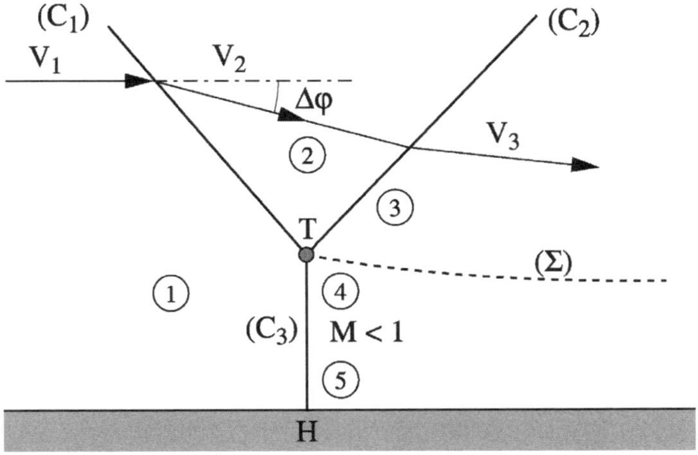

Figure 14.16. *Singular reflection of an oblique shock wave, Mach phenomenon, and image in the physical plane*

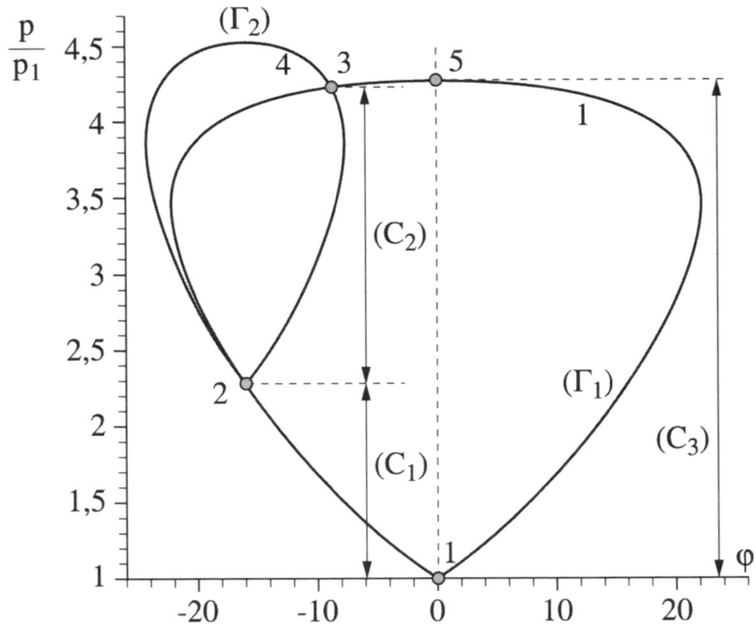

Figure 14.17. *Singular reflection of an oblique shock wave, Mach phenomenon for $M_1 = 2$ and $\varphi = -16°$, and image in the polar plane*

Downstream of (C_3) the flow is subsonic, thus the solution depends on conditions imposed downstream. It is therefore no longer possible to construct a flow in as simple a way as the previous case. In particular, if we can discuss the organization of the shock system that has to form, it is impossible to determine the position of the point T. In addition, as the slip line (Σ) is generally curved, the flow in region (3) is not uniform; thus the shock (C_2) is not rectilinear. The calculation of the solution requires the use of numerical methods that will allow taking into account the existence of a region with subsonic flow.

Shock intersection and reflection. The problem of the reflection of a shock wave on a rectilinear wall is identical to that of the intersection of two symmetric oblique waves, as drawn in Figure 14.18. The plane of symmetry of this situation plays the role of the wall in the case that has been analyzed. This is true in a non-viscous fluid, since the slip condition imposed on a wall is here identical to the condition of symmetry. This would not be true in a viscous fluid where the presence of a boundary layer on the wall may significantly alter the phenomenon of shock reflection.

The schlieren picture in Figure 14.19 shows a regular intersection of two symmetric shocks induced by deflections of +10° and −10° in a flow at Mach number 1.95. The slip line starting from the intersection point (a line that should not exist) is due to a slight asymmetry in the configuration.

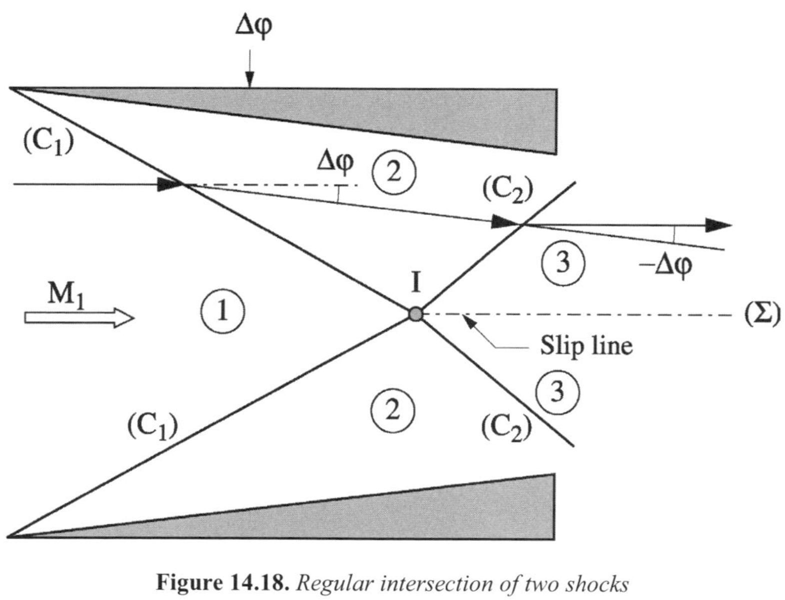

Figure 14.18. *Regular intersection of two shocks*

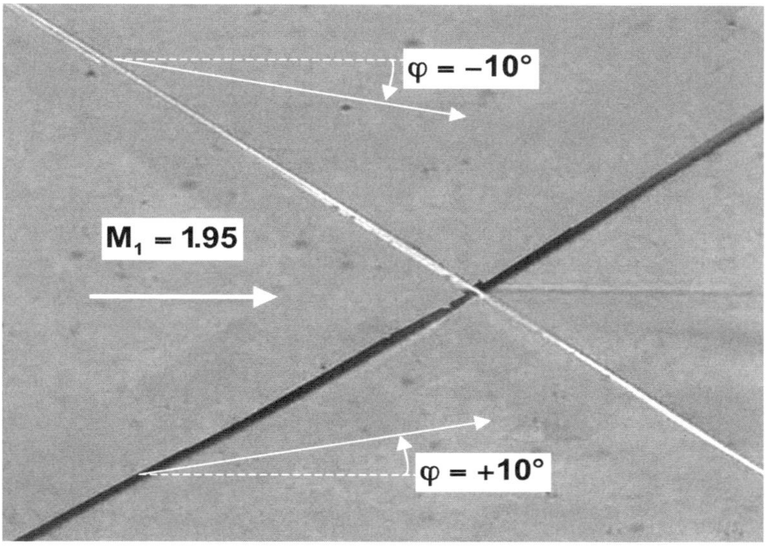

Figure 14.19. *Schlieren visualization of the regular intersection of two symmetric shock waves: $M_1 = 1.95, \varphi = -10°$ and $\varphi = +10°$ (document Onera)*

Figure 14.20 shows the intersection in the plane of shock polars. The two polars associated with the diffracted shocks (C_2) intersect here on the axis $\varphi = 0$ at point 3 where the upper and lower flows are compatible.

If we increase the deflection across the shock (C_1), or if the upstream Mach number is decreased, the intersection cannot be regular any longer and the Mach phenomenon occurs as shown in Figure 14.21. For an axisymmetric flow, such as a propulsion jet, it is called *Mach disk*. Mach disks occur in the non-adapted supersonic jet where shock waves are formed (see Chapter 20).

If we assume that the flow downstream of the shock (C_3) is one-dimensional, which it practically the case, it is possible to calculate the flow downstream of the singular reflection by determining step by step the internal subsonic flow (*I*) and the contiguous supersonic flow (*E*). Flow (*I*) is calculated by the one-dimensional theory and the flow (*E*) by the method of characteristics (see Chapter 18). This calculation is conducted by ensuring in each station the equality of the pressures on both sides of the slip lines (Σ_1) and (Σ_2) starting from the triple points T_1 and T_2 (here, the configuration is assumed symmetric). In most cases, the observed structure is similar to what has been described above, the direction of the velocity downstream of the triple point T_1 is negative and positive downstream of T_2. The channel occupied by the subsonic flow downstream of the shock (C_3) thus begins to contract and the flows (*I*) and (*E*) expand in concert, separated by the fluidic boundaries (Σ_1) and (Σ_2) that define for the subsonic flow (*I*) a channel in the shape of a convergent fluid nozzle. If, during this expansion, the Mach number $M_{(I)}$ in the internal channel becomes sonic, it faces the problem of singularity at the throat of a convergent–divergent nozzle (see section 9.4).

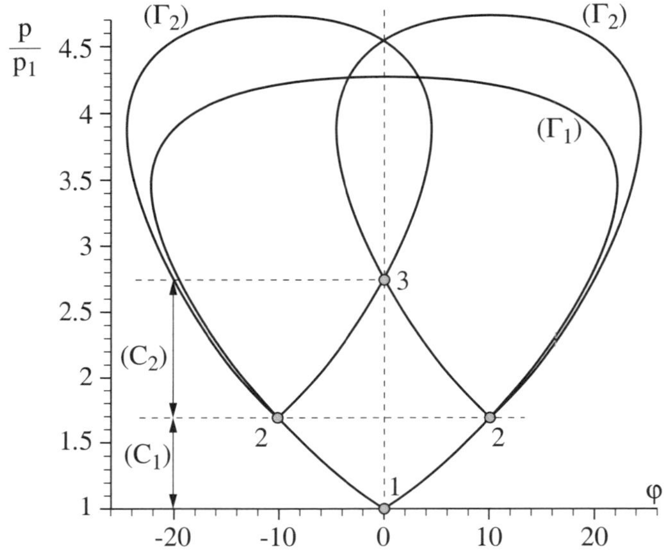

Figure 14.20. *Regular intersection in the plane of shock polars; $M_1 = 1.95$, $\varphi = -10°$, and $\varphi = +10°$*

The solution can be regular only if section *A* available to the flow (*I*) is stationary, i.e. if $dA = 0$ in the section where $M_{(I)} = 1$. In other words, in the section where $M_{(I)} = 1$, it is necessary that the velocities on the slip lines (Σ_1) and (Σ_2) are parallel, as shown in Figure 14.21. We face a two-point boundary value problem where the solution depends on a condition to be met during integration. The eigenvalue here is the location of the triple points. There is only one position of these points where the conditions at the beginning of the calculation of the flows (*I*) and (*E*) meet the conditions of regularity of the solution in the sonic section. The position of T_1 and T_2, hitherto unknown, is fixed by this condition, which closes the problem. In this situation, the shock structure does not depend on conditions far downstream, the throat and the subsequent supersonic flow preventing any downstream influence.

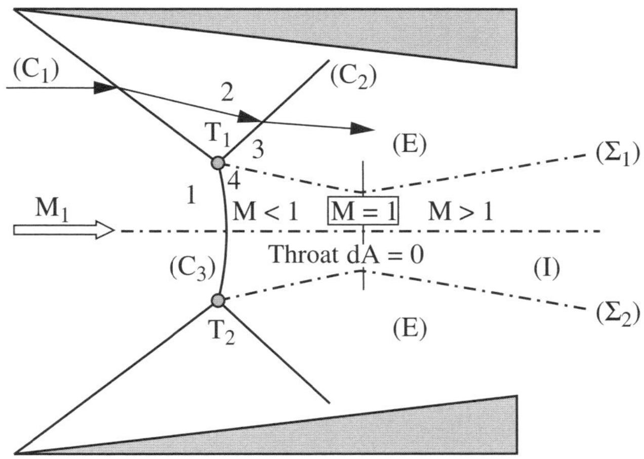

Figure 14.21. *Singular intersection of two shocks or Mach phenomenon*

The schlieren picture in Figure 14.22 shows the Mach phenomenon that occurs when the upstream flow is at Mach number 1.95, with deflections of $\varphi = -14°$ *and* $\varphi = +14°$. The situation is represented in Figure 14.23 in the plane of shock polars where we see that the shock polar associated with the shock (C_2) does not intersect the axis $\varphi = 0$.

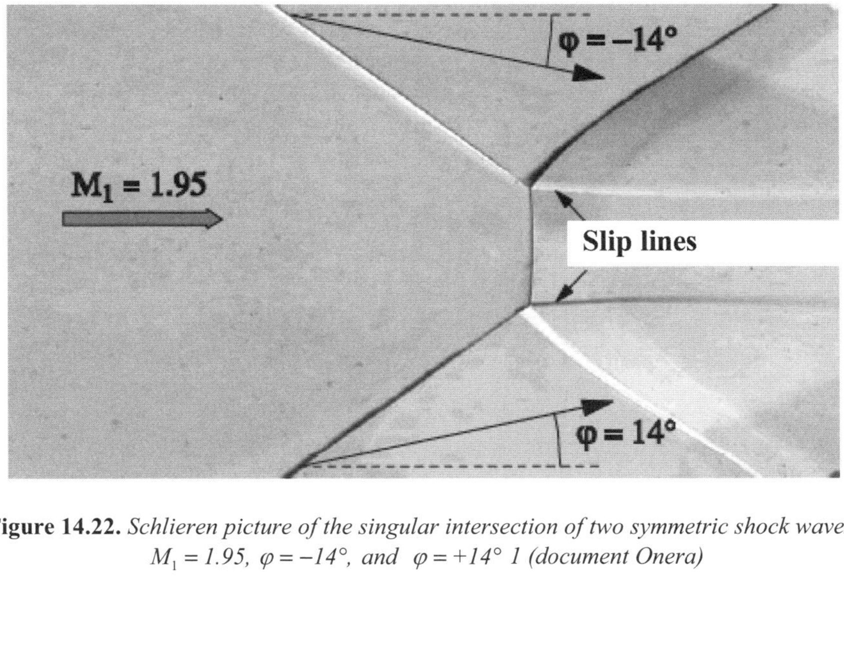

Figure 14.22. *Schlieren picture of the singular intersection of two symmetric shock waves;* $M_1 = 1.95$, $\varphi = -14°$, *and* $\varphi = +14°$ *1 (document Onera)*

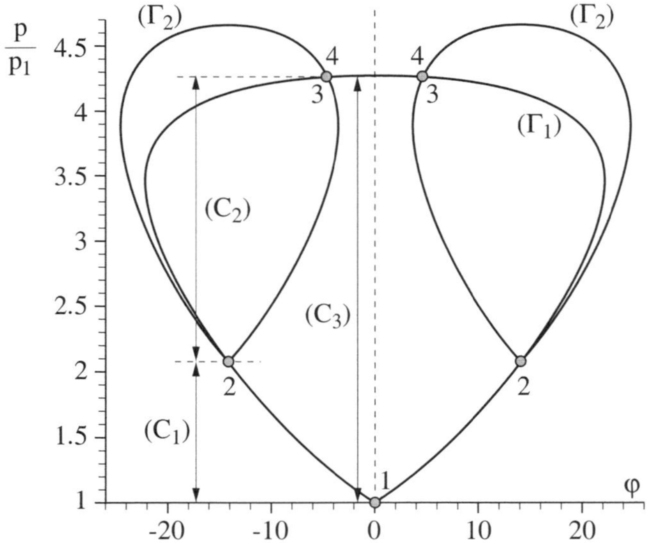

Figure 14.23. *Singular intersection or Mach phenomenon in the plane of the shock polars; $M_1 = 1.95$, $\varphi = -14°$, and $\varphi = +14°$*

At large Mach numbers, as has been observed in strongly overexpanded jets of rocket engines, a shock reflection may happen so that the polars associated with the separation shocks (C_1) and (C_2) have the configuration drawn in Figure 14.24. We note that then a Mach reflection occurs, represented by a point 3.4 for which the velocity angle φ is positive (with, in addition, two possible regular reflections). In this configuration, the Mach disk is bulging forward, as drawn in Figure 14.25, and the slip line (Σ) diverges from the axis at the triple point *T*. We talk of the *inverse Mach reflection*.

The channel occupied by the internal subsonic flow (*I*) being divergent, the flow slows down and its pressure increases correspondingly. However, in a non-viscous fluid, the static pressure cannot exceed the stagnation pressure; at most it can equal it when the velocity is zero. As it is mainly the external supersonic field (*E*) which imposes pressure, a situation can be attained when the internal flow is no longer able to slow down, the imposed pressure becoming greater than its stagnation pressure. Thus, a "breakdown" of this flow occurs with the formation of a recirculation bubble where the opposing pressure field makes the fluid flow back upstream. The complete interpretation of this phenomenon cannot be done in the framework of the assumption of a non-viscous fluid.

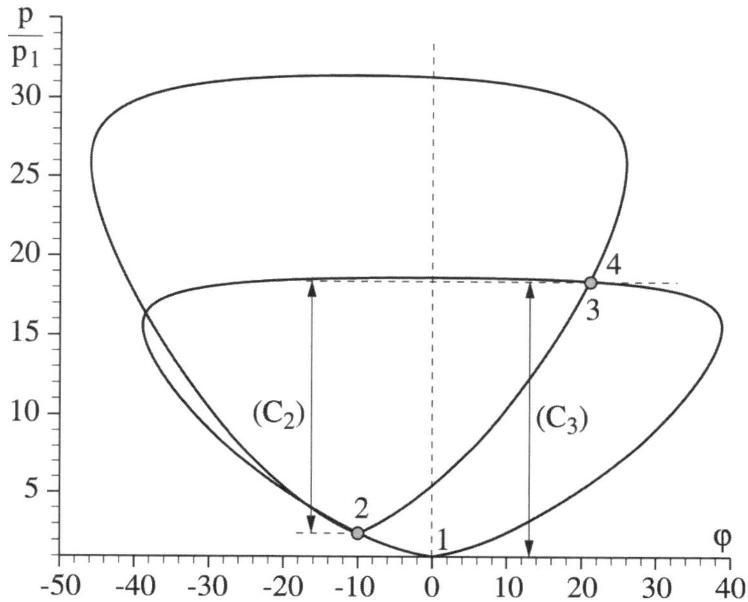

Figure 14.24. *Shock intersection with inverse Mach reflection and situation in the plane of the shock polars* $(M_1 = 4, \varphi = -10°)$

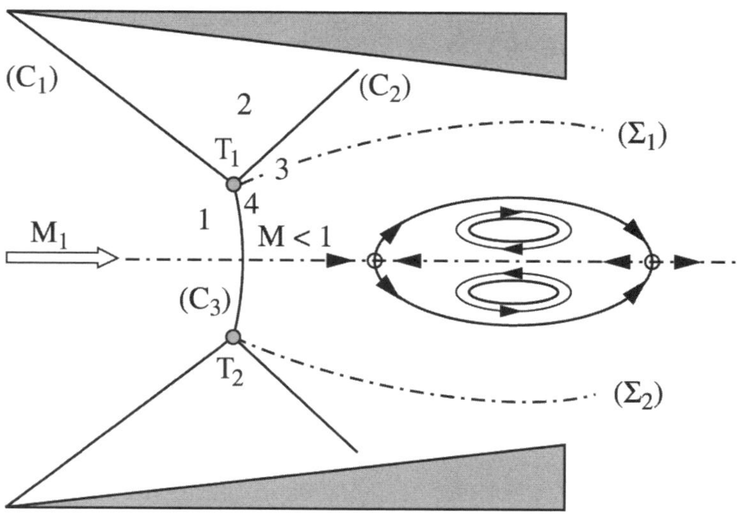

Figure 14.25. *Shock intersection with inverse Mach effect*

To conclude this paragraph, the photograph of Figure 14.26 shows the Mach disk formed in the propulsion jet of the *Vulcain 2* engine of the European launcher *Ariane 5* in a ground test facility. The intense brightness of the jet at the level and downstream of the disk is the result of the heating of the gas crossing the shock, the static temperature becoming again almost equal to the stagnation temperature which is close to 3,500 K. Under these conditions, the water produced by the combustion of the engine propellants (Hydrogen + Oxygen) radiates strongly.

Figure 14.26. *Mach disk in the jet of the Vulcain2 engine tested in a ground facility (document Snecma)*

14.5. Transition from regular reflection to Mach reflection

At a large Mach number, the problem of transition from regular reflection to Mach reflection is more complex. To examine the question, we have to consider the regular reflection represented in the plane of shock polars in Figure 14.27a. Given its shape, the polar (Γ_2) attached to the state (2) intersects the axis $\varphi = 0$ at two points, labeled *RR1* and *RR2*, corresponding to regular reflections. The downstream state of (C_2) being subsonic the solution *RR2* leads to a reflected strong shock (C_2). This solution is unstable. There is also the solution *RM*, intersection of (Γ_1) and (Γ_2), which leads to a Mach reflection. In this case, we see that the angle φ_3 is positive, leading to a forward bulging Mach disk. As we have seen in the preceding paragraph, such a configuration has been called inverse Mach reflection. In this case, we focus on the regular reflection *RR1*.

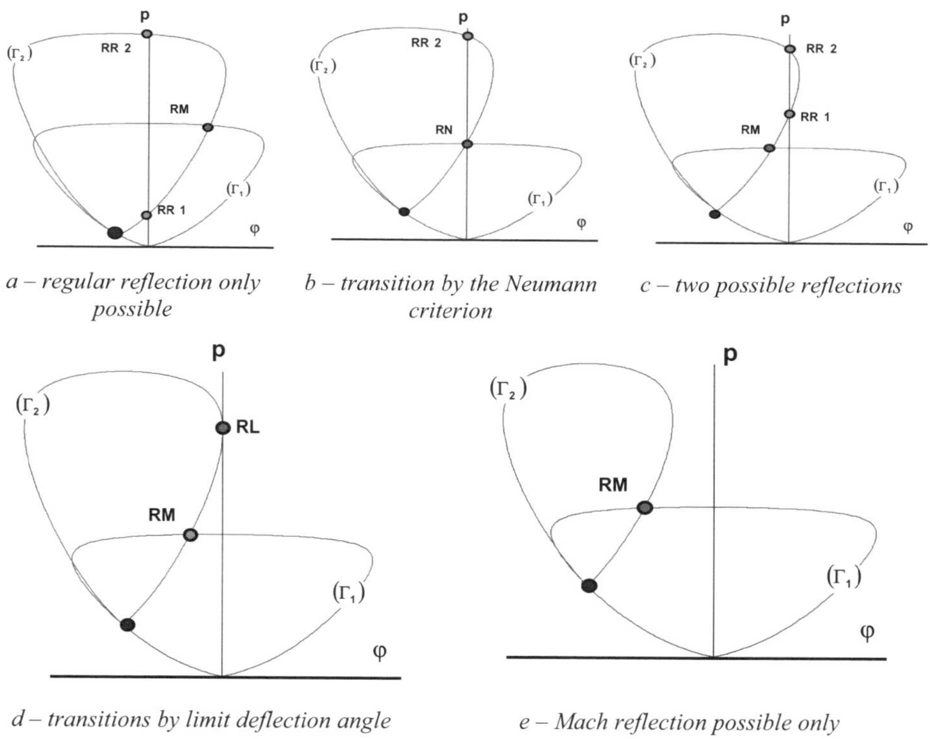

Figure 14.27. *Transition from regular reflection to Mach reflection and representation in the polar plane (case where $M_1 = 4$)*

The upstream Mach number M_1 being constant, we gradually increase the deflection φ_1 across the incident shock (C_1). A situation will be reached where the points *RR1* and *RM* merge in *RN* located on the image point of the normal shock of (Γ_1) (see Figure 14.27b). This limit situation corresponds to a condition for transition from regular reflection to Mach reflection, called the *Neumann criterion*.

If we increase φ_1 further, we arrive at the situation of Figure 14.27c, again with two regular reflections and a Mach reflection *RM*. The reflection *RR2* is excluded for the same reasons as above. In this situation, two solutions are possible: the Mach reflection *RM* and the regular reflection *RR1* which gives a pressure jump higher than the Mach reflection. Both reflection types are actually observed, one or other of the solutions establishing depending on the circumstances (sense of the variation of the Mach number or of the deflection φ_1). Here, the velocity angle φ_3 at the triple point is negative, the stream tube downstream of the Mach disk being convergent at the beginning.

If we continue to increase φ_1, we reach a second extreme situation, represented in Figure 14.27d, where the polar (Γ_2) is tangent to the axis $\varphi = 0$ at the point *RL* which corresponds to the limit deflection φ_{lim}. If φ_1 still rises, then only the Mach reflection is possible (see Figure 14.27e).

Note that, according to the Neumann criterion, the transition to Mach reflection takes place in a continuous manner; in particular, there is no pressure discontinuity. However, exceeding the limit deflection angle causes a "jump" of the representative point from *RL* to *RM* in Figure 14.27e, leading to a pressure discontinuity.

The various situations encountered in the reflection of a shock are represented in the plane $\left(M_1, \varphi_1\right)$ in Figure 14.28. Between the curves (*N*) and (*L*), there is a region called *dual domain* where two reflections are possible and observed. Either one establishes when we vary a parameter in a certain direction, such as the intensity of the incident shock. If we interrupt the change to resume it in the other direction, we often observe a hysteresis.

14.6. External, internal and isentropic compressions

The compression through a normal shock can slow down a supersonic flow to a subsonic Mach number all by increasing its static pressure. The process is *a priori* advantageous as it is very simple and not accompanied by any deviation of the flow. However, any compression through a shock leads to a loss of stagnation pressure which is maximum for the normal shock. We can avoid such a loss by causing an isentropic compression through a ramp that compresses the flow through a progressive wave (see Chapter 17). But it causes a deflection of the flow and, if we impose a final pressure equal to the one caused by a normal shock, a still supersonic Mach number at the end of the ramp. An intermediate solution is to replace the single normal shock with a series of oblique shocks each causing elementary compression until the level of the normal shock is reached. The loss of stagnation pressure after passing through these successive shocks is smaller when the elementary shocks. But to each of these shocks are more numerous, a deflection is attached so that the final flow can be strongly deflected.

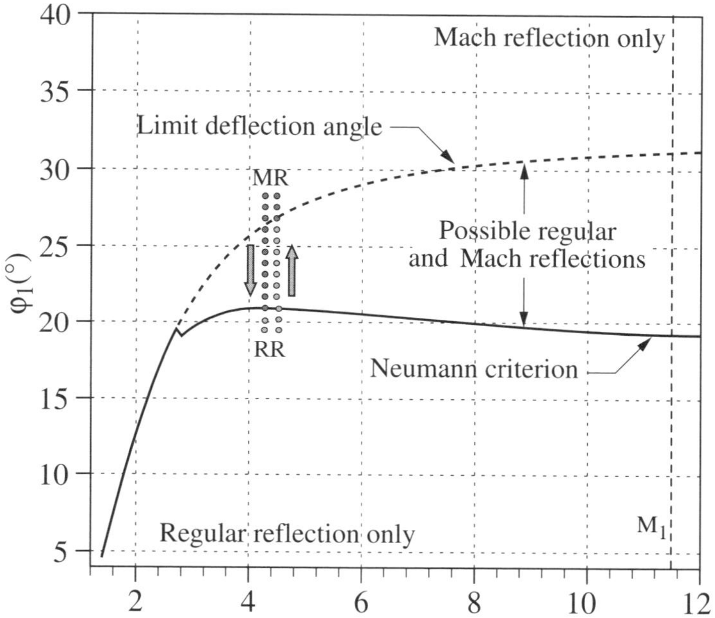

Figure 14.28. *Transition from regular reflection to Mach reflection in the plane: Mach number – deflection*

These compression modes are represented in Figure 14.29a which shows the various compression mechanisms in the plane of polars for an initial Mach number $M_1 = 6$ in the case of air ($\gamma = 1.4$). In this example, the shock compression is done by seven successive shocks, each causing a deflection of 5°, except the last one which ensures the compression to the final level. Note that the images corresponding to the oblique shocks are almost on the isentropic compression polar. The coincidence would be even better if we increased the number of elementary shocks, the isentropic evolution being the limit of a succession of an infinite number of shocks whose intensity tends to zero. Table 14.1 provides information on the change of certain quantities during compression.

single normal shock	$\varphi(°)$	0							0
	Mach	6							0.404
	p/p_0	1							41.48
	p_{st}/p_{st_0}	1							0.029
isentropic compression	$\varphi(°)$	0							34.85
	Mach	6							3.018
	p/p_0	1							41.83
	p_{st}/p_{st_0}	1							1
successive oblique shocks	$\varphi(°)$	0	5	10	15	20	25	30	35.38
	Mach	6	5.316	4.757	4.288	3.890	3.541	3.236	2.941
	p/p_0	1	2.010	3.755	6.605	11.036	17.670	27.2250	41.83
	p_{st}/p_{st_0}	1	0.9663	0.9427	0.9257	0.9131	0.9034	0.8961	0.8889

Table 14.1. *Compression through successive oblique shocks*

The last column highlights the dilemma for the design of hypersonic air intakes (see Chapter 16). A compression through a normal shock does not deflect the flow and ensures a final subsonic Mach number which facilitates the operation of the combustion chamber of the engine. However, in this case, the loss of stagnation pressure, or efficiency η, is catastrophic (in this example, the final stagnation pressure is only about 3% of the upstream stagnation pressure, the efficiency being $\eta = 0.0297$). For an isentropic compression, the efficiency is 1 but the final Mach number is 3.018 and the flow has been deflected by 34.85°. These are essentially the same results for the compression by multiple oblique shocks, the efficiency here being 0.889, i.e. to say fairly close to 1. If we want an acceptable efficiency, we have to resolve to have a final flow, still supersonic and with an important deflection. Hence, the current research on supersonic combustion scramjet engines (the propulsion system for hypersonic flight) where combustion takes place at a supersonic Mach number.

Figure 14.29b shows the ramp system designed so that the isentropic compression waves and oblique shocks converge at a point which is the leading edge of the air intake duct. We see that the isentropic ramp is longer than the multiramp for oblique shocks. It is the difficulty to maintain the convergence at a point of the compression waves which makes this air intake layout unusable. In practice, we use multiramps, with a smaller number of elementary shocks, to minimize the hazards of operation and also for reasons of technological simplicity.

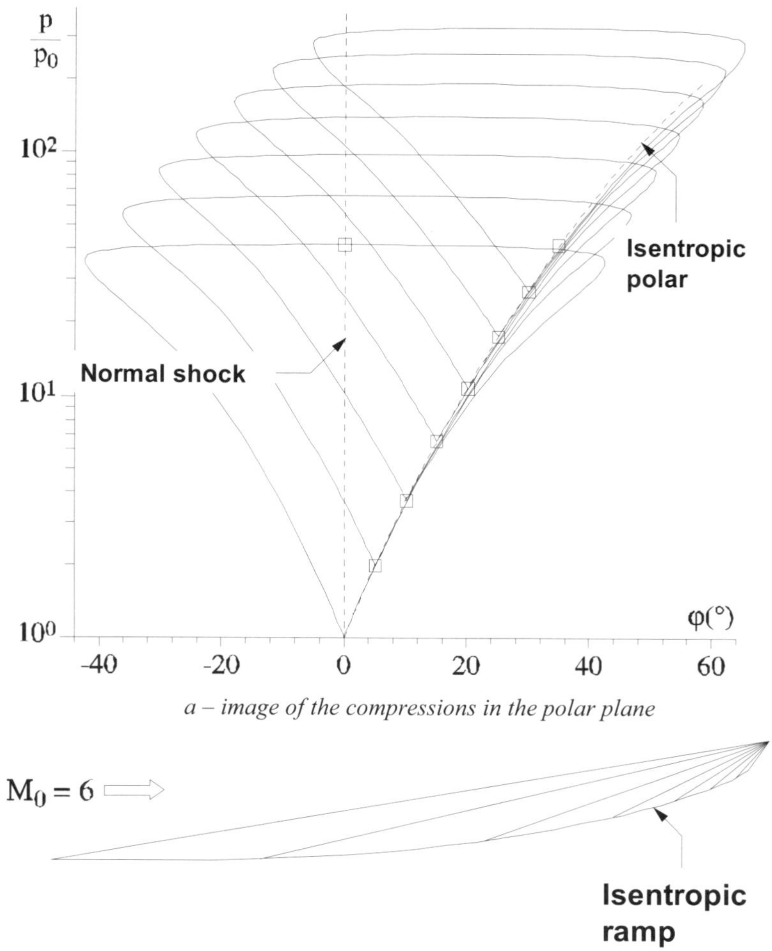

b – physical plane

Figure 14.29. *Air intake for flight at Mach number 6. External compression by seven deflections of 5° and isentropic ramp*

It is possible, as shown in Figure 14.30b, to achieve compression by successive shocks, with an identical efficiency and to avoid excessive deflection of the flow, through the process of shock reflections within the duct. At each reflection, there is a compression similar to the previous case, but at the end of the process the flow deflection it is no more than 5.38° in this example (see Table 14.2). This layout offers, at a first glance, great advantages but, with each shock reflection on the ramp or in the interior of the duct, occurs an interaction with the boundary layer forming on the walls and the risk of boundary layer separation.

single normal shock	$\varphi(°)$	0							0
	Mach	6							0.404
	p/p_0	1							41.48
	p_{st}/p_{st_0}	1							0.029
isentropic compression	$\varphi(°)$	0							34.85
	Mach	6							3.018
	p/p_0	1							41.83
	p_{st}/p_{st_0}	1							1
Successive oblique shocks	$\varphi(°)$	0	5	0	5	0	5	0	5.38
	Mach	6	5.316	4.757	4.288	3.890	3.541	3.236	2.941
	p/p_0	1	2.010	3.755	6.605	11.036	17.670	27.2250	41.83
	p_{st}/p_{st_0}	1	0.9663	0.9427	0.9257	0.9131	0.9034	0.8961	0.8889

Table 14.2. *Air intake for flight at Mach number 6 and compression through shock reflection*

Figure 14.30a shows the image of the flow in the polar plane. The points representing the states after each oblique shock oscillate between $\varphi = 5°$ and $\varphi = 0$, the ramp inducing a deflection of 5° and the internal straight wall of the duct straightens the flow in a way that $\varphi = 0$. The performance of the system is summarized in Table 14.2.

14.7. Conical shock wave

Often projectiles or missiles are fitted with a conical warhead. Also, some types of air intakes for supersonic aircraft are equipped with a conical ramp compressing the flow across a conical shock wave. It is the case of the SR71 *Blackbird* and the *Mirage* family of Dassault Aviation, whose air intakes are semi-circular (see Chapter 16). In supersonic the flow around such a body has, as we shall see, remarkable properties.

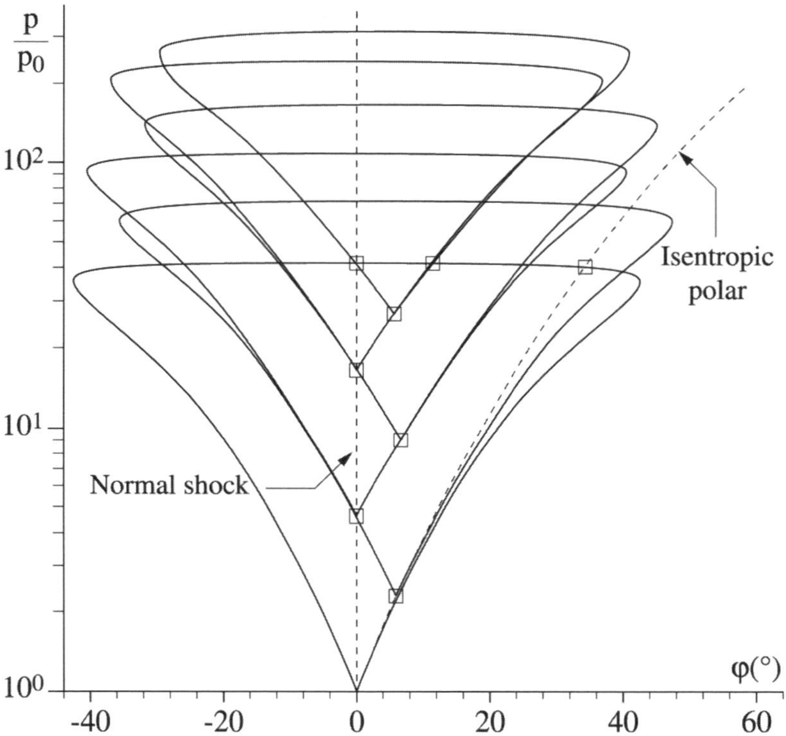

a – image of the compressions in the polar plane

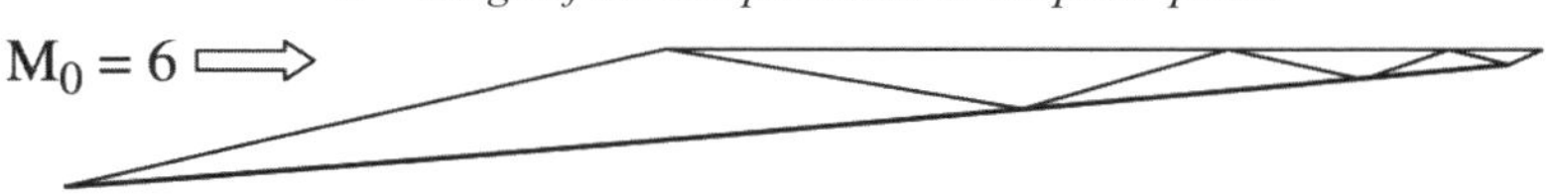

b – physical plane

Figure 14.30. *Air intake for flight at Mach number 6. Internal compression by six reflections of 6°*

The approach is to seek a solution for an axisymmetric flow which is invariant along straight lines, or radii, all issuing from an origin, or apex, *O*. It is therefore an axisymmetric flow independent of the distance r to the apex and only a function of the angle θ of the radius vector with respect to the axisymmetry axis (see Figure 14.31).

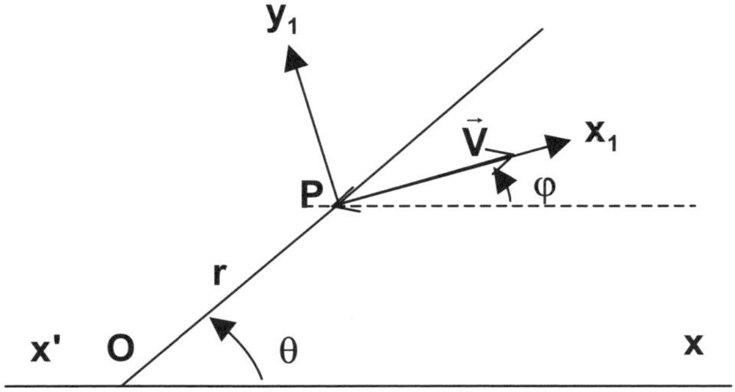

Figure 14.31. *Axis system for conical flow*

We begin with the Euler equations written in an intrinsic system linked to the direction $\vec{x}_1$ of the local velocity vector and the perpendicular direction $\vec{y}_1$ (these equations are established in Chapter 17):

– *Motion equation along the direction* $\vec{x}_1$:

$$\left(\frac{V^2}{a^2}-1\right)\frac{\partial p}{\partial x_1}+\rho V^2\frac{\partial \varphi}{\partial y_1}=-\varepsilon\rho V^2\frac{\sin\varphi}{y}.$$

– *Motion equation along the normal* $\vec{y}_1$ *(radial equilibrium)*:

$$\rho V^2\frac{\partial \varphi}{\partial x_1}+\frac{\partial p}{\partial y_1}=0.$$

– *Energy equation*:

$$\frac{\partial s}{\partial x_1}=0.$$

The flow is invariant along the radii vectors and the entropy is the same on the streamlines that cross any radius vector; consequently, the flow is isentropic in its total or *homentropic*.

The derivatives with respect to r are zero by assumption, and it follows that:

$$\frac{\partial}{\partial x_1}=-\sin(\theta-\varphi)\frac{\partial}{r\partial\theta}\qquad\frac{\partial}{\partial y_1}=\cos(\theta-\varphi)\frac{\partial}{r\partial\theta}.$$

And the above equations become:

$$-\sin(\theta-\varphi)\left(\frac{V^2}{a^2}-1\right)\frac{\partial p}{r\partial\theta}+\rho V^2\cos(\theta-\varphi)\frac{\partial\varphi}{r\partial\theta}=-\varepsilon\rho V^2\frac{\sin\varphi}{r\sin\theta}$$

$$-\sin(\theta-\varphi)\rho V^2\frac{\partial\varphi}{r\partial\theta}+\cos(\theta-\varphi)\frac{\partial p}{r\partial\theta}=0.$$

We verify that the distance r is eliminated from these equations, which justifies the assumption made. Thus, we can find solutions depending only on the angle θ. We state:

$$z=\ln\ p\rightarrow dz=\frac{dp}{p}.$$

In addition, for a calorically perfect gas, $\rho V^2=\gamma pM^2$. The above equations can be solved as:

$$\frac{\partial z}{\partial\theta}=\gamma M^2\frac{\sin\varphi\sin(\theta-\varphi)}{\left[M^2\sin(\theta-\varphi)^2-1\right]}\frac{1}{\sin\theta}$$

$$\frac{\partial\varphi}{\partial\theta}=\frac{\sin\varphi\cos(\theta-\varphi)}{\left[M^2\sin^2(\theta-\varphi)-1\right]}\frac{1}{\sin\theta}.$$

We construct a conical flow which connects to a uniform upstream flow with the Mach number M_0 across a shock wave (C) which, in the framework of the current assumption, will be rectilinear and starting from O; i.e. conical. By giving the angle of (C) relative to the axis, i.e. the shock angle σ, the conditions (1) downstream of (C) are calculated from oblique shock relations. Downstream of the shock, the normal Mach number $M\sin(\theta-\varphi)$ is subsonic (see section 13.5.2), and hence the derivatives of z and φ, compared to θ, are negative. If we integrate the above system from the shock in the direction of the axis, thus by negative $d\theta$ steps, the pressure p and the direction of the velocity φ increase. Starting from the conditions (1), integration of the system will go up to a value θ_s of θ so that the velocity is aligned with the radius vector $(\varphi = \theta)$, which corresponds to the slip condition on the surface of a physical cone with the half-angle θ_s, on which we have:

$$\varphi(\theta_s) = \theta_s.$$

Figures 14.32 and 14.33 show plots giving the angle of the conical shock and the Mach number on the cone, as a function of the cone half-angle θ_s for increasing values of the upstream Mach number M_0. For each value, there are two solutions; the Mach number on the cone can be supersonic or subsonic.

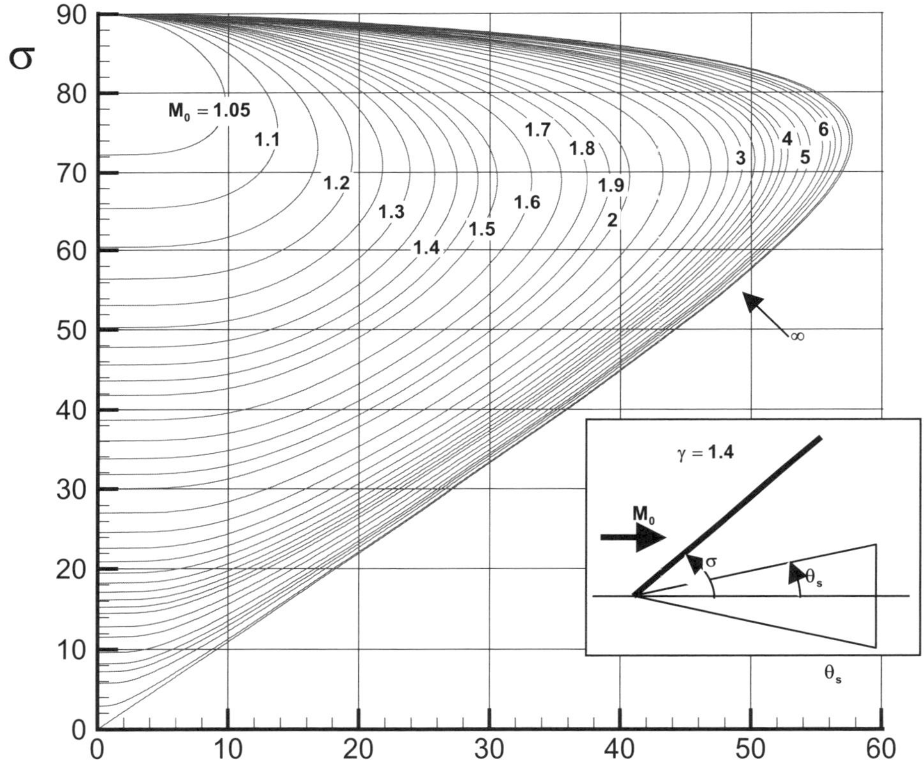

Figure 14.32. *Conical shock. Shock angle as a function of the cone half-angle for different values of the upstream Mach number (case of air γ = 1.4)*

14.8. Indication on the effects of relaxation or non-equilibrium

As we have seen in section 14.1, this case reduces to that of the normal shock wave with the effect of non-equilibrium, discussed in section 13.5.3. We consider the velocity component normal to the shock wave, equal to $V_1 \sin\sigma$ if σ is the shock angle relative to the upstream velocity. This enables us to determine the after-shock conditions, in particular, the velocity normal component V'_{n_2} (see Figure 14.34).

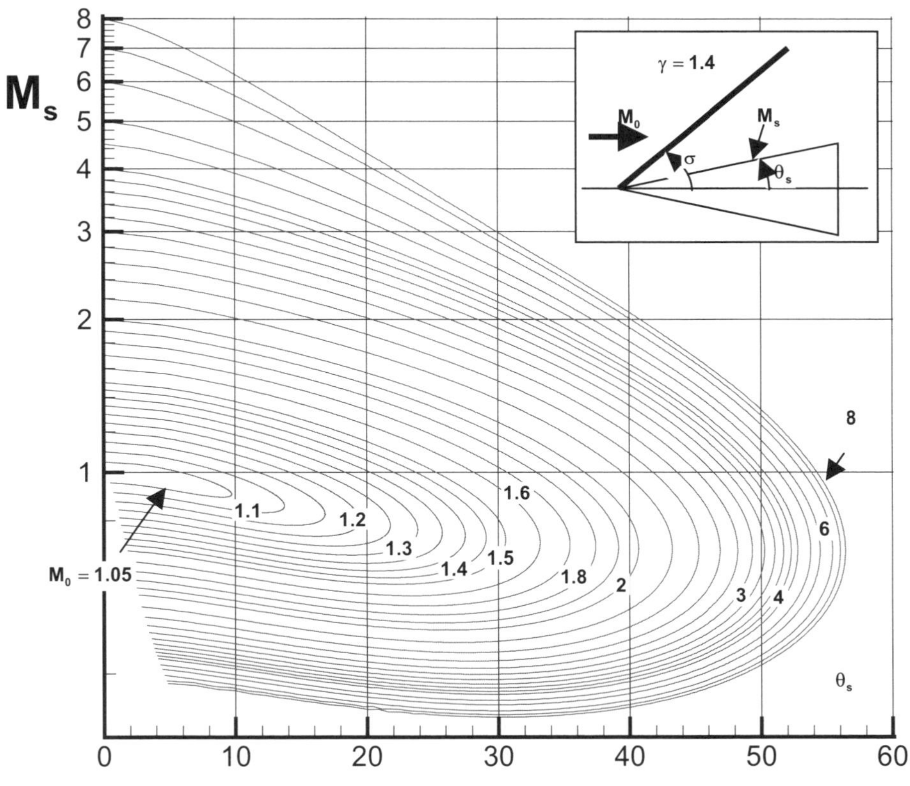

Figure 14.33. *Conical shock. Mach number on the cone as a function of the cone half-angle for different values of the upstream Mach number (case of air $\gamma = 1.4$)*

The change of quantities in the relaxation region is then calculated by the just indicated method. The downstream velocities are obtained by adding the velocity component along the plane of the shock to all normal velocities. We know that this component, whose value is $V_1 \cos\sigma$, is conserved across the shock. Because the normal velocity varies in the relaxation domain between V'_{n_2} and V_{n_2}, it follows that the angle φ of the velocity downstream of the shock varies from the value φ'_2 on the downstream side of the shock until the value φ_2 reached in the equilibrium region. Thus, as shown in Figure 14.35a, the streamlines of the flow downstream of the planar oblique shock are not straight.

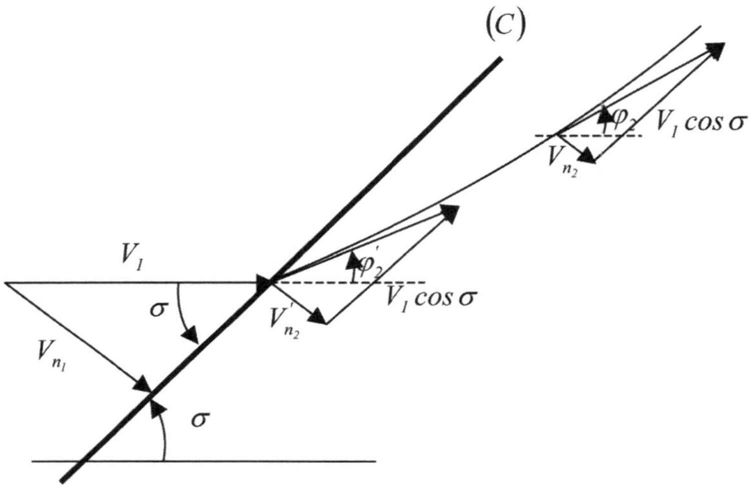

Figure 14.34. *Relaxation zone downstream of an oblique shock wave*

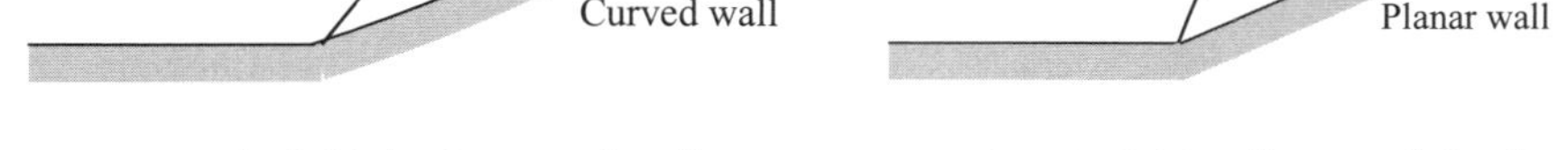

a – straight shock: curved wall *b – straight wall: curved shock*

Figure 14.35. *Effect of chemical kinetics on the shape of an oblique shock wave*

Accordingly, to obtain an oblique planar shock by wedge effect in a uniform upstream flow and if relaxation phenomena occur after the shock, it requires that the downstream wall be curved along a profile depending on the chemical kinetic properties of the fluid (see Figure 14.35b). Conversely, if a shock occurs with relaxation on a wedge with straight sides, then the shock will be curved. Its calculation will have to use more sophisticated methods, particularly based on the method of characteristics (see Chapter 18).

Chapter 15

Shock Intersections or Shock–Shock Interferences

15.1. Introduction

Shock–shock interference phenomena are the result of the crossing or intersection of shock waves. This intersection causes the formation of waves whose presence affects the aerodynamic field around a body. If the crossing occurs near the vehicle, these waves can influence the aerothermodynamic loads to which its surface is subjected, or degrade its aerodynamic performance. Shock–shock interferences can occur in any supersonic flow where shock waves are generated either by a deflection produced by a wall (airfoil, control surface, compression ramp of air intake, and cowl) or a back pressure (interblade channel of a transonic cascade, separated nozzle). The shock–shock interferences to be feared are those in hypersonics where the shocks are, first, extremely intense, and second, highly inclined with respect to the upstream velocity (the shock angle σ is small). Thus, their intersection occurs near the vehicle and the resulting waves will strike almost inevitably its surface. In section 14.4, we have studied the intersection of two shock waves by considering the case where the configuration is symmetric; the problem is then equivalent to the reflection of a shock on a planar wall. Here, we will be investigating more complex situations where shocks have different intensities.

The first example of shock interference, presented in Figure 15.1a, is a vehicle of the *Space Shuttle* type. In the imagined situation, the head shock of the main tank meets the shock at the nose of the shuttle, the latter intersecting, in turn, the shock that forms ahead of the cockpit. This last shock can interfere with the shock caused by the vertical stabilizer. The second case (see Figure 15.1b) relates to a space launcher on which the head shock of the main body (*Cryogenic main stage*, or *Etage Propulseur Cryotechnique* (*EPC*) for the European launcher *Ariane 5*) intersects the shock at the nose of a lateral booster (*Etage d'Accélération à Poudre* (EAP), for *Ariane 5*). The third example concerns an air intake where shocks induced by compression ramps meet the detached shock in front of the rounded leading edge of the cowl (see Figure 15.1c). On an air-breathing hypersonic vehicle, that is, using atmospheric air as an oxidizer, these different interactions can occur simultaneously. As shown in Figure 15.1d, interferences exist within an interblade channel of a turbomachine, where shock waves generated by two neighboring blades intersect. Shock–shock interferences also occur in overexpanded supersonic nozzles when the boundary layer separates due to a too intense adaptation shock at the exit.

Many other cases of interference can be encountered in practice. Interferences are associated with shock wave/boundary layer interactions when there is separation. Then, the resulting thickening of the boundary layer induces typical shock structures (*shock patterns*) in the contiguous supersonic or transonic flow. The shock–shock interferences can be devastating at high Mach numbers due to local thermal loads they generate. As such, they were most often studied in flows with Mach numbers higher than 5. However, interferences have also side effects that can be very damaging at moderate Mach numbers where the created disturbances can affect the aerodynamic functioning of the device and trigger violent instabilities, such as air intake buzz.

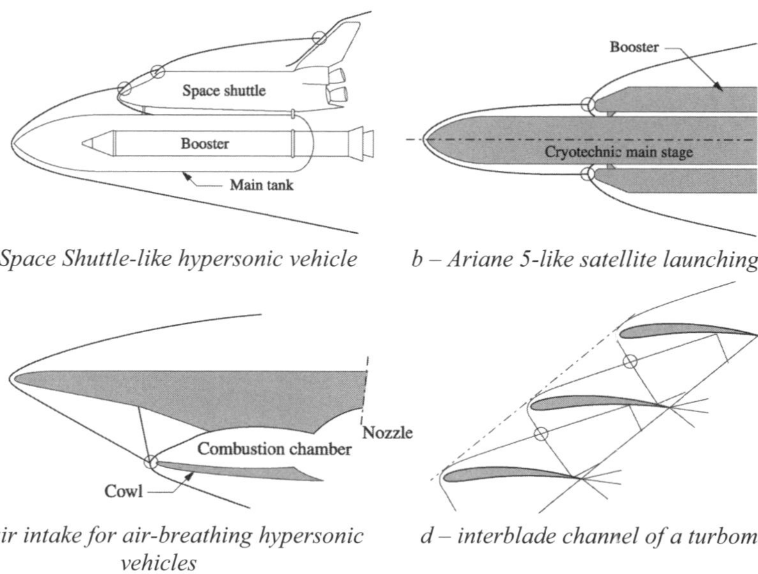

a – Space Shuttle-like hypersonic vehicle   *b – Ariane 5-like satellite launching vehicle*

c – air intake for air-breathing hypersonic vehicles   *d – interblade channel of a turbomachine*

Figure 15.1. *Examples of shock–shock interferences*

Shock–shock interferences are one of the most spectacular phenomena of gas dynamics. The intersection of two shock waves can indeed generate wave systems (shocks, expansion) and slip surfaces, forming structures with identifiable properties, revealed by optical visualization methods (schlieren, shadowgraph, interferometry, and electron beam fluorescence). As we shall see, the theory of discontinuities allows an interpretation of these remarkable phenomena governed mainly by the non-viscous part of the flow, that is, the Euler equations. The viscosity, essential for transfer processes (heat flux in this case), occurs only secondarily to the overall organization of the flow field. Its contribution can be superimposed on the structure of the non-viscous field as most often thin mixing layers.

According to a classification proposed by Edney in 1968, we should distinguish six types of interferences that occur according to the orientation and relative positions of the two intersecting shock waves. We can introduce these types considering the situation shown in Figure 15.2, where a "weak" oblique shock wave (C_1), as defined by the weak shock solution of the Rankine-Hugoniot equations, intersects the curved shock (C_2). This situation corresponds to the intersection of the shock induced by a control surface and the shock present in front of a blunt nose or a rounded leading edge. Also, this is what happens when the shock caused by a compression ramp crosses the shock created by the leading edge of an air intake cowl.

For simplicity, the supersonic upstream flow is assumed uniform and the geometry two-dimensional. We imagine that the shock wave (C_1) comes from below. When (C_1) meets (C_2) far away from the nose, the situation corresponds to the intersection of two weak oblique shocks with opposite directions or families. This means that (C_1) and (C_2) are very inclined to the upstream flow direction and the angles identifying these shocks are of opposite signs. When the intersection point goes back up to the nose, (C_2) becomes more intense so that, for the same shock intensity (C_1), the intersection involves shocks with different intensities. When the intersection is above the nose region, the shocks (C_1) and (C_2) belong to the same family. This means that their inclination angles with respect to the upstream flow direction have the same sign. Then, the nature of the interaction changes radically.

Depending on the relative position of the shocks (C_1) and (C_2), we observe the six types of flow patterns shown in the diagrams I-VI in Figure 15.2. To simplify the presentation, the flow upstream of the interference will be assumed uniform and we consider the two-dimensional case, which allows us to easily obtain exact solutions in some cases. The made analyses also apply to more complex situations of non-uniform and axisymmetric flows; but then they have a more qualitative character.

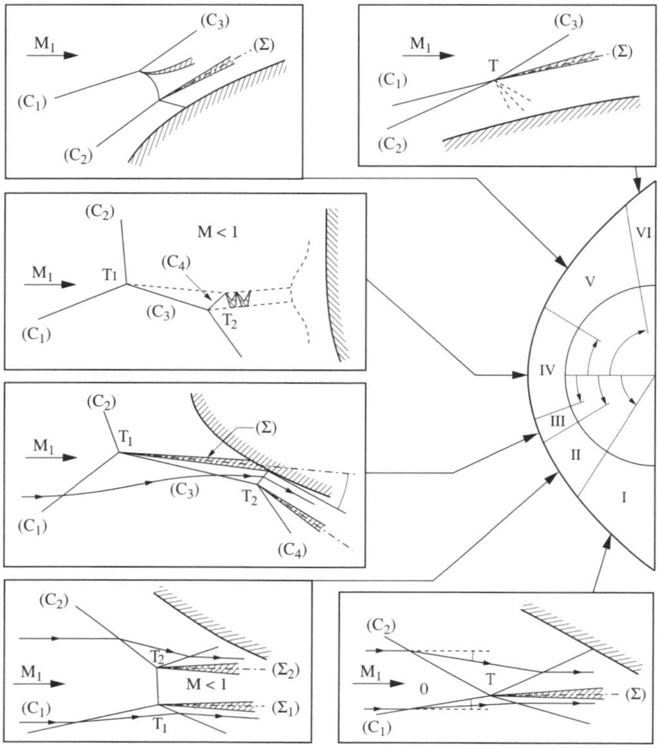

Figure 15.2. *The different types of shock interferences (according to Edney)*

15.2. Type I interference

This interference corresponds to the case of a shock reflection analyzed in detail in section 14.4 when the crossing shock waves are symmetric. Here, we will only repeat the essential results of this analysis.

We consider the situation shown in Figure 15.3, where a uniform supersonic flow (Mach number M_1, pressure p_1) is subjected in A_1 and A_2 to two deviations φ_2 and φ_3 with opposite signs. This leads to two shock waves, one ascending (C_1) and the other descending (C_2), which intersect in the point I. The problem is to determine the flow resulting from this intersection. For this, we will consider the situation in the plane of the shock polars $[\varphi, p]$. At high Mach numbers, we often trace the polars in semi-log coordinates because of very large pressure ratios (see Figure 15.4).

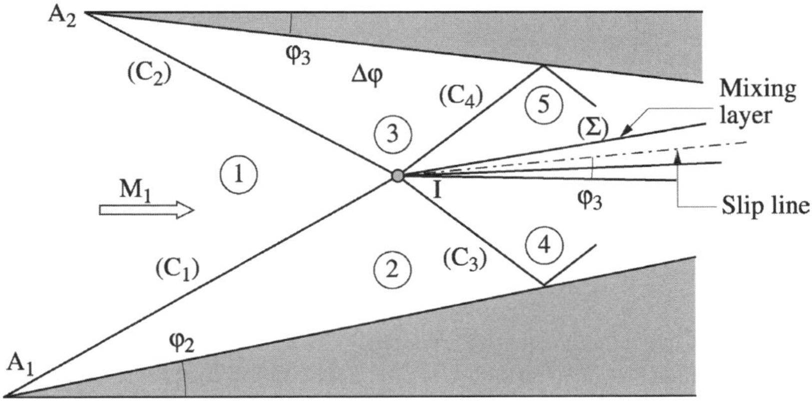

Figure 15.3. *Type I interference. Physical plane*

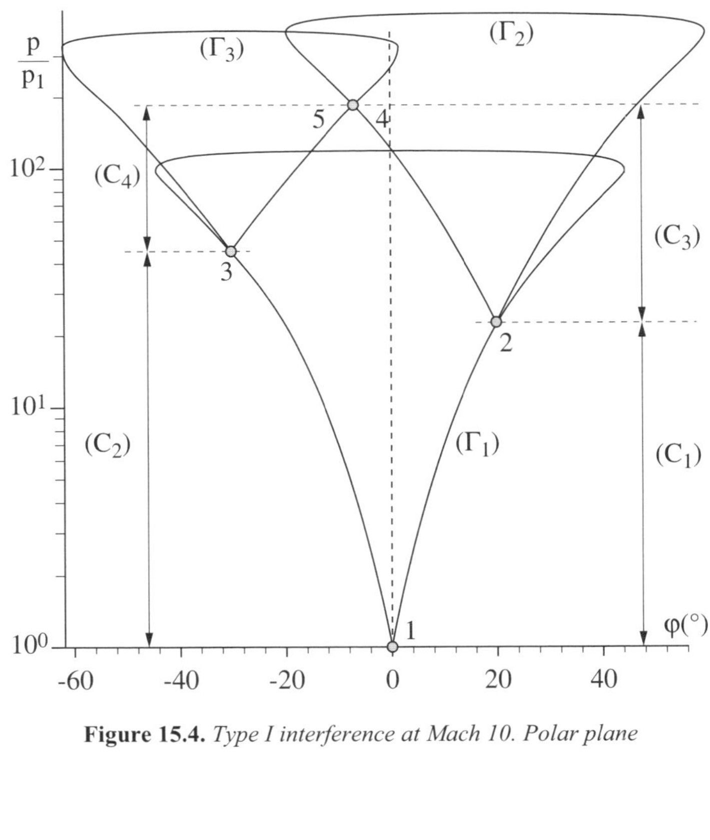

Figure 15.4. *Type I interference at Mach 10. Polar plane*

Points 2 and 3 of this plane, images of the flows downstream of the shocks (C_1) and (C_2), are located on the polar (Γ_1) related to the upstream flow (1) through the deflections φ_2 and φ_3. It is clear that states (2) and (3) are not compatible, since both the directions of the velocity vectors φ_2 and φ_3 as well as the pressures p_2 and p_3 are different. The pressures are only equal if the deflections are the same in absolute values. The incident shocks (C_1) and (C_2) should therefore be extended by the shocks (C_3) and (C_4) that adjust the flows (2) and (3) in such a way that they take a common direction $\varphi_4 = \varphi_5$. At the same time, the pressures pass from p_2 to p_4 and p_3 to p_5, the compatibility condition $p_4 = p_5$ having to be ensured. The dual condition $\varphi_4 = \varphi_5$, $p_4 = p_5$ is satisfied at the intersection of the polars (Γ_2) and (Γ_3) associated with the shocks (C_3) and (C_4). Thus, the flows (4) and (5) of the physical plane have, as images, the two coincident points 4 and 5 in the plane of the shock polars. The construction of the polars shows that two solutions are possible, since (Γ_2) and (Γ_3) intersect in two points, the right solution being determined by downstream conditions.

Downstream of *I*, the solution consists of flows (4) and (5), separated by a slip line (Σ) emerging from the intersection *I*. Indeed, the flows (4) and (5) generally have different properties, except pressure, since they result from the crossing of different shocks.

Figure 15.5 shows an Electron Beam Fluorescence (EBF) visualization of the intersection of two shocks in a flow at Mach 10. The oblique shock caused by a wedge intersects the detached shock forming ahead of a cylindrical body, according to a type I interference.

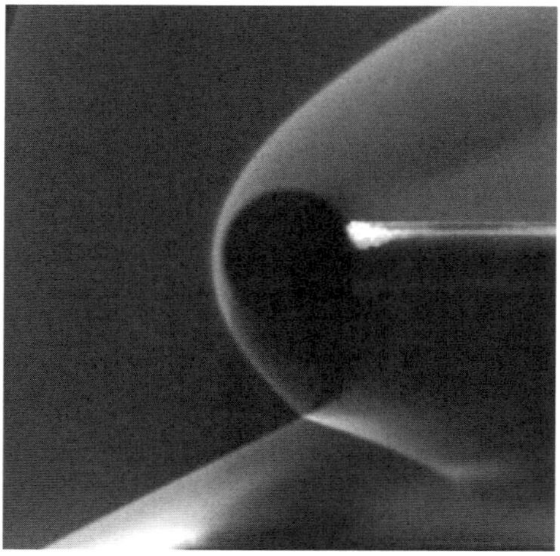

Figure 15.5. *Type I interference between two shocks in a flow at Mach 10 (document Onera)*

We show that the fluid velocities are unequal on both sides of (Σ), which has the property of being a slip line. We denote by Δ the *assumed weak* variation of any quantity when we cross (Σ) from (4) to (5). The invariance of the stagnation enthalpy across a shock allows us to write:

$$\Delta h + V\,\Delta V = 0$$

or if s is the specific entropy:

$$\Delta h = T\,\Delta s + \frac{\Delta p}{\rho}.$$

In the above expressions, V, T, and ρ are averages of the values in (4) and (5). The compatibility of the flows downstream of I requires:

$$\Delta p = p_4 - p_5 = 0$$

from which follows by eliminating Δh:

$$\frac{\Delta V}{V} = -\frac{T}{V^2}\Delta s\,.$$

For a calorically perfect gas then:

$$\frac{\Delta V}{V} = -\frac{1}{\gamma M^2}\frac{\Delta s}{r}\,.$$

If the shocks crossed by the flows (4) and (5) do not have the same intensity, then, $\Delta s \neq 0$, and consequently ΔV too. In addition, the stagnation temperature T_{st} being invariant, we have:

$$\Delta s = -r\frac{\Delta p_{st}}{p_{st}}$$

which leads to:

$$\frac{\Delta V}{V} = \frac{1}{\gamma M^2}\frac{\Delta p_{st}}{p_{st}}$$

or:

$$\frac{V_5 - V_4}{V_{average}} = \frac{1}{\gamma\left(M^2\right)_{average}} \frac{p_{st_5} - p_{st_4}}{\left(p_{st}\right)_{average}}.$$

Such a slip line also corresponds to a temperature discontinuity defined by:

$$\frac{\Delta T}{T} = -\frac{V\,\Delta V}{C_p T}$$

and since $\Delta p = 0$, to a density discontinuity:

$$\frac{\Delta \rho}{\rho} = -\frac{\Delta T}{T} = -\frac{V\,\Delta V}{C_p T}.$$

Therefore, these lines are easily seen by schlieren of shadowgraphic visualization of the flow. Because of the viscosity of the fluid, the discontinuities ΔV, ΔT, etc., cannot be maintained and a mixing layer develops along (Σ). This layer ensures a continuous transition of the flow properties between the states (4) and (5).

15.3. Type II interference

If the initial shocks (C_1) and (C_2) become more intense, because of an increase in the deflections φ_2 and φ_3, points 2 and 3 become more distant on the polar (Γ_1). In addition, the polars (Γ_2) and (Γ_3) are increasingly less developed with diminishing Mach numbers M_2 and M_3. They end up being tangential and no longer cross. Thus, as we have seen in section 14.4, a type I solution is no longer possible. The flow reorganizes into a new structure that obeys the compatibility rules on the deflections and pressures (see Figure 15.6). As it is now impossible to find the contiguous and compatible states (4) and (5), we have to introduce other states between (4) and (5). The solution that establishes can be constructed in the following way in the polar plane (see Figure 15.7):

– States (2) and (3) are always located on the polar (Γ_1).

– From (2) and (3), we pass to (4) and (5) across the shocks (C_3) and (C_4) which have the polars (Γ_2) and (Γ_3), respectively.

– Points 4 and 5 are on the intersection of (Γ_1) and (Γ_2) and (Γ_3), respectively.

– As the states (4) and (5) are incompatible, the flows on the physical plane are disjoint. The point of intersection I of the previous case splits into two distinct points T_1 and T_2, which are triple points (see Figure 15.6).

– Exists a third flow (6) in T_1 whose image in the polar plane is at the intersection of (Γ_1) and (Γ_2) and which, therefore, is compatible with (4). In the physical plane, (4) and (6) are separated by a slip line (Σ_1), the entropy change resulting from the passage across (C_1), then (C_3), to arrive at (4) being less than the crossing of the single shock (C_5) from (1) to (6). The flow in (6) is mostly subsonic.

– The situation in T_2 is analogous: the state (7) is compatible with (5) at the intersection of (Γ_1) and (Γ_3). Similarly, a slip line (Σ_2) originates from T_2.

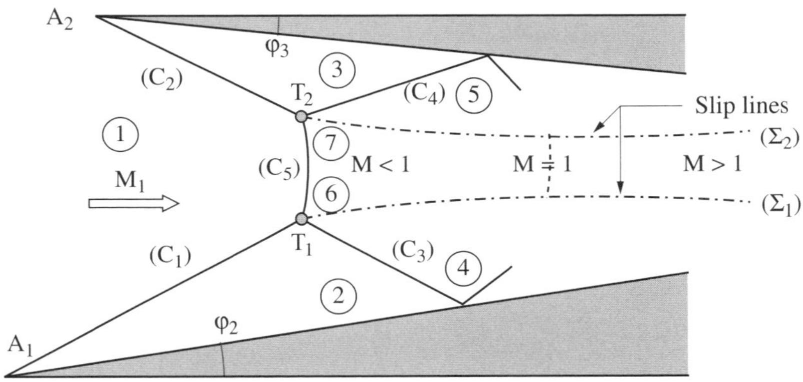

Figure 15.6. *Type II interference. Physical plane*

The nearly normal shock (C_5), which forms between points T_1 and T_2 within flow (1), has as image the arc of (Γ_1) comprised between 4 and 5. It is, therefore, a *strong oblique shock*, except in the point whose image in the plane $[p, \varphi]$ is located on the axis $\varphi = 0$; then, as we know: $\sigma = \dfrac{\pi}{2}$.

The flow being subsonic in the area delimited by (C_5), (Σ_1), and (Σ_2), the whole phenomenon, called *Mach effect* or *Mach phenomenon*, is influenced by the downstream conditions (this configuration has been discussed in detail in section 14.4). If the upstream flow is uniform and the ramps inducing the shocks (C_1) and (C_2) are straight, then the states (2) and (3) are also uniform. In contrast, the flows downstream of the shocks (C_3) and (C_4) are not uniform because of the curved shape of the slip lines (Σ_1) and (Σ_2). Waves emanating from these lines affect the flows, and consequently, the shocks (C_3) and (C_4) are curved because the conditions on their downstream side are variable.

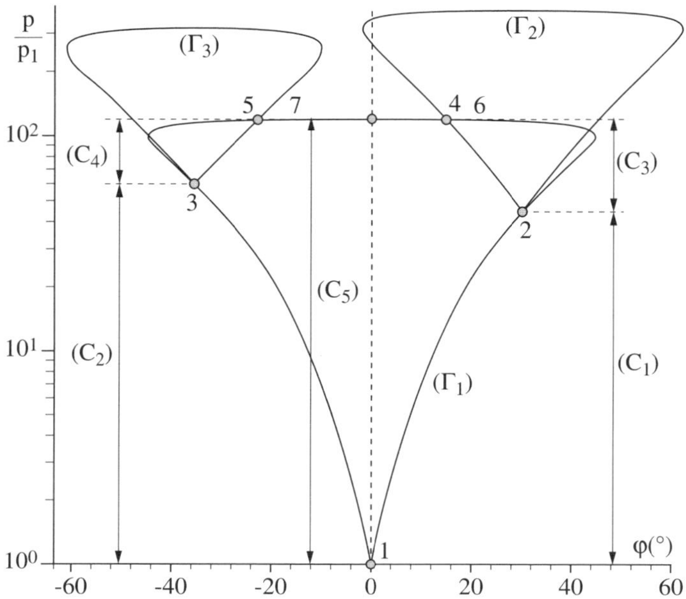

Figure 15.7. *Type II interference at Mach 10. Polar plane*

The flows behind (C_3) and (C_4) are, therefore, rotational because entropy there is variable (see section 8.2.1). Similarly, the flow downstream of (C_5) is variable. Consequently, discussion of the field structure and shock waves applies to local conditions at points T_1 and T_2. This comment does not alter the substance of the problem but we should be aware of this fact: the states (2), (3), (4), etc., may be related to points and not to extended fields. Hereof, this will not always be stated explicitly, but will be obvious. Figure 15.8 shows an EBF visualization of a type II interference in a flow at Mach 10.

15.4. Type III interference

We consider the case where the intensities of the two shocks, intersecting at a point T_1, are very different, the shock (C_1) being a *weak* oblique shock and (C_2) a *strong* oblique shock (see Figure 15.9). The situation is represented in polar plane in Figure 15.10. The flow downstream of (C_2) being subsonic, a type II solution is no longer possible. In contrast, a shock (C_3) causing the flow to pass downstream of (C_1) from (2) to (4), so that the image of (4) is located at the intersection of the polars (Γ_1) and (Γ_2), leads to a situation where the states (3) and (4) are compatible on both sides of the slip line (Σ_1). Downstream of the weak oblique shocks (C_1) and (C_3), the flow (4) is still supersonic, while being subsonic after crossing (C_2).

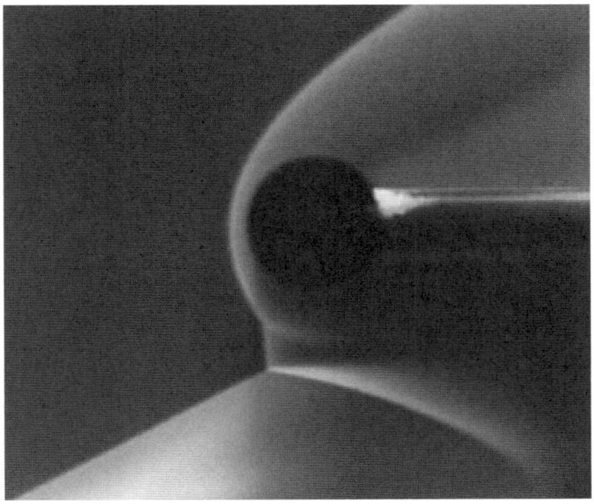

Figure 15.8. *Type II interference between two shocks at Mach 10 (document Onera)*

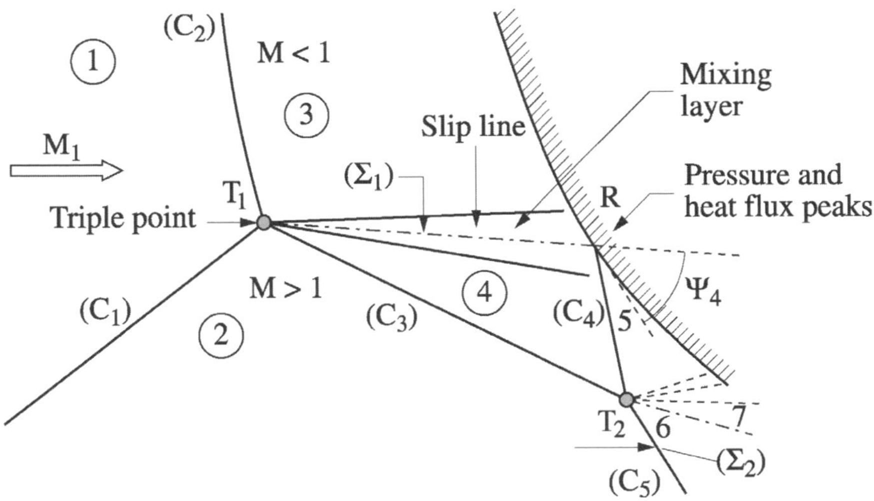

Figure 15.9. *Type III interference. Physical plane*

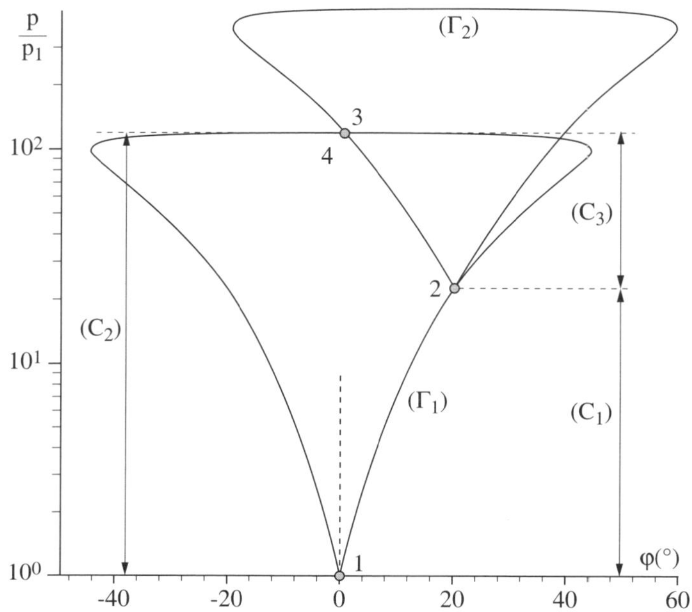

Figure 15.10. *Type III interference at Mach 10. Polar plane*

In a non-viscous fluid, the discontinuities on (Σ_1) are important, with the result that a viscous layer through which the velocity and density vary in large proportions develops rapidly along (Σ_1). The situation is similar to that of a separated flow where a mixing zone separates an external supersonic field from a quasi-stagnant region.

We look at the case where, at the level of point R, the line (Σ_1) impacts on a nearby surface (see Figure 15.9). Flow (4) undergoes a deflection ψ_4 to which a shock wave (C_4) (polar (Γ_3)) is associated. This shock intersects (C_3) at a second triple point T_2. A new interference of the type VI, analyzed in section 15.7, is formed in T_2. It leads downstream to two flows separated by another slip line (Σ_2). The situation in the shock polar plane is shown in Figure 15.11. The shock (C_4) causes transition from state (4) to state (5), incompatible with (2). In (2), a shock wave (C_5) is formed causing the transition on the polar (Γ_2) from the state (2) to the state (6), which is not compatible with (5). A state (7) compatible with (6) has to form so that: $\varphi_7 = \varphi_6$, $p_6 = p_7$. The state (7) is connected to (5) by an expansion represented by the isentropic polar (Δ). In Figure 15.11, corresponding to an upstream Mach number $M_1 = 10$, the shock (C_5) has a very small intensity.

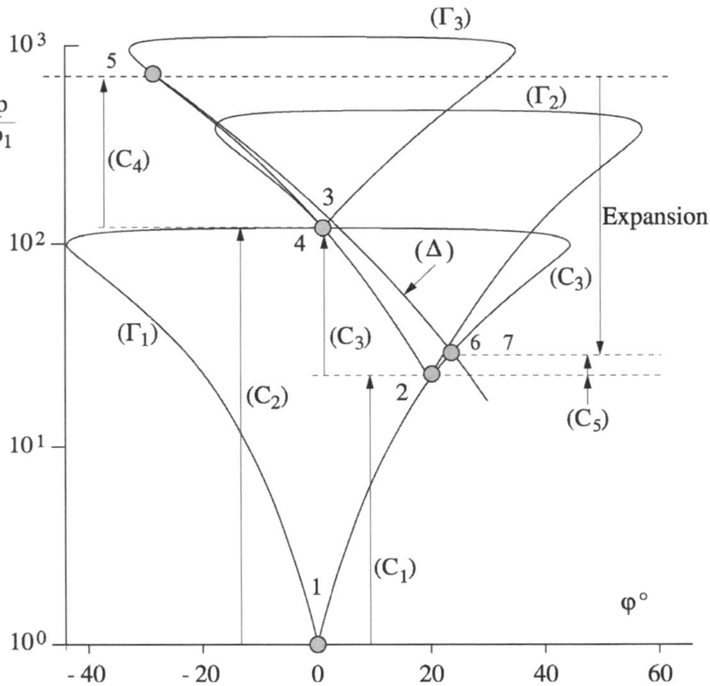

Figure 15.11. *Type III interference at Mach 10 and slip line impact. Polar plane*

In hypersonics, the heat transfer in the vicinity of *R* can be 10–20 times higher than the heat transfer existing at the nose of the vehicle. As we know, the heat flux $\dot{q}_w$ (in *W/m*²) between a wall with the temperature T_w and a flow with the velocity U_e, the density ρ_e and the stagnation temperature T_{st} , is given by an equation of the form (see section 2.4.3):

$$\dot{q}_w = C_h \, \rho_e \, U_e \, C_p \left(T_{st} - T_w\right)$$

where C_h is the heat transfer coefficient (Stanton number) and C_p the specific heat at constant pressure. We assume that the coefficient C_h varies little with the configurations. Knowing that the stagnation temperature T_{st} does not change across a shock wave, we see that the heat flux will be higher the more elevated the velocity and density are. The velocity is related to the local Mach number, which is supersonic in region (3), whereas it is weakly subsonic behind the shock forming in front of the nose. Above all, the density ρ_e is proportional to the stagnation pressure of the flow. The compression experienced by flow (3) contiguous to the mixing layer is, in terms of stagnation pressure loss, much more efficient than the one experienced by the flow stagnating at a front stagnation point after crossing an intense shock. This fact allows us to achieve pressure levels well above the Pitot pressure recovered after the shock forming at the nose. This explains (qualitatively) the high heat transfer levels at the impact of the mixing layer developing along (Σ_1).

Such interferences, thus, may have a local destructive effect on the surface of the vehicle. For this reason, these risks must be carefully evaluated, and if unavoidable, local thermal protections will have to be installed. Figure 15.12 shows an EBF visualization of a type III interference in a flow at Mach 10.

Figure 15.12. *Type III interference between two shocks in a flow at Mach 10 (document Onera)*

15.5. Type IV interference

When the shock (C_2) is nearly normal, a situation is reached where the flow (4) cannot reattach regularly on the body. Thus, as shown in Figure 15.13, a more complex flow establishes, composed of a supersonic jet imbedded in the subsonic domain located between the front shock and the obstacle.

Until the region (4), the field is the same as in the case of a type III interference. Thus, such a flow can also be regarded as a special case of this interference, but with significant differences in the organization of the downstream field. We again find a second triple point T_2, but now the shock (C_5) is more intense, so that the flow downstream of (C_5) is subsonic, which means that the pressure there will not vary much compared to a supersonic condition. Downstream of shock (C_4), flow (4) is supersonic. We consider a supersonic jet between two subsonic, almost isobaric, flows (6) and (3). This jet is limited by two boundaries, (f_1) on the side of (6) and (f_2) on the side of (3), the pressure being nearly constant along (f_1) and (f_2). In order to compensate for the pressure jump across (C_5), a centered expansion arises at the point J, where (C_4) meets the boundary (f_2) (see section 17.5). As a result, (f_2) is deflected upward (according to the drawing in Figure 15.13), this positive deflection adding itself to the one produced by (C_5) (see Figure 15.14).

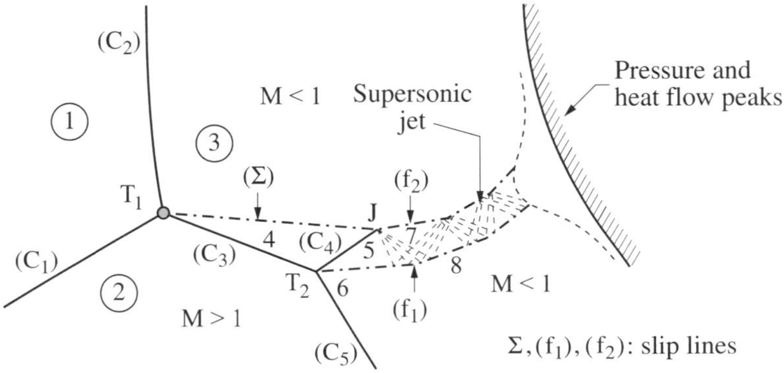

Figure 15.13. *Type IV interference. Physical plane*

The structure of a supersonic jet is discussed in greater detail in section 20.8. The image of the expansion in J is the isentropic polar (Δ_1). Image point 7 of the state after expansion is found on this polar at the point where the pressure is equal to p_2. As the pressure in 7 is incompatible with the pressure of the flow (6) $(p_3 < p_6)$, compression waves must originate in the region where the expansion waves coming from J meet the boundary (f_1) in order to restore on (f_1) the pressure p_6. The image of these waves is the part of the isentropic polar (Δ_2), uniform state (8) having as image the point (Δ_2), where $p = p_6$. This expansion has the effect of increasing the deflection of the flow, the velocity angle φ being larger. Thus, as shown in Figure 15.13, the supersonic jet is more and more curved upward. The images of the successive expansions and compressions are parts of the isentropic polars (Δ_1), (Δ_2), etc., plotted in Figure 15.14.

The above scenario is repeated until the jet reaches a close vicinity to the body surface. Then it is the origin of a quasi-normal shock, which is followed by a subsonic region with a stagnation point at the impact on the body where we experience pressure and heat flux peaks that may be much higher than in a type III interference.

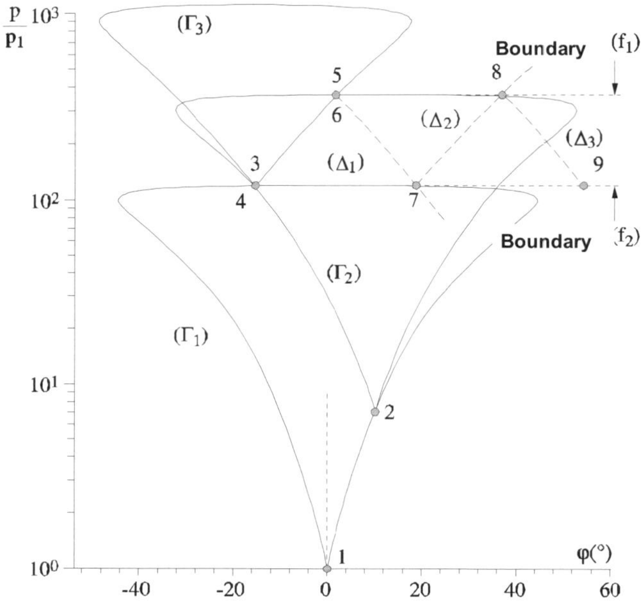

Figure 15.14. *Type IV interference at Mach 10. Polar plane: situation a*

Figure 15.15 shows an enlarged view of the jet, which consists of a series of triangular areas. If the flow downstream of the shock (C_4) is uniform and the pressures in the subsonic regions (3) and (6) are constants, then the zones (5), (7) (8), etc., are uniform and the compressions and expansions are of the simple wave type, except for regions where waves intersect (see Chapter 17).

Figure 15.16 shows an alternative to the previous scenario. As the compression waves are converging, it is possible that they intersect. Thus, a focalization shock forms within the supersonic jet, the impact of this shock with the boundary (f_2) being the origin of a centered expansion (the situation is analogous to that at the point *J*). As the focalization shock is curved, the jet will contain a rotational region (see section 20.7). History may well repeat itself until the impact with the body. In this case, the structure of the jet will be more complex since it contains rotational regions, so that it is no longer a succession of uniform and simple wave regions. Figure 15.17 illustrates the structure of the supersonic jet developing between two regions with constant pressure (3) and (8). This calculation was performed by the rotational method of characteristics presented in Chapter 18. In Figure 15.17a is plotted the mesh of the characteristics highlighting focalizations of compression waves giving rise to shocks. Figure 15.17b gives the iso-pressure contours in the jet. Figure 15.18 shows an EBF visualization of a type IV interference in a flow at Mach 10.

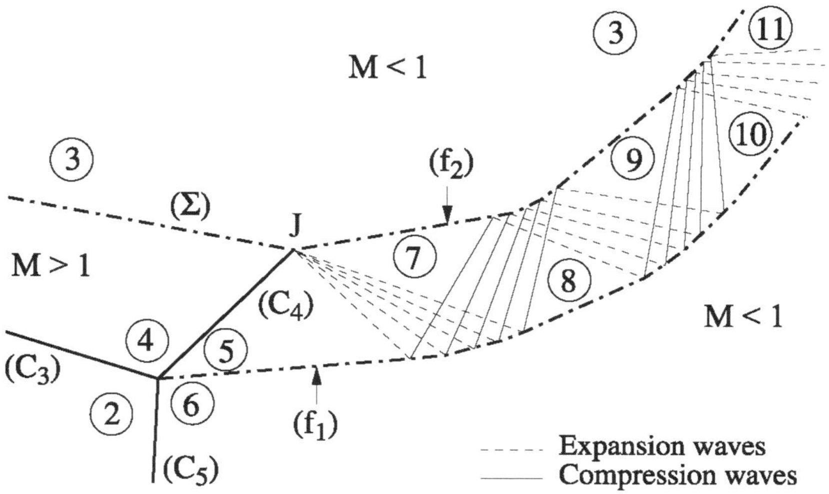

Figure 15.15. *Type IV interference.*
Detail of the supersonic jet structure: situation a

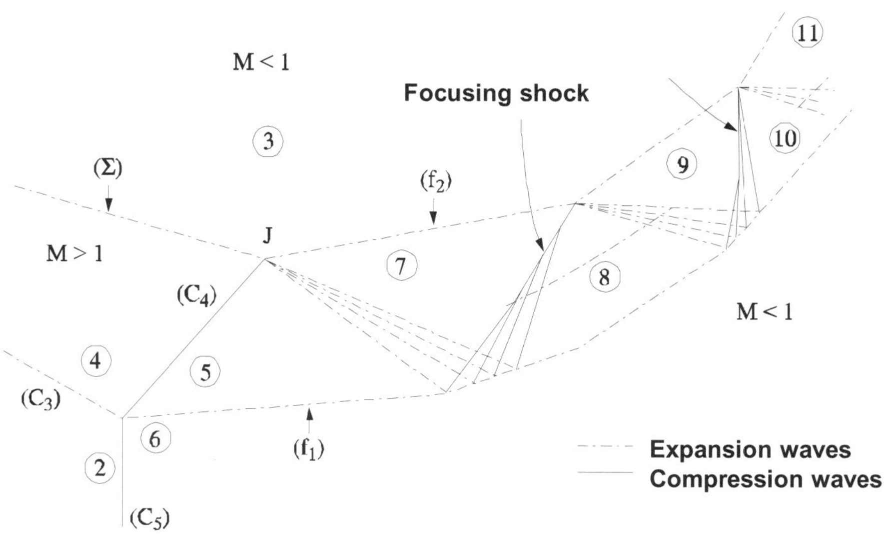

Figure 15.16. *Type IV interference.*
Detail of the supersonic jet structure: situation b

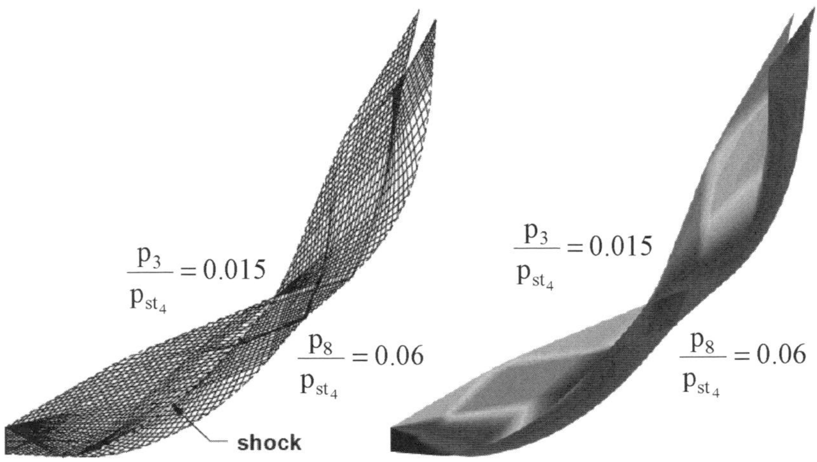

a – characteristics mesh *b – iso-pressure contours*

Figure 15.17. *Supersonic jet between two regions with constant pressure. Calculation by the rotational method of characteristics*

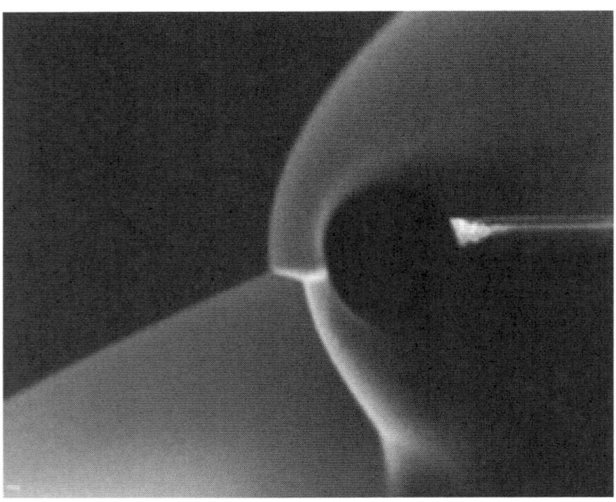

Figure 15.18. *Type IV interference between two shocks in a flow at Mach 10 (document Onera)*

15.6. Type V interference

This type of interference occurs when a weak oblique shock (C_1) encounters an intense shock (C_2) according to the drawing shown in Figure 15.19. In this case, the flow in front of (C_2) is the flow (2) downstream of (C_1). This kind of interference results from the intersection of shocks of the *same family* (two upward shocks in the present situation, whereas in previous situations, one shock was upward and the other downward). As shown in Figure 15.19, the structure of the field resulting from the intersection of (C_1) and (C_2) is rather complex. Here, the presence of a wall in the vicinity was taken into account.

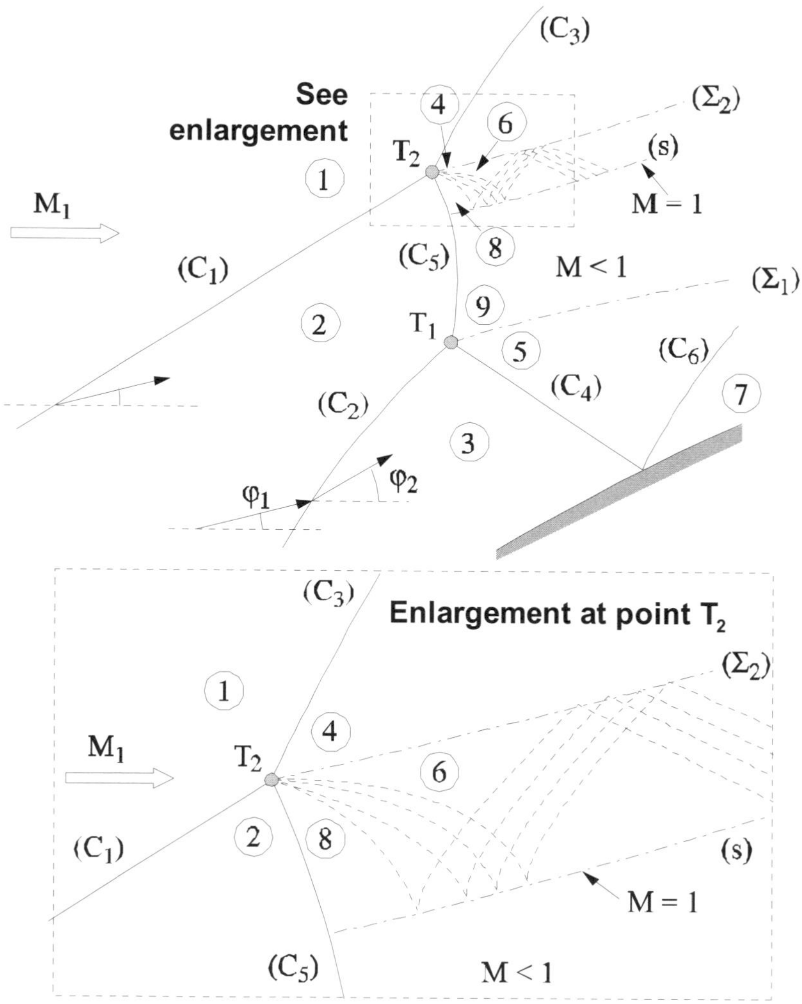

Figure 15.19. *Type V interference. Physical plane*

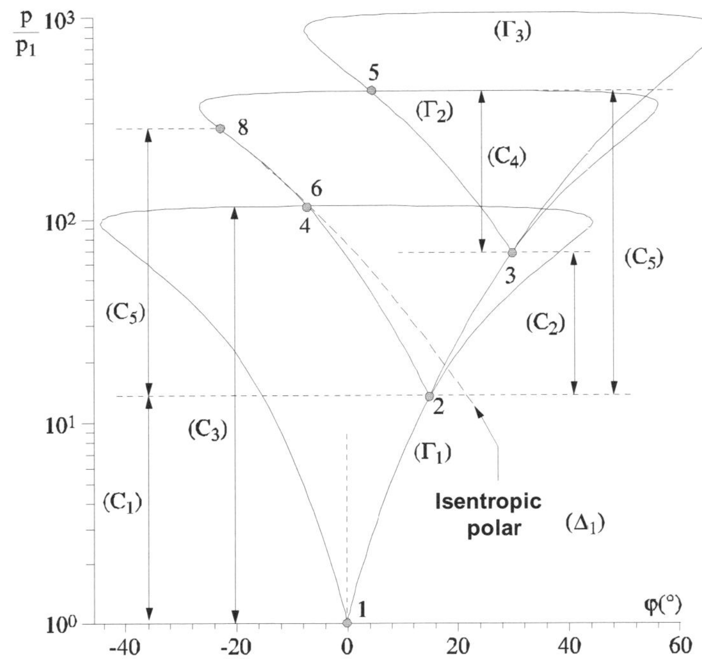

Figure 15.20. *Type V interference at Mach 10. Polar plane*

The field contains two triple points T_1 and T_2. The structure of the flow near these points is established by considering the polar plane in Figure 15.20. We start with point T_1 where the situation is simpler. The uniform upstream flow (1) going through the shocks (C_1) and (C_2) with successive positive deflections, the corresponding states (2) and (3) have the image points 2 and 3 located, respectively, on the polar (Γ_1) for the shock (C_1) and on the polar (Γ_2) for shock (C_2). Shock (C_5), also represented by polar (Γ_2), and shock (C_4), whose image is the polar (Γ_3), originate from the point T_1. States (5) and (9) with compatible pressures and directions are at the intersection of polars (Γ_2) and (Γ_3).

For the same reasons as above, a slip line (Σ_1) starting from T_1 separates two flows of which one is supersonic, downstream of (C_4), and the other subsonic, downstream of (C_4).

We look at the situation at point T_2 where the shocks (C_1), (C_5), and (C_3) converge. The solution, which establishes here and which has been observed, is as follows (other solutions may be possible): through shock (C_5), the flow passes from (2) to a downstream state (8) located on the polar (Γ_2), the Mach number at 8 is supersonic (the jumps from 2 to 9 and 2 to 8 are due to the same shock (C_5) whose properties in T_1 and T_2 are different). The shock (C_5) between the points T_1 and T_2 has as image the portions 8 to 9 of the polar (Γ_2). Point 8 is located below the point where the Mach number after the shock is just equal to 1. Thus, the flow is supersonic in 8. Moving along the shock (C_5) from 8 to 9, the flow on the downstream side of (C_5) passes thus from supersonic to subsonic.

Shock (C_3) makes state (1) pass to a downstream subsonic state (4) whose image is on the polar (Γ_1). It is clear that the states (4) and (8) are incompatible, the pressure in 8 being greater than the pressure in 4: flow (8) must expand. This expansion takes place through a centered expansion wave emerging from the point T_2, as shown in Figure 15.19. The plot of this expansion is the isentropic polar (Δ_1), which meets the shock polar (C_3) at the image point of state (6) compatible with state (4). As in the case of the type IV interference, a supersonic jet forms between the slip line (Σ_2), emerging from T_2, and the sonic line (S) of the flow downstream of the shock (C_5). The expansion waves coming from T_2 are reflected on (S) as compression waves, which, in turn, reflect on (Σ_2) as expansion waves, etc. A close-up of the flow near T_2 is given in Figure 15.19.

In types III, IV, and V interferences, the position of the interference region cannot be determined from the mere knowledge of the shocks (C_1) and (C_2) since the flow contains subsonic regions. A calculation of the interference requires the use of numerical methods based on solving the Euler equations and taking into account the whole field. If we wish to represent the heat transfer effects and have a quantitative prediction, we have to use the Navier-Stokes equations.

Figure 15.21 shows an EBF visualization of a type IV interference in a flow at Mach 10. The phenomena here are too diffuse and the visualization technique is not sensitive enough for us to distinguish the previously described details of the flow.

15.7. Type VI interference

This problem is one of the intersection of two shocks of the same family, upward shocks, for example, whose intensities are not too different, and moderate (weak oblique shocks). This situation, shown in Figure 15.22, corresponds to the intersection of two shocks generated by an air intake ramp with two successive changes of direction.

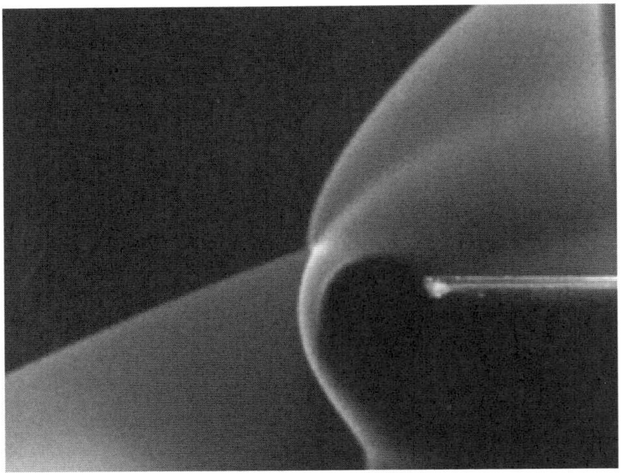

Figure 15.21. *Type V interference between two shocks in a flow at Mach 10 (document Onera)*

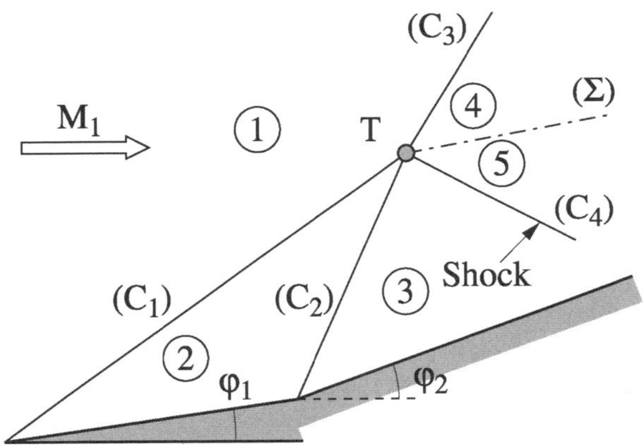

Figure 15.22. *Type VI interference. Physical plane: situation VIa*

The structure of the flow is simpler here than for the previous problems. As shown in Figure 15.23, the first shock (C_1) [polar (Γ_1)] causes the flow of the upstream state (1) to pass to the state (2) to which correspond a direction φ_2 and a pressure p_2. A second deflection $\Delta\varphi = \varphi_3 - \varphi_2$ gives rise to a shock (C_2) leading to condition (3) located on the polar (Γ_2). The two shocks (C_1) and (C_2) intersect at the triple point T from which the shock (C_3) emerges, causing state (1) to pass to state (4) whose image point is on the polar (Γ_1). The plot of the polars shows that flows (3) and (4) are incompatible. We must introduce an additional state (5) connected to (3) either by a shock (situation VIa) or by an expansion (situation VIb) according to the relative positions of the polars (Γ_1) and (Γ_2).

Figure 15.23 is the plot of the polars for a situation VIa represented in Figure 15.22, which occurs for low or moderate Mach numbers (in this case, $M_1 = 2$).

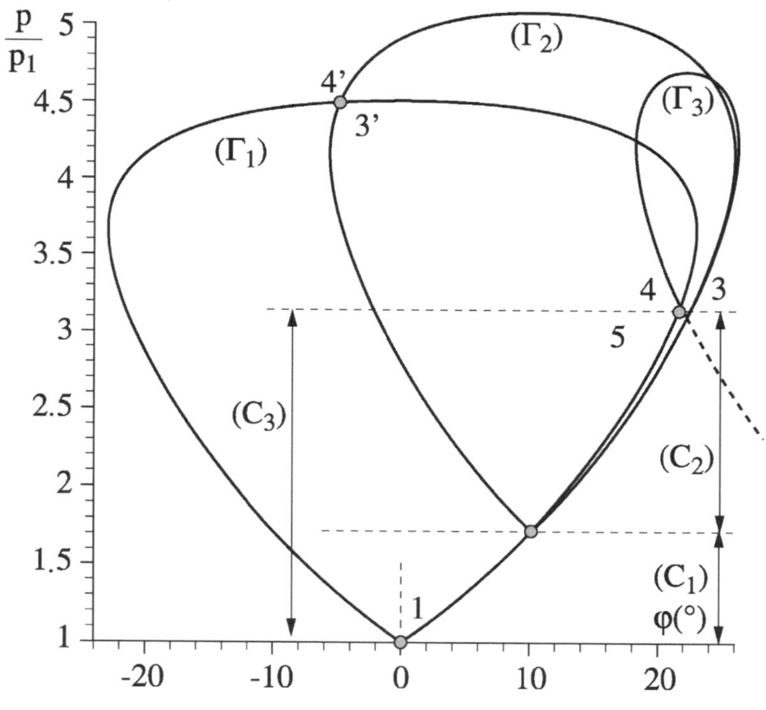

Figure 15.23. *Type VIa interference at Mach 2. Polar plane*

We see that the polar (Γ_1) passes over (Γ_2), resulting in pressure p_3 lower than p_4. State (5) compatible with (4) must therefore be on the shock polar (Γ_3) attached to (3), a shock wave (C_4) starting from the triple point *T*. Flows (4) and (5) are separated by a slip line (Σ). We note that if the upstream state is uniform and the ramps straight, the flows (2) to (5) are uniform in a two-dimensional flow and we built an exact solution for the entire field (in the considered region). Figure 15.24 shows a close-up of the situation at the triple point where we see that the states (3) to (5) are very close. As the polars (Γ_2) and (Γ_3) are close to each other, the adjustment shock (C_4) has a very small intensity.

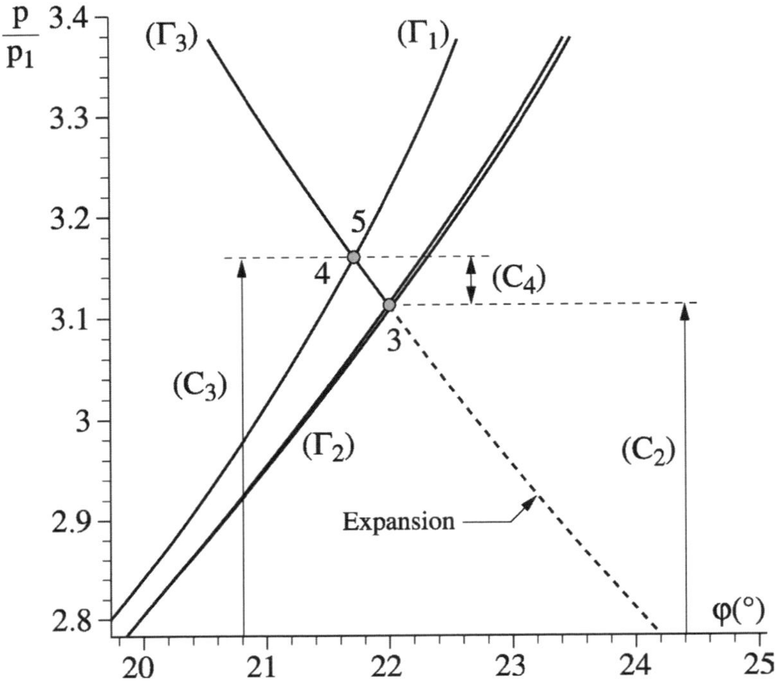

Figure 15.24. *Type VIa interference at Mach 2. Polar plane (close-up)*

Situation VIb is represented in the plane of polars in Figure 15.25. In the analyzed case, the upstream Mach number is equal to 10 and then the polar (Γ_2) passes over (Γ_1). Thus, if the pressure p_3 is greater than p_4, state (5), compatible with (4), is obtained by a centered expansion emerging out of T (see Figure 15.26). The image of (5) is at the intersection of (Γ_1) and the expansion polar (Δ_1). Figure 15.25 shows a variant of this solution in the case where the polar (Γ_2) is sufficiently developed to intersect the polar (Γ_1) in 3'. We then have in 3' and 4' compatible states. However, as 3' is on the strong shock branch of (Γ_1), the flow (3') is generally subsonic, thus depending on downstream conditions. If these conditions are such that the condition (3') must be supersonic, then there will be a jump to the solution of Figure 15.25.

This remark is very general. In the case of multiple solutions, it is the solution with weak shocks that establishes; unless the downstream flow must be subsonic, as in the case of types III and IV interferences, where the proximity of the body causes stagnation in the flow.

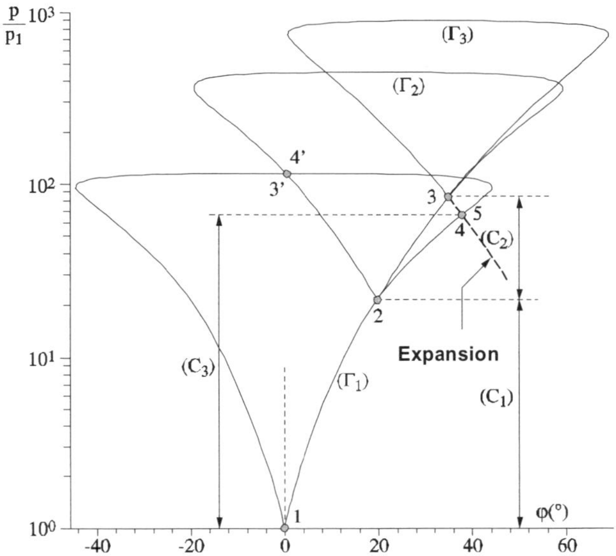

Figure 15.25. *Type VIb interference at Mach 10. Polar plane*

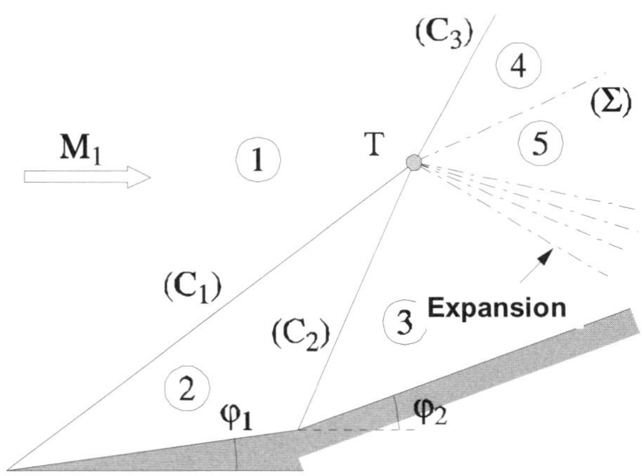

Figure 15.26. *Type VI interference. Physical plane. Situation VIb*

15.8. Lambda shock structure

In flows with low Mach numbers (less than about 2), we frequently observe the following interference of shocks, called *lambda shock*, although it involves three shocks. For example, a deflection induced by a wall generates an oblique shock (C_1), which intersects a normal, or near-normal, shock (C_3) (see Figure 15.27). The situation, shown in the plane of shock polars in Figure 15.28, is for an upstream flow with a Mach number of 1.5 and a deflection of 5°. Since the states (2) and (4), respectively, downstream of (C_1) and (C_3) are incompatible, a shock wave (C_2) forms, imposing the passage from (2) to (3) on the polar (Γ_1). The states (3) and (4) are compatible and separated by a slip line (Σ), which is usually made visible by optical visualization techniques. The situation is analogous to the situation at points 3' and 4' shown in Figure 15.25. The considered solution here is driven by the flow downstream of the shocks (C_2) and (C_3), which is subsonic and thereby dependent on downstream conditions. The lambda shock structures are frequently observed in transonics, especially during the separation of the boundary layer caused by a shock wave.

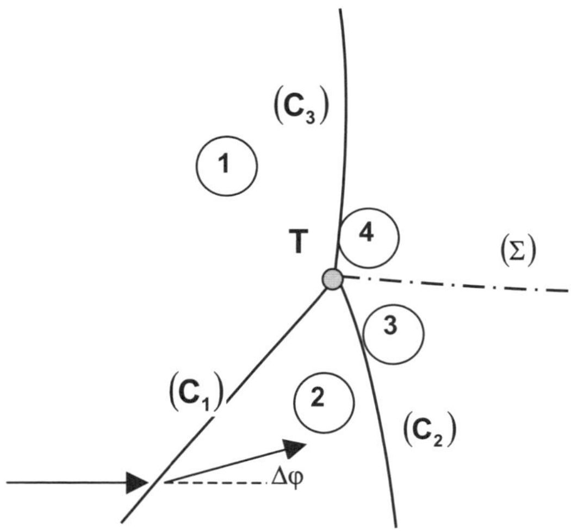

Figure 15.27. *Lambda shock structure. Physical plane*

The schlieren visualization in Figure 15.29 shows a lambda shock structure observed in a transonic flow. We note the slip line starting at the triple point where the three shocks converge.

15.9. Transition between type VI and type V interferences

If the conditions are such that the polar (Γ_2) is well developed and if the deflection $\Delta\varphi = \varphi_3 - \varphi_2$ is large, it may happen that point 3 is above the polar (Γ_1) as shown in Figure 15.30. Then the shock polar (Γ_3), attached to the condition (3), no longer intersects (Γ_1) and the branch (Δ_1) of the isentropic polar intersects (Γ_1) in 4' on its strong shock part. This represents a possible solution. Similarly, we can imagine the formation of an intermediate shock (C_5) within the flow, whose image is the polar portion between 5.9 and 4''.8', point 4'' being on the strong shock branch of (Γ_1). However, states such as (4') and (4), mostly subsonic, may be incompatible with downstream conditions that force the weak shock solution. Thus, the solution "flips" toward the type V structure where 4 is on the weak shock branch of (Γ_1), the states (8) and (4) being connected by an isentropic expansion (see figures relating to the type V interference).

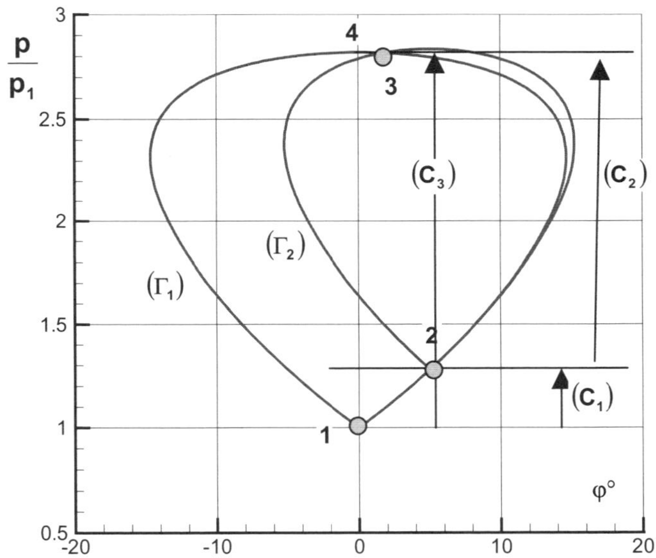

Figure 15.28. *Lambda shock structure. Shock polar plane* $(M_1 = 1.5, \varphi = 5°)$

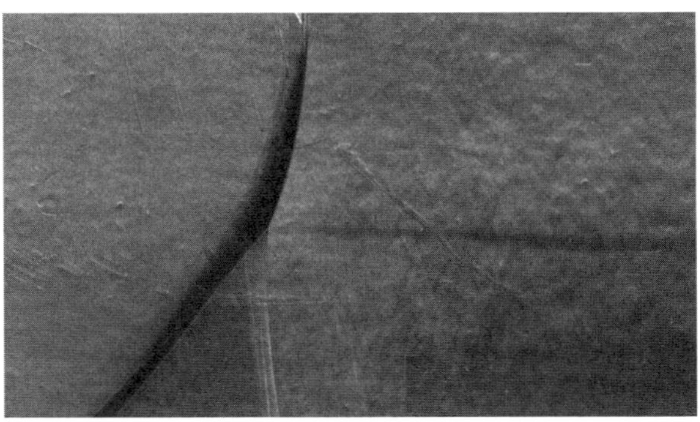

Figure 15.29. *Schlieren picture of a lambda shock in a transonic flow (document Onera)*

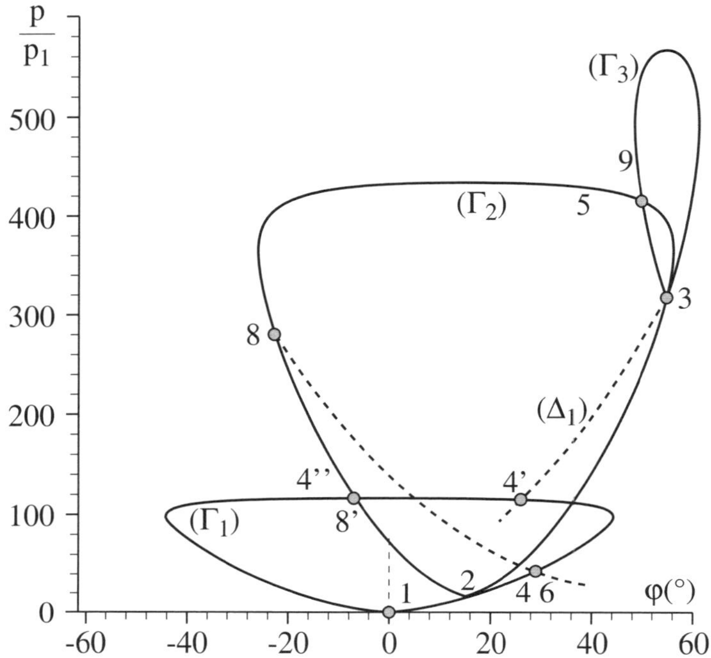

Figure 15.30. *Transition from type VI interference to type V interference at Mach 10*

Thus, type V interference is a special case of "normal" type VI interference, the flow associated with type V being similar to the Mach phenomenon of the type II interference. In this discussion, however, we preferred to adopt the usual order in which the interferences are numbered by considering, as shown in Figure 15.2, a transition of the intersection point from the bottom to the upper side of the body.

Chapter 16

Application of the Shock Wave Theory to Supersonic Air Intakes

16.1. The role of air intake

Air intakes for supersonic aircraft presents an excellent example for the application of the theory of shock waves in steady flow. This chapter describes the operation of such devices given that it is only an introduction to the aerodynamics of air intakes whose design has many aspects that will not be mentioned here. We know that the propulsion of aircraft is provided by the so-called *propulsion nacelle*, or more simply the nacelle, which brings together the various elements contributing to the propulsion (see section 5.2), namely:

– the air intake, capturing air to feed the engine;

– the engine that can be a turbojet, a ramjet, or even a scramjet engine for air breathing hypersonic vehicles;

– a possible reheat duct or afterburner, where kerosene is burnt downstream of the engine in order to provide extra thrust; and

– the exhaust nozzle whose aerodynamics is presented in Chapter 10.

The engines of current airliners additionally have a fan, arranged in front of the compressor and drawing an airflow that does not cross the compressor–turbine set. The operation of the fan is characterized by the bypass ratio, a ratio of the total drawn airflow and the airflow through the compressor. This bypass ratio is around 10 for modern engines. Engines for combat aircraft and missiles have no fan: it is this type of engine that we will consider here.

By convention, we consider the following sections, numbered from 0 to 4 (see Figure 16.1):

– Section (0), upstream of the nacelle where the flow is considered to be undisturbed by the presence of the aircraft. There, we identify the upstream conditions, corresponding to flight conditions: Mach number M_0, pressure p_0, velocity V_0, density ρ_0, etc.

– Section (1) located at the inlet section of the air intake.

– Section (2) or engine entry level located at the entrance to compressor.

– Section (3) or engine exit level at the turbine outlet.

– Section (4) containing the exhaust nozzle exit plane.

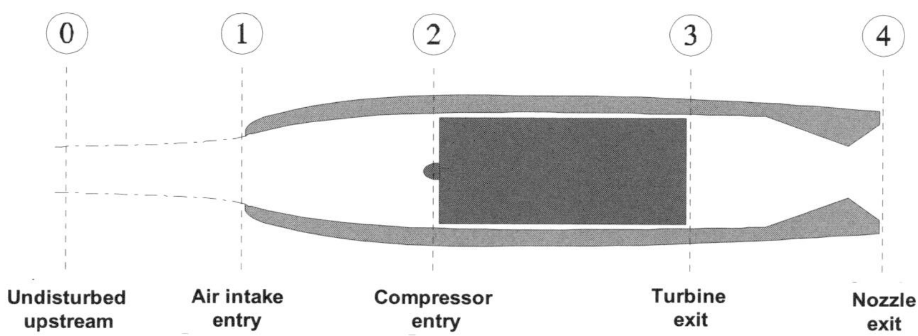

Figure 16.1. *Main sections of a propulsion nacelle*

The role of the air intake is to capture the air that feeds the engine while slowing it down so that the Mach number in the entry level of the engine is close to 0.3-0.5. Taking into account that the functioning, or performance, of the engine will be better the higher the pressure at the entry of the compressor is, the capture should happen with the least loss, that is, with as little decrease of the captured flow stagnation pressure as possible. However, the decrease of the stagnation pressure of a flow is directly related to the production of entropy, which is through the action of the viscosity in the boundary layers and shock waves. In the case of air intakes, the processes are adiabatic because there is no exchange of heat or combustion, and the stagnation temperature is always conserved.

The architecture of air intake is thus made up of a captation part, which can either be in the form of a simple Pitot tube, with a carefully shaped leading edge to avoid separation, or have a more complex geometry in supersonics. Downstream of the captation part, if the flow is subsonic, the inside of the air intake comprises a channel with a cross section ensuring a slowdown of the flow to the desired Mach number in section (2).

For scramjet engines where the combustion takes place in a still supersonic flow, the air intake only slows down the Mach numbers to about 2 to 3 (this type of air intake will not be discussed here).

16.2. Pitot type air intake

16.2.1. *Arrangements and definitions*

It is formed like a Pitot tube placed facing the flow, capturing a stream tube whose section in the upstream undisturbed flow at Mach number M_0 is A_0 (see Figure 16.2). We recall the expression of the mass flow q_m in a one-dimensional, non-viscous, and steady flow (see section 9.5.2):

$$\boxed{q_m = \frac{\delta(\gamma)}{\sqrt{C_p \;\; T_{st}}} p_{st} \;\; A_c}$$

where:

$$\delta(\gamma) = \frac{\gamma}{\sqrt{\gamma - 1}} \left(\frac{\gamma + 1}{2} \right)^{-\frac{\gamma+1}{2(\gamma-1)}}$$

is a function of γ with the value $\delta(\gamma) = 1.281$ in the case of air.

In the case of an air intake, the captured flow undergoes an adiabatic evolution, its stagnation temperature remaining constant, even if it passes through shock waves. If the mass flow is constant (except in special cases), then the product:

$$p_{st} \; A_c$$

is an invariant of the flow. This relation is frequently used in the design and analysis of the operation of air intake. Furthermore, the air intake provides a slowdown of the captured flow to a subsonic Mach number (close to 0.3-0.5) in the section of the compressor entry, a value imposed by the engine operation.

During this slowdown, the flow suffers losses, that is, the (average) stagnation pressure decreases. This decrease is caused by:

– the crossing of shock waves in supersonic flight; and

– the boundary layers developing in the diffusing part of the air intake.

The losses are related to the entropy production, which occurs here by the action of the viscosity (no combustion or heat transfer). Thus, if p_{st_0} denotes the stagnation pressure in the upstream flow, then in the section of the engine inlet:

$$p_{st_2} < p_{st_0}.$$

We denote by A_1 the frontal section of the air intake and A_0 the section of the stream tube captured by the air intake in the part where the flow is uniform with the Mach number M_0.

The performances of an air intake are characterized by two parameters, namely:

– *the* mass flow coefficient:

$$\boxed{\varepsilon = \frac{A_0}{A_1}}$$

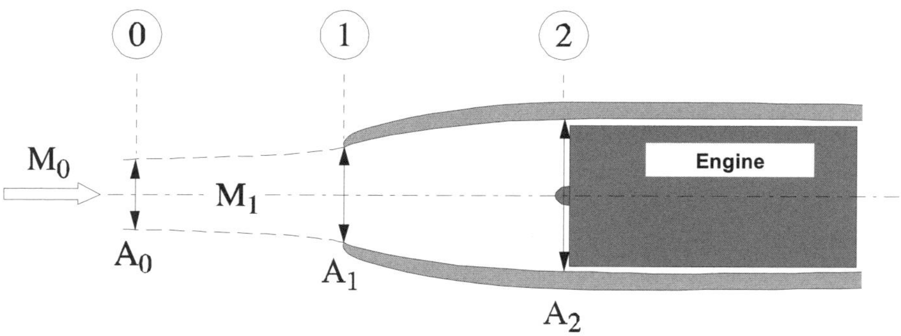

Figure 16.2. *Definition of a Pitot type air intake*

The mass flow coefficient represents the ratio between the section of the stream tube of the upstream flow that is captured by the air intake and the geometric section of the air intake entrance. We can understand intuitively that the optimum is achieved if $\varepsilon = 1$. The entry of the air intake is then dimensioned at its best given the airflow required for engine operation. In this case, the air intake is said to be *adapted*. Adaptation is only fulfilled for conditions corresponding to cruise flight.

Apart from this case, and if for various reasons, the regime of the engine varies, then there will be no adaptation.

The mass flow expressed from quantities relative to the uniform upstream flow with Mach number M_0 is expressed:

$$q_m = \varepsilon\, A_1\, \rho_0\, V_0$$

– *the efficiency*:

$$\boxed{\eta = \frac{p_{st_2}}{p_{st_0}}.}$$

This number represents the capacity of the air intake to slow the flow without significantly degrading the level of the maximum recoverable pressure. Indeed, the maximum pressure that could be achieved in front of the engine is the isentropic stagnation pressure p_{st_0} of the upstream flow. If it were so, η would be equal to 1. For the reasons discussed above, η is always less than 1.

One of the objectives of air intake designers is to achieve maximum efficiency, that is, to ensure that η is as close as possible to 1. We know that the ideal slowdown is the isentropic stagnation, which conserves the stagnation pressure. In reality, this condition cannot be achieved due to the presence of boundary layers, and in supersonics, shock waves.

16.2.2. *Operation in subsonic flight*

In cruise flight, the captured flow is slowed not only in the internal diffusing area of the air intake (a duct with increasing section), but also upstream of it, the section of the stream tube of the captured flow expanding from A_0 to the value A_1. We recall that in subsonics, a slowdown increases the section of the stream tube (see section 9.3, Hugoniot theorem). It is therefore necessary (see Figure 16.3a) that:

$$A_0 < A_1$$

and consequently:

$$\varepsilon = \frac{A_0}{A_1} < 1.$$

However, at ground level where the velocity of the aircraft is low or zero, the engine sucks in outside air that is thus accelerated so that in this case (see Figure 16.3b):

$$\varepsilon = \frac{A_0}{A_1} > 1.$$

At a fixed point on the ground, the definition of ε becomes tricky because the section of the captured stream tube current tends towards infinity.

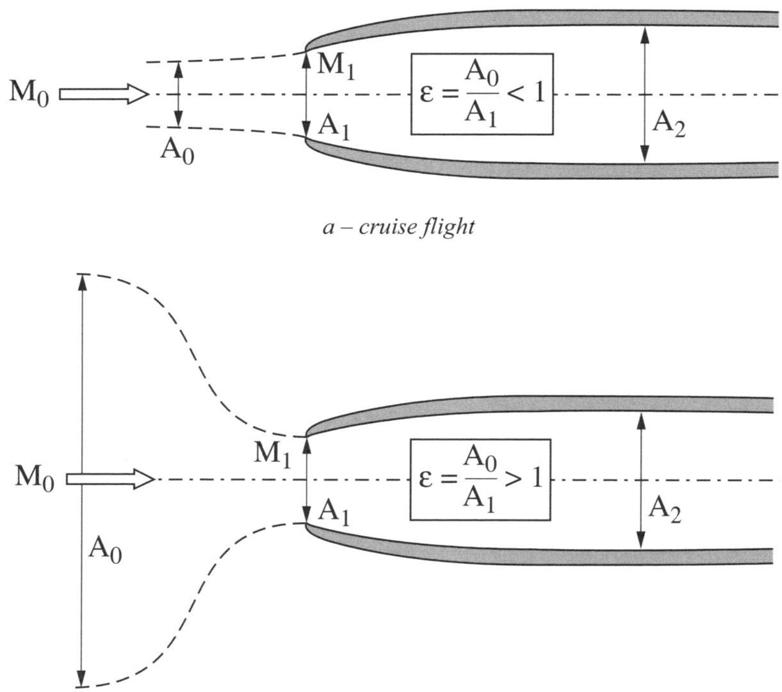

b – flight close to the ground (taxiing, take off, landing)

Figure 16.3. *Pitot air intake in subsonic flow*

16.2.3. *Operation in supersonic flight*

The air intakes considered here are of the Pitot type, the deceleration of the flow occurring primarily through a normal, or nearly normal, shock wave that forms before the air intake.

We distinguish three operation types.

a – Subcritical regime *(see Figure 16.4a)*

This is established when the engine is in a reduced regime, which causes a reduction of the captured mass flow. As:

$$\overline{q}_m = A_1 \, \rho_0 \, V_0$$

is the maximum mass flow that the air intake can capture, it is a requirement that the effective mass flow:

$$q_m = \varepsilon \, A_1 \, \rho_0 \, V_0$$

is lower than $\overline{q}_m$, which requires: $\varepsilon < 1$. Then, the upstream stream tube has a section A_0 smaller than the inlet section A_1. In this case, a detached shock forms in front of the air intake. The section of the captured stream tube, which is constant and equal to A_0 upstream of the shock, increases downstream of the shock where the flow slows down to subsonic.

b – Critical regime (see Figure 16.4b)

If the engine requires an increased captured mass flow, then the quantity:

$$q_m = \varepsilon \, A_1 \, \rho_0 \, V_0$$

increases, which imposes an increase in ε and thus in A_0.

The shock will be closer to the inlet section of the air intake and a situation will be reached, where:

$$A_0 = A_1.$$

Then the captured mass flow has a value:

$$q_m = A_1 \, \rho_0 \, V_0$$

and:

$$\boxed{\varepsilon = \overline{\varepsilon} = 1.}$$

c – Supercritical regime (see Figure 16.4c)

We assume that for some reason, the Mach number M_2 increases. Since section A_2 is fixed due to construction, the critical section A_{c_2} of the flow in section (2) must increase. Consequently, according to the area-Mach number relation in subsonic flow, the ratio $\frac{A_2}{A_{c_2}} = \Sigma(M_2)$ has to decrease (see section 9.5.2.3).

As the mass flow conservation requires that:

$$p_{st_0} A_{c_0} = p_{st_2} A_{c_2}$$

and the upstream supersonic flow is invariant, it follows that p_{st_2} must decrease.

Therefore, the normal shock must be more intense, thus occurring at a higher Mach number. Consequently, the shock penetrates inside of the air intake, where the diffusing section is increasing (in supersonics, under the Hugoniot theorem, an increase in the Mach number must be accompanied by an increase in the section).

It is noteworthy that, since the stream tube entering by the air intake is fixed, the captured mass flow now remains constant, i.e.:

$$\boxed{\varepsilon = \overline{\varepsilon} = 1.}$$

In subcritical and critical regimes, the efficiency is practically only defined by the loss of stagnation pressure across the normal shock. Thus, if the Mach number M_0 is constant, this ratio p_{st_1}/p_{st_0} does not vary (the loss through friction at the wall in the diffusing duct varies little and is much less important than the loss through the shock when the Mach number becomes clearly supersonic). Thus, for these regimes, if ε decreases, η remains virtually constant.

In contrast, $\varepsilon = \overline{\varepsilon}$ remains constant in supercritical regime while η drops due to the increased intensity of the shock wave. Hence, the *operating characteristic* of the air intake has the form shown in Figure 16.5.

If we only consider the loss of stagnation pressure across the shock forming in front of the air intake, efficiency is given by the normal shock relation:

$$\eta = \Pi(M_0, \gamma)$$

where the function $\Pi(M_0, \gamma)$ is the ratio of the stagnation pressures on both sides of a normal shock (see section 13.5.2.3).

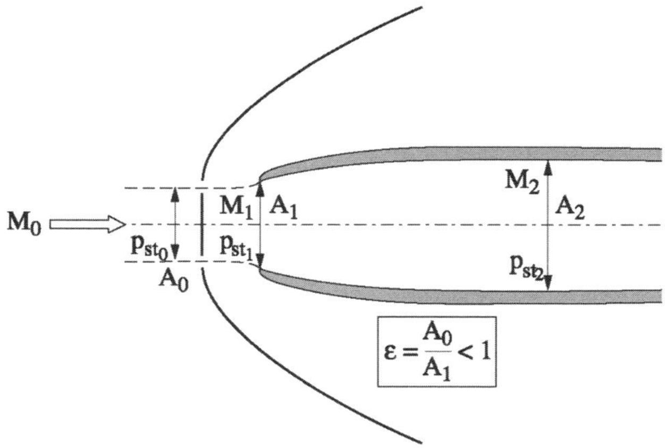

a – subcritical regime

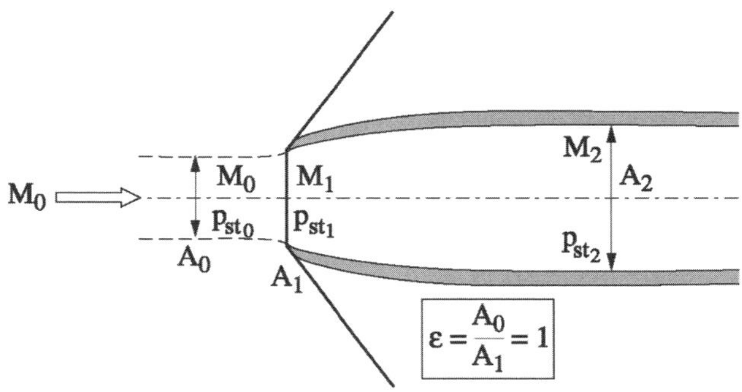

b – critical regime

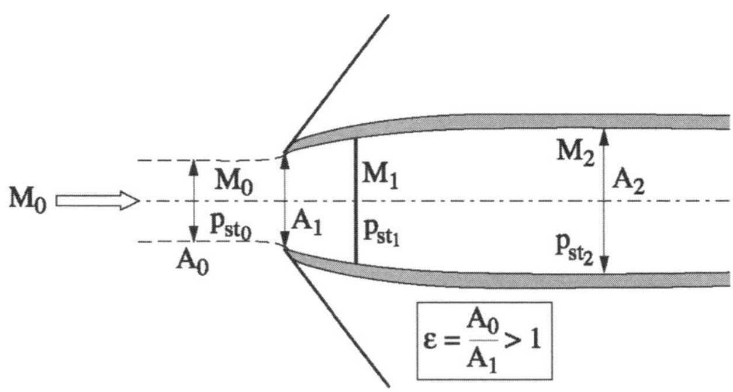

c – supercritical regime

Figure 16.4. *Pitot type air intake in supersonic flow*

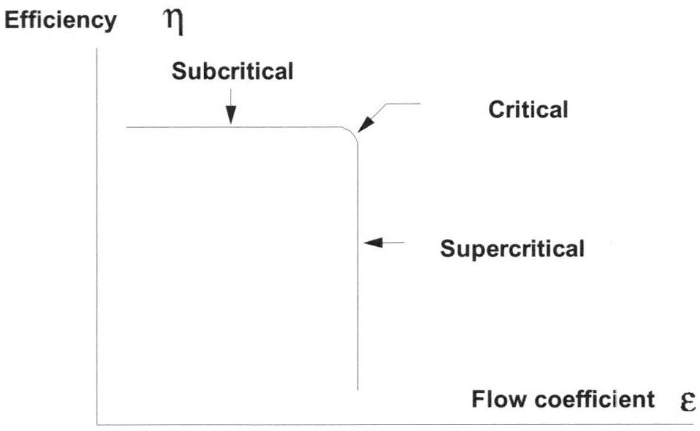

Figure 16.5. *Operating characteristic of a supersonic air intake*

Table 16.1 below gives the stagnation pressure loss through a normal shock as a function of the flight Mach number:

M_0	1	2	3	4	5	6
η	1.0000	0.7210	0.3280	0.1390	0.0617	0.0297

Table 16.1. *Loss of stagnation pressure across a normal shock wave*

We see that the efficiency of Pitot air intake deteriorates quickly when the Mach number increases. In practice, it is unacceptable from Mach 2 on.

16.2.4. *An important relation between the efficiency and the mass flow coefficient*

As in air intake flows the stagnation temperature is conserved, it has been demonstrated in section 16.2.1 that *the product* $p_{st}A_c$ *is an invariant*, then:

$$\boxed{p_{st_0}\ A_{c_0} = p_{st_2}\ A_{c_2}}$$

from which follows:

$$A_{c_0} = A_{c_2}\frac{p_{st_2}}{p_{st_0}} = \frac{A_{c_2}}{A_2}\frac{p_{st_2}}{p_{st_0}}\ A_2$$

then:

$$\varepsilon = \frac{A_0}{A_1} = \frac{A_0}{A_{c_0}} \frac{A_{c_2}}{A_2} \frac{p_{st_2}}{p_{st_0}} \frac{A_2}{A_1}$$

which gives:

$$\boxed{\varepsilon = \frac{\Sigma(M_0)}{\Sigma(M_2)} \eta \frac{A_2}{A_1}}$$

where $\Sigma(M) = A/A_c$ is the area-Mach number relation (see section 9.5.2.3).

The Pitot type air intakes equip the propulsion nacelles of airliners whose cruise Mach number is in the range of 0.75-0.9. Thus, the operation is of the subsonic type. They also equip combat aircraft operating at Mach numbers up to 1.6: Dassault Aviation *Rafale*, *F16*, *F18,* for example, despite a non-negligible loss of efficiency through shock ($p_{st_1} / p_{st_0} = 0.895$) Under these conditions, their simplicity, coupled with the fact that fighter jets only operate in supersonics for short periods, is an advantage that makes them preferable to ramp air intakes whose efficiency is better, but the design more complex.

16.3. Two-dimensional supersonic air intake

16.3.1. *Design and operation regimes*

Pitot type air intakes have poor efficiency in supersonics because almost all of the compression is done through a single normal shock wave. We know that for a transition from the pressure p_0 to the pressure p_1, it is the normal shock that produces the maximum of entropy, thus causing the greatest stagnation pressure loss (see section 13.5). In practice, Pitot air intakes are not used in supersonics beyond Mach numbers 1.6-1.8. For high flight Mach numbers, we will seek to define an air intake where the compression is split into several shocks leading to a lower entropy production and a higher compression ratio p_2/p_0 .

In a two-dimensional flow, the most simple air intake consists of a straight ramp that has an angle φ with respect to the uniform upstream flow with the Mach number M_0 (see Figure 16.6). The ramp induces an oblique shock wave, inclined at an angle σ , which causes an initial compression of the flow. On the ramp, the flow is uniform with the Mach number M_1, usually supersonic. A second deceleration is achieved through an almost normal shock, called *terminal shock*, which ensures the transition to the subsonic flow. The rest of the compression is done inside the air intake whose duct is diffusing (increasing section).

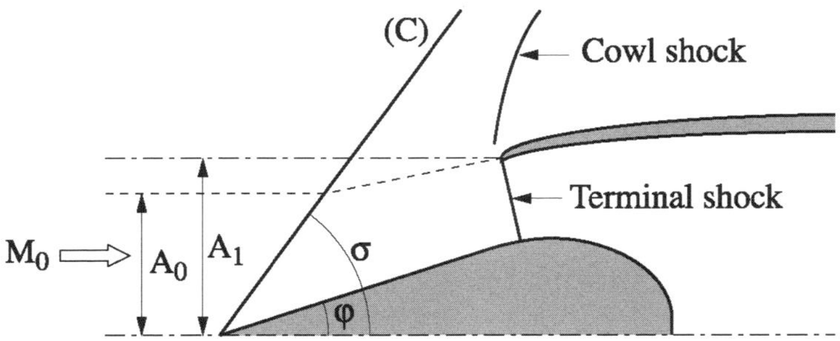

Figure 16.6. *Simple two-dimensional supersonic air intake*

If A_1 designates the front surface of the inlet section and A_0 the section of the upstream flow stream tube captured by the air intake, we define, as for the Pitot air intake, the mass flow coefficient by:

$$\varepsilon = \frac{A_0}{A_1}.$$

Note that downstream of the shock (*C*), caused by the compression ramp, we consider a situation similar to that of a Pitot air intake, the compression of the flow with the Mach number M_1 being carried out through a single shock. Thus, depending on the engine regime, we find again the three cases: subcritical, critical, and supercritical, shown in Figure 16.4.

Note the following points:

– The flow on the ramp, downstream of (*C*), depends only on the Mach number M_0 and the angle φ of the ramp.

– According to the sketches in Figure 16.7, it is clear that the mass flow coefficient ε is less than 1, even for *critical and supercritical regimes*.

This mass flow coefficient will only be equal to 1 if the air intake is *adapted*. In this case, the shock wave (*C*) passes exactly through the leading edge of the cowl and the regime is critical (see Figure 16.8). The adaptation corresponds to the regime of normal operation of air intake in cruise flight. We can represent the change from the upstream state (0) to the downstream state (2) (entry section of the engine), in the hodographic plane $[p, \varphi]$ (see Figure 16.9). The state (1) lies on the shock polar (Γ_0) attached to the upstream state (0) and the state (2) on the polar (Γ_1) attached to the state (1) (normal shock solution).

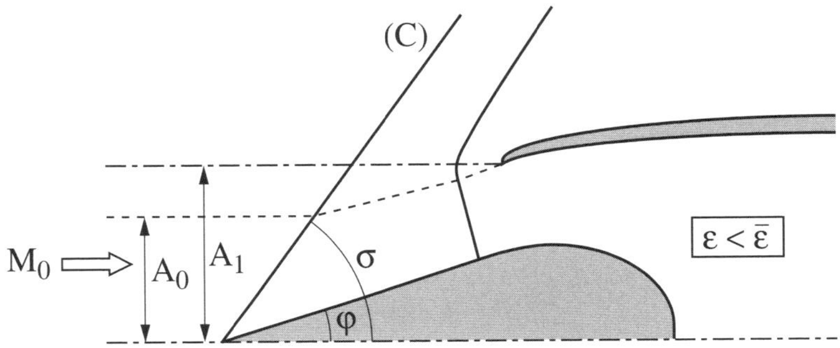

a – subcritical regime

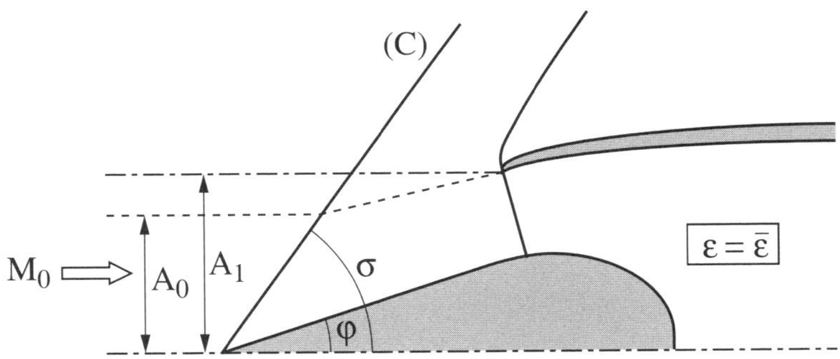

b – critical regime

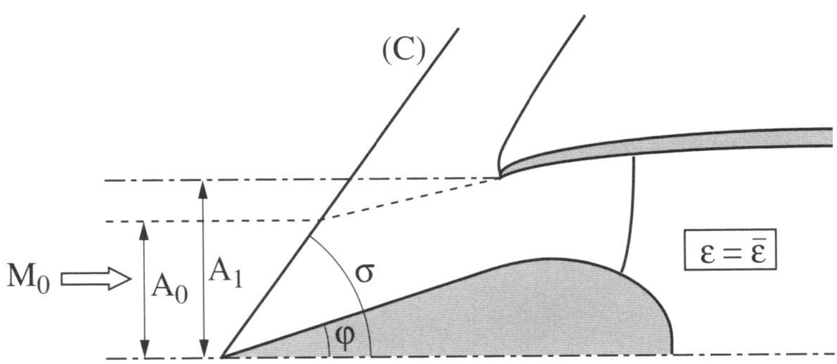

c – supercritical regime

Figure 16.7. *Operation regimes of supersonic air intake*

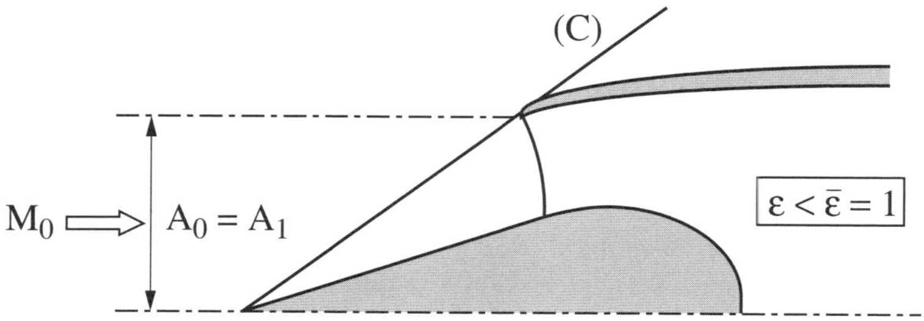

Figure 16.8. *Adapted supersonic air intake*

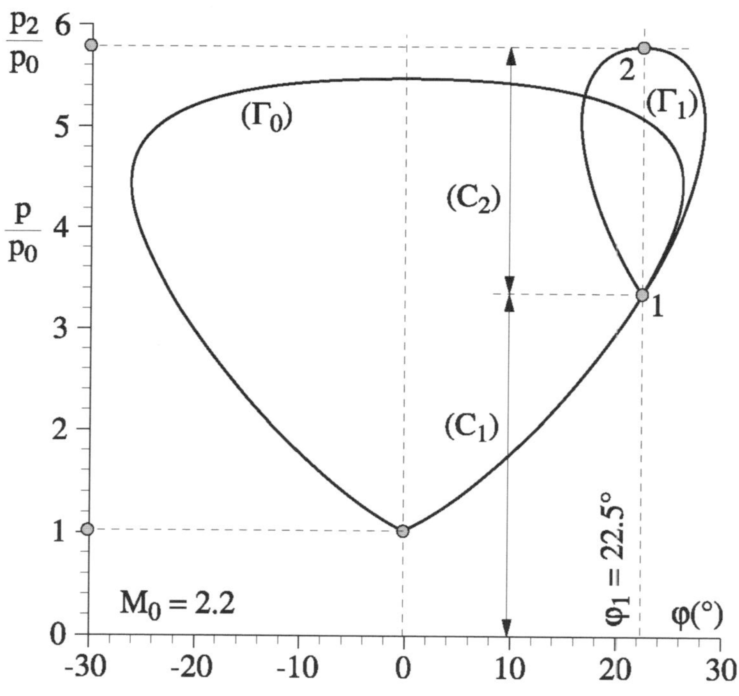

Figure 16.9. *Operation of a simple ramp air intake. Representation in the polar plane*

The transition from (0) to (2), which will cause a single normal shock, happens now through the intermediate state (1). It should be noted that the flow downstream of the terminal shock (considered here as normal) has an angle φ compared to the direction of the upstream flow, whereas φ is zero for a single normal shock. Thus, we will have to deviate the flow in such a way that its direction is along the axis $X'X$, that is, oriented along the axis of the engine. This condition imposes a suitable shape of the internal duct and of the cowl. In Figure 16.7, we see that the cowl will have a certain frontal area facing the flow. As this cowl causes a deflection with a shock wave, the pressure exerted on it is higher than the pressure p_0 of the upstream flow. The resulting a drag is called the *cowl drag*.

16.3.2. *Two-dimensional supersonic air intake with multiple ramps*

It is still possible to split the compression by carrying out a dual ramp air intake, as shown in Figure 16.10, where the air intake is represented for adapted operation. Thus, the system is designed so that both shocks (C_1) and (C_2), as well as the terminal shock, converge on the leading edge of the cowl or very close to it. The situation in the hodographic plane is shown in Figure 16.11, where the polars (Γ_1) and (Γ_2) are associated with the shocks (C_1) and (C_2). In the definition of the compression ramps, we impose, in general, the Mach number on the upstream side of the terminal shock. A value of 1.3-1.4 is usually chosen because it corresponds to incipient shock-induced boundary layer separation by a normal shock.

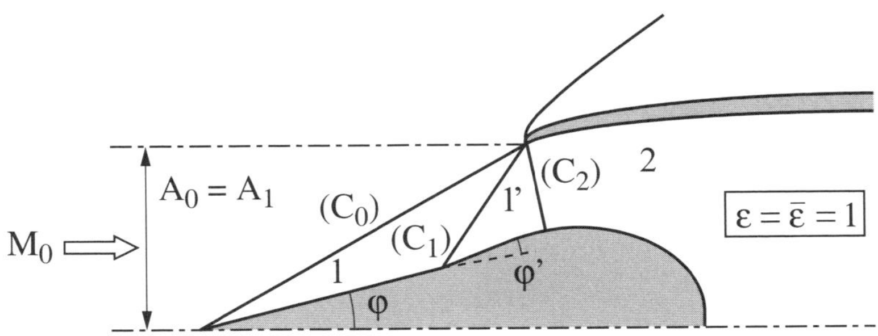

Figure 16.10. *Double ramp supersonic air intake*

It is possible to reduce the Mach number before the terminal shock by extending the ramps with a curved surface providing a progressive isentropic compression (see section 17.5.4). The contour of this ramp is calculated so that the compression waves converge at a point coinciding with the cowl leading edge. The gain produced by a double ramp is important at relatively high flight Mach numbers. Thus, for the *Concorde*, at a flight Mach number of 2, an efficiency of 0.95 was achieved with two ramps followed by an isentropic compression. The Pitot air intake at $M_0 = 2$ has an efficiency of 0.721.

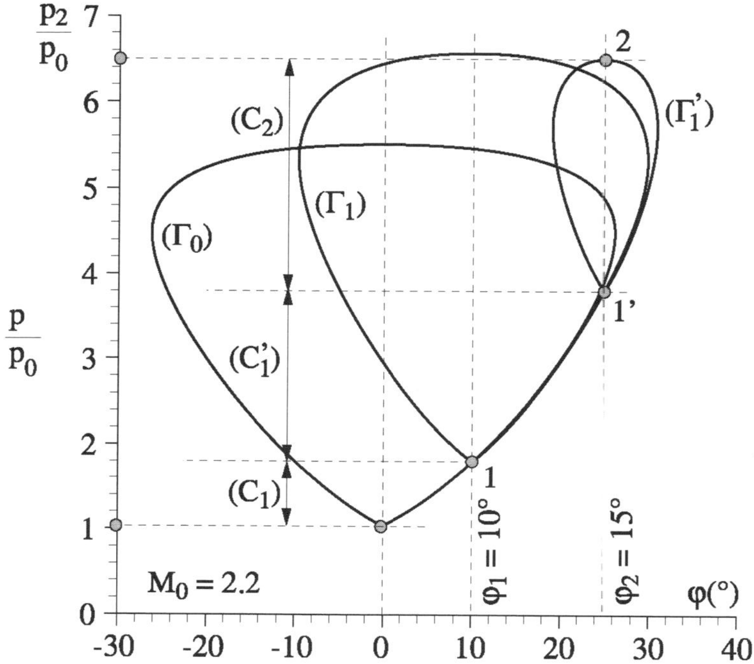

Figure 16.11. *Operation of the double ramp air intake. Representation in the polar plane*

We can ask ourselves why the slowdown of flow could not be ensured in an isentropic manner through the use of a ramp with an appropriately shaped progressive curvature so that all compression waves converge at the cowl leading edge. Then the efficiency would be maximum.

In reality, it would be very difficult to ensure and maintain the convergence of the compression waves at a point. The crossing of waves would be unavoidable, leading to the formation of unwanted shock waves. In addition, for Mach numbers less than 2.4, the isentropic ramp causes a deflection of the flow greater than that produced by a multishock system, resulting in an increased cowl drag (see section 14.2). The Pitot air intake, on the other hand, has the lowest cowl drag, but a very poor efficiency. The different situations are illustrated in Figure 16.12 and Figure 16.13 shows their operation in the plane of the shock and isentropic evolutions polars.

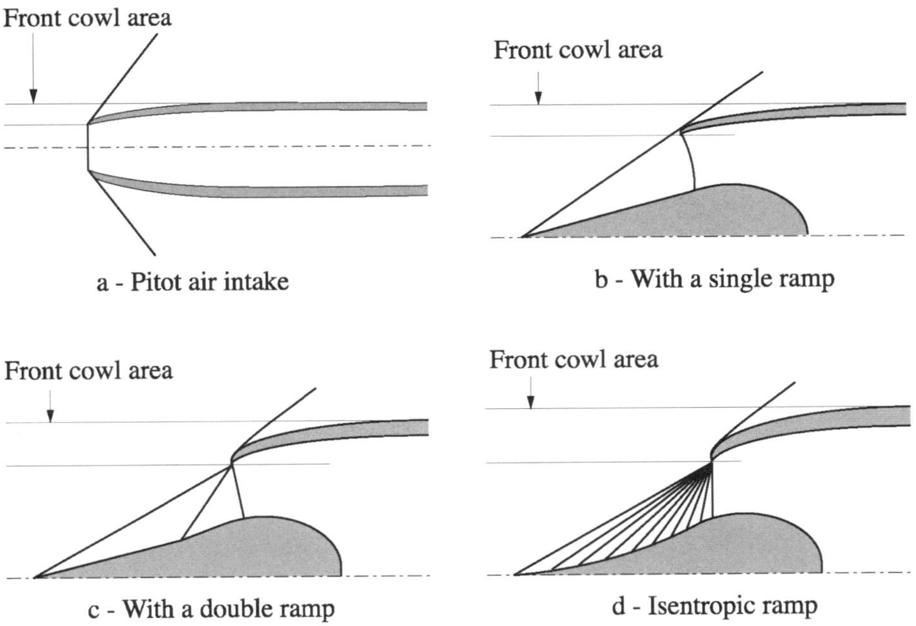

Figure 16.12. *Different types of two-dimensional air intakes*

Table 16.2 qualitatively summarizes the conclusions of the previous review. Each configuration must, of course, be optimized as a function of the flight Mach number. In particular, not only the efficiency and cowl drag but also the ease of installation and robustness should be taken into account. Thus, for a Mach number of 1.6 and less than this, the Pitot type air intake seems obvious because of its simplicity despite a (relatively) poor efficiency. We often take advantage of a pre-compression on the part of the fuselage before the air intake to reduce the Mach number of the flow entering the air intake, which helps increasing its efficiency. Thus, combat aircraft of the *Rafale*, *F16*-type and others are equipped with Pitot air intakes. It is true that these aircraft fly supersonic over a short duration of their mission. The problem must be studied more closely for airliners, which will fly in supersonic for several hours.

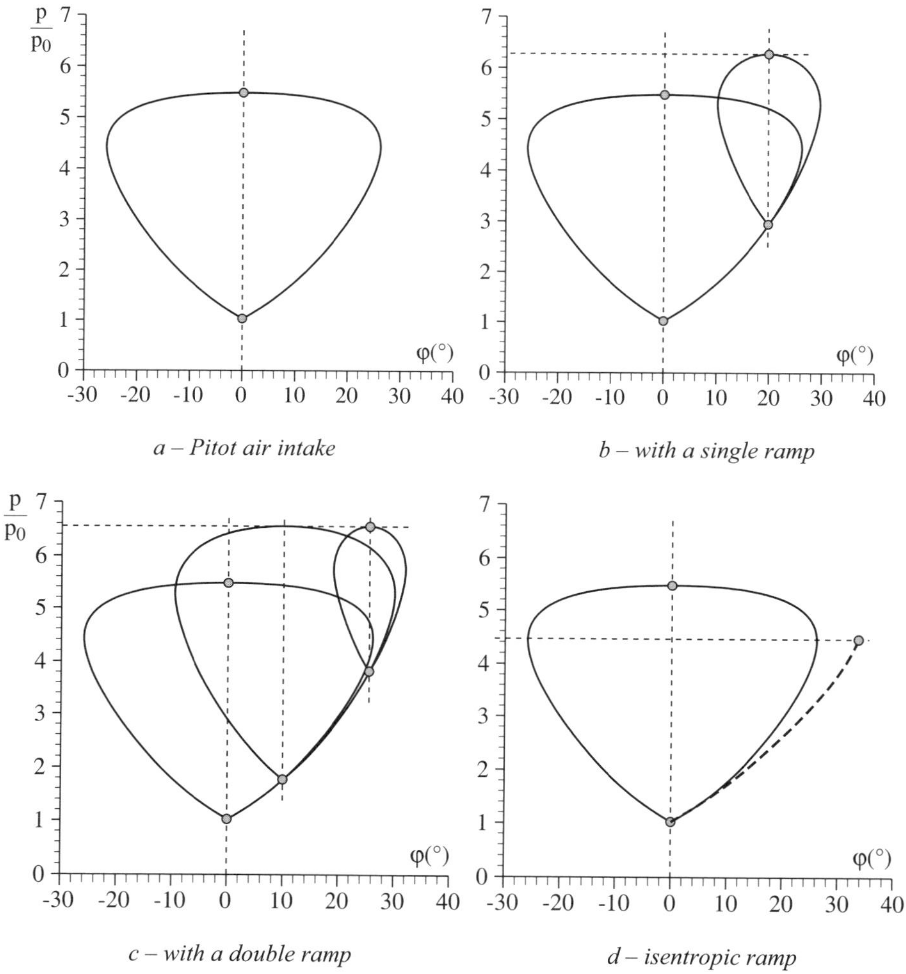

Figure 16.13. *Operation of different air intakes. Representation in the polar plane (upstream Mach number: $M_0 = 2.2$)*

Type	Pitot	One ramp	Two ramps	Isentropic
Efficiency	bad	average	very good	maximum
Cowl drag	weak	appreciable	noticeable	important

Table 16.2. *Quality comparison of different air intakes*

The supersonic air intakes where the system of oblique shock waves is located in front and outside of the captation channel are said to be of *supersonic external compression type*. This design, which is most often used up to flight Mach number of 2.5, has the disadvantage, as we have seen, of causing a significant cowl drag due to the need to compensate for the deviation produced by the compression ramps. Such a design becomes impractical for high Mach numbers, which leads to the design of air intakes where the compression is achieved, as shown in Figure 16.14, through shock reflections (see also section 14.6). With each reflection, the flow is both straightened and compressed, which seems a double advantage. In fact, at each shock impact, there is an interaction with the boundary layer of the wall whose effect can be catastrophic if the interaction causes boundary layer separation. Therefore, local control devices should be provided (often by local fluid suction) in order to avoid such a separation. Figure 16.15 shows the operation of the air intake in the polar plane. The first shock induced by the ramp causes a compression from p_0 to p_1 with a positive deflection. The shock reflecting on the cowl leads to a compression from p_1 to p_1' and a negative deflection that restores the flow direction approximately along the engine axis. The residual compression from p_1' to p_2 is achieved with the terminal shock.

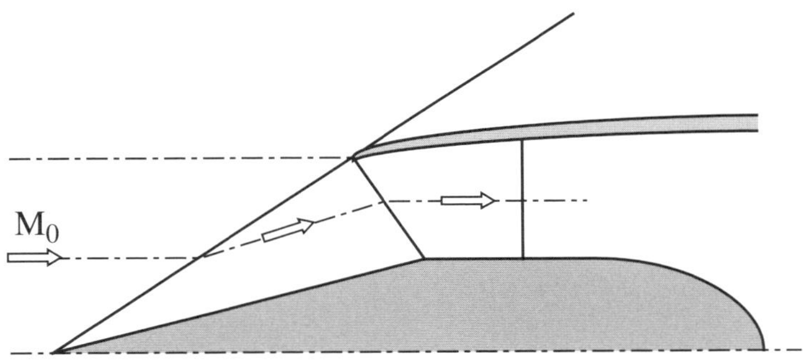

Figure 16.14. *Air intake with mixed external–internal compression. One reflection of the internal shock*

The air intakes where the shock system is entirely within the channel are of *supersonic internal compression type*. We often combine external and internal shocks to construct a supersonic mixed external–internal compression *type* air intake.

Figures 16.16 and 16.17 relate to air intake with three oblique internal shocks. The example corresponds to a capture at Mach 4, the ramp having an angle of 13.6°. The different reflections cause, up to a Mach number of 1.32, deflections of +13.6° and -13.6° in the flow (4), where the terminal shock occurs. We leave it to the reader to interpret the phenomena in the polar plane.

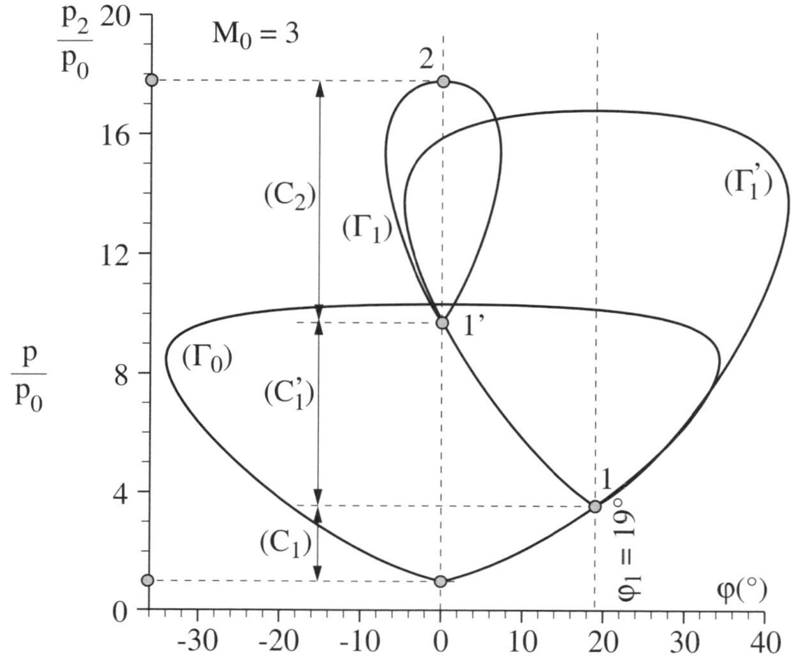

Figure 16.15. *Air intake with mixed external–internal compression. One reflection of the internal shock. Representation in the polar plane*

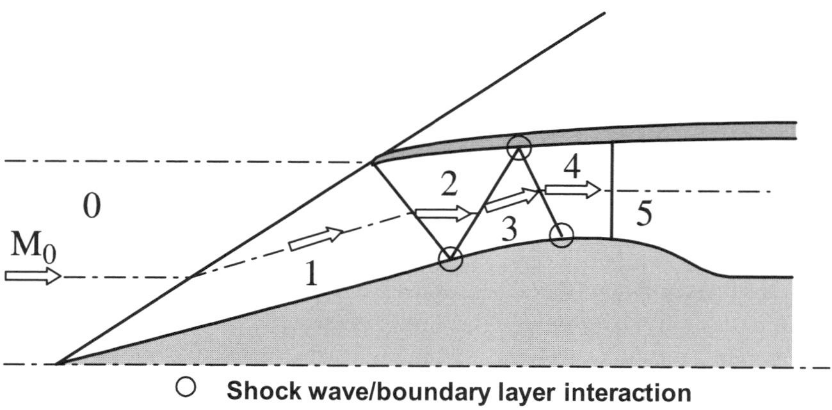

Figure 16.16. *Air intake with mixed external–internal compression. Three internal shocks reflections*

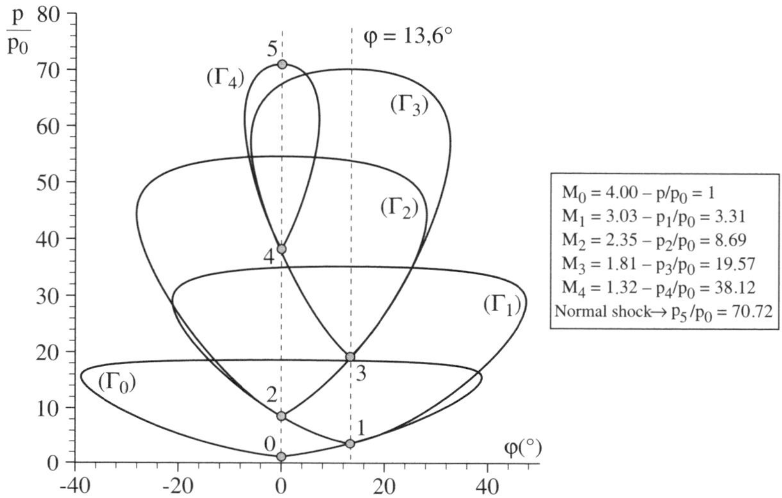

Figure 16.17. *Air intake with mixed external–internal compression. Three internal shock reflections. Representation in the polar plane*

16.4. Air intake examples

Previously, we considered two-dimensional configurations of supersonic ramp air intakes. This geometry corresponds to many supersonic air intakes, for example, those of the *Concorde*, the American fighter jets *F14 Tomcat* and *F15 Eagle*, and the *Tornado* (see Figures 16.18 and 16.19 for the *F15* and the *Concorde*).

In reality, the air intakes have a finite transverse dimension and are limited by sidewalls. Lateral interactions and corner flows occur, affecting the operation of the air intake. Figure 16.20 shows a model used at Onera to develop the air intake of the *Concorde*. Windows arranged at the side of the model allow us to visualize the flow inside the air intake, as the shadowgraph in Figure 16.21 shows.

Supersonic compression is achieved by two ramps followed by a continuous curved wall providing an isentropic compression before the terminal shock. The visualization shows the oblique shock waves associated with the two ramps as well as the terminal shock. In Figure 16.21, three shocks converge on the cowl leading edge from where the external shock emerges, the regime being adapted. A vast cavity, connected to a vacuum system and placed at the level of the terminal shock, is used to suck off the boundary layer that forms and thickens over the compression ramps. This is done to improve the quality of the flow before the deceleration in the internal diffusing duct. This device, called a *boundary layer bleed system*, also serves to regulate the mass flow captured by the air intake in order to avoid potential air intake unstart. As the air intake is adapted in cruise flight, a drop in engine regime leads to a reduction of the captured flow. Hence, there is a transition to the subcritical regime and an upstream shift of the shock system to the front of the air intake (see section 16.2.3). If this flow reduction is significant, a single normal shock forms before the air entry. To avoid this circumstance, the captured flow is maintained constant by sucking through the boundary layer bleed system a mass flow compensating the reduction of the mass flow traversing the engine.

Figure 16.18. *Two-dimensional air intakes of the F15 "Eagle"*

Other supersonic air intakes are of circular geometry. In this case, the planar front shock becomes a conical shock attached to the apex of the air intake. As seen in section 14.7, the supersonic flow downstream of this shock has also conical properties. The organization of the flow is qualitatively the same as for two-dimensional air intakes; the calculation being done by using the conical flow properties. The air intake may be entirely circular, which is the case of the air intakes of the reconnaissance aircraft *SR71 Blackbird* (see Figure 16.22), which flew at Mach 3, or of a semi-circular shape, as in the family of *Mirage* aircraft (see Figure 16.23). Two-dimensional or circular air intakes have both their advantages and disadvantages: easy adaptation of the geometry in the first case and more compact form in the second. This point will not be discussed here.

Figure 16.19. *Two-dimensional air intakes of the Concorde*

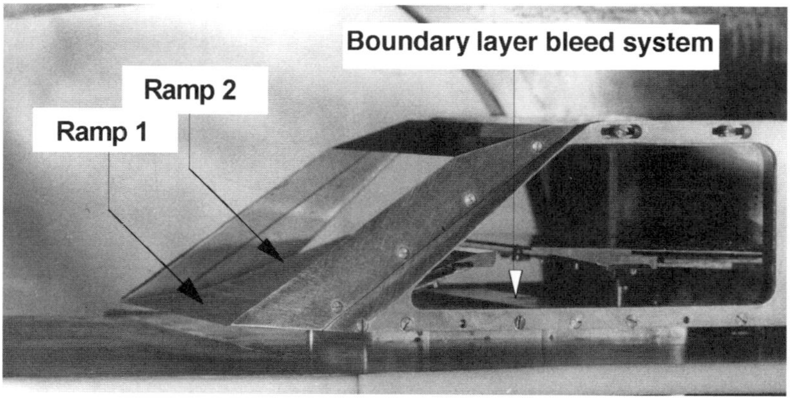

Figure 16.20. *Model of the air intake of the supersonic airliner Concorde (document Onera)*

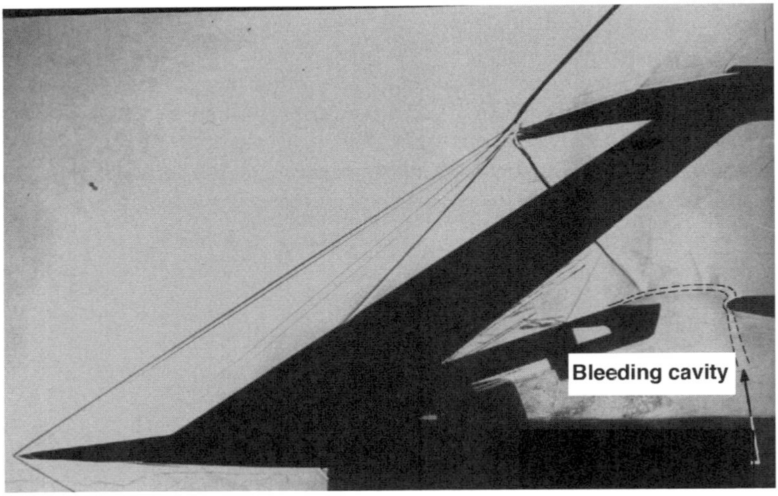

Figure 16.21. *Shadowgraph of the flow in the air intake of the Concorde (document Onera)*

Figure 16.22. *Circular air intakes of the SR71 Blackbird*

Figure 16.23. *Semi-circular air intakes of the Dassault Aviation Mirage F1*

Figure 16.24 shows the air intake of the *Rafale* fighter jet, which is of the Pitot type (as on the *F16 Falcon,* the *F18 Hornet,* and the *Eurofighter*). The entrance to the air intake is mounted on a plate, which lifts it off the fuselage to avoid swallowing the fuselage boundary layer.

Figure 16.24. *Pitot type air intake of the fighter jet Dassault Aviation Rafale*

Chapter 17

Supersonic, Steady, Two-Dimensional Flows and the Theory of Characteristics

17.1. The theory of characteristics: introduction

In this chapter, we will present the key results of the theory of characteristics applied to a steady, two-dimensional planar or axisymmetric flow of a non-viscous fluid. We shall see that this flow must also be supersonic. The theory of characteristics also applies to the three-dimensional, but then its presentation and interpretation become more difficult. In addition, this theory, which leads to a very efficient calculation method in two dimensions, becomes, in the three dimensions, an implementation complex in the plane of algorithms, which makes it preferred over other methods based on direct numerical solution of Euler equations. The method of characteristics based on the theory of characteristics is actually a numerical solution technique of the Euler equations. The practice has, however, established itself in phraseology to distinguish the method of characteristics from the (numerical) resolution of the Euler equations. This is actually a misunderstanding.

There are several ways to introduce the theory of characteristics, either by considering the framework of the wave theory or mathematically from the study of the properties of pseudo-linear differential systems whose equations do not contain partial derivative products. Here, we will adopt a more direct approach by specializing in the two-dimensional case to avoid too heavy a formalism that is not useful for our purpose. We specify the basic assumptions, which should be kept in mind during the applications; the studied flow is:

– without body forces,

– non-viscous and non-heat conducting,

– steady, and

– adiabatic.

In addition, the fluid is in thermal and chemical equilibrium, divariant from the thermodynamic point of view. The motion then obeys the steady Euler equations, which are written by adopting the tensor notation (see section 7.1):

– *continuity equation:*

$$\frac{\partial}{\partial x_i}(\rho u_i) = 0$$

– *equation of motion:*

$$\frac{\partial}{\partial x_j}\left(\rho u_i u_j\right) = -\frac{\partial p}{\partial x_i}$$

– *energy equation:*

$$u_i \frac{\partial s}{\partial x_i} = 0.$$

The energy equation shows the particle (material) derivative of the entropy is zero (here, we recall that $\partial / \partial t = 0$). It follows that the entropy is constant along a streamline.

17.2. Intrinsic equations of the planar or axisymmetric flows

In most cases, we consider the Euler equations for a steady axisymmetric flow written in cylindrical coordinates (r, θ, z), where r is the radial coordinate, θ the angular coordinate, and z the axial coordinate (see Figure 17.1), the corresponding components of the velocity being written as V_r, V_θ, V_z.

Then, Euler equations take the following form:

– *continuity equation:*

$$\frac{\partial(\rho V_z)}{\partial z} + \frac{\partial(\rho V_r)}{\partial r} + \frac{\partial(\rho V_\theta)}{r\partial\theta} + \frac{\rho V_r}{r} = 0$$

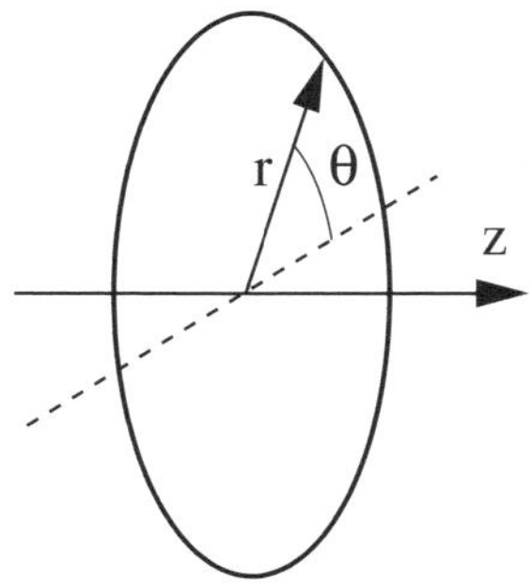

Figure 17.1. *Cylindrical coordinates*

– *equation of motion, split into three scalar equations*:

$$V_z \frac{\partial V_r}{\partial z} + V_r \frac{\partial V_r}{\partial r} + \frac{V_\theta}{r}\frac{\partial V_r}{\partial \theta} - \frac{V_\theta^2}{r} = -\frac{1}{\rho}\frac{\partial p}{\partial r}$$

$$V_z \frac{\partial V_\theta}{\partial z} + V_r \frac{\partial V_\theta}{\partial r} + \frac{V_\theta}{r}\frac{\partial V_\theta}{\partial \theta} + \frac{V_r V_\theta}{r} = -\frac{1}{\rho}\frac{\partial p}{\partial \theta}$$

$$V_z \frac{\partial V_z}{\partial z} + V_r \frac{\partial V_z}{\partial r} + \frac{V_\theta}{r}\frac{\partial V_z}{\partial \theta} = -\frac{1}{\rho}\frac{\partial p}{\partial z}$$

– *energy equation*:

Given the assumption that has been made, the energy equation reduces to (see section 4.7):

$$\frac{ds}{dt} = \frac{\partial s}{\partial t} + \vec{V}.\overrightarrow{grad}\, s = \vec{V}.\overrightarrow{grad}\, s = 0$$

which expresses the constancy of the entropy along the streamlines. This property may interest all of the flow, which is then called homentropic or only the streamlines whose level of the entropy can be variable from one streamline to another. This is what happens downstream of a curved shock due to the variable entropy. The flow is then rotational under the Crocco relation (see section 8.2.1).

Here, we will consider axisymmetric flows of the first kind whose velocity vector is contained in a plane (r,z) (i.e. $V_\theta = 0$) and which are invariant on the parallels, (derivatives with respect to θ equal to zero). Hence, the new set of equations:

– *continuity equation:*

$$\frac{\partial(\rho V_z)}{\partial z}+\frac{\partial(\rho V_r)}{\partial r}+\frac{\rho V_r}{r}=0$$

– *equation of motion:*

$$V_z\frac{\partial V_r}{\partial z}+V_r\frac{\partial V_r}{\partial r}=-\frac{1}{\rho}\frac{\partial p}{\partial r}$$

$$V_z\frac{\partial V_z}{\partial z}+V_r\frac{\partial V_z}{\partial r}=-\frac{1}{\rho}\frac{\partial p}{\partial z}.$$

As these equations are very close to those of the two-dimensional planar flow, we rewrite them using the notations x and y instead of z and r, understanding that in an axisymmetric flow y is the *distance to the axis of revolution*:

– *continuity equation*:

$$\frac{\partial(\rho V_x)}{\partial x}+\frac{\partial(\rho V_y)}{\partial y}+\frac{\rho V_y}{y}=0$$

– *equation of motion*:

$$V_x\frac{\partial V_x}{\partial x}+V_y\frac{\partial V_x}{\partial y}=-\frac{1}{\rho}\frac{\partial p}{\partial x}$$

$$V_x\frac{\partial V_y}{\partial x}+V_y\frac{\partial V_y}{\partial y}=-\frac{1}{\rho}\frac{\partial p}{\partial y}.$$

With these notations, we find that the equations are almost identical for the two-dimensional planar and the axisymmetric cases, with the exception of an additional term in the equation of continuity in the second case. We come quickly to the *so-called characteristic* form of these equations by transposing them first in an intrinsic coordinate system, where $\vec{x}_1$ is the unit vector directed along the velocity vector $\vec{V}$ and $\vec{y}_1$ the unit vector directly perpendicular to $\vec{x}_1$, with the vectors $\vec{x}_1$ and $\vec{y}_1$ forming a system of local axes linked to the velocity (see Figure 17.2).

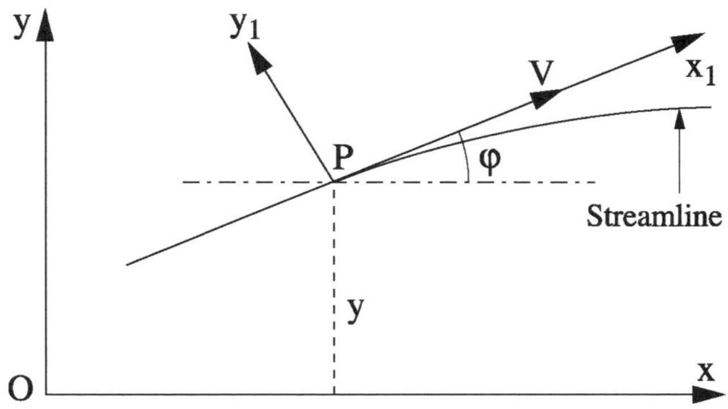

Figure 17.2. *Intrinsic coordinates system*

Under these conditions:

$$\vec{V} = V\,\vec{x}_1$$

where V denotes the modulus of velocity vector. If $\vec{x}$ and $\vec{y}$ are fixed axes and if φ is the angle of the velocity $\vec{V}$ with respect to $\vec{x}$, the velocity components in the fixed frame of reference are:

$$V_x = V\cos\varphi$$
$$V_y = V\sin\varphi$$

and we have the relations of the axis transformation:

$$\frac{\partial}{\partial x_1} = \cos\varphi\ \frac{\partial}{\partial x} + \sin\varphi\ \frac{\partial}{\partial y}$$
$$\frac{\partial}{\partial y_1} = -\sin\varphi\ \frac{\partial}{\partial x} + \cos\ \varphi\frac{\partial}{\partial y}$$

or for an inverse transformation:

$$\frac{\partial}{\partial x} = \cos\varphi\ \frac{\partial}{\partial x_1} - \sin\varphi\ \frac{\partial}{\partial y_1}$$
$$\frac{\partial}{\partial y} = \sin\varphi\ \frac{\partial}{\partial x_1} + \cos\varphi\ \frac{\partial}{\partial y_1}.$$

In the fixed frame, the continuity equation is written as follows:

$$\frac{\partial}{\partial x}(\rho V \cos\varphi) + \frac{\partial}{\partial y}(\rho V \sin\varphi) + \varepsilon \frac{\rho V \sin\varphi}{y} = 0.$$

The factor ε before the term of the axisymmetry effect is such that:

– $\varepsilon = 0$ in a two-dimensional planar flow; and

– $\varepsilon = 1$ in an axisymmetric flow.

Its introduction gives a unified formulation for both cases.

The change of frame of reference $(x, y) \rightarrow (x_1, y_1)$ gives the following expression for the continuity equation:

$$\cos\varphi \frac{\partial}{\partial x_1}(\rho V \cos\varphi) - \sin\varphi \frac{\partial}{\partial y_1}(\rho V \cos\varphi) +$$
$$\sin\varphi \frac{\partial}{\partial x_1}(\rho V \sin\varphi) + \cos\varphi \frac{\partial}{\partial y_1}(\rho V \sin\varphi) = -\varepsilon \frac{\rho V \sin\varphi}{y}.$$

By developing, and after elementary calculations that are unnecessary to specify, we obtain the equation:

$$\boxed{\frac{\partial \rho V}{\partial x_1} + \rho V \frac{\partial \varphi}{\partial y_1} = -\varepsilon \frac{\rho V \sin\varphi}{y}.} \qquad [17.1]$$

The same change of frame of reference is applied to the two components of the equation of motion:

– *first equation:*

$$V_x \frac{\partial V_x}{\partial x} + V_y \frac{\partial V_x}{\partial y} = -\frac{1}{\rho}\frac{\partial p}{\partial x}$$

which gives:

$$V\cos\varphi\left(\cos\varphi \frac{\partial(V\cos\varphi)}{\partial x_1} - \sin\varphi \frac{\partial(V\cos\varphi)}{\partial y_1}\right) +$$
$$V\sin\varphi\left(\sin\varphi \frac{\partial(V\cos\varphi)}{\partial x_1} + \cos\varphi \frac{\partial(V\cos\varphi)}{\partial y_1}\right) = -\frac{1}{\rho}\frac{\partial p}{\partial x}.$$

Hence, after development and simplifications:

$$V\cos\varphi\frac{\partial V}{\partial x_1}-V^2\sin\varphi\frac{\partial\varphi}{\partial x_1}=-\frac{1}{\rho}\left(\cos\varphi\frac{\partial p}{\partial x_1}-\sin\varphi\frac{\partial p}{\partial y_1}\right).$$

– *second equation:*

$$V_x\frac{\partial V_y}{\partial x}+V_y\frac{\partial V_y}{\partial y}=-\frac{1}{\rho}\frac{\partial p}{\partial y}$$

which gives:

$$V\cos\varphi\left(\cos\varphi\frac{\partial(V\sin\varphi)}{\partial x_1}-\sin\varphi\frac{\partial(V\sin\varphi)}{\partial y_1}\right)+$$
$$V\sin\varphi\left(\sin\varphi\frac{\partial(V\sin\varphi)}{\partial x_1}+\cos\varphi\frac{\partial(V\sin\varphi)}{\partial y_1}\right)=-\frac{1}{\rho}\frac{\partial p}{\partial y}$$

and after elementary calculations:

$$V\sin\varphi\frac{\partial V}{\partial x_1}+V^2\cos\varphi\frac{\partial\varphi}{\partial x_1}=-\frac{1}{\rho}\left(\sin\varphi\frac{\partial p}{\partial x_1}+\cos\varphi\frac{\partial p}{\partial y_1}\right).$$

By simple combination of the above equations, we obtain the so-called intrinsic equations of motion:

– *equation for the axial motion:*

$$\boxed{\rho V\frac{\partial V}{\partial x_1}+\frac{\partial p}{\partial x_1}=0} \qquad [17.2]$$

– *equation for the normal motion:*

$$\boxed{\rho V^2\frac{\partial\varphi}{\partial x_1}+\frac{\partial p}{\partial y_1}=0.} \qquad [17.3]$$

If $\Re=\left(\dfrac{\partial\varphi}{\partial x_1}\right)^{-1}$ denotes the local radius of curvature of the streamline, the above equation is written:

$$\frac{\rho V^2}{\Re} = -\frac{\partial p}{\partial y_1}.$$

This relation expresses the radial equilibrium, the rotation of the fluid equilibrating the pressure gradient normal to the streamline.

If we observe that the derivative $\partial V/\partial x_1$ is only present in the first two equations, the combination $V \times [17.1] - [17.2]$ allows us to eliminate this derivative, which gives:

$$V^2 \frac{\partial \rho}{\partial x_1} + \rho V^2 \left(\frac{\partial \varphi}{\partial y_1} + \varepsilon \frac{\sin \varphi}{y} \right) - \frac{\partial p}{\partial x_1} = 0.$$

Since the change along a streamline is isentropic, the velocity of sound is given by:

$$a^2 = \left(\frac{\partial p}{\partial \rho} \right)_s = \frac{(\partial p / \partial x_1)}{(\partial \rho / \partial x_1)}.$$

The energy equation in a steady flow is written: $\frac{ds}{dt} = \vec{V}.\overrightarrow{grads} = 0$, which shows the entropy is constant along a streamline, or:

$$\boxed{\frac{\partial s}{\partial x_1} = 0.} \qquad [17.4]$$

Thus, the system of Euler equations can be written as:

$$\left(\frac{V^2}{a^2} - 1 \right) \frac{\partial p}{\partial x_1} + \rho V^2 \frac{\partial \varphi}{\partial y_1} = -\varepsilon \rho V^2 \frac{\sin \varphi}{y} \qquad [17.5]$$

$$\rho V^2 \frac{\partial \varphi}{\partial x_1} + \frac{\partial p}{\partial y_1} = 0 \qquad [17.6]$$

$$\frac{\partial s}{\partial x_1} = 0. \qquad [17.7]$$

These equations have a great character of generality since we have not made very restrictive assumptions on the gas thermodynamics.

There is a great connection between the steady two-dimensional flows (x, y) and the unsteady flows (x, t), discussed in Chapter 21. In both cases, we deal with pseudo-linear differential systems involving two independent variables, if we limit ourselves to two dimensions. The approach followed here is similar to the one adopted in Chapter 21. Similarly, the addressed problems and analyzed solutions will have a great resemblance. However, we should be careful not to make a simple substitution of the space dimension y with the time t in order to derive immediate results. The equations have similar forms, but are not reducible to a common writing. In addition, the analogy only exists for two-dimensional steady and supersonic flows.

17.3. The Cauchy problem and characteristic relations

The Cauchy problem can be formulated in the following way. Knowing a solution along a curve (C), called the initial curve, is it possible to extend the solution outside of (C) using an analytical expansion? In this case, (C) is a curve on which all the properties of the flow are known (p, V, φ, s) and satisfy the Euler equations (see Figure 17.3).

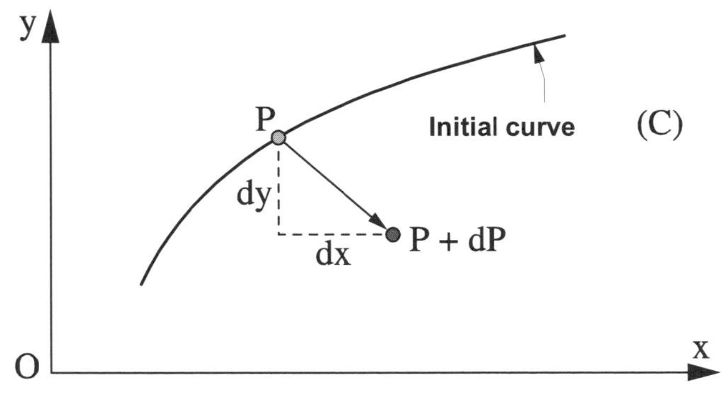

Figure 17.3. *Cauchy problem: extension of the solution from P to P+dP*

COMMENT.– We will see that for solving Euler equations, the entropy must be specified on each streamline. It is thus a piece of information whose knowledge implies that of the stagnation conditions: temperature T_{st} and pressure p_{st}. Under these circumstances, it is possible to determine V, as well as all thermodynamic properties, from the local value of the pressure p (see the theory of one-dimensional steady flows in Chapter 9). The flow is, therefore, fully defined by the variables p, φ, and s.

To analytically extend the solution from a point P of (C) to a neighboring point $P+dP$, we must know the partial derivatives of p, φ, s as a function of x_1 and y_1. For this, we have three equations [17.5]–[17.7] and the variations $d\overline{p}$, $d\overline{\varphi}$, $d\overline{s}$ of p, φ, and s for an elementary displacement along (C). Thus, we have three additional equations:

$$d\overline{p} = \left(\frac{\partial p}{\partial x_1}\right)d\overline{x}_1 + \left(\frac{\partial p}{\partial y_1}\right)d\overline{y}_1 \qquad [17.8]$$

$$d\overline{\varphi} = \left(\frac{\partial \varphi}{\partial x_1}\right)d\overline{x}_1 + \left(\frac{\partial \varphi}{\partial y_1}\right)d\overline{y}_1 \qquad [17.9]$$

$$d\overline{s} = \left(\frac{\partial s}{\partial x_1}\right)d\overline{x}_1 + \left(\frac{\partial s}{\partial y_1}\right)d\overline{y}_1. \qquad [17.10]$$

Since p, φ, and s are known on (C), the differences $d\overline{p}$, $d\overline{\varphi}$, and $d\overline{s}$ are also known quantities. We thus have six equations [17.5]–[17.10] to calculate the six partial derivatives:

$$\frac{\partial p}{\partial x_1}, \frac{\partial p}{\partial y_1}, \frac{\partial \varphi}{\partial x_1}, \frac{\partial \varphi}{\partial y_1}, \frac{\partial s}{\partial x_1}, \frac{\partial s}{\partial y_1}.$$

Equations [17.5]–[17.10] form a linear algebraic system (remember that p, φ, and s are known on (C)), which has a unique solution, unless it's principal determinant vanishes, that is, if:

$$\begin{vmatrix} \left(\frac{V^2}{a^2}-1\right) & 0 & 0 & \rho V^2 & 0 & 0 \\ 0 & 1 & \rho V^2 & 0 & 0 & 0 \\ 0 & 0 & 0 & 0 & 1 & 0 \\ d\overline{x}_1 & d\overline{y}_1 & 0 & 0 & 0 & 0 \\ 0 & 0 & d\overline{x}_1 & d\overline{y}_1 & 0 & 0 \\ 0 & 0 & 0 & 0 & d\overline{x}_1 & d\overline{y}_1 \end{vmatrix} = 0.$$

The development of this determinant leads to the equation:

$$\boxed{d\overline{y}_1 \left[\left(\frac{V^2}{a^2} - 1 \right) \rho V^2 d\overline{y}_1^2 - \rho V^2 d\overline{x}_1^2 \right] = 0.}$$

This determinant is zero if dy_1 is zero, that is, if (*C*) is a streamline. In this case, it is not possible to determine the entropy *s* at a point outside the streamline. We already know that *s* is constant along a streamline since $\partial s / \partial x_1 = 0$. In other words, in the calculation of a flow of a non-viscous fluid, the entropy must be specified for each streamline since there is no entropy creating (or diffusion) process. If the streamline crosses a shock wave, then entropy is created at the level of the shock wave (see the shock wave theory in Chapter 13). Locally, the entropy of the streamline will undergo a jump, its value then remains constant until the eventual crossing of another shock.

The cancellation of the determinant may also come from the term in square brackets, i.e. after simplification with ρV^2:

$$\left(\frac{V^2}{a^2} - 1 \right) d\overline{y}_1^2 - d\overline{x}_1^2 = 0.$$

We see that for a subsonic flow, where $V < a$, the term cannot vanish because $(V^2 / a^2 - 1)$ is always negative. In a subsonic flow, the Cauchy problem has a solution for any curve (*C*).

However, in a supersonic flow, where $V > a$, the above expression can vanish. This circumstance occurs if the curve (*C*) is such that its local slope $d\overline{y}_1 / d\overline{x}_1$ is:

$$\frac{d\overline{y}_1}{d\overline{x}_1} = \pm \frac{1}{\sqrt{\frac{V^2}{a^2} - 1}}.$$

We recall that $d\overline{y}_1 / d\overline{x}_1$ is the local slope of (*C*) relative to the streamline. We introduce the Mach number:

$$M = \frac{V}{a}$$

and the *Mach angle* α so that:

$$\tan \alpha = \left| \frac{d\bar{y}_1}{d\bar{x}_1} \right| = \frac{1}{\sqrt{\frac{V^2}{a^2} - 1}} = \frac{1}{\sqrt{M^2 - 1}}.$$

We show without difficulty:

$$\boxed{\sin \alpha = \frac{1}{M} \quad or \quad \alpha = \sin^{-1} \frac{1}{M}.}$$

Thus, at any point P of a supersonic flow, we can define two specific directions with the angles $+\alpha$ and $-\alpha$ with respect to the velocity vector. Any curve (C), tangential at every of its points to one or the other of these directions, is called *characteristic curve* (see Figure 17.4). If the initial curve coincides with a characteristic curve, then the Cauchy problem cannot be solved: it is impossible to analytically extend the solution outside of (C).

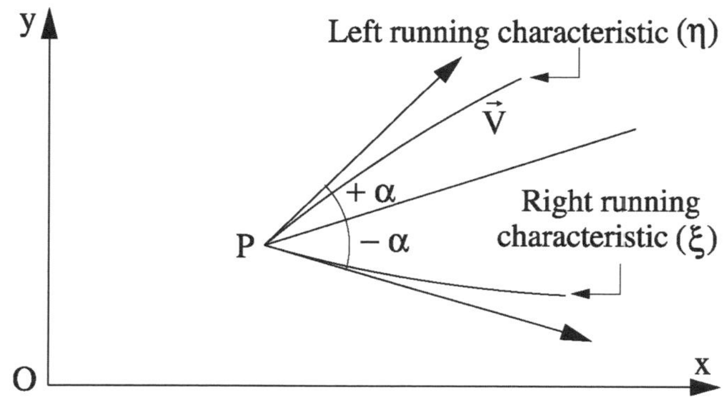

Figure 17.4. *Characteristic lines at a point P*

In fluid mechanics, the characteristic curves are also called *Mach lines* or even *waves* (here, the waves are stationary). By solving the Cauchy problem limited to the four equations for the four unknowns:

$$\frac{\partial p}{\partial x_1}, \frac{\partial p}{\partial y_1}, \frac{\partial \varphi}{\partial x_1}, \frac{\partial \varphi}{\partial y_1}$$

(we have eliminated the entropy whose equation simply leads to s = *constant* along a streamline) we obtain after an elementary calculation:

$$\frac{\partial p}{\partial x_1} = \frac{d\overline{p}d\overline{x}_1 + \rho V^2 d\overline{\varphi} \;\; d\overline{y}_1 + \varepsilon \rho V^2 \left(\frac{\sin\varphi}{y} \right) d\overline{y}_1^2}{\left(M^2 - 1\right) \rho V^2 d\overline{y}_1^2 - \rho V^2 d\overline{x}_1^2}$$

and similar expressions for other derivatives.

If the curve (C) is characteristic, the two following situations can occur:

– The numerator of the previous expression is non-zero; then we are led to an absolute impossibility, the derivatives cannot be calculated.

– The numerator is also zero, thus, the derivatives exist, but the Cauchy problem cannot be solved as it leads to indeterminacy.

Thus, for the problem to be physically possible (i.e. such that the field does not contain infinite derivatives), it is necessary that the following condition is satisfied on a characteristic curve:

$$d\overline{p} \;\; dx_1 + \rho V^2 d\overline{\varphi} \;\; dy_1 = -\varepsilon \frac{\sin\varphi}{y} \rho V^2 dy_1^2 .$$

In order to simplify the writing, we now remove the bar over the quantities whose variations are to be taken along the characteristic curve.

As we are on a characteristic curve:

$$\frac{dy_1}{dx_1} = \pm \frac{1}{\sqrt{M^2 - 1}} .$$

Thus, after dividing the above equation by $\rho V^2 dy_1$, we must have along the characteristic with the angle $+\alpha$ to the velocity vector, which we will call *left running characteristic* (η), the compatibility relation:

$$\frac{\sqrt{M^2 - 1}}{\gamma \;\; pM^2} dp + d\varphi = -\varepsilon \frac{\sin\varphi}{y} dy_1 .$$

If the gas is calorically perfect, then:

$$\rho \ V^2 = \gamma \ p \ M^2.$$

As the variations dp and $d\varphi$ result from a displacement $d\eta$ on the characteristic (η):

$$dy_1 = d\eta \ \sin\alpha.$$

Hence, the relation:

$$\frac{\sqrt{M^2-1}}{\gamma M^2}\frac{dp}{p} + d\varphi = -\varepsilon\frac{\sin\varphi\sin\alpha}{y}d\eta \ \ along \ \ (\eta).$$

If we consider the characteristic (ξ) with the angle $-\alpha$ to the velocity vector, called *right running characteristic*, we obtain the relation:

$$\frac{\sqrt{M^2-1}}{\gamma M^2}\frac{dp}{p} - d\varphi = -\varepsilon\frac{\sin\varphi\sin\alpha}{y}d\xi \ \ along \ \ (\xi).$$

We can make a passage to the limit $d\eta, d\xi \rightarrow 0$, η, and ξ denoting the coordinates along (η) and (ξ), respectively, to reach the characteristic equations:

$$\boxed{\begin{array}{ll} \dfrac{\sqrt{M^2-1}}{\gamma p M^2}\dfrac{\partial p}{\partial \eta} + \dfrac{\partial \varphi}{\partial \eta} = -\varepsilon\dfrac{\sin\varphi\sin\alpha}{y} & along \ \ (\eta) \\ \dfrac{\sqrt{M^2-1}}{\gamma p M^2}\dfrac{\partial p}{\partial \xi} - \dfrac{\partial \varphi}{\partial \xi} = -\varepsilon\dfrac{\sin\varphi\sin\alpha}{y} & along \ \ (\xi). \end{array}} \qquad [17.11]$$

These relations must be satisfied by a supersonic flow on the characteristic lines. We recall that in subsonics, the characteristic lines do not exist; the Cauchy problem has always a unique solution. The equations of motion are then elliptical; they are hyperbolic for a steady flow in supersonics.

Equations [17.11] can also be written in the equivalent form, which we will easily establish:

$$\boxed{\begin{array}{ll} \dfrac{\sin\alpha \ \ \cos\alpha}{\gamma p}\dfrac{\partial p}{\partial \eta} + \dfrac{\partial \varphi}{\partial \eta} = -\varepsilon\dfrac{\sin\varphi\sin\alpha}{y} & on \ \ (\eta) \\ \dfrac{\sin\alpha \ \ \cos\alpha}{\gamma p}\dfrac{\partial p}{\partial \xi} - \dfrac{\partial \varphi}{\partial \xi} = -\varepsilon\dfrac{\sin\varphi\sin\alpha}{y} & on \ \ (\xi). \end{array}}$$

COMMENT.–

Having introduced the Mach angle α through $\alpha = sin^{-1}(1/M)$, it is possible to find the characteristic equations by projecting the equations of motion along the two directions of angles α and $-\alpha$. Using the characteristic coordinates η and ξ, we have:

$$\frac{\partial}{\partial\eta} = \cos\alpha\frac{\partial}{\partial x_1} + \sin\alpha\frac{\partial}{\partial y_1}$$

$$\frac{\partial}{\partial\xi} = \cos\alpha\frac{\partial}{\partial x_1} - \sin\alpha\frac{\partial}{\partial y_1}.$$

A simple calculation leads to equations [17.11], which is another more direct but less revealing way to obtain the characteristic relations.

The characteristic equations have very specific properties. As the initial equations, they are pseudo-linear, meaning that they are linear with respect to partial derivatives. Moreover, each involves only the derivatives with respect to only one of the independent variables η or ξ. Thus, starting from a system where the equations contain simultaneously derivatives with respect to x_1 and y_1, we obtain system [17.11] where the first equation contains only derivatives with respect to η and the second only derivatives with respect to ξ. This is a property typical of a first-order pseudo-linear differential system, which only exists under certain conditions. In the case of steady Euler equations, it is necessary that the Mach number is supersonic.

If we again consider the characteristic equations as relations between the variations dp and $d\varphi$ along a characteristic line, we write:

$$\frac{\sqrt{M^2-1}}{\gamma M^2}\frac{dp}{p} + d\varphi = -\varepsilon\frac{\sin\varphi\sin\alpha}{y}d\eta \;\; along \;\; (\eta)$$

$$\frac{\sqrt{M^2-1}}{\gamma M^2}\frac{dp}{p} - d\varphi = -\varepsilon\frac{\sin\varphi\sin\alpha}{y}d\xi \;\; along \;\; (\xi). \qquad [17.12]$$

These relationships can be put in compact form by stating:

$$dz = \frac{dp}{p} \;\; where \;\; z = \ln p, \quad Z = \frac{\sqrt{M^2-1}}{\gamma \;\; M^2}, \;\; N = -\varepsilon\frac{\sin\varphi\sin\alpha}{y}$$

which gives:

$$Zdz + d\varphi = Nd\eta \quad on \quad (\eta) \qquad [17.13]$$

$$Zdz - d\varphi = Nd\xi \quad on \quad (\xi).$$

17.4. Characteristic relations for the two-dimensional planar flow of a calorically perfect gas

In the case of a planar flow $(\varepsilon = 0)$, relations [17.12] become:

$$\frac{\sqrt{M^2-1}}{\gamma M^2}\frac{dp}{p} + d\varphi = 0 \quad along \quad (\eta)$$

$$\frac{\sqrt{M^2-1}}{\gamma \ M^2}\frac{dp}{p} - d\varphi = 0 \quad along \quad (\xi).$$

If the flow is isentropic (see section 9.5):

$$\frac{p}{p_{st}} = \left(1 + \frac{\gamma-1}{2}M^2\right)^{-\frac{\gamma}{\gamma-1}}.$$

Hence, since the total pressure p_{st} is constant:

$$\frac{dp}{p} = -\frac{\gamma \, MdM}{1 + \frac{\gamma-1}{2}M^2}$$

and:

$$\frac{\sqrt{M^2-1}}{\gamma \ M^2}\frac{dp}{p} = -\frac{\sqrt{M^2-1}}{1+\frac{\gamma-1}{2}M^2}\frac{dM}{M}.$$

We introduce the function $\omega(M,\gamma)$ such that:

$$d\omega = -\frac{\sqrt{M^2-1}}{1+\frac{\gamma-1}{2}M^2}\frac{dM}{M}.$$

Its primitive is:

$$\omega(M,\gamma)=\sqrt{\frac{\gamma+1}{\gamma-1}}\tan^{-1}\sqrt{\frac{\gamma-1}{\gamma+1}\left(M^2-1\right)}-\tan^{-1}\sqrt{M^2-1}.$$

As defined here, the function $\omega(M,\gamma)$ is an increasing function of the Mach number M (see Figure 17.5). It is zero for $M=1$, its derivative $d\omega/dM$ being zero for $M=1$. It tends toward a finite limit when the Mach number tends to infinity, which is:

$$\omega_{\max}(\gamma)=\left(\sqrt{\frac{\gamma+1}{\gamma-1}}-1\right)\times 90°=130.45° \quad \textit{for} \quad \gamma=1.4.$$

This function, called the *Busemann pressure number* or *Prandtl–Meyer function,* is sometimes given as tabulated according to the Mach number *M.*

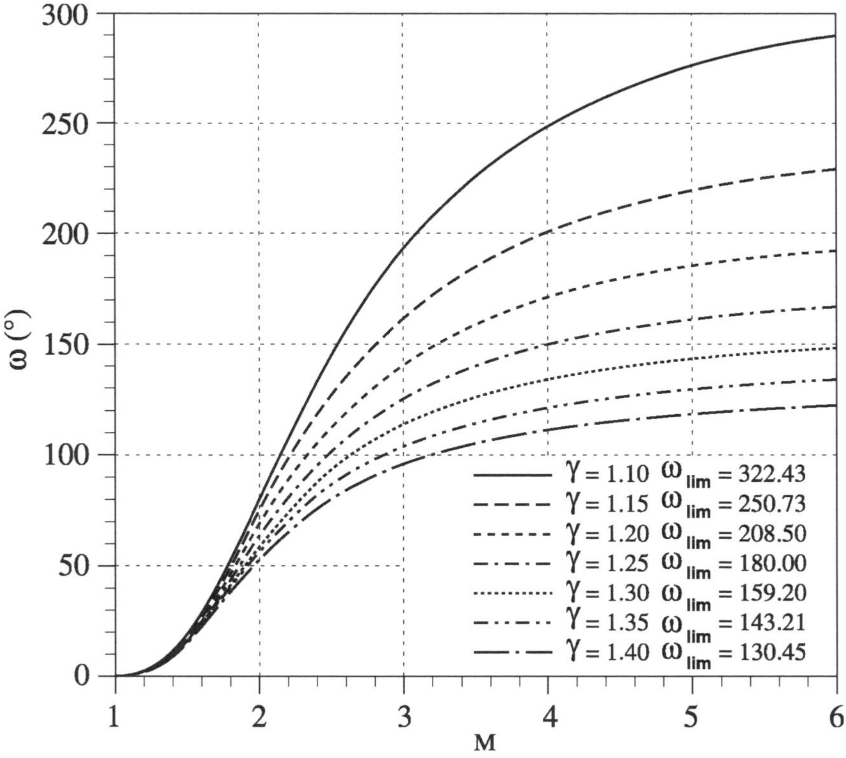

Figure 17.5. *Busemann pressure number*

Ultimately, the characteristic relations for the planar case and a calorically perfect gas can be integrated to give the two equations:

$$\boxed{\begin{array}{l}\omega(M,\gamma)-\varphi=\text{constant} \ \ \textit{along a curve } (\eta)\\ \omega(M,\gamma)+\varphi=\text{constant} \ \ \textit{along a curve } (\xi).\end{array}}$$

We should focus on the sign change in the characteristic relations that comes from the definition made for the function $\omega(M,\gamma)$. Except in special cases (uniform flow or simple wave flow), the constant varies from one characteristic to another.

17.5. Applications of the theory of characteristics

17.5.1. *The progressive expansion wave*

Consider a two-dimensional planar, uniform flow with a supersonic Mach number M_0, pressure p_0, and direction φ_0, which we take, for convenience, to be equal to zero. This flow is along a straight wall with the direction φ_0. From point *A*, the wall curves, its angle φ being negative, as shown in Figure 17.6, but without discontinuity of the derivative $(d\varphi/dx)$. At a point *B,* φ reaches the value φ_1 and stops to change, the wall again being straight downstream of *B*. The problem is to determine the flow that establishes itself along this wall.

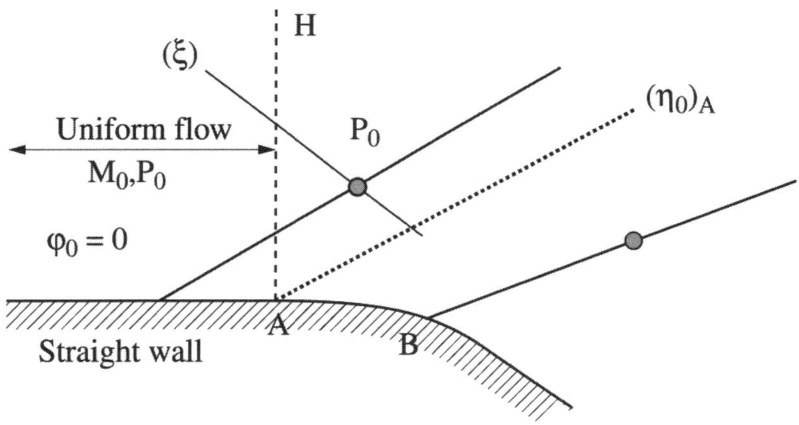

Figure 17.6. *Extension of the uniform flow zone*

If the flow is isentropic, in the light of the made assumptions, the equations of motion reduce to the two following characteristic equations:

$$\omega(M,\gamma)-\varphi = cste. \text{ on any left running characteristic } (\eta)$$

$$\omega(M,\gamma)+\varphi = cste. \text{ on any right running characteristic } (\xi).$$

We consider a point upstream of the left running characteristics (η_0) passing through A where the wall begins to curve. Through P_0 pass:

– a left running characteristic (η); and

– a right running characteristic (ξ).

The left running characteristic (η) comes from a point on the wall, located upstream of A, where the direction of the velocity is:

$\varphi = \varphi_0 =$ constant (*for convenience, we set* $\varphi_0 = 0$).

Consequently:

$$\omega\left(M_{P_0},\gamma\right)-\varphi_{P_0} = \omega\left(M_0\right)-\varphi_0 = \omega\left(M_0\right).$$

On the (ξ) coming from the upstream region where the flow is uniform:

$$\omega\left(M_{P_0},\gamma\right)+\varphi_{P_0} = \omega\left(M_0\right)+\varphi_0 = \omega\left(M_0\right).$$

By adding, then subtracting, the relations above, we obtain:

$$\omega\left(M_{P_0},\gamma\right) = \omega\left(M_0\right)$$

$$\varphi_{P_0} = \varphi_0 = 0.$$

Thus, the flow is uniform and identical to the state prior to the point P_0, this property being true as long as P_0 is located upstream of (η_0), i.e. as long as the left running characteristic (η) passes through a point on the wall located upstream of A. In other words, if a flow is uniform in a section HA, perpendicular to the wall in A, it is also uniform down to the left running characteristics coming from A. Physically, this fact reflects the phenomenon of *propagation*. A disturbance at a point A (here the beginning of a change in the direction of the velocity) is only felt in the adjacent flow downstream of the characteristics (here left running), emanating from the point

A where the disturbance occurs. As long as it does not meet the wave (η_0), the flow "ignores" that something has happened on the wall.

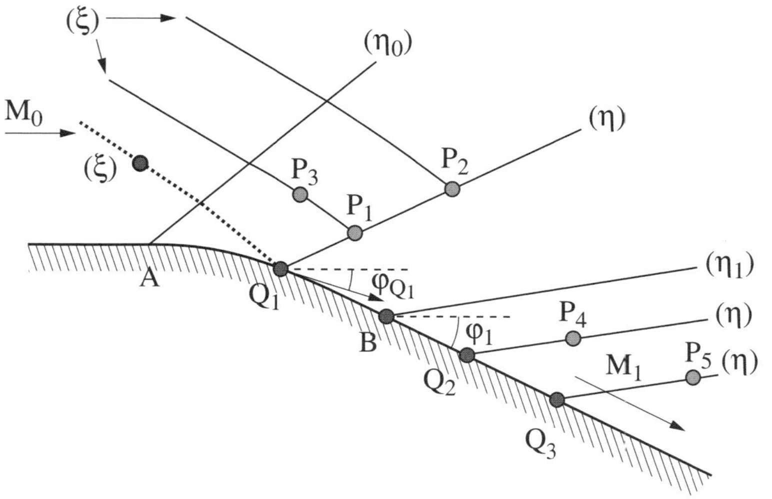

Figure 17.7. *Progressive expansion on a convex wall*

According to its nature, we know that a characteristic line locally has the angle $+\alpha$ or $-\alpha$ with respect to the velocity vector. Its slope relative to a reference axis will thus be:

$$\boxed{\begin{array}{ll}\psi = \varphi + \alpha & \textit{for a curve } (\eta) \\ \psi = \varphi - \alpha & \textit{for a curve } (\xi).\end{array}}$$

In a uniform flow, the left and right running characteristics are thus rectilinear and parallel to each other since α and φ have everywhere the same value.

We consider a point P_1 so that the characteristic (η) passing through P_1 meets the wall at a point Q_1 in its curved part (see Figure 17.7). We can write on this (η):

$$\omega(M_{P_1}, \gamma) - \varphi_{P_1} = \omega(M_{Q_1}, \gamma) - \varphi_{Q_1}. \qquad [17.14]$$

If we consider the characteristic (ξ) passing through P_1, this characteristic comes from a region where the flow is uniform, so that:

$$\omega(M_{P_1}, \gamma) + \varphi_{P_1} = \omega(M_0, \gamma) + \varphi_0 = \omega(M_0, \gamma). \qquad [17.15]$$

If P_2 is a second point located on the same characteristic curve (η), we have:

$$\omega(M_{P_2},\gamma)-\varphi_{P_2}=\omega(M_{Q_1},\gamma)-\varphi_{Q_1} \qquad [17.16]$$

$$\omega(M_{P_2},\gamma)+\varphi_{P_2}=\omega(M_0,\gamma). \qquad [17.17]$$

Comparing the relations [17.14]–[17.17], we see that:

$$\omega(M_{P_1},\gamma)=\omega(M_{P_2},\gamma)=\omega(M_{Q_1},\gamma)$$

$$\varphi_{P_1}=\varphi_{P_2}=\varphi_{Q_1}.$$

We deduce that the flow has *constant properties along the characteristics* (η) *meeting the wall between the points A and B*. These properties change from one (η) to another. If the Mach number is constant on an (η), hence, also the Mach angle: thus, the slope $\psi=\varphi+\alpha$ is constant. *The left running characteristics* (η) *are straight*, which is not the case for the (ξ). In particular, the direction of the velocity at any point (η) is equal to the slope of the wall at the point where this (η) meets the wall.

Such a flow, where a family of characteristics is made of straight lines, is called a *simple wave*, or is said to be made of simple waves. The demonstration that has been made shows the following general property:

Any flow adjacent to a uniform flow is a simple wave.

In the application of interest here, the simple wave region is limited by the upstream characteristic (η_0) passing through the point *A*. The slope of this initial wave is:

$$\psi_0=\varphi_0+\alpha_0=\alpha_0.$$

The slope of the final wave (η_1), emerging from the point *B*, is:

$$\psi_1=\varphi_1+\alpha_1.$$

Moreover, if we consider the characteristic (ξ) ending in Q_l and coming from the upstream uniform region, we have:

$$\omega(M_{Q_1},\gamma)+\varphi_{Q_1}=\omega(M_0,\gamma)+\varphi_0=\omega(M_0,\gamma)$$

$$\omega(M_{Q_1},\gamma)=\omega(M_0,\gamma)-\varphi_{Q_1}.$$

As the angle of the wall φ_{Q_1} is negative here, it follows that:

$$\omega(M_{Q_1},\gamma) > \omega(M_0,\gamma).$$

Hence, considering the behavior of the function $\omega(M,\gamma)$ (see Figure 17.5):

$$M_{Q_1} > M_0.$$

The flow expands continuously along the contour *AB*.

Now let a point P_3 be on that same (ξ) as P_1, but located, as shown in Figure 17.7, on a more upstream (η). We have the relation:

$$\omega(M_{P_3},\gamma) + \varphi_{P_3} = \omega(M_{P_1},\gamma) + \varphi_{P_1}$$

from which follows:

$$\omega(M_{P_3},\gamma) = \omega(M_{P_1},\gamma) + \varphi_{P_1} - \varphi_{P_3}.$$

As $(\varphi_{P_1} - \varphi_{P_3})$ is negative (the negative angle φ increases in absolute value between *A* and *B*), it follows that:

$$\omega(M_{P_1},\gamma) > \omega(M_{P_3},\gamma).$$

Knowing that $\omega(M,\gamma)$ is an increasing function of the Mach number, we deduce that:

$$M_{P_1} > M_{P_3}.$$

Thus, when we move along a (ξ) in the upstream–downstream direction, the pressure decreases: *we are dealing with an expansion wave.* We can also demonstrate that the slope $\psi = \varphi + \alpha$ of the waves (η) decreases between (η_0) and (η_1): the characteristics (η) form a *set, or fan, of diverging waves.*

If we consider two points P_4 and P_5, located within the flow downstream of the characteristic (η_1) emerging from point B, we can write:

– on one hand:

$$\omega(M_{P_4},\gamma) + \varphi_{P_4} = \omega(M_0,\gamma) \qquad [17.18]$$

$$\omega(M_{P_5},\gamma) + \varphi_{P_5} = \omega(M_0,\gamma) \qquad [17.19]$$

by considering the (ξ) passing through P_4 and P_5 and coming from the upstream region of uniform flow.

– on the other hand:

$$\omega(M_{P_4},\gamma)-\varphi_{P_4}=\omega(M_{Q_2},\gamma)-\varphi_{Q_2}=\omega(M_{Q_2},\gamma)-\varphi_1 \qquad [17.20]$$

$$\omega(M_{P_5},\gamma)-\varphi_{P_5}=\omega(M_{Q_3},\gamma)-\varphi_{Q_3}=\omega(M_{Q_3},\gamma)-\varphi_1 \qquad [17.21]$$

by considering the (η) passing through P_4 and P_5 and meeting the wall downstream of B at the points Q_2 and Q_3.

From relations [17.18]–[17.21], it follows that:

$$M_{P_4}=M_{Q_2},\ \varphi_{P_4}=\varphi_1$$

$$M_{P_5}=M_{Q_3},\ \varphi_{P_5}=\varphi_1.$$

These conclusions were predictable because the results relative to the simple wave can also be applied to the region downstream of (η_1).

Hence, considering relations [17.18] and [17.19]:

$$M_{P_4}=M_{P_5},\ \varphi_{P_4}=\varphi_{P_5}.$$

Downstream of the final wave (η_1), the flow is uniform if the wall downstream of B is straight. Its Mach number M_1 can be calculated according to the deflection $\Delta\varphi=\varphi_1-\varphi_0$, through:

$$\omega(M_1,\gamma)=\omega(M_0,\gamma)-\varphi_1+\varphi_0. \qquad [17.22]$$

17.5.2. *The limit case of the Prandtl–Meyer expansion*

If the length of arc AB is reduced, the deflection φ_1 remaining unchanged, the expansion wave narrows at its base, the waves (η_0) and (η_1) maintaining constant directions. Indeed, M_0 is invariable by hypothesis, and M_1 is unchanged, according to relation [17.22]. If M_1 does not change, the Mach angle α_1 does not change either, thus also the slope of the wave (η_1).

It is possible to consider the passage to the limit where the length of arc *AB* becomes zero. The wall then has a sudden change of its direction $\Delta\varphi = \varphi_1 - \varphi_0$ in *A*. The resulting flow is called the *Prandtl–Meyer expansion*. It is shown schematically in Figure 17.8. The characteristic curves (η), constituting the simple expansion waves, all emanate from the slope discontinuity at *A* to form a fan of centered waves. In *A*, the flow undergoes a deflection with the angle $\Delta\varphi = \varphi_1 - \varphi_0$, the Mach number passing from M_0 to $M_1 > M_0$. Point *A* is a particular point where all the intermediate states between the upstream and downstream flows are merged. Otherwise, all results relating to the progressive expansion can be applied.

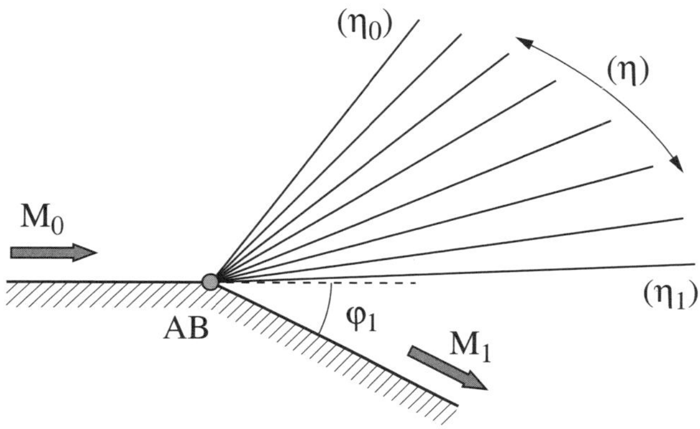

Figure 17.8. *The centered Prandtl-Meyer expansion*

Figure 17.9 is a visualization by interferometric schlieren of a Prandtl–Meyer expansion from a flow at Mach 3. The streak visible upstream of the slope change of the wall is due to a slight misalignment between two parts of the upstream wall. We see that the effect of this disturbance propagates in the supersonic flow as a wave whose slope is equal to the Mach angle for a Mach number of 3. In the vicinity of the wall, the expansion phenomenon is disturbed due to the presence of a boundary layer.

Figure 17.10 shows the two types of expansion, with the corresponding pressure distributions at the wall.

In the foregoing, we considered a wall located “below”, therefore, the expansion waves (η) went to “the top”. It is obvious that we treat in the same way, *mutatis mutandis*, the case where the wall is “above”. Then, the expansion waves are (ξ) characteristics (see Figure 17.11).

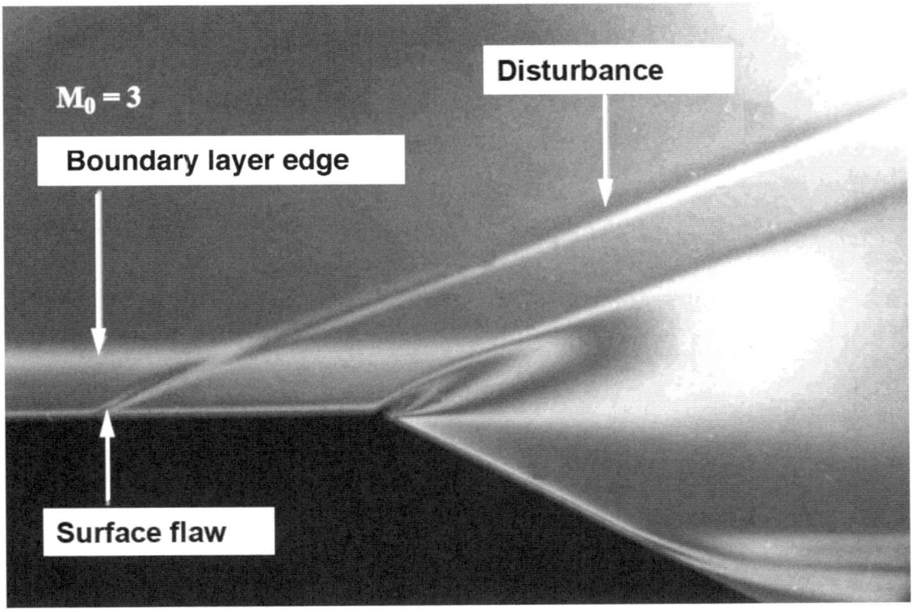

Figure 17.9. *Interferometric schlieren photo of a Prandtl-Meyer expansion (document Onera)*

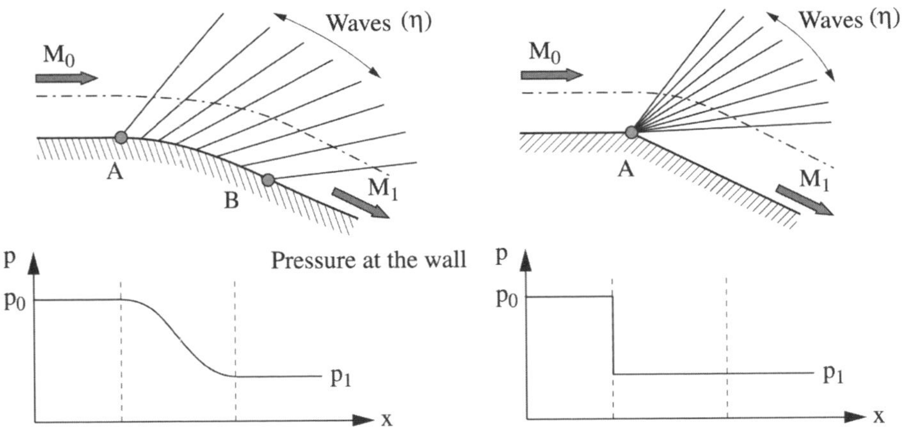

Figure 17.10. *Progressive and centered expansion wave*

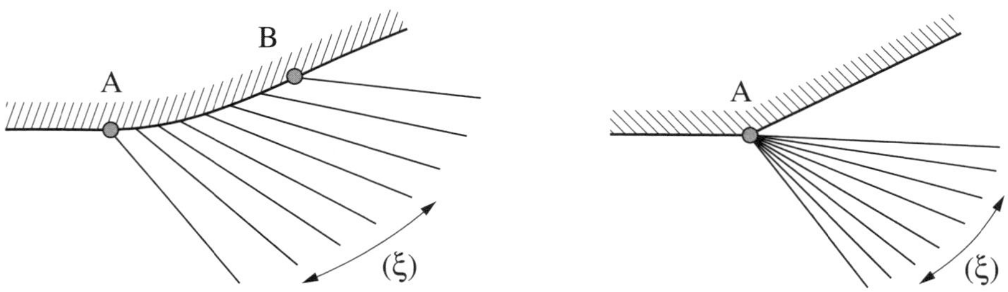

Figure 17.11. *"Downward" expansion waves*

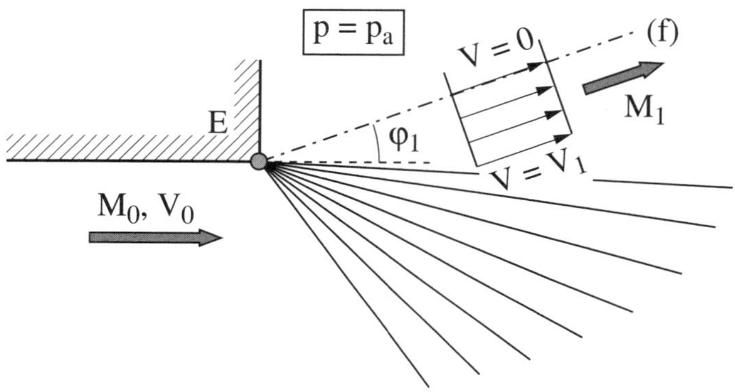

Figure 17.12. *Centered expansion at the origin of a supersonic jet*

A centered expansion is also formed when the wall stops at a point *E*, the downstream being an atmosphere at rest (or ambient) with a constant pressure p_a. The flow with the Mach number M_0 then leaves the wall at *E* and continues in the atmosphere as a jet confined by a fluid (or jet boundary) (*f*) isobaric at the pressure p_a, straight if the upstream flow is planar and uniform (see Figure 17.12). The boundary (*f*) is a line across which the fluid properties are discontinuous, with the exception of the pressure. Thus, the velocity passes from V_1 to 0, the density from ρ_1 to the value ρ_a, etc. The existence of a discontinuity surface such as (*f*) is compatible with the Euler equations: it is a slip line (see section 13.4).

If the pressure p_a is such that $p_a < p_0$, the flow in *E* undergoes a sudden expansion, allowing its pressure to adjust from p_0 to $p_1 = p_a$. The point *E* is the origin of an expansion fan whose properties are similar to those of the Prandtl–Meyer expansion. The difference is that in this case, the downstream Mach number

M_1 is given (as the expansion is isentropic, it can be deduced from the pressure p_a). We deduce the deflection by:

$$\Delta\varphi = \varphi_1 - \varphi_0 = \omega(M_1,\gamma) - \omega(M_0,\gamma).$$

It is the "mirror" problem of the wall deflection studied above.

– change of direction $\Delta\varphi$ → pressure variation Δp,

– pressure variation Δp → change of direction $\Delta\varphi$.

Figure 17.13 is an interferogram of the expansion of a flow at a wall interruption *D*. The interference fringes are lines of equal density value, thus in regions where the flow is isentropic, there are also iso-Mach lines, isobars, etc. The Mach number before the expansion is very close to 1. The document highlights the centered expansion from *D* as well as the mixing zone developing along the slip line separating the supersonic flow from the region without flow at constant pressure.

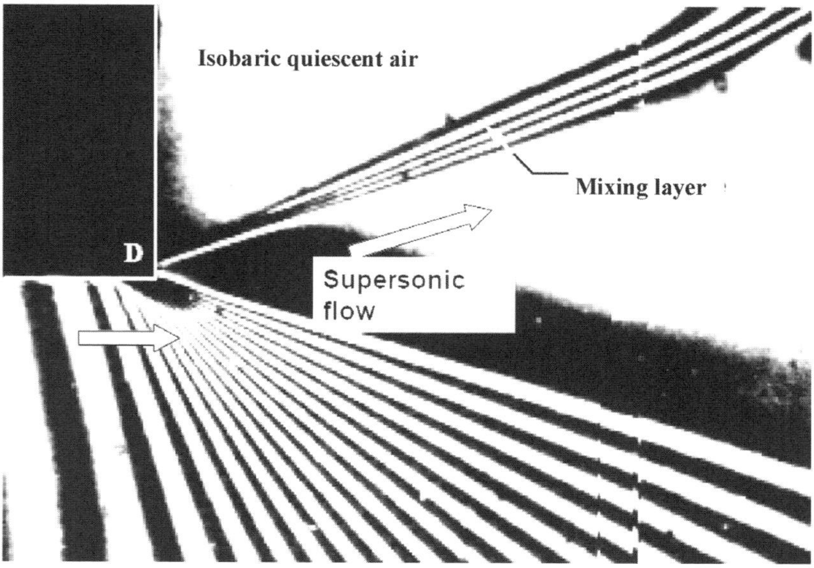

Figure 17.13. *Interferogram of a centered expansion at a wall interruption (document Onera)*

COMMENT.– A centered expansion can also occur from a non-uniform state, or in a rotational flow, where the characteristics (η) are no longer straight. Such a flow is sometimes called a Prandtl–Meyer expansion, which is incorrect. Strictly speaking, a Prandtl–Meyer expansion is an expansion, through centered waves, from a uniform state in a two-dimensional planar flow.

17.5.3. *The limit expansion angle*

We consider a Prandtl–Meyer expansion from the sonic state $M_0 = 1$ and, by convention, state $\varphi_0 = 0$. For a deflection φ_1, we have:

$$\omega(M_1,\gamma)+\varphi_1=\omega(M_0,\gamma)+\varphi_0=0$$

from which follows:

$$\varphi_1=-\omega(M_1,\gamma).$$

We have seen in section 17.4 that the function $\omega(M,\gamma)$ tends toward a finite limit when the Mach number tends toward infinity, i.e. when the flow expands to the vacuum (zero pressure). Thus, during a Prandtl–Meyer expansion undertaken to the vacuum, the deflection does not exceed a limit value such that:

$$\varphi_{\max}=\left|\omega_{\max}(\gamma)\right|.$$

If the angle of the wall is greater than $\varphi_{\max}$, the flow cannot remain in contact with the wall: a vacuum zone forms between the flow and the wall (see Figure 17.14). Such a phenomenon is largely illusory, since this result implies that the gas behaves as a continuous medium. In reality, when the pressure decreases, a threshold is reached beyond which the gas will either be liquefied or switch from the continuous to the molecular condition for which the above calculations do not apply. A similar phenomenon occurs during unsteady expansion of one-dimensional flow (see section 21.3.6).

In the case where the flow expands from a supersonic Mach number $M_0 > 1$, the maximum deflection is:

$$\Delta\varphi_{\max}=\left|\varphi_1-\varphi_0\right|=\omega_{\max}(\gamma)-\omega(M_0,\gamma).$$

This limit angle phenomenon is observed on space launcher vehicles at very high altitudes. Then, the external pressure being almost zero, the propulsion jets leaving the nozzles take the shape of plumes with very large sizes whose boundary has an initial slope approximately equal to the limit deflection.

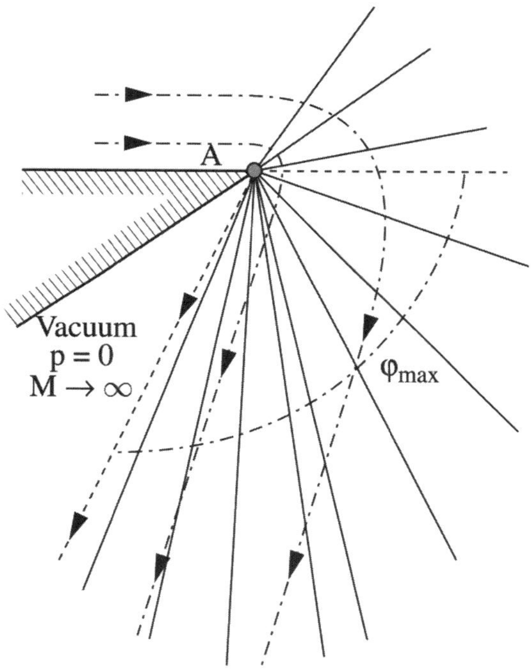

Figure 17.14. *Limit expansion angle*

17.5.4. *The progressive compression wave*

We imagine a wall, first straight, which presents from A a part where its slope φ rises $(\varphi > 0)$ to the point B beyond which $\varphi = \varphi_1$ is constant (see Figure 17.15). To address this problem, we can repeat the same previous reasoning and reach the following conclusions:

– The arc AB is the origin of the straight left running characteristics (η) forming a simple wave.

– Downstream of the final wave (η_1), the flow is uniform with the Mach number M_1.

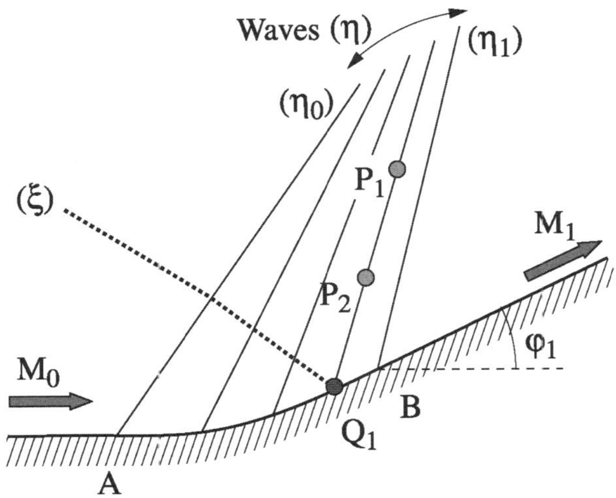

Figure 17.15. *Progressive compression wave*

This flow, however, presents a major difference compared to the previous case. To calculate the Mach number M at the wall at a point where the direction is φ, we use the characteristic relation (see above):

$$\omega(M,\gamma)=\omega(M_0,\gamma)-\varphi+\varphi_0=\omega(M_0,\gamma)-\varphi \quad (\textit{if} \quad \varphi_0=0).$$

As here $\varphi>0$, it follows that $\omega(M,\gamma)<\omega(M_0,\gamma)$. Thus, the pressure increases gradually as the deflection increases because $M<M_0$. We are dealing with a *compression wave.*

From upstream to downstream, from one (η) to another, since the Mach number decreases, the Mach angle α increases and as φ also increases, so does the slope $\psi=\varphi+\alpha$ of the characteristics. *The compression waves* (η) *are converging and form a convergent fan.*

This result leads to a remarkable consequence. Nothing prevents, mathematically and in the context of the problem addressed here, the converging waves (η) from intersecting. This leads to the flow diagram shown in Figure 17.16a. If we consider the point P located in the intersection region of the waves (η), this point is found:

– on one hand, on the (η) coming from Q_1, upstream of A, thus belonging to the uniform flow, and

– on the other hand, on the (η) coming from Q_2, located between A and B.

Since P belongs to two characteristics of the same family, the calculation of the flow in P will lead to two different values of the Mach number and the direction φ. Thus, there are two distinct velocities in P. This is possible from a purely mathematical point of view, but it is in contradiction with the postulate of mechanics according to which there can be only one velocity value in the same point of the space (realizability condition). Moreover, in such a region, the streamlines would have, as shown in Figure 17.16a, two turn backs (cusp points).

The solution obtained in the part of the plane where the characteristics intersect is physically not admissible. This does not mean that the problem of fluid mechanics has no solution but a solution made of a progressive wave type is not acceptable; we must find something else. In this case, we will have to introduce *a shock wave* from the place where the compression waves intersect (see Figure 17.16b). This shock wave will ensure the transition from the upstream uniform state to a downstream flow, satisfying the boundary conditions.

In general, the calculation of the shock resulting from the crossing, or *focalization*, of the characteristics can be performed with numerical techniques. We have to detect the first intersection point F of the characteristics, which will be the origin of the shock, then calculate it step-by-step from the Rankine–Hugoniot equations (see Chapter 20). The downstream flow is no longer uniform, nor isentropic; it is rotational.

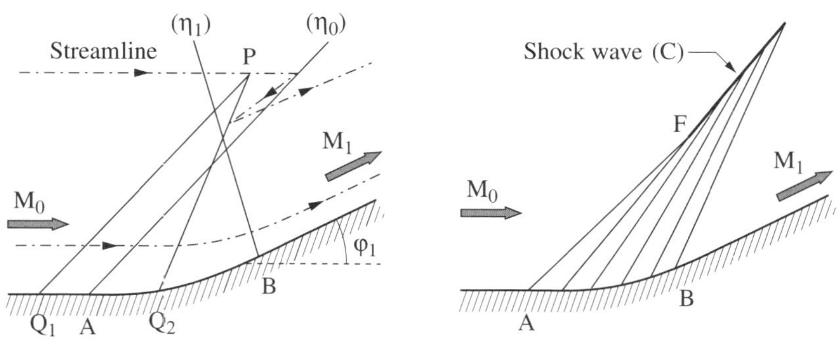

a – intersection of compression waves *b – focalization of compression waves*

Figure 17.16. *Formation of a shock wave through focalization*

When the length of arc AB tends to zero, the wall presents in A a discontinuity in the slope $\Delta\varphi = \varphi_1 - \varphi_0$ (see Figure 17.17). As in the case of the negative deflection $\Delta\varphi$, the mathematical solution consists of a fan of straight waves centered in A. But since for a deflection $\Delta\varphi > 0$ the slope of the waves (η) forming the fan increases, all (η) move in front of the initial wave (η_0). In particular, the final wave (η_1) is now further upstream. For the previously mentioned reasons, the solution in the domain defined by (η_0) and (η_1) is not unique, and therefore, unacceptable.

In this case, the introduction of an oblique planar shock wave, attached to *A*, allows us to achieve a uniform downstream flow, with the direction φ_1, satisfying the equations of motion and boundary conditions. The deflection φ_1 being known, the downstream flow 1 is calculated from the upstream flow 0 by solving the oblique shock equations. This problem is dealt with in Chapter 14.

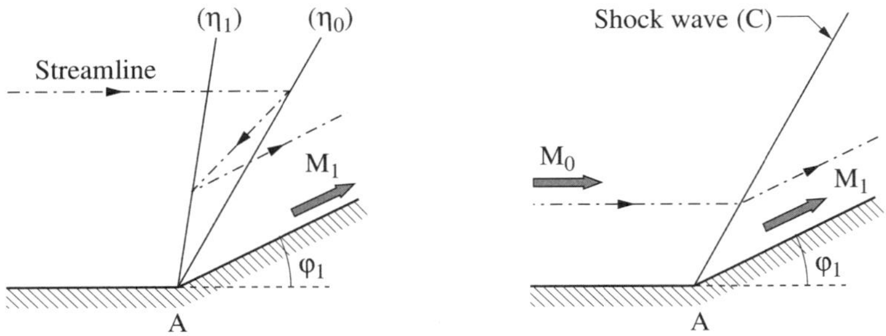

Figure 17.17. *Shock wave caused by a slope discontinuity*

We will find again the focusing phenomenon of compression waves in the theory of unsteady flows, where, in a compression, waves emitted moments later catch those emitted earlier (see section 21.4). There results a situation leading to the formation of a shock wave. These focalization phenomena are also the origin of the formation of a sonic boom when an aircraft maneuvers under conditions leading to the emission of compression waves. This phenomenon should not be confused with the classic supersonic boom related to shock waves attached to parts of the aircraft and which exist during steady flight.

Chapter 18

The Numerical Method of Characteristics

18.1. Introductory remarks

The integration technique of Euler equations by the method of characteristics is based on the use of equations [17.13] of the previous chapter that have been linearized and then discretized. Different operators are used depending on whether the point is calculated within the fluid or located on a boundary (wall, axis of symmetry or centerline, and fluid boundary). The procedure is similar in spirit to that used for one-dimensional unsteady flows (see section 24.1). As in any numerical method, the solution to the equations is sought in discrete points, which together form a mesh, chosen close enough to each other so that the flow properties vary only slightly when we pass from one point to one of its neighbors. This condition justifies the linearization of the equations and ensures the convergence of the calculation under certain conditions that we will not discuss here.

The method of characteristics is to calculate the flow step-by-step, progressing along the characteristic lines (η) and (ξ). In this integration technique, the meshing depends on the solution itself since (η) and (ξ) are functions of local properties of the flow. Thus, this meshing is constructed as the calculation progresses, which introduces some algorithmic complexity into the structure of a computer program, especially when shock waves are formed within the flow. However, the method ensures correct transmission of information in agreement with the rules governing the system of hyperbolic equations, which leads to a high accuracy, even in regions with high gradient, and great execution velocity because of a very fast convergence.

In conventional finite difference methods, the meshing is most often defined once for all before starting the calculation, which tends to substantially simplify the resolution algorithm. However, the rules of information transmission are not well respected and regions with rapid changes can be miscalculated, for example, the origin of an expansion fan. Despite these drawbacks, finite difference methods have the advantage that they can handle complex flows, with shock waves and subsonic regions. In addition, finite difference methods (as well as finite volume methods) generalize easily to three-dimensions, which is not the case with the method of characteristics.

Nevertheless, despite its limitations, the method of characteristics is a tool well suited to the calculation of two-dimensional supersonic flows, planar or axisymmetric, containing a limited number of shock waves. It is a very precise method of handling discontinuities, such as shock waves and slip lines, as well as regions with rapid change, such as the origin of a centered expansion. The integration technique enables a rigorous treatment of these discontinuities without having to stabilize the calculation by adding an artificial viscosity, as the shocks are *fitted* and not *captured*. It is also very fast, the calculation of the flow in a nozzle requiring only a fraction of a second on a PC. Finally, its knowledge is of great educational value because, in its operation, it reproduces the mechanism of disturbance propagation in supersonic flows. From this perspective, it is a powerful means of understanding the physics of supersonic flows whose mastering is essential for the aerodynamicist of high velocities.

18.2. The basic operators

18.2.1. *Normal point in a flow: operator N*

We remember the characteristic equations established in the previous chapter that connect variations of pressure δp (or δz) and direction $\delta\varphi$ of the velocity for the displacements $\delta\eta$ and $\delta\varphi$ along the characteristics (η) and (ξ):

$$Z\delta z + \delta\varphi = N\delta\eta \quad on \quad (\eta) \qquad [18.1]$$

$$Z\delta z - \delta\varphi = N\delta\xi \quad on \quad (\xi).$$

As the flow is known at points 1 and 2 (see Figure 18.1), we propose to calculate the pressure p_3 (or $z_3 = \ln p_3$), the direction φ_3 of the velocity, and the entropy s_3 at point 3, located at the intersection of the right running characteristic (ξ), emerging from 1, and the left running characteristic (η) going through 2. As Z and N are functions of p, φ, and s, the system [18.1] above is nonlinear, and we must proceed for its resolution by linearization and iteration.

Very often, the pressure p is related to a reference pressure p_r ; thus, we take as a working variable:

$$z = \ln\left(\frac{p}{p_r}\right).$$

In the case of a calorically perfect gas, the entropy will be given by (see section 9.5.2.1):

$$s = C_v \ln\left(\frac{T_{st}}{T_r}\right) - r \ln\left(\frac{p_{st}}{p_r}\right)$$

where T_r is a reference temperature. In an adiabatic flow, which represents almost all treated cases, the stagnation temperature T_{st} is constant throughout the flow field. It is then unnecessary to consider this quantity as it is a constant and we agree to call entropy the quantity:

$$s = -\ln\left(\frac{p_{st}}{p_r}\right)$$

which is actually the total pressure assigned to each streamline.

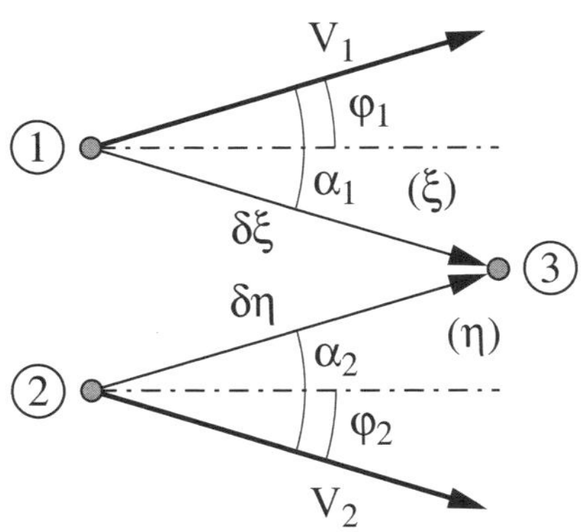

Figure 18.1. *Normal point in a flow: operator N*

The Mach number M is calculated from the relation:

$$M = \sqrt{\frac{2}{\gamma-1}\left[\left(\frac{p}{p_{st}}\right)^{-\frac{\gamma-1}{\gamma}} - 1\right]}$$

where the ratio p / p_{st} will be deducted from the variables z and s by:

$$\frac{p}{p_{st}} = \exp(z+s).$$

We assume that we are calculating the iteration with the rank (n). Having the values of p, φ, and s of the previous iteration $(n-1)$, we can calculate:

$$M_3^{(n-1)}, \quad \alpha_3^{(n-1)}, \quad Z_3^{(n-1)}.$$

The characteristic (ξ) passing through 1 and 3 is approximated by a straight line with the direction:

$$\psi_{13} = \frac{1}{2}\left(\varphi_1 - \alpha_1 + \varphi_3^{(n-1)} - \alpha_3^{(n-1)}\right)$$

and (η) passing through 2 and 3 in a straight line with the direction:

$$\psi_{23} = \frac{1}{2}\left(\varphi_2 + \alpha_2 + \varphi_3^{(n-1)} + \alpha_3^{(n-1)}\right).$$

Thus, we can calculate the coordinates of point 3:

$$x_3^{(n-1)} \qquad and \qquad y_3^{(n-1)}$$

which leads to $N_3^{(n-1)}$, and thus, to the lengths of the characteristic segments:

$$\delta\xi^{(n-1)} \qquad and \qquad \delta\eta^{(n-1)}.$$

We state:

$$Z_{13} = \frac{1}{2}\left(Z_1 + Z_3^{(n-1)}\right), \qquad N_{13} = \frac{1}{2}\left(N_1 + N_3^{(n-1)}\right)$$

$$Z_{23} = \frac{1}{2}\left(Z_2 + Z_3^{(n-1)}\right), \qquad N_{23} = \frac{1}{2}\left(N_2 + N_3^{(n-1)}\right).$$

The characteristics relations [18.1] are discretized in the form of:

$$Z_{13}\left(z_3^{(n)}-z_1\right)-\left(\varphi_3^{(n)}-\varphi_1\right)=N_{13}\delta\xi^{(n-1)}$$

$$Z_{23}\left(z_3^{(n)}-z_2\right)+\left(\varphi_3^{(n)}-\varphi_2\right)=N_{23}\delta\eta^{(n-1)}.$$

These two equations form a linear algebraic system for the unknowns $z_3^{(n)}$ and $\varphi_3^{(n)}$. Its resolution gives the new pair:

$$\left(z_3^{(n)},\quad \varphi_3^{(n)}\right).$$

The entropy in 3 must be calculated in another way by exploiting the fact that it remains constant along a streamline (which is a characteristic direction resulting from the energy equation). As the entropy is known in the points 1 and 2, where it has the values s_1 and s_2, respectively, the entropy s_3 in 3 is determined by interpolation between the streamlines through 1 and 2 (see Figure 18.2). As these streamlines are treated as straight lines, locally tangential to the velocity vector, we use the interpolation formula:

$$s_3=\frac{s_1\sin\left(\alpha_{23}\right)\Delta\eta+s_2\sin\left(\alpha_{13}\right)\Delta\xi}{\sin\left(\alpha_{23}\right)\Delta\eta+\sin\left(\alpha_{13}\right)\Delta\xi}.$$

As shown in Figure 18.2, $\Delta\eta$ and $\Delta\xi$ represent the distances between streamlines passing, on one hand, through 2 and 3, and, on the other hand, through 1 and 3, and α_{ij} denotes the arithmetic average of the Mach angle α between the points (*i*) and (*j*).

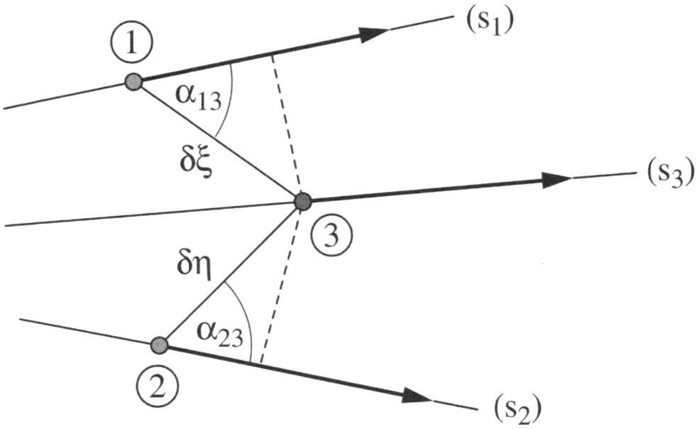

Figure 18.2. *Entropy calculation at the normal point*

At this level, we conduct a convergence test that involves comparing the values of z_3 and φ_3 with those of the previous iteration. If ε is a preset accuracy threshold, the calculation is considered converged at the point 3 if we simultaneously have:

$$\left|z_3^{(n)} - z_3^{(n-1)}\right| < \varepsilon \qquad and \qquad \left|\varphi_3^{(n)} - \varphi_3^{(n-1)}\right| < \varepsilon.$$

If these conditions are not met, a new iteration loop is started. The iterative process is summarized in Figure 18.3.

The iterative process is initialized by putting, for example:

$$z_3^{(0)} = z_1, \quad \varphi_3^{(0)} = \varphi_1, \quad s_3^{(0)} = s_1.$$

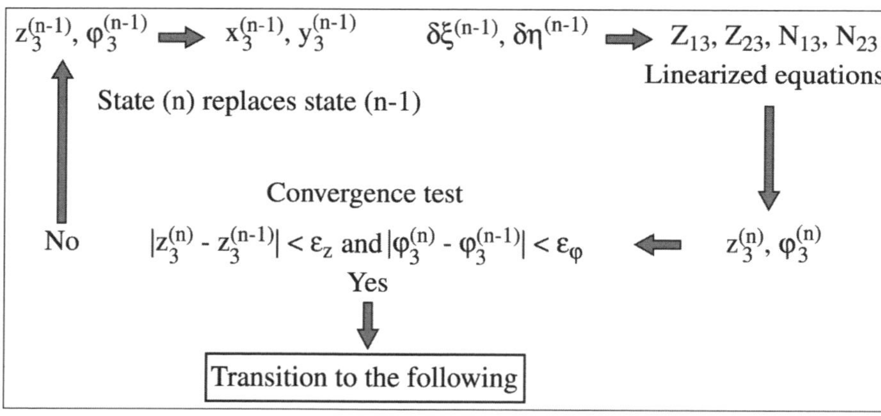

Figure 18.3. *Calculation procedure for the normal point: operator N*

18.2.2. *Point on a wall: operator W*

We consider the situation in Figure 18.4. The flow being known in the point 2, we have to calculate it in point 3 located at the intersection of the wall (W) and the characteristic (η) emerging from 2. The shape of the wall is given either in analytical form or via a series of points.

The slip condition to be met in 3 requires that the direction of the velocity φ_3 is equal to the slope φ_w of the wall in 3, φ_w resulting from the wall definition. In addition, as the wall is a streamline, the entropy s_w on (W) is known, $s_3 = s_w$. Therefore, there is only one unknown quantity in 3, the pressure p_3. As for the normal point, the solution is obtained by iteration, the point $W \equiv 3$ moving during the calculation.

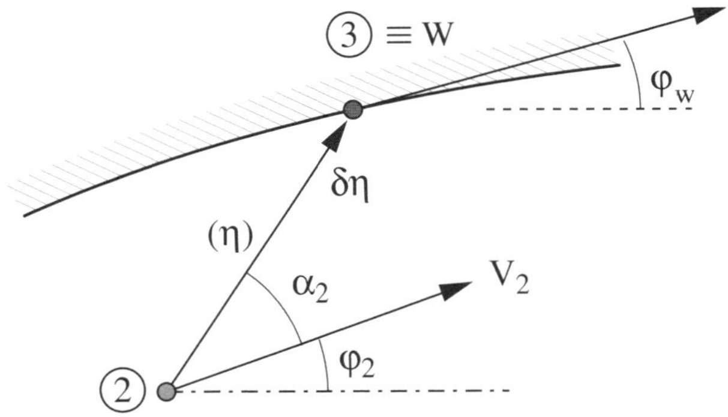

Figure 18.4. *Point on a wall: operator W*

The coordinates:

$$x_w^{(n)} \quad and \quad y_w^{(n)}$$

of W in the iteration number (n) are calculated by determining the intersection of the wall (W) with the left running characteristic (η) emerging from 2, which is approximated by a straight line of slope:

$$\psi_{23} = \frac{1}{2}\left(\varphi_2 + \alpha_2 + \varphi_w^{(n-1)} + \alpha_w^{(n-1)}\right).$$

We deduce the value of $\varphi_w^{(n)}$ for the new position of W. Then, $z_3^{(n)}$ is calculated by the characteristic relation on (η):

$$Z_{23}\left(z_3^{(n)} - z_2\right) + \left(\varphi_w^{(n)} - \varphi_2\right) = N_{23}\delta\eta^{(n-1)}.$$

We deduce $M_3^{(n)}$, $\alpha_3^{(n)}$, then $Z_3^{(n)}$, $N_3^{(n)}$, etc., the process is repeated until the convergence test for z_3 is satisfied.

18.2.3. *Point on a jet boundary: operator J*

This concerns the calculation of a point $J \equiv 3$ located on a jet boundary (*J*), which can be the boundary (jet line) of a jet issuing from a nozzle into an ambient with the pressure p_a. In this case, one of the boundary conditions imposed on *J* is the value of the pressure, $p = p_a$, the other condition being that (*J*) is a streamline.

We will calculate the point $3 \equiv J$ at the intersection of (*J*) and the left running characteristic (η), passing through the point 2 (see Figure 18.5). We denote with 1 the point of (*J*) calculated immediately before 3 and located on the right running (ξ) passing through 2. We still proceed iteratively. At the iteration (n), the coordinates of 3 are calculated at the intersection:

– of the characteristic (η) approximated by a segment of the straight line with the slope:

$$\psi_{23} = \frac{1}{2}\left(\varphi_2 + \alpha_2 + \varphi_3^{(n-1)} + \alpha_3^{(n-1)}\right)$$

– of the boundary (*J*) approximated by a straight line passing through 1 with the slope:

$$\varphi_j = \frac{1}{2}\left(\varphi_1 + \varphi_3^{(n-1)}\right).$$

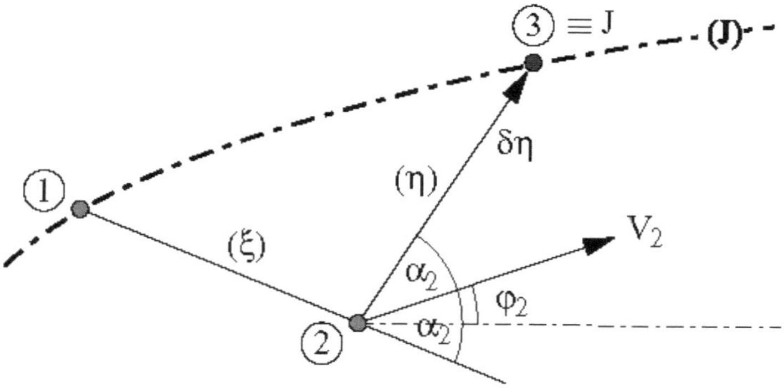

Figure 18.5. *Point on a jet boundary: operator J*

We are thus able to calculate the coordinates $x_3^{(n)}$ and $y_3^{(n)}$ of 3. We deduce the pressure $p_3^{(n)}$, which, in the general case of a non-isobaric jet boundary (*J*), is given as a function $p_f = p(x)$, or a sequence of values $[x_n, p_n]$. The direction of the velocity in 3 is calculated by the characteristic relation:

$$Z_{23}\left(z_3^{(n)} - z_2\right) + \left(\varphi_3^{(n)} - \varphi_2\right) = N_{23}\delta\eta^{(n-1)}.$$

As the boundary (*J*) is a streamline, the entropy is known in 3: $s_3 = s_1$. The above loop is repeated until the convergence test on φ_3 is satisfied. In the case of an isobaric boundary, $p_3 = p_a$ does not vary from one iteration to another.

18.2.4. *Point on an axisymmetry axis: operator A*

In the planar two-dimensional case, if there is an axis of symmetry $X'X$, the velocity vector must be aligned with $X'X$, a condition amounting to imposing $\varphi = 0$ if the angles are evaluated with respect to $X'X$. Hence, the calculation of a point *A* located on the axis of symmetry $X'X$ is identical to that of a point on a wall.

In an axisymmetric flow, a difficulty arises for the calculation of the term:

$$N = -\frac{\sin\alpha \sin\varphi}{y}$$

as on the axis φ and y simultaneously tend to zero. We must resolve the indetermination 0/0 before proceeding with the discretization (the same problem occurring in conventional numerical methods). We calculate the point $3 \equiv A$ at the intersection of the axis $X'X$ and the right running characteristic (ξ), passing through the point 1 (see Figure 18.6). For a displacement near the axis on the characteristic (ξ), we have if δ denotes a variation along (ξ):

$$\frac{\sin\varphi}{y} \cong \frac{\delta\varphi}{\delta y} \cong \left(\frac{\partial\varphi}{\partial y}\right)_\xi.$$

However, as on the axis (ξ) has the slope $-\alpha$ (since $\varphi = 0$), we have:

$$\delta y = -\sin\alpha\, \delta\xi.$$

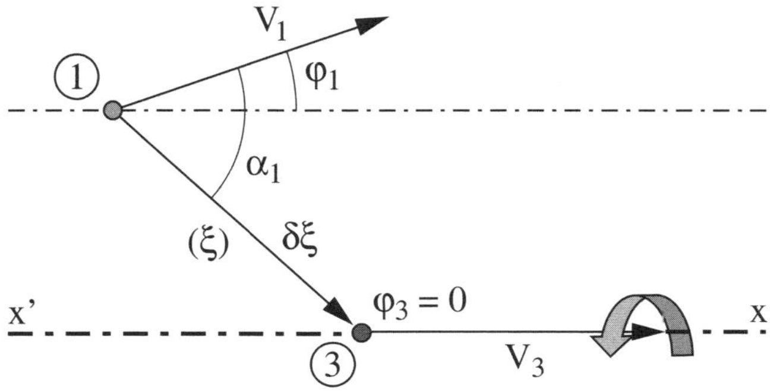

Figure 18.6. *Point on an axisymmetry axis: operator A*

Therefore, when $y \to 0$:

$$-\frac{\sin\alpha\sin\varphi}{y}\delta\xi \cong -\left(\frac{\partial\varphi}{\partial y}\right)_{\xi}\delta\xi\sin\alpha = \frac{\delta\varphi}{\delta y}\frac{\delta y}{\sin\alpha}\sin\alpha = \delta\varphi$$

so that the second characteristic relation [17.13] is written on an axisymmetry axis:

$$\boxed{Z\ \delta z - 2\delta\varphi = 0.}$$

Knowing that $\varphi_3 = 0$, we have the following two relations to calculate z_3:

– the characteristic relation written in 1:

$$Z_1(z_3 - z_1) - (\varphi_3 - \varphi_1) = Z_1(z_3 - z_1) + \varphi_1 = N_1\delta\xi$$

– the limit form written in 3:

$$Z_3(z_3 - z_1) + 2\varphi_1 = 0.$$

Taking the average value of z_3, given by the two relations above, we get:

$$z_3 = z_1 - \frac{1}{2}\left[\varphi_1\left(\frac{1}{Z_1} + \frac{2}{Z_3}\right) - \frac{N_1}{Z_1}\delta\xi\right].$$

This combination allows us to keep the same accuracy as for the previous operators. Despite this difference in determining z_3, the calculation procedure is the same as for the point on a wall. The entropy in 3 is known as the axis is a streamline.

18.2.5. *Origin of a centered expansion: operator Q*

When a wall has a discontinuity in the slope $\Delta\varphi$ at a point *Q*, due to the change of direction $\Delta\varphi$ relative to the upstream velocity, *Q* is the origin, either of a shock wave or an expansion fan formed of waves emanating from *Q*. Here, we examine the second case. The situation in *Q* is illustrated in Figure 18.7a, where we consider the case where *Q* is the origin of expansion waves (ξ). The point *Q* is a particular point where coexist an infinite number of states that correspond to the transition from an upstream state 0 (such that $\varphi = \varphi_0$ and $p = p_0$) to a downstream state 1 (such that $\varphi = \varphi_1$ and $p = p_1 < p_0$). The point *Q* is the limit position of the segment of the characteristic (η) joining the flows 0 and 1 when we tend toward point *Q*.

Since $\Delta\eta \to 0$, the first of relations [17.12], valid for a characteristic (η), takes the limit form in *Q*:

$$\frac{\sqrt{M^2-1}}{\gamma M^2}\frac{dp}{p} + d\varphi = 0.$$

In other words, in *Q*, the relation pressure–deflection is the same in planar and axisymmetric flows. If we are dealing with an isentropic flow (which is the case here), as *Q* is a streamline, and if the gas is calorically perfect, for a deflection $\Delta\varphi$, the downstream pressure p_1 is calculated from:

$$\omega(M_1,\gamma) = \omega(M_0,\gamma) + \varphi_1 - \varphi_0 = \omega(M_0,\gamma) + \Delta\varphi$$

where $\omega(M,\gamma)$ is the Busemann function, defined in section 17.4.

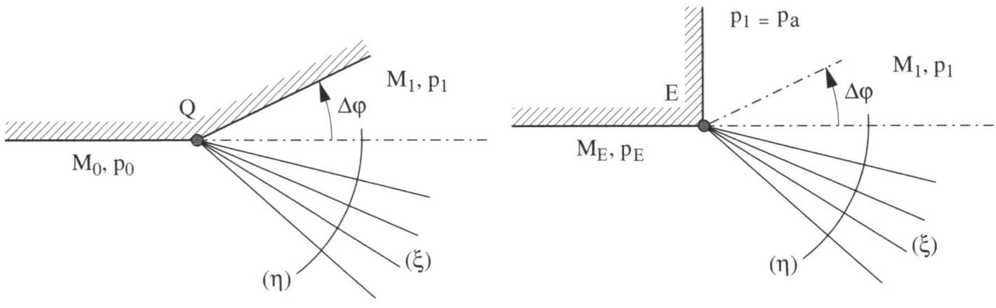

a – centered expansion caused by a wall discontinuity

b – centered expansion caused by a pressure discontinuity

Figure 18.7. *Origin of a centered expansion wave: operator Q*

We face a similar problem when a flow, initially guided by a wall to a point E, continues after E into an atmosphere whose pressure p_a is lower than the pressure p_E in E (see Figure 18.7b). Here, E is also the origin of an expansion fan across which the pressure passes from $p_0 = p_E$ to $p_1 = p_a$. A deflection $\Delta\varphi = \varphi_1 - \varphi_0$ of the velocity vector results, the unknown being the direction φ_1. It is the "mirror" problem of the previous case. Assuming a calorically perfect gas, φ_1 will be obtained by:

$$\varphi_1 = \varphi_0 + \omega(M_1, \gamma) - \omega(M_0, \gamma).$$

The Mach number M_1 is deduced from the pressure p_1 and the entropy s_1 is one of the flows along the wall before the expansion in E.

To apply the method of characteristics, in one case as well as in the other, we divide the finite expansion associated with the deflection $\Delta\varphi = \varphi_1 - \varphi_0$ into N_Q elementary deflections:

$$\delta\varphi = \frac{\Delta\varphi}{N_Q}.$$

Hence, at point Q, there are N_Q states, each characterized by:

– the direction of the velocity : $\varphi_n = \varphi_0 + n\,\delta\varphi$ and

– the pressure p_n, or z_n, calculated from:

$$\omega(M_n, \gamma) = \omega(M_0, \gamma) + n\,\delta\varphi$$

where n varies from $n = 0$ to $n = N_Q$. In practice, the splitting N_Q is chosen by ensuring that $\delta\varphi \approx 1°$.

18.3. Calculation of a supersonic field

18.3.1. *Domains of dependence and influence*

Let us assume that we know the flow on a curve (C) of finite length, non-characteristic, and limited by the points A and B (see Figure 18.8). The flow on AB is the only data we have, as A and B are located neither on a wall, or axis of symmetry, nor on a jet boundary. We propose to calculate the flow downstream of AB by solving the motion equations in the characteristic form, which are equivalent to the equations written in the conventional Cartesian form. Thus, any solution to the system of characteristic equations [17.12] is the solution of the problem.

Using the described numerical method, we choose a distribution of N discrete points on the curve AB (five in Figure 18.8) numbered: 01, 02, 03, 04, and 05. Applying the normal point operator N, we can calculate the flow in point 11, at the intersection of the right running characteristic from 01 and the left running characteristic rising from 02. Then, we move to point 12 that is calculated from 02 and 03, then to 13, and finally, to 14. We note that we cannot go further because we do not have data beyond B. The above process can be repeated to calculate 21 from 11 and 12, then 22 and 23. We then move to 31 and 32, and finally, the calculation stops at $41 \equiv P$ for want of data.

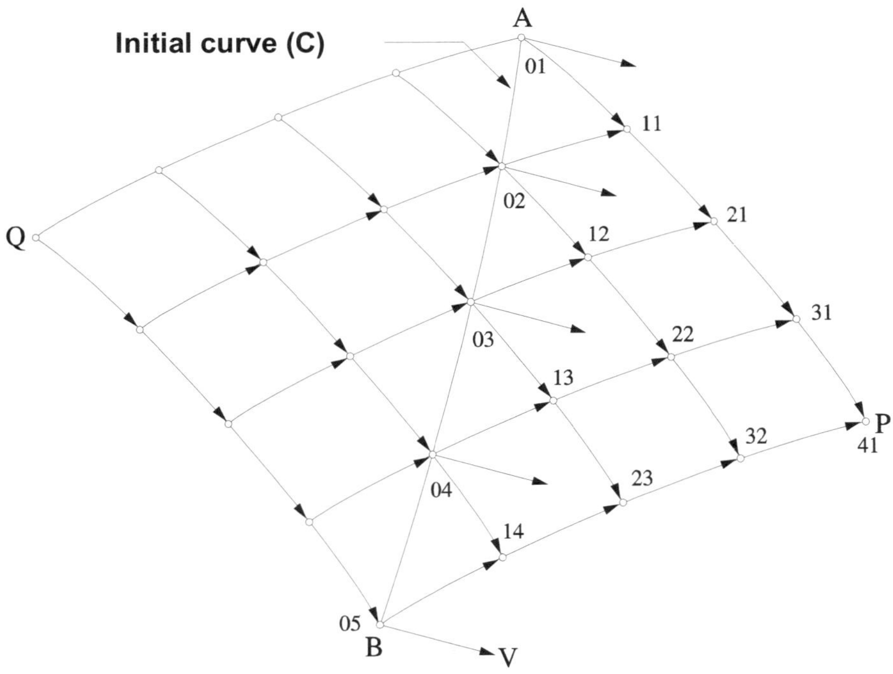

Figure 18.8. *Calculable domain starting from an initial non-characteristic curve (C)*

Thus, from AB, we can only calculate the flow inside the curvilinear triangle ABP, two of whose sides being the segment of the right running characteristic AP springing from A and the segment of the left running characteristic BP springing from B. We see that the flow in ABP is independent of conditions outside of the segment AB. It is possible to connect arbitrary flows along BP and AP without affecting the field inside of *ABP*. This result is only true if there is no formation of shock waves, in other words, the segments *AP* and *BP* can only support discontinuities affecting the derivatives of dependent variables. If discontinuities of order 1 form in *A* or *B*, affecting the variables themselves, then the formation of a shock wave can occur which will penetrate the interior of the triangle *ABP*.

In the absence of shock waves, the flow in the point *P* depends only on the conditions in *AB*, that is, more generally, the conditions inside the roughly triangular area limited by the left and right running characteristics passing through *P*. This area is called the *domain of dependence of point P* (see Figure 18.9a).

A similar reasoning can be made for the triangle *ABQ*, where *Q* is upstream of *AB* (see Figure 18.8). We see that the conditions in *Q* do not influence the solution inside the triangle *ABQ*. Of course, this solution does not depend on *Q* only, but it is influenced by the situation in *Q*. For this reason, the area defined by the two characteristics passing through *Q* is called *domain of influence of Q* (see Figure 18.9b).

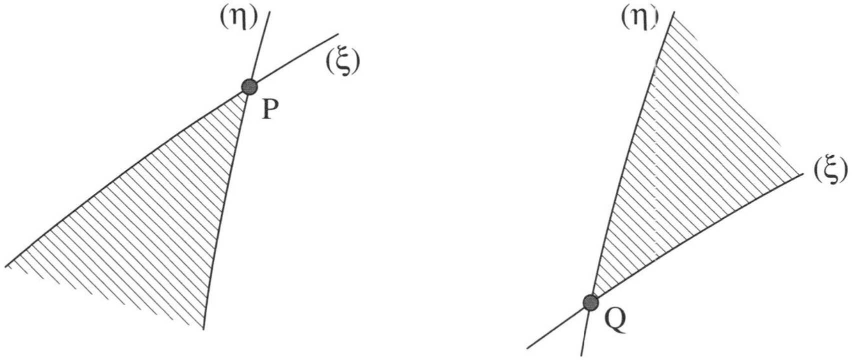

a – domain of dependence of P *b – domain of influence of Q*

Figure 18.9. *Domains of dependence and influence*

The reasoning we have done by relating to the numerical method of characteristics, to be more specific, is obviously true for a continuous field.

We will note that it is possible, starting from the segment *AB*, to calculate the flow to the left of *AB* to the point *Q* by "climbing back up" along the characteristics. This ability follows from the reversibility of the Euler equations mentioned in Chapter 7.

We return to the Cauchy problem, in light of what we have learned about the method of characteristics, by considering the situation where the segment AB is a part of a characteristic line. We imagine the previous calculation procedure to extend the calculation outside of AB (see Figure 18.10). We begin by trying to calculate the point 11 at the intersection of the right running (ξ) coming from 01 and the left running (η) coming from 02. However, as this left running characteristic is precisely the segment AB, the intersection of (η) passing through 02 and the (ξ) coming from 01 will be the point 01 itself: it is impossible to "take off" the calculation from the characteristic segment AB.

The Cauchy problem is undetermined. We have, thus, found the theoretical result in an "empirical" way.

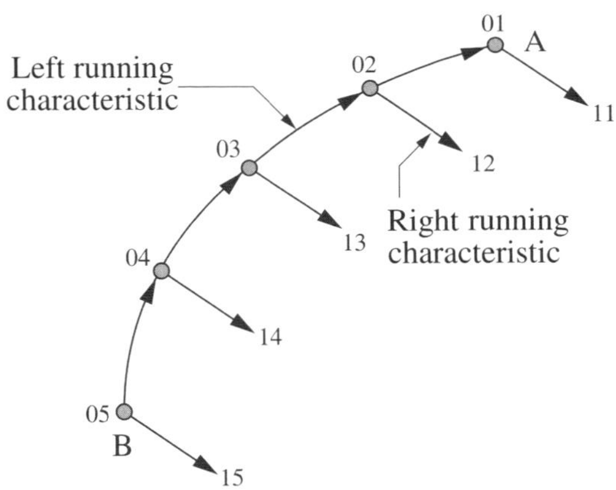

Figure 18.10. *Situation where the initial curve (C) is characteristic*

18.3.2. *Extension of the calculation: taking into account a boundary condition*

We consider the situation downstream of B and impose a boundary condition. This act provides additional information that will help extend the calculation in a larger domain than the triangle ABP. The condition imposed downstream of B can be either wall-like or jet-boundary-like. Also, B may be on a symmetry or axisymmetry axis, which is actually equivalent to a wall condition, as seen in section 18.2.4. Here, to clarify, we assume that downstream of B the flow must follow a wall that does not impose a velocity discontinuity at B, which would be the cause of a shock wave.

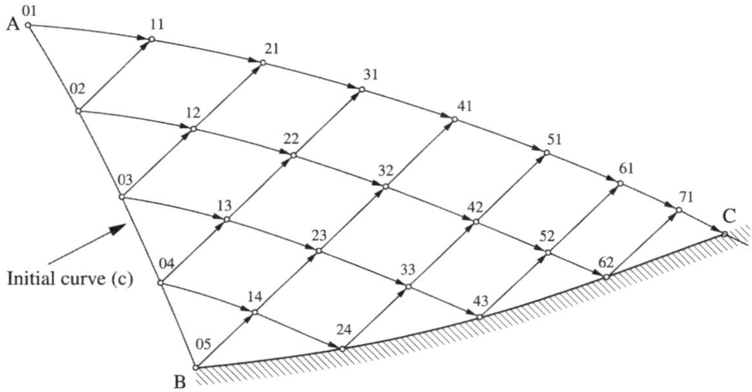

Figure 18.11. *Calculable domain starting from an initial curve (C) and a wall*

Referring to Figure 18.11, the calculation will be performed in the following way. Applying the previous algorithm, from the points 01, 02, 03, 04, and 05, we calculate the points 11, 12, 13, and 14. Applying the operator W (J or A in other cases) allows us to calculate the point 24 located on the wall (either on the fluid boundary or the symmetry/axisymmetry axis). Returning to the original algorithm, we calculate 21, 22, 23, 33, then again through the operator W the point 43, and so on.

The calculation may be continued until the end of the wall is reached or the right running characteristic coming from A has met the wall. Then it is no longer possible to continue the calculation, the entire flow field defined by the data of the segment AB and the wall having been determined.

As shown in the following example, the inclusion of a boundary condition allows us to perform a calculation from data located on a characteristic line (see Figure 18.12). We know the flow on the initial characteristic (η_0) starting point B, which is also the origin of the wall (W). The starting points are numbered 01, 02, 03, 04, etc. Applying the operator W (or J or A), starting from 02, we calculate 12 on the wall, then with the operator N we can determine 13, and then 23 through the operator W. Then we move to the starting point 04 and calculate with the same method the right running characteristic ending at the wall in the point 34. The calculation continues until the depletion of points on the starting characteristic (η_0) or until the end of the wall has been reached. Thus, the calculation algorithm consists of constructing, step-by-step, the right running characteristics starting from the points 01, 02, 03, etc. given on the initial characteristic starting from the wall (W). In this case, we can start the calculation from data located on a characteristic curve as the boundary condition to satisfy (here ship condition on a wall) provides the information needed to extend the solution.

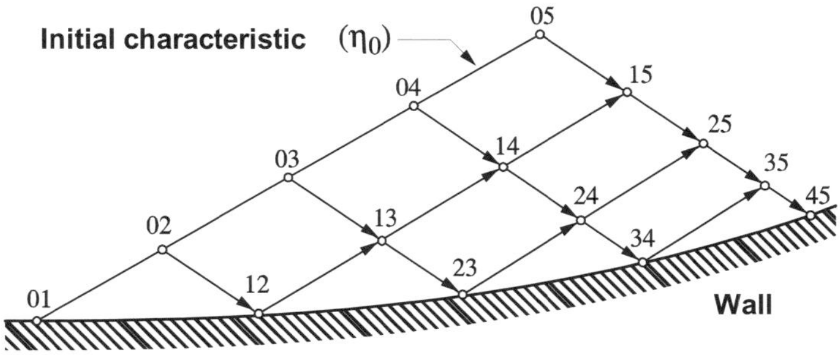

Figure 18.12. *Calculable domain between a starting characteristic and a wall*

18.4. Some examples of calculation

The following calculations illustrate the application of the method of characteristics to the calculation of simple flows. The first case is a progressive expansion of a flow at Mach number $M_0 = 2$ along a convex wall between the points *A* and *B*, the final deflection having an angle of $-20°$. We observe in Figure 18.13 the expansion waves emanating from the wall that form a set of divergent straight lines as well as the establishment of a new uniform state downstream of the last wave of the expansion rising from *B*.

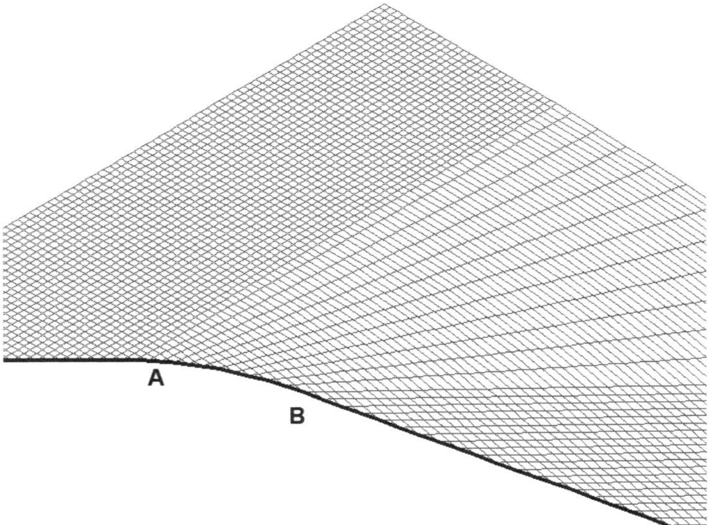

Figure 18.13. *Continuous expansion along a convex wall AB. Initial Mach number $M_0 = 2$, final deflection $\Delta\varphi = -20°$*

The second calculation corresponds to an expansion caused by a sudden deviation of the wall at an angle of −20° (see Figure 18.14). The fan of centered waves from point *A* forms a Prandtl–Meyer expansion (see section 17.5.2).

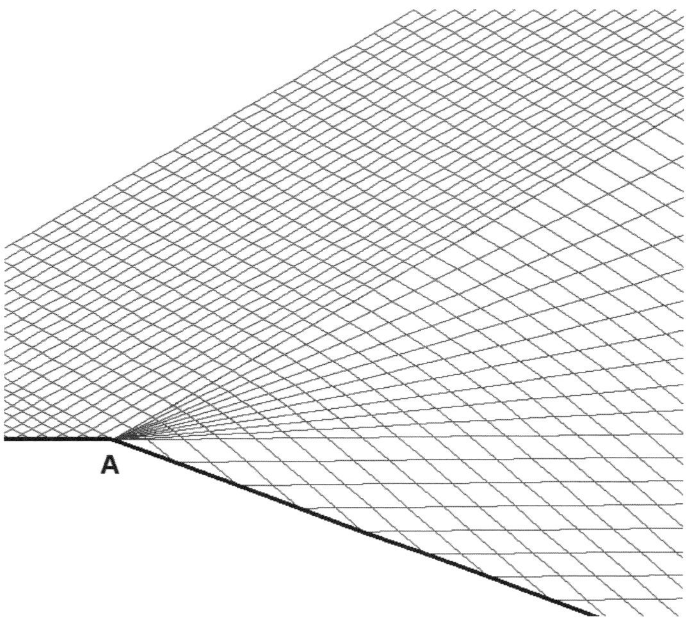

Figure 18.14. *Discontinuous expansion through centered waves. Initial Mach number* $M_0 = 2$ *, deflection* $\Delta\varphi = -20°$

The following example shows the calculation of the jet issuing from a two-dimensional planar nozzle, which produces a uniform flow with the Mach number $M_0 = 2$. As shown in Figure 18.15, the jet, emerging from an environment with the constant pressure p_1, is limited by a rectilinear wall. The expansion ratio is such that $\frac{p_1}{p_0} = 0.313$. We see the expansion fan attached to the nozzle lip and the reflection of those waves as expansion waves on the straight wall. The latter is reflected from the isobaric jet boundary (*J*) as compression waves, the reflection inducing a bending of (*J*). Up to the impact of the expansion waves reflected from the wall, the isobaric boundary is straight with the direction $\varphi_1 = 22.21°$, calculable from the Busemann number (see section 17.5.1). At the end of the reflection process of the expansion waves, (J) has taken a direction equal to $\varphi_1 = -22.21°$. Hence, the conclusion: in a two-dimensional planar flow, the reflection of an expansion wave from an isobaric boundary causes a deflection of this boundary equal to twice the deflection through the incident expansion wave. The compression waves reflected from the isobaric boundary converge and focalize to give rise to a shock wave. This phenomenon is analyzed in section 20.5.

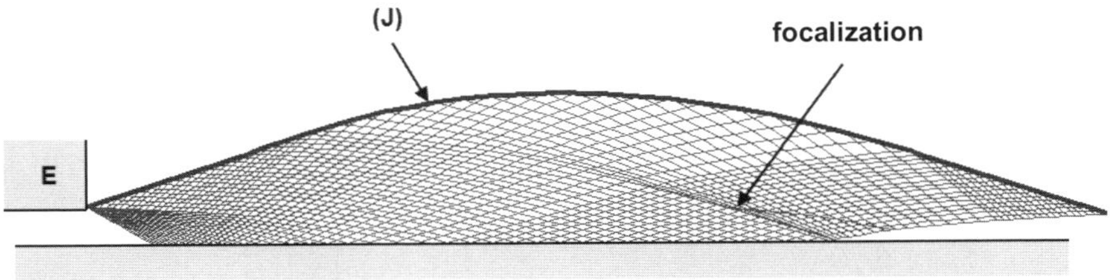

Figure 18.15. *Isobaric two-dimensional planar jet emerging from a nozzle at Mach number* $M_0 = 2$ *, expansion ratio* $\frac{p_1}{p_0} = 0.313$

Chapter 19

Application of the Method of Characteristics to the Calculation of Supersonic Nozzles

19.1. General principle

Because of its precision and low cost in computation time, the method of characteristics is well suited to determine the flow in supersonic nozzles, planar or axisymmetric, aimed to equip wind tunnels or rocket engines. As we know, a supersonic nozzle has a convergent–divergent shape with a throat (see Chapter 10). It is supplied by a gas (with the stagnation conditions: pressure p_{st_j}, temperature T_{st_j}) which accelerates in the convergent part in subsonic, becomes sonic in the throat, and continues its acceleration to supersonic in the divergent part. A supersonic nozzle thus comprises three parts:

– a subsonic region where the flow is governed by an elliptic differential system,

– a transonic region,

– a supersonic region where the system is hyperbolic and which can be treated by the method of characteristics.

We see that the method of characteristics cannot be applied to the entire flow. The calculation procedure proposed here is to adopt a particular treatment for the transonic domain. This will enable us to determine the initial part of the supersonic region from which the method of characteristics can be used.

19.2. Calculation of the transonic domain

19.2.1. *The potential equation in compressible flow*

There are various methods, including finite difference-like numerical techniques, to calculate the flow in the throat region. Here, we will present two methods based on the potential equation which give good results and have the advantage of leading to analytical expressions for velocity components, which can be of interest for practical implementation. The methods apply to nozzles, planar or axisymmetric, whose meridian section profile has at the throat a large radius of curvature $\Re$ compared to the radius r_c of the throat section. In the planar case r_c is to be replaced by the half-height h_c of the throat. Then, it is possible to find an approximate solution to the motion equations in the form of analytical developments.

We consider a steady flow, irrotational and two-dimensional, planar or axisymmetric, which is the case of a conventional supersonic nozzle. The hypothesis of irrotationality allows us to introduce a potential function of velocities designated as $\Phi(x,y)$ such that (remembering that $\overrightarrow{curl}\left(\overrightarrow{grad f}\right)=0$, see Appendix 1):

$$\vec{V}=\overrightarrow{grad}\ \Phi\ (x,y).$$

The components u and v of $\vec{V}$ along the Cartesian axes (x,y) are given by:

$$u=\frac{\partial\Phi}{\partial x}\qquad v=\frac{\partial\Phi}{\partial y}.$$

The equation for the function $\Phi(x,y)$, or potential equation, is obtained by writing that the flow also satisfies the continuity equation. In compressible flow, the energy equation must also be considered.

We established in section 17.2 the following form of the continuity equation for a two-dimensional planar ($\varepsilon=0$) or axisymmetric ($\varepsilon=1$) flow:

$$\frac{\partial}{\partial x}(\rho V\cos\varphi)+\frac{\partial}{\partial y}(\rho V\sin\varphi)+\varepsilon\frac{\rho V\sin\varphi}{y}=0$$

which leads to:

$$\frac{\partial}{\partial x}(\rho u)+\frac{\partial}{\partial y}(\rho v)+\varepsilon\frac{\rho v}{y}=0.$$

By developing and introducing the potential function:

$$\rho\frac{\partial^2\Phi}{\partial x^2}+\frac{\partial\Phi}{\partial x}\frac{\partial\rho}{\partial x}+\rho\frac{\partial^2\Phi}{\partial y^2}+\frac{\partial\Phi}{\partial y}\frac{\partial\rho}{\partial y}=-\varepsilon\frac{\rho v}{y}$$

or:

$$\rho\left(\frac{\partial^2\Phi}{\partial x^2}+\frac{\partial^2\Phi}{\partial y^2}\right)+\frac{\partial\Phi}{\partial x}\frac{\partial\rho}{\partial x}+\frac{\partial\Phi}{\partial y}\frac{\partial\rho}{\partial y}=-\varepsilon\frac{\rho v}{y} \qquad [19.1]$$

for a steady, irrotational flow of a non-viscous fluid, the motion equation is written as:

$$dp=-\rho V\,dV=-\rho\,d\left(\frac{V^2}{2}\right)=-\frac{\rho}{2}d\left(u^2+v^2\right)$$

or:

$$dp=-\frac{\rho}{2}d\left[\left(\frac{\partial\Phi}{\partial x}\right)^2+\left(\frac{\partial\Phi}{\partial y}\right)^2\right].$$

As the flow is isentropic, the sound velocity a is such that:

$$a^2=\left(\frac{\partial p}{\partial\rho}\right)_s=\frac{dp}{d\rho}$$

from which follows:

$$dp=a^2d\rho.$$

Thus:

$$d\rho=-\frac{\rho}{2\;a^2}d\left[\left(\frac{\partial\Phi}{\partial x}\right)^2+\left(\frac{\partial\Phi}{\partial y}\right)^2\right].$$

Or by developing:

$$d\rho = -\frac{\rho}{a^2}\left[\frac{\partial\Phi}{\partial x}\frac{\partial^2\Phi}{\partial x^2}+\frac{\partial\Phi}{\partial y}\frac{\partial^2\Phi}{\partial x\partial y}\right]dx-\frac{\rho}{a^2}\left[\frac{\partial\Phi}{\partial x}\frac{\partial^2\Phi}{\partial x\partial y}+\frac{\partial\Phi}{\partial y}\frac{\partial^2\Phi}{\partial y^2}\right]dy$$

but as:

$$d\rho = \frac{\partial\rho}{\partial x}dx+\frac{\partial\rho}{\partial y}dy$$

by identifying:

$$\frac{\partial\rho}{\partial x} = -\frac{\rho}{a^2}\left[\frac{\partial\Phi}{\partial x}\frac{\partial^2\Phi}{\partial x^2}+\frac{\partial\Phi}{\partial y}\frac{\partial^2\Phi}{\partial x\partial y}\right] \qquad \frac{\partial\rho}{\partial y} = -\frac{\rho}{a^2}\left[\frac{\partial\Phi}{\partial x}\frac{\partial^2\Phi}{\partial x\partial y}+\frac{\partial\Phi}{\partial y}\frac{\partial^2\Phi}{\partial x\partial y}\right]$$

and substituting these expressions into equation [19.1], it becomes:

$$\rho\left(\frac{\partial^2\Phi}{\partial x^2}+\frac{\partial^2\Phi}{\partial^2 y}\right)-\frac{\rho}{a^2}\left[\left(\frac{\partial\Phi}{\partial x}\right)^2\frac{\partial^2\Phi}{\partial x^2}+\frac{\partial\Phi}{\partial x}\frac{\partial\Phi}{\partial y}\frac{\partial^2\Phi}{\partial x\partial y}\right]$$

$$-\frac{\rho}{a^2}\left[\frac{\partial\Phi}{\partial y}\frac{\partial\Phi}{\partial x}\frac{\partial^2\Phi}{\partial x\partial y}+\left(\frac{\partial\Phi}{\partial y}\right)^2\frac{\partial^2\Phi}{\partial y^2}\right]=-\varepsilon\frac{\rho v}{y}.$$

After division by ρ and by clustering the terms, we obtain the equation:

$$\boxed{\left[a^2-\left(\frac{\partial\Phi}{\partial x}\right)^2\right]\frac{\partial^2\Phi}{\partial x^2}+\left[a^2-\left(\frac{\partial\Phi}{\partial y}\right)^2\right]\frac{\partial^2\Phi}{\partial y^2}-2\left(\frac{\partial\Phi}{\partial x}\right)\left(\frac{\partial\Phi}{\partial y}\right)\left(\frac{\partial^2\Phi}{\partial x\partial y}\right)=-\varepsilon\frac{a^2}{y}\left(\frac{\partial\Phi}{\partial y}\right)}$$

[19.2]

which is the equation sought for the function $\Phi(x,y)$. The velocity of sound a is calculated from the energy equation:

$$h_{st} = h+\frac{V^2}{2}$$

or for a calorically perfect gas:

$$C_p\ T_{st} = C_p\ T+\frac{V^2}{2}.$$

But knowing that:

$$a^2 = \gamma\left(C_p - C_v\right)T$$

it follows, by denoting a_{st} the sound velocity for the stagnation temperature T_{st} :

$$a_{st}^2 = a^2 + \frac{\gamma-1}{2}V^2$$

and hence the following equation allows expressing a^2 from Φ and a_{st} which constitutes data pertaining to the problem:

$$\boxed{a_{st}^2 = a^2 + \frac{\gamma-1}{2}\left[\left(\frac{\partial\Phi}{\partial x}\right)^2 + \left(\frac{\partial\Phi}{\partial y}\right)^2\right].}$$

Equation [19.2] is the general equation for the velocity potential, valid for a two-dimensional planar or axisymmetric, steady and compressible non-viscous flow. This is a partial differential equation, strongly non-linear, and hence difficult to solve. There are no exact analytical solutions. However, we can solve it with a good accuracy and short computation time by using numerical methods or, as we shall see, by approximate analytical methods. We will again find the potential equation for compressible flow in Chapter 25.

19.2.2. *The Carrière development method*

At this stage, we usually normalize the velocities at the throat with the velocity of sound, or critical velocity, a_c by stating:

$$\overline{a} = \frac{a}{a_c}, \quad \overline{u} = \frac{u}{a_c}, \quad \overline{v} = \frac{v}{a_c}$$

from which follows the energy equation:

$$\frac{2}{\gamma-1}\overline{a}^2 + \overline{u}^2 + \overline{v}^2 = \frac{\gamma+1}{\gamma-1}$$

and the potential equation where we reintroduce $\overline{u}$ and $\overline{v}$ to arrive at the equation:

$$\left(\overline{u}^2 - \overline{a}^2\right)\frac{\partial^2\overline{\Phi}}{\partial\overline{x}^2} + 2\overline{u}\overline{v}\frac{\partial^2\overline{\phi}}{\partial\overline{x}\partial\overline{y}} + \left(\overline{v}^2 - \overline{u}^2\right)\frac{\partial^2\overline{\Phi}}{\partial\overline{y}^2} = \varepsilon\overline{a}^2\frac{\overline{v}}{\overline{y}}.$$

The lengths x and y are normalized by a scale which will be specified later.

We adopt as the origin of the coordinate system the sonic point where $M = 1$ on the axis, and we seek solutions satisfying in the transonic domain a velocity distribution on the axis $\overline{y} = 0$ in the form of:

$$\overline{u}(x,0) = 1 + \beta x.$$

We adopt the dimensionless coordinates such that:

$$\overline{x} = \beta x \quad \text{and} \quad \overline{y} = \beta y.$$

The reference length $1/\beta$ will be defined later. Thus, we impose on the axis:

$$\overline{u}(\overline{x},0) = 1 + \overline{x}.$$

A solution to the potential equation is sought in the form of a limited expansion in the vicinity of the sonic point on the axis of the form:

$$\overline{\Phi}(\overline{x},\overline{y}) = \sum_{m,n} \alpha_{m,n} \overline{x}^m \overline{y}^n.$$

Hence, for the velocity components:

$$\overline{u} = \frac{\partial \overline{\varphi}}{\partial \overline{x}} = \sum_{m,n} m\, \alpha_{m,n}\ \overline{x}^{(m-1)} \overline{y}^n$$

$$\overline{v} = \frac{\partial \overline{\varphi}}{\partial \overline{y}} = \sum_{m,n} n\ \alpha_{m,n}\ \overline{x}^m \overline{y}^{(n-1)}.$$

As the change of $\overline{y}$ to $-\overline{y}$ ought to change $\overline{v}$ to $-\overline{v}$ (symmetry condition), all coefficients $\alpha_{m,n}$ are zero if n is odd. On the other hand, the condition to be met on the axis requires:

$$\alpha_{1,0} = 1,\ 2\alpha_{2,0} = 1,\ \alpha_{3,0} = \alpha_{4,0} = \cdots = 0.$$

Substituting the above developments in the potential equation, in the light of the energy equation, it is possible to determine the still unknown coefficients $\alpha_{m,n}$ by identifying to zero the terms of the same power $x^m y^n$. This operation leads to tedious calculations that are not reproduced here.

For an *axisymmetric flow*, we obtain for $\overline{u}$ and $\overline{v}$ the expressions:

$$\overline{u} = 1 + \overline{x} + \frac{\gamma+1}{4}\overline{y}^2\left[1 + (2\gamma-1)\overline{x} + 3\gamma(\gamma-1)\overline{x}^2 + (\gamma-1)\left(4\gamma^2 - 2\gamma - 1\right)\overline{x}^3\right]$$
$$+ \frac{(\gamma+1)^2}{64}\overline{y}^4\left[6\gamma - 1 + \left(36\gamma^2 - 19\gamma - 4\right)\overline{x}\right] + \cdots$$

$$\overline{v} = \frac{\gamma+1}{2}\overline{y}\left[\overline{x} + \frac{2\gamma-1}{2}\overline{x}^2 + \gamma(\gamma-1)\overline{x}^3 + \frac{(\gamma-1)\left(4\gamma^2 - 2\gamma - 1\right)}{4}\overline{x}^4\right]$$
$$+ \frac{(\gamma+1)^2}{16}\overline{y}^3\left[1 + (6\gamma-1)\overline{x} + \frac{36\gamma^2 - 19\gamma - 4}{2}\overline{x}^2\right]$$
$$+ \frac{(\gamma+1)^3(6\gamma-1)}{128}\overline{y}^5 + \cdots$$

In the case of a *planar flow* we obtain for $\gamma = 1.4$:

$$\overline{u} = 1 + \overline{x} + \overline{y}^2\left[1.2 + 2.16\overline{x} + 2.016\overline{x}^2 + 1.9392\overline{x}^3\right]$$
$$+ \overline{y}^4\left[2.256 + 13.622\overline{x}\right]$$

$$\overline{v} = \overline{y}\left[2.4\overline{x} + 2.16\overline{x}^2 + 1.344\overline{x}^3 + 0.9696\overline{x}^4\right]$$
$$+ \overline{y}^3\left[0.96 + 9.24\overline{x} + 27.245\overline{x}^2\right] + 3.871\overline{y}^5.$$

Knowing the components $\overline{u}$ and $\overline{v}$ of the velocity as a function of $\overline{x}$ and $\overline{y}$, we are able to determine all the flow properties in the transonic domain. In particular, it is possible to trace:

– the *sonic line*, defined by: $\overline{u}^2 + \overline{v}^2 = 1$,

– the *line of the throats*, locus of the points where the velocity is parallel to the axis, which is such that: $\overline{v} = 0$.

The curvature of a streamline is given by:

$$\frac{1}{\overline{\Re}} = \frac{\partial}{\partial \overline{x}}\left(\frac{\overline{v}}{\overline{u}}\right) = \frac{1}{\overline{u}^2}\left(\overline{u}\frac{\partial \overline{v}}{\partial \overline{x}} - \overline{v}\frac{\partial \overline{u}}{\partial \overline{x}}\right).$$

On a line of the throats $\overline{v} = 0$, which leads to:

$$\frac{1}{\overline{\mathfrak{R}}} = \frac{\partial}{\partial \overline{x}}\left(\frac{\overline{v}}{\overline{u}}\right) = \frac{1}{\overline{u}}\left(\frac{\partial \overline{v}}{\partial \overline{x}}\right)_{\overline{v}=0}.$$

The line of the throats $\overline{v} = 0$ can be graduated with the values of $\frac{\overline{\mathfrak{R}}}{\overline{r}_c}$, recalling that $\overline{\mathfrak{R}}$ and $\overline{r}_c$ are normalized by $1/\beta$ (in the planar case, the radius $\overline{r}_c$ is replaced by the throat half-height $\overline{h}_c$).

In practice, to calculate the transonic domain of a nozzle, we give:

– the radius of curvature of the meridian contour $\mathfrak{R}$,

– the radius of the throat r_c (or its half-height h_c).

The transposition of the solution in the physical plane is then accomplished as follows, knowing the metric values of the radius of curvature $\mathfrak{R}$ and of the radius r_c (or half-height h_c) of the throat. The line of the throats can be graduated in values of $\frac{\overline{\mathfrak{R}}}{\overline{r}_c}$, where r_c is the ordinate of the point corresponding to the line of the throats. As:

$$\frac{\mathfrak{R}}{r_c} = \frac{\overline{\mathfrak{R}}}{\overline{r}_c}$$

we define the contour in the calculation plane corresponding to the given value of $\frac{\overline{\mathfrak{R}}}{\overline{r}_c}$. This defines the value of $\overline{r}_c$ expressed in metric units. If appropriate, we can renormalize by the radius (or half-height) of the throat. Now we only have to transpose all the lengths in metric units. We can do the same for the velocities if the total temperature T_{st} is given, thus:

$$a_c = \sqrt{\gamma r T_c}$$

with:

$$T_c = \frac{2}{(\gamma+1)} T_{st}.$$

The Mach number is deduced from:

$$M = \frac{\sqrt{u^2 + v^2}}{a} = \frac{a_c \sqrt{\overline{u}^2 + \overline{v}^2}}{a}.$$

Now:

$$\frac{a_c}{a} = \sqrt{\frac{T_c}{T}} = \sqrt{\frac{\left(\frac{\gamma+1}{2}\right)^{-1}}{\left(1+\frac{\gamma-1}{2}M^2\right)^{-1}}}.$$

Hence, after simple calculations:

$$\boxed{M = \sqrt{\frac{\left(\overline{u}^2 + \overline{v}^2\right)}{\left(\frac{\gamma+1}{2} - \frac{\gamma-1}{2}\left(\overline{u}^2 + \overline{v}^2\right)\right)}}.}$$

Figure 19.1 shows the structure of the transonic domain of an axisymmetric nozzle, the results being qualitatively the same in a planar flow. We find that the sonic line is curved, the sonic state being reached at the wall upstream of the geometric throat and downstream of it on the centerline. Knowledge of the transonic domain enables us to determine a *starting characteristic* to initiate the calculation of the supersonic flow by the method of characteristics.

It is customary to take as a starting line the right running characteristic $(\xi)_0$. beginning from the point C of the wall located at the geometric throat where the Mach number is already supersonic.

This line is defined by:

$$\frac{dy}{dx} = \tan\left(\alpha - \varphi\right)$$

with:

$$\varphi = \tan^{-1}\left(\frac{\overline{v}}{\overline{u}}\right)$$

and:

$$\alpha = \sin^{-1}\left(\frac{1}{M}\right) = \sin^{-1}\left(\frac{\overline{a}^2}{\sqrt{\overline{u}^2 + \overline{v}^2}}\right).$$

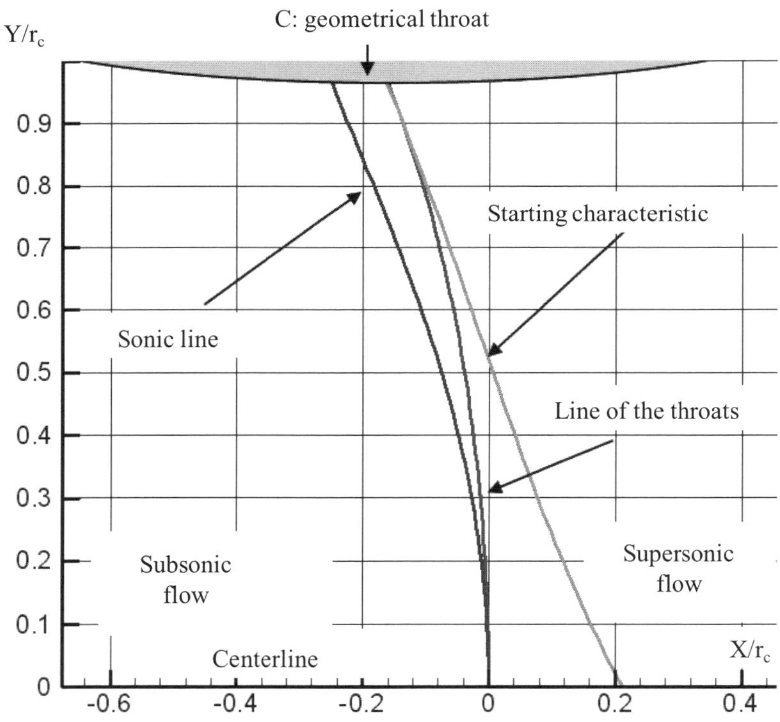

Figure 19.1. *Transonic domain of an axisymmetric nozzle*

The starting characteristic is built step-by-step starting from the point *C*, the flow properties (pressure, Mach number, and velocity direction) being calculated at each point of $(\xi)_0$. In practice, we determine about 20 points, evenly spaced between the wall and axis.

This expansion method is only applicable if the ratio $\frac{\overline{\Re}}{\overline{r}_c}$ is large enough: in practice, $\frac{\overline{\Re}}{\overline{r}_c} \geq 4$. As this is an expansion in the vicinity of a uniform sonic state, the accuracy of such a calculation can only be good if the flow does not deviate too much from this state. In particular, the streamlines must be not too curved close to the wall, which would lead to small values of $\Re / r_c$.

19.2.3. *The Sauer small perturbation method*

Another analytical method was proposed by Sauer from equation [19.1], written under the assumption of small perturbations (see section 25.2). For this, the velocity components are set as:

$$u = a_c + \hat{u}$$

$$v = \hat{v}.$$

Thus, we assume that the perturbations $\hat{u}$ and $\hat{v}$ of the velocity are small compared to a_c. This assumption is justified in the throat region of a nozzle whose section law varies slowly, the flow being almost one-dimensional and the section area stationary at the throat $(dA = 0)$. Under these conditions:

$$\frac{\hat{u}}{a_c}, \frac{\hat{v}}{a_c} << 1 \quad \text{and,} \quad \textit{a fortiori}: \quad \frac{\hat{u}^2}{a_c^2}, \frac{\hat{v}^2}{a_c^2} << \frac{\hat{u}}{a_c}, \frac{\hat{v}}{a_c} << 1.$$

As the flow is irrotational, we introduce the potential function Φ:

$$\Phi = a_c\left(x + \Phi'\right)$$

whose above expression shows the pertubation potential Φ'. Consequently:

$$u = a_c + \hat{u} = a_c\left(1 + u'\right) = a_c\left(1 + \frac{\partial \Phi'}{\partial x}\right)$$

$$v = \hat{v} = a_c v' = a_c \frac{\partial \Phi'}{\partial y}$$

and hence the equation for the perturbation potential Φ' which can be deduced from the general equation following calculations that we will not detail here (see Chapter 25):

$$(\gamma+1)\frac{\partial\Phi'}{\partial x}\frac{\partial^2\Phi'}{\partial x^2}-\frac{\partial^2\Phi'}{\partial y^2}-\varepsilon\frac{1}{y}\frac{\partial\Phi'}{\partial y}=0. \qquad [19.3]$$

We seek a solution to this equation in the form of a polynomial development in y whose coefficients are functions of x. Due to the symmetry with respect to y, the polynomial contains only even powers in y. Thus:

$$\Phi'(x,y)=\sum_{n=0}^{\infty}f_{2n}(x)y^{2n}=f_0(x)y^0+f_2(x)y^2+f_4(x)y^4+....$$

knowing that $y^0=1$. Thus, we have the following expressions:

$$\frac{\partial\phi'}{\partial x}=f_0'(x)+f_2'(x)y^2+f_4'(x)y^4+....$$

$$\frac{\partial^2\phi'}{\partial x^2}=f_0''(x)+f_2''(x)y^2+f_4''(x)y^4+....$$

$$\frac{\partial\phi'}{\partial y}=2f_2(x)y+4f_4(x)y^3+....$$

$$\frac{\partial^2\phi'}{\partial y^2}=2f_2(x)+12f_4(x)y^2+.....$$

By inserting these expressions into the equation for the perturbation potential above, we obtain:

$$y^0\left[(\gamma+1)f_0'f_0''-2f_2-2\varepsilon f_2\right]+y^2\left[(\gamma+1)\left(f_0'f_2''+f_0''f_2'\right)-12f_4-4\varepsilon f_4\right]+y^4[....]+....=0.$$

Since this polynomial equation must be satisfied for any value of x, it follows that the coefficients of powers of y must be zero, which leads by limiting the polynomial to the 2nd degree:

$$f_2(x) = \frac{(\gamma+1) f_0' f_0''}{2(1+\varepsilon)}$$

$$f_4(x) = \frac{(\gamma+1)\left(f_0' f_2'' + f_0'' f_2'\right)}{4(3+\varepsilon)}.$$

The functions $f_2(x)$ and $f_4(x)$ can be identified if $f_0'(x)$ is known. By writing the development on the axis, we have:

$$\frac{\partial \phi'}{\partial x}(x,0) = u'(x,0) = f_0'(x).$$

Thus, $f_0'(x,0)$ is the perturbation velocity on the axis. The problem is solved by setting a linear law for the variation of u' on the axis. Thus:

$$u'(x,\ 0) = f_0'(x) = \alpha x.$$

By substituting this law in the expressions of $f_2(x)$ and $f_4(x)$, we obtain:

$$\Phi'(x,y) = f_0(x) + \frac{(\gamma+1)\alpha^2 x y^2}{2(1+\varepsilon)} + \frac{(\gamma+1)^2 \alpha^3 y^4}{8(1+\varepsilon)(3+\varepsilon)}.$$

The velocity components, normalized with a_c, are deduced without difficulty:

$$\boxed{\begin{aligned} \overline{u} &= 1 + \alpha x + \frac{(\gamma+1)\alpha^2 y^2}{2(1+\varepsilon)} \\ \overline{v} &= \frac{(\gamma+1)\alpha^2 x y}{(1+\varepsilon)} + \frac{(\gamma+1)^2 \alpha^3 y^3}{2(1+\varepsilon)(3+\varepsilon)}. \end{aligned}}$$

If we take $\beta = \dfrac{1}{\alpha}$ as a length scale, then we get in the plane of dimensionless variables:

$$\overline{x} = \frac{x}{\beta},\ \overline{y} = \frac{y}{\beta}.$$

The velocity components become:

$$\overline{u} = 1 + \overline{x} + \frac{(\gamma+1)\overline{y}^2}{2(1+\varepsilon)}$$

$$\overline{v} = \frac{(\gamma+1)\overline{x}\ \overline{y}}{(1+\varepsilon)} + \frac{(\gamma+1)^2\overline{y}^3}{2(1+\varepsilon)(3+\varepsilon)}.$$

These expressions are used to calculate the velocity at any point in the transonic domain, especially:

– the sonic line: locus of points such that: $\overline{u}^2 + \overline{v}^2 = 1$,

– the line of the throats or position of points where the velocity is parallel to the axis: $\overline{v} = 0$,

– the limit characteristic which is the characteristic ending on the sonic point of the axis.

The radius of curvature of the line of the throats is expressed as:

$$\frac{1}{\overline{\Re}} = \frac{\partial}{\partial \overline{x}}\left(\frac{\overline{v}}{\overline{u}}\right) = \frac{1}{\overline{u}}\left(\frac{\partial \overline{v}}{\partial \overline{x}}\right)_{\overline{v}=0} = \frac{1}{\overline{u}}\frac{(\gamma+1)\overline{y}}{(1+\varepsilon)}.$$

The transposition of the solution in the physical plane then is done, as for the Carrière method, from the radius of curvature imposed at the throat of the nozzle and the radius r_c (or half-height h_c) of the throat. The developments are identical for the calculation of the local Mach number.

19.3. Calculation of the flow in a nozzle with a given contour: direct method

In the direct problem, we have to calculate the flow inside a nozzle of a given shape to determine, for example, its thrust. The first step is to calculate the transonic domain, using either of the methods of the preceding paragraph. This step provides a distribution of discrete points on the starting characteristic $(\xi)_0$. These points are denoted 00, 01, 02, 03, 04, their number being limited to five for clarity of the explanations (see Figure 19.2).

The calculation of flow in the nozzle is organized in the following way:

– applying the operator *W*, we calculate, from the point 01 and the wall geometry, the point 11 on the wall,

– applying the operator N from 11 and 02, we determine the point 12, and then the same operator is used to calculate 13 and 14,

– applying the operator A from 14, the point 15 is calculated on the axis or centerline.

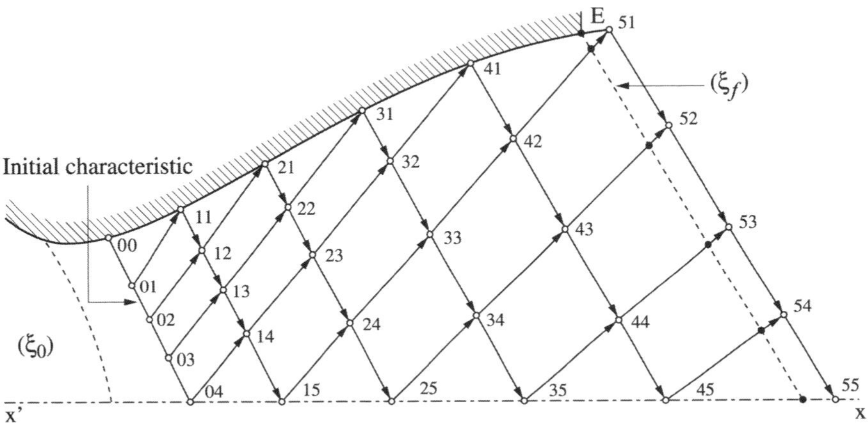

Figure 19.2. *Calculation of the flow in a nozzle with a given contour: first method*

At the beginning of this calculation step, we have the right running characteristic $(\xi)_1$ consisting of the five points: 11, 12, 13, 14, and 15. This sequence is repeated, $(\xi)_1$ playing the role of $(\xi)_0$, to determine the characteristic $(\xi)_2$ whose points are: 21, 22, 23, 24, and 25. We can thus continue the calculation of the characteristics (ξ) to a point, here 51, located beyond the end of the wall. It is then easy to define, by an appropriate interpolation, the final characteristic $(\xi)_f$ passing exactly through the nozzle extremity E. We note that the calculation of the supersonic flow cannot be continued beyond $(\xi)_f$, which is the boundary of the influence domain of the nozzle.

It is also possible to arrange the calculation by progressing along the left running characteristics (η) as shown in Figure 19.3. To do this, from 01 and the definition of the wall, we determine 11, which gives the characteristic $(\eta)_1$ carrying the points 01 and 11. Then, we calculate the point 12 from 02 and 11, and 21 on the wall from 12, which provides the characteristic $(\eta)_2$ composed of 02, 12, and 21. Then, we move to the calculation of $(\eta)_3$, made with the points 03, 13, 22, and 31. The process is repeated until the starting point 04, located on the centerline or the symmetry plane. Here, we determine the point 15 on axis from 14; then, we calculate the characteristic $(\eta)_5$ composed of the points 15, 24, 33, 42, and 51. The process is interrupted when the last calculated point is beyond E.

The second approach saves computational time, and the calculated field is less extensive than the one given by the first technique. In fact, the pressure distribution on the nozzle wall, which gives directly the thrust of the divergent (without the friction losses), depends only on the field upstream of the characteristic $(\eta)_f$ ending in E. Thus, it is not useful to continue the calculation beyond $(\eta)_f$.

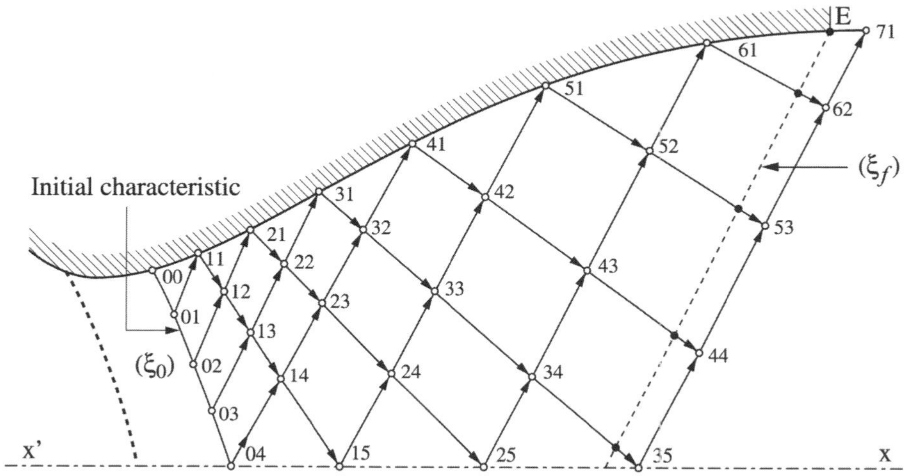

Figure 19.3. *Calculation of the flow in a nozzle with a given contour: second method*

19.4. Calculation of the contour of a nozzle producing a given flow: inverse or design method

In practice, most often the problem is to define the contour of a nozzle producing a flow whose properties are imposed. For example, to generate a uniform supersonic flow in the exit plane of a nozzle intended to equip:

– either a wind tunnel to simulate a uniform flow with a given Mach number,

– or the engine of a missile or satellite launcher whose thrust we want to optimize.

To achieve a uniform flow with the Mach number M_E in the exit section whose area A_E is imposed, we first calculate the area A_c of the throat with the formula:

$$A_c = \frac{A_E}{\Sigma(M_E, \gamma)}$$

where the function $\sum(M,\gamma) = A/A_c$ is given in section 9.5.2.3. We deduce the radius of the throat r_c and we choose for the throat region a circular meridian contour with the radius $\mathfrak{R}$ so that the ratio $\mathfrak{R}/r_c$ is at least equal to 4.

COMMENT.– A value $\mathfrak{R}/r_c > 4$ is recommended to define a nozzle equipping a wind tunnel where the quality of the flow is the main constraint. Such large values of $\mathfrak{R}/r_c$ lead to rather long nozzle which would be too big and too heavy to equip a rocket engine. In this case, values as low as $\mathfrak{R}/r_c = 0.5$ are adopted. Consequently, the flow contains regions of overacceleration leading to the formation of focalization shock waves (often call *internal shocks*). There results some loss of performance which is outbalanced by the gain of weight. In addition, even with a so low value of $\mathfrak{R}/r_c$, a nozzle giving a uniform flow in its exit sections still would be too long. Then, the designed nozzle is truncated with the consequence that there is a new loss of thrust. A nozzle defined in this way is sometimes called the *Truncated Ideal Contour* nozzle (or TIC nozzle).

When the wall radius of curvature is too small to allow the use of analytical solutions, the flow in the nozzle transonic domain can be computed by a numerical solution of the full Euler equations. Then, the starting characteristic is extracted from the field thus calculated.

The method of defining the contour of the nozzle is divided into five steps:

1. In the first step, we perform a calculation of the transonic domain according to one of the methods described above. We deduce a starting characteristic $(\xi)_0$ intersecting the axis at a point of abscissa x_0 where we know the Mach number M_0 as well as the derivative:

$$\left(\frac{dM}{dx}\right)_0 .$$

2. In a second step, we define a distribution of Mach number $M(x)$ along the centerline (axis). Very often, $M(x)$ is defined by an analytic curve satisfying the following conditions:

– going at x_0 through the point $M = M_0$,

– has in x_0 the slope : $\dfrac{dM}{dx} = \left(\dfrac{dM}{dx}\right)_0$,

– reaches a constant level $M = M_E$ at an abscissa: $x = x_1$,

– has at x_1 a slope equal to zero.

These four conditions define a cubic distribution, the chosen distribution has the shape presented in Figure 19.4. The conditions to be met in x_0 and x_1 ensure a good continuity of the distribution $M(x)$. These precautions allow avoiding discontinuities, or too rapid accelerations, which could cause disturbances or shock waves.

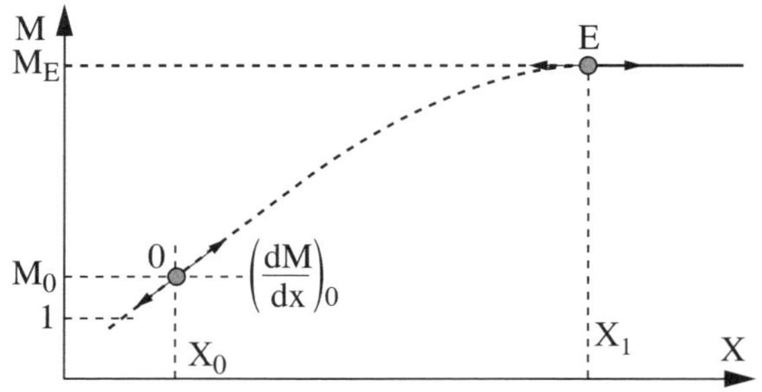

Figure 19.4. *Distribution of the Mach number prescribed on the axis*

The choice of x_1 leads to a compromise between the wish to ensure the most progressive acceleration possible, and thus an excellent quality of the flow, and to limit the length of the nozzle, whose size and weight are important parameters in propulsion.

3. In the third step, we calculate the mass flow $\overline{q}_m$ of the nozzle which is determined by calculating the flow through the starting characteristic $(\xi)_0$. If the nozzle is axisymmetric the elementary mass flow $d\overline{q}_m$ crossing an element $d\xi$, is expressed (see Figure 19.5) as:

$$d\overline{q}_m = 2\pi\ \rho \left(\vec{V}.\vec{n}\right) y\ d\xi$$

which gives:

$$d\overline{q}_m = 2\pi\ \rho\, V\ \sin\alpha\ y\ d\xi$$

where $\alpha = \sin^{-1}(1/M)$ is the Mach angle.

We see that: $d\overline{q}_m = 2\pi\ \rho\ a\ y\ d\xi$ where a is the velocity of sound.

Ultimately, by integrating from the axis to point P_0 on the wall:

$$\boxed{\overline{q}_m = 2\pi \int_0^{(\xi)_{P0}} \rho\ a\ y\ d\xi.}$$

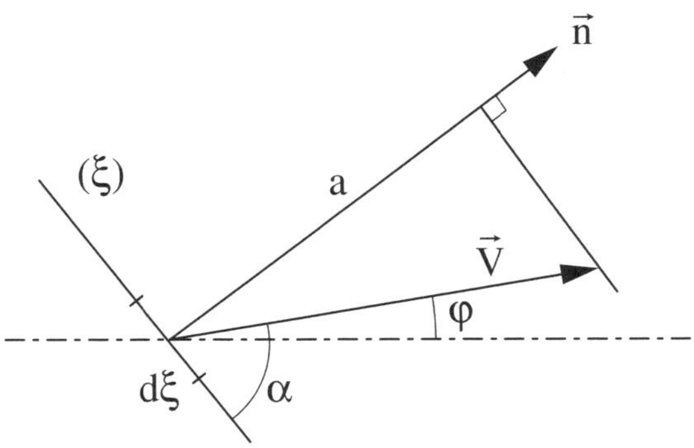

Figure 19.5. *Mass flow across a characteristic*

COMMENT.– The mass flow of a two-dimensional planar nozzle, whose width is taken as the unit, is simply:

$$\overline{q}_m = \int_0^{(\xi)_{P0}} \rho \ a \ d\xi.$$

4. In the fourth step, we calculate the flow downstream of $(\xi)_0$, taking into account the distribution $M(x)$ imposed on the axis (see Figure 19.6). Fixing a small step dx_A on the axis, we determine the Mach number at the point 10, where $\varphi = 0$. Thus the conditions in 10 are known.

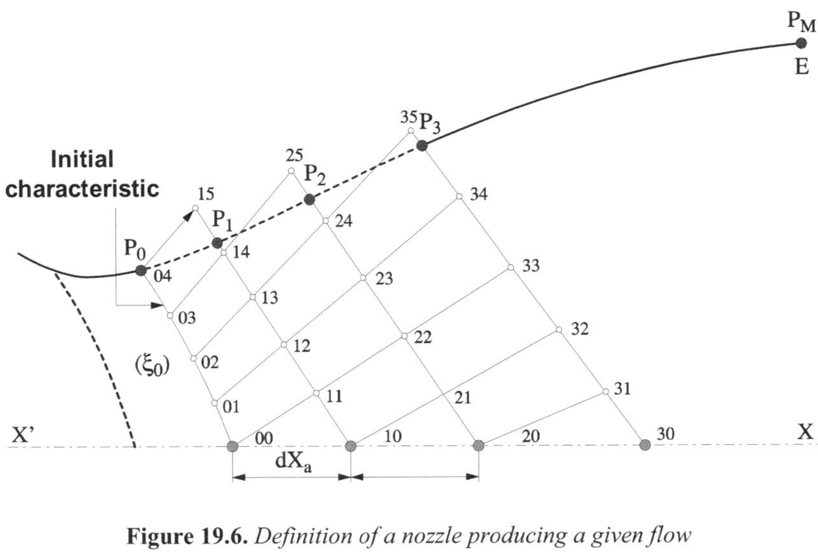

Figure 19.6. *Definition of a nozzle producing a given flow*

Applying the operator N, the point 11 is calculated from 00 and 10, then 12 from 11 and 01, and so on until point 15. We have thus determined the characteristic $(\xi)_1$. Then, we move to the characteristic $(\xi)_2$ passing through the point 20 on the axis, located at the distance dx_A from 10. The calculation of (ξ) is continued until the characteristic $(\xi)_f$ on which the flow is uniform with the Mach number M_E.

5. In the fifth stage, the wall of the nozzle $P(x)$ is determined. As $P(x)$ is a streamline, points P_1, P_2, P_3, etc., which define it, are determined on each (ξ), calculated by writing that the mass flow between the axis and the point P_n in question, is equal to $\overline{q}_m$, hence for an axisymmetric nozzle:

$$2\pi \int_0^{(\xi)_{Pn}} \rho \ a \ y \ d\xi = \overline{q}_n .$$

The point P_n is interpolated between the two points of the computed characteristic which surround it.

The contour of the nozzle is defined by a circular arc in the throat region and a succession of discrete points downstream. Before manufacturing, we proceed to a fairing of these points using suitable numerical techniques (polynomial curves or spline functions).

19.5. Example of the definition of the contour of a two-dimensional planar nozzle

The method above is applied to define the contour of a two-dimensional planar nozzle producing a uniform flow with the Mach number $M_0 = 2.5$. The length of the nozzle is set to 500 mm and the wall radius or curvature at the throat is such that $\Re / h = 10$; the gas is air in ambient conditions $(\gamma = 1.4)$.

Figure 19.7a shows the starting characteristic on which were defined 11 points and Figure 19.7b shows the Mach number distribution prescribed on the nozzle symmetry axis. The application of the procedure leads to the contour plotted in Figure 19.8. The mesh of calculated characteristics is plotted in Figure 19.9 with a close-up of the throat region in Figure 19.10.

Figure 19.11 is an image of the iso-Mach number contours determined from the characteristics calculation. A check of the calculation is to inject the contour thus defined in a calculation in direct method and to verify that the two calculations give the same flow. The pressure distributions on the nozzle wall plotted in Figure 19.12 show that there is a good agreement between the two modes of calculations.

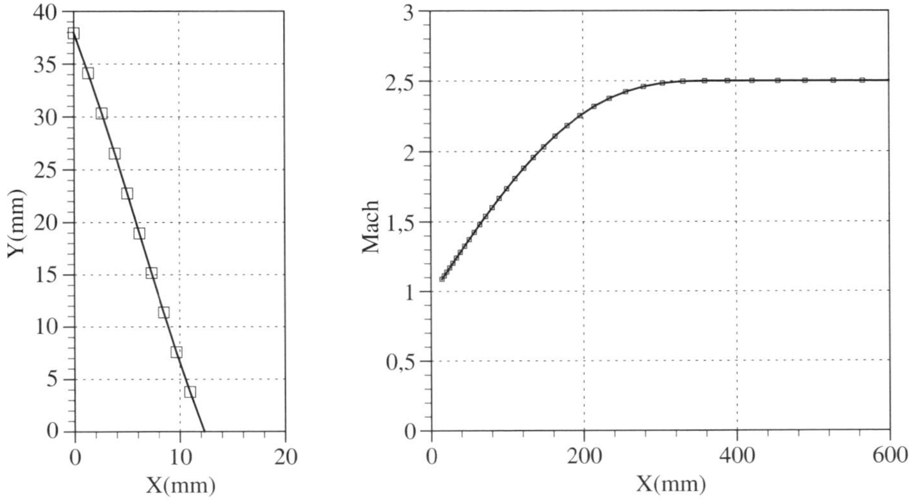

a – starting characteristic *b – Mach number law imposed on the axis*

Figure 19.7. *Definition conditions of a two-dimensional planar nozzle giving a uniform flow with Mach number* $M_0 = 2.5$

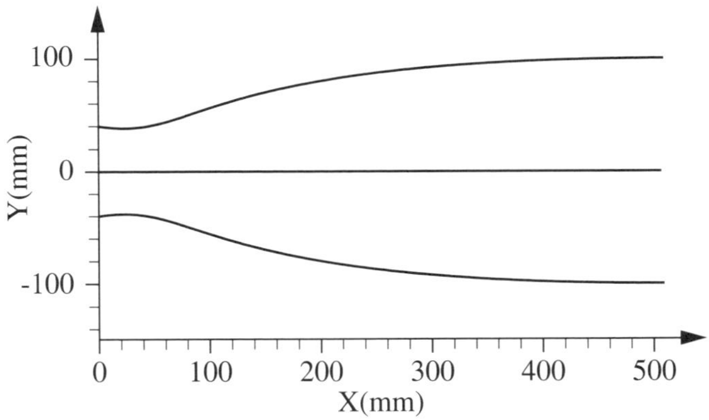

Figure 19.8. *Contour of a planar nozzle with Mach number* $M_0 = 2.5$

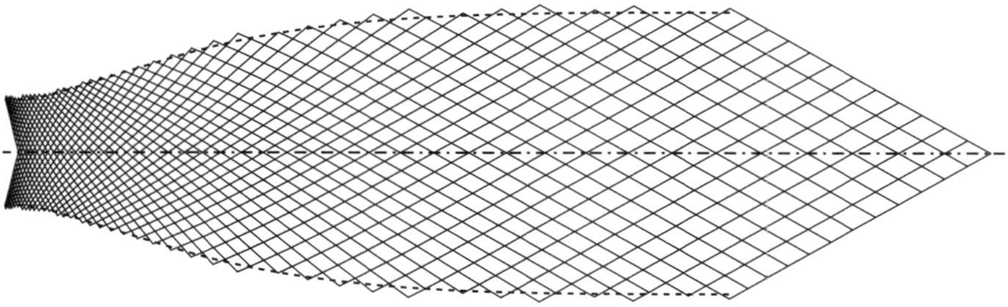

Figure 19.9. *Two-dimensional planar nozzle with Mach number* $M_0 = 2.5$*: Characteristics mesh*

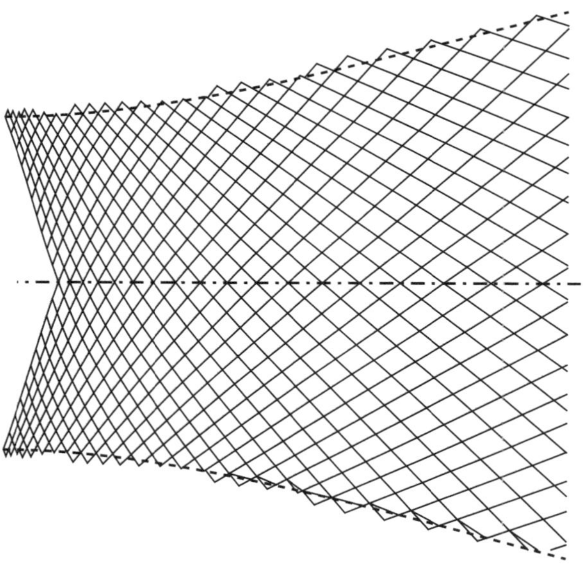

Figure 19.10. *Two-dimensional planar nozzle with Mach number* $M_0 = 2.5$*: Mesh in the throat region*

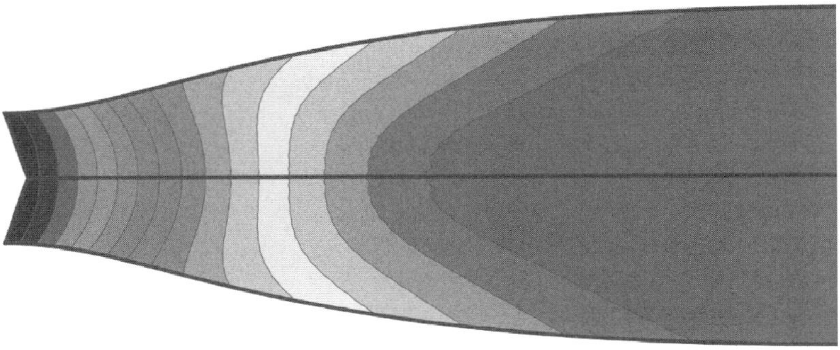

Figure 19.11. *Two-dimensional planar nozzle of Mach number $M_0 = 2.5$: Iso-Mach number contours*

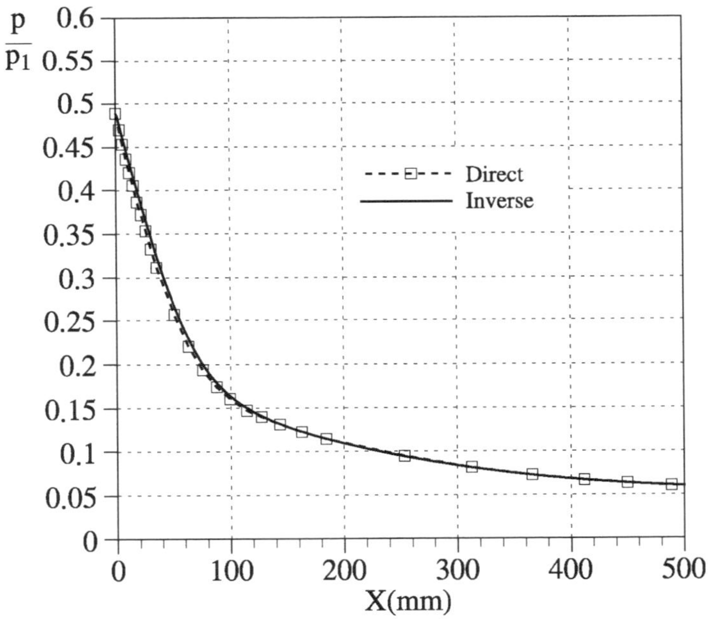

Figure 19.12. *Two-dimensional planar nozzle at $M_0 = 2.5$, wall pressure distribution, and comparison of the inverse and direct methods*

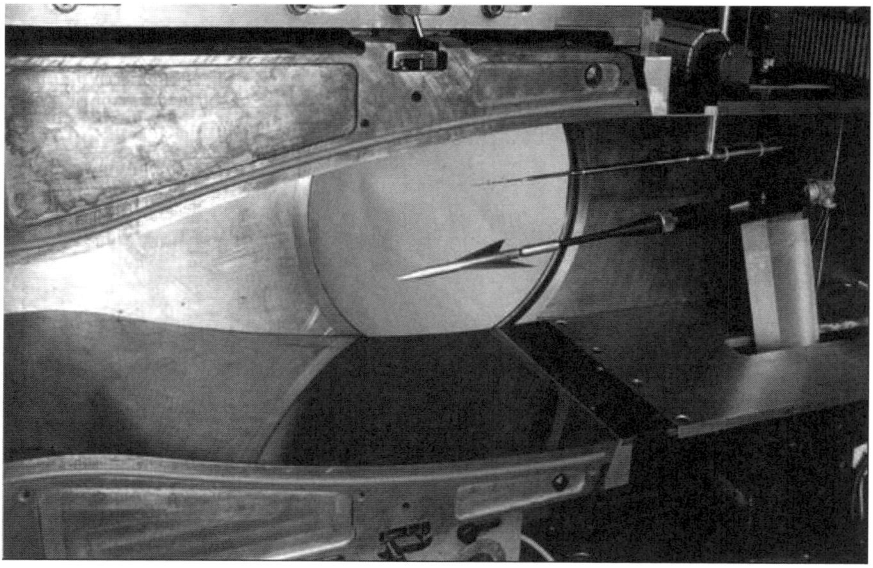

Figure 19.13. *Mach 2 nozzle of the wind tunnel S5 of the Onera Meudon center*

The photograph of Figure 19.13 shows a contoured supersonic nozzle installed in one of the Onera supersonic wind tunnels.

19.6. Mach rhombus of a nozzle

We consider a nozzle designed to produce a uniform flow with the Mach number M_0 in the exit section *EE′* (see Figure 19.14). Using the results of section 17.5, it is easy to show that if the flow is uniform in *EE'*, so it is in *EAE'* limited upstream by the left running (η_E) and right running $(\xi_{E'})$ characteristics going through the points *E* and *E'*, respectively. If α_0 denotes the Mach angle $\alpha_0 = \sin^{-1}\left(\dfrac{1}{M_0}\right)$, the angles of these characteristics relative to the nozzle axis are equal to α_0 for (η_E) and $-\alpha_0$ for $(\xi_{E'})$.

If the nozzle is interrupted in *E* and *E'*, for the same reasons, the flow will be uniform toward downstream in the area *EBE'* limited by the left running $(\eta_{E'})$ and the right running (ξ_E) characteristics, emerging from *E'* and *E,* respectively. The volume defined by the shape *E'AEBE'*, which has a transverse dimension equal to the width of the nozzle, is sometimes called the *Mach rhombus*. In an axisymmetric flow, it has the shape of two cones with as common basis the circle coincident with the nozzle exit section. The Mach rhombus is the volume within which the supersonic flow is uniform and where we can place the test model. The higher the Mach number, the greater this volume is.

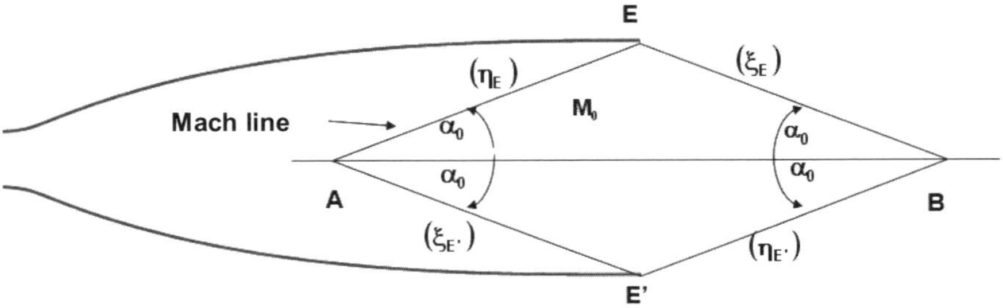

Figure 19.14. *Mach rhombus of a supersonic wind tunnel nozzle*

If the nozzle opens into a plenum chamber maintained at a pressure p_c less than the pressure p_0 in the exit plane of the nozzle, the domain where to place the model coincides with the Mach rhombus since then the expansion waves at E and E' propagate downstream of EE'. If p_c exceeds p_0, then oblique shock waves start from E and E'. As the angle of these shocks is higher than the Mach angle, they propagate upstream of (ξ_E) and $(\eta_{E'})$, which will restrict the useful test section. In this case, in addition, if strong enough, the shocks are likely to separate the boundary layer of the nozzle causing its rapid unstart.

Chapter 20

Flows with Shock Waves: Rotational Method of Characteristics

20.1. Shock wave and rotational flow

When calculating a supersonic flow by the method of characteristics, it may be necessary to take into account the presence of shock waves caused by discontinuities of direction (a wall with a sudden slope change), of pressure (flow at the exit of an overexpanded nozzle), or even focalization of compression waves. To address this problem, we must introduce *shock operators* which enable us to calculate the shock during the construction during the construction of the characteristic mesh.

As we know, crossing a shock wave (C) causes an increase in entropy s. If (C) is rectilinear, the downstream entropy s is constant, and thus we can apply the operators defined in section 18.2 as such, taking into account that the stagnation pressure has become $p_{st} < p_{st_0}$. In fact, the shock waves encountered in practical problems are almost always curved. Locally, the shock equations remain the same and the relations of Chapter 13 apply. However, as the shock angle σ varies along (C), the entropy increase also varies from one streamline to another after the crossing of (C). According to the Crocco relation (see section 8.2.1), a non-zero vorticity value is associated with the entropy gradient: the flow downstream of a curved shock is rotational. Consequently, the quantities characterizing the flow will be three, namely: the pressure, the direction of the velocity, and the local entropy. Denoting with r an arbitrary reference state, in the case of a calorically perfect gas, the specific entropy s has the expression:

$$s - s_r = C_v \ln \frac{T}{T_r} - r \ln \frac{p}{p_r} = C_v \ln \frac{T_{st}}{T_r} - r \ln \frac{p_{st}}{p_r}.$$

As the stagnation temperature remains constant across shock wave, we can replace s by the local total pressure p_{st} or its logarithm. Thus, we use as working variables:

$$z = \ln\left(\frac{p}{p_r}\right), \ \varphi \ , \ s = -\ln\left(\frac{p_{st}}{p_r}\right).$$

The variables z and p are calculated from the characteristic relations, the entropy, or stagnation pressure, as determined by the method described in section 18.2, which takes into account invariance of entropy along a streamline (except across a shock).

20.2. The origin of a shock caused by a wall discontinuity

We consider the situation shown in Figure 20.1: the right running characteristic (*I*) was calculated to the point $(I, N-1)$, located on the wall (*W*), the index I denoting the considered characteristic and the index N identifying the position of the point on this characteristic. We conduct the calculation of the next (ξ), indexed *(J)*, the point $(J, N-1)$ being calculated from $(J, N-2)$ and $(I, N-1)$ by the operator N with $(J, N-2) \equiv 1$ and $(I, N-1) \equiv 2$ (see section 18.2.1).

At point A, the wall has a change of direction $\Delta\varphi$, giving rise to a shock wave (*C*), attached to A where two states coincide:

– the first, (A_1), corresponds to the conditions *upstream* of (*C*),

– the second, (A_2), to the conditions *downstream* of (*C*).

Progressing along (J) and wanting to calculate the point (J, N) by applying the operator *W*, we find that (J, N) lies downstream of the wall discontinuity. As shown in Figure 20.1, the point (J, N) is then calculated by extrapolating the upstream wall beyond A. Therefore, (J, N) is a fictitious point which allows us to calculate the portion of the segment $\overline{(J, N-1)(J, N)}$ which is not influenced by the shock (C). Also, (J, N) allows us to determine the conditions at A on the upstream side of (C) by linear interpolation between (J, N) and $(I, N-1)$. Knowing the state (A_1) as well as the deflection $\Delta\varphi$ across (C), we are able to solve the shock equations given in Chapter 14 and thus determine the downstream conditions (A_2).

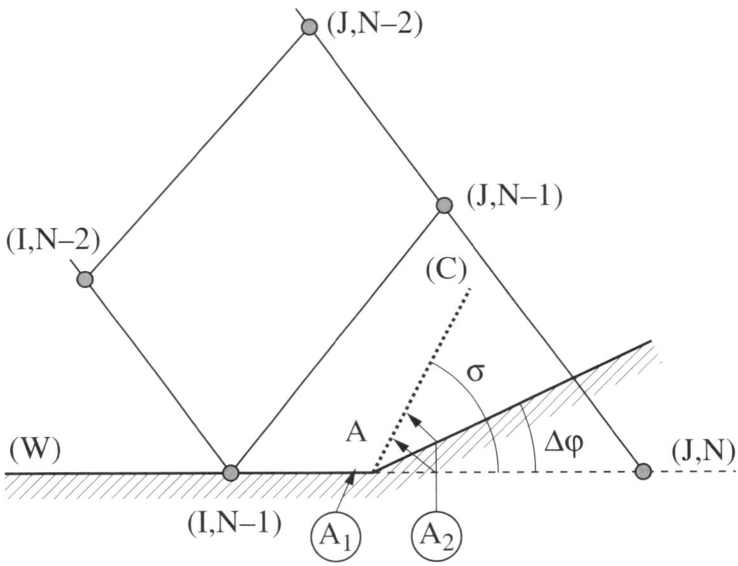

Figure 20.1. *Origin of a shock caused by a wall discontinuity*

The equation giving the shock angle σ as a function of the deflection and the conditions upstream of (A_1) is easily solved by a dichotomy technique with search of the root comprised between the Mach angle α_{A_1} and the angle $\sigma_{\max}$, corresponding to the maximum deviation $\varphi_{\max}$ (this angle is given explicitly as a function of M_{A_1} and γ, section 14.2.1). This approach enables us to capture the weak shock solution located on the lower part of the shock polar. If no root is found, it means that σ is higher than $\sigma_{\max}$. Then, the shock in A is detached and the method of characteristics must be abandoned, as a subsonic pocket forms downstream of (*C*). Knowing σ, the oblique shock relations give the properties of the flow downstream of the shock, i.e. the state (A_2).

20.3. Origin of a shock caused by a pressure discontinuity

A problem similar to the previous case is that of the shock created by a pressure discontinuity, such as at the exit lip of an overexpanded nozzle. We assume a wall (W) ending at a point E after which an ambient pressure p_a prevails.

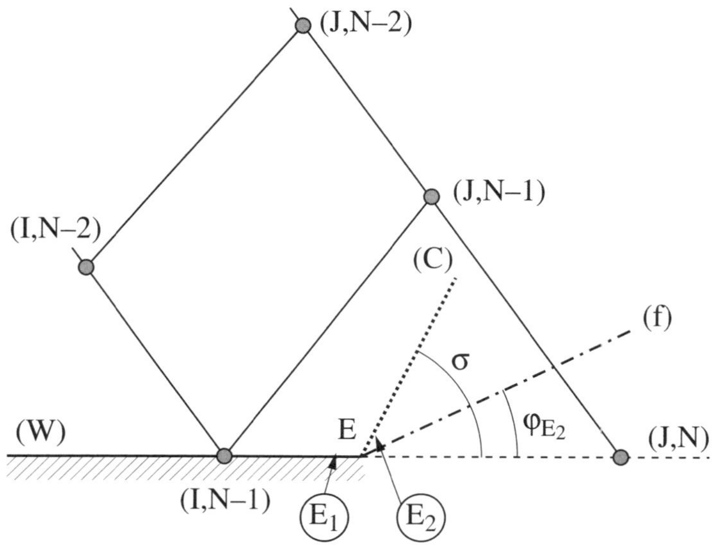

Figure 20.2. *Origin of a shock caused by a pressure discontinuity*

The calculation method corresponding to the situation in the vicinity of *E* is represented in Figure 20.2. The point (J,N), calculated at the wall from $(J,N-1)$ by the operator *W*, is found downstream of the point *E*, beyond which the wall (W) ends. Thus, we proceed to calculate (J,N) by extrapolating (W) beyond *E*, which provides the segment $\overline{(I,N-1)(J,N)}$. Then, the upstream state (E_1) in *E* is determined by linear interpolation between $(I,N-1)$ and (J,N). The pressure p_{E_1} is compared to p_a:

– if $p_{E_1} > p_a$, *E* is the origin of an expansion fan which we calculate by applying the operator Q (see section 18.2),

– if $p_{E_1} < p_a$, a shock wave (*C*) arises in *E*.

Here, we will only consider the second case, as the first is discussed in section 18.2. Knowing the ratio $\dfrac{p_a}{p_{E_1}}$, and the Mach number M_{E_1}, the relation giving the pressure jump across an oblique shock allows calculation of the shock angle σ. All other downstream conditions are deduced, in particular, the initial direction φ_{E_2} of the jet boundary (f) leaving *E*.

20.4. The point on a shock in the vicinity of its origin

Calculating the first shock point (*C*) located on the characteristic (J) requires a special procedure. We will only examine in detail the case of the shock caused by a wall discontinuity, as the problem of the shock created by a pressure jump is processed almost identically.

The meshing structure in the vicinity of *A* is represented in Figure 20.3a. From (A_1) and $(J,N-1)$, we calculate the fictitious point (J,N) located behind (*C*), the shock angle being always greater than the characteristic angle (relative to the velocity vector). The problem is to determine the point *C* of the shock at the intersection of (*C*) and $\overline{(J,N-1)(J,N)}$. We denote by (C_1) and (C_2) the states upstream and downstream of (*C*). The unknowns are: the coordinates of *C*, the downstream conditions $z_{C_2}, \varphi_{C_2}, s_{C_2}$, and the shock angle σ_C in *C*. The three shock equations do not allow calculation of the four unknown quantities: we must involve additional relations. On the left running characteristic (η) going through *C*, and ending on the wall at *W*, we have:

$$\bar{Z}_{WC_2}\left(z_{C_2}-z_W\right)+\left(\varphi_{C_2}-\varphi_W\right)=\bar{N}_{WC_2}\Delta\eta. \qquad [20.1]$$

Moreover, considering the right running characteristic (ξ) going through *W* and coming from the point *T*, located on the *downstream* side of the shock, we have the relation:

$$\bar{Z}_{TW}\left(z_W-z_T\right)-\left(\varphi_W-\varphi_T\right)=\bar{N}_{TW}\Delta\xi\,. \qquad [20.2]$$

The conditions in *T* are determined by linear interpolation between the downstream shock points A_2 and C_2:

$$\begin{aligned}\varphi_T &= \varphi_{A_2}+\left(\varphi_{A_2}-\varphi_{C_2}\right)\frac{\overline{AT}}{\overline{AC}}\\ z_T &= z_{A_2}+\left(z_{A_2}-z_{C_2}\right)\frac{\overline{AT}}{\overline{AC}}\\ s_T &= s_{A_2}+\left(s_{A_2}-s_{C_2}\right)\frac{\overline{AT}}{\overline{AC}}.\end{aligned} \qquad [20.3]$$

The shock relations plus the two characteristic equations [20.1] and [20.2] plus the interpolation formulas [20.3] form a set of eight equations for the eight unknowns:

$$\sigma_C \ , \ \varphi_{C_2} \ , \ z_{C_2} \ , \ s_{C_2} \ , \ z_W \ , \ \varphi_T \ , \ z_T \ , \ s_T \, .$$

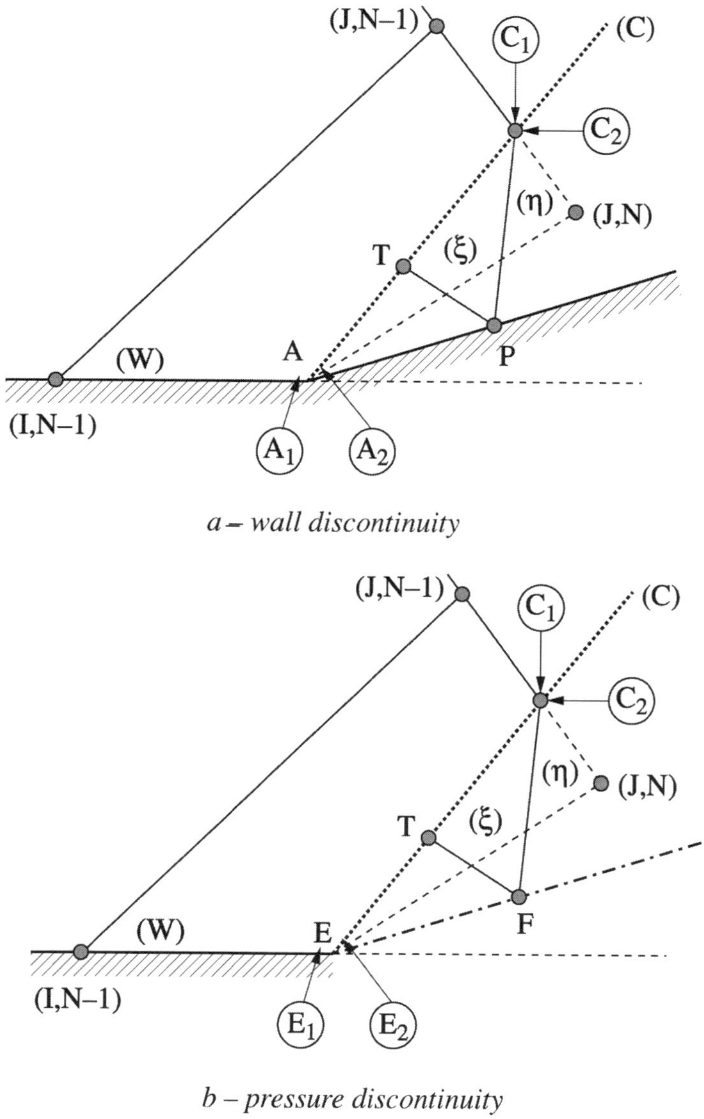

Figure 20.3. *Point on a shock in the vicinity of its origin*

The problem is thus "closed". The direction φ_W is given by the definition of the wall and s_W is equal to s_{A_2} as the entropy is constant along the wall which is a streamline. Knowing that the pressure jump across a shock is given by an equation of the form (see Chapter 14):

$$\frac{p_{C_2}}{p_{C_1}} = F(M_{C_1} \sin\sigma, \gamma)$$

we deduce from [20.1] and [20.2] that:

$$\varphi_{C_2} = \varphi_W + \overline{N}_{WC_2}\Delta\eta - \overline{Z}_{WC_2}\left[\ln F\left(M_{C_1} \sin\sigma, \gamma\right) + z_{C_1} - z_W\right].$$

Thus, the equation allowing calculation of the shock angle σ is written as (see section 10.1):

$$\begin{aligned} &ctg\left\{\varphi_W + \overline{N}_{WC_2}\Delta\eta - \overline{Z}_{WC_2}\left[\ln F\left(M_{C_1} \sin\sigma, \gamma\right) + z_{C_1} - z_W\right] - \varphi_{C_1}\right\} \\ &= \tan\sigma\left(\frac{\gamma+1}{2}\frac{M_{C_1}^2}{M_{C_1}^2 \sin^2\sigma - 1} - 1\right). \end{aligned} \qquad [20.4]$$

This equation is solved without difficulty by dichotomy. The Mach number after the shock M_{C_2} is then deduced from the pressure p_{C_2} and the entropy calculated as specified in section 18.2. The determination of the point C is performed by an iterative method, the calculation algorithm being organized as follows.

We assume we are in the iteration with rank (n). The point $C^{(n)}$ on the shock (*C*), which moves during the iteration, is calculated at the intersection of $\overline{(J,N-1)(J,N)}$ and the shock segment passing through A with the slope (we recall that σ is the shock angle relative to the upstream velocity vector):

$$\psi_{shock} = \frac{1}{2}\left(\varphi_{A_1} + \sigma_{A_1} + \varphi_{C_1}^{(n-1)} + \sigma_{C_1}^{(n-1)}\right).$$

Knowing the position of $C^{(n)}$, the state $(C_1)^{(n)}$ is deduced by interpolating between $(J,N-1)$ and (J,N). The point $W^{(n)}$ is calculated at the intersection of the wall and the straight line passing through $C^{(n)}$ with the slope:

$$\psi_{WC_2} = \frac{1}{2}\left(\varphi_{C_2}^{(n-1)} + \alpha_{C_2}^{(n-1)} + \varphi_W^{(n-1)} + \alpha_W^{(n-1)}\right).$$

Knowing the coordinates of $C^{(n)}$, the new value $\varphi_W^{(n)}$ is calculated from the definition of the wall. The point T is defined at the intersection of the shock segment $\overline{AC^{(n)}}$ and the straight line going through $W^{(n)}$ and with the slope:

$$\psi_{WT} = \frac{1}{2}\left(\varphi_W^{(n)} - \alpha_W^{(n-1)} + \varphi_T^{(n)} - \alpha_T^{(n-1)}\right).$$

Knowing the coordinates of T, we deduce by [20.3]:

$$z_T^{(n)}, \quad \varphi_T^{(n)}, \quad s_T^{(n)}$$

leading to $z_W^{(n)}$ through relation [20.2] where $\overline{Z}_{TW}$ and $\overline{N}_{TW}$ are evaluated with the values at the iteration $(n-1)$; similarly for $\overline{Z}_{WC_2}$ and $\overline{N}_{WC_2}$. At this level, the implicit equation [20.4] is solved, leading to $\sigma^{(n)}$, of which follow the actual values:

$$z_{C_2}^{(n)}, \quad \varphi_{C_2}^{(n)}, \quad s_{C_2}^{(n)}.$$

Then, we start a new iteration loop. If ε is an accuracy threshold fixed in advance, the process is considered converged when $\left|\sigma^{(n)} - \sigma^{(n-1)}\right| < \varepsilon$.

The case of a shock caused by a pressure discontinuity is solved practically by the same algorithm. The only difference lies in determining the downstream point, as W is replaced by a point F, located on the jet boundary (*f*). Placing ourselves at the iteration (n), J is determined at the intersection (see Figure 20.3b):

– of the characteristic (η) emerging from $C_2^{(n-1)}$, assimilated to a straight line passing through $C_2^{(n-1)}$ with the slope:

$$\psi_{FC_2} = \frac{1}{2}\left(\varphi_{C_2}^{(n-1)} + \alpha_{C_2}^{(n-1)} + \varphi_F^{(n-1)} + \alpha_F\right)$$

– of the boundary (f) assimilated to a straight line passing through E with the slope:

$$\psi_{EF} = \frac{1}{2}\left(\varphi_{E_2} + \varphi_F^{(n-1)}\right).$$

As the pressure is known along (*f*) (which is a streamline), hence z_F and s_F are constant, and therefore also the Mach number, the Mach angle, etc. Apart from these differences, the calculation procedure is the same as in the previous case.

COMMENT.– If the jet boundary (*f*) is not at constant pressure, the calculation procedure is nearly the same. In this case, since the location of the point *F* moves during the iteration process, one has to calculate at each iteration the value of the pressure at *F*. The pressure distribution along (*f*) can be defined either by a set of points or analytically. As (*f*) is a streamline, the entropy s_F is constant.

20.5. Origin of a focalization shock

During a calculation, it can happen that characteristics belonging to the *same family* intersect. This circumstance may occur when compression waves are generated by a continuous concave ramp *AB* or in a strongly expanded axisymmetric jet where the waves reflected by the jet boundary focus. We have seen in section 17.5.4 that the solution with a progressive wave then becomes unacceptable and a shock wave must form in the flow. Figure 20.4 shows an intersection of waves (ξ) occurring during the construction of the characteristic (J). Having calculated the point (J,N) from (I,N) and $(J,N-1)$, we see that the segments $\overline{(I,N-1)(J,N-1)}$ and $\overline{(I,N)(J,N)}$ intersect at a point *K*. In this circumstance, a shock wave (*C*) starts from the point *K*. We assume that in *K*, the shock (*C*) is directed along the *bisector* of the angle formed by $\overline{(I,N-1)(J,N-1)}$ and $\overline{(I,N)(J,N)}$. Thus, the direction of (*C*) is imposed at its origin.

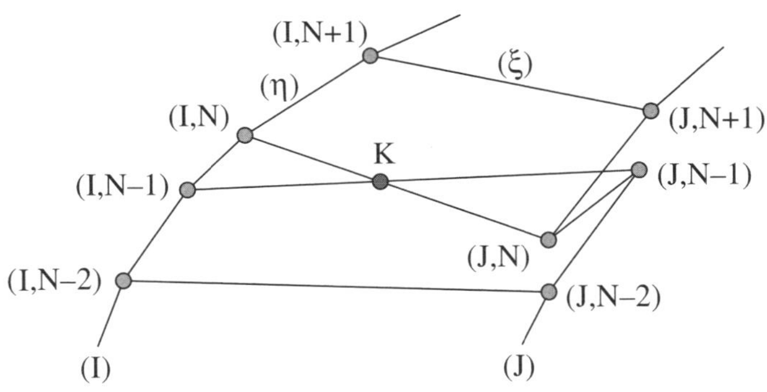

Figure 20.4. *Formation of a focalization shock through the crossing of characteristics*

The intersection of (*C*) with the segment of the left running characteristic $\overline{(J,N-2)(J,N-1)}$ gives the point on the shock *C* (see Figure 20.5). The upstream state (C_1) is determined by interpolation between $(J,N-2)$ and $(J,N-1)$. As the direction of (*C*) is imposed, the angle σ of the shock relative to the velocity in (C_1) results and the shock relations allow the calculation of state (C_2) on the downstream side. Performing the substitution of the points shown in Figure 20.5, the progression is continued downstream by calculating $(J,N+1)$ from (J,N), coinciding with (C_2), and of $(I,N+1)$, etc.

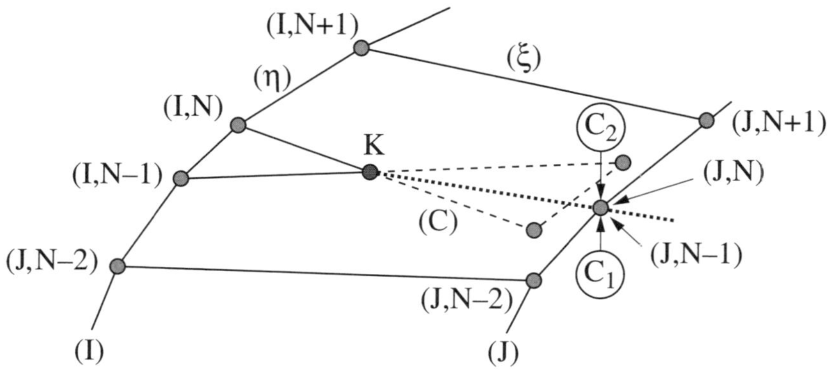

Figure 20.5. *Start of a focalization shock*

20.6. Normal point of a shock

The characteristic (I), crossing a shock (*C*) at the point *B*, has been fully calculated. In *B*, the flow has two states, namely (see Figure 20.6):

– an upstream state (B_1) located in (I,N),

– a downstream state (B_2) located in $(I,N1)$.

Because of the shock, the numbering *N* undergoes a discontinuity since the shock "absorbs" characteristics. The characteristic (J) is being calculated. Thus, from (I,N) and $(J,N-1)$, the application of the operator *N* has given the point (J,N). The search for the crossing of the shock (*C*) (extrapolated from *B* with a direction equal to its direction in *B*) with (J) indicates that the intersection point *C* falls between $(J,N-1)$ and (J,N).

The determination of the point *C*, belonging to (*C*) and located on the characteristic (J), is done in the following way. The unknowns of the problem are the shock angle σ and the flow conditions downstream of (C_2): $z_{C_2}, \varphi_{C_2}, s_{C_2}$. As there are three shock equations, we have to include an additional equation obtained by considering the left running characteristic (η) passing through *C* and meeting

(I) at the point *T*. In Figure 20.6, *T* is between $(I, N1)$ and $(I, N1+1)$. In fact, (η) of *C*, can cross the characteristic (I) beyond $(I, N1+1)$. If this is the case, we should look for the points of (I) surrounding *T*. We have between *T* and the point downstream of the shock (C_2):

$$\bar{Z}_{TC_2}\left(z_{C_2} - z_T\right) + \left(\varphi_{C_2} - \varphi_T\right) = \bar{N}_{TC_2}\Delta\eta. \quad [20.5]$$

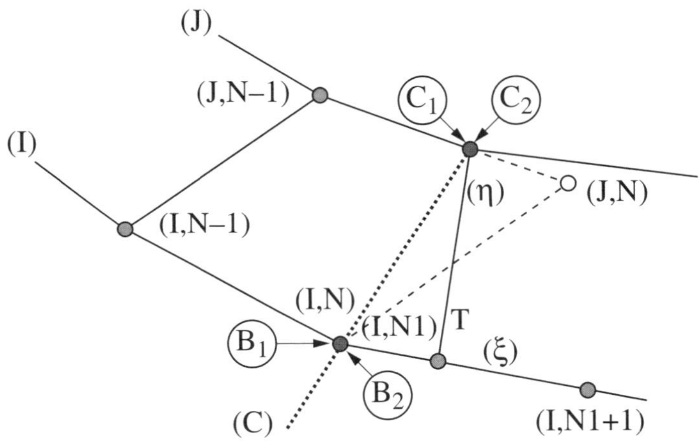

Figure 20.6. *Normal point on a shock*

The quantities in *T* are linearly interpolated between the two points of (I) on each side of *T*; i.e. $(I, N1)$ and $(I, N1+1)$ in the case of Figure 20.6. Thus z_T , φ_T , s_T are not really unknowns. The problem is then "closed" and the algorithm is organized in the following way.

As we are at the iteration with the rank (n): the position of $C^{(n)}$ is found by taking the intersection:

– of the characteristic segment $\overline{(J, N-1)(J, N)}$,

– and of the shock (C) approximated by a straight line passing through *B* with the direction:

$$\psi_{shock} = \frac{1}{2}\left(\varphi_{B_1} + \sigma_B + \varphi_{C_1}^{(n-1)} + \sigma_C^{(n-1)}\right).$$

Knowing $C^{(n)}$, the conditions relative to the upstream state $\left(C_1^{(n)}\right)$ are determined by linearly interpolating between (J, N) and $(J, N-1)$. The point *T* is then determined at the intersection of (I) and the characteristic (η) going through $\left(C_2^{(n)}\right)$ and having the direction:

$$\psi_{TC_2} = \frac{1}{2}\left(\varphi_{C_2}^{(n-1)} + \alpha_{C_2}^{(n-1)} + \varphi_T^{(n)} + \alpha_T^{(n)}\right).$$

Then, as indicated above, the conditions T are determined by linear interpolation. The fundamental equation of the oblique shock, allowing to calculate σ_C, is written by taking into account [20.5]:

$$ctg\left\{\varphi_T^{(n)} + \bar{N}_{TC_2}\Delta\eta - \bar{Z}_{TC_2}\left[\ln F\left(M_{C_1}\sin\sigma_C,\gamma\right) + z_{C_1}^{(n)} - z_T^{(n)}\right] - \varphi_{C_1}^{(n)}\right\}$$

$$= \tan\ \sigma_C\left(\frac{\gamma+1}{2}\frac{M_{C_1}^2}{M_{C_1}^2\sin^2\sigma_C - 1} - 1\right).$$

This equation is solved by a dichotomy technique, which gives $\sigma_C^{(n)}$. Then, we calculate the conditions relative to the state $\left(C_2^{(n)}\right)$:

$$z_{C_2}\ ,\ \varphi_{C_2}\ ,\ s_{C_2}.$$

The above procedure is repeated until the error $|\sigma^n - \sigma^{(n-1)}|$ is less than a precision threshold ε set in advance.

20.7. Calculation examples of flows with shock formation

To illustrate, Figure 20.7 shows an application of the method of rotational characteristics for calculating a uniform flow with the Mach number $M_0 = 2.5$ in which a shock wave is caused by a deflection $\Delta\varphi = 10°$ from the wall. In the case of Figure 20.7a, the deflection is abrupt, which leads to the formation of a shock attached to the apex of the wedge. In the case of Figure 20.7b, the deflection is a gradual process, which induces compression waves that focus to give a shock wave.

Figure 20.8, relative to the focalization shock, shows the distribution of the Mach number and the stagnation pressure (equivalent to the entropy). Near the wall, below the focalization shock, the flow is isentropic. After focalization, entropy is created, gradually increasing its level in the region where the shock is curved. Other examples are given hereafter.

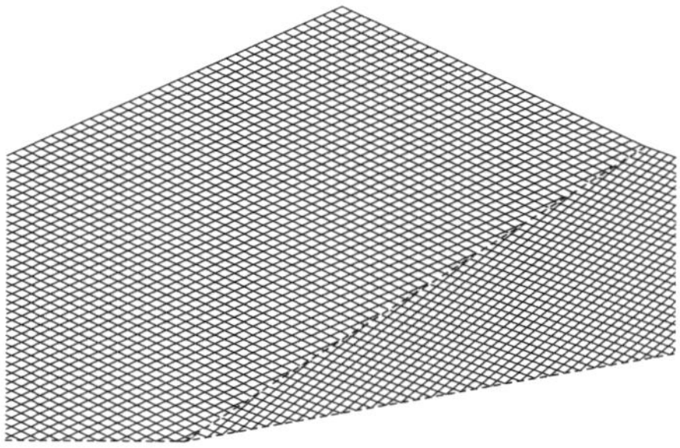

a – shock attached to the discontinuity

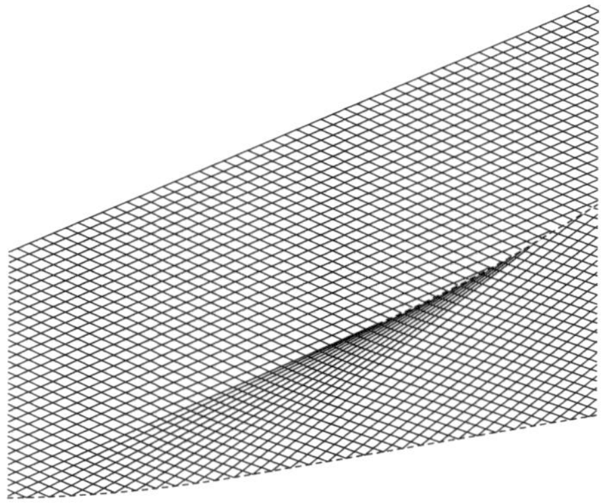

b – focalization of waves

Figure 20.7. *Formation of a shock wave in a flow. Upstream Mach number $M_0 = 2.5$, deflection $\Delta\varphi = 10°$*

a – iso-Mach number contours

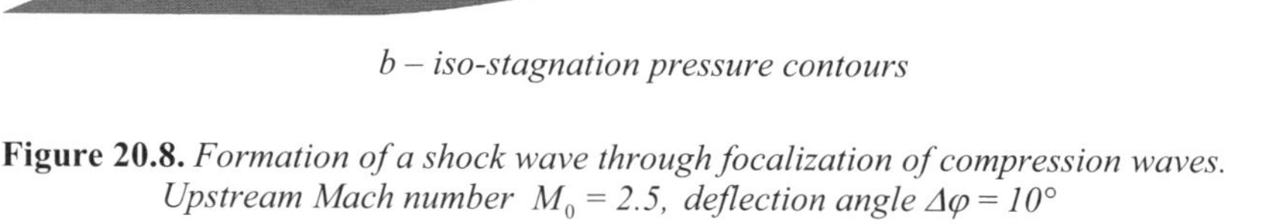

b – iso-stagnation pressure contours

Figure 20.8. *Formation of a shock wave through focalization of compression waves. Upstream Mach number $M_0 = 2.5$, deflection angle $\Delta\varphi = 10°$*

The following two examples are related to the impact of a shock wave. Figure 20.9a shows a calculation by the method of characteristics of the impact of an oblique shock wave on a planar wall, causing a deflection $\Delta\varphi = -10°$ in a flow with upstream Mach number $M_0 = 2.5$. The impact causes the formation of a reflected shock that deflects the flow of an angle $\Delta\varphi = +10°$, giving it back the direction of the wall.

Figure 20.9b shows the impact of the same incident shock on an isobaric jet boundary (J). Upon the impact, the boundary experiences a deflection $\Delta\varphi = -20.03°$, which is approximately twice the deflection across the shock wave. The Mach number on (J) after the impact is 2.484, rather than 2.5 before the impact, due to stagnation pressure loss across the shock.

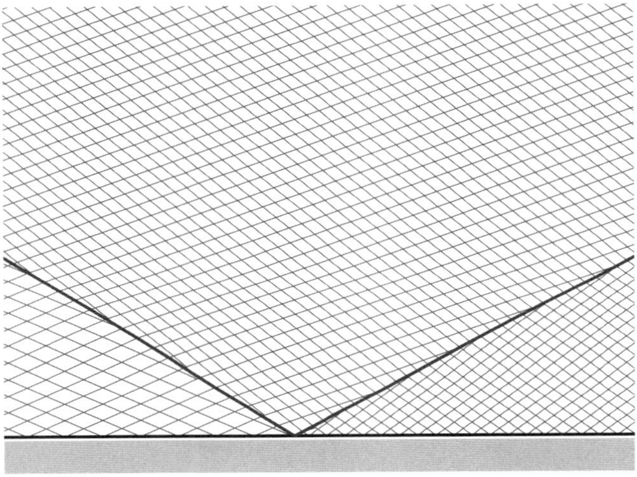

a – on a straight wall

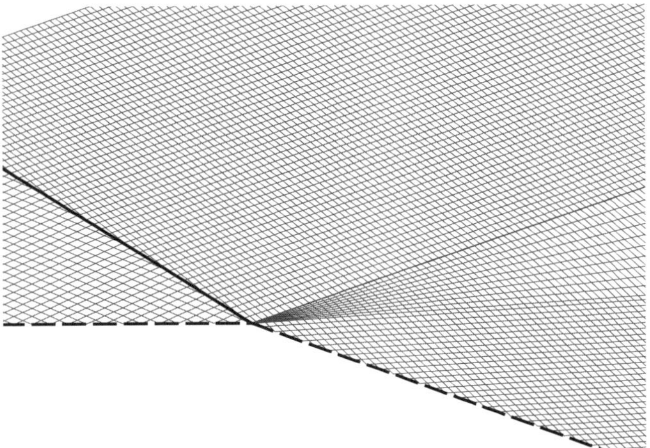

b – on an isobaric jet boundary

Figure 20.9. *Reflection of an oblique shock wave. Upstream Mach number* $M_0 = 2.5$, *deflection across the shock* $\Delta\varphi = -10°$

20.8. Application to the analysis of the structure of a supersonic jet

First, we will show some properties related to the reflection of a wave. The pressure and direction variations along the characteristic lines or waves are related by the following relations in a *two-dimensional planar* flow:

$$\frac{\sqrt{M^2 - 1}}{\gamma M^2} \frac{\delta p}{p} + \delta\varphi = 0 \quad \text{on a characteristic } (\eta)$$

$$\frac{\sqrt{M^2 - 1}}{\gamma M^2} \frac{\delta p}{p} - \delta\varphi = 0 \quad \text{on a characteristic } (\xi).$$

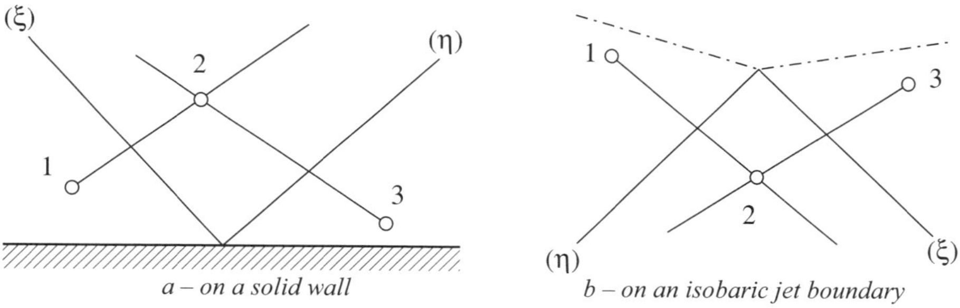

Figure 20.10. *Reflection of a wave in two-dimensional planar flow*

To fix ideas, we consider an expansion wave (ξ) reflecting on a wall with the local direction $\varphi = 0$. To simplify the notation, we write:

$$Z = \frac{\sqrt{M^2 - 1}}{\gamma M^2}.$$

The states (1) and (2) on either side of (ξ) and located on a wave (η) are related by the following relation, where δ_{21} denotes the variations across the wave (ξ) (see Figure 20.10a):

$$Z \frac{\delta p_{21}}{p} + \delta\varphi_{21} = 0.$$

If the wave is an expansion, then we must have $\delta\varphi_{21} > 0$. As the direction φ_1 in (1) is such that $\varphi_1 = 0$ (slip condition on the wall), (2) is not compatible with the wall direction. A wave (η) must be created to lead to a state (3), such that: $\varphi_3 = 0$. As the states (2) and (3) are on the same wave (ξ):

$$Z\frac{\delta p_{32}}{p} - \delta\varphi_{32} = 0$$

from which follows:

$$Z\frac{\delta p_{32}}{p} = \delta\varphi_{32} = -\,\delta\varphi_{21}$$

because $\delta\varphi_{32} = -\,\delta\varphi_{21}$ to restore the flow parallel to the wall. Since $\delta\varphi_{21} > 0$, it follows that $\delta p_{32} < 0$: the flow expands across (η). The same reasoning can be applied to a compression wave. If the wall is a symmetry plane, it also amounts to impose φ and the situations are identical.

Now we consider a compression wave (η) reflecting from the isobaric jet boundary (f) (see Figure 20.10b). The states (1) and (2) on both sides of (η) and on (ξ) are such that:

$$Z\ \frac{\delta p_{21}}{p} - \delta\varphi_{21} = 0.$$

Since, by hypothesis $\delta p_{21} > 0$, it follows that $\delta\varphi_{21} > 0$. As the pressure in (2) is no longer compatible with the pressure on the boundary (f), a wave (ξ) must be created to produce a state (3) where $p_3 = p_a$. The states (2) and (3) are related through the equation:

$$Z\ \frac{\delta p_{32}}{p} + \delta\varphi_{32} = 0.$$

As the pressure p_1 is equal to p_a and $p_2 > p_1 = p_a$, δp_{32} has to be negative: the compression wave is reflected in an expansion wave. A similar argument can be made if (η) is an expansion wave. It should be noted that:

$$\delta\varphi_{32} = -\ Z\ \frac{\delta p_{32}}{p} = Z\ \frac{\delta p_{21}}{p} = \delta\varphi_{21}.$$

On the impact of the wave (η), the boundary (f) undergoes a deflection equal to twice the deflection across the incident wave.

In conclusion, in a two-dimensional planar flow:

A wave, expansion or compression, is reflected on a wall, or symmetry plane, as a wave of a similar type.

A wave, expansion or compression, is reflected on an isobaric jet boundary as a wave of opposite type.

We find the observed behaviors during the impact of a shock wave on, either a wall or on an isobaric boundary, despite the fact that the shock is a strong first-order disturbance.

20.8.1. *Case of the under-expanded jet*

We consider a two-dimensional planar flow and study the structure, or organization, of the jet leaving an *under-expanded* supersonic nozzle, i.e. operating under such conditions that the pressure in the exit plane is greater than the assumed uniform external pressure p_a (see section 10.5).

The nozzle produces a uniform supersonic flow with Mach number M_0 and pressure p_0 in its exit plane so that $p_0 > p_a$. To ensure the adaptation of the supersonic jet to the external pressure, an expansion must occur at the nozzle exit so that the pressure on the boundary (f) of the jet, which is a slip line, is equal to the external pressure p_a. This expansion occurs via an expansion fan centered on the exit lip *E* of the nozzle.

Figure 20.11 shows the structure of the supersonic jet after leaving the nozzle. The expansion fan (ξ) is reflected on the symmetry plane as an expansion wave made of characteristics (η). These waves propagate toward the isobaric boundary (f) of the jet where they are reflected as compression waves (ξ) which propagate in the direction of the symmetry axis. These waves are reflected in turn on the symmetry axis as compression waves (η), etc. The process of alternating between compression and expansion waves in principle repeats itself indefinitely since in a non-viscous fluid there is no mechanism to dissipate the kinetic energy of the flow into heat.

We will see further down that another phenomenon may amend this structure. When an expansion wave is reflected on the boundary (f), it deflects it as shown in the figure. The boundary (f) is therefore bent toward the symmetry axis. If the impacting wave is a compression, the deflection occurs in the opposite direction.

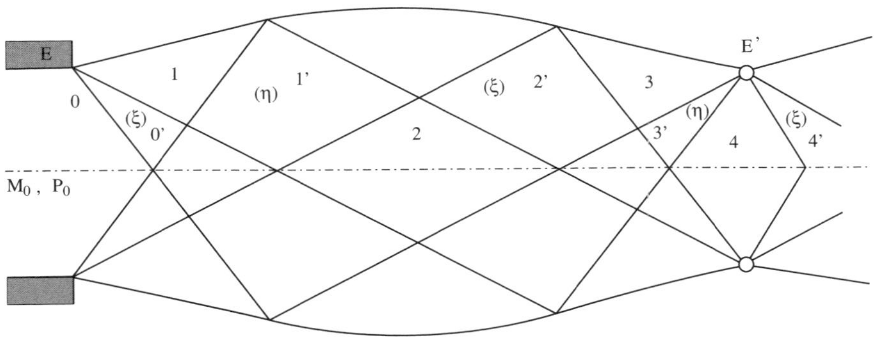

Figure 20.11. *General structure of the supersonic planar under-expanded jet*

We analyze the structure of the supersonic jet in detail. As the flow (0) in the exit of the nozzle is by design uniform, the contiguous area *0'* is a simple wave up to the first wave reflected by the axis. As the boundary (f) is at the constant pressure p_a, the flow (Mach number, stagnation pressure, etc.) there is uniform, the process being isentropic (no shock). The domain (1) following (0') is uniform. Indeed, the left running characteristics (η) of this region come all from the uniform region (0) and they end on (f) where the flow is constant: the property is then deduced without difficulty. The region (1') is again a simple wave as it is contiguous to the uniform area (1). As the velocity on the axis has an imposed direction $\varphi = 0$, it follows that (2) is uniform, etc. We demonstrate without difficulty that the flow in (4) is the same as in the domain (1). The compression waves of (3) converge at a point *E'* where the situation is identical to that in *E*. Thus, in the absence of viscosity, the structure of the planar jet repeats itself in a periodic undulating form to downstream infinity. These results are summarized below in Table 20.1.

Uniform regions	0		1		2		3		4	
Simple wave regions		0'		1'		2'		3'		4'

Table 20.1. *Particular flow regions in a supersonic planar jet*

Figure 20.12a shows the characteristics calculated in a supersonic planar jet with the total pressure p_{st_j} expanding from a uniform state at Mach 2 in an atmosphere where the pressure p_a is such that $p_a / p_{st_j} = 0.08$. This plot highlights the expansion at the exit (E) of the nozzle which is reflected on the symmetry plane as an expansion which is returned from the isobaric boundary as a compression wave whose characteristics are converging. This compression is reflected by the symmetry plane as a compression whose waves in theory converge at a point (E') on the isobaric boundary. The intersection of characteristics which we observe, are due to a lack in resolution of the meshing. At point (E'), the situation is identical to that in (E) and, in the absence of any real or numerical viscosity, the structure of the jet should be repeated in theory to infinity. Figure 20.12b is a display of the jet by iso-Mach contours which highlights uniform regions and the achieved pressure levels. We verify that the situation in (E') is identical to that in (E) so that the story can repeat itself indefinitely.

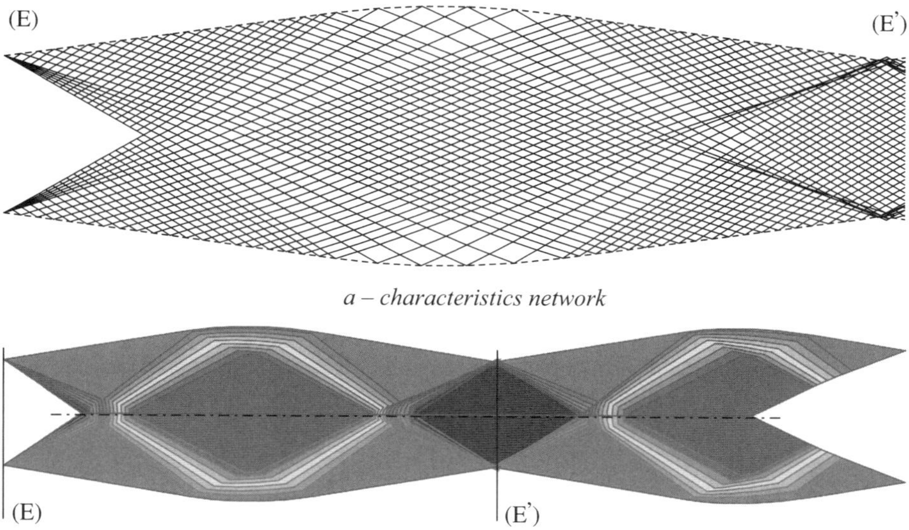

a – characteristics network

b – isovalue contours of the Mach number

Figure 20.12. *Supersonic planar jet expanding from $M_0 = 2$ in an isobaric atmosphere with the pressure p_a such that $p_a / p_{st_j} = 0.08$*

We verify that the waves returned from the isobaric boundary (*f*) are compression waves, and thus these waves converge and can intersect in particular if the flow in the exit of the nozzle is not uniform. The progressive wave solution is then unacceptable and a shock wave will form in the flow, as shown in Figure 20.13 (only one-half of the assumed symmetric jet is represented).

At the impact on the isobaric (f), the shock is reflected as a centered expansion wave to ensure continuity of the pressure (see calculation in Figure 20.9b). Then, the boundary (f) is suddenly deflected as indicated in the Figure. Downstream of the impact point, the pressure on (f) is equal to p_a but the stagnation pressure has decreased because of loss across the shock. Thus, the Mach number is not equal to M_1 any longer, but has become smaller.

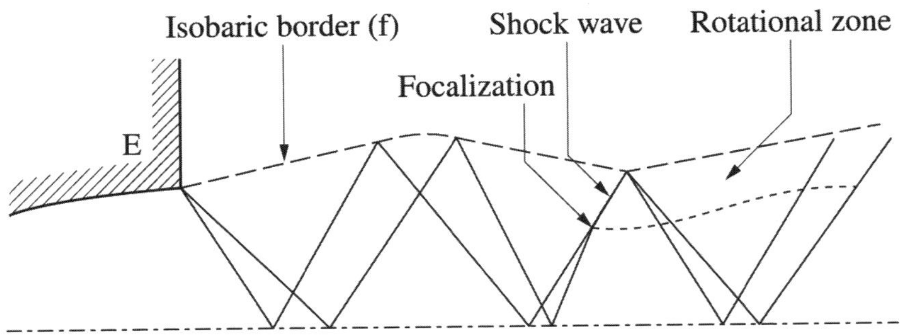

Figure 20.13. *Focalization of compression waves in a supersonic jet*

20.8.2. *Case of the over-expanded jet*

We analyze the case where the nozzle exit pressure is lower than outside pressure. Since $p_0 < p_a$, the adaptation of the jet pressure to the external pressure must be through a shock wave (C_1) that forms at the lip of the nozzle exit so as to ensure the compatibility of the pressures downstream of (E), leading to the diagram shown in Figure 20.14. In this case, the jet contracts instead of expanding.

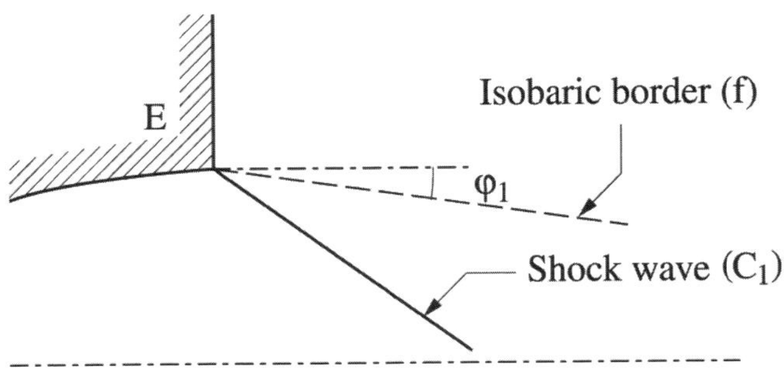

Figure 20.14. *Contraction of the over-expanded jet at the nozzle exit*

As seen in Figure 20.15, arriving on the axis of symmetry, the shock wave (C_1) is reflected as a shock wave (C_2) in order to deviate the flow (1) whose direction is incompatible with that of the axis of symmetry $(\varphi = 0)$. We have therefore a compression from p_1 to p_2. Thus, the pressure in (2) downstream of (C_2) will be incompatible with the pressure on the boundary (f) at the point I where (C_2) impacts (f). The pressure jump $p_1 \rightarrow p_2$ must be compensated by an expansion restoring the pressure from p_2 to p_1, which is done by an expansion centered in I. Beyond, the structure of the jet is comparable to what we have seen for the under-expanded jet with the probable formation of shock waves through focalization of compression waves.

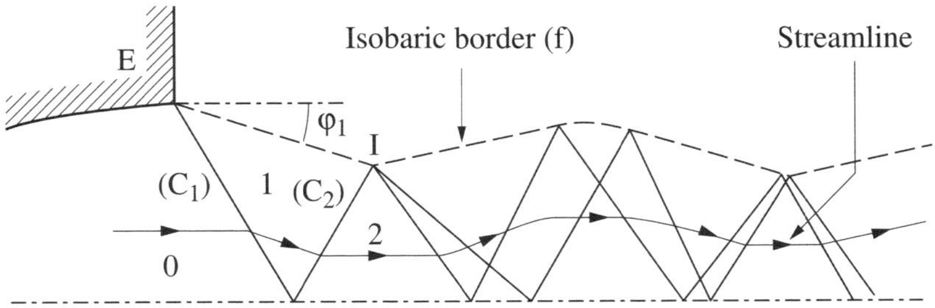

Figure 20.15. *General structure of the over-expanded supersonic jet*

If we maintain a constant jet stagnation pressure (and therefore the pressure p_0), the external pressure is increased, the shock (C_1) at the nozzle exit intensifies. It will be the same for the reflected shock (C_2), as the deflection through (C_1) will increase. A situation will be reached where it will be impossible to find an oblique shock (C_2), the deflection to be achieved on the axis being greater than the maximum deflection angle: we pass from a *regular reflection* to a *singular reflection* or *Mach reflection*. This phenomenon is studied in detail in section 14.5, a symmetry plane behaving like a straight wall in a non-viscous fluid. Therefore, the Mach phenomenon appears with the formation on the axis of a nearly-normal shock wave (see Figure 20.16). A triple point T exists at the confluence of the shocks (C_1), (C_2) and the nearly-normal shock (C_3). A slip line (Σ) emanates from T separating the flows (2) and (3) which have different levels of velocity, density, stagnation, pressure, etc., only the pressures in (2) and (3) having to be identical on both sides of (Σ).

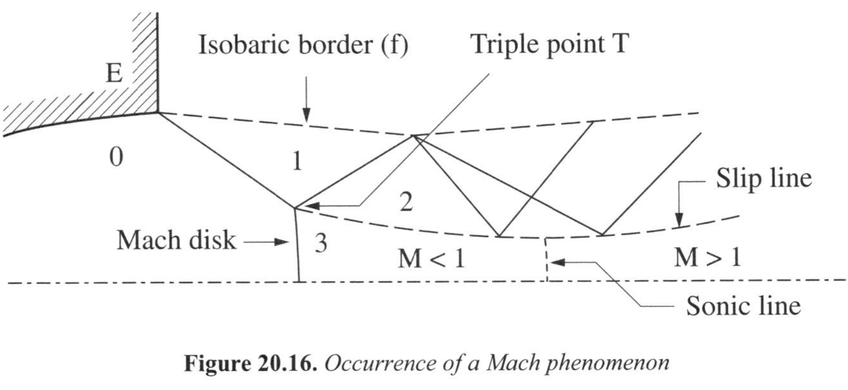

Figure 20.16. *Occurrence of a Mach phenomenon*

The phenomena described above are also observed in axisymmetric jets, the most common case in practice. We also observe a pseudoperiodic structure of the jet with alternating crests and troughs (the so-called barrel-type organization). The construction of the solution is less simple and requires the use of numerical techniques. Thus, the isobaric boundary curves out of the nozzle and the jet is not formed by assembly of uniform and simple wave regions. In addition, in an axisymmetric flow, we observe a phenomenon, resulting from the reflection of expansion waves (η) as compression waves when they reach the isobaric jet boundary (f). These converging waves tend to intersect and form a shock wave by focalization as soon as the expansion ratio is significant. An example is given in Figure 20.17 which shows the calculation of a supersonic axisymmetric jet with the formation of a focalization shock.

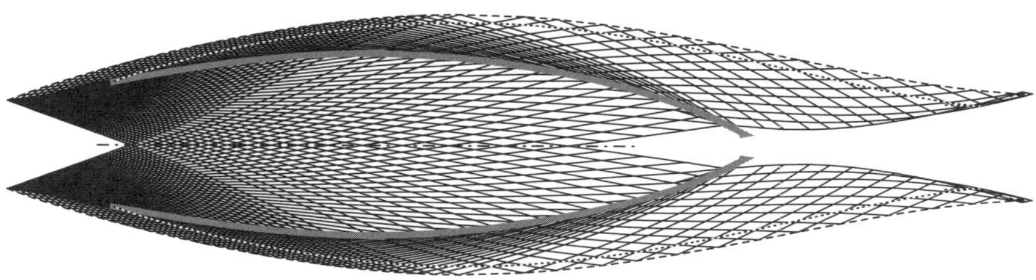

a – characteristics mesh

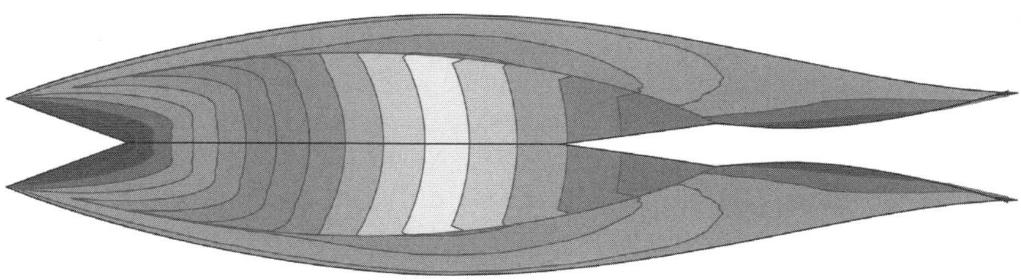

b – iso-Mach number contour

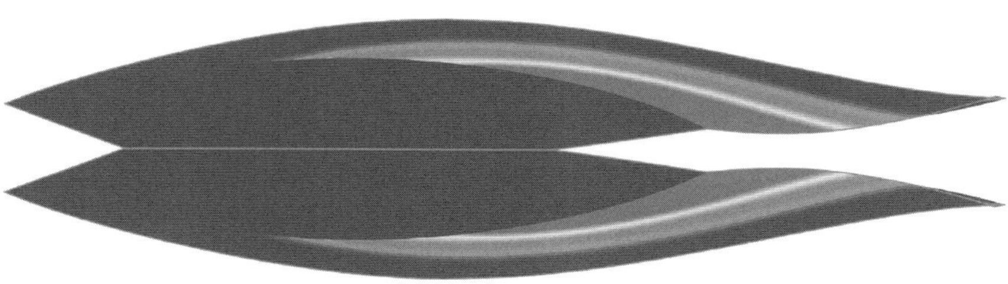

c – iso-stagnation pressure contours (entropy)

Figure 20.17. *Supersonic axisymmetric jet with focalization shock*

This "barrel" shock converges toward the centerline where it can give rise to a Mach phenomenon and the formation of what is called a *Mach disk*. The static temperature, which significantly decreased during the expansion in the nozzle and then in the jet, undergoes a jump when crossing the Mach disk which almost brings it to the level of the stagnation temperature. Thus, in a propulsion jet (of a rocket or jet engine), the Mach disks are locally very hot regions where the gas strongly radiates (see Figure 14.26). The Mach disks forming in a jet engine are intense infrared sources that can make a fighter very vulnerable to an infrared-guided missile.

20.9. Drag of the symmetric bump in supersonic flow

A sometimes advanced argument to question the generality of the Oswatitsch relation, established in section 5.3, is the drag of a symmetric bump, or profile, in a non-viscous supersonic flow. The apparent paradox of a drag with no entropy source is illustrated by a calculation using the method of characteristics for the flow around a bump with sinusoidal shape placed on a straight wall in a uniform supersonic flow of Mach number $M_0 = 2.5$. The result obtained assuming a totally isentropic (homentropic) flow is represented as iso-Mach contours in Figure 20.18.

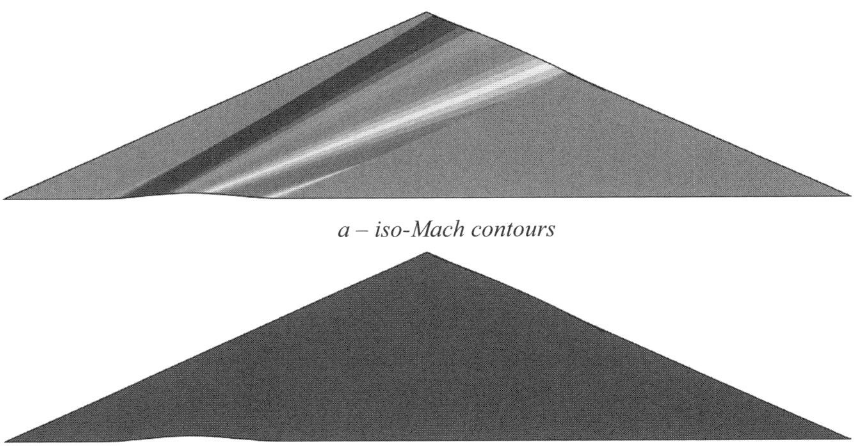

a – iso-Mach contours

b – iso-stagnation pressure contours

Figure 20.18. *Flow past a symmetric bump. Upstream Mach number $M_0 = 2.5$. Calculation for a homentropic flow. Characteristics mesh*

The field is clearly unsymmetrical, the flow undergoing a compression on the front part of the bump and an expansion on the rear part. This result is interpreted from the theoretical analysis of Chapter 17. The compression and expansion are simple waves because the upstream flow is uniform. A compression occurs in the region of the trailing edge of the bump, the flow is deflected to follow the straight downstream wall. As shown in Figure 20.19, the Mach number distribution on the wall is antisymmetric. Hence, the existence of a pressure drag which can be calculated by integrating the wall pressure distribution (the pressure is deduced from the Mach number by the classic isentropic relation). In this case, we find a drag coefficient such that: $C_D = 0.003839$. Therefore, there is an obvious contradiction with the Oswatitsch theorem, since there seems to be drag without entropy production.

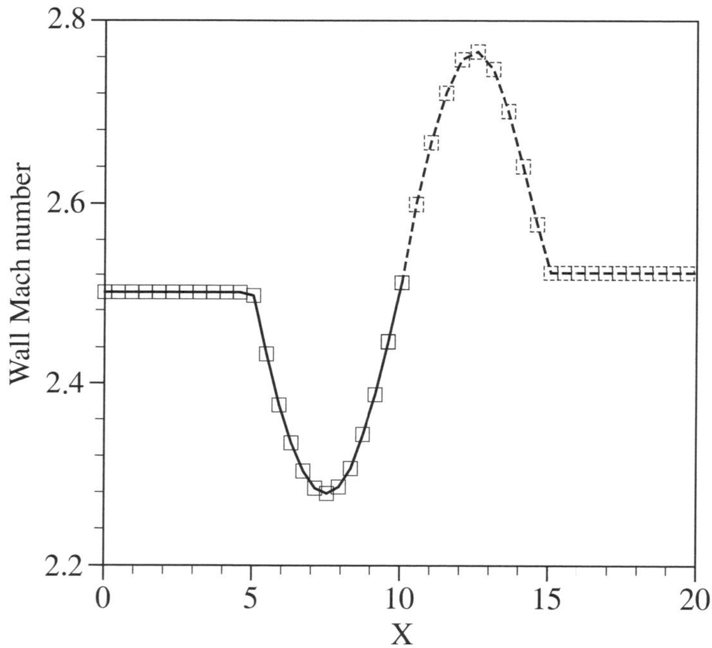

Figure 20.19. *Flow past a symmetrical bump, upstream Mach number $M_0 = 2.5$, calculation for a homentropic flow. Wall Mach number distribution*

However, if we plot the characteristics mesh (see Figure 20.20), we see that the compression waves generated at the front and rear of the bump intersect causing a folding of the network in regions where the solution is not unique any longer. Such a situation is physically not acceptable: we have to treat the focalizations by introducing shock waves (see Figure 20.21). This does not alter the pressure distribution over the bump, but the field is not homentropic any longer as shown in Figure 20.22b representing the iso-stagnation pressure contours. The entropy produced across shock waves forms a "wake" containing the wave drag of the bump.

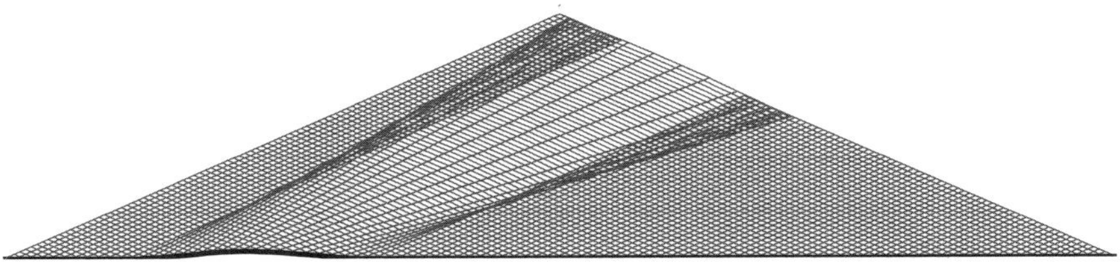

a – characteristics mesh

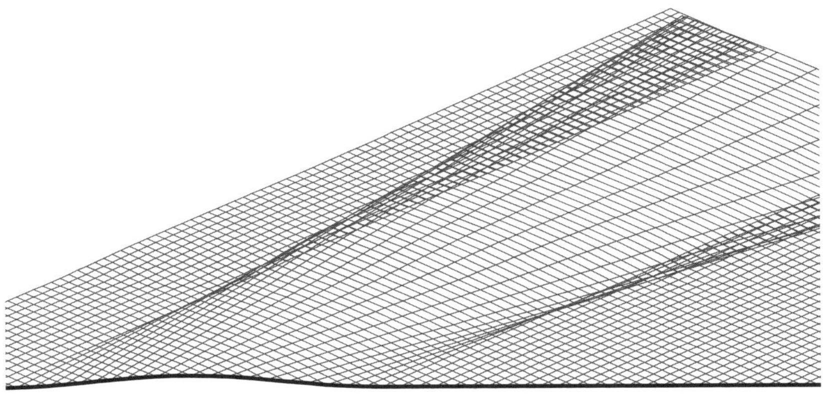

b – characteristics mesh: close-up of the bump region

Figure 20.20. *Flow past a symmetrical bump, upstream Mach number* $M_0 = 2.5$*, and calculation for a homentropic flow*

This calculation highlights the link between drag and entropy, the drag of a body in a flow without a source of entropy being zero (paradox of d'Alembert). The source of entropy in flows without chemical reaction is mainly the viscosity and we know that shock waves are viscous phenomena (see section 13.6).

Therefore, this result is consistent with the Oswatitsch relation, whose application here is difficult as we have to calculate the entropy flux through a surface on which the pressure is again constant, which creates a difficulty because, in theory, the shock waves propagate to infinity. In the real viscous world, shock waves are attenuated and eventually disappear (which is lucky for the neighborhood!), but not the entropy which is preserved forever.

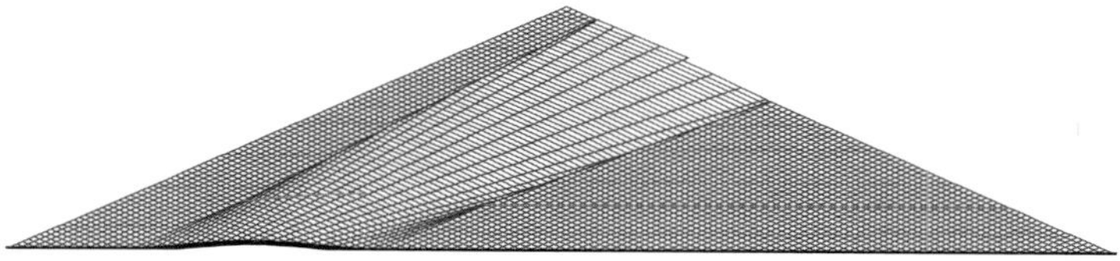

a – characteristics mesh

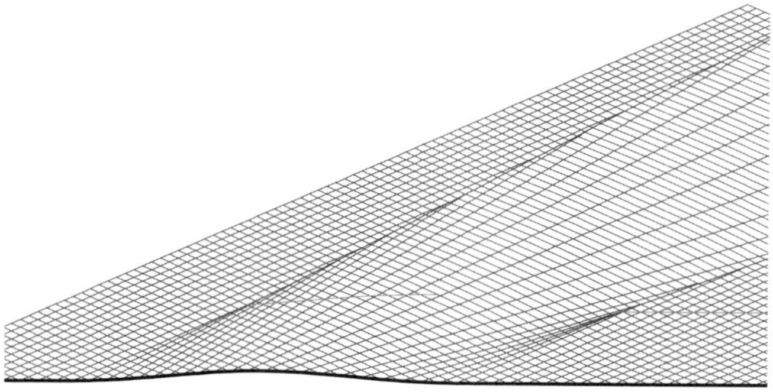

b – characteristics mesh: close-up of the bump region

Figure 20.21. *Flow past a symmetrical bump. Upstream Mach number $M_0 = 2.5$. Taking into account the focalizations and shock waves. Characteristics mesh*

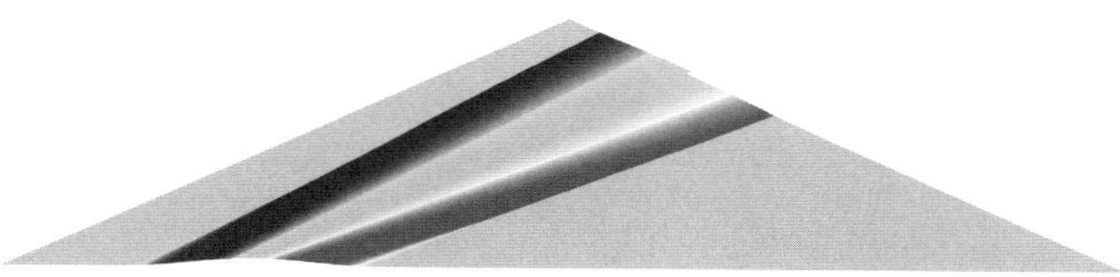

a – iso-Mach contour for an isentropic flow

b – iso-stagnation pressure contours

Figure 20.22. *Flow past a symmetrical bump. Upstream Mach number $M_0 = 2.5$. Non-isentropic field*

The incompressible flow around the same bump was calculated using the *FreeFem* code, developed by the University Pierre-et-Marie Curie of Paris, which solves the Laplace equation by a finite-element method. The plot of isopressure lines in Figure 20.23 confirms the fact that in the assumption of an incompressible non-viscous irrotational flow, the flow field is symmetric. It is clear that the integration of the pressure distribution plotted in Figure 20.24 gives a drag coefficient of zero. An identical result would be obtained for a non-symmetrical bump, the regions with overpressure being compensated by underpressure regions to achieve a zero drag.

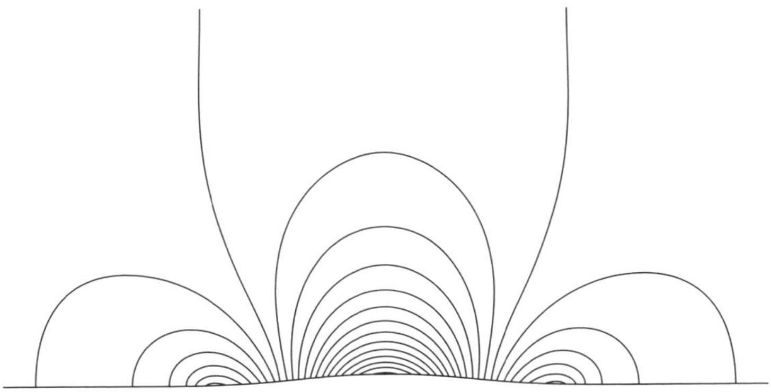

Figure 20.23. *Flow past a symmetrical bump in an incompressible irrotational flow. Iso-static pressure curves (calculation by FreeFem code)*

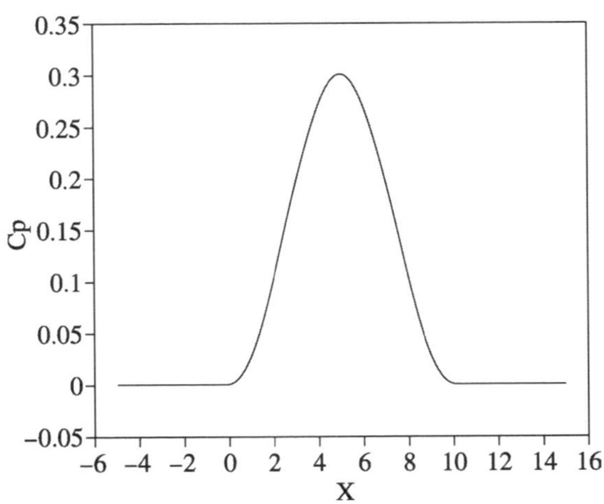

Figure 20.24. *Flow past a symmetrical bump in an incompressible irrotational flow. Pressure on the wall (calculation by FreeFem code)*

Chapter 21

One-Dimensional, Non-Viscous and Adiabatic Unsteady Flows

21.1. Introduction

This chapter examines the propagation of pressure disturbances, or waves, in space and time. They may be small disturbances of the acoustic type, but also discontinuities such as shock waves. In Chapters 9–20, we have considered steady flows, i.e. independent of time. But the full Euler equations contain derivatives with respect to time; they reflect the temporal change of a spatial field (see Chapters 4 and 7). If we are given a certain initial state at a given instant in the form of velocity, pressure, and density distributions, the Euler equations govern the subsequent change of this field over time. These equations have the property of being hyperbolic as a function of time, which means they can be integrated progressing in time without any influence on the solution at a time point t of the conditions that may prevail at a future time $t+\Delta t$. This property is exploited by the numerical integration methods to find a steady solution to the Navier–Stokes or Euler equations. Starting from a more or less arbitrary initial state, we progress in time (*time marching integration method*) due to the Euler, or Navier–Stokes, equations to reach a steady state, that is, until the derivatives with respect to time vanish. Of course, it is necessary for a steady solution to exist, which is compatible with the boundary conditions to be satisfied.

The study of unsteady behavior is fundamental to understand the phenomena of wave propagation, especially in acoustics, in detonics but also in aerodynamics, where the flow may be affected by violent unsteady phenomena such as air intakes buzz, buffeting, flutter, surging of turbomachines, formation and propagation of a supersonic boom, etc. In addition, the unsteady approach is essential to understand steady behaviors because in the real world, things always have a beginning, then a settling time (transient state) before (if possible) a steady state is established. We shall see that only the unsteady analysis allows, for example, to understand the mechanisms of starting and unstarting of supersonic nozzles (see section 24.5).

Here, we shall restrict the analysis to the case of one-dimensional flows, the purpose being primarily an introduction to phenomena the understanding of which is sometimes difficult even though they are governed by simple equations.

21.2. General equations

The present theory applies under the following assumptions:

– The flow is non-viscous and non-heat conducting.

– It is one-dimensional, meaning that all properties are function of only one space variable x. Here, we study the flow in a slender tube with a slowly varying section $A(x)$ and a straight axis Ox. The properties of the flow (velocity u, pressure p, density ρ, etc.) are assumed to be constant throughout the section $A(x)$ (see Figure 21.1). We will study in section 24.4 the case of a spherical flow, which is also a one-dimensional flow whose properties depend only on the radial distance r to the origin O. In this situation, the properties of the flow are constant on spheres centered on O.

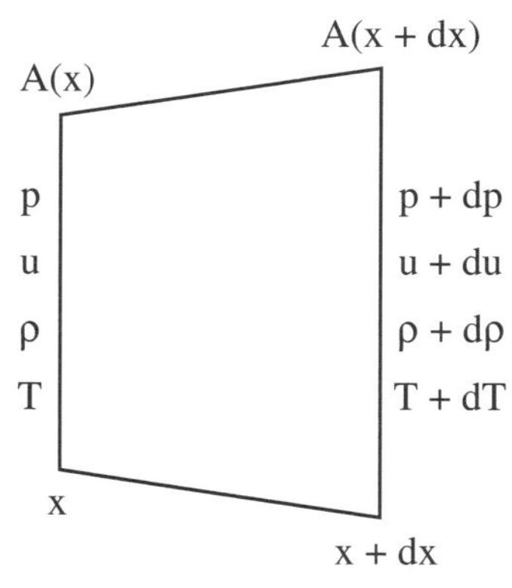

Figure 21.1. *Control volume for establishing the equations of motion*

The equations governing the flow can be deduced from the general equations established in Chapter 4. In order to give this chapter some autonomy, we will, here, deduce them with a direct approach. To do this, we consider a slice of the tube with the thickness *dx* and apply the fundamental laws to the fluid in this slice. The approach follows that for one-dimensional steady flows used in Chapter 9. Thus, we will be more concise.

– *Continuity equation:*

During a time interval *dt*, the increase in mass in the slice *dx* is the difference between the outgoing and incoming masses, hence to the second order:

$$\left[\rho uA-\left(\rho uA+\frac{\partial}{\partial}\left(\rho uA\right)\right)dx\right]dt.$$

This difference must be equal to the mass change in the slice *dx* at the same time interval:

$$\left[\frac{\partial}{\partial t}\left(\rho Adx\right)\right]dt$$

which leads to the continuity equation:

$$\boxed{\frac{\partial}{\partial t}\left(\rho A\right)+\frac{\partial}{\partial x}\left(\rho uA\right)=0.}$$

– *Equation of motion (Euler equation):*

The fundamental equation of mechanics (Newton's law) applied to the mass $\left(A\rho dx\right)$, contained in the slice *dx*, gives us:

$$\rho Adx\frac{du}{dt}=pA-\left[pA+\frac{\partial}{\partial x}\left(pA\right)dx\right]+p\frac{\partial A}{\partial x}dx$$

i.e.:

$$\rho Adx\frac{du}{dt}=pA-pA-\frac{\partial}{\partial x}\left(pA\right)dx+p\frac{\partial A}{\partial x}dx$$

or:

$$\rho Adx\frac{du}{dt}=pA-pA-p\frac{\partial A}{\partial x}dx-A\frac{\partial p}{\partial x}dx+p\frac{\partial A}{\partial x}dx.$$

Hence:

$$\boxed{\rho\frac{du}{dt}=-\frac{\partial p}{\partial x}.}$$

– *Energy equation:*

The energy equation is reduced here to the entropy invariance condition, since the terms of viscosity and heat conduction are ignored. Therefore, we have:

$$\boxed{\frac{ds}{dt}=0.}$$

21.3. Unsteady isentropic flow in a cylindrical tube

21.3.1. *Equation of motion*

In this section, we will consider the case of a tube with a constant section (i.e. $A(x)=constant$), which will enable interesting developments. The basic equations take the following form:

– *continuity equation:*

$$\frac{\partial\rho}{\partial t}+\frac{\partial}{\partial x}(\rho u)=0$$

or by developing:

$$\frac{1}{\rho}\frac{\partial\rho}{\partial t}+\frac{u}{\rho}\frac{\partial\rho}{\partial x}+\frac{\partial u}{\partial x}=0.$$

– *equation of motion (Euler equation):*

$$\frac{\partial u}{\partial t}+u\frac{\partial u}{\partial x}+\frac{1}{\rho}\frac{\partial p}{\partial x}=0.$$

The following calculations are intended to eliminate the density so that only the velocity u and pressure p are involved in the equations. To do this, we are led to state:

$$a^2=\left(\frac{\partial p}{\partial\rho}\right)_s.$$

We will see that quantity a is the propagation velocity of pressure disturbances, or pressure waves, which we have identified to the sound velocity.

The flow is isentropic:

$$\frac{\partial \rho}{\partial t}=\left(\frac{\partial \rho}{\partial p}\right)_s \frac{\partial p}{\partial t}+\left(\frac{\partial \rho}{\partial s}\right)_p \frac{\partial s}{\partial t}=\left(\frac{\partial \rho}{\partial p}\right)_s \frac{\partial p}{\partial t}=\frac{1}{a^2}\frac{\partial p}{\partial t}$$

and for the same reason:

$$\frac{\partial \rho}{\partial x}=\left(\frac{\partial \rho}{\partial p}\right)_s \frac{\partial p}{\partial x}=\frac{1}{a^2}\frac{\partial p}{\partial x}$$

which leads to the equations:

$$\frac{1}{\rho a^2}\frac{\partial p}{\partial t}+\frac{u}{\rho a^2}\frac{\partial p}{\partial x}+\frac{\partial u}{\partial x}=0 \qquad [21.1]$$

$$\frac{1}{\rho}\frac{\partial p}{\partial x}+u\frac{\partial u}{\partial x}+\frac{\partial u}{\partial t}=0. \qquad [21.2]$$

As the entropy is constant, the variables p and u are sufficient to define the flow as ρ and a can be calculated as a function of the pressure p alone. The system of partial differential equations [21.1] and [21.2] is a pseudo-linear system in that it is linear with regard to partial derivatives. Let us consider the plane (x,t), where the distance x is plotted along the horizontal axis and time t along the vertical axis. Such a representation is called the *wave diagram*. If we assume a fluid element located on the abscissa x_0 at time t_0, over time, this element will move in the tube, its abscissa x changing with time t. The movement of the fluid particle can be represented in the plane (x,t) by a curve $x(t)$, which is the *trajectory* of the element in question.

21.3.2. *Cauchy problem and characteristic curves*

We assume the flow known on a certain curve (C) of the space (x,t) and seek to determine this flow at a point Q close to the point $\bar{Q}$ located on (C) (see Figure 21.2). This is the Cauchy problem already considered in Chapter 17. The point $\bar{Q}$ has $\bar{x}$ and $\bar{t}$ as coordinates, the pressure and velocity in $\bar{Q}$ have the values $\bar{p}$

and $\bar{u}$. To calculate p and u in Q, we perform a first order Taylor expansion in the vicinity of $\bar{Q}$, i.e.:

$$u = \bar{u} + (x - \bar{x})\frac{\partial u}{\partial x} + (t - \bar{t})\frac{\partial u}{\partial t}$$

$$p = \bar{p} + (x - \bar{x})\frac{\partial p}{\partial x} + (t - \bar{t})\frac{\partial p}{\partial t}.$$

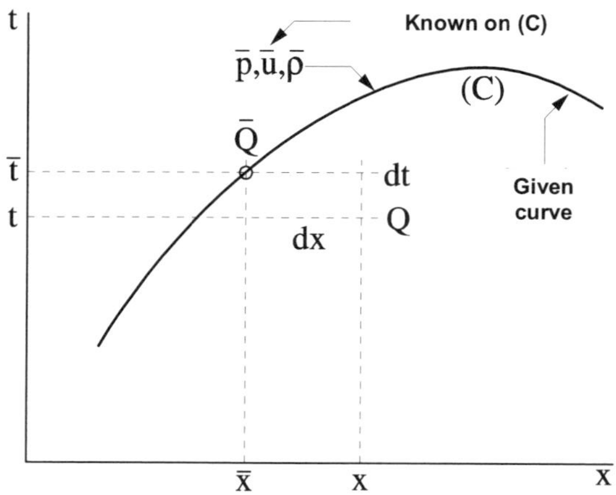

Figure 21.2. *Cauchy problem in the plane of the curve (C)*

To calculate u and p, we need to know the four derivatives:

$$\frac{\partial u}{\partial x}, \quad \frac{\partial u}{\partial t}, \quad \frac{\partial p}{\partial x}, \quad \frac{\partial p}{\partial t}$$

at the point $\bar{Q}$. For this, we have two equations, [21.1] and [21.2], and relations expressing the variations $d\bar{u}$ and $d\bar{p}$ of u and p when we move along (C). This elementary displacement has the components $d\bar{x}$ and $d\bar{t}$. As the flow is known on (C), the variations $d\bar{u}$ and $d\bar{p}$ are known and we have:

$$d\bar{u} = \frac{\partial u}{\partial x} d\bar{x} + \frac{\partial u}{\partial t} d\bar{t} \tag{21.3}$$

$$d\bar{p} = \frac{\partial p}{\partial x} d\bar{x} + \frac{\partial p}{\partial t} d\bar{t}. \tag{21.4}$$

We have four relations [21.1]–[21.4] to calculate the four sought derivatives. We rewrite this system as:

$$\frac{1}{\bar{\rho}\bar{a}^2}\frac{\partial p}{\partial t}+\frac{\bar{u}}{\bar{\rho}\bar{a}^2}\frac{\partial p}{\partial x}+0\times\frac{\partial u}{\partial t}+\frac{\partial u}{\partial x}=0$$

$$0\times\frac{\partial p}{\partial t}+\frac{1}{\bar{\rho}}\frac{\partial p}{\partial x}+\frac{\partial u}{\partial t}+\bar{u}\frac{\partial u}{\partial x}=0$$

$$0\times\frac{\partial p}{\partial t}+0\times\frac{\partial p}{\partial x}+d\bar{t}\,\frac{\partial u}{\partial t}+d\bar{x}\,\frac{\partial u}{\partial x}=d\bar{u}$$

$$d\bar{t}\,\frac{\partial p}{\partial t}+d\bar{x}\,\frac{\partial p}{\partial x}+0\times\frac{\partial u}{\partial t}+0\times\frac{\partial u}{\partial x}=d\bar{p}.$$

In general, the above system admits a solution and only one, unless its principal determinant is zero. We write this determinant and express its nullity, i.e.:

$$\begin{vmatrix} \dfrac{1}{\bar{\rho}\bar{a}^2} & \dfrac{\bar{u}}{\bar{\rho}\bar{a}^2} & 0 & 1 \\ 0 & \dfrac{1}{\bar{\rho}} & 1 & \bar{u} \\ 0 & 0 & d\bar{t} & d\bar{x} \\ d\bar{t} & d\bar{x} & 0 & 0 \end{vmatrix} = 0.$$

By expanding:

$$d\bar{x}^2+\left(\bar{u}^2-\bar{a}^2\right)d\bar{t}^2-2\bar{u}\,d\bar{x}\,d\bar{t}=0.$$

The roots of this algebraic equation of the 2nd degree in $d\bar{x}/d\bar{t}$ are:

$$\frac{d\bar{x}}{d\bar{t}}=\bar{u}+\bar{a} \quad \text{and} \quad \frac{d\bar{x}}{d\bar{t}}=\bar{u}-\bar{a}.$$

The interpretation is that if at the point $\bar{Q}$ of an initial curve (C), the slope $d\bar{x}/d\bar{t}$ of (C) has one or other of the above values, the Cauchy problem is either impossible or indeterminate.

21.3.3. *Characteristic relation: Riemann invariants*

If the Cauchy problem is indeterminate (the only situation that we are interested in), it is necessary that the determinants, associated with the system written above, are also equal to zero. Writing this condition leads to *characteristic relations* that can be obtained more directly in the following way.

We denote with (η) and (ξ) the singular directions of slope (see Figure 21.3):

$$\frac{d\overline{x}}{d\overline{t}} = \overline{u} + \overline{a} \quad for \quad (\eta)$$

$$\frac{d\overline{x}}{d\overline{t}} = \overline{u} - \overline{a} \quad for \quad (\xi).$$

Overscoring the magnitudes will be removed to simplify the writing.

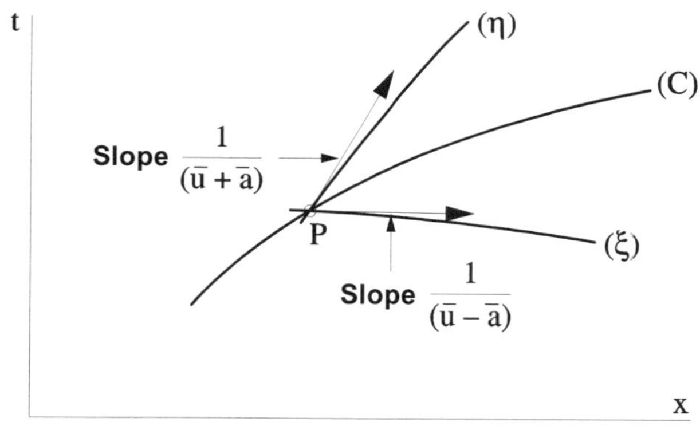

Figure 21.3. *Characteristic directions* (η), (ξ) *in the plane* $[x, t]$

We denote with δ^+ and δ^- the variations of a quantity *f as we move along* (η) *and* (ξ), respectively. We have the following relations for the temporal variations of *f*:

$$\frac{\delta^+ f}{\delta t} = \frac{\partial f}{\partial t} + (u + a)\frac{\partial f}{\partial x} \quad along \quad (\eta)$$

$$\frac{\delta^- f}{\delta t} = \frac{\partial f}{\partial t} + (u - a)\frac{\partial f}{\partial x} \quad along \quad (\xi).$$

Solving this system gives us:

$$\frac{\partial f}{\partial t}=\frac{1}{2a}\left[(u+a)\frac{\delta^- f}{\delta t}-(u-a)\frac{\delta^+ f}{\delta t}\right]$$

$$\frac{\partial f}{\partial x}=\frac{1}{2a}\left(\frac{\delta^+ f}{\delta t}-\frac{\delta^- f}{\delta t}\right).$$

By substituting these expressions in the system, [21.1] and [21.2], it follows that:

– for the continuity equation [21.1]:

$$\frac{1}{\rho a^2}\left[\frac{u+a}{2a}\frac{\delta^- p}{\delta t}-\frac{u-a}{2a}\frac{\delta^+ p}{\delta t}\right]+\frac{u}{\rho a^2}\left[\frac{1}{2a}\frac{\delta^+ p}{\delta t}-\frac{1}{2a}\frac{\delta^- p}{\delta t}\right]+\frac{1}{2a}\left[\frac{\delta^+ u}{\delta t}-\frac{\delta^- u}{\delta t}\right]=0.$$

After development and simplification:

$$\frac{1}{\rho a}\frac{\delta^+ p}{\delta t}+\frac{1}{\rho a}\frac{\delta^- p}{\delta t}+\frac{\delta^+ u}{\delta t}-\frac{\delta^- u}{\delta t}=0. \qquad [21.5]$$

– for the equation of motion [21.2]:

$$\frac{1}{\rho}\frac{1}{2a}\left[\frac{\delta^+ p}{\delta t}-\frac{\delta^- p}{\delta t}\right]+\frac{u}{2a}\left[\frac{\delta^+ u}{\delta t}-\frac{\delta^- u}{\delta t}\right]+\frac{1}{2a}\left[(u+a)\frac{\delta^- u}{\delta t}-(u-a)\frac{\delta^+ u}{\delta t}\right]=0$$

i.e.:

$$\frac{1}{\rho a}\frac{\delta^+ p}{\delta t}-\frac{1}{\rho a}\frac{\delta^- p}{\delta t}+\frac{\delta^+ u}{\delta t}+\frac{\delta^- u}{\delta t}=0. \qquad [21.6]$$

By carrying out [21.5] + [21.6], then [21.5] – [21.6], we obtain:

$$\boxed{\begin{aligned}&\frac{1}{\rho a}\frac{\delta^+ p}{\delta t}+\frac{\delta^+ u}{\delta t}=0\\ &\frac{1}{\rho a}\frac{\delta^- p}{\delta t}-\frac{\delta u^-}{\delta t}=0.\end{aligned}} \qquad [21.7]$$

These relations mean that if in one of its points (C) has a slope $(u+a)$ or $(u-a)$, then the Cauchy problem is indeterminate. This means that for a solution to exist, the variations of p and u along (C) must satisfy one or the other of the above equations [21.7].

These conditions, called characteristic relations, were first given by Riemann. The method by which we have obtained them shows that these conditions are the motion equations projected onto (η) and (ξ), respectively.

If the gas is, from the thermodynamic point of view, divariant, then:

$$\rho = \rho(p, s)$$

and:

$$a = a(p, s).$$

Since the flow is also isentropic, the density ρ and velocity of sound a vary only as a function of the pressure p. Then the entropy s is a known constant, and we can introduce a function P that depends only on p so that:

$$P(p) = \int^{p} \frac{dp}{\rho a}.$$

Therefore, it is possible to integrate the relations [21.7] to arrive at:

$$\boxed{\begin{aligned} P + u &= 2\mu = constant \quad on \quad (\eta) \\ P - u &= 2\lambda = constant \quad on \quad (\xi) \end{aligned}}$$

where λ and μ are called *Riemann invariants*. They are also characteristic coordinates, in the sense that they are numbers which are constant along the corresponding characteristics (η) or (ξ).

21.3.4. *Case of a calorically perfect gas*

If the gas is calorically perfect, (γ constant), the velocity of sound a is given by:

$$a^2 = \gamma \frac{p}{\rho}$$

which leads to:

$$2\frac{da}{a}=\frac{dp}{p}-\frac{d\rho}{\rho}.$$

The flow being isentropic:

$$\frac{d\rho}{\rho}=\frac{1}{\gamma}\frac{dp}{p}$$

it follows that:

$$\frac{dp}{p}=\frac{2\gamma}{\gamma-1}\frac{da}{a}$$

and:

$$\frac{dp}{\rho a}=\frac{p}{\rho a}\frac{2\gamma}{\gamma-1}\frac{da}{a}=\frac{a^2}{\gamma a}\frac{2\gamma}{\gamma-1}\frac{da}{a}.$$

Therefore:

$$\frac{dp}{\rho a}=\frac{2}{\gamma-1}da$$

so that:

$$\int^{p}\frac{dp}{\rho a}=\frac{2}{\gamma-1}a+constant$$

which leads to the characteristic equations:

$$\boxed{\begin{aligned}&\frac{2}{\gamma-1}a+u=2\mu=constant\quad on\quad(\eta)\\&\frac{2}{\gamma-1}a-u=2\lambda=constant\quad on\quad(\xi).\end{aligned}}$$

These relations express conditions that must satisfy the quantities u and a when we move along a characteristic, either (η) or (ξ). As we highlighted, these are the motion equations projected on (η) and (ξ), which take a remarkably simple integrated form. We found similar relations in the theory of steady supersonic flows in Chapter 17.

We will see the characteristics (η) and (ξ) are also called *waves*.

The wave (η) whose slope is $(u+a)$ propagates a disturbance with the velocity $+a$ with regard to the flow and we call it a *progressive wave.*

The wave (ξ) whose slope is $(u-a)$ propagates a disturbance with the velocity $-a$ with regard to the flow and we call it a *regressive wave.*

21.3.5. *Initial and boundary conditions*

Solving a fluid mechanics problem is seeking a state satisfying the equations of motion as well as the initial and boundary conditions. These are those conditions which dictate the form of the solution. In most of the problems we consider here, the initial conditions correspond to a resting state in the tube characterized by velocity $u = 0$ and uniform thermodynamics conditions (pressure, density, and temperature). It is clear that such a state is the solution of the motion equations. Thus, it is the change of boundary conditions that will produce a change in this state, i.e. a flow in the tube.

When the problem has a wall, we express that the gas must remain in contact with the wall. This means that it can neither penetrate the wall nor a vacuum can form between the gas and the wall (this condition is similar to the slip condition for a steady state). We will find this boundary condition in the case of a piston moving inside the tube or when the tube is closed at its ends. If V_W denotes the velocity of the wall, we write that the gas velocity on the wall must be such that:

$$u = V_W .$$

If at the abscissa x_f the tube has an open end, where the pressure is kept constant and equal to p_f, then we impose the condition:

$$p = p_f \quad for \quad x = x_f .$$

Later, we will return to the question of boundary conditions, which is, in fact, trickier; but before we do this, it is necessary to have the knowledge of a number of results to understand the information transmission mechanism. We will apply the above theory to the construction of flows that can occur in a tube with a cylindrical section (see Figure 21.4).

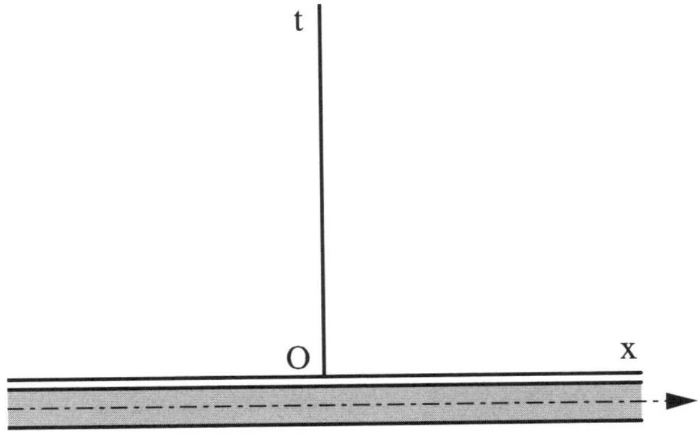

Figure 21.4. *Wave diagram for solving unsteady problems*

21.3.6. *The expansion wave*

21.3.6.1. *General properties*

We consider a cylindrical tube containing a fluid at rest $(u_1 = 0)$, where the pressure is p_1 and the velocity of sound a_1. A more immediate thermodynamic datum, directly accessible from the measurement, is the temperature T_1. For a calorically perfect gas, the sound velocity can be deduced from T_1 with the formula $a_1 = \sqrt{\gamma r T_1}$.

A piston placed at the abscissa $x = 0$ is set in motion to the left of the tube at time $t = 0$ (i.e. it moves to the negative side of the abscissa, see Figure 21.4). This displacement of the piston causes the increase in the volume occupied by gas. The piston is moved *progressively*, meaning that its velocity V increases starting from a value of zero at $t = 0$ and then increases steadily. In other words, we say that initially we create at $x = 0$ a change of the derivative of the velocity; V is continuous but its derivative dV/dt (in this case an acceleration) undergoes a discontinuity from zero to a non-zero value. It is a *second-order disturbance* affecting the derivative(s), as opposed to a disturbance of the first order which affects the quantities themselves.

The trajectory of the piston in the plane (x,t) is a curve $X(t)$ called the *piston trajectory*. The tangent at the origin to this curve is vertical because we start from a value $V = 0$ at $t = 0$ [we keep in mind that in the wave diagram (x,t), the slopes of the curves dt/dx are the inverse of the velocity].

We seek to describe the "events" that occur in the *right-hand side* of the tube after the start of the piston. For this, we use the wave diagram of Figure 21.5, where we have represented the piston motion law (or trajectory) $X(t)$.

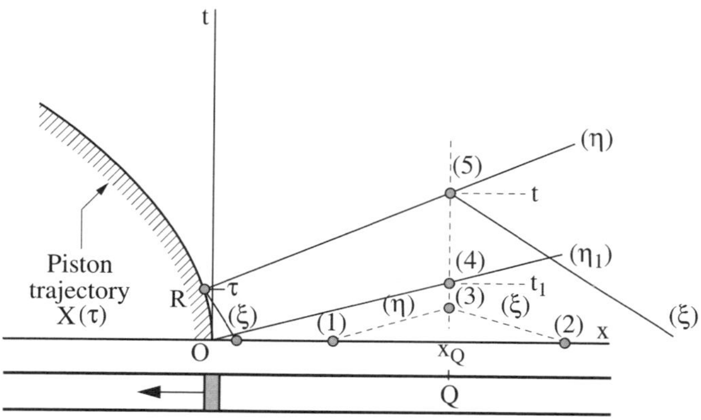

Figure 21.5. *Expansion wave*

We consider a point Q of the tube located at the abscissa x_Q and examine the situation in Q at a small but finite time $t > 0$. In the wave diagram, Q is represented by a point labeled (3) with the coordinates x_Q and *t*. To determine the situation in Q, that is, to know the state of the fluid and its velocity at time *t*, we trace the characteristics (η) and (ξ) going through the point (x_Q, t). These characteristics are straight with the slopes $(u+a)$ and $(u-a)$, respectively. If *t* is small enough, the curves (η) and (ξ) intersect the axis $t = 0$, corresponding to the initial state in the tube at the points (1) and (2) for which:

$$u_{(1)} = u_{(2)} = 0$$

and:

$$a_{(1)} = a_{(2)} = a_1.$$

The characteristic relations give:

$$P_{(3)} + u_{(3)} = P_{(1)} = P_1 \quad on \quad (\eta)$$

$$P_{(3)} - u_{(3)} = P_{(2)} = P_1 \quad on \quad (\xi).$$

It follows that:

$$P_{(3)} = P_1$$

$$u_{(3)} = 0.$$

We recall that P is a function of the pressure p and the entropy s, hence, also of a and p. Knowing P, and the entropy which is a given constant, we deduce p and a. We find here:

$$p_{(3)} = p_1 \quad and \quad a_{(3)} = a_1.$$

Conclusion: the conditions in Q at the considered time t are identical to the initial conditions in the tube before the start of the piston: the point Q has not yet experienced the movement of the piston. This situation will persist until the characteristic (η) going through (3) intersects the axis $t=0$ to the right of the origin O, that is, the side of positive x. The conditions in Q will start to change from the moment t_1 where (η) comes from O (point (4) of the wave diagram). We denote by (η_1) this characteristic whose slope is:

$$\frac{dx}{dt} = a_1.$$

We consider a time t after t_1. It corresponds to the point (5) in the wave diagram. The conditions in (5) are determined by considering (η) and (ξ) passing through this point. However, the wave (η) with the slope $(u+a)$ intersects the trajectory of the piston at a time τ prior to t. At the moment τ, the piston velocity is $V(\tau)$ while, due to the contact conditions, the velocity of the gas on the piston surface is $u = V(\tau)$. The corresponding sound velocity being $a(\tau)$, the function P has the value $P(\tau)$. In contrast, the wave (ξ) with the slope $(u-a)$ always meets the axis $t=0$. Thus, we have the relations:

$$P_{(5)} + u_{(5)} = P(\tau) + u(\tau) = P(\tau) + V(\tau)$$

$$P_{(5)} - u_{(5)} = P_1$$

leading to:

$$2P_{(5)} = P(\tau) + P_1 + V(\tau)$$

and:

$$2u_{(5)} = P(\tau) + V(\tau) - P_1.$$

These relations are true for any point (5) located on the wave (η) encountering the trajectory of the piston at the point R corresponding to the time τ. It follows that P and u *are constant along the* (η), since the right-hand side of the above relations is invariable: it depends only on initial conditions and the situation on the trajectory of the piston at the time τ.

Conclusions

At a point Q of the tube, the acceleration of the piston is only experienced after the leading wave (η_1) with the slope a_1, emitted from $x = 0$ and at the time $t = 0$ of the acceleration of the piston, arrives in Q. The passage of the wave (η_1) causes an acceleration of the fluid located in Q, therefore, induces a flow in the tube.

The "events" happening in the tube are caused by propagation phenomena with the velocities $(u + a)$ and $(u - a)$ according to the waves (η) and (ξ). In particular, the departure of the piston propagates to the right side of the tube with the velocity a_1 (u being zero here); hence, the name of sound velocity given to the quantity $a = \sqrt{(\partial p / \partial \rho)_s}$.

The flow is the same on any wave (η) on which u and a are constant. As the slope of (η) is $(u + a)$, the waves (η) are *straight lines*. Thus, the waves (η) meeting the trajectory of the piston are straight lines. In particular, on any wave (η), the velocity u is equal to the velocity $V(\tau)$ the piston had at the time τ, where (η) intersects the trajectory of the piston.

Such a flow whose properties are invariant along one family of characteristics is called a simple wave. We can easily convince ourselves of the following general property (see section 17.5.1):

Any flow adjacent to a uniform state is a simple wave.

COMMENT.– In everyday language, we often call a wave the characteristic that is related to the propagation of a punctual pressure disturbance as well as the extended phenomenon caused by the continuous succession of a set of elementary disturbances.

The wave (ξ) going through the point *R* on the piston (see Figure 21.5) ends on the axis $(t=0)$ at a point where $u=0$ and $P=P_1$. We have on this wave (ξ):

$$V(\tau)-P(\tau)=0-P_1$$

or:

$$V(\tau)=P(\tau)-P_1.$$

We have shown above:

$$2u_{(5)}=P(\tau)+V(\tau)-P_1.$$

Hence, after simple manipulations, the results:

$$u_{(5)}=V(\tau)$$

$$P_{(5)}=P_1+V(\tau).$$

As defined, the function $P(p)$ increases with the pressure *p*, which means that *P* and *p* vary in the same sense. In the problem studied here, the velocity $V(\tau)$, thus also $u_{(5)}$, is *negative* because the motion of the piston is in the direction of decreasing *x*. It follows that:

$$\boxed{P_{(5)}<P_1}$$

leading to a pressure in (5) lower than p_1. As during its motion (as shown in Figure 21.5) the velocity of the piston $V(\tau)$ continuously decreases (increases in absolute value), the pressure *p* decreases as a result on the waves (η) generated by the acceleration of the piston. All these waves form an *expansion wave*, the characteristic (η_1) being called the front wave.

If we consider the point *Q*, the pressure *p* will remain there constant until the moment t_1 of the front wave (η_1) arrival. Then, as time elapses, the pressure decreases at the same time as the fluid starts moving to *the left* with a velocity increasing correlatively with the decrease of the pressure.

As the slope of the wave (η) is:

$$\frac{dx}{dt} = u + a$$

we see that $\frac{dx}{dt}$ decreases as u (which is negative) increases in absolute value and that a decreases with the pressure. Therefore, the slope $\frac{dt}{dx}$ of (η) in the plane (x,t) increases:

The expansion waves (η) form a family of divergent straight lines.

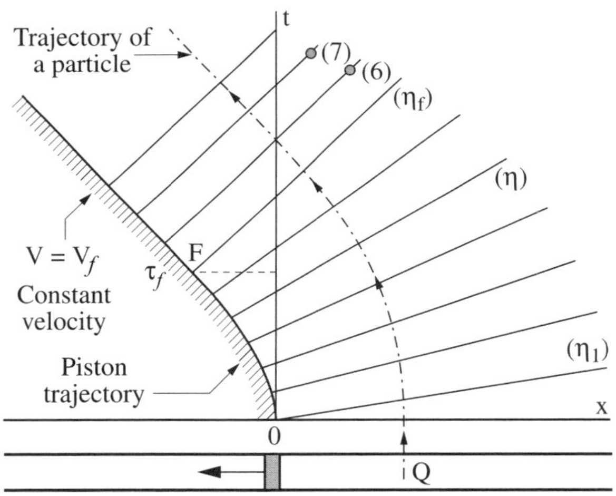

Figure 21.6. *Expansion wave caused by the acceleration of the piston to the constant velocity* V_f

We assume that at a time $\tau = \tau_f$ the acceleration phase of the piston ends, its velocity now remaining constant; and thus, $V_f = V(\tau_f)$, as shown in Figure 21.6. We denote by (η_f) the wave (η) emitted from the trajectory of the piston at time τ_f, and by taking into account Q, we consider the points (6) and (7) of the wave diagram located "above" of (η_f), that is, corresponding to the time after the passage of the wave (η_f) at Q. The waves (η) going through (6) and (7) meet the trajectory of the piston in the region where it travels at constant speed.

Consequently, the velocities in (6) and (7) (and therefore, the pressures) are the same. Thus, after the passage of the final wave (η_f), the point Q experiences a uniform flow with the velocity $u = V_f$ and constant pressure p_f given by:

$$P(p_f) = P_1 + V_1.$$

This property is true for the entire region of the plane (x,t) located to the left of the wave (η_f) and limited by the trajectory of the piston. The change with time of the pressure at point Q located at the abscissa x in the tube is drawn in Figure 21.7 (p is plotted horizontally and t vertically like in the wave diagram). Until $t = t_1$, the pressure p is constant and equal to p_1. From t_1 onward, the pressure continuously decreases until the value p_f is reached at $t = t_f$. This change of $p(t)$ would be the signal delivered by a pressure transducer placed in Q. Such a curve is called the *expansion wave signature*. A similar diagram can be traced for the velocity u in Q.

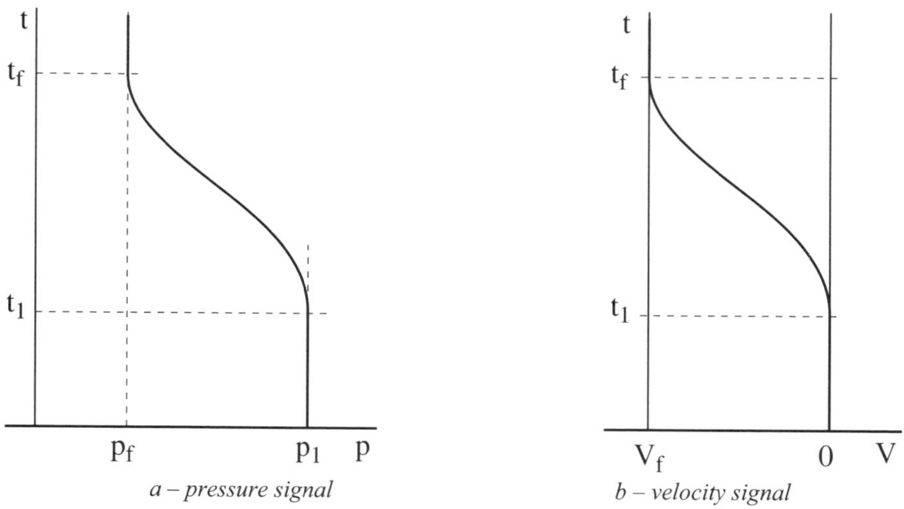

Figure 21.7. *Signature of an expansion wave*

The signal of a pressure transducer placed at a greater distance x_2 from the origin will be delayed, compared to that of the first, by the time used by the front wave to travel the distance $x_2 - x_1$. In addition, as the expansion wave spreads with time, the response curve of the sensor will expand a while longer before showing the final level p_f (see Figure 21.8).

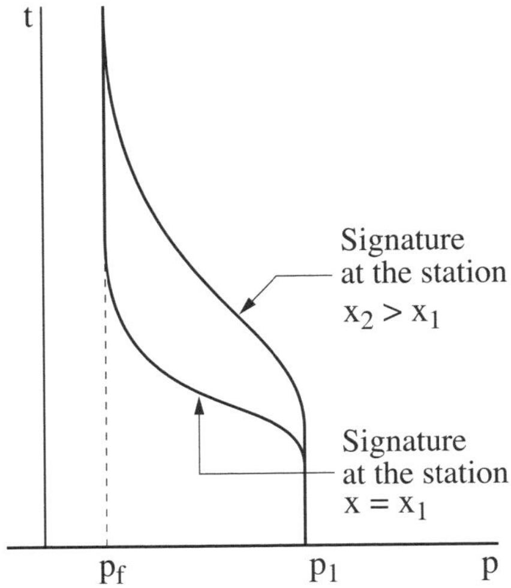

Figure 21.8. *Progressive spreading of an expansion wave*

Figure 21.9a shows a calculation by the method of characteristics (see Chapter 24) of the flow generated in a cylindrical tube by the progressive acceleration of a piston to the left until a constant velocity of –200 m/s. We observe the following: the divergent expansion waves (η) generated by the acceleration of the piston, the state of rest remaining undisturbed until the arrival of the front wave of the expansion, and the uniform state with the velocity –200 m/s after the passage of the last expansion wave. The phenomena are made more clear by tracing the isobaric contours (see Figure 21.9b). What happens after the arrival of the expansion waves on the fixed right wall has not been calculated.

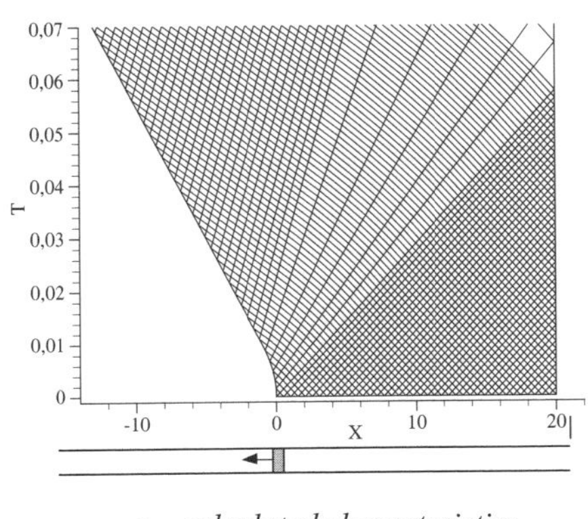

a – calculated characteristics

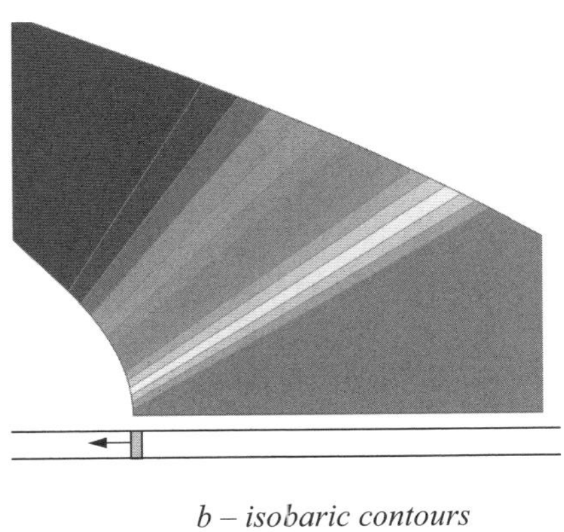

b – isobaric contours

Figure 21.9. *Flow generated in a cylindrical tube by the progressive acceleration of a piston up to the constant velocity of –200 m/s (air $\gamma = 1.4$)*

Conclusions

> The leftward movement of the piston causes an acceleration of the fluid in the tube. This effect arises through a *propagation* mechanism of disturbances, or waves, generated by the acceleration of the piston. It causes an expansion wave, limited by initial (η_1) and final (η_f) characteristics. The wave gradually spreads in such a way that it occupies the larger a time interval $\Delta t = t_f - t_1$, the greater the distance x to the initial position of the piston is.

Intuitively, the gas initially contained in the tube tends to fill the volume created by the displacement of the piston; hence, its acceleration and pressure decrease, as the volume to be occupied is greater than the one at the start.

When studying unsteady flows, it is useful to trace the trajectory of a *fluid element, or fluid particle*, in the wave diagram (x,t). In the framework of one-dimensional approximation, it is rather the fluid contained in a section of the tube and which will move collectively. The trajectory of the fluid element initially located at Q is drawn in Figure 21.6. Until the time t_1, the element is immobile, as its trajectory is a straight segment parallel to the time axis. The arrival of the expansion wave in Q causes its acceleration to the left in the direction of piston displacement. Its trajectory is then a curve passing through the expansion fan. After the passage of the final wave (η_f), the trajectory is a straight line parallel to the trajectory of the piston that became straight from $t = \tau_f$.

Mathematically, the trajectory of a fluid particle is defined as the integral curve of the differential equation:

$$\frac{dx}{dt} = u$$

i.e.:

$$x = x_1 + \int_{t_1}^{t} u dt.$$

This is the locus of positions in the plane (x,t) of the fluid particle located at the abscissa $x = x_1$ at the time $t = t_1$. (Lagrangian representation of the fluid motion, see Chapter 8.)

21.3.6.2. *Case of the calorically perfect gas: limit velocity*

If the gas is calorically perfect, the characteristic relations are written as:

$$\frac{2}{\gamma-1}a+u=constant \quad on\ a\ wave\ (\eta)$$

$$\frac{2}{\gamma-1}a-u=constant \quad on\ a\ wave\ (\xi)$$

which leads to the relations:

$$\frac{2}{\gamma-1}a_{(5)}+u_{(5)}=\frac{2}{\gamma-1}a(\tau)+u(\tau)$$

and:

$$\frac{2}{\gamma-1}a_{(5)}-u_{(5)}=\frac{2}{\gamma-1}a_1.$$

We deduce:

$$\frac{2}{\gamma-1}a=\frac{2}{\gamma-1}a_1+V(\tau)$$

dropping the index (5), which has no special meaning.

As $V(\tau)$ is negative, the sound velocity a in the wave is lower than a_1 and decreases more when $V(\tau)$ is higher. This behavior is correlated with the decrease of the pressure p as, in an isentropic flow, if p decreases, the temperature T decreases too, therefore in the same way as the velocity of sound.

A physical limit is reached when $V(\tau)$ is such that $a=0$. Then the pressure and temperature of the gas will also be zero. If $V(\tau)$ exceeds this limit, the gas *can no longer follow the piston*. The corresponding gas velocity is called *limit velocity* $U_{\lim}$; its (absolute) value is:

$$\boxed{U_{\lim}=\frac{2}{\gamma-1}a_1.}$$

The existence of a limit velocity for the gas is explained due to the fact that its kinetic energy cannot grow indefinitely, its total energy remaining bounded. The maximum value of $|u|$ is determined by the initial energy level of the fluid. It should be noted that in a steady state, the limit value of the velocity is given by (see section 10.2):

$$V_{\lim} = \sqrt{\frac{2}{\gamma - 1}}\, a_1.$$

As $\left(\frac{2}{\gamma-1}\right) > 1$, $U_{\lim} > V_{\lim}$. In the case of air $(\gamma = 1.4)$, the ratio $U_{\lim}/V_{\lim}$ is equal to $\sqrt{\frac{2}{(1.4-1)}} = \sqrt{5} = 2.236$ so that, when starting from an initial ambient condition for which $a_1 = 340\,\text{m/s}$, the limit velocity is $U_{\lim} \approx 1700\,\text{m/s}$.

The limit velocity concept is partly fictional, since the gas will become liquid below a certain temperature so that it will not be possible to apply perfect gas thermodynamics. Nevertheless, the value $U_{\lim}$ thus calculated gives a fairly accurate idea of the maximum velocity that can be achieved.

21.3.6.3. *The centered expansion wave*

We analyze the case where the piston is accelerated *suddenly* to the left, which means that at the initial time $t = 0$, its velocity passes abruptly from $V = 0$ to $V = V_f$ (supposed constant). The solution to this problem can be obtained by considering a passage to the limit of the previous case where the portion of the trajectory *OF*, corresponding to the gradual acceleration of the piston, tends to zero (see Figure 21.6). The set of spread out waves $(\eta_1), (\eta_f)$ then tends toward a fan of expansion waves centered in *O*, all the constituent waves (η) which have *O* as a common origin (see Figure 21.10). The results obtained above can be applied directly to this situation and will not be repeated.

COMMENT.– We can ask ourselves about the physical significance of the solution at the origin *O* of a centered expansion where an infinite number of states of the fluid are concentrated. All these states, previously distributed on a finite segment of the characteristic (ξ) crossing the expansion wave, are now coincident in *O*, where there is discontinuous change from the state (1) to the state (f). In reality, this question does not really arise, as a sudden acceleration of the piston, which would imply infinite acceleration, is mechanically not possible. In a real device, the acceleration of the piston can be done very quickly, but always for a finite time (a few milliseconds) which brings us back to the previous problem. However, if we consider a certain distance from the origin *O*, either in time or in space, it is as if the fan was actually centered in *O*. The "microstructure" of the phenomenon in the immediate vicinity of *O* is indistinguishable. The same applies to the centered expansion of a steady two-dimensional supersonic flow in the space (x, y) (see section 17.5.2), as the discontinuous change of the direction of a physical wall does not exist. At a certain microscale, the wall has a shape where there are irregularities causing that the expansion there is gradual. Moreover, in this case, the presence of a boundary layer complicates the phenomenon.

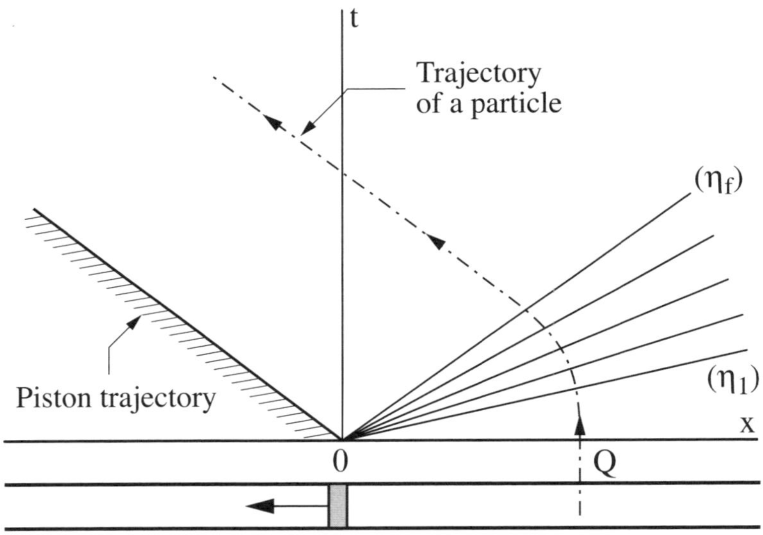

Figure 21.10. *Sudden acceleration of the piston. Centered expansion wave*

We would examine in a similar manner the case where, when the piston moves to *the right* (either with a gradual or discontinuous motion law), an expansion wave forms propagating along the left side of the tube. We reach, of course, the same conclusions, *mutatis mutandis*, the waves (ξ) playing the role of the waves (η) and vice versa.

Figure 21.11 shows the calculation of the expansion caused by the sudden acceleration of a piston to the velocity of –200 m/s.

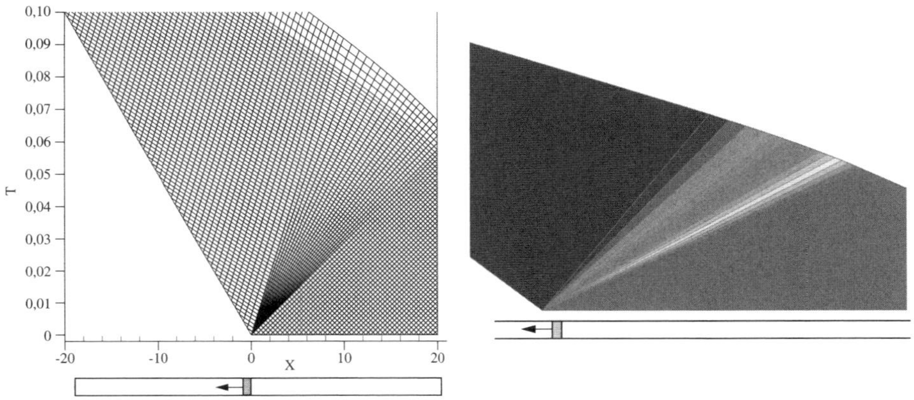

a – calculated characteristics

b – isobaric contours

Figure 21.11. *Flow generated in a cylindrical tube by the sudden acceleration of a piston up to the constant velocity of –200 m/s (air $\gamma = 1.4$)*

21.4. The compression wave

We analyze what happens in the right side of the tube when the piston is accelerated to *the right* along increasing *x*, the situation being shown in Figure 21.12.

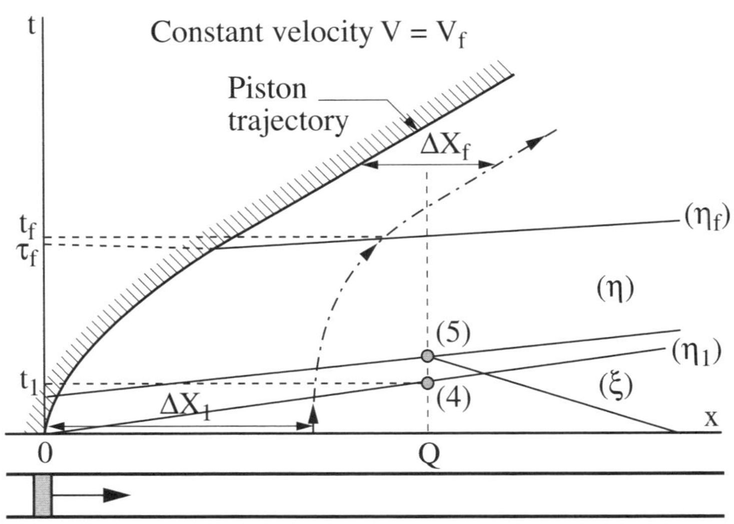

Figure 21.12. *Progressive compression wave*

Initially, the tube is filled with a gas at rest (velocity $u_1 = 0$) with the pressure p_1, the velocity of sound being a_1. The piston is gradually accelerated, its velocity V starting from $V = 0$ at the time point $t = 0$ to grow by becoming positive. For a reasoning identical to that for the expansion wave, we show that at a point Q of the tube, located on the abscissa x_Q, the motion of the piston will only be felt from the moment t_1 when the front wave (η_1) reaches the position Q. We know that (η_1) has a slope $dx/dt = a_1$, i.e. $dt/dx = 1/a_1$ in the wave diagram.

We look at a later time $t > t_1$ and determine the situation in Q, corresponding to point (5) in the wave diagram (x,t). The conditions in (5) are obtained by considering (ξ) intersecting the axis $t = 0$ and (η) meeting the trajectory of the piston at the time $\tau < t$, leading to the relations:

$$P_{(5)} + u_{(5)} = P(\tau) + u(\tau) = P(\tau) + V(\tau) \quad \textit{on the wave } (\eta)$$

$$P_{(5)} - u_{(5)} = P_1 \qquad \textit{on the wave } (\xi)$$

which leads to:

$$2P_{(5)} = P(\tau) + P_1 + V(\tau)$$

$$2u_{(5)} = P(\tau) + V(\tau) - P_1.$$

We draw the same conclusions as for an expansion wave, namely:

– The flow is constant on all the waves (η).

– The waves (η) are straight lines.

However, the following results are different. Considering a characteristic (ξ) joining the trajectory of the piston at the time point τ to the initial rest area, we can demonstrate (see section 21.3.6.1):

$$u_{(5)} = V(\tau)$$

$$P_{(5)} = P_1 + V(\tau).$$

As $V(\tau)$ is positive here, it follows that $P_{(5)} > P_1$. Thus, the pressure in (5) is greater than the initial pressure p_1. Since, by assumption, the velocity $V(\tau)$ increases with time, therefore the pressure in the section Q will grow. Thus, as the pressure is increasing more and more on the successive waves (η), the waves (η) form a *compression wave*. If we assume that $V(\tau)$ increases to a value V_f, reached at the time τ_f and then kept constant, then p increases to $p = p_f$, the value reached at the time point t_f in Q. Then, the pressure and the velocity of the gas remain constant, u taking the value $u = V_f$. The piston pushes the gas and transfers its velocity V_f after an acceleration phase perceived *via* a propagation process.

The slope of the compression waves (η) has the values:

$$\frac{dx}{dt} = V(\tau) + a.$$

Now, as $V(\tau)$ is here an increasing function of τ and as the velocity of sound increases with pressure, dx/dt increases from one (η) to another. In the plane (x,t) of the wave diagram, the successive (η) of the compression wave has slope dt/dx that decreases: the waves (η) are straight *converging* lines.

If we plot the signature of the wave $p(t)$ for two abscissas x_1 and x_2 so that $x_2 > x_1$, the signature at x_2 will be more "condensed", or shorter, as at x_1 (see Figure 21.13). In an expansion wave, the elementary waves propagate slower and slower, and consequently, dx/dt decreases. On the contrary, in a compression wave, the elementary waves propagate faster and faster so that dx/dt increases, one wave racing after the other.

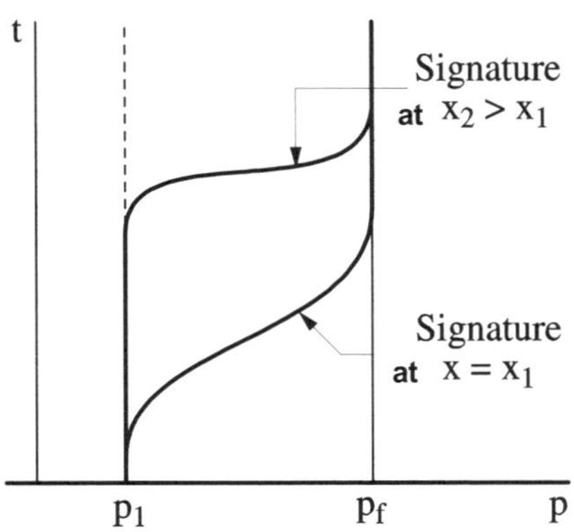

Figure 21.13. *Signature of a compression wave*

We have drawn in the wave diagram the trajectory of a fluid particle located in the section Q at the initial moment. As the gas mass contained between this particle and the piston is constant at any time, it follows that:

$$\rho_1 A \Delta x_1 = \rho_f A \Delta x_f$$

where Δx_1 and Δx_f designate the position of the particle relative to the piston and A the (constant) section of the tube. Since in this case $\rho_f > \rho_1$, it follows that $\Delta x_f < \Delta x_1$. The fluid particle is closer to the piston.

21.5. Intersection of compression waves: formation of a shock wave

As the waves (η) are converging straight lines, there will be a region of the plane (x,t) where these waves will cross. This will happen in a particular section F (see Figure 21.14). The leading wave (η_1) will thus pass "before" the final wave (η_f). In other words, in the wave diagram (η_1) will be "above" (η_f). If we consider a point (5) of this overlapment region, then (5) belongs to:

– the region located "below" (η_1), that is, not yet reached by the compression wave and where the velocity is zero and the pressure equal to p_1;

– the region affected by the compression wave when an (η) meeting the trajectory of the piston passes through (5); and

– the region following the passage of the wave.

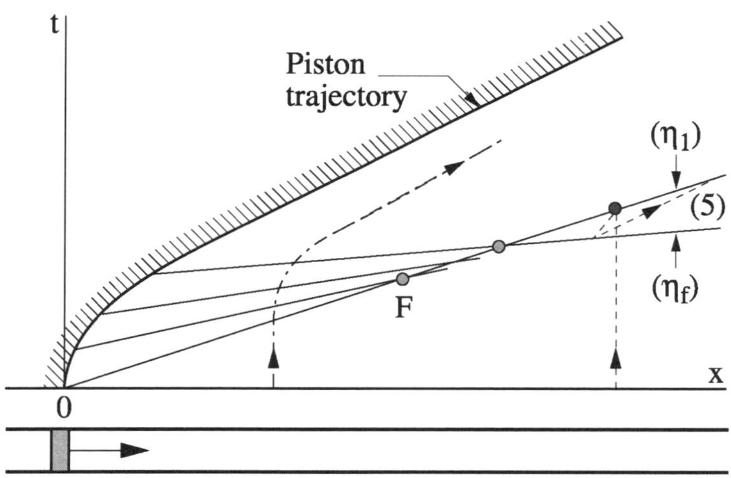

Figure 21.14. *Intersection of characteristics in a compression wave*

There are *three states* in (5): one uniform, another having felt the wave induced by the acceleration of the piston, and the third corresponding to the uniform state after passage of the terminal wave (η_f). Such a situation is excluded for obvious physical reasons. From the section F where the compression waves (η) begin to intersect, the solution consisting of waves causing a progressive compression is no longer acceptable. We have to seek a different kind of solution, which will include a *shock wave*.

Furthermore, we can see that the trajectory of a particle belonging to this region will contain two switchback points: the particle, initially motionless, moves first toward the origin O after the arrival of the leading wave (η_1), and then away from O after the passage of (η_f). Such a movement is unrealistic from a physical point of view. Finally, in the region of the intersection of the characteristics (η), we consider the signature of the wave at three positions located at increasing distances from the origin O with the abscissas x_1, x_2, and x_3, the latter being beyond F (see Figure 21.15). As we have seen, when the distance x increases, the signature becomes shorter and shorter. At the position x_3, the signal will be inverted with, for the same time t, three values for the pressure: p_1, p_f, and an intermediate value $p_1 > p > p_f$.

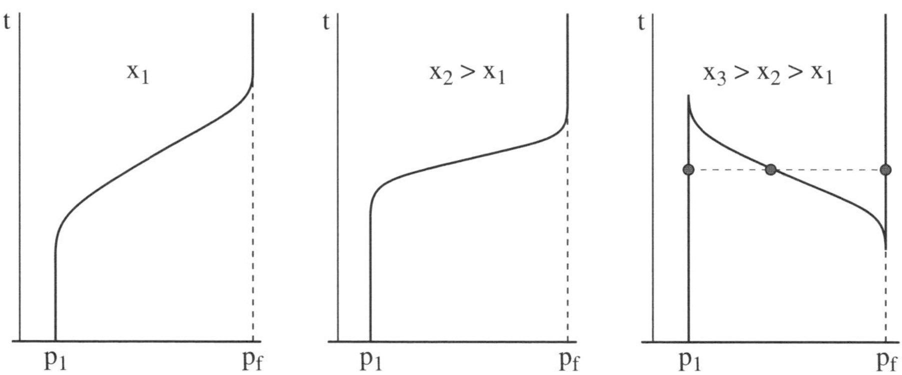

Figure 21.15. *Change of the pressure signature in a compression wave*

Figure 21.16 shows the calculation of the flow caused by the acceleration of the piston to the right until it reaches a velocity of 150 m/s. The movement induces a compression through waves (η) that eventually intersect. The further calculation after the first meeting of waves needs to take account of the formation of so-called focalization shock waves.

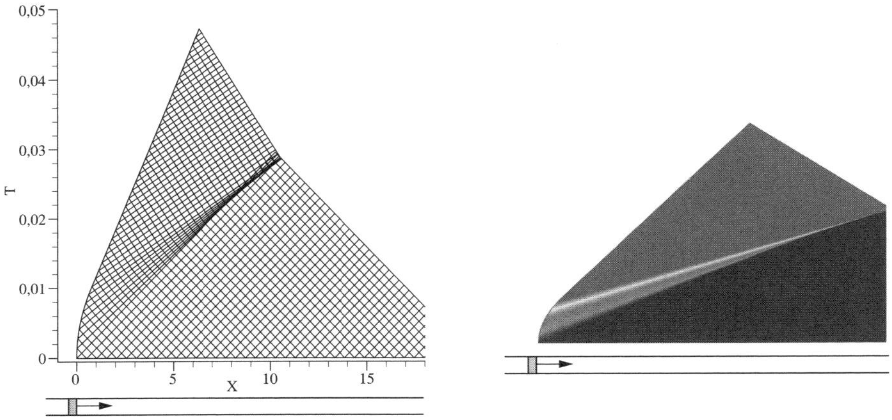

a – calculated characteristics *b – isobaric contours*

Figure 21.16. *Progressive acceleration of a piston to the right until the velocity 150 m/s. Formation of a shock through focalization of pressure waves (air $\gamma = 1.4$)*

We encounter the same difficulty in the case where the piston is accelerated *suddenly* to the right. Without the need to repeat calculations similar to those already made, we find that the solution consists of a fan centered in O whose final wave (η_f) has in the wave diagram a lower slope than the first wave (η_1), leading to a region in the plane (x,t) where the solution is not unique, which is ruled out. The situation is shown in Figure 21.17.

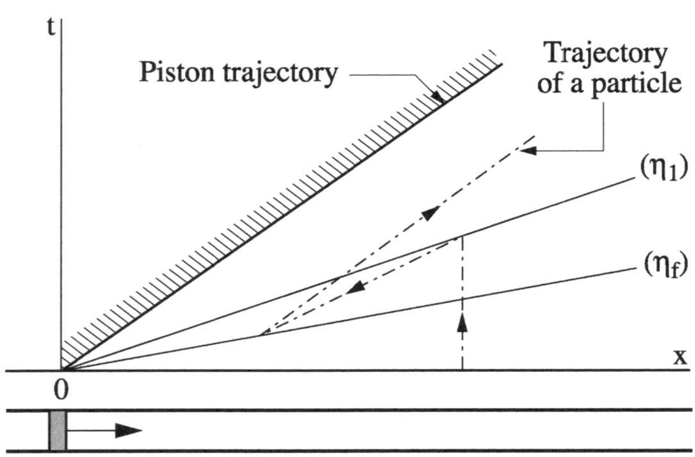

Figure 21.17. *Sudden acceleration of a piston in the direction of a compression*

Conclusions

The compression waves (η) form a family of converging straight lines.

The rightward movement of the piston causes an acceleration of the fluid in the tube. This effect arises through a *propagation* mechanism of disturbances, or waves, generated by the acceleration of the piston. The result is a compression wave, limited by the initial (η_1) and final (η_f) characteristics. The compression gradually shrinks in such a way that it occupies the shorter a time interval $\Delta t = t_f - t_1$, the greater the distance x to the initial position of the piston is.

In the case of an intersection of compression waves (η), or a sudden acceleration of the piston in a direction tending to compress the gas, a progressive wave-like solution is physically not real anymore. To solve the problem, we have to introduce *discontinuities* or *shock waves*.

21.6. Distortion of a periodic wave and nonlinear acoustics

The fact that a compression wave propagates faster than an expansion wave is the cause of the progressive distortion of an initially sinusoidal pressure fluctuation. As shown in Figure 21.18a, we assume that we create in the origin O a sinusoidal variation of the pressure p around the value p_1. Considering x in the tube where the pressure maxima (or overpressures) propagate faster than the minima (or depressions), the observed wave will be distorted as shown in Figure 21.18b. As the distortion amplifies as a function of the distance, the situation shown in Figure 21.18c can be reached where the overpressures pass before the depressions. Thus, there is in the period of the signal an interval to which, at a given time t, correspond three pressure values, which is unacceptable.

In reality, as shown schematically in Figure 21.18d, the fluctuation will eventually include shock wave-like discontinuities. Such phenomena occur in *nonlinear acoustics*, that is, in circumstances where the pressure fluctuations are so strong that they affect the properties of the environment in which they propagate. This is the case, for example, of the strong instabilities occurring in combustion chambers of rocket engines.

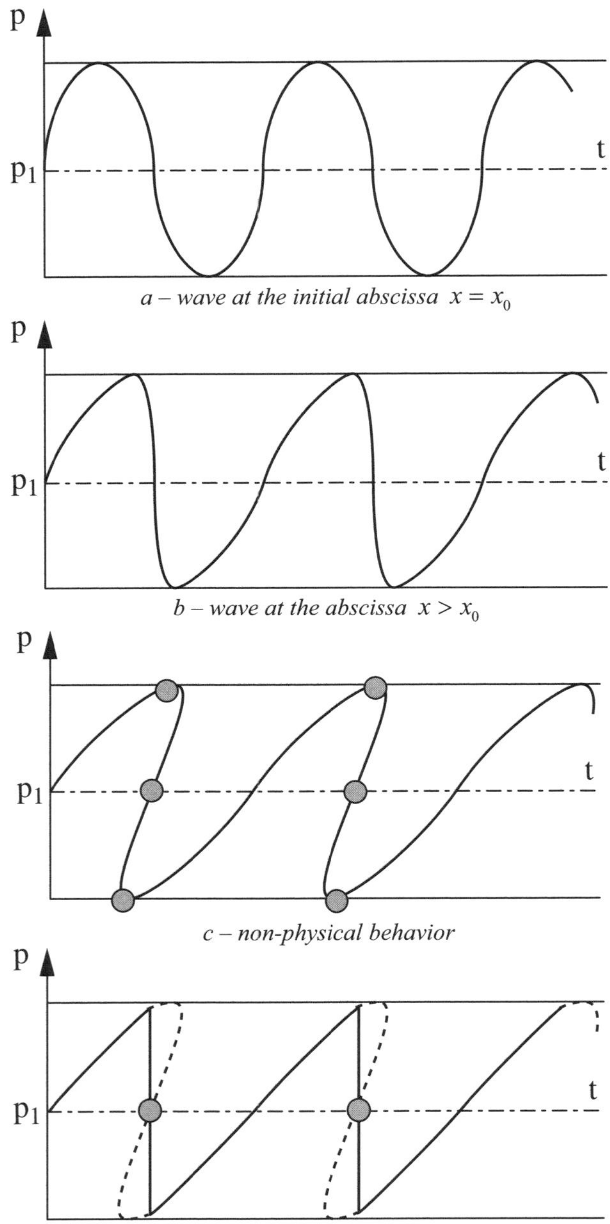

a – wave at the initial abscissa $x = x_0$

b – wave at the abscissa $x > x_0$

c – non-physical behavior

d – formation of shock waves in the periodic signal

Figure 21.18. *Distortion of a sinusoidal pressure wave*

21.7. Reflection–refraction of a wave on a contact surface: acoustic impedance

We show in section 22.3 that two flows (1) and (2) moving in a tube can be separated by a contact surface (Σ) on each side of which the properties of the gases are different. The compatibility conditions require equality of velocities and pressures on both sides of (Σ), i.e.:

$$u_1 = u_2 \qquad \textit{and} \qquad p_1 = p_2.$$

Consider a disturbance, or wave, with a small amplitude propagating in the medium (1) along a characteristic (η) (see Figure 21.19). This wave causes the transition of the environment (1) from a state (1) to a state (3) and we have between the states (1) and (3) connected by a wave (ξ) (Riemann's relations):

$$\frac{1}{\rho a}\delta p - \delta u = 0 \;\rightarrow\; \frac{1}{\rho_1 a_1}(p_3 - p_1) - (u_3 - u_1) = 0 \;\rightarrow\; \frac{1}{\rho_1 a_1}\delta p_{31} - \delta u_{31} = 0.$$

It is understood that changes in pressure and velocity are weak (it may be an acoustic disturbance). As the state (3) no longer is compatible with the prevailing condition in (2), it is necessary that disturbances occur in order to create in each medium compatible states that have the same pressure and same velocity.

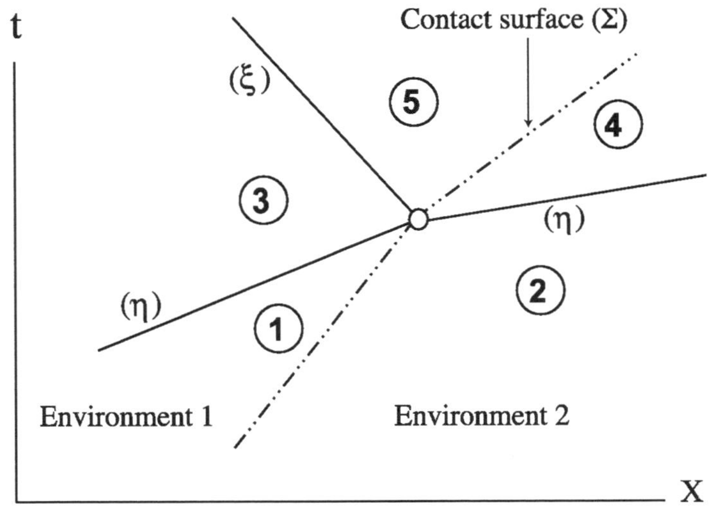

Figure 21.19. *Reflection-refraction of a wave on a contact surface*

We therefore introduce a reflected wave in medium 1 causing a change from the state (3) to a state (5), and therefore:

$$\frac{1}{\rho_1 a_1}(p_5 - p_3) + (u_5 - u_3) = 0.$$

Simultaneously, a wave (η) is transmitted into medium 2 causing the transition from the state (2) to a state (4) so that:

$$\frac{1}{\rho_2 a_2}(p_4 - p_2) - (u_4 - u_2) = 0 \quad \rightarrow \frac{1}{\rho_2 a_2}\delta p_{42} - \delta u_{42} = 0.$$

The two states resulting from the impact of the wave are compatible if:

$$p_5 = p_4 \qquad \textit{and} \qquad u_5 = u_4.$$

We deduce:

$$\frac{1}{\rho_1 a_1}(p_5 - p_3) + (u_5 - u_3) = \frac{1}{\rho_1 a_1}(p_4 - p_3) + (u_4 - u_3).$$

But also:

$$p_5 = p_4,\, p_2 = p_1 \quad \textit{and} \quad u_5 = u_4,\, u_2 = u_1:$$

$$p_4 - p_3 = p_4 - p_2 - (p_3 - p_1) = \delta p_{42} - \delta p_{31} \quad \textit{and} \quad u_4 - u_3 = \delta u_{42} - \delta u_{31}.$$

It follows that:

$$\frac{1}{\rho_1 a_1}(\delta p_{42} - \delta p_{31}) + (\delta u_{42} - \delta u_{31}) = 0.$$

By introducing the *acoustic impedances*:

$$Z_1 = \rho_1 a_1 \qquad \textit{and} \qquad Z_2 = \rho_2 a_2$$

we can write:

$$\delta p_{31} = Z_1 \delta u_{31}$$

$$\delta p_{42} = Z_2 \delta u_{42}$$

$$\delta p_{42} - \delta p_{31} = Z_1 (\delta u_{31} - \delta u_{42}).$$

This leads to the relations:

$$\boxed{\frac{\delta u_{42}}{\delta u_{31}} = \frac{2Z_1}{Z_2 + Z_1} \quad and \quad \frac{\delta p_{42}}{\delta p_{31}} = \frac{2Z_2}{Z_2 + Z_1}.}$$

Conclusions

– If $Z_2 = Z_1$, then $\delta p_{42} = \delta p_{31}$ and $\delta u_{42} = \delta u_{31}$; there is no need for a reflected wave in medium 1. Both environments are said to be acoustically adapted.
– If $Z_1 > Z_2$, then $\delta p_{42} < \delta p_{31}$; the transmitted wave is less intense than the incident wave. In this case: $\delta p_{53} = \delta p_{42} - \delta p_{31} < 0 \rightarrow \frac{\delta p_{53}}{\delta p_{31}} = \frac{\delta p_{42}}{\delta p_{31}} - 1 < 0$; the wave reflected in medium 1 is of opposite nature to the incident wave.
– If $Z_1 < Z_2$, then $\delta p_{42} > \delta p_{31}$; the transmitted wave is more intense than the incident wave: $\delta p_{53} = \delta p_{42} - \delta p_{31} > 0 \rightarrow \frac{\delta p_{53}}{\delta p_{31}} = \frac{\delta p_{42}}{\delta p_{31}} - 1 > 0$; the wave reflected in medium 1 has the same nature as the incident wave.

21.8. Back on the boundary conditions

We consider a cylindrical tube, open at both ends, with the fluid entering from the left end (L) and exiting through the right end (R). First, we look at the expression of the solution near the inlet section (L). The solution is calculated by advancing in time, for example, by a numerical method. As the flow is known everywhere in the tube at time *t*, a new state at time point $t + \Delta t$ is determined using the Riemann relations. As the calculated point (3) is close to the section (L), two situations are possible.

First case

This situation corresponds to the diagram in Figure 21.20a. The slope to be calculated of the right running characteristic (ξ) going through the point (3) on the border (L) is given by:

$$(\xi) \quad \rightarrow \quad \frac{dt}{dx} = \frac{1}{u - a}.$$

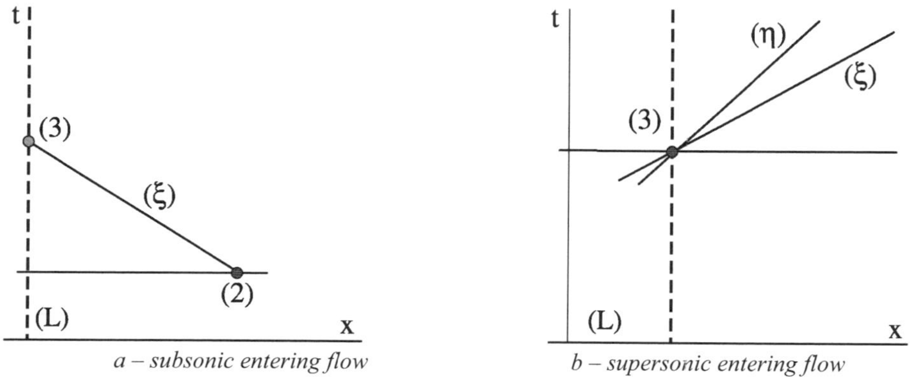

Figure 21.20. *Boundary conditions in an entry section of the fluid*

If the velocity u is assumed as positive and is such that $u < a$ (incoming subsonic flow), then this slope is negative and the (ξ) going through (3) starts from a point (2) inside the tube. We have the following relation between (2) and (3):

$$\frac{2}{\gamma-1}a_3 - u_3 = \frac{2}{\gamma-1}a_2 - u_2.$$

If the pressure p_L is imposed on (L), then a_3 takes the value corresponding to $p_L = p_3$. In this situation, the solution in (3) involves a condition imposed on (L) as well as the conditions inside the tube through the intermediate of the point (2). In other words, we can only impose one condition on (L), here, the pressure. The velocity u_3 of the gas entering the tube results from the calculation.

Second case

In the plane $[x,t]$, the characteristics (η) and (ξ) have the slopes:

$$(\eta) \quad \rightarrow \quad \frac{dt}{dx} = \frac{1}{u+a}$$

$$(\xi) \quad \rightarrow \quad \frac{dt}{dx} = \frac{1}{u-a}.$$

We assume that the gas enters the tube $(u > 0)$. Then the slope of (η) is always positive and finite. In contrast, the slope of (ξ) changes sign if $u > a$, that is, if the gas entering the tube is *supersonic*. We will consider the situation of Figure 21.20b: (ξ) can come no longer from a point inside the tube, which would allow calculating

u_3. In this case, we must know both the pressure and velocity in the inlet section since no information can be transmitted from inside of the tube.

Now we look at the conditions at the exit section.

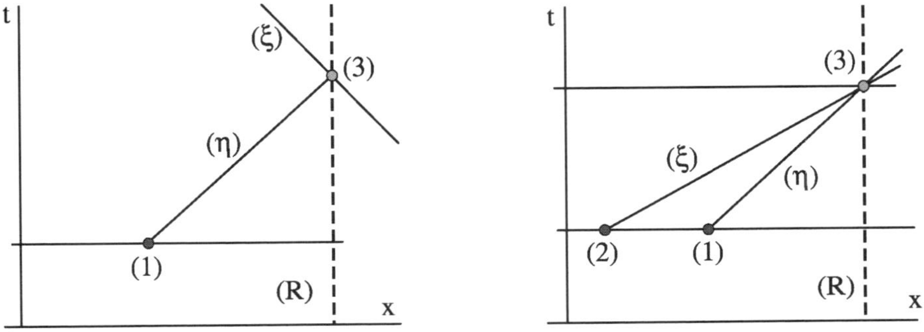

a – subsonic outgoing flow *b – supersonic outgoing flow*

Figure 21.21. *Boundary conditions in an exit section of the fluid*

First case

If the velocity is subsonic in the section (R), then the slope of (η) is positive and that of (ξ) negative, the corresponding diagram being shown in Figure 21.21a. The calculation of the flow in (3) involves the point (1) located in the tube, (η) going through (1) allowing to calculate a condition in (3), for example, the velocity u_3. We have to impose the missing information, namely the pressure $p_3 = p_R$ in (3).

Second case

If the velocity in section (R) is supersonic (see Figure 21.21b), then for the same reasons as above, the slope of (ξ) is positive and the point (2) is inside of the tube. The calculation of the conditions in (3) is done by using the two characteristic relations applied between the points (1) and (3), on one side, and (2) and (3) on the second, i.e.:

$$\text{on } (\eta) \quad \frac{2}{\gamma-1}a_3 + u_3 = \frac{2}{\gamma-1}a_1 + u_1$$

$$\text{on } (\xi) \quad \frac{2}{\gamma-1}a_3 - u_3 = \frac{2}{\gamma-1}a_2 - u_2.$$

The conditions in the exit section (L) are derived entirely from conditions inside the tube: no conditions have to be imposed on the exit section. The flow in the tube is independent of conditions outside of the tube.

Table 21.1 below summarizes these results.

Entry section (flow enters through the left or right section)	
Subsonic velocity: $u < a$	One condition to impose
Supersonic velocity: $u > a$	Two conditions to impose
Exit section (the flow exits through the left or right section)	
Subsonic velocity: $u < a$	One condition to impose
Supersonic velocity: $u > a$	No condition to impose

Table 21.1. *Expression of the boundary conditions*

The considerations we have developed in the simple one-dimensional case are generalized for unsteady flows with two and three dimensions. They play a key role in the expression of boundary conditions during the numerical solution of Euler and Navier–Stokes equations by numerical time marching methods.

Chapter 22

Unsteady Shock Wave, Contact Surface, and Wave Reflections

22.1. The shock wave equations and relations

We examine the problem of a shock wave (Ω) propagating parallel to the velocity $\vec{V}_1$ of the upstream fluid in a fixed reference frame (A) (see Figure 22.1). We denote with D the propagation velocity of the shock wave *in relation to the upstream fluid* counted positive upstream: D is called the shock wave *celerity*.

The absolute velocity of the wave in the reference frame (A) is:

$$V_1 + D$$

(since the flow is one-dimensional, we omit the vector sign above V_1). The velocity V_1 undergoes a discontinuity across (Ω) to take a value V_2 and we write:

$$V_2 = V_1 + w$$

The quantity $w = V_2 - V_1$ represents the *velocity difference* experienced by the fluid across the shock (Ω). To make the shock wave *steady*, we consider a reference frame (R) moving with the velocity of the shock; i.e. $V_1 + D$ (see Figure 22.2). In (R), the upstream flow (1) encounters the shock wave with a velocity:

$$V_1 - (V_1 + D) = -D$$

and becomes a state (2) with the velocity in (R):

$$V_2 - (V_1 + D) = w - D$$

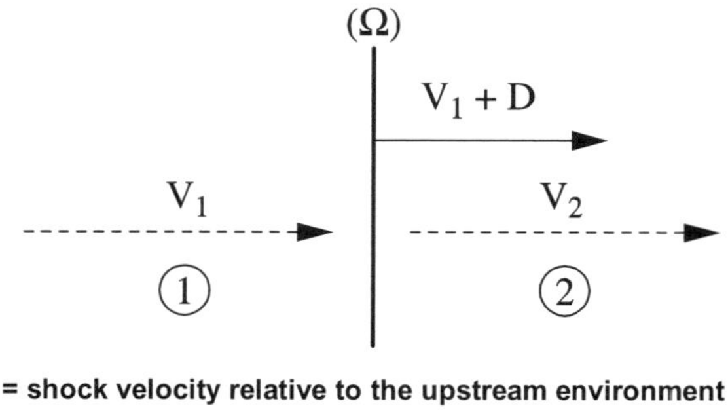

Figure 22.1. *Shock wave in the absolute reference frame*

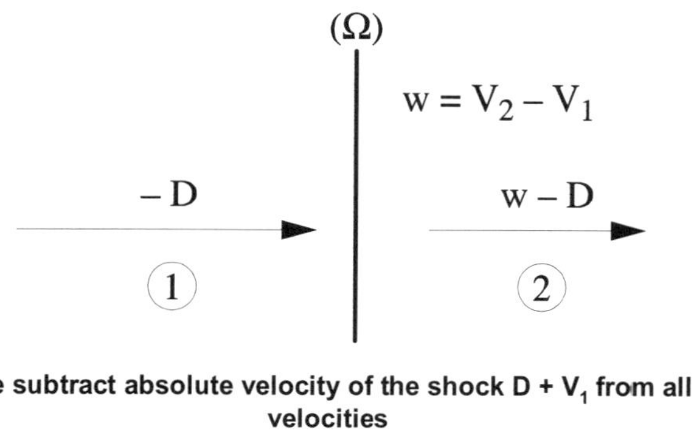

Figure 22.2. *Shock wave in the reference frame linked to the shock*

By considering the reference frame (R), we can apply the equations of the steady normal shock, i.e. (see Chapter 13):

– *continuity equation:*

$$-\rho_1 D = \rho_2 (w - D)$$

or:

$$\rho_1 D = \rho_2 (D - w).$$

– *equation of motion:*

$$p_1 + \rho_1 D^2 = p_2 + \rho_2 (D - w)^2 .$$

But, given the continuity equation, it becomes:

$$p_1 + \rho_1 D^2 = p_2 + \rho_1 D (D - w)$$

i.e.:

$$p_2 - p_1 = \rho_1 D w.$$

– *energy equation:*

$$h_1 + \frac{D^2}{2} = h_2 + \frac{1}{2}(D - w)^2 .$$

We introduce the unsteady shock Mach number M_s by stating:

$$\boxed{M_s = \frac{D}{a_1}}$$

where a_1 is the sound velocity in the *upstream medium.*

We recapitulate the equations of the unsteady shock wave:

$$\boxed{\begin{array}{l} \textit{continuity:}\quad \rho_1 D = \rho_2 (D - w) \\ \textit{motion:}\quad p_2 - p_1 = \rho_1 D w \\ \textit{energy:}\quad h_1 + \dfrac{D^2}{2} = h_2 + \dfrac{1}{2}(D - w)^2 . \end{array}}$$

If the upstream state (1) is known and if the velocity D of the shock wave is fixed, the unknown quantities in the above equations are:

$$w,\ \rho_2,\ p_2\ \textit{and}\ h_2 .$$

The system must be supplemented by the thermodynamic relation:

$$h = h(p, \rho) \quad or \quad \rho = \rho(p, h)$$

which implies that the gas is divariant, i.e. in equilibrium. The relation for calculating the enthalpy h as a function of p and ρ can be either explicit, for a calorically perfect gas, or provided in the form of a chart or tables.

The general method of solving unsteady shock equations is similar to that used for normal steady shocks (see section 13.5). Taking as parameter $\varepsilon = \rho_1 / \rho_2$, which can be fixed at the start of the calculation, the continuity equation gives w, and thus the motion equation p_2 and that of energy h_2. Hence, we get a new value of ρ_2 from the thermodynamic relation as well as the calculation of a new ε.

The velocity jump across the shock is obtained by:

$$w = D(1-\varepsilon).$$

We note that if the upstream state is given and if a quantity downstream of the shock is known, then we can infer the velocity D of the shock and therefore all other downstream quantities.

In the case of a calorically perfect gas, we can transpose the explicit relations obtained for the steady normal shock by substituting the shock Mach number M_s with M_1, which gives us:

$$\boxed{\frac{p_2}{p_1} = 1 + \frac{2\gamma}{\gamma+1}\left(M_s^2 - 1\right) = \frac{2}{\gamma+1}\left(\gamma\ M_s^2 - \frac{\gamma-1}{2}\right)}$$

$$\frac{\rho_1}{\rho_2} = \frac{\gamma-1}{\gamma+1}\left[1 + \frac{2}{(\gamma-1)M_s^2}\right] = \frac{2+(\gamma-1)M_s^2}{(\gamma+1)M_s^2}$$

i.e.:

$$\boxed{\frac{\rho_2}{\rho_1} = \frac{(\gamma+1)M_s^2}{2+(\gamma-1)M_s^2} = \left(\frac{\gamma+1}{2}\right)\frac{M_s^2}{1+\frac{\gamma-1}{2}M_s^2}}$$

$$\boxed{\frac{T_2}{T_1} = \left[1 + \frac{2\gamma}{\gamma+1}\left(M_s^2 - 1\right)\right]\left[\frac{2+(\gamma-1)M_s^2}{(\gamma+1)M_s^2}\right].}$$

The jump of the velocity w is calculated from the continuity equation, i.e.:

$$w = D\left(1-\frac{\rho_1}{\rho_2}\right)$$

leading to:

$$\frac{w}{a_1}=\frac{D}{a_1}\left\{1-\frac{\gamma-1}{\gamma+1}\left[1+\frac{2}{(\gamma-1)M_s^2}\right]\right\}$$

or:

$$\frac{w}{a_1}=M_s\left\{1-\frac{\gamma-1}{\gamma+1}\left[1+\frac{2}{(\gamma-1)M_s^2}\right]\right\}.$$

After elementary calculations, we have:

$$\boxed{\frac{w}{a_1}=\frac{2}{(\gamma+1)}\left(M_s-\frac{1}{M_s}\right).}$$

If the medium (1) is at rest $(u_1=0)$, the Mach number M_2 in the flow after the shock is given by:

$$M_2=\frac{u_2}{a_2}=\frac{w}{a_1}\frac{a_1}{a_2}=\frac{w}{a_1}\sqrt{\frac{T_1}{T_2}}.$$

Hence, by substituting the already known expressions:

$$M_2=\frac{2}{(\gamma+1)}\left(M_s-\frac{1}{M_s}\right)\left[1+\frac{2\gamma}{\gamma+1}\left(M_s^2-1\right)\right]^{-\frac{1}{2}}\left[\frac{(\gamma+1)M_s^2}{2+(\gamma-1)M_s^2}\right]^{-\frac{1}{2}}$$

and after some calculations:

$$\boxed{M_2=\left(M_s^2-1\right)\left[\left(1+\frac{\gamma-1}{2}M_s^2\right)\left(\gamma\ M_s^2-\frac{\gamma-1}{2}\right)\right]^{-\frac{1}{2}}.}$$

One should not confuse the Mach number corresponding to the absolute velocity u_2 and the Mach number downstream of the shock associated with the downstream velocity *relative to the shock*. This second Mach number has to be subsonic (see the Prandtl relation in section 13.5.2.1).

For applications, it may be interesting to use the ratio of pressures as a parameter:

$$\Pi = \frac{p_2}{p_1}.$$

Thus, we find successively:

$$M_s = \sqrt{\frac{1}{2\gamma}\left[(\gamma+1)\Pi+(\gamma-1)\right]}$$

$$\frac{w}{a_1} = \sqrt{\frac{2}{\gamma}}\frac{\Pi-1}{\sqrt{(\gamma+1)\Pi+(\gamma-1)}}$$

$$\frac{\rho_1}{\rho_2} = \frac{(\gamma-1)\Pi+(\gamma+1)}{(\gamma+1)\Pi+(\gamma-1)}$$

$$\frac{T_2}{T_1} = \frac{p_2}{p_1}\frac{\rho_1}{\rho_2} = \Pi\ \frac{(\gamma-1)\Pi+(\gamma+1)}{(\gamma+1)\Pi+(\gamma-1)}.$$

We calculate the entropy jump across the shock. For a calorically perfect gas, the specific entropy s is given by:

$$s = C_p \ln\ T - r\ln\ p.$$

Hence, using the equation of state $\frac{p}{\rho} = rT$, the expression (for calculation details, see section 13.5.2.3):

$$\frac{s_2-s_1}{C_v} = \ln\ \left[\frac{p_2}{p_1}\left(\frac{\rho_1}{\rho_2}\right)^{\gamma}\right].$$

Thus:

$$\boxed{\frac{s_2 - s_1}{C_v} = \ln \left\{ \Pi \left[\frac{(\gamma-1)\Pi + (\gamma+1)}{(\gamma+1)\Pi + (\gamma-1)} \right]^{\gamma} \right\}.}$$

If the shock becomes very intense, then $\Pi \to \infty$ and we have the following limit behaviors:

$$M_s \approx \sqrt{\frac{\gamma+1}{2\gamma}\Pi} \quad \to \quad \infty$$

$$\frac{w}{a_1} \approx \sqrt{\frac{2}{\gamma(\gamma+1)}}\,\Pi \quad \to \quad \infty$$

$$\frac{\rho_2}{\rho_1} \approx \frac{\gamma+1}{\gamma-1} \quad \to \quad \textit{finite limit}$$

$$\frac{T_2}{T_1} \approx \frac{\gamma-1}{\gamma+1}\Pi \quad \to \quad \infty$$

$$M_2 = \frac{w}{a_2} \approx \sqrt{\frac{2}{\gamma(\gamma-1)}} \to \textit{finite limit.}$$

22.2. Formation of a shock wave

We will reconsider the problem of the sudden acceleration of the piston *to the right of the tube*; i.e. to the side occupied by the gas (see Figure 22.3). We have seen that in this case a progressive wave-type solution led to physical impossibilities. The problem of building an acceptable solution will be resolved by considering that the sudden acceleration of the piston at the time $t = 0$ generates a shock wave that, if the velocity V_p of the piston is constant, propagates through the right side of the tube with a constant velocity D_1. When passing through a section Q of the abscissa x, the shock wave causes an instantaneous acceleration of the gas whose velocity changes from $u_1 = 0$ to $u_2 = V_p$.

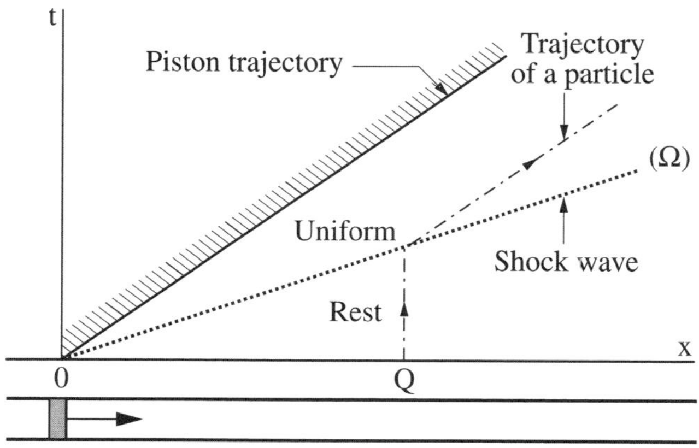

Figure 22.3. *Formation of a shock wave by sudden acceleration of a piston*

The condition (2) downstream of the shock is here a uniform state (velocity, pressure, density, etc., independent of x and t), thus satisfying the motion equations. The velocity u_2 is equal to that of the piston, and therefore satisfying the contact condition on the piston. It is a solution admissible to the stated problem, the condition (2) having to be derived from (1) in such a way as to satisfy the unsteady shock equations, knowing that downstream of the shock we must have $u_2 = V_p$.

The difference in the velocity w across the shock is a factor important to the problem, since:

$$w = V_p$$

(the velocity of upstream state (1) is assumed to be zero).

The Mach number M_s obeys the relation:

$$\frac{w}{a_1} = \frac{2}{(\gamma+1)}\left(M_s - \frac{1}{M_s}\right).$$

This leads to the second-order algebraic equation for M_s:

$$\boxed{M_s^2 - \frac{\gamma+1}{2}\left(\frac{V_P}{a_1}\right)M_s - 1 = 0}$$

whose roots are:

$$M_s = \left(\frac{\gamma+1}{4}\right)\left(\frac{V_p}{a_1}\right) \pm \frac{1}{2}\sqrt{\left(\frac{\gamma+1}{2}\right)^2\left(\frac{V_P}{a_1}\right)^2 + 4}..$$

According to the shock theory (see Chapter 13), we know that the root to consider must be such that $|M_s| > 1$ to satisfy the second law of thermodynamics. As the product of the roots is equal to 1, one of them is necessarily higher than the unit. However, the root to retain can be either positive or negative, depending on whether *D* is positive (shock propagating to the right) or negative (shock propagating to the left). In the problem that we are considering here, M_s is positive. Hence, knowing M_s, we deduce:

– the shock velocity: $D_1 = a_1 M_s$

– the pressure ratio: $\Pi = \frac{p_2}{p_1} = 1 + \frac{2\gamma}{\gamma+1}\left(M_s^2 - 1\right)$

– thus: $\frac{a_2}{a_1} = \sqrt{\frac{T_2}{T_1}} = \sqrt{\Pi \frac{(\gamma-1)\Pi + (\gamma+1)}{(\gamma+1)\Pi + (\gamma-1)}}$

– by definition: $u_2 = V_p$

– hence, the Mach number of flow behind the shock:

$$M_2 = \frac{u_2}{a_2} = \frac{V_P}{a_2}$$

etc.

Conclusions

The sudden acceleration of the piston generates a shock wave propagating in the gas at a velocity exceeding that of the piston. The passage of the shock in a section *x* of the tube causes an instantaneous acceleration of the fluid in the travel direction of the shock at a velocity equal to that of the piston. The passage of the shock causes a compression and heating of the gas at the same time as an increase in its entropy.

As the Mach number of the shock is greater than 1, it follows that a shock wave always propagates faster than an acoustic disturbance whose velocity is equal to the velocity of sound.

22.3. Contact surface

The shock wave is a surface across which all the properties of the fluid, with the exception of its stagnation enthalpy, undergo discontinuities. The shock wave moves with a velocity different from that of the fluid, thus it is crossed by a mass flow.

We may ask whether there are not other surfaces through which certain properties of the fluid would suffer discontinuities. To examine this problem, and therefore for a study of the shock, we consider a reference frame linked to the discontinuity surface, thus moving at the velocity D of the latter. The motion equations impose the following relations:

– *continuity equation:*

$$\rho_1 D = \rho_2 (D - w)$$

– *equation of motion:*

$$p_1 + \rho_1 D^2 = p_2 + \rho_2 (D - w)^2$$

which we write as:

$$p_1 + (\rho_1 D) D = p_2 + \rho_2 (D - w)(D - w).$$

– *energy equation:*

$$(\rho_1 D)\left(h_1 + \frac{D^2}{2}\right) = \rho_2 (D - w)\left[h_2 + \frac{(D - w)^2}{2}\right].$$

In the equations of motion and energy, we have restored the mass flow that occurs in the balance relations. In the case of the shock wave theory, we divided these equations by this mass flow *implicitly assumed that this mass flow was not zero*.

However, the continuity equation can be satisfied by taking $D = 0$ and $w = 0$, i.e. considering a situation where the mass flow through the discontinuity is zero. Knowing that D is the velocity of the discontinuity surface (Σ) in relation to the upstream medium, $D = 0$ means that (Σ) moves at the velocity V_1 of the upstream fluid. And since $w = 0$, the velocity V_2 is equal to V_1. The considered discontinuity surface *follows the fluid in its movement.*

If $D = w = 0$, it follows that the densities ρ_1 and ρ_2 may differ. Similarly, the energy equation shows that the specific enthalpies h_1 and h_2 may be different. However, the motion equation requires:

$$\boxed{p_1 = p_2.}$$

Conclusion

For a non-viscous fluid, there may be a discontinuity surface moving at the velocity of the fluid on both sides of which the fluid properties are discontinuous, *with the exception of the pressure*. Such a discontinuity is called *contact surface*, and it separates two different gases moving inside the tube.

A contact surface is of the same type as the slip line introduced in section 22.5.4.

22.4. Reflection of a wave

We have already addressed the problem of the piston accelerated either to right or left, which gives rise to the formation of a compression or expansion wave. We have also analyzed the sudden acceleration of the piston which generates either a shock wave or a centered expansion. Now we will consider other basic problems that can be solved simply through this theory.

We analyze the following problem: the tube, of finite length, has *its right end* (E) *open*, which puts it in contact with the atmosphere where the pressure p_a is constant. The left end is closed by a piston initially at rest. The initial state of the gas in the tube is characterized by $p_1 = p_a$, a_1, the value of the velocity of sound corresponding to p_a, and $u_1 = 0$ (see wave diagram in Figure 22.4).

At the time $t = 0$, the piston is suddenly accelerated *to the left*. We know that a centered expansion wave propagates to the right of the tube, associated with the acceleration of the gas which follows the piston. This expansion wave is formed by wave (η) with the slope $dx/dt = u + a$. The states (1) and (2) on both sides of the wave are linked by the relation:

$$\frac{2}{\gamma-1}a_2 - u_2 = \frac{2}{\gamma-1}a_1$$

and thus:

$$a_2 = a_1 + \frac{\gamma-1}{2}u_2.$$

Hence, $a_2 < a_1$ since $u_2 < 0$. However, at the exit (*E*) of the tube, the pressure is imposed to be equal to p_a: therefore, the velocity of sound must be equal to a_1. There is an incompatibility between the state (2) and the conditions imposed in (E). Consequently, it is necessary that waves form to establish a state (3) where $a_3 = a_1$, so that $p_3 = p_a$. These waves are the waves (ξ) moving left, as shown in Figure 22.4. The states (3) and (2) are linked by the relation:

$$\frac{2}{\gamma-1}a_3 + u_3 = \frac{2}{\gamma-1}a_2 + u_2$$

leading to:

$$u_3 = u_2 + \frac{2}{\gamma-1}(a_2 - a_3) = u_2 + \frac{2}{\gamma-1}(a_2 - a_1).$$

As $a_2 < a_1$, $u_3 < u_2$, the gas velocity algebraically decreases toward the left: it is negative and its absolute value increases. This fact is represented by the plot of the trajectory of a particle fluid. Such a movement corresponds to the effect of the external atmosphere whose pressure p_a is greater than p_2 and which tends to push the gas inside the tube. The wave (ξ) is a *compression wave*, since $p_3 > p_2$.

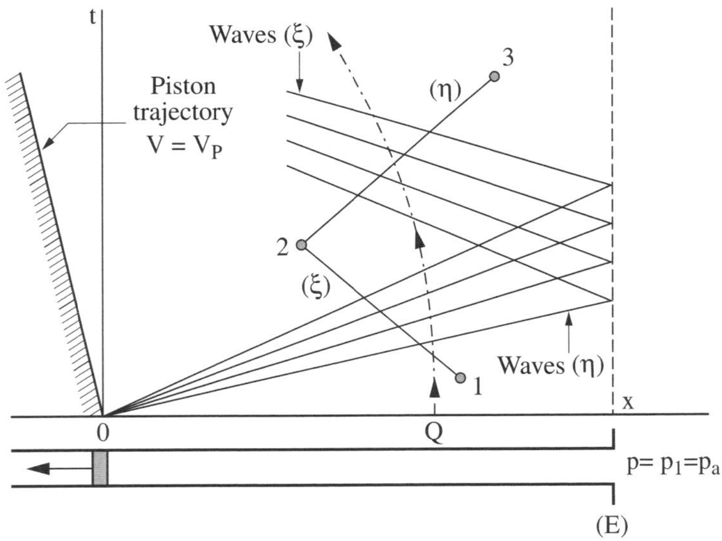

Figure 22.4. *Reflection of a wave on an open constant pressure end*

Conclusions

An expansion wave is reflected by an open extremity, where the pressure is imposed, as a compression wave.

We could show in the same way that a compression wave is reflected as an expansion wave.

We look at the case where the right end (E) of the tube is *closed by a fixed bottom* (see Figure 22.5). As we know, the wave generated by the acceleration of the piston to the left causes (since $u_2 < 0$) an acceleration of the gas to the left. Therefore, there is incompatibility between the state (2) and the condition of zero velocity $u = 0$ to satisfy in (E). It is necessary that a wave forms leading to a state (3) so that $u_3 = 0$. Between states (2) and (3), we have:

$$\frac{2}{\gamma - 1} a_2 + u_2 = \frac{2}{\gamma - 1} a_3 + u_3 = \frac{2}{\gamma - 1} a_3 .$$

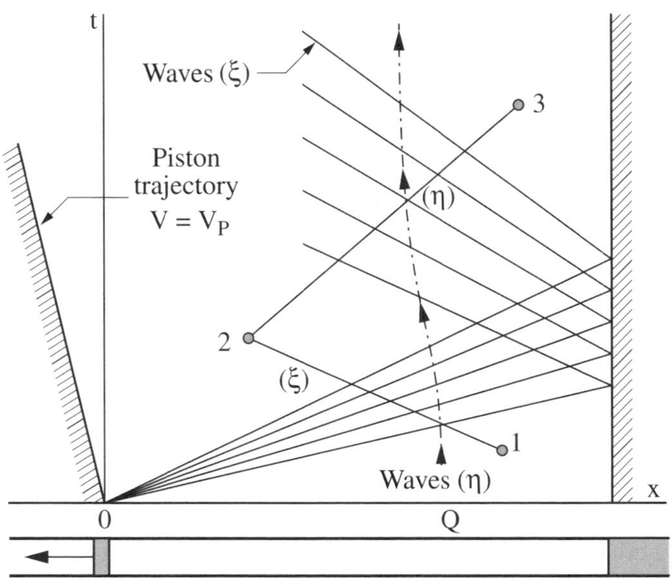

Figure 22.5. *Reflection of a wave from a closed fixed end*

Thus:

$$a_3 = a_2 + \frac{\gamma - 1}{2} u_2.$$

As $u_2 < 0$, it follows that $a_3 < a_2$.

Conclusions

An expansion wave is reflected from a fixed wall as an expansion wave.

A compression wave is reflected from a fixed wall as a compression wave.

22.5. Multiple reflections

We reconsider the problem of the tube open at its right end (E). We assume that the piston, after a sudden acceleration, moves to the left retaining its initial velocity that we denote by V_P (see Figure 22.6).

We have seen that the velocity in (3), after reflection of the centered wave on the open end of the tube, was such that:

$$u_3 = u_2 + \frac{2}{\gamma - 1}(a_2 - a_1)$$

but as:

$$a_2 - a_1 = \frac{\gamma - 1}{2} u_2$$

it follows that:

$$u_3 = 2u_2 = 2V_P.$$

Thus, after the passage of the wave (ξ) reflected from the end (E), the gas in (3) moves to the left with a velocity equal to twice the velocity of the piston.

Thus, there is incompatibility between the state (3) and the condition $u = V_P$ to satisfy on the piston. A wave has to form in order to produce a state (4) where $u_4 = V_P$. Considering a characteristic (ξ) joining (3) to (4), we have between (3) and (4):

$$\frac{2}{\gamma - 1} a_3 - u_3 = \frac{2}{\gamma - 1} a_4 - u_4 = \frac{2}{\gamma - 1} a_4 - V_P$$

leading to:

$$a_4 = a_3 + \frac{\gamma - 1}{2}\left(V_P - u_3\right).$$

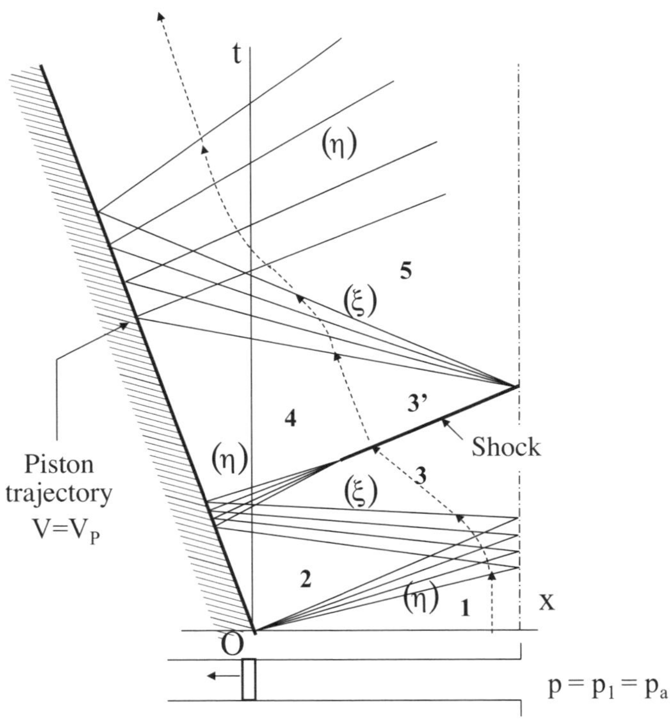

Figure 22.6. *Multiple reflections between a moving piston and an open end*

As u_3 is negative and equal to $2V_P$, it follows that $a_4 > a_3$. The wave causing the transition from (3) to (4) is a compression wave, a result consistent with what we found for the reflection of a wave on a fixed wall. We note that this conclusion would not be true if the velocity began to decrease (increase in absolute value) so that $|V_P|$ become higher than $|u_3|$. We can also imagine that at the impact of the first wave (ξ) returned by the open end on the piston, the piston trajectory is modified to cause a reduction δV_p of V_P equal to δu across this (ξ) and this is continued for all following waves (ξ). The trajectory will be such that there will be progressive cancellation of the wave (ξ). At the end of the process, the piston now has a constant velocity $V_p = u_3$. There will be no return to the right of a compression wave.

Let us leave this particular case and return to our problem. For compression waves forming converging families (see section 21.4), it may be that the waves (η) returned by the piston intersect as the waves (ξ) already converge. We know that beyond the point of intersection the progressive wave solution is no longer acceptable. A shock wave will form in the tube between the state (3) and a state (3'), which is different from (3), since in (3') the entropy is not the same as in (3). Across the shock wave, the pressure jumps from $p_3 = p_a$ to $p_{3'} > p_a$, and therefore is incompatible with the value p_a imposed at the open end of the tube. Upon its arrival in (E), the pressure jump $p_3 \rightarrow p_{3'}$ must be compensated by a local variation $p_{3'} \rightarrow p_a$, which is caused by a centered expansion wave formed by waves (ξ) propagating inside the tube. These waves are reflected by the piston as expansion waves (η), etc. The process can thus, under the assumption of a *non-viscous fluid*, continue indefinitely since there is no mechanism to dissipate the mechanical energy of the fluid as heat. In the real viscous fluid, dissipation will be felt by extinguishing the waves little by little.

We can imagine the same exercise where the end (E) is blocked by a fixed bottom. Thus, if the piston keeps the same motion law $V_P = constant$, only expansions will form in the tube which is normal since the gas then occupies larger and larger volume.

Figure 22.7a shows a calculation by the method of characteristics of the sudden discharge into the atmosphere (pressure $p_1 = 10^5\,\text{Pa}$) of a tube where the pressure is initially $p_2 = 1.5 \times 10^5\,\text{Pa}$. The sudden opening of tube at the right creates a centered expansion wave after whose passage the pressure in the tube is $p_3 = 10^5\,\text{Pa}$. The expansion wave reflects off the blocked left end into an expansion wave, lowering the pressure again. This wave is returned by the right end, where the atmospheric pressure is imposed, as a compression wave which is reflected from the fixed left bottom as a new compression wave. As shown in the figure, the latest pressure waves intersect, which, in turn, generates a shock wave.

Such a phenomenon occurs if a valve sealing a pipeline fails abruptly, connecting with the atmosphere a part of the pipeline sealed with another closed valve. As the initial pressure in the pipeline is well above that of the previous calculation (up to 100 bars), an intense shock wave forms that can cause the rupture of the pipeline, the contained methane igniting on contact with the outside air.

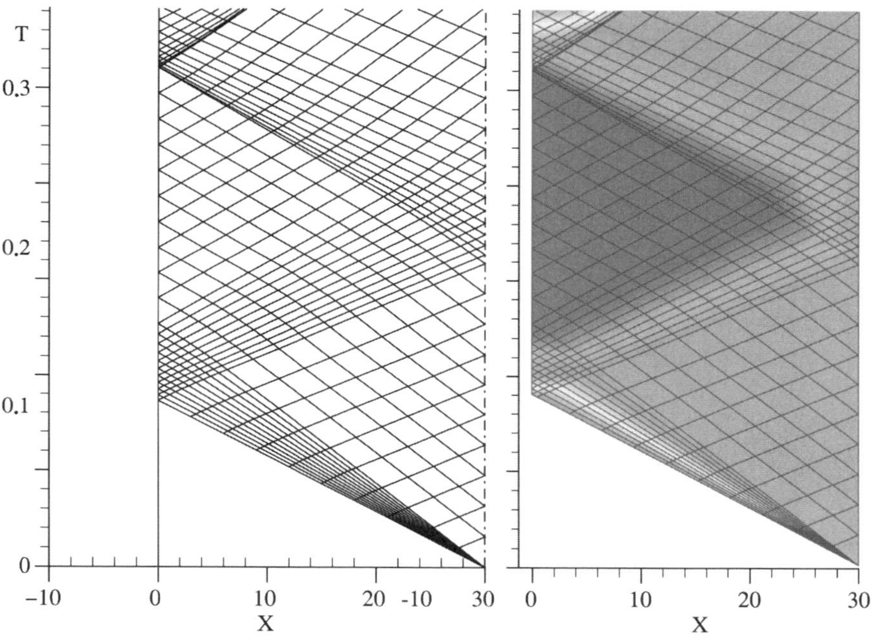

a – calculated characteristics *b – isobaric contours*

Figure 22.7. *Flow caused by the abrupt discharge of a tube into the atmosphere* ($p_1 = 10^5$ *Pa*, $p_2 = 1.5 \times 10^5$ *Pa*)

22.6. Collision of two shock waves

In the problem considered here, the piston moves suddenly to the right, by taking a velocity V_{P_1} at the initial time $t = 0$. We know that this movement generates a shock wave (Ω_1) propagating in the tube with the velocity D_1. The wave (Ω_1) causes the acceleration of the initially at rest gas which takes the velocity V_{P_1} of the piston. Thus, the fluid passes from the state (1) to the state (2).

After a time t_1, the piston undergoes an instantaneous velocity increment to reach a new velocity $V_{P_2} > V_{P_1}$ (see Figure 22.8).

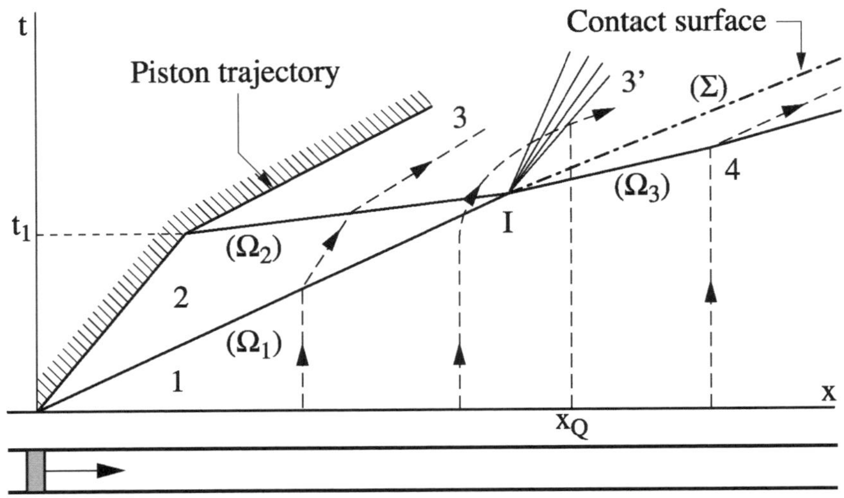

Figure 22.8. *Collision of two shock waves*

As the velocity of the fluid downstream of (Ω_1) is $u_2 = V_{P_1} < V_{P_2}$, it follows that the situation at the moment of the velocity change is comparable to that of the time $t = 0$. From t_1 on a shock wave (Ω_2) forms and propagates in the moving gas (2).

We have the following elements:

– velocity of the shock (Ω_1): D_1,

– velocity difference across (Ω_1): $w_1 = u_2 - u_1 = u_2 = V_{P_1}$,

– Mach number of the shock (Ω_1): $M_{s_1} = \dfrac{D_1}{a_1}$.

Satisfying the equation:

$$M_{s_1}^2 - \frac{\gamma+1}{2}\left(\frac{V_{P_1}}{a_1}\right)M_{s_1} - 1 = 0$$

– velocity of the shock (Ω_2): D_2,

– velocity difference across (Ω_2): $w_2 = u_3 - u_2 = u_3 - w_1$.

This leads to:

$$u_3 = w_1 + w_2.$$

Also:

$$M_{s_2} = \frac{D_2}{a_2}$$

is given by the equation:

$$M_{s_2}^2 - \frac{\gamma+1}{2}\left(\frac{w_2}{a_2}\right)M_{s_2} - 1 = 0$$

or:

$$M_{s_2}^2 - \frac{\gamma+1}{2}\left(\frac{V_{P_2} - V_{P_1}}{a_2}\right)M_{s_2} - 1 = 0.$$

We evaluate the absolute velocities of the waves (Ω_1) and (Ω_2). If $\left(\frac{dx}{dt}\right)_\Omega$ is the velocity of the shock, we have:

$$\left(\frac{dx}{dt}\right)_\Omega = V + D$$

where V is the (absolute) velocity of the medium in which the shock propagates.

In this case:

$$\left(\frac{dx}{dt}\right)_{\Omega_1} = D_1$$

and:

$$\left(\frac{dx}{dt}\right)_{\Omega_2} = u_2 + D_2 = w_1 + D_2$$

or:

$$\left(\frac{dx}{dt}\right)_{\Omega_2} = w_1 + a_2 M_{s_2}.$$

But the shock (Ω_1) has a subsonic velocity relative to the downstream medium (2), this velocity is:

$$u_1 + D_1 - u_2 = D_1 - w_1.$$

Thus:

$$|D_1 - w_1| < a_2 \quad or \quad |w_1 + a_2| > |D_1|.$$

As (Ω_2) is *supersonic* relative to the upstream environment (2):

$$|D_2| > a_2.$$

Thus:

$$|D_2 + w_1| > |a_2 + w_1| > |D_1|.$$

We see that the velocity of the shock wave (Ω_2) *exceeds the velocity of the shock wave* (Ω_1). The shock (Ω_2) will catch up with the shock (Ω_1) in a certain section of the tube *(I)* of abscissa x_I. This result was predictable, since we have seen that compression waves tend to catch up.

The problem is to determine the flow beyond (*I*), i.e. the answer to the question: considering a section of the tube of abscissa x so that $x > x_I$, what will "see" this section as a function of the time?

It is possible to construct a solution with the following form: the collision of the two shocks (Ω_1) and (Ω_2) gives birth to a single shock (Ω_3) which propagates with a velocity D_3 relative to the upstream environment (1). If we consider two sections on both sides of (*I*) where the two shocks meet, one, as $x > x_I$, will see passing a single shock (Ω_3), the other, as $x < x_I$, two successive shocks: (Ω_1) and (Ω_2). The result is that the downstream states (3) and (4) have different properties, because there is no reason for the passage of two successive shocks leading to the same state as the passage of a single shock. The states (3) and (4) are separated by a contact surface (Σ) that allows such a discontinuity. The contact surface (Σ) propagates with a velocity u_Σ equal to the velocity u_4 of the fluid downstream of shock (Ω_3).

The surface (Σ) separates the states (4) of a state (3') where the gas has been crossed by the two shocks, i.e. (Ω_1) and (Ω_2). The conditions with respect to both sides of the contact surface require:

$$u_3^{'} = u_4 = u_\Sigma$$

$$p_3^{'} = p_4.$$

In general, these conditions are not met by the states (3) and (4), hence the necessity of the state (3'), the transition from (3) to (3') being done through a wave leaving from (*I*) and propagating in the fluid that has undergone a succession of shocks $(\Omega_1)+(\Omega_2)$. Most often, this is an expansion wave centered at (*I*).

The problem has no simple explicit solution; to solve, it we must proceed by iterations. For example, by assuming a value of u_4, we can infer the Mach number M_{s_3} of the shock wave (Ω_3), and therefore the properties of the gas downstream of (Ω_3), namely, the pressure p_4. Moreover, the state (3) is known since it follows from the calculation of the shocks (Ω_1) and (Ω_2). Knowing that, we must have:

$$u_3^{'} = u_4 = u_\Sigma$$

by considering a characteristic (ξ) linking the states (3) and (3'). Then:

$$\frac{2}{\gamma-1}a_3^{'} - u_3^{'} = \frac{2}{\gamma-1}a_3 - u_3 = \frac{2}{\gamma-1}a_3^{'} - u_4$$

which allows us to calculate $a_3^{'}$, and hence the pressure $p_3^{'}$. The solution is obtained if $p_3^{'} = p_4$.

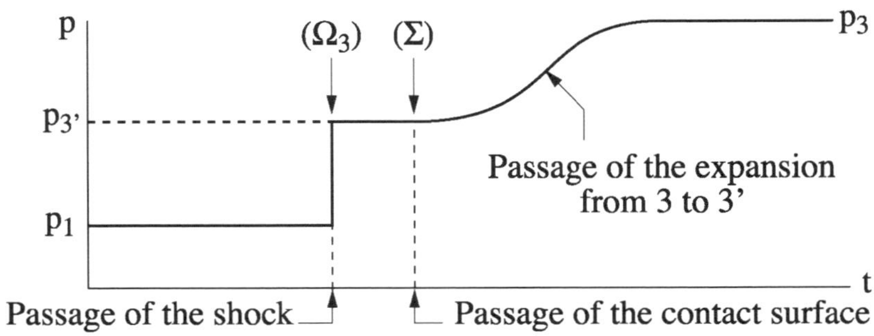

Figure 22.9. *Signature of the phenomena at the abscissa* $x_Q > x_I$

The signature of phenomena for a section located at an abscissa $x > x_I$ has the form shown schematically in Figure 22.9. The passage of the contact surface will not be detected by a pressure sensor, while it would be by a heat flux sensor. Indeed, the heat flux, which depends on the temperature of the medium, must change when crossing (Σ).

The problem we just solved is similar to the problem posed by the type VI interference resulting from the encounter of two shock waves in steady flow (see section 15.7).

22.7. Impact of a shock wave on a fixed wall

We analyze the following problem: the tube is now obturated at its right end by a fixed bottom. It is initially filled with a gas at rest with the following characteristics: pressure p_1, velocity of sound a_1, velocity $u_1 = 0$. At the time $t = 0$, the piston positioned at the left end of the tube suddenly moves toward the right with a constant velocity V_P. The acceleration of the piston causes the formation of a shock wave (Ω_1) propagating in the tube with a velocity D_1 (here, it equals its absolute velocity). Across (Ω_1), the gas velocity changes abruptly from $u_1 = 0$ to $u_2 = V_P$. The above given relations can fully assess the downstream condition (2).

Problem: what happens when the shock impacts the fixed bottom of the tube? As the gas downstream of (Ω_1) has a velocity $u_2 = V_P > 0$, there is incompatibility with the zero velocity condition to be imposed on the fixed bottom. Therefore, the gas undergoes an abrupt deceleration causing its velocity to pass from V_P to $u_3 = 0$. This is achieved through a so-called *reflected* shock wave which propagates toward the piston, as shown in the wave diagram of Figure 22.10.

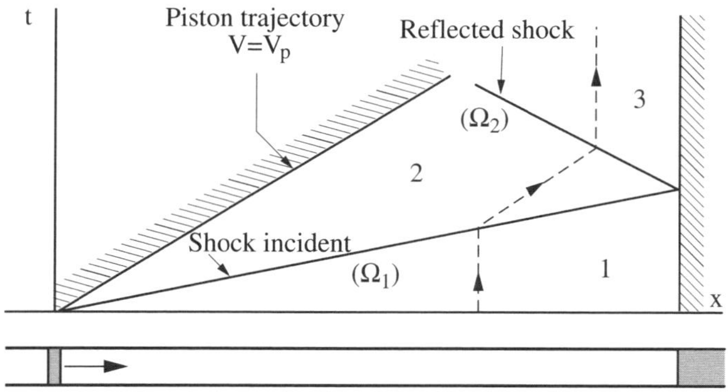

Figure 22.10. *Reflection of a shock wave from a fixed wall*

In the case of a calorically perfect gas, the solution to this problem can be obtained in analytical form. The pressure jump across the incident wave $\Pi_1 = \dfrac{p_2}{p_1}$ is deduced from the initial state and V_P.

We state:

$$\Pi_2 = \frac{p_3}{p_2}.$$

If $w_1 = V_P$ is the velocity difference across the incident shock (Ω_1), we have (see above):

$$\frac{w_1}{a_1} = \sqrt{\frac{2}{\gamma}}\frac{\Pi_1 - 1}{\sqrt{(\gamma+1)\Pi_1 + (\gamma-1)}}$$

and:

$$\frac{a_2}{a_1} = \sqrt{\Pi_1 \frac{(\gamma-1)\Pi_1 + (\gamma+1)}{(\gamma+1)\Pi_1 + (\gamma-1)}}.$$

Similarly, we have across (Ω_2):

$$\frac{w_2}{a_2} = \sqrt{\frac{2}{\gamma}}\frac{\Pi_2 - 1}{\sqrt{(\gamma+1)\Pi_2 + (\gamma-1)}}.$$

However, we must have $w_2 = -w_1$, since $u_3 = u_2 + w_2 = 0$ and $u_2 = w_1$.

Thus:

$$\frac{w_2}{a_2} = -\frac{w_1}{a_2} = -\frac{w_1}{a_1}\frac{a_1}{a_2}$$

giving:

$$\sqrt{\frac{2}{\gamma}}\frac{\Pi_2 - 1}{\sqrt{(\gamma+1)\Pi_2 + (\gamma-1)}} = -\sqrt{\frac{2}{\gamma}}\frac{\Pi_1 - 1}{\sqrt{(\gamma+1)\Pi_1 + (\gamma-1)}}\sqrt{\frac{1}{\Pi_1}\frac{(\gamma+1)\Pi_1 + (\gamma-1)}{(\gamma-1)\Pi_1 + (\gamma+1)}}$$

which leads to the explicit relation:

$$\boxed{\Pi_2 = \frac{(3\gamma-1)\Pi_1 - (\gamma-1)}{(\gamma-1)\Pi_1 + (\gamma+1)}.}$$

Limit cases

1. If the incident shock wave (Ω_1) is very intense, $\Pi_1 \to \infty$, so that:

$$\Pi_2 \to \frac{3\gamma-1}{\gamma-1} = 8 \quad \text{for} \quad \gamma = 1.4$$

and:

$$\frac{p_3}{p_1} = \frac{p_3}{p_2}\frac{p_2}{p_1} \to \frac{3\gamma-1}{\gamma-1}\Pi_1.$$

2. If the wave (Ω_1) is very weak, we can state:

$$\Pi_1 = 1+\eta \quad \textit{with} \quad \eta \ll 1.$$

Substituting this expression into the formula giving Π_2, it follows that:

$$\Pi_2 = \frac{(3\gamma-1)(1+\eta)-(\gamma-1)}{(\gamma-1)(1+\eta)+(\gamma+1)} = \frac{2\gamma+\eta(3\gamma-1)}{2\gamma+\eta(\gamma-1)}$$
$$\cong \frac{1}{2\gamma}\left[2\gamma+\eta(3\gamma-1)\right]\left[1-\eta\frac{\gamma-1}{2\gamma}\right]$$

which gives to the first order in η:

$$\Pi_2 \cong 1+\eta.$$

Then:

$$\frac{p_3}{p_1} = \Pi_1\ \Pi_2 \cong (1+\eta)(1+\eta) \cong 1+2\eta.$$

In the case of a weak shock wave, the pressure jump doubles in a normal reflection.

We can also calculate the Mach number M_{s_2} of the reflected shock wave which is defined by:

$$M_{s_2} = \frac{D_2}{a_2}$$

where D_2 is the velocity of the reflected shock wave. The absolute velocity of (Ω_2) is given by:

$$\left(\frac{dx}{dt}\right)_{\Omega_2} = D_2 + u_2 = D_2 + V_P.$$

We note that this velocity must be negative since the shock moves to the left, and hence $D_2 < 0$ as V_P is positive.

The Mach number M_{s_2} is the solution of the equation:

$$M_{s_2}^2 - \frac{\gamma+1}{2}\left(\frac{u_3 - u_2}{a_2}\right)M_{s_2} - 1 = 0.$$

In this case:

$$\frac{u_3 - u_2}{a_2} = -\frac{V_P}{a_2} = -M_2$$

where M_2 is the Mach number of the flow (2) which we count as positive.

This leads to the equation for M_{s_2} :

$$M_{s_2}^2 + \left(\frac{\gamma+1}{2}M_2\right)M_{s_2} - 1 = 0$$

of which we retain the root greater than 1 in the modulus, i.e.:

$$M_{s_2} = -\left(\frac{\gamma+1}{4}\right)M_2 - \sqrt{\left[\left(\frac{\gamma+1}{4}\right)^2 M_2^2 + 1\right]}.$$

This root is negative, which corresponds to a shock wave traveling to the left, i.e. according to the decreasing abscissas x.

Knowing that (see the shock relations given above):

$$M_2 = \left(M_{s_1}^2 - 1\right)\left[\left(1 + \frac{\gamma-1}{2} M_{s_1}^2\right)\left(\gamma M_{s_1}^2 - \frac{\gamma-1}{2}\right)\right]^{-\frac{1}{2}}$$

we deduce:

$$\boxed{M_{s_2} = \sqrt{\frac{\gamma \ M_{s_1}^2 - \frac{\gamma-1}{2}}{1 + \frac{\gamma-1}{2} M_{s_1}^2}}.}$$

We can then calculate the various ratios: $\frac{p_3}{p_1}, \frac{T_3}{T_1}$, etc.

Thus:

$$\frac{p_3}{p_1} = \frac{p_3}{p_2}\frac{p_2}{p_1} = \left[1 + \frac{2\gamma}{\gamma+1}\left(M_{s2}^2 - 1\right)\right]\left[1 + \frac{2\gamma}{\gamma+1}\left(M_{s1}^2 - 1\right)\right]$$

which after substitution gives:

$$\frac{p_3}{p_1} = \frac{2}{\gamma+1}\left[\frac{\gamma \ M_{s_1}^2 - \frac{\gamma-1}{2}}{1 + \frac{\gamma-1}{2} M_{s_1}^2}\right]\left[\frac{3\gamma-1}{2} M_{s_1}^2 - (\gamma-1)\right].$$

In a similar way we find:

$$\frac{T_3}{T_1} = \left(\frac{2}{\gamma+1}\right)^2 \frac{1}{M_{s_1}^2}\left[(\gamma-1) M_{s_1}^2 + \frac{3-\gamma}{2}\right]\left[\frac{3\gamma-1}{2} M_{s_1}^2 - (\gamma-1)\right].$$

Numerical example

The shock wave (Ω_1) moves in the air $(\gamma = 1.4)$ in the condition:

$$p_1 = 0.1 \times 10^5 \,\text{Pa} \quad \text{and} \quad T_1 = 300\,\text{K}.$$

The corresponding velocity of sound has the value:

$$a_1 = \sqrt{\gamma r T_1} = 347 \,\text{m/s} \qquad (\text{knowing that: } r = 287\,\text{J/kg/K.})$$

If this shock wave moves with a velocity $D_1 = 3{,}470$ m/s, its Mach number is therefore:

$$M_{s_1} = 10.$$

For large values of M_s (roughly $M_s > 7$), we can use the approximations:

$$\frac{p_2}{p_1} \cong \frac{2\gamma}{\gamma+1} M_s^2$$

and:

$$\frac{T_2}{T_1} \cong \frac{2\gamma(\gamma-1)}{(\gamma+1)^2} M_s^2$$

which gives in this case:

$$\frac{p_2}{p_1} \cong 117 \qquad \text{leading to: } p_2 = 11.7\times10^5\,\text{Pa}$$

$$\frac{T_2}{T_1} \cong 19.4 \qquad \text{leading to: } T_2 = 5{,}833\,\text{K}.$$

If the wave is reflected from the fixed bottom, we can use the approximations:

$$\frac{p_3}{p_1} \cong \frac{2\gamma(3\gamma-1)}{(\gamma+1)(\gamma-1)} M_s^2$$

$$\frac{T_3}{T_1} \cong \frac{2(\gamma-1)\,(3\gamma-1)}{(\gamma+1)^2} M_s^2.$$

This gives:

$$\frac{p_3}{p_1} \cong 933 \qquad \text{leading to: } p_3 = 93\times10^5\,\text{Pa}$$

$$\frac{T_3}{T_1} \cong 44 \qquad \text{leading to: } T_3 = 13{,}300\,\text{K}.$$

We note that the shock and its reflection cause considerable heating of the air. In fact, these results are not realistic because at these pressure and temperature levels, the assumption of a calorically perfect gas is no longer justified. Before reaching these conditions, the so-called real gas effects occur (see sections 3.7 and 15.5.3). In particular, there is a dissociation of molecules of oxygen and nitrogen so that the temperatures reached are, in fact, lower (about 3,500 K behind the incident shock). The assumption $\gamma = constant$ is better justified for a monatomic gas such as argon, for which $\gamma = 5/3 = 1.667$.

Chapter 23

Shock Tube

23.1. General theory

The *shock tube* is a simple way to achieve supersonic flows with elevated pressure and temperature levels (several thousands of kelvins for the temperature). The shock tube and its variants, such as the shock tunnel, are used to simulate hypersonic flows, which are distinguished by high stagnation enthalpy. The energy to expend to operate a shock tube is modest or low. Accordingly, and in return, the time during which a useful flow is established is very short: a few milliseconds to a few hundred microseconds. The shock tube is also used in detonics (to study conditions of shock wave formation during an explosion) and to investigate the phenomena resulting from the propagation and reflection of shocks, for example, the creation and propagation of the shock produced by a high-speed train entering a tunnel. Shock tubes are also used for the analysis of fast processes involved in chemical reactions, in particular, chemical kinetics.

A shock tube consists of a cylindrical tube, with a circular or square section, closed at both ends, and divided into two compartments initially isolated from each other (see Figure 23.1). The low-pressure compartment contains a gas with the pressure p_1 and temperature T_1. This gas is called *driven gas*. The high-pressure compartment is filled with a gas at high pressure p_4 (several hundreds or even thousands of atmospheres), and the temperature T_4 (often close to T_1). This gas is called the *driver gas*. The low-pressure part is much longer than the high-pressure part (5-10 times).

Initially, for $t < 0$, the two gases are separated by a diaphragm, located at the abscissa $x = 0$. At the time $t = 0$, we suddenly remove the diaphragm (by breaking it): the problem is to determine the phenomena occurring in the tube for $t > 0$.

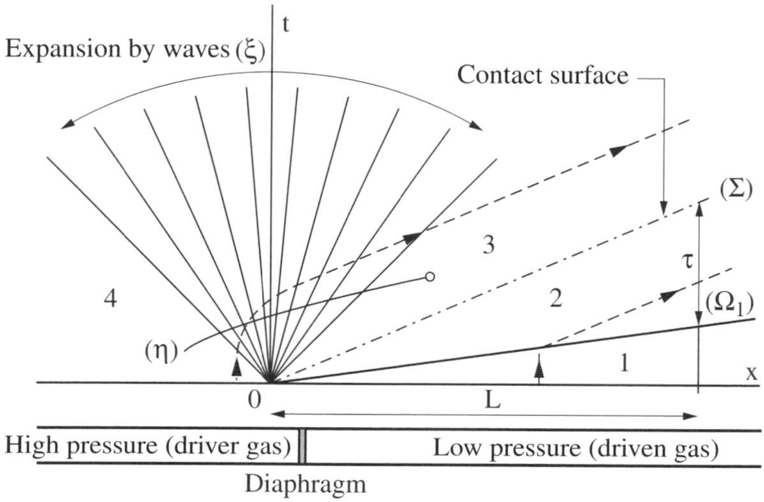

Figure 23.1. *Operation principle of a shock tube*

For $t \leq 0$, the driver gas and the driven gas, usually of different nature, are separated by a physical surface on both sides of which the pressures are very different. The removal of this wall at $t = 0$ causes the two gases to become separated by a contact surface (Σ), which cannot be maintained in that state for $t \geq 0$ as $p_4 >> p_1$. The surface (Σ) begins to move toward the low pressure and acts like a piston causing:

– a shock wave (Ω_1) in the driven gas (travel to the "right"), and

– a centered expansion wave in the driver gas (travel to the "left").

From $t = 0$ onward, a shock wave (Ω_1) propagates in the driven gas, causing it to pass from the state (1) to a state (2), which we have to determine. Correspondingly, an expansion fan propagates in the driver gas that passes from the state (4) to a state (3). The fluids in the states (2) and (3) are separated by the contact surface (Σ), also called interface. The conditions to satisfy on both sides of (Σ) are:

– pressure equality: $p_2 = p_3$ and

– velocity equality: $u_2 = u_3$.

The phenomena are illustrated in Figure 23.1. The shock wave (Ω_1) propagates in the medium (1) before the interface (Σ).

The data of the problem for $t \leq 0$ are the following:

– conditions in the driven gas: p_1, T_1 (giving the velocity of sound a_1), $u_1 = 0, \gamma_1$ and

– conditions in the driver gas: p_4, T_4 (giving a_4), $u_4 = 0, \gamma_4$.

We have to determine the states (2) and (3). To do this, we will first seek a relation between the ratio p_4 / p_1 and the Mach number M_{s_1} of the shock wave (Ω_1). Knowing M_{s_1} the state (2) will therefore be fully determined.

We have:

$$\frac{p_4}{p_1} = \frac{p_4}{p_3}\frac{p_3}{p_2}\frac{p_2}{p_1} = \frac{p_4}{p_3}\frac{p_2}{p_1}$$

because:

$$p_3 = p_2 .$$

The states (3) and (4) are connected, as shown in Figure 23.1, by the following relation on a characteristic (η):

$$u_3 + \frac{2}{\gamma_3 - 1} a_3 = u_4 + \frac{2}{\gamma_4 - 1} a_4 .$$

We assume that γ is the same in the states (3) and (4), which are relative to the same fluid. Since $u_4 = 0$, it follows that:

$$\frac{a_3}{a_4} = 1 - \frac{\gamma_4 - 1}{2}\frac{u_3}{a_4} .$$

But as: $a^2 = \gamma \dfrac{p}{\rho}$:

$$\frac{a_3}{a_4} = \left(\frac{p_3}{p_4} \frac{\rho_4}{\rho_3} \right)^{\frac{1}{2}}.$$

As the transition of (3) to (4) is done isentropically:

$$\frac{\rho_4}{\rho_3} = \left(\frac{p_3}{p_4} \right)^{-\frac{1}{\gamma_4}}$$

we have:

$$\frac{a_3}{a_4} = \left(\frac{p_3}{p_4} \right)^{\frac{1}{2}} \left(\frac{p_3}{p_4} \right)^{-\frac{1}{2\gamma_4}} = \left(\frac{p_3}{p_4} \right)^{\frac{\gamma_4 - 1}{2\gamma_4}}.$$

Moreover, as $u_3 = u_2$:

$$\frac{u_3}{a_4} = \frac{u_2}{a_2} \frac{a_2}{a_4} = M_2 \frac{a_2}{a_4}.$$

The characteristic relation between (3) and (4) thus leads to:

$$\left(\frac{p_3}{p_4} \right)^{\frac{\gamma_4 - 1}{2\gamma_4}} = 1 - \frac{\gamma_4 - 1}{2} M_2 \frac{a_2}{a_4} = 1 - \frac{\gamma_4 - 1}{2} \left(M_2 \frac{a_2}{a_1} \right) \frac{a_1}{a_4}$$

but:

$$M_2 \frac{a_2}{a_1} = \frac{u_2}{a_1} = \frac{2}{\gamma_1 + 1} \left(M_{s_1} - \frac{1}{M_{s_1}} \right)$$

from which follows:

$$\left(\frac{p_3}{p_4} \right)^{\frac{\gamma_4 - 1}{2\gamma_4}} = 1 - \frac{\gamma_4 - 1}{\gamma_1 + 1} \left(M_{s_1} - \frac{1}{M_{s_1}} \right) \frac{a_1}{a_4}.$$

Moreover, according to the shock relations:

$$\frac{p_2}{p_1} = 1 + \frac{2\gamma_1}{\gamma_1 + 1}\left(M_{s_1}^2 - 1\right) = \frac{\gamma_1 - 1}{\gamma_1 + 1}\left(\frac{2\gamma_1}{\gamma_1 - 1} M_{s_1}^2 - 1\right)$$

from which follows finally:

$$\boxed{\frac{p_4}{p_1} = \frac{\dfrac{\gamma_1 - 1}{\gamma_1 + 1}\left(\dfrac{2\gamma_1}{\gamma_1 - 1} M_{s_1}^2 - 1\right)}{\left[1 - \dfrac{\gamma_4 - 1}{\gamma_1 + 1}\dfrac{a_1}{a_4}\left(M_{s_1} - \dfrac{1}{M_{s_1}}\right)\right]^{\frac{2\gamma_4}{\gamma_4 - 1}}}.}$$

The above relation allows us to calculate the shock wave Mach number M_{s_1} based on the data of the ratio p_4 / p_1. We can verify that for gas with a given nature, M_{s_1} increases when the ratio p_4 / p_1 increases. Knowing M_{s_1} we can entirely calculate the state (2), i.e. p_2, T_2, ρ_2, as well as the velocity u_2 through:

$$\frac{u_2}{a_1} = \frac{2}{\gamma_1 + 1}\left(M_{s_1} - \frac{1}{M_{s_1}}\right).$$

We deduce $u_3 = u_2$, $p_3 = p_2$, then a_3, p_3, and T_3 for the isentropic relations between the states (3) and (4), i.e.:

$$\frac{a_3}{a_4} = \left(\frac{p_3}{p_4}\right)^{\frac{\gamma_4 - 1}{2\gamma_4}}$$

$$\frac{\rho_3}{\rho_4} = \left(\frac{p_3}{p_4}\right)^{\frac{1}{\gamma_4}}$$

$$\frac{T_3}{T_4} = \left(\frac{p_3}{p_e}\right)^{\frac{\gamma_4 - 1}{\gamma_4}}.$$

When the pressure ratio p_4 / p_1 becomes very large, the shock Mach number M_{s_1} tends toward a limit value $M_{s(\lim)}$, which is the value canceling the denominator of the expression of p_4 / p_1 given above, i.e.

$$1-\frac{\gamma_4-1}{\gamma_1+1}\frac{a_1}{a_4}\left[M_{s(\lim)}-\frac{1}{M_{s(\lim)}}\right]=0.$$

If we state that:

$$\alpha=\frac{\gamma_4-1}{\gamma_1+1}\frac{a_1}{a_4}$$

we obtain the equation:

$$\alpha M_{s(\lim)}^2-M_{s(\lim)}-\alpha=0$$

which gives us:

$$M_{s(\lim)}=\frac{1+\sqrt{1+4\alpha^2}}{2\alpha}.$$

We verify that the larger $M_{s(\lim)}$ is, the smaller α is. The driven gas has properties most often dictated by the nature of the studied phenomena, for example, the thermochemistry of the air during an atmospheric reentry. To create intense shocks, we must play with the properties of the driver gas by assuming a gas where the velocity of sound is large, which tends to decrease α . As:

$$a_4=\sqrt{\gamma_4 r_4 T_4}=\sqrt{\gamma_4\frac{\Re}{M}T_4}$$

where $\Re$ is the universal perfect gas constant, we will use a gas whose molar mass M is small. Table 23.1 gives the limit Mach numbers for the following conditions: $T_1=T_4=300$ K, $\gamma_1=1.4$, and different driver gas.

Driver Gas	Molar Mass (kg/mole)	γ	$M_{s(\lim)}$
Air	0.029	1.4	6.16
Helium	0.004	1.667	10.67
Hydrogen	0.002	1.407	22.84

Table 23.1. *Limit Mach number for a pressure ratio tending toward infinity*

Figure 23.2 shows the change of the shock Mach number as a function of the pressure ratio p_4 / p_1 for the combinations air–air, air–helium, and air–hydrogen, the temperature being in all cases 300 K in the driver and driven gases.

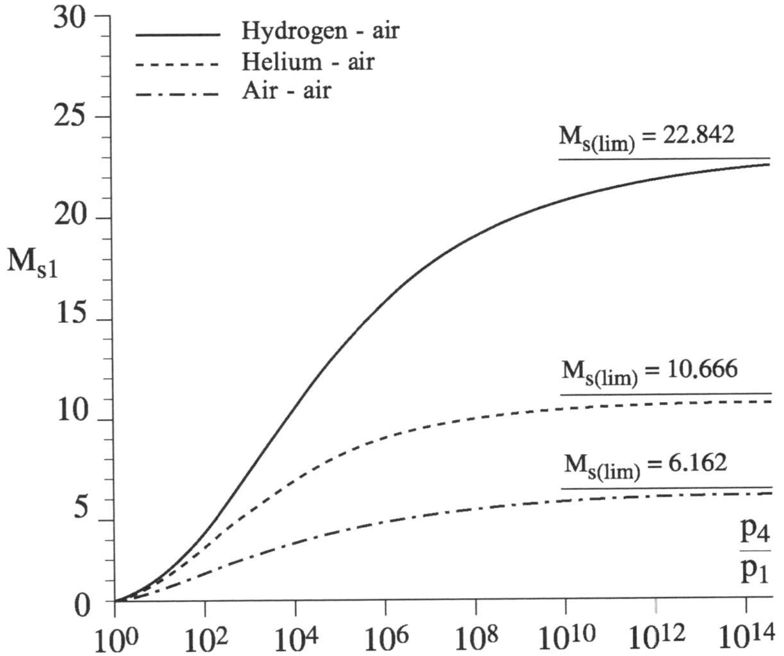

Figure 23.2. *Shock Mach number as a function of the upstream/downstream pressure ratio*

It is possible to further increase the shock Mach number by heating the driver gas, i.e. by acting on T_4. In combustion shock tube, the high-pressure compartment is filled with a mixture of oxygen and hydrogen, which is ignited. At the time of ignition, the diaphragm burst and the resulting shock is produced by a high-temperature driver gas.

Figure 23.3 schematizes the pressure distribution in the tube at a time $t > 0$. In the part of the tube farthest to the right, the pressure is still equal to the initial value p_1. On the abscissa x_2, the pressure passes from p_1 to p_2 through the shock (Ω_1) and then remains constant until the abscissa x_4 from which onward it grows steadily to reach the level of the initial pressure of the driver gas p_4. This gradual growth corresponds to the expansion wave propagating in the driver gas. The contact surface is located at an abscissa x_3, where the pressure is continuous and equal to the plateau value $p_3 = p_2$.

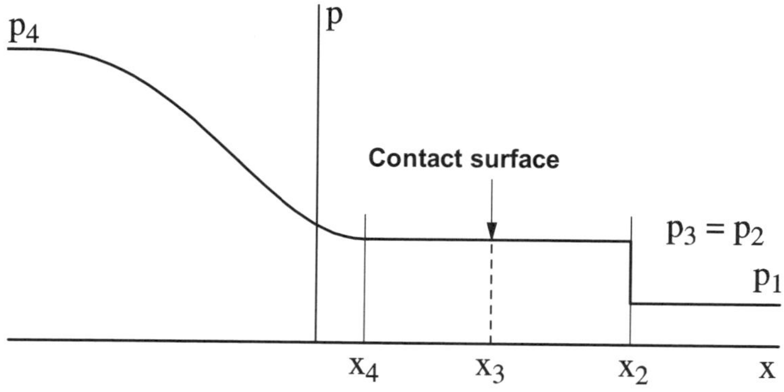

Figure 23.3. *Pressure in the tube at a given time*

In Figure 23.1, two trajectories of fluid particles are plotted. In the driven gas, the particle remains motionless until the passage of the wave (Ω_1) causing its acceleration from $u = u_1 = 0$ to $u = u_2$. The trajectory of the particle is then parallel to that of the contact surface (Σ) moving with the velocity u_2. The particle located in the driving gas gradually accelerates at the arrival of the first wave of the expansion fan. Across the wave, its velocity increases continuously to reach the value $u = u_3$, which is the velocity of the contact surface. Driven and driver gases move toward the right with the same velocity while being separated by the contact surface. We note that the shock moves faster than (Σ), the time interval separating the passage of the shock and the contact surface increasing with the distance x from the origin O where the diaphragm is located.

Useful testing time

The measuring device, which may include a scale model, is placed at the abscissa $x = L$. The duration of the test time τ is the time interval between the passage of the wave (Ω_1) and the arrival of the contact surface (Σ) in $x = L$. This interval corresponds to the flow of the driven gas after its compression and heating by the shock (Ω_1) with the velocity D_1. The duration τ is thus given by:

$$\tau = \frac{L}{u_2} - \frac{L}{D_1}.$$

To give orders of magnitude, we consider a length $L = 10\ \text{m}$ and a shock propagating in the air where $T_1 = 300\ \text{K}$ (leading to a velocity of sound $a_1 = 347\ \text{m/s}$). Table 23.2 gives the values of τ for the shock Mach numbers 5, 9, and 20 corresponding, respectively, to a driver gas consisting of air, helium, and hydrogen.

Driver gas	M_{s_1}	$D_1(m/s)$	$u_2(m/s)$	$\tau(\mu s)$
Air	5	1,736	1,388	1,450
Helium	9	3,123	2,570	862
Hydrogen	20	6,940	5,769	292

Table 23.2. *Useful testing time for different driving gases*

We see that the duration of the test for useful data varies between 1.45 ms and 290 μs.

23.2. Reflected shock tube

We seek to determine the flow in a tube closed at both ends. Figure 23.4 shows the corresponding wave diagram. The basic problems addressed above provide us with the information to construct the flow. Arriving at the right bottom, the shock (Ω_1) is reflected as a shock (Ω_2) in such a way that the velocity u_5 downstream of (Ω_2) is zero (contact condition on the fixed wall).

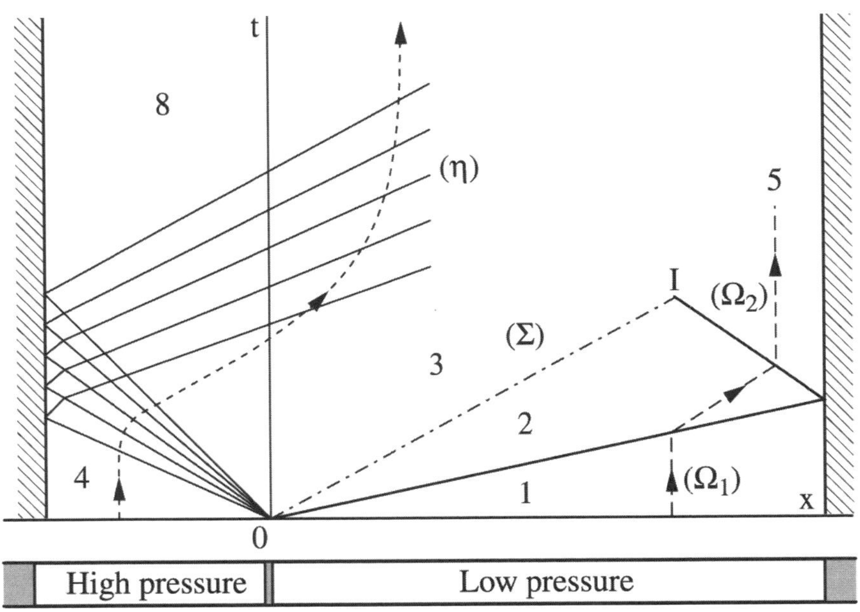

Figure 23.4. *Shock tube closed at the two ends*

The state (2) is fully defined: we know p_2, ρ_2, T_2, a_2 and u_2, the velocity difference being $w_1 = u_2$, since $u_1 = 0$. The velocity difference w_2 across the reflected shock (Ω_2) should be such that:

$$w_2 = u_5 - u_2 = -u_2 = -w_1.$$

The Mach number of the reflected shock is provided by the equation:

$$M_{s_2}^2 - \frac{\gamma+1}{2}\left(\frac{w_2}{a_2}\right) M_{s_2} - 1 = 0.$$

Knowing that $w_2 = -w_1$, the solution of this equation gives M_{s_2}. We will find here a negative Mach number since the reflected wave (Ω_2) moves to the left. We deduce all properties of state (5). This calculation can also be done using as parameter the pressure ratio $\Pi_1 = p_2 / p_1$ (see section 22.7).

To give orders of magnitude, consider the case where $T_1 = 300$ K, $p_1 = 0.1 \times 10^5$ Pa, and the Mach number of the incident shock (Ω_1) is $M_{s_1} = 9$ (driver gas is helium). The conditions in (2) can be calculated using the approximate relations if the shock Mach number is large (see section 22.7), which gives for $M_{s_1} = 9$:

$$\frac{p_2}{p_1} \cong \frac{2\gamma_1}{\gamma_1 + 1} M_{s_1}^2 = 1.167 M_{s_1}^2 \rightarrow \frac{p_2}{p_1} = 94.5 \text{ leading to: } p_2 = 9.45 \times 10^5 \text{ pa}$$

$$\frac{T_2}{T_1} \cong \frac{2(\gamma_1 - 1)}{(\gamma_1 + 1)^2} M_{s_1}^2 = 0.194 M_{s_1}^2 \rightarrow \frac{T_2}{T_1} = 15.75 \text{ leading to: } T_2 = 5{,}830 \text{ K}$$

thus:

$$\frac{p_5}{p_1} \cong \frac{2\gamma_1(3\gamma_1 - 1)}{(\gamma_1 + 1)(\gamma_1 - 1)} M_{s_1}^2 = 9.33 M_{s_1}^2 \rightarrow \frac{p_5}{p_1} = 756 \text{ leading to: } p_5 = 75.6 \times 10^5 \text{ Pa}$$

$$\frac{T_5}{T_1} \cong \frac{2(\gamma_1 - 1)(3\gamma_1 - 1)}{(\gamma_1 + 1)^2} M_{s_1}^2 = 0{,}444 M_{s_1}^2 \rightarrow \frac{T_5}{T_1} = 36 \text{ leading to: } T_5 = 10{,}800 \text{ K}.$$

We note that the successive passages of the shocks (Ω_1) and (Ω_2) allow strong heating of the driven gas whose temperature passes from 300 to 10,800 K. As we noted before, such a calculation is partly fictitious since at these temperature levels the gas dissociates, thus invalidating the assumption of a calorically perfect gas. The temperature reached in reality will be closer to 6,000 K.

23.3. Tailoring the contact surface

We examine what happens in the section *I,* where the reflected shock (Ω_2) meets the contact surface (Σ) (see Figures 23.4 and 23.5). As there is an incompatibility between the states (3) and (5), disturbances have to form in *I* to create compatible states on both sides of the contact surface after it has been crossed by the shock (Ω_2). As the states (2) and (3) are different (only the velocities and pressures are equal), the shock (Ω_2) will propagate in the medium (3) as a *transmitted shock* (Ω_3), which induces a pressure jump $p_3 \to p_6$ and a velocity jump $u_3 \to u_6$. In general, it is not possible to find p_6 and u_6 so that:

$$p_6 = p_5$$

$$u_6 = u_5$$

thus, ensuring the compatibility of states (5) and (6) on both sides of (Σ). Two circumstances may occur (see Figures 23.5a and 23.5b).

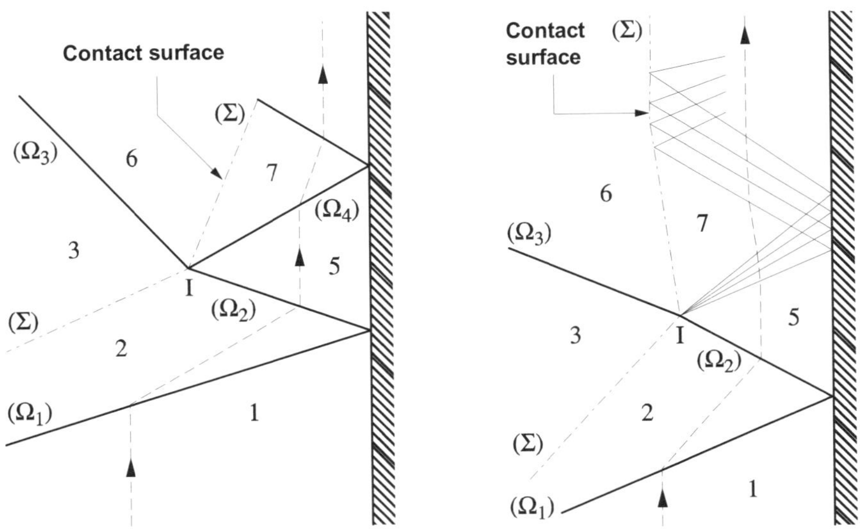

a – case A *b – case B*

Figure 23.5. *Interaction of the reflected shock with the contact surface*

Case A: if $p_6 > p_5$, a shock wave (Ω_4) has to propagate in the environment (5), which is the driven gas, to ensure a compression p_5 to p_7 so that $p_6 = p_7$ (with $u_6 = u_7$).

Case B: if $p_6 < p_5$, an expansion wave propagates in the environment (5) so that $p_6 = p_7$.

In both cases, the solution is obtained by iterating on the velocity of the contact surface (Σ) after its crossing of the reflected shock (Ω_2). Setting this velocity amounts to imposing $u_6 = u_7$.

In case A, knowing the velocity differences $u_6 - u_3$ and $u_7 - u_5$, we are able to calculate the shocks (Ω_3) and (Ω_4), leading to p_6 and p_7 which should be such that $p_6 = p_7$.

In case B, p_7 is calculated from p_5 with the characteristic relation written on a wave (ξ):

$$u_7 - \frac{2}{\gamma_1 - 1} a_7 = u_5 - \frac{2}{\gamma_1 - 1} a_5.$$

As u_7 is fixed, we deduce a_7, thus p_7 by the isentropic relations.

If we consider the case A, we see that the shock (Ω_4) causes an acceleration of the driven gas with $u_7 > 0$. Thus, a reflected shock (Ω_5) is formed when (Ω_4) impacts the bottom of the tube, causing the stop of the gas. In turn, (Ω_5) interferes with the contact surface (Σ) and the above scenario repeats itself with the formation of a series of shocks that compress more and more the driven gas.

In case B, the expansion wave is reflected as an expansion wave from the static bottom (see section 22.4). The reflected wave will interfere with the contact surface (Σ) through a mechanism generating transmitted and reflected (compression) waves, etc. (see section 21.7).

It is possible to define the operating conditions of the tube such that the shock wave (Ω_2) is fully transmitted in the environment (3) without formation of disturbances in the driven gas. Then, there is neither shock wave (Ω_4) nor expansion fan. The gas remains at rest in contact with the bottom of the tube, the velocity of the contact surface (Σ) being zero after meeting (Ω_2). This situation is illustrated in Figure 23.6. If these conditions are fulfilled, we say that the *contact surface is tailored.* Case A corresponds to an *overtailored* contact surface and case B to an *undertailored* contact surface.

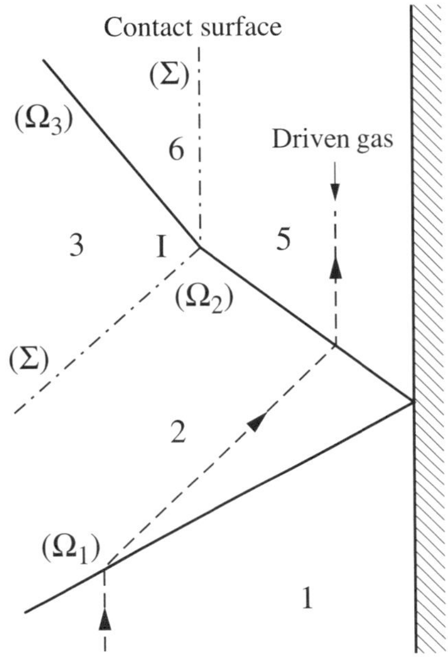

Figure 23.6. *Tailored contact surface*

We examine the phenomena in the *left-hand side of the tube*. When the expansion wave, forming at the rupture of the diaphragm, reaches the bottom of the tube, it is reflected as an expansion wave. This wave, formed by characteristics (η), propagates to the right side of the tube, where it meets the contact surface (Σ), either before *I* or after *I*. This leads to increasingly complex wave interferences, which we will not analyze here. The solution must be found by numerical integration methods, for example, based on the method of characteristics (see section 24.1). The arrival of the waves (η) in the right side of the tube will affect the phenomena discussed above by modifying the different cases A, B, or tailored. After the passage of the reflected wave (η), the driver gas is in a state (8) with the velocity $u_8 = 0$ and the pressure p_8 so that $p_8 < p_3 < p_4$. This state (8) will be perturbed by the arrival of waves from the right side of the tube. Thus, little by little, the pressure and velocity in the tube will become uniform.

In principle, the construction of the previous waves can continue indefinitely in a non-viscous fluid. In reality, viscosity rapidly occurs due to the development of boundary layers along the walls and of mixing layer at the contact surfaces. Gradually, the viscous effects invade the whole tube, transforming the kinetic energy into heat until the flow is completely at rest.

To illustrate, Figure 23.7 shows the wave diagram of an air–air shock tube, where $p_1 = 10^5\,\text{Pa}$ and $p_4 = 100\times 10^5\,\text{Pa}$. The diagram shows the shock wave produced by the rupture of the diaphragm, the expansion fan propagating in the driver gas, and the contact surface separating the driver and driven gases. The first shock heats the driven gas up to 604 K and the reflected shock up to 971 K.

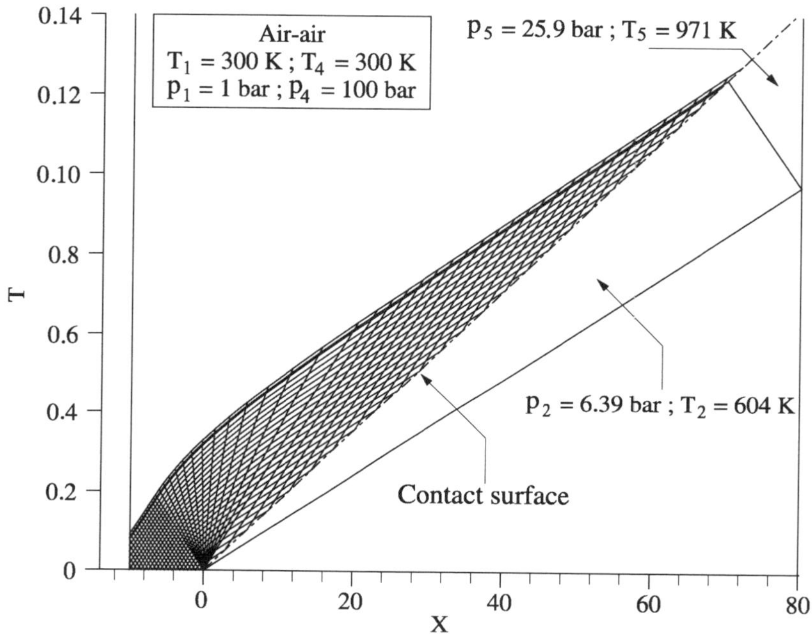

Figure 23.7. *Wave diagram of an air-air shock tube*

Figure 23.8 shows the wave diagram of a hydrogen–air shock tube for the same initial pressure ratio ($p_1 = 10^5\,\text{Pa}$ and $p_4 = 100\times 10^5\,\text{Pa}$). In this case, the shock heats the driven gas up to 1,461 K and the reflected shock up to 2,927 K; the maximum pressure is such that $p_5 = 148.4\times 10^5\,\text{Pa}$.

23.4. Numerical simulation of the functioning of a shock tube

In this section, we examine the functioning of a shock tube by a finite difference solution of the conservation equations using the methodology described in section 24.2. The tube length is 10 m, the diaphragm separating the high- and low-pressure compartments is in the middle of the tube, the pressure p_4 of the driving gas is 10, and that of the driven gas p_1 is 1 (units are irrelevant). The two gases have identical temperatures and the same γ.

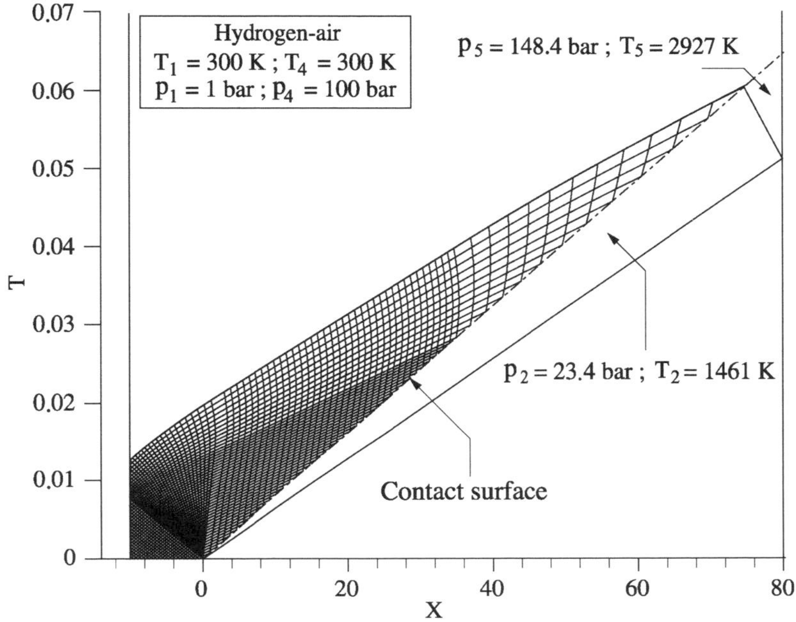

Figure 23.8. *Wave diagram of a hydrogen-air shock tube*

The velocity of sound in each medium is 347 m/s, the data of the length and the velocity of sound being necessary to define the timescale. The tube is closed at both ends.

Figure 23.9 shows the pressure distribution in the tube at different instants. At $t = 0.00144$ s, we are at the beginning of the phenomena: a shock (Ω_1), causing the transition of the pressure from 1 to 2.8, propagates in the driven gas, while an expansion spreads in the high-pressure side. At $t = 0.0043$ s, the shock is at the abscissa $x = 0.73$. At $t = 0.0086$ s, it has almost reached the end on the right. At $t = 0.0115$ s, the shock (Ω_1) has been reflected by the right bottom, the intensity of the reflected shock (Ω_2) being roughly twice that of the incident shock.

At later times, the shock (Ω_2) propagates to the left and meets the expansion wave that was reflected on the left end of the tube, which weakened the shock. Shortly after $t = 0.0418$ s, the shock (Ω_2) impacts the left bottom and a reflected shock moves to the right (last image). The reflection of the expansion wave on the left bottom leads to a pressure decrease to a level of about 1.3 before the impact of the shock (Ω_2). At times $t = 0.0158$ s and $t = 0.0202$ s, a second shock is visible following the reflected shock. It is caused by the encounter of (Ω_2) with the contact surface separating the states (4) and (1) (see section 23.4).

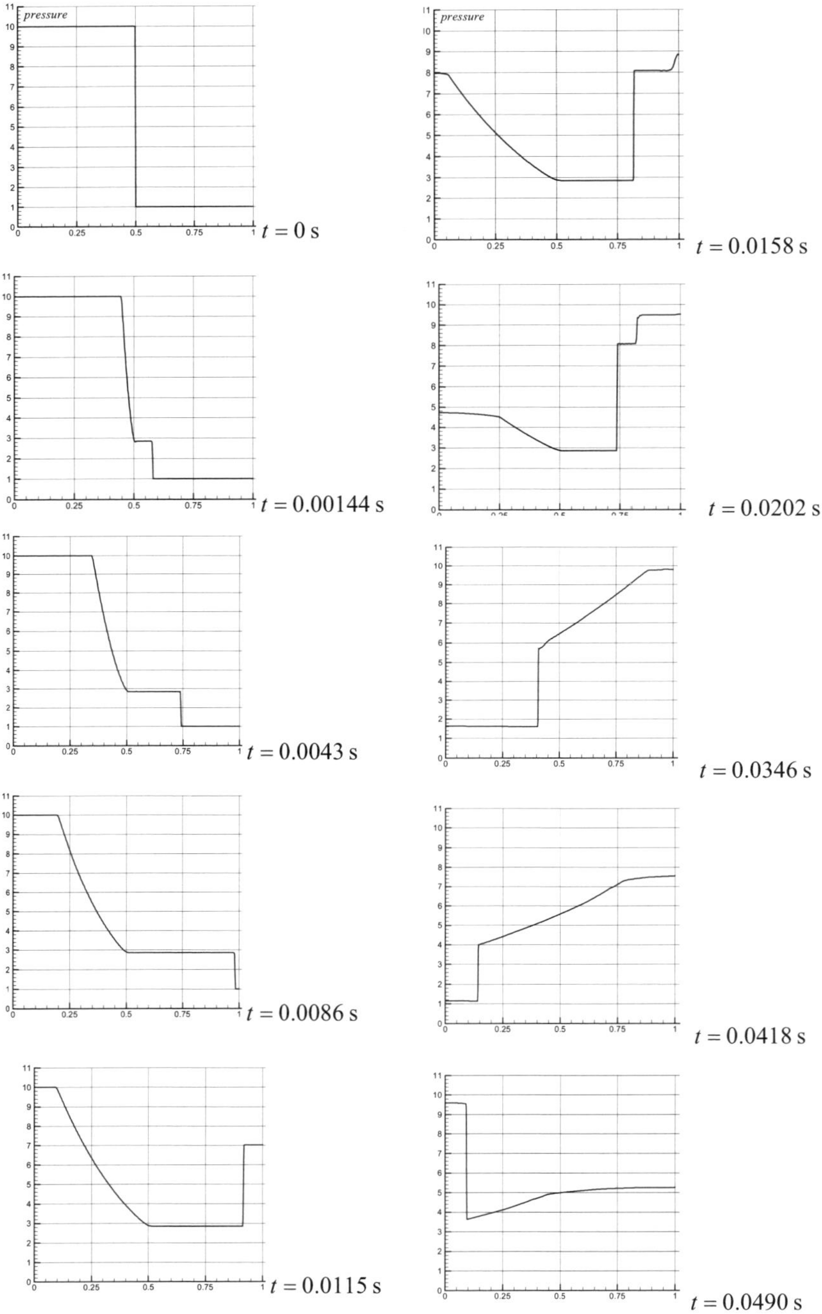

Figure 23.9. *Pressure evolution in an air-air shock tube with closed ends*

Figure 23.10 shows the Mach number distributions, which can be interpreted in conjunction with those of the pressure. The fixed bottoms impose Mach numbers of zero at both ends.

In Figure 23.11 are plotted the distributions of the reduced entropy $\overline{s}$, calculated with the relation (see Chapter 3):

$$\overline{s} = \frac{s}{C_v} = \gamma \ln T - (\gamma - 1)\ln p.$$

The discontinuity of $\overline{s}$ at the time $t = 0$ corresponds to differences between the states (1) and (4) separated by the diaphragm. After rupture of the latter, the contact surface (Σ) separating (1) and (4) moves to the right. The smaller entropy jump at the right of (Σ), and moving faster than (Σ), corresponds to the shock (Ω_1). We have seen that the useful testing time at a given section x in the tube is the time interval between the passage of the shock and that of (Σ). The contact surface (Σ) stabilizes near the right end, where it oscillates around an average position as shown in the examination of phenomena over a longer time. The medium (1) is therefore compressed into a small volume in contact with the right end of the tube.

Firing conditions leading to the immobilization of (Σ) correspond to those that we call contact surface tailoring (see section 23.3). We also observe the jump in entropy associated with the reflected shock (Ω_2) moving to the right and crossing the contact surface.

The calculation can be continued, as progressive waves and shock waves that interfere with each other and reflect on the ends of the tube travel through the tube. In a non-viscous fluid, these phenomena continue indefinitely. In the case of a real viscous fluid, boundary layers develop along the walls of the tube and mixing layers on the contact surfaces. The dissipative regions gradually invade the entire flow and help to dissipate the kinetic energy into heat. After a finite time, the tube is filled with a gas mixture at uniform pressure and temperature (care should be taken if the driver and driven gases are hydrogen and air). The situation is analogous to the non-adapted supersonic jet in a steady flow whose organization can develop in a pseudo-periodic manner in a theoretically infinite length for a non-viscous fluid (see section 20.8).

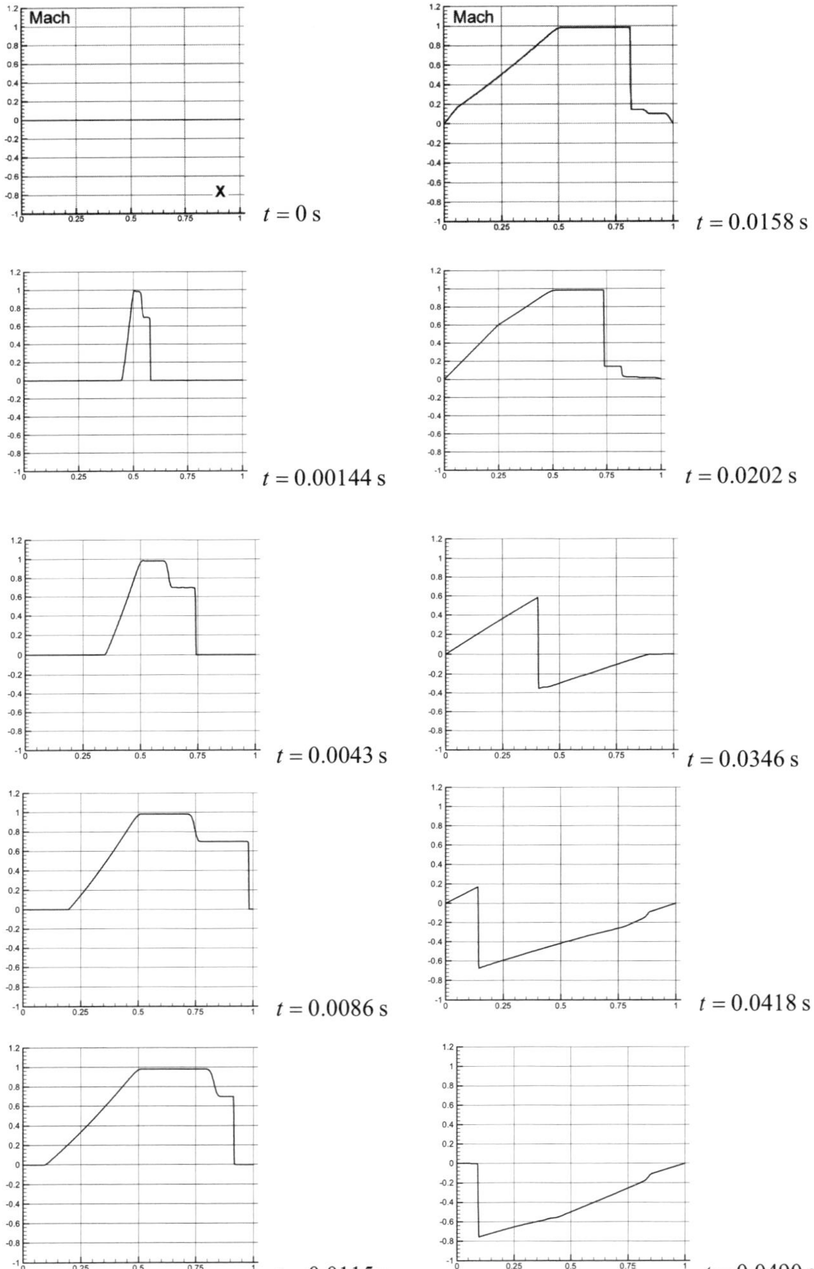

Figure 23.10. *Mach number evolution in an air-air shock tube with closed ends*

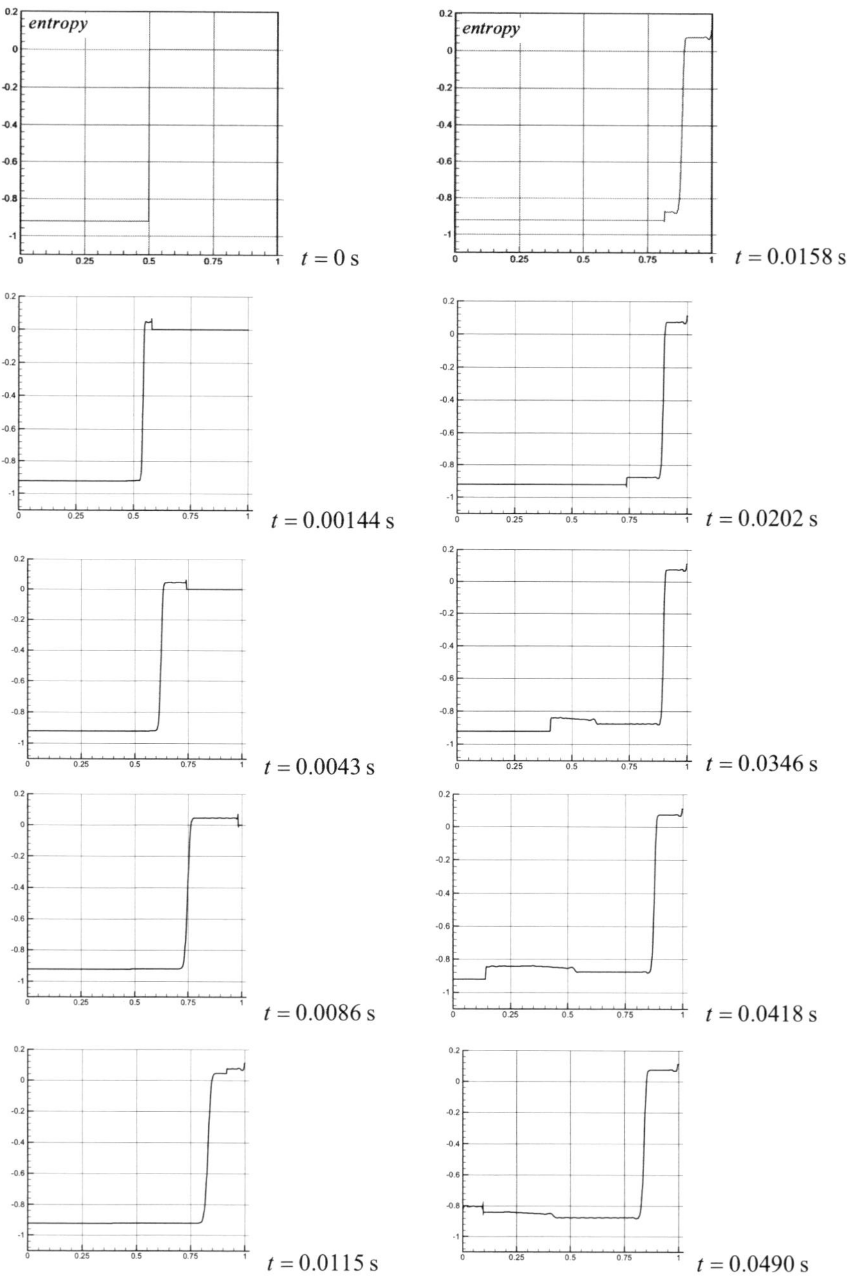

Figure 23.11. *Entropy evolution in an air-air shock tube with closed ends*

23.5. Shock tube arrangement: the shock tunnel

Figure 23.12 shows a schematic layout of a conventional research shock tube. The right compartment is filled with the driver gas under high pressure (up to 2,000 atmospheres). In the most advanced facilities, the gas is hydrogen which imposes security rules and stringent testing procedures due to the risk of explosion from mixing with oxygen in the air. The low-pressure side is the tube itself, length and diameter varying according to facilities. As the test time is proportional to the length of the tube, it appears that the optimum is to build a tube as long as possible. In reality, the boundary layers that develop in the tube downstream of the shock limit the useful operating time, as these boundary layers completely fill the tube after a certain travel distance. The optimum, dictated by experience, is a length not exceeding 100 times the diameter of the tube. Thus, a tube 60 mm in diameter will have a length of about 6 m.

The test chamber is placed at the end of the tube. It is often equipped with glass windows for the observation of phenomena and to allow measurements with optical techniques. Downstream of the test chamber is an exhaust pipe connecting the shock tube to the expansion tank collecting the gases after a shot. For safety reasons, the total volume of the facility, including the expansion tank, is calculated so that the pressure in the facility after a shot is not greater than the atmospheric pressure. The expansion tank is connected to a pumping unit for the removal of the gases and adjustment of the initial pressure in the low-pressure part. Before test, the tube is filled with the driven gas under preset pressure and temperature conditions.

Before shooting, the high- and low-pressure parts are separated by a diaphragm formed by a metal disk whose material and thickness are set to burst at the aimed pressure difference $(p_4 - p_1)$. In order to have a clean burst and to avoid sending metal fragments caused by the rupture, predetermined breaking notches are cut into the diaphragm (see Figure 23.12). These notches are cross-shaped causing an opening in petals, completely clearing the path for the driving gas. The disadvantage of this single diaphragm device is that the rupture moment is not precisely known, the rupture conditions being not reproducible from one diaphragm to another. The test procedure is to gradually fill the high-pressure side until the rupture of the diaphragm takes place.

In order to control the starting time of the shot better, double diaphragm devices are used (see Figure 23.12). Two identical diaphragms, placed in series, enclose a cavity with a small volume, which is initially filled to a pressure $p_4^{'}$ intermediate between p_1 and p_4, for example, $p_4^{'} = 1/2(p_4 + p_1) \approx p_4/2$. The diaphragms are calculated to withstand the pressure difference $(p_4 - p_4^{'})$, but not the application of the pressure $(p_4 - p_1)$. The shot is fired by suddenly emptying the cavity that submits the two diaphragms to the pressure $(p_4 - p_1)$, causing them to burst almost instantly in chain.

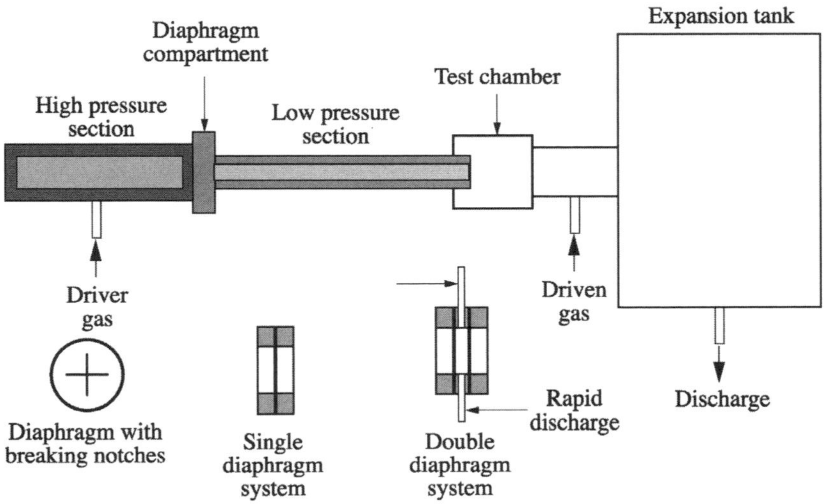

Figure 23.12. *Layout of a shock tube*

Both devices are used, the one with one diaphragm has the merit of being simpler and consuming less diaphragms which may be difficult to manufacture and often expensive. In fact, the precise moment of firing is irrelevant in practice, as the rupture moment of the diaphragm cannot serve as a starting signal for the phenomena in the tube because of duration of the rupture (as we have seen, the characteristic times of the phenomena in a shock tube are close to the millisecond). The events following the rupture of the diaphragm (or diaphragms) are measured by timing the travel of the shock in the tube. To do this, fast response transducers are placed at regularly spaced and known abscissas along the tube. The passage of the shock over a sensor gives rise to a signal whose onset time is recorded by a suitable device. Thus, we record the instants of the shock passage on several transducers, and knowing their spacing, we can deduce the velocity of the shock. Having also measured the initial conditions in the driven gas (temperature, pressure), we then calculate the shock Mach number and deduce the working conditions p_2, T_2, ρ_2, etc. with the unsteady shock theory.

Figure 23.13 is an image of a research shock tube with 60 mm diameter.

Figure 23.14 is an image with very short exposure time showing a shock wave emerging from a shock tube with a circular cross section. Because of the low pressure in the test chamber, the supersonic flow following the shock expands through an expansion fan accelerating the flow. This leads to the formation of a shock closely following the main shock that incurves due to the expansion at the exit of the tube.

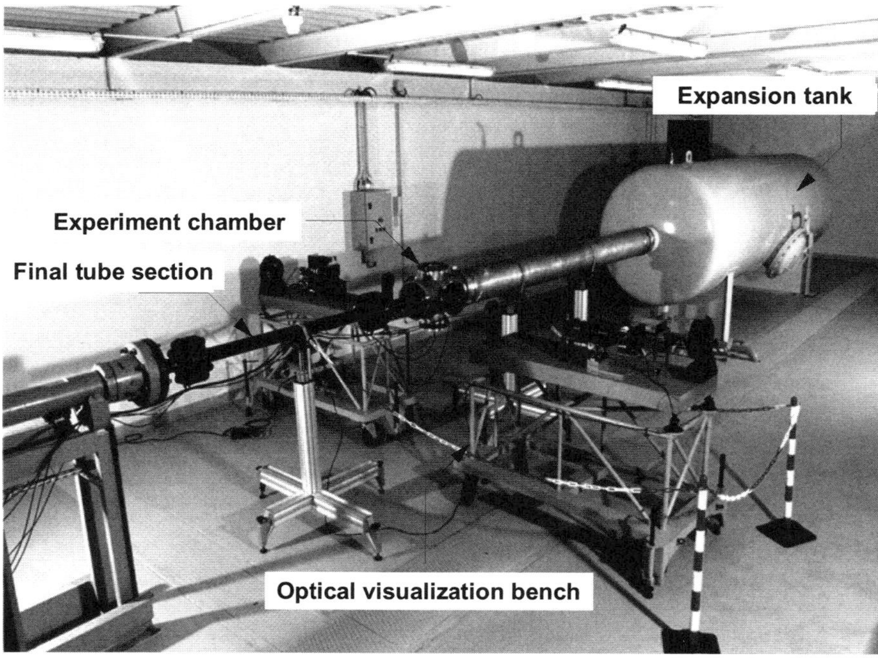

Figure 23.13. *Research shock tube (diameter 60 mm, document Onera)*

As shown by numerical examples, the shock tube is a means of obtaining, downstream of the shock, a flow with high pressure and temperature. Taking into account the real gas effects, the temperatures behind the shock (Ω_1) are much lower than those given by the calorically perfect gas model; yet they can exceed 8,000 K in tubes using hydrogen as a driver gas. In *shock tunnels*, the driven gas obtained under these conditions is used to supply a nozzle with a large area ratio to achieve a flow that is both hypersonic (high Mach number) and hyperenthalpic (high stagnation enthalpy) for studies of atmospheric reentry and hypersonic vehicles. The nozzle is placed at the end of the tube, where, after the shock passage, it is fed for a short time (a few milliseconds) by the driven gas in state (2). It is possible to increase the level of stagnation enthalpy by working with a reflected shock. As shown in Figure 23.15, a second diaphragm D_2 is then placed at the end of the tube, upstream of the nozzle. After reflection of the shock (Ω_1) from the diaphragm D_2, the gas is in the state (5) whose properties were calculated in section 22.7. Immediately after the reflection, the diaphragm D_2 bursts and gas in the state (5) feeds the nozzle. A tailoring of the device optimizes the useful testing time.

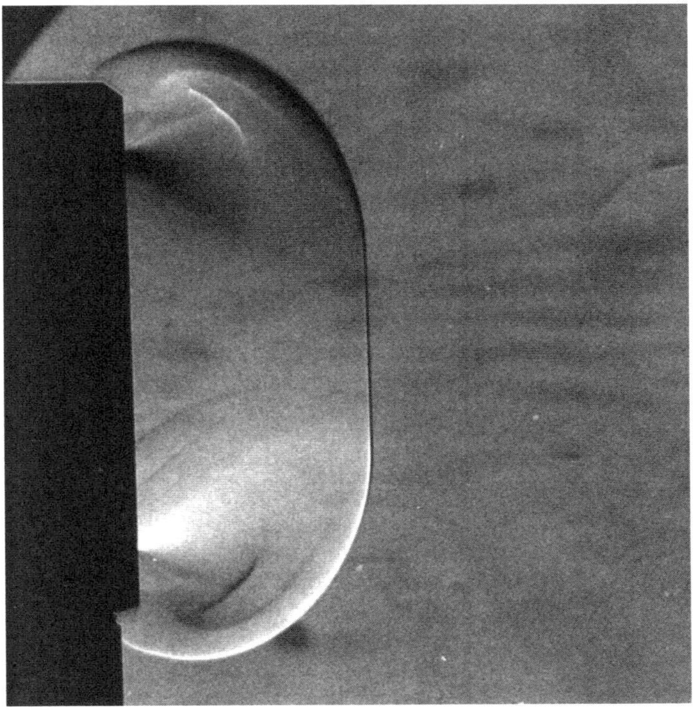

Figure 23.14. *Shock wave leaving an air-air shock tube (document Onera)*

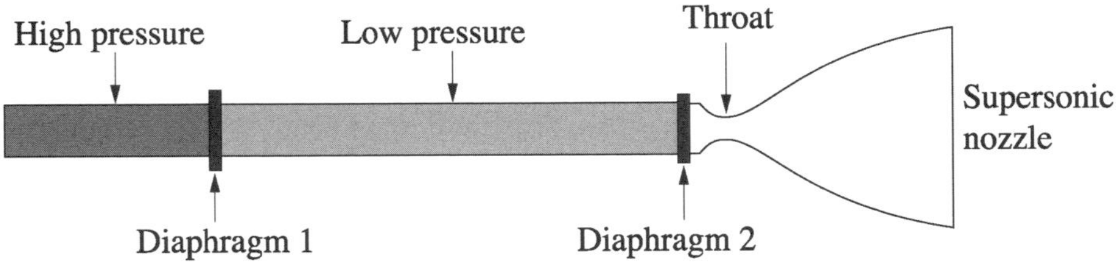

Figure 23.15. *Schematic lay out of a hypersonic wind tunnel with reflected shock*

Chapter 24

Numerical Methods for Calculating Unsteady Flows

24.1. Integration by the method of characteristics

There are circumstances where the solution to the problem cannot be obtained as simply as in the cases that have been analyzed, for example, if the boundary conditions vary with time, or if the piston moves at a variable speed. It is then necessary to resort to numerical integration techniques by determining the solution at discrete points of the space $[x,t]$ and proceeding step by step in the calculation domain. Under these conditions, the solution is obtained only at the calculation points and has an approximated character, since the equations are not solved exactly. This is a shortcoming of all numerical methods, widely used today, but it is possible to get a high enough accuracy by refining the calculation grid; i.e. by considering calculation points very close to each other and using accurate discretization schemes. The method of characteristics, described at length in Chapter 18 dedicated to steady supersonic flows, builds on the mathematical properties of unsteady flows which we have just seen.

We assume that in the plane $[x,t]$, the flow is, as shown in Figure 24.1a, known at two points (1) and (2). The intersection of the characteristic (η) through (1) and of the characteristic (ξ) through (2) is a point (3) whose properties are related to those in (1) and (2) by the Riemann relations:

$$on\ (\eta)\ \frac{2}{\gamma-1}a_3+u_3=\frac{2}{\gamma-1}a_1+u_1$$

$$on\ (\xi)\ \frac{2}{\gamma-1}a_3-u_3=\frac{2}{\gamma-1}a_2-u_2.$$

These two relations allow the calculation of u_3 and a_3 through:

$$u_3=\frac{1}{\gamma-1}(a_1-a_2)+\frac{1}{2}(u_1-u_2)$$

$$a_3=\frac{1}{2}(a_1+a_2)+\frac{\gamma-1}{4}(u_1+u_2).$$

The position of the point (3) in the plane $[x,t]$ is calculated at the intersection of the characteristics (η) and (ξ) which are approximated by straight segments passing through (1) and (2) with the slopes:

$$for\ \ (\eta)\rightarrow\frac{1}{2}\left[(u_1+a_1)+(u_3+a_3)\right]$$

$$for\ \ (\xi)\rightarrow\frac{1}{2}\left[(u_2-a_2)+(u_3-a_3)\right].$$

If the point (3) to calculate is, according to the diagram in Figure 24.1b, on the trajectory of the piston, then the velocity is imposed in (3) by the boundary condition:

$$u_3=V_P(\tau).$$

The Riemann relation along the (ξ) between (3) and (2) is enough to calculate the velocity of sound a_3. In the case of an open end (E), the conditions to be imposed depend on the subsonic or supersonic nature of the velocity in (E). This issue has been discussed in detail in section 21.8 and will not be repeated here. The various operators of the method of characteristics are summarized in Figure 24.1.

Let us consider an initial discretization of the data on a portion *AB* of a non-characteristic curve (C_0) of the plane $[x,t]$. As drawn in Figure 24.2, these points are numbered 00, 01, 02, 03, 04, 05, limiting ourselves to six points for more clarity. By using the above procedure, we calculate 10 from 00 and 01, then 11 from 01 and 02, and so on until point 14 calculated from 04 and 05. It is not possible to go further due to the lack of data beyond B. We calculate 20 from 10 and 11, then 21 from 11 and 12 until 23. The process continues through 30 up to 32, 40 up to 41, and finally 50. It is impossible to go farther than 50 due to the lack of information outside of the arc AB of the starting curve $\left(C_0\right)$.

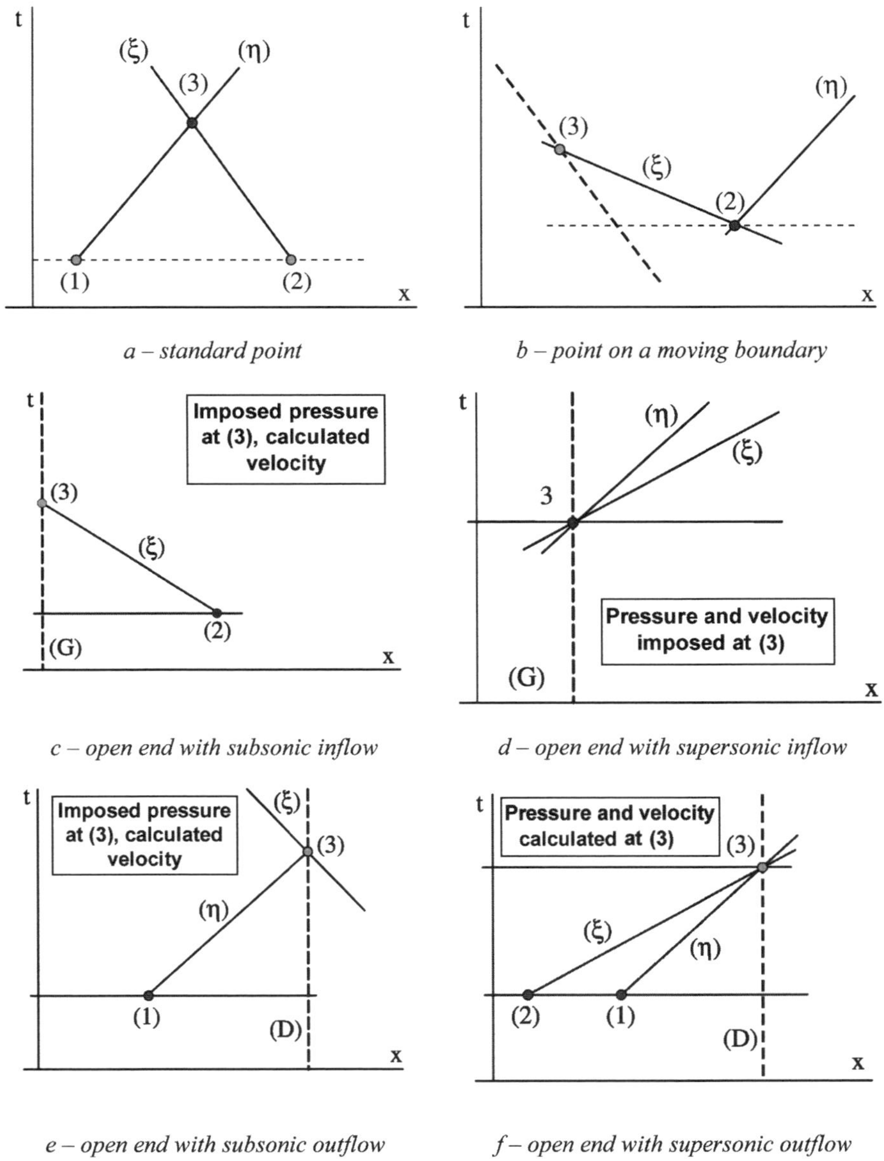

a – standard point

b – point on a moving boundary

c – open end with subsonic inflow

d – open end with supersonic inflow

e – open end with subsonic outflow

f – open end with supersonic outflow

Figure 24.1. *Operators for the method of characteristics*

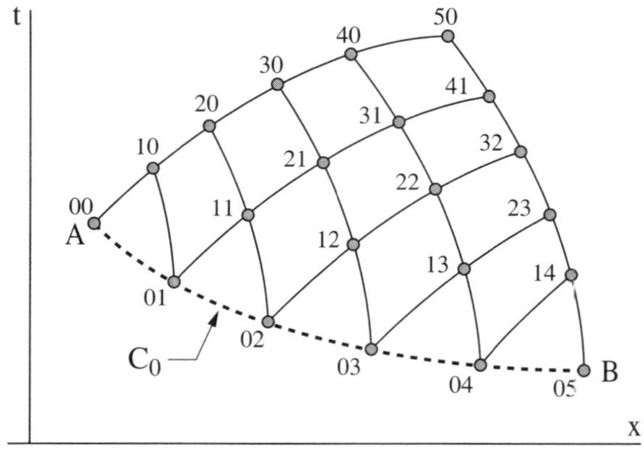

Figure 24.2. *Domain calculable from the data on the portion of a non-characteristic curve*

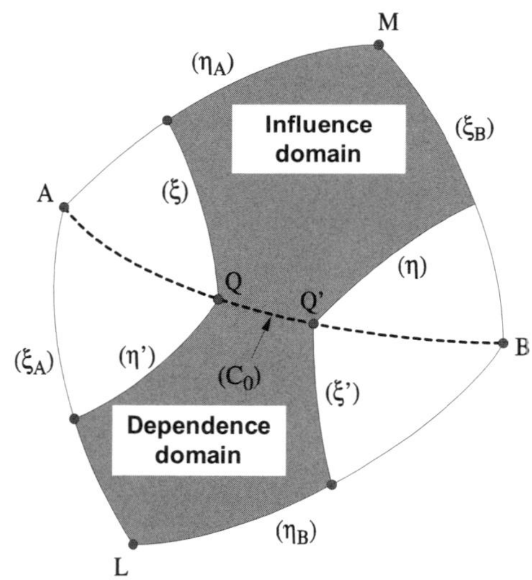

Figure 24.3. *Domain of dependence and domain of influence*

This result highlights important conclusions, referring to Figure 24.3:

> The distribution data on the arc AB of the non-characteristic (C_0) in the plane $[x,t]$ determines the flow in the quadrilateral $LAMB$ formed by the characteristics of the two families going through A and B. We are free to extend the initial data on both sides of AB without modifying the solution constructed in $LAMB$, provided that the continuity of data is preserved in A and B.

The state of flow at the point L depends exclusively on the data on AB: we say that this arc belongs to the *domain of dependence* of point L.

Conversely, if we slightly modify the data on a small arc QQ' of AB, this change is only perceived in the *domain of influence* defined by the extreme characteristics through Q and Q'.

It is easy to convince ourselves that if the curve (C_0) is characteristic, then it is impossible to extend the calculation beyond (C_0) (see section 18.3). The Cauchy problem is well undetermined. However, if one end is a curve on which a boundary condition must be satisfied, then this condition allows, as shown in Figure 24.4, "taking off" from (C_0) and extending the calculation. Data are provided on the curve (C_0) which is here a left running characteristic (η_0). As the point 00 of (η_0) belongs to the trajectory of the piston, it is possible to calculate, by the process described above, point 11 on this trajectory, and hence the left running characteristic rising from 11 to 17. The reader will easily understand the calculation procedure for calculating the domain limited by the trajectory of the piston and the left running characteristic passing through the end 07 of (η_0).

These calculable domains issues were discussed in more detail in Chapter 18, devoted to the theory of characteristics for two-dimensional supersonic non-viscous steady flows.

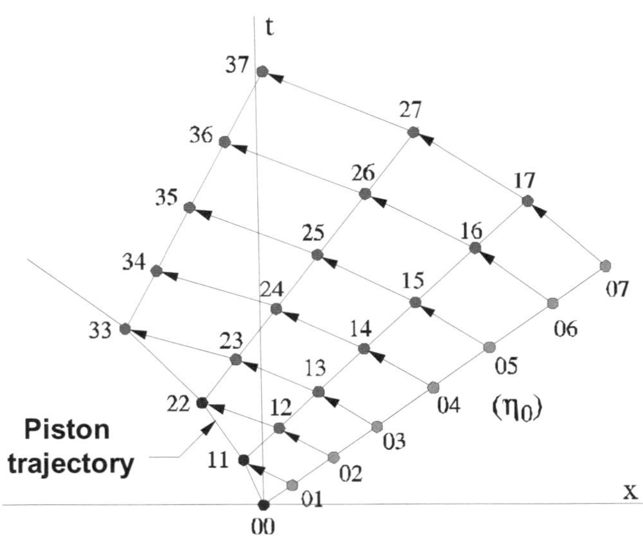

Figure 24.4. *Domain calculable from a characteristic curve and a boundary condition*

24.2. Integration by finite differences

24.2.1. *Dimensionless equations in conservative form*

The integration technique based on the theory of characteristics is extremely fast and accurate; however, it quickly leads to serious programming difficulties. The management of the calculation grid becomes difficult when multiple wave reflections occur or when focalizations waves occur with formation of shocks. In the case of complex flows, we may prefer a finite differences technique where the different phenomena, including shock waves, are automatically captured. The calculation accuracy is smaller than with the characteristics but it is often sufficient for the intended applications and the programming of the method is much easier. We present in this section a finite difference method using the MacCormack predictor–corrector scheme of which the merit is to be conceptually simple and give good results (provided the shocks are not too intense). It can be easily encoded and therefore useful for exercise applications. This scheme has been supplanted by more elaborate schemes allowing the capture of discontinuities more accurately and to calculate very intense shocks, but their implementation is more complex. For more information, the reader is referred to books on numerical analysis.

We reconsider the equations for a one-dimensional, non-viscous, and unsteady flow written in section 21.1. The notations are identical; we recall that A denotes the area of the section of flow which, here, can vary with both x and time t. For example, if it is a one-dimensional supersonic nozzle with a second adjustable throat.

– *continuity equation:*

$$\frac{\partial(\rho A)}{\partial t}+\frac{\partial(\rho u A)}{\partial x}=0.$$

– *equation of motion:*

We reach in section 21.1 the equation:

$$\rho A dx\left(\frac{\partial u}{\partial t}+u\frac{\partial u}{\partial x}\right)=pA-\left[pa+\frac{\partial(pA)}{\partial x}dx\right]+p\frac{\partial A}{\partial x}dx$$

i.e.:

$$\rho A\frac{\partial u}{\partial t}+\rho A u\frac{\partial u}{\partial x}+\frac{\partial(pA)}{\partial x}=p\frac{\partial A}{\partial x}.$$

But as:

$$\frac{\partial}{\partial x}\left[A\left(p+\rho u^2\right)\right]=\frac{\partial (pA)}{\partial x}+\rho A u\frac{\partial u}{\partial x}+u\frac{\partial}{\partial x}(\rho u A).$$

Hence, given the continuity equation, the equation of motion becomes:

$$\frac{\partial(\rho u A)}{\partial t}+\frac{\partial}{\partial x}\left[A\left(p+\rho u^2\right)\right]=p\frac{\partial A}{\partial x}.$$

This expression is called conservative or quasi-conservative form for the quantity $A\left[p+\rho u^2\right]$ which is the dynalpy of the fluid (see Chapter 5).

– *energy equation:*

This equation should be reconsidered because the flow cannot be assumed isentropic if we want to be able to calculate shocks. The first principle is applied by considering the control volume drawn in Figure 24.5 and by using the internal energy *e* as the state function.

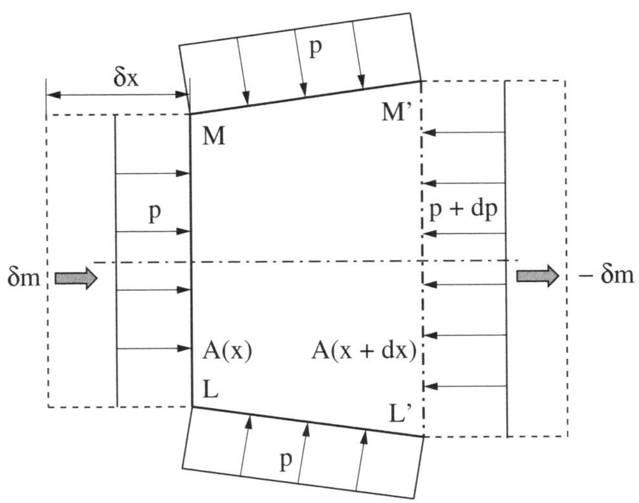

Figure 24.5. *Control volume for the energy equation*

The total energy contained in the control volume *MM'L'L* is:

$$\rho A dx\left(e+\frac{u^2}{2}\right).$$

The variation of this energy during the time interval δt is:

$$\frac{\partial}{\partial t}\left[\rho A dx\left(e+\frac{u^2}{2}\right)\right]\delta t+\left\{\frac{\partial}{\partial x}\left[\rho Au\left(e+\frac{u^2}{2}\right)\right]\right\}dx\delta t.$$

No heat exchange (adiabaticity hypothesis).

Work of external forces. The term expressing the work of the pressure forces during introduction and extraction of the fluid contained in the control volume is written:

$$pAu\ \delta t-\left[pAu+\frac{\partial(pAu)}{\partial x}dx\right]\delta t.$$

The work of the lateral pressure forces due to the section variation as a function of time is given by:

$$-p\frac{\partial A}{\partial t}dx\delta t.$$

The balance equation is:

$$\frac{\partial}{\partial t}\left[\rho A dx\left(e+\frac{u^2}{2}\right)\right]\delta t+\left\{\frac{\partial}{\partial x}\left[\rho Audx\left(e+\frac{u^2}{2}\right)\right]\right\}dx\delta t=-\frac{\partial(pAu)}{\partial x}dx\delta t-p\frac{\partial A}{\partial t}dx\delta t$$

$$\frac{\partial}{\partial t}\left[\rho A\left(e+\frac{u^2}{2}\right)\right]+\frac{\partial}{\partial x}\left[\rho Au\left(e+\frac{u^2}{2}\right)+pAu\right]=-p\frac{\partial A}{\partial t}.$$

This leads to the energy equation:

$$\frac{\partial}{\partial t}\left[\rho A\left(e+\frac{u^2}{2}\right)\right]+\frac{\partial}{\partial x}\left[\rho Au\left(e+\frac{u^2}{2}+\frac{p}{\rho}\right)\right]=-p\frac{\partial A}{\partial t}.$$

By introducing the specific total internal energy:

$$e_T=e+\frac{u^2}{2}$$

we obtain another form of the energy equation:

$$\frac{\partial(\rho e_T A)}{\partial t}+\frac{\partial}{\partial x}\left[\rho A u\left(e_T+\frac{p}{\rho}\right)\right]=-p\frac{\partial A}{\partial t}.$$

COMMENT.– If the flow is steady and section A is constant with respect to x, the equations above become:

– *continuity:* $\frac{\partial(\rho u A)}{\partial x}=0 \rightarrow \rho u A = constant$

– *motion:* $\frac{\partial}{\partial x}\left[A\left(p+\rho u^2\right)\right]=0 \rightarrow A\left(p+\rho u^2\right)=constant$

– *energy:* $\frac{\partial}{\partial x}\left[\rho A u\left(e_T+\frac{p}{\rho}\right)\right]=0 \rightarrow \rho A u\left(e_T+\frac{p}{\rho}\right)=constant.$

We recognize the Rankine–Hugoniot equations (see section 13.2). The adopted formulation contains these equations, which will allow a more accurate capture of shock waves.

Before integration, the equations are put into dimensionless form by introducing reference quantities chosen in reference to the shock tube problem (see Figure 24.6):

– density in the medium 2: ρ_2,

– pressure in the medium 2: p_2,

– velocity in the medium 2: a_2,

– length of the tube (or nozzle for other applications): L,

– reference section of the tube: A^*.

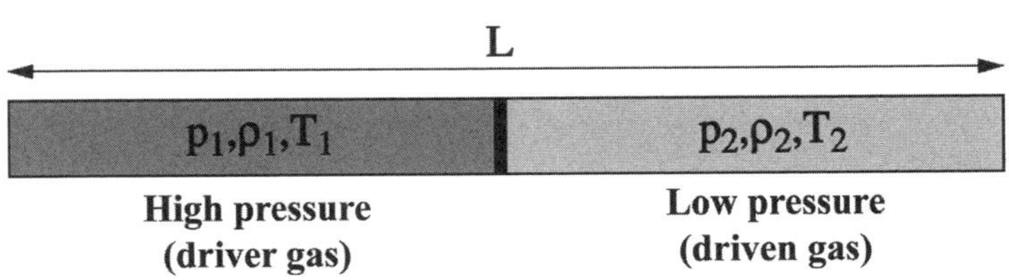

Figure 24.6. *Initial conditions for the shock tube*

Hence, the dimensionless variables are:

$$\bar{\rho}=\frac{\rho}{\rho_2},\quad \bar{p}=\frac{p}{p_2},\quad \bar{u}=\frac{u}{a_2},\quad \bar{x}=\frac{x}{L},\quad \bar{A}=\frac{A}{A^*}.$$

– *continuity equation:*

It becomes:

$$\frac{\rho_2 A^*}{t^*}\frac{\partial\left(\bar{\rho}\bar{A}\right)}{\partial\bar{t}}+\frac{\rho_2 a_2 A^*}{L}\frac{\partial\left(\rho\bar{u}A\right)}{\partial\bar{x}}=0.$$

This leads us to adopt as a time scale:

$$t^*=\frac{L}{a_2}$$

$$\frac{\rho_2 a_2 A^*}{L}\frac{\partial\left(\bar{\rho}\bar{A}\right)}{\partial\bar{t}}+\frac{\rho_2 a_2 A^*}{L}\frac{\partial\left(\bar{\rho}\bar{u}\bar{A}\right)}{\partial\bar{x}}=0.$$

Hence, the dimensionless continuity equation:

$$\boxed{\frac{\partial\left(\bar{\rho}\bar{A}\right)}{\partial\bar{t}}+\frac{\partial\left(\bar{\rho}\,\bar{u}\,\bar{A}\right)}{\partial\bar{x}}=0}$$

– *equation of motion:*

It takes the form:

$$\frac{\rho_2 a_2^2 A^*}{L}\frac{\partial\left(\bar{\rho}\bar{u}\bar{A}\right)}{\partial\bar{t}}+\frac{\rho_2 a_2^2 A^*}{L}\frac{\partial}{\partial\bar{x}}\left[\bar{A}\left(\frac{p}{\rho_2 a_2^2}+\bar{\rho}\bar{u}^2\right)\right]=p\frac{A^*}{L}\frac{\partial\bar{A}}{\partial\bar{x}}.$$

Knowing that:

$$a_2^2=\gamma\frac{p_2}{\rho_2}\quad\rightarrow\quad \rho_2 a_2^2=\gamma p_2$$

and by introducing the dimensionless pressure:

$$\overline{p} = \frac{p}{\rho_2}$$

it follows that:

$$\frac{p}{\rho_2 a_2^2} = \frac{\overline{p}}{\gamma}.$$

Hence, the dimensionless motion equation:

$$\boxed{\frac{\partial\left(\overline{\rho}\,\overline{u}\,\overline{A}\right)}{\partial\overline{t}} + \frac{\partial}{\partial\overline{x}}\left[\overline{A}\left(\frac{\overline{p}}{\gamma} + \overline{\rho}\,\overline{u}^2\right)\right] = \frac{\overline{p}}{\gamma}\frac{\partial\overline{A}}{\partial\overline{x}}}$$

– *energy equation:*

Expression of the dimensionless total energy:

$$\overline{e}_T = \frac{e}{a_2^2} + \frac{1}{2}\frac{u^2}{a_2^2}.$$

Hence, the equation:

$$\frac{\rho_2 a_2^3 A^*}{L}\frac{\partial\left(\overline{\rho}\overline{e}_T\overline{A}\right)}{\partial\overline{t}} + \frac{\rho_2 a_2^3 A^*}{L}\frac{\partial}{\partial\overline{x}}\left[\overline{\rho}\overline{A}\overline{u}\left(\overline{e}_T + \frac{\overline{p}}{\gamma\overline{\rho}}\right)\right] = -\frac{p_2 a_2 A^*}{L}\overline{p}\frac{\partial\overline{A}}{\partial\overline{t}} = -\frac{\rho_2 a_2^3 A^*}{L}\frac{\overline{p}}{\gamma}\frac{\partial\overline{A}}{\partial\overline{t}}$$

and the dimensionless energy equation:

$$\frac{\partial\left(\overline{\rho}\,\overline{e}_T\overline{A}\right)}{\partial\overline{t}} + \frac{\partial}{\partial\overline{x}}\left[\overline{\rho}\overline{A}\overline{u}\left(\overline{e}_T + \frac{\overline{p}}{\gamma\overline{\rho}}\right)\right] = -\frac{\overline{p}}{\gamma}\frac{\partial\overline{A}}{\partial\overline{t}}.$$

The expression of the specific internal energy as a function of p and ρ :

$$e = C_v T = C_v\frac{p}{r\rho} = \frac{C_v}{C_p - C_v}\frac{p}{\rho} = \frac{1}{\gamma - 1}\frac{p}{\rho}.$$

Hence:

$$\bar{e} = \frac{e}{a_2^2} = \frac{1}{\gamma - 1} \frac{p_2}{\rho_2 a_2^2} \frac{\bar{p}}{\bar{\rho}} = \frac{1}{\gamma(\gamma - 1)} \frac{\bar{p}}{\bar{\rho}}$$

and:

$$\boxed{\bar{e} = \frac{1}{\gamma(\gamma - 1)} \frac{\bar{p}}{\bar{\rho}}}$$

Another form of the energy equation is:

$$\boxed{\frac{\partial}{\partial \bar{t}}\left[\bar{A}\left(\frac{\bar{p}}{\gamma(\gamma-1)} + \frac{1}{2}\bar{\rho}\bar{u}^2\right)\right] + \frac{\partial}{\partial \bar{x}}\left[\bar{A}\bar{u}\left(\frac{\bar{p}}{\gamma-1} + \frac{1}{2}\bar{\rho}\bar{u}^2\right)\right] = -\frac{\bar{p}}{\gamma}\frac{\partial \bar{A}}{\partial \bar{t}}.}$$

24.2.2. *Explicit numerical integration method*

We define the "vector" of the unknowns to be calculated by:

$$Q(x_j, t) = \begin{cases} \bar{\rho}\bar{A} \\ \bar{\rho}\ \bar{u}\ \bar{A} \\ \bar{A}\left(\dfrac{\bar{p}}{\gamma(\gamma-1)} + \dfrac{1}{2}\bar{\rho}\ \bar{u}^2\right) \end{cases} \rightarrow \textit{physical variables} \begin{cases} \bar{\rho} \\ \bar{u} \\ \bar{p}. \end{cases}$$

Then, the "vector" of the flux terms:

$$F(x_j, t) = \begin{cases} \bar{\rho}\bar{u}\bar{A} \\ \bar{A}\left(\dfrac{\bar{p}}{\gamma} + \bar{\rho}\bar{u}^2\right) \\ \bar{A}\bar{u}\left(\dfrac{\bar{p}}{\gamma-1} + \dfrac{1}{2}\bar{\rho}\bar{u}^2\right). \end{cases}$$

Finally that of the source terms:

$$G(x_j,t) = \begin{cases} 0 \\ \dfrac{\overline{p}}{\gamma}\dfrac{\partial \overline{A}}{\partial \overline{x}} \\ -\dfrac{\overline{p}}{\gamma}\dfrac{\partial \overline{A}}{\partial \overline{t}}. \end{cases}$$

The equations are written in compact form as:

$$\frac{\partial}{\partial t}\left[Q(x,t)\right] + \frac{\partial}{\partial x}\left[F(x,t)\right] = G(x,t).$$

The length of the pipe (or nozzle) is divided into N steps Δx which we assume to be equal to simplify and the integration in time is done by time steps Δt. The steps Δx and Δt cannot be arbitrary; they must satisfy a stability criterion, the so-called Courant–Friedrich–Lewy or CFL criterion (see section 24.3), under penalty of seeing the calculation diverging. Knowing the state of the flow at the time $t^{(n)}$, we intend to advance this solution in time from $t^{(n)}$ to $t^{(n+1)}$ (see Figure 24.7).

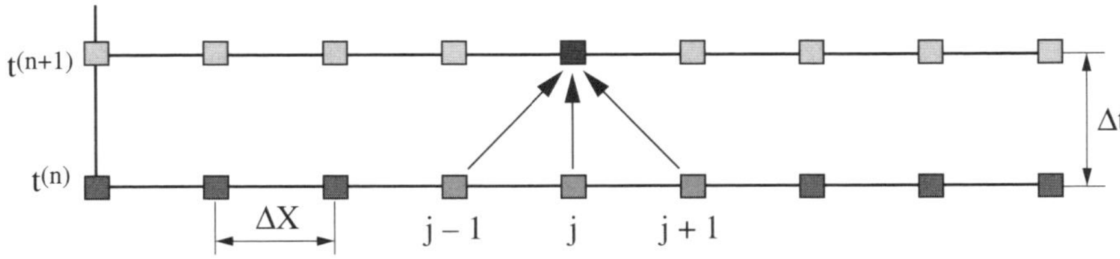

Figure 24.7. *Advancement of the solution in time*

To do this, we write for the calculation point at the abscissa x_j (to first order in time):

$$\boxed{Q\left(x_j,t^{(n+1)}\right) = Q\left(x_j,t^{(n)}\right) + \left(\frac{\partial Q}{\partial t}\right)^{(n)} \Delta t}$$

According to the motion equations:

$$\left(\frac{\partial Q}{\partial t}\right)^{(n)} = -\left(\frac{\partial F}{\partial x}\right)^{(n)} + G\left(x_j,t^{(n)}\right).$$

The term $\left(\dfrac{\partial F}{\partial x}\right)^{(n)}$ is calculated from the values of the solution known at the time $t^{(n)}$.

According to the MacCormack scheme, accuracy is improved by proceeding in two steps.

– *Predictor step:*

We advance the solution by the already given relation:

$$\tilde{Q}\left(x_j, t^{(n+1)}\right) = Q\left(x_j, t^{(n)}\right) + \left(\frac{\partial Q}{\partial t}\right)^{(n)} \Delta t.$$

In this step, the flux and source terms are calculated by using an upwind discretization scheme between x_j and x_{j+1} (see Figure 24.8).

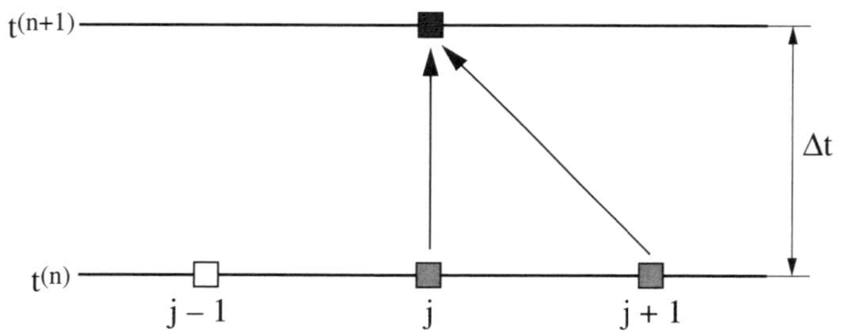

Figure 24.8. *Predictor step and upwind discretization scheme*

Thus:

$$\left(\frac{\partial F}{\partial x}\right)^{(n)} = \frac{1}{\Delta x}\left[F\left(x_{j+1}, t^{(n)}\right) - F\left(x_j, t^{(n)}\right)\right].$$

The source term is calculated by taking the average between x_j and x_{j+1}:

$$\rightarrow \quad \frac{1}{2}\left[G\left(x_j, t^{(n)}\right) + G\left(x_{j+1}, t^{(n)}\right)\right].$$

Hence, the unknown vector $t^{(n)}$ is calculated from:

$$\left(\frac{\partial Q}{\partial t}\right)^{(n)} = -\frac{1}{\Delta x}\left[F\left(x_{j+1}, t^{(n)}\right) - F\left(x_j, t^{(n)}\right)\right] + \frac{1}{2}\left[G\left(x_{j+1}, t^{(n)}\right) + G\left(x_j, t^{(n)}\right)\right].$$

This calculation provides an estimate of the solution or provisional values:

$$\tilde{Q}\left(x_j,t^{(n+1)}\right) \rightarrow \tilde{F}\left(x_j,t^{(n+1)}\right),\ \tilde{G}\left(x_j,t^{(n+1)}\right).$$

– *Corrector step:*

The advance between $t^{(n)}$ and $t^{(n+1)}$ is repeated by considering for the derivative $\left(\frac{\partial Q}{\partial t}\right)$ the arithmetic mean of its value at time $t^{(n)}$ and the estimate made at time $t^{(n+1)}$ with the provisional solution.

Thus:

$$Q\left(x_j,t^{(n+1)}\right)=Q\left(x_j,t^{(n)}\right)+\frac{1}{2}\left[\left(\frac{\partial Q}{\partial t}\right)^{(n)}+\left(\frac{\partial \tilde{Q}}{\partial t}\right)^{(n+1)}\right]\Delta t.$$

This time, the flow and source terms are calculated by using a downwind discretization scheme between x_{j-1} and x_j (see Figure 24.9).

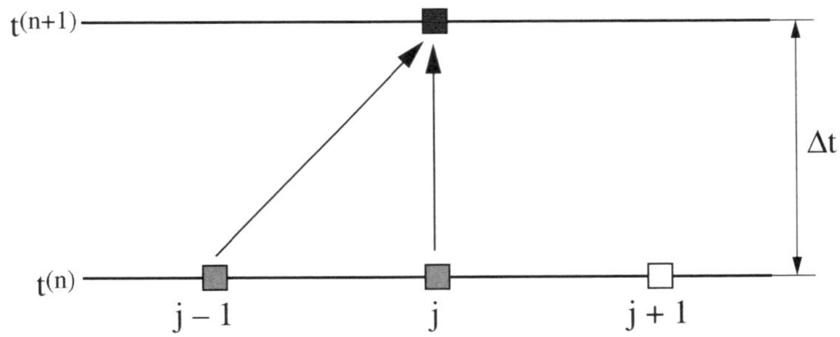

Figure 24.9. *Corrector step and downwind discretization scheme*

Hence:

$$\left(\frac{\partial \tilde{F}}{\partial x}\right)^{(n+1)}=\frac{1}{\Delta x}\left[\tilde{F}\left(x_j,t^{(n+1)}\right)-\tilde{F}\left(x_{j-1},t^{(n+1)}\right)\right]$$

and:

$$\frac{1}{2}\left[\tilde{G}\left(x_j,t^{(n+1)}\right)+\tilde{G}\left(x_{j-1},t^{(n+1)}\right)\right].$$

This approach improves the calculation accuracy of derivatives with respect to x.

Hence:

$$\left(\frac{\partial \tilde{Q}}{\partial t}\right)^{(n+1)} = -\frac{1}{\Delta x}\left[\tilde{F}\left(x_j, t^{(n+1)}\right) - \tilde{F}\left(x_{j-1}, t^{(n+1)}\right)\right] + \frac{1}{2}\left[\tilde{G}\left(x_j, t^{(n+1)}\right) + \tilde{G}\left(x_{j-1}, t^{(n+1)}\right)\right].$$

The expression of the corrected or improved value is:

$$Q\left(x_j, t^{(n+1)}\right) = Q\left(x_j, t^{(n)}\right) - \frac{\Delta t}{2\Delta x}\left[F\left(x_{j+1}, t^{(n)}\right) - F\left(x_j, t^{(n)}\right) + \tilde{F}\left(x_j, t^{(n+1)}\right) - \tilde{F}\left(x_{j-1}, t^{(n+1)}\right)\right] + \frac{\Delta t}{2}\left[\frac{G\left(x_{j+1}, t^{(n)}\right) + G\left(x_j, t^{(n)}\right)}{2} + \frac{\tilde{G}\left(x_j, t^{(n+1)}\right) + \tilde{G}\left(x_{j-1}, t^{(n+1)}\right)}{2}\right].$$

But:

$$-\frac{\Delta t}{2\Delta x}\left[F\left(x_{j+1}, t^{(n)}\right) - F\left(x_j, t^{(n)}\right)\right] + \frac{\Delta t}{2}\left[\frac{G\left(x_{j+1}, t^{(n)}\right) + G\left(x_j, t^{(n)}\right)}{2}\right] = \frac{1}{2}\left[\tilde{Q}\left(x_j, t^{(n+1}\right) - Q\left(x_j, t^{(n}\right)\right]$$

which leads to:

$$Q\left(x_j, t^{(n+1)}\right) = Q\left(x_j, t^{(n)}\right) - \frac{\Delta t}{2\Delta x}\left[\tilde{F}\left(x_j, t^{(n+1)}\right) - \tilde{F}\left(x_{j-1}, t^{(n+1)}\right)\right] + \frac{\Delta t}{2}\left[\frac{\tilde{G}\left(x_j, t^{(n+1)}\right) + \tilde{G}\left(x_{j-1}, t^{(n+1)}\right)}{2}\right] + \frac{1}{2}\left[\tilde{Q}\left(x_j, t^{(n+1}\right) - Q\left(x_j, t^{(n}\right)\right].$$

Ultimately:

$$\boxed{Q\left(x_j, t^{(n+1)}\right) = \frac{1}{2}\left[Q\left(x_j, t^{(n)}\right) + \tilde{Q}\left(x_j, t^{(n+1)}\right)\right] - \frac{\Delta t}{2\Delta x}\left[\tilde{F}\left(x_j, t^{(n+1)}\right) - \tilde{F}\left(x_{j-1}, t^{(n+1)}\right)\right] + \frac{\Delta t}{2}\left[\frac{\tilde{G}\left(x_j, t^{(n+1)}\right) + \tilde{G}\left(x_{j-1}, t^{(n+1)}\right)}{2}\right]}$$

24.3. Stability condition and CFL criterion (Courant–Friedrich–Lewy)

The just described numerical method uses a so-called explicit discretization in the sense that the space derivatives (along x) are evaluated at the time $t^{(n)}$ for a calculation of the solution at the time $t^{(n+1)}$. To lead to a stable calculation, such a scheme must satisfy a constraint on the time and space steps which, simply stated, is that the time step Δt should be smaller than the space step Δx is. This constraint is expressed as a stability criterion, *called Courant–Friedrich–Lewy criterion* or simply *CFL*. Here, we give an intuitive presentation of this criterion, referring the reader to the literature on numerical analysis for a rigorous proof.

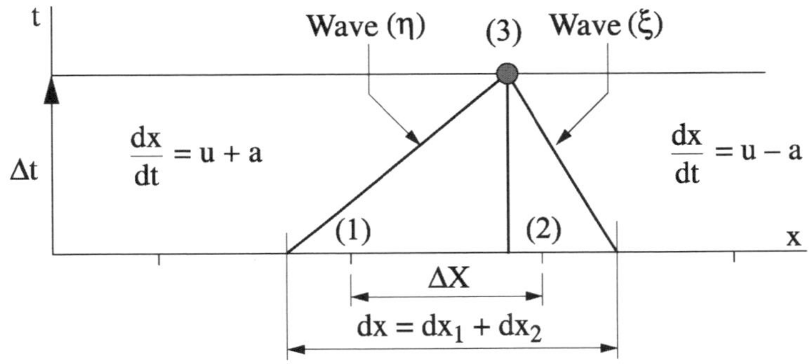

Figure 24.10. *Numerical and physical domains of dependence*

We consider a discretization in space with the constant step Δx and a stage where the solution, known at the time $t^{(n)}$, is advanced by the time step Δt (see Figure 24.10). We calculate the solution in (3) by estimating the space derivatives from the points (1) and (2) separated by Δx. Through (3) pass the characteristics (η) (slope: $\dfrac{dx}{dt} = u + a$) and (ξ) (slope: $\dfrac{dx}{dt} = u - a$) intersecting at the time level $t^{(n)}$ at points separated by a distance dx such that:

$$dx = dx_1 + dx_2 = \left(|u+a| + |u-a|\right)\Delta t.$$

The distance dx defines the *domain of physical dependence* of the solution (3), the state in (3) depending only on the conditions in the region delimited by the characteristics (η) and (ξ) passing through (3) (see above).

The distance Δx defines the *domain of numerical dependence* since the numerical solution (3) involves only the conditions in (1) and (2) separated by Δx. For a MacCormack-like scheme, the situation is somewhat different.

The *CFL* criterion states that the calculation is only stable if the numerical domain of dependence includes the domain of physical dependence, namely:

$$\Delta x > dx$$

i.e.:

$$\Delta x > \left(|u+a| + |u-a|\right)\Delta t.$$

Hence, the constraint of the time step:

$$\boxed{\frac{\Delta t}{\Delta x} < \frac{1}{\left(|u+a| + |u-a|\right)}}$$

The limit condition for which $\Delta x = dx$ is such that:

$$(\Delta t)_{\lim} = \frac{\Delta x}{\left(|u+a| + |u-a|\right)}.$$

We call *CFL* number, or simply *CFL*, the ratio of the time step used in the calculation to the maximum allowed time step:

$$\boxed{CFL = \frac{\Delta t}{(\Delta t)_{\lim}}}$$

Figure 24.11 summarizes these findings.

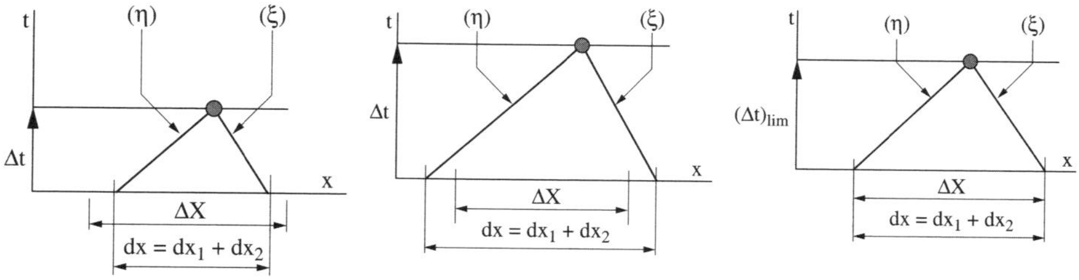

a – CFL<1 stable *b – CFL>1 unstable* *c – CFL=1 marginal*

Figure 24.11. *CFL stability criterion of an explicit calculation scheme*

In terms of computing time economy, it would be advantageous to adopt a *CFL* equal to 1. However, the stability analysis from which this criterion is derived being based on a simplified form of the equations, experience shows that the stability of the calculation is most often provided for *CFL* smaller than 1. We note that the method of characteristics ensures a rigorous physical dependence of the solution at the calculated point: the *CFL* is always equal to 1. The constraint related to *CFL* does not exist for implicit methods where we can use *CFL* well above 1, which enables very quick advancement in time. But there is a limit under the penalty of degrading the accuracy of the calculation.

24.4. Equations in spherical coordinates: spherical shock wave

Some problems of wave propagation are discussed in spherical coordinates, such as the case of an explosion caused by a sudden energy supply in a quasi-punctual region or the rupture of a spherical tank containing a gas initially at high pressure. These two processes, similar to what happens in a shock tube, cause the formation of a spherical shock wave that spreads as it expands into the surrounding environment, the shock being followed by expansion and compression waves.

This kind of problem is discussed with spherical coordinates by numerical methods. Here we give, without proof, the equations to be solved, the notations being conventional.

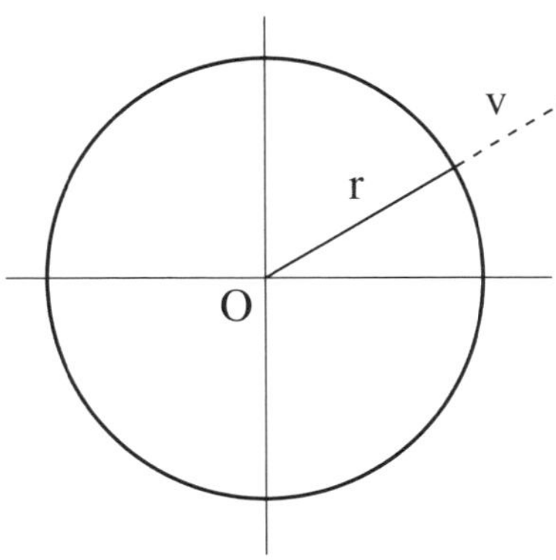

Figure 24.12. *Coordinate system for a spherical wave*

The flow is one-dimensional, its properties being the same on spheres centered on the origin *O* and depending only on the distance *r* to *O* (see Figure 24.12). Here, *V* is the radial component of the velocity (only non-zero component).

– *continuity equation:*

$$\boxed{\frac{\partial \rho}{\partial t}+\frac{\partial(\rho V)}{\partial r}+\frac{2\rho V}{r}=0}$$

– *equation of motion:*

$$\boxed{\frac{\partial(\rho V)}{\partial t}+\frac{\partial}{\partial r}\left(p+\rho V^2\right)+\frac{2\rho V^2}{r}=0}$$

– *energy equation:*

$$\boxed{\frac{\partial(\rho e_T)}{\partial t}+\frac{\partial}{\partial r}\left[\rho V\left(e_T+\frac{p}{\rho}\right)\right]+\frac{2(\rho e_T V+pV)}{r}=0}$$

If the gas is calorically perfect, the total internal energy e_T has the expression:

$$e_T=e+\frac{V^2}{2} \quad\rightarrow\quad e_T=\frac{1}{\gamma-1}\frac{p}{\rho}+\frac{V^2}{2}$$

leading to the equation:

$$\frac{\partial}{\partial t}\left(\frac{p}{\gamma-1}+\frac{1}{2}\rho V^2\right)+\frac{\partial}{\partial r}\left[V\left(\frac{p}{\gamma-1}+\frac{1}{2}\rho V^2+p\right)\right]+\frac{2}{r}\left[V\left(\frac{p}{\gamma-1}+\frac{1}{2}\rho V^2+p\right)\right]=0$$

i.e.:

$$\boxed{\frac{\partial}{\partial t}\left(\frac{p}{\gamma-1}+\frac{1}{2}\rho V^2\right)+\frac{\partial}{\partial r}\left[V\left(\frac{\gamma}{\gamma-1}p+\frac{1}{2}\rho V^2\right)\right]+\frac{2}{r}\left[V\left(\frac{\gamma}{\gamma-1}p+\frac{1}{2}\rho V^2\right)\right]=0}$$

Before resolution, these equations are rewritten in dimensionless form by referring to the situation in Figure 24.13. Before the initial time, the two domains (1) and (2) are separated by a sphere with the radius $R_{\min}$, the calculation domain being limited by the sphere of radius $R_{\max}$. The reference quantities are those concerning the medium (2) in which the shock wave will propagate, i.e.:

– density: ρ_2,

– pressure: p_2,

– velocity of sound: a_2.

We introduce the dimensionless variable:

$$\bar{\rho} = \frac{\rho}{\rho_2} \quad , \quad \bar{p} = \frac{p}{p_2} \quad , \quad \bar{V} = \frac{V}{a_2} \quad , \quad \bar{r} = \frac{r}{R_{\max}}$$

and obtain the dimensionless equations

– *continuity equation:*

$$\rho_2 \frac{\partial \bar{\rho}}{\partial t} + \frac{\rho_2 a_2}{R_{\max}} \frac{\partial \left(\bar{\rho}\bar{V}\right)}{\partial \bar{r}} + \frac{\rho_2 a_2}{R_{\max}} \frac{2\bar{\rho}\bar{V}}{\bar{r}} = 0.$$

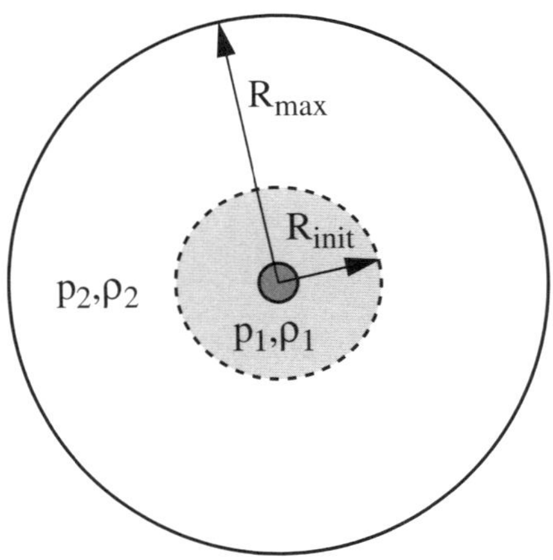

Figure 24.13. *Initial state for the spherical shock wave*

By adopting as time scale $t^* = \dfrac{R_{\max}}{a_2}$, the continuity equation reads:

$$\boxed{\frac{\partial \bar{\rho}}{\partial \bar{t}} + \frac{\partial \left(\bar{\rho}\bar{V}\right)}{\partial \bar{r}} + \frac{2\bar{\rho}\bar{V}}{\bar{r}} = 0}$$

– *equation of motion:*

$$\frac{\partial (\rho V)}{\partial t} + \frac{\partial}{\partial r}\left(p + \rho V^2\right) + \frac{2\rho V^2}{r} = 0$$

$$\frac{\rho_2 a_2^2}{R_{\max}}\frac{\partial(\bar{\rho}\bar{V})}{\partial\bar{t}}+\frac{\rho_2 a_2^2}{R_{\max}}\frac{\partial}{\partial\bar{r}}\left(\frac{\bar{p}}{\gamma}+\bar{\rho}\bar{V}^2\right)+\frac{\rho_2 a_2^2}{R_{\max}}\frac{2\bar{\rho}\bar{V}^2}{\bar{r}}=0$$

i.e.:

$$\boxed{\frac{\partial(\bar{\rho}\bar{V})}{\partial\bar{t}}+\frac{\partial}{\partial\bar{r}}\left(\frac{\bar{p}}{\gamma}+\bar{\rho}\bar{V}^2\right)+\frac{2\bar{\rho}\bar{V}^2}{\bar{r}}=0}$$

– *energy equation:*

$$\frac{\partial(\rho e_T)}{\partial t}+\frac{\partial}{\partial r}\left[\rho V\left(e_T+\frac{p}{\rho}\right)\right]+\frac{2(\rho e_T V+pV)}{r}=0$$

$$\frac{\rho_2 a_2^3}{R_{\max}}\frac{\partial(\bar{\rho}\bar{e}_T)}{\partial\bar{t}}+\frac{\rho_2 a_2^3}{R_{\max}}\frac{\partial}{\partial\bar{r}}\left[\bar{\rho}\bar{V}\left(\bar{e}_T+\frac{\bar{p}}{\bar{\rho}}\right)\right]+\frac{\rho_2 a_2^3}{R_{\max}}\frac{2(\bar{\rho}\bar{e}_T\bar{V}+\bar{p}\bar{V})}{\bar{r}}=0$$

i.e.:

$$\boxed{\frac{\partial(\bar{\rho}\bar{e}_T)}{\partial\bar{t}}+\frac{\partial}{\partial\bar{r}}\left[\bar{\rho}\bar{V}\left(\bar{e}_T+\frac{\bar{p}}{\bar{\rho}}\right)\right]+\frac{2(\bar{\rho}\bar{e}_T\bar{V}+\bar{p}\bar{V})}{\bar{r}}=0.}$$

Another form is:

$$\boxed{\frac{\partial}{\partial\bar{t}}\left[\frac{\bar{p}}{\gamma(\gamma-1)}+\frac{1}{2}\bar{\rho}\bar{V}^2\right]+\frac{\partial}{\partial\bar{r}}\left[\bar{V}\left(\frac{\bar{p}}{\gamma-1}+\frac{1}{2}\bar{\rho}\bar{V}^2\right)\right]+\frac{2}{\bar{r}}\left[\bar{V}\left(\frac{\bar{p}}{\gamma-1}+\frac{1}{2}\bar{\rho}\bar{V}^2\right)\right]=0.}$$

Integration method

The internal radius being given, the initial conditions in 1 and 2 are set as:

$$p_1,\quad \rho_1,\quad a_1=\gamma\frac{p_1}{\rho_1},\quad V_1=0$$

$$p_2,\quad \rho_2,\quad a_2=\gamma\frac{p_2}{\rho_2},\quad V_2=0.$$

If $p_1 > p_2$, we simulate an explosion; if $p_1 < p_2$, it is an implosion.

The equations can be solved by a conventional numerical technique, the already described MacCormack method for example.

Figure 24.14 presents eight views taken from an animation visualizing an explosion caused by a sudden and local energy input (deposition) by a laser flash. The images show the evolution with time of the pressure distributions in a three-dimensional representation, the base plane being a section of the sphere by a meridian plane.

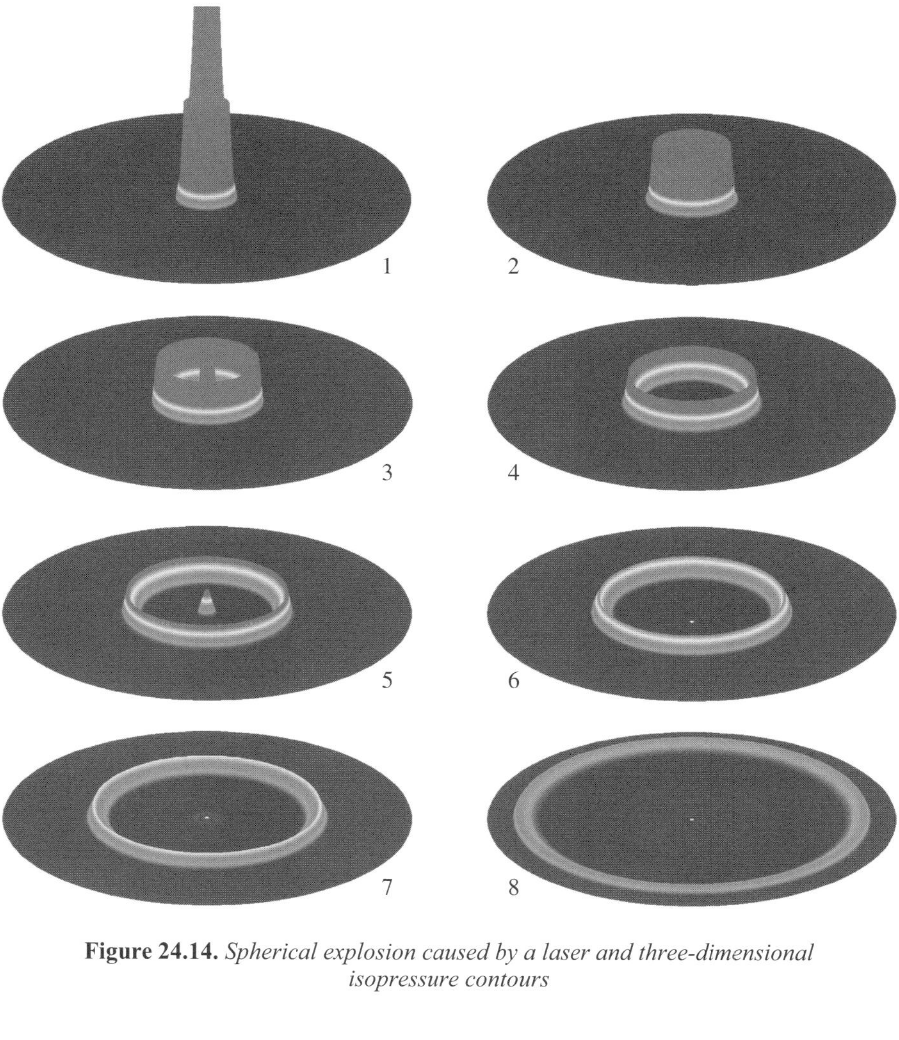

Figure 24.14. *Spherical explosion caused by a laser and three-dimensional isopressure contours*

24.5. Starting and unstarting a supersonic nozzle

The unsteady approach enables an understanding of the starting process of a supersonic nozzle whose operation was examined under the steady aspect in section 9.4. To illustrate, we imagine a symmetric planar nozzle whose profile is shown in Figure 24.15. It includes a converging part (C_1), a throat of half section $A_{c_1} = 0.6$ located at the abscissa $x = 0.5$, a divergent followed by a channel whose half section is increasing slowly from $A_1 = 1$ to $A_2 = 1.03$. A second throat (C_2) is located at the abscissa $x = 2.5$ whose section can be changed with time. A diaphragm is placed at $x = 1$ between an upstream compartment with the pressure $p_4 = 4$ and a downstream compartment with the pressure $p_1 = 1$ (*units* are irrelevant). At the initial time $t = 0$, the diaphragm is broken releasing the air at high pressure, which reproduces the rapid opening of a valve placed at the end of the divergent. The channel is open at both ends where we impose the pressures $p_4 = 4$ at the left and $p_1 = 1$ at the right. The operation of the system is similar to that of a shock tube (see Chapter 23), but here section A of the tube is variable in x and the tube is open on both sides.

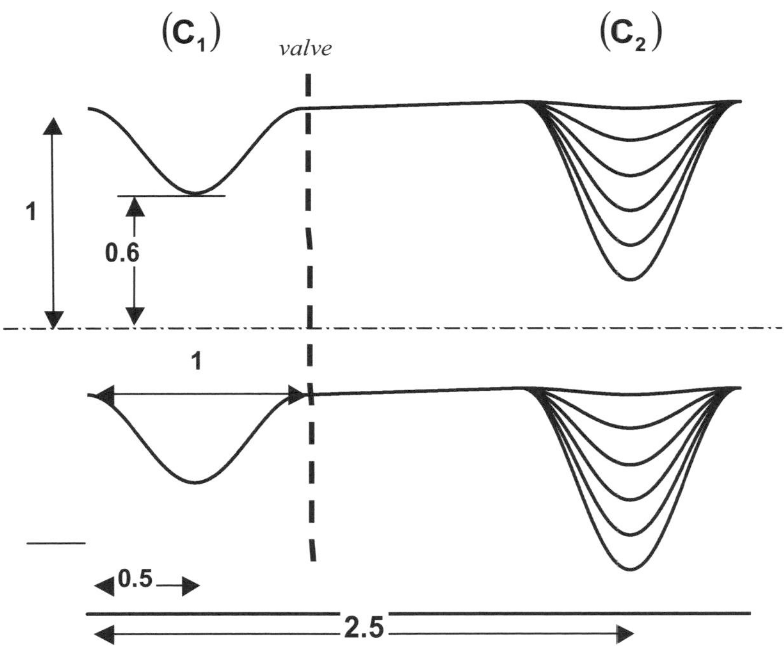

Figure 24.15. *Nozzle with a first fixed throat and a second throat with variable section*

In the first case, the second throat is wide open. An unsteady, one-dimensional calculation allows simulating the phenomena occurring in the nozzle after opening the valve. To give an idea of the duration of the phenomena, we consider a system with a total length of 3 m and a velocity of sound, identical in both initial environments, equal to 340 m/s. Figure 24.16 shows the Mach number distribution along the nozzle at different times following the opening of the valve. As in a shock tube after diaphragm rupture, a shock wave propagates to the right of the diaphragm and an expansion wave to the left. It results in an acceleration of the gas to the right.

We observe at $t = 0.0018$ s the beginning of the propagation of the shock and the expansion. The initial shock is followed by a second, less intense, shock, caused by waves generated by the change in the diverging section. At the time $t = 0.0088$ s, the initial shock is at the abscissa $x \approx 2.4$ and the induced flow is accelerated in the convergent part of the nozzle and decelerated in the divergent part, the Mach number being less than 1 (see Hugoniot theorem in section 9.4). At $t = 0.0176$ s the flow, having become sonic in the throat (C_1), accelerates in the nozzle where a shock wave forms, imposed by the downstream pressure level which is much higher than the pressure caused by the expansion of the supersonic flow in the divergent. From the time where $M = 1$ is reached in (C_1), the upstream flow is invariant: there is choking of the first throat. The initial shock is now outside of the calculation domain. At later times, the process continues, the said starting shock is moving toward the exit section which it reaches at $t \approx 0.1400$ s. Then, the nozzle is fully started, the Mach number being close to 2. The thus established situation has become steady, provided that the pressure imposed on the right end of the system is maintained at a low enough level.

Now we imagine that the second throat (C_2) is closed with a section $A_{c_2} = 0.8$, larger than the area A_{c_1} of the first throat. The events following the rupture of the diaphragm are shown in Figure 24.17. The beginning of the process is similar to the previous case, until $t \approx 0.0088$ s: there is choking of the first throat with the formation in the nozzle diverging part of the starting shock which travels downstream. From $t \approx 0.0132$ s on, the choking of the second throat begins. The flow induced by the initial shock is accelerated in the convergent, becoming sonic in (C_2) at $t \approx 0.0150$ s. For later times, the second throat is choked and the subsequent flow in the divergent is supersonic, up to a Mach number of 1.5 at the end of the system.

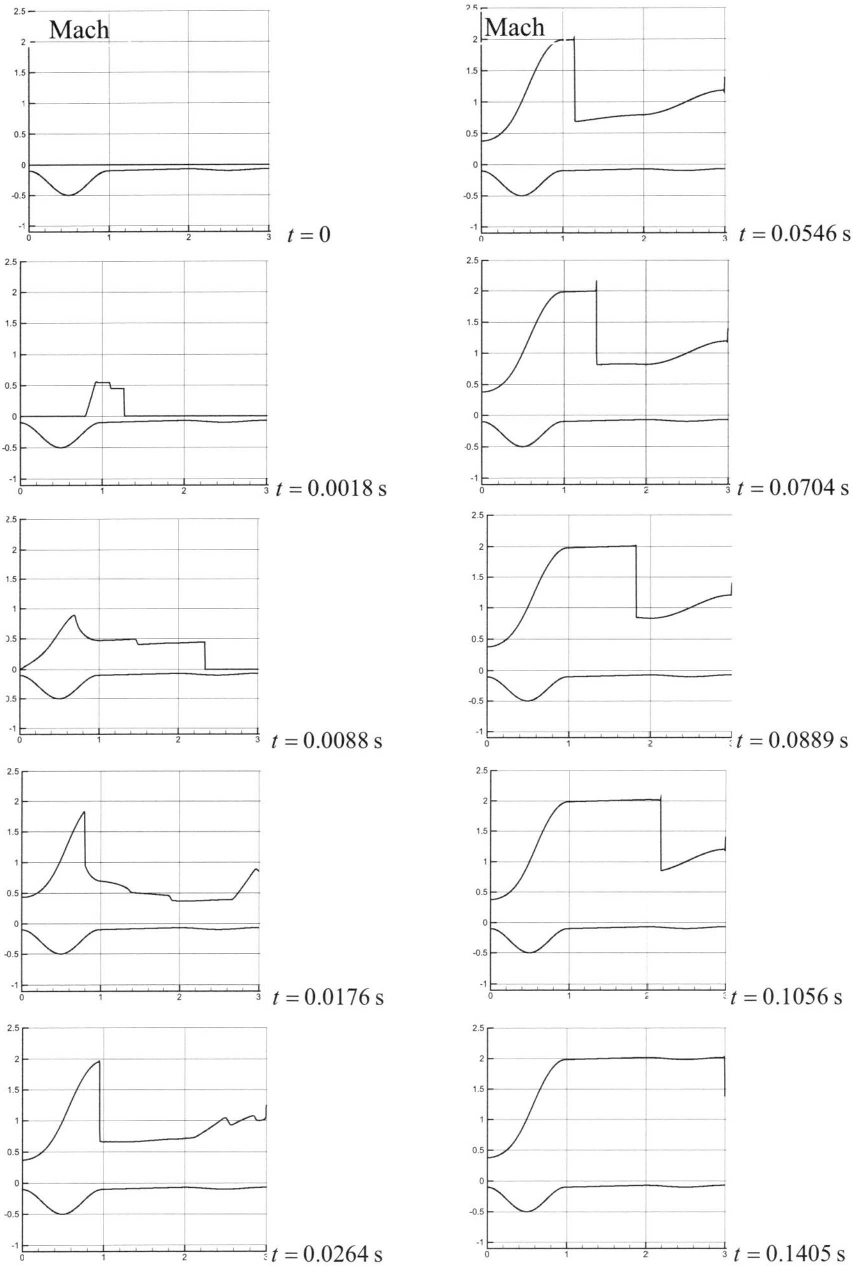

Figure 24.16. *Starting phase of a Mach 2 nozzle. Time in seconds for a total length of 3 m and a velocity of sound of 340 m/s*

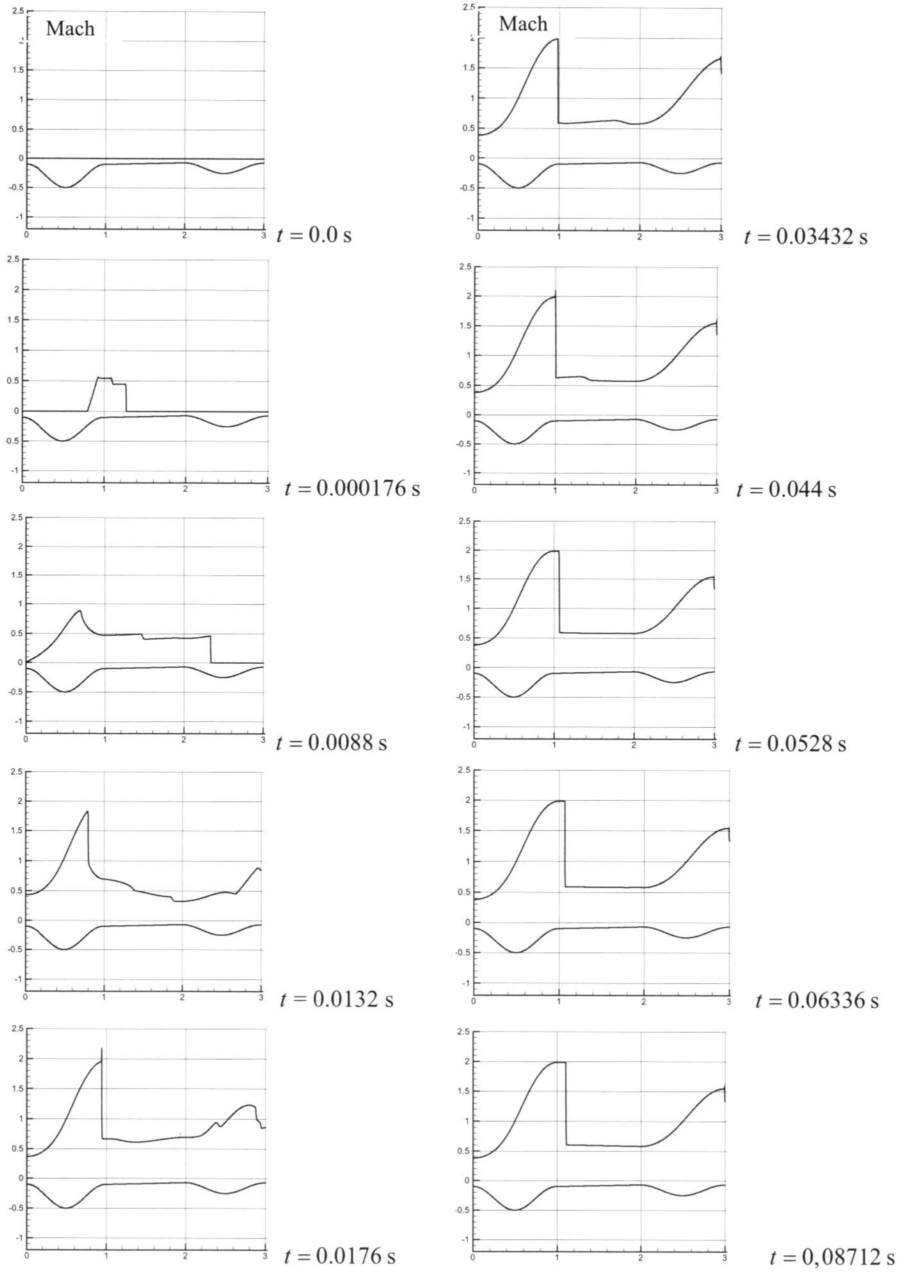

Figure 24.17. *Starting phase of a Mach 2 nozzle with second throat effect. Time in seconds for a total length of 3 m and a velocity of sound of 340 m/s*

In this situation, we must satisfy mass flow condition in the two throats which is expressed by the relation (see section 16.2.4):

$$p_{st_1} A_{c_1} = p_{st_2} A_{c_2} \qquad [24.1]$$

where p_{st_1} and p_{st_2} are the stagnation pressures upstream of A_{c_1} and A_{c_2}, respectively, and where we assumed that the stagnation temperature is constant.

As the second throat has an area larger than the first $\left(A_{c_2} > A_{c_1}\right)$, it follows that the stagnation pressure of the flow through the second throat must be less than that upstream of the first throat $\left(p_{st_2} < p_{st_1}\right)$. The drop of the stagnation pressure is ensured by the shock forming here at the end of the divergent of the first nozzle. We are led to a situation in a steady system with two throats where a shock occurs to ensure the mass flow conservation condition.

The just made reasoning is true only for the steady state because the relation is established for a steady flow. We therefore can only apply it strictly when the flow in the system no longer changes. Similarly, the stagnation temperature is constant in an adiabatic *steady* flow (see section 4.6).

If, from the previous situation, we open the throat (C_2), relation [24.1] requires a reduction of the stagnation pressure, thus increasing the strength of the shock which has to happen in a larger section where the Mach number is higher. As the section of the part following the first divergent is almost constant, the shock cannot find an equilibrium section. Thus, it travels quickly downstream, passing through the throat (C_2), the flow becoming supersonic in (C_2). In the example of a nozzle with the Mach number $M_1 = 2$, the establishment of a supersonic flow in the whole system happens for an area of the second throat given by:

$$A_{c_2} = \frac{A_{c_1}}{\Pi(M_1, \gamma)} = \frac{0.6}{0.72087} = 0.8323$$

where $\Pi(M_1, \gamma)$ is the loss of stagnation pressure through a normal shock at Mach 2 (see Chapter 13).

If, conversely, we close the throat (C_2), the shock moves upstream, travels in the divergent of the first nozzle, and reaches the throat (C_1) where its intensity becomes zero and then vanishes. In the limit situation, the two throats (C_1) and (C_2) are sonic. A subsequent slight decrease of (C_2) causes the unchoking of (C_1) where the flow will therefore become subsonic. The sonic state has moved from (C_1) to (C_2).

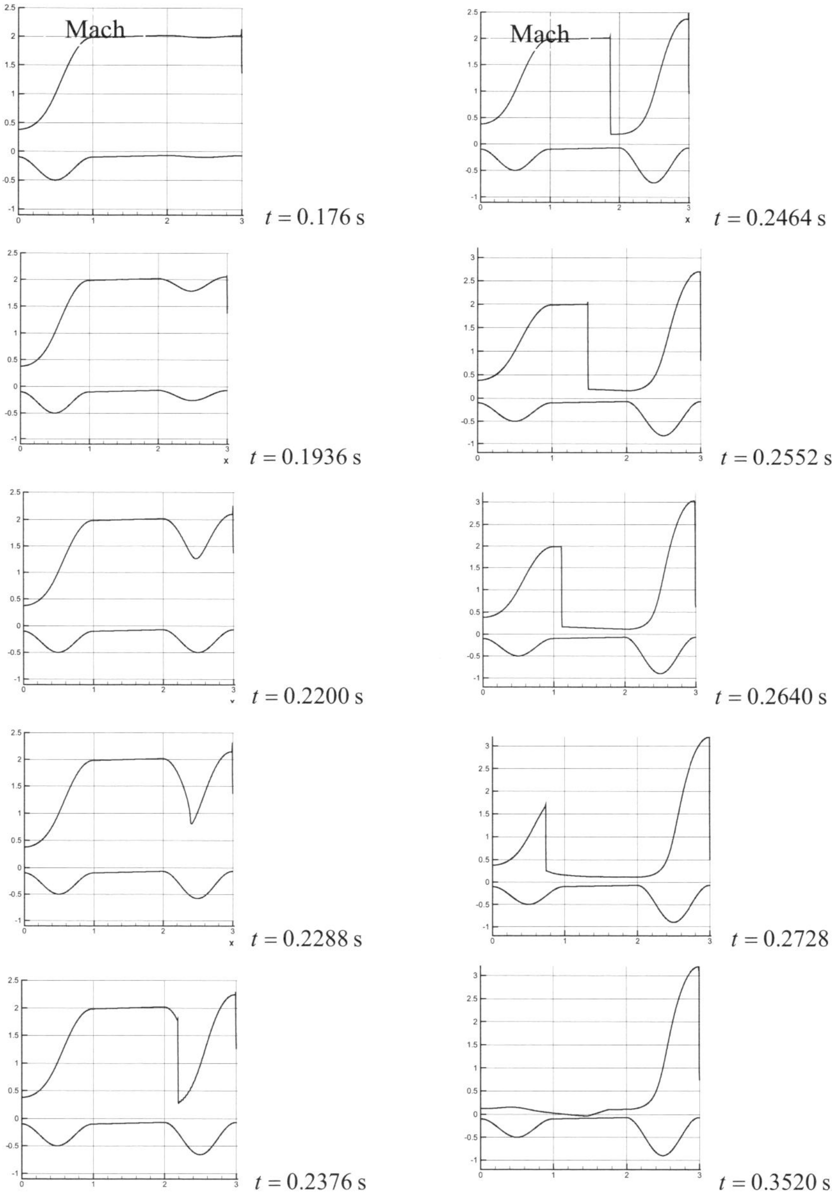

Figure 24.18. *Unstarting of a Mach 2 nozzle by closure of a second throat. Time in seconds for a total length of 3 m and a velocity of sound of 340 m/s*

We return to the first situation and examine in detail the unstarting process of the nozzle by closing the second throat. The nozzle being fully started, from the time $t = 0.0176$ s on, we gradually close the second throat (C_2). The resulting change of the Mach number distribution is shown in Figure 24.18. Until the time $t = 0.220$ s, the closure of the throat (C_2) causes a decrease of the Mach number in its converging part, followed by an increase in the divergent, according to the theorem of Hugoniot. At the time $t = 0.2288$ s, the section of the second throat has become smaller than that of the first one, the Mach number in the second throat being sonic. There is an incompatibility between the flows through (C_1) and (C_2). A shock wave forms in the flow due to the choking effect of (C_2). This shock, which cannot find an equilibrium position because the satisfaction of [24.1] would require $p_{st_2} > p_{st_2}$, travels the flow up to the first throat where it vanishes. After its passage, the throat is unchoked and the flow in the nozzle is subsonic (time $t = 0.352$ s). Downstream of the second throat, the flow expands to supersonic up to about Mach 3.2.

Chapter 25

Some Properties of Transonic Flows

25.1. Introduction

A flow is called transonic if the field contains both subsonic and supersonic regions. If the flow is retarded (undergoes a compression), these regions are usually separated by a normal or nearly-normal shock wave, whereas in the case of an expansion, the transition from subsonic to supersonic happens continuously with as boundary a sonic line which is a locus of points where the critical velocity is reached (Mach number equal to 1). Transonic flows are met:

– at the throat of a convergent–divergent nozzle accelerating the flow to supersonic velocities (see Figure 25.1a),

– in a channel or interblade passage, where a counter pressure causes the formation of a shock wave (see Figure 25.1b),

– before the nose of a projectile where a detached shock wave forms downstream of which the subsonic flow accelerates and becomes supersonic (see Figure 25.1c),

– on a wing profile where, at high subsonic flight Mach numbers, the acceleration created by the shape of the airfoil causes the supersonic passage of the flow, the supersonic pocket being ended by a shock wave (see Figure 25.1d).

The prediction of transonic flow in a non-viscous fluid has long faced the difficulty of the coexistence of subsonic regions, governed by elliptic equations, and supersonic regions where the equations are hyperbolic. This raises difficult theoretical problems due to the change of the disturbance propagation mechanisms which, in the first case, eventually affect the whole field, while, in the second case, they are transmitted in a limited domain of influence. This mixed nature gives the transonic flows specific physical properties, often causing dangerous unsteady phenomena. Indeed, the presence of a subsonic domain in the flow around an airfoil allows downstream disturbances to propagate upstream and thus exciting the shock wave terminating the supersonic pocket. It may result in a coupling mechanism that causes a periodic oscillation with large amplitude of the whole flow, known as *transonic buffeting*. A similar phenomenon is at work in air intake buzz.

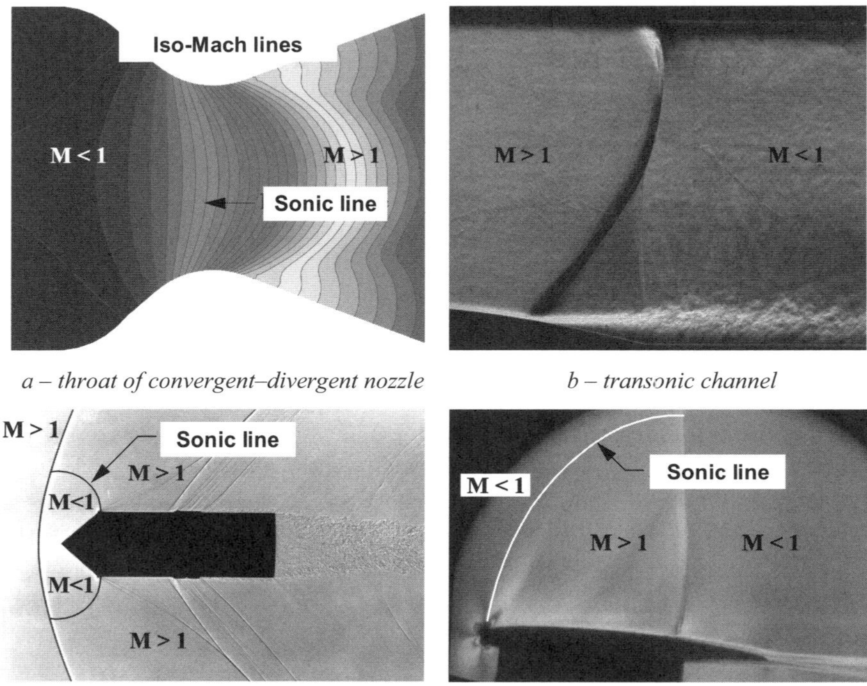

a – throat of convergent–divergent nozzle

b – transonic channel

c – projectile in flight at Mach 1.24 (ISL document)

d – supercritical transonic airfoil

Figure 25.1. *Examples of transonic flows*

In addition, the changing nature of the equations leads to solutions with different properties which dramatically change the pressure distributions, and hence aerodynamic loads, on an airfoil when the flow goes supersonic. This causes a shift of the pressure center, affecting the trimming conditions of the aircraft, and the appearance of a wave drag linked to the shock wave. The transonic disturbances, causing what was called the sound barrier, have been a serious obstacle to high velocity flight in the early 1950s.

25.2. Potential equation for small disturbances

As we have seen in section 19.2, there is a compressible version of the potential equation which has largely been used to calculate subsonic flows at high velocities, particularly in the transonic domain which became strategical in the early 1950s. Linearized forms of this equation have been deduced to examine the effects of compressibility at high subsonic or frankly supersonic velocities (i.e. outside the transonic domain). One of the earliest methods of calculating transonic flows, from the early 1970s, was based on the numerical solution of the potential equation for transonic small disturbances. Advances in numerical aerodynamics have caused a loss of interest in the potential equation which only applies to irrotational flows. However, it retains some advantages because of the low cost of its resolution in terms of computing time, either by approximate analytical methods (see section 19.2) or by numerical techniques.

Due to the lack of high calculation power, simplified versions of the potential equation have been established under the assumption of small disturbances of which we have already seen an application for the transonic region of a nozzle. Here, we consider the initially uniform flow with the velocity $\vec{V}_\infty$ around a thin airfoil with a small incidence. We can then assume that the profile weakly disturbs the flow so that the components u and v of the velocity in a system of orthogonal axes, where Ox is aligned with the upstream velocity, can be written as:

$$u = V_\infty + \hat{u} = V_\infty\left(1+u'\right), \quad v = \hat{v} = V_\infty v'.$$

Under the assumption of small disturbances, we have $u', v' << 1$. With these notations, the potential equation is written as:

$$\left[1-\frac{V_\infty^2}{a^2}(1+u')^2\right]\frac{\partial^2\Phi}{\partial x^2}+\left[1-\frac{1}{a^2}(V_\infty v')^2\right]\frac{\partial^2\Phi}{\partial y^2}-\frac{2}{a^2}\left(\frac{\partial\Phi}{\partial x}\right)\left(\frac{\partial\Phi}{\partial y}\right)\left(\frac{\partial^2\Phi}{\partial x\partial y}\right)=0$$

or:

$$\left[\frac{a^2}{a_\infty^2}-M_\infty^2(1+u')^2\right]\frac{\partial^2\Phi}{\partial x^2}+\left[\frac{a^2}{a_\infty^2}-M_\infty^2 v'^2\right]\frac{\partial^2\Phi}{\partial y^2}-\frac{2}{a_\infty^2}\left(\frac{\partial\Phi}{\partial x}\right)\left(\frac{\partial\Phi}{\partial y}\right)\left(\frac{\partial^2\Phi}{\partial x\partial y}\right)=0.$$

The velocity of sound is calculated from the energy equation, i.e.:

$$\frac{a_\infty^2}{\gamma-1}+\frac{V_\infty^2}{2}=\frac{a^2}{\gamma-1}+\frac{V_\infty^2\left[(1+u')^2+v'^2\right]}{2}$$

leading to:

$$a^2 = a_\infty^2 + \frac{\gamma-1}{2} V_\infty^2 \left[1-(1+u')^2 - v'^2\right]$$

or, by introducing the Mach number:

$$\frac{a^2}{a_\infty^2} = 1 + \frac{\gamma-1}{2} M_\infty^2 \left[1-(1+u')^2 - v'^2\right] = 1 + \frac{\gamma-1}{2} M_\infty^2 \left[1-1-2u'-u'^2-v'^2\right] \approx 1-(\gamma-1) M_\infty^2 u'.$$

Also:

$$\frac{2}{a_\infty^2}\left(\frac{\partial\Phi}{\partial x}\right)\left(\frac{\partial\Phi}{\partial y}\right)\left(\frac{\partial^2\Phi}{\partial x\partial y}\right) = \frac{2}{a_\infty^2} V_\infty (1+u') V_\infty v' V_\infty \left(\frac{\partial u'}{\partial y}+\frac{\partial v'}{\partial x}\right) = 2M_\infty^2 (1+u') v' \left(\frac{\partial u'}{\partial y}+\frac{\partial v'}{\partial x}\right)$$

hence the equation:

$$\begin{aligned}&\left[1-M_\infty^2 - M_\infty^2 u'(\gamma+1+u')\right]\frac{\partial^2\Phi}{\partial x^2} + \left\{1-M_\infty^2\left[(\gamma-1)u'+v'^2\right]\right\}\frac{\partial^2\Phi}{\partial y^2} \\ &= 2M_\infty^2(1+u')v'\left(\frac{\partial u'}{\partial y}+\frac{\partial v'}{\partial x}\right).\end{aligned} \qquad [25.1]$$

First we consider the case where the Mach number is not very close to 1; i.e. outside of the transonic domain. Then, an inspection of the magnitude of its terms shows that equation [25.1] reduces to:

$$\boxed{\left(1-M_\infty^2\right)\frac{\partial^2\Phi}{\partial x^2}+\frac{\partial^2\Phi}{\partial y^2}=0}$$

or, equivalently, by introducing the perturbation potential Φ' so that:

$$u = V_\infty(1+u') = V_\infty\left(1+\frac{\partial\Phi'}{\partial x}\right) \quad , \quad v = V_\infty v' = V_\infty \frac{\partial\Phi'}{\partial y}$$

$$\left(1-M_\infty^2\right)\frac{\partial^2\Phi'}{\partial x^2}+\frac{\partial^2\Phi'}{\partial y^2}=0.$$

In a three-dimensional flow where the flow is little affected by the presence of the body according to both directions Oy and Oz, the equation for the perturbation potential is written as:

$$\boxed{\left(1-M_\infty^2\right)\frac{\partial^2 \Phi'}{\partial x^2}+\frac{\partial^2 \Phi'}{\partial y^2}+\frac{\partial^2 \Phi'}{\partial z^2}=0.}$$

This situation corresponds to an aircraft wing where we can assume that the velocity field is unaffected by the normal to the chord Oy and the wingspan Oz. We thus end up with a linear equation much simpler than the complete equation. But it should be borne in mind that this equation:

– will only give an approximate solution of the problem,

– ceases to be valid for thick bodies or thin bodies placed at high incidence,

– can neither be applied in the transonic range (roughly $0.8 < M_\infty < 1.2$) nor in hypersonic flows (approximately $M_\infty > 5$).

Expression of the pressure coefficient

The objective of the theories mentioned here is the prediction of the pressure distribution on the body to deduce the lift. Therefore, we look at the pressure coefficient:

$$C_p = \frac{p - p_\infty}{q_\infty}$$

where $q_\infty = 1/2\rho_\infty V_\infty^2$ is the upstream flow dynamic pressure. For a calorically perfect gas:

$$q_\infty = \frac{\gamma}{2} p_\infty M_\infty^2$$

and hence the expression of the pressure coefficient:

$$\boxed{C_p = \frac{2}{\gamma M_\infty^2}\left(\frac{p}{p_\infty}-1\right).}$$

As the studied flows are isoenergetic, the stagnation enthalpy is everywhere constant and the energy equation allows us to express the temperature with:

$$\frac{T}{T_\infty} = 1 + \frac{\gamma-1}{2}\frac{V_\infty^2 - V^2}{a_\infty^2}.$$

If we introduce the perturbation velocities, we have:

$$V^2 = V_\infty^2 \left(1+u'\right)^2 + V_\infty^2 v'^2 = V_\infty^2 \left(1+2\,u'+u'^2+v'^2\right)$$

so that:

$$\frac{T}{T_\infty} = 1+\frac{\gamma-1}{2}\frac{V_\infty^2}{a_\infty^2}\left(-2u'-u'^2-v'^2\right) = 1-\frac{\gamma-1}{2}M_\infty^2\left(2u'+u'^2+v'^2\right).$$

Knowing that in an isentropic flow:

$$\frac{p}{p_\infty} = \left(\frac{T}{T_\infty}\right)^{\frac{\gamma}{\gamma-1}}$$

we have:

$$\frac{p}{p_\infty} = \left[1-\frac{\gamma-1}{2}M_\infty^2\left(2u'+u'^2+v'^2\right)\right]^{\frac{\gamma}{\gamma-1}}.$$

Introducing the hypothesis of small perturbations and developing, it follows that:

$$\frac{p}{p_\infty} \approx 1-\gamma\, M_\infty^2 u'$$

which gives:

$$C_p \approx \frac{2}{\gamma\, M_\infty^2}\left(1-\gamma\, M_\infty^2 u'-1\right)$$

i.e.:

$$\boxed{C_p = -2u'.}$$

Under the small perturbations assumption, the pressure coefficient is expressed solely as a function of the dimensionless perturbation component u'. In a non-viscous fluid, the condition imposed on the body is a slip condition expressing that the velocity vector is tangent to the wall. If θ denotes the local angle of the wall relative to the upstream velocity, we write on the body:

$$\tan\theta = \frac{v}{u} = \frac{v'}{1+u'}.$$

In the case where $u' << 1$, it follows that:

$$\boxed{v' = \tan\theta}$$

or, considering the perturbation potential Φ':

$$\boxed{\frac{\partial \Phi'}{\partial y} = \tan\ \theta.}$$

25.3. Compressibility correction: the Prandtl–Glauert rule

The aim of the compressibility correction is to use incompressible results for velocities where compressibility effects are felt; i.e. values of M_∞ that are no longer negligible compared to 1. Thus, we could transpose results obtained in subsonic wind tunnels with low velocities $\left(V_\infty \leq 100 \text{ m/s}\right)$ to the case of flight at Mach numbers of 0.5–0.6, avoiding much more expensive tests in high-velocity wind tunnels (which are not always available!). The oldest and best known compressibility correction is the so-called Prandtl–Glauert rule.

Under the already made assumptions, we consider a thin airfoil placed with a small incidence in a flow with the velocity V_∞ and the Mach number M_∞. The shape of this airfoil is defined by a curve $y=f(x)$ where x and y are coordinates in the physical plane. As the flow is subsonic $\left(M_\infty < 1\right)$, we state:

$$\beta = \sqrt{1 - M_\infty^2}.$$

The equation for the perturbation potential Φ' is then written as:

$$\beta^2 \frac{\partial^2 \Phi'}{\partial x^2} + \frac{\partial^2 \Phi'}{\partial y^2} = 0.$$

We proceed to the change of variables:

$$\xi = x, \qquad \eta = \beta y$$

that maps the physical plane (x, y) into a transformed plane (ξ, η) and we introduce a new potential $\bar{\Phi}$ defined by:

$$\bar{\Phi}\ (\xi, \eta) = \beta \Phi'\ (x, y).$$

The chain derivation rules give:

$$\frac{\partial \Phi'}{\partial x} = \frac{\partial \Phi'}{\partial \xi}\frac{\partial \xi}{\partial x} + \frac{\partial \Phi'}{\partial \eta}\frac{\partial \eta}{\partial x} \qquad \frac{\partial \Phi'}{\partial y} = \frac{\partial \Phi'}{\partial \xi}\frac{\partial \xi}{\partial y} + \frac{\partial \Phi'}{\partial \eta}\frac{\partial \eta}{\partial y}$$

or:

$$\frac{\partial \xi}{\partial x} = 1, \qquad \frac{\partial \eta}{\partial x} = 0, \qquad \frac{\partial \xi}{\partial y} = 0, \qquad \frac{\partial \eta}{\partial y} = \beta.$$

Thus:

$$\frac{\partial \Phi'}{\partial x} = \frac{\partial \Phi'}{\partial \xi}, \qquad \frac{\partial \Phi'}{\partial y} = \beta \frac{\partial \Phi'}{\partial \eta}.$$

By introducing the potential $\bar{\Phi}(\xi, \eta)$, we have:

$$\frac{\partial \Phi'}{\partial x} = \frac{1}{\beta}\frac{\partial \bar{\Phi}}{\partial \xi}, \qquad \frac{\partial \Phi'}{\partial y} = \frac{\partial \bar{\Phi}}{\partial \eta}.$$

We now calculate the second derivatives:

$$\frac{\partial^2 \Phi'}{\partial x^2} = \frac{1}{\beta}\left(\frac{\partial^2 \bar{\Phi}}{\partial \xi^2}\frac{\partial \xi}{\partial x} + \frac{\partial^2 \bar{\Phi}}{\partial \xi \partial \eta}\frac{\partial \eta}{\partial x}\right) = \frac{1}{\beta}\frac{\partial^2 \bar{\Phi}}{\partial \xi^2}$$

$$\frac{\partial^2 \Phi'}{\partial y^2} = \frac{\partial^2 \bar{\Phi}}{\partial \eta \partial \xi}\frac{\partial \xi}{\partial y} + \frac{\partial^2 \bar{\Phi}}{\partial \eta^2}\frac{\partial \eta}{\partial y} = \beta \frac{\partial^2 \bar{\Phi}}{\partial \eta^2}.$$

Thus:

$$\beta^2 \frac{\partial^2 \Phi'}{\partial x^2} + \frac{\partial^2 \Phi'}{\partial y^2} = \beta^2 \frac{1}{\beta}\frac{\partial^2 \bar{\Phi}}{\partial \xi^2} + \beta \frac{\partial^2 \bar{\Phi}}{\partial \eta^2} = \beta\left(\frac{\partial^2 \bar{\Phi}}{\partial \xi^2} + \frac{\partial^2 \bar{\Phi}}{\partial \eta^2}\right) = 0.$$

The potential function $\overline{\Phi}$ satisfies the Laplace equation for incompressible flows, since:

$$\frac{\partial^2 \overline{\Phi}}{\partial \xi^2} + \frac{\partial^2 \overline{\Phi}}{\partial \eta^2} = 0.$$

The shape of the airfoil in the physical plane is given by $y = f(x)$. Its shape in the transformed plane (ξ, η) is a curve $\eta = g(\xi)$. The slip condition on the airfoil is expressed by:

$$\tan \theta = \frac{\partial \Phi'}{\partial y}$$

or:

$$V_\infty \frac{df}{dx} = \frac{\partial \Phi'}{\partial y} = \frac{\partial \overline{\Phi}}{\partial \eta}.$$

In the transformed plane, this equation is written as:

$$\frac{dg}{d\xi} = \frac{\partial \overline{\Phi}}{\partial \eta}.$$

It follows that $g(\xi) = f(x)$. In other words, the applied transformation makes an incompressible flow around a certain airfoil correspond to a compressible flow around the *same airfoil*.

We have seen that the pressure coefficient is given by:

$$C_p = -2u' = -2\frac{\partial \Phi'}{\partial x} = -2\frac{1}{\beta}\frac{\partial \overline{\Phi}}{\partial \xi}.$$

But $-2\dfrac{\partial \overline{\Phi}}{\partial \xi}$ is the pressure coefficient $C_{p(inc)}$ on the same profile in incompressible. Hence:

$$C_p = \frac{C_{p(inc)}}{\beta}$$

i.e.:

$$C_p = \frac{C_{p(inc)}}{\sqrt{1-M_\infty^2}}.$$

The preceding expression is called Prandtl–Glauert rule. It allows transposing the results obtained in incompressible to compressible. It is therefore a compressibility correction. As the lift and moment coefficients result from the integration of the pressure distribution over the airfoil, it is easy to see that:

$$C_L = \frac{C_{L(inc)}}{\sqrt{1-M_\infty^2}}$$

$$C_m = \frac{C_{m(inc)}}{\sqrt{1-M_\infty^2}}.$$

We note:

– that in subsonic $C_p \propto \frac{1}{\sqrt{1-M_\infty^2}}$, and therefore increases with the Mach number M_∞;

– that in supersonic $C_p \propto \frac{1}{\sqrt{M_\infty^2 - 1}}$, and therefore decreases when the Mach number M_∞ increases.

The change of C_p is illustrated in Figure 25.2. The transonic domain, for which according to the above formulas $C_p \to \infty$ when $M_\infty \to 1$, must be excluded. In this domain, the potential equation with which we work is false, since it implies that $\left|M_\infty^2 - 1\right|$ is not small. In section 25.5, we will see the transonic form of the potential equation for small disturbances.

The Prandtl–Glauert rule gives good results up to $M_\infty \sim 0.8$. Beyond that, its approximate character becomes apparent, the correction factor $1/\beta$ tending to infinity if $M_\infty \to 1$. There are more precise alternatives to the Prandtl–Glauert rule which will not be given here.

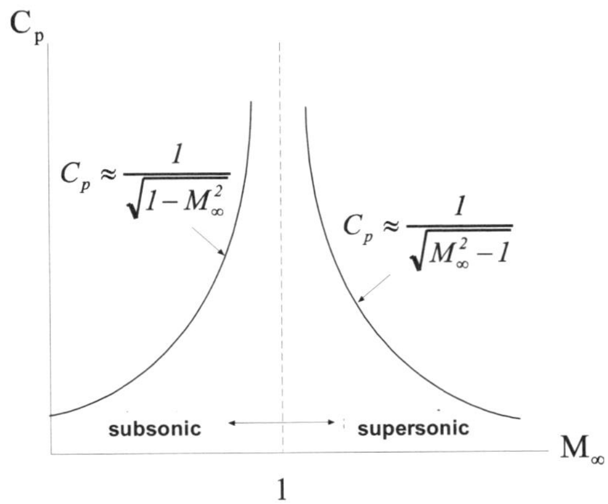

Figure 25.2. *Prandtl–Glauert rule for the compressibility effects*

25.4. Linearized theory for supersonic flows

We return to the equation for the potential Φ' which is valid both in supersonic and subsonic, the transonic domain being excluded (roughly the range $0.8 < M_\infty < 1.2$):

$$\left(1-M_\infty^2\right)\frac{\partial^2\Phi'}{\partial x^2}+\frac{\partial^2\Phi'}{\partial y^2}=0.$$

If $M_\infty < 1$ (subsonic case), this equation is called *elliptic*; if $M_\infty > 1$,(supersonic case), it is *hyperbolic*. As we will see, the behavior of solutions dramatically changes, according to the elliptical or hyperbolic character of the equation. We consider supersonic and write the above equation in the form:

$$\left(M_\infty^2-1\right)\frac{\partial^2\Phi'}{\partial x^2}-\frac{\partial^2\Phi'}{\partial y^2}=0.$$

We state:

$$\lambda=\sqrt{M_\infty^2-1}.$$

Then:

$$\lambda^2 \frac{\partial^2 \Phi'}{\partial x^2} - \frac{\partial^2 \Phi'}{\partial y^2} = 0. \qquad [25.2]$$

We seek a solution to the equation [25.2] of the form:

$$\Phi'(x,y) = f(x - \lambda\, y).$$

In other words, the desired solution depends on x and y through the intermediate of the variable $z = x - \lambda y$. Let us check if f satisfies equation [25.2]. We have:

$$\frac{\partial \Phi'}{\partial x} = \frac{df}{dz}\frac{\partial z}{\partial x} = \frac{df}{dz} \quad \textit{then} \quad \frac{\partial^2 \Phi'}{\partial x^2} = \frac{d^2 f}{dx^2}$$

$$\frac{\partial \Phi'}{\partial y} = \frac{df}{dz}\frac{\partial f}{\partial y} = -\lambda \frac{df}{dz} \quad \textit{then} \quad \frac{\partial^2 \Phi'}{\partial y^2} = \lambda^2 \frac{d^2 f}{dz^2}.$$

Hence:

$$\lambda^2 \frac{d^2 f}{dz^2} - \lambda^2 \frac{d^2 f}{dz^2} = 0.$$

Thus, any function f depending on x and y through the intermediate of $z = x - \lambda\, y$ satisfies the linearized potential equation in supersonic. At this level, the function f is arbitrary; it will be determined by the boundary conditions that it must satisfy. The obtained result shows that Φ' is constant along the straight lines of equation $x - \lambda y = constant$ whose slope is:

$$\frac{dy}{dx} = \frac{1}{\lambda} = \frac{1}{\sqrt{M_\infty^2 - 1}}.$$

If α denotes the Mach angle (see section 17.3):

$$\alpha = \sin^{-1}\left(\frac{1}{M_\infty}\right)$$

and we see that the slope of these lines is precisely the tangent of the Mach angle since:

$$\boxed{\frac{dy}{dx} = \tan\alpha = \frac{1}{\sqrt{M_\infty^2 - 1}}.}$$

The function Φ' is constant along the Mach lines. Any disturbance, for example, introduced by a small change in the wall shape at a point P, induces a variation of the function Φ'. However, since according to the relation given above, Φ' is a function of x and y through the intermediate of z, this change will be felt (or will propagate) in the flow along the line with the slope $dy/dx = \left(M_\infty^2 - 1\right)^{-\frac{1}{2}}$ coming from the point P. As this slope is positive, the disturbance does not propagate upstream of the point where it is applied. We find again the essential difference between subsonic flows, where any disturbance affects the whole field, and the supersonic flows, where disturbances propagate upstream along characteristics or Mach lines.

To stay within the linearized theory, we consider a uniform supersonic flow with the Mach number M_∞ flowing along a wall with a bump of small local angle θ. We seek to express the local pressure coefficient C_p given by:

$$C_p = -2u'.$$

In this case, we have:

$$u' = \frac{\partial \Phi'}{\partial x} = \frac{df}{dz}$$

but on the other hand:

$$v' = \frac{\partial \Phi'}{\partial y} = -\lambda \frac{df}{dz} = -\lambda\, u'.$$

As the angle θ is small, the slip condition is written as:

$$v' = \tan\theta \cong \theta.$$

Finally:

$$C_p = -2u' = \frac{2\,v'}{\lambda} = \frac{2\,\theta}{\sqrt{M_\infty^2 - 1}}$$

i.e.:

$$C_p = \frac{2\,\theta}{\sqrt{M_\infty^2 - 1}}.$$

As, according to the usual conventions, the angle θ is positive on the part ahead of the bump and negative on the other part, we have:

– an overpressure $(C_p > 0)$ on the front part of the bump,

– a depression $(C_p < 0)$ on the rear part.

Hence, the existence of a drag called *wave drag*, which is typical of supersonic flows. The existence of this drag, in apparent contradiction to the d'Alembert paradox, is linked to the presence of a shock wave formed in the actual flow by focalization of waves coming from the compression part of the bump (see Figure 25.3). The mechanism of the shock wave formation is analyzed in section 20.9. It is interesting to note that the linearized theory, which ignores the shock waves, also predicts the existence of a wave drag.

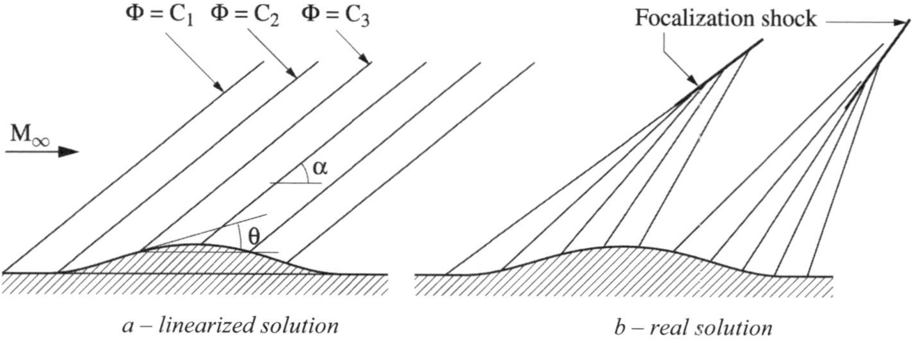

Figure 25.3. *Linearized theory in supersonic flows*

COMMENT.– If δp denotes a small pressure variation compared to the upstream state (∞), the pressure coefficient is expressed by:

$$C_p = \frac{2\delta p}{\rho_\infty V_\infty^2} = \frac{2\delta p}{p_\infty M_\infty^2}.$$

But, according to the relation expressing δp in the context of the linearized theory, we have:

$$C_p\sqrt{M_\infty^2 - 1} - 2\theta = 0$$

i.e.:

$$\frac{\sqrt{M_\infty^2 - 1}}{\gamma M_\infty^2}\frac{\delta p}{p_\infty} - \theta = 0.$$

The non-linearized theory of supersonic flows presented in section 17.3 establishes the following relation between the pressure variation δp and the direction $\delta\varphi$ for a wave (ξ):

$$\frac{\sqrt{M^2 - 1}}{\gamma \; M^2}\frac{\delta p}{p} - \delta\varphi = 0.$$

If we linearize this relation by expressing that the flow remains close to an upstream state (M_∞, p_∞), we obtain:

$$\frac{\sqrt{M_\infty^2 - 1}}{\gamma \; M_\infty^2}\frac{\delta p}{p_\infty} - \delta\varphi = 0.$$

This relation is identical to that derived from the linearized potential theory, knowing that the deflections θ are small. Within the framework of the linearized theory, one considers a family of rectilinear characteristics of constant slope.

Under the supersonic linearized theory, we examine the case of an airfoil consisting of an infinitely thin flat plate with the length L at an incidence ε (small) in a uniform flow with the Mach number M_∞ (see Figure 25.4).

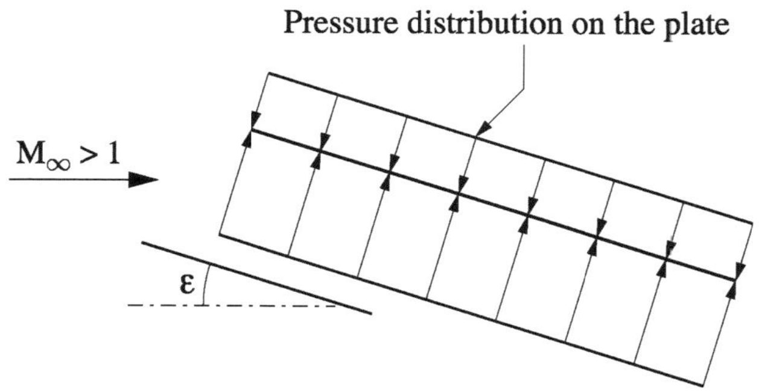

Figure 25.4. *Flat plate at incidence in a supersonic flow*

According to the previous results, the pressure coefficient C_p at the lower surface has the value:

$$C_{p_e} = -\frac{2\varepsilon}{\sqrt{M_\infty^2 - 1}}$$

and at the upper surface:

$$C_{p_e} = \frac{2\varepsilon}{\sqrt{M_\infty^2 - 1}}$$

it being understood that the angle ε is counted positive for an effective incidence (i.e. a negative value counterclockwise). If the incidence is positive $(\varepsilon > 0)$, the upper surface is in depression and the lower side in overpressure. The normal force coefficient results from the integration (see Chapter 2) are:

$$C_N = \frac{1}{L}\int_0^L \left(C_{p_l} - C_{p_u}\right) dx$$

i.e:

$$C_N = \frac{4\varepsilon}{\sqrt{M_\infty^2 - 1}} \frac{1}{L}\int_0^L dx = \frac{4\varepsilon}{\sqrt{M_\infty^2 - 1}}.$$

As the plate has zero thickness and as the pressure force is perpendicular to the plate, the axial force C_A is zero. The lift and drag coefficients are given by the relations (see section 2.1):

$$C_L = C_N \cos\varepsilon - C_A \sin\varepsilon$$
$$C_D = C_N \sin\varepsilon + C_A \cos\varepsilon.$$

As the incidence angle ε is small:

$$C_L = C_N - C_A\varepsilon$$
$$C_D = C_N\varepsilon + C_A$$

which gives:

$$\boxed{C_L = \frac{4\varepsilon}{\sqrt{M_\infty^2 - 1}} \qquad C_D = \frac{4\varepsilon^2}{\sqrt{M_\infty^2 - 1}}.}$$

Figure 25.5a represents the organization of the flow as a result of the linearized theory.

Upstream of the plate, the waves are straight lines with the angle α_0 relative to the velocity. On the upper side of the plate, these waves have the angle $\alpha_0 - \varepsilon$ relative to a horizontal reference. Here, as the inclination angle of the plate is negative (the incidence ε being in fact positive), they are more inclined than the waves upstream of the airfoil. The angle of the waves leaving the airfoil upper side at the trailing edge is $\alpha_0 - \varepsilon$ and there is a compression, the angle α_0 of the final wave being higher than the initial wave. This situation is physically impossible since the final state precedes the initial state. We would do the same analysis at the lower side, the compression occurring this time at the leading edge and the expansion at the trailing edge. The prediction paradox of a drag in a non-viscous fluid is at the cost of the violation of a basic physical principle. As drawn in Figure 25.5b, to find an acceptable physical solution, we must introduce shock waves; which restores the influence of viscosity. This issue is discussed in greater detail in section 20.9.

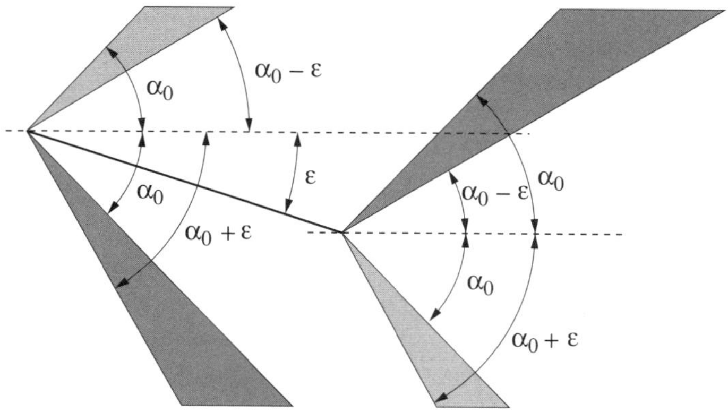

a – linearized solution

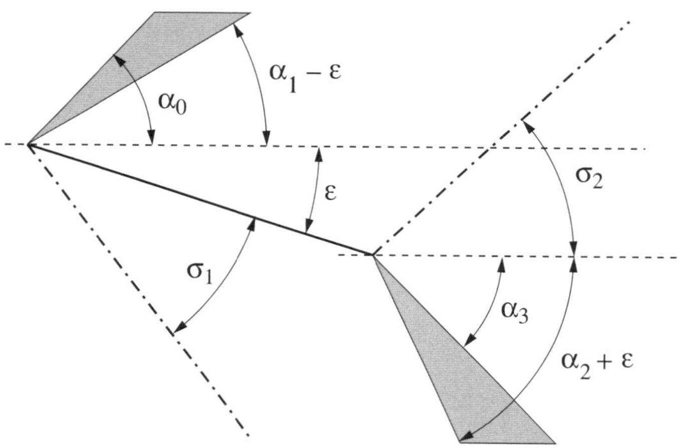

b – solution with shock waves

Figure 25.5. *Solutions for the flow past a flat plate at incidence*

25.5. The small perturbation potential equation in transonic

We consider the case of a thin airfoil, placed at low incidence, in a uniform transonic flow. The goal is to find a simplified form of the potential equation valid when the Mach number is close to 1. The establishment of a rigorous linearized equation in transonics requires an assessment of the magnitudes of its terms by the introduction of parameters, small compared to 1. This requires prior normalization of the different quantities by representative scales, for example, the velocity V_∞, the chord of the profile or wing, the dynamic pressure q_∞, etc. This kind of analysis can lead far enough, but we must ensure that the simplified form is consistent (have terms been wrongly neglected?). As this is a very specialized discipline, we content ourselves here with a more intuitive approach that will lead us quickly to the desired result and that will suffice for our purpose. However, we should be aware that this approach is not without risk, because it uses expressions like "small compared to", "large compared to", "slowly varying depending on such or such direction", having vague and imprecise character, etc.

Given the assumptions made above, we can estimate that the presence of the airfoil has little effect on the flow in the transverse direction, i.e. that the induced vertical velocities v are small, not only compared to V_∞, but also compared to the disturbance of the longitudinal velocity. Experience shows that the pressure varies greatly on the surface of the airfoil (upper and lower sides), and thus the velocity u must also vary significantly. In any event, the disturbance $\hat{u}$ of u remains small compared to V_∞, which is expressed by $\frac{\hat{u}}{V_\infty} = u' << 1$. For similar reasons, derivatives with respect to y are considered small compared to the derivatives with respect to x, the flow varying much faster along x than along y. In the vicinity of $M_\infty = 1$, the factor $\left(1 - M_\infty^2\right)$ is small, its magnitude becoming comparable to the terms previously regarded as negligible. Considering equation [25.1], the right-hand side (RHS) is always negligible and the coefficient of $\frac{\partial^2 \Phi}{\partial y^2}$ is still equal to 1. However, it is not possible to neglect $(\gamma+1) M_\infty^2 u'$ any further compared to $\left(1 - M_\infty^2\right)$, as these two terms have now the same order of magnitude.

The potential equation for the small transonic perturbations is therefore written as:

$$\boxed{\left[1 - M_\infty^2 - (\gamma+1) M_\infty^2 u'\right] \frac{\partial^2 \Phi'}{\partial x^2} + \frac{\partial^2 \Phi'}{\partial y^2} = 0} \qquad [25.3]$$

or:

$$\boxed{\left[1 - M_\infty^2 - (\gamma+1) M_\infty^2 \frac{\partial \Phi'}{\partial x}\right] \frac{\partial^2 \Phi'}{\partial x^2} + \frac{\partial^2 \Phi'}{\partial y^2} = 0.}$$

For similar arguments in a three-dimensional setup, we arrive at the equation:

$$\left[1 - M_\infty^2 - (\gamma+1) M_\infty^2 u'\right] \frac{\partial^2 \Phi'}{\partial x^2} + \frac{\partial^2 \Phi'}{\partial y^2} + \frac{\partial^2 \Phi'}{\partial z^2} = 0.$$

According to the energy equation:

$$a_\infty^2 + \frac{\gamma - 1}{2} V_\infty^2 = a^2 + \frac{\gamma - 1}{2} V^2.$$

At the sonic point $V = a_c$, leading to:

$$a_\infty^2 + \frac{\gamma - 1}{2} V_\infty^2 = \frac{\gamma + 1}{2} a_c^2.$$

But in this case:

$$V^2 = V_\infty^2 (1 + u')^2 \cong V_\infty^2 + 2\, u' V_\infty^2$$

as the term u'^2 is negligible. If u'_c denotes the value u' corresponding to the sonic condition, we find the relation:

$$u'_c = \frac{1 - M_\infty^2}{(\gamma + 1) M_\infty^2}.$$

Referring to equation [25.3], we see that the coefficient of $\dfrac{\partial^2 \Phi'}{\partial x^2}$ vanishes for exactly:

$$u' = \frac{1 - M_\infty^2}{(\gamma + 1) M_\infty^2}.$$

Thus, there is a transition from an elliptical equation to a hyperbolic equation when the local velocity changes from subsonic to supersonic, i.e. when the local Mach number is equal to 1. The linearized equation preserves the mixed elliptic–hyperbolic nature of transonic flows. The linearized small perturbation potential equation in transonics has historical significance as its resolution by numerical methods in the early 1970s allowed for the first time to calculate the flow around a transonic airfoil.

25.6. Final comments on the potential equation

The small perturbation potential equation, in its various forms, can lead to remarkably simple results and can be solved economically, even if we resort to numerical methods. However, the accuracy of its solutions is deteriorating rapidly when we move away from the conditions justifying the assumption of small perturbations, namely: small local slope of the airfoil relative to the upstream velocity, thin airfoil, and small incidence. The last two conditions are well verified for a wing of an airliner at cruise flight, where the incidence is only a few degrees. In contrast, at the rounded leading edge of the wing, the slope of the profile becomes large, so that calculations will be incorrect in this region. However, it is at the leading edge where the most important pressure variations occur. An incorrect prediction of the leading edge flow will cause an inaccuracy in the calculation of the lift which may be deemed unacceptable.

Thus, we often resort to solve the full potential equation, even in three dimensions. For this, we have very efficient numerical methods that lead to reduced computing time. With the restriction of a non-viscous fluid, the full potential equation has a great character of generality since it applies to the three conditions sub-, trans-, and supersonic. However, we should not forget that it assumes an irrotational flow, which excludes the presence of shock waves in the flow. We can therefore not use it in supersonics, except for very slender bodies producing very weak shocks. Its use in high subsonics is also limited; its accuracy may become critical in *off-design* flight conditions where relatively intense transonic shocks form on the wing.

What follows is dedicated to the presentation of some properties of transonic flows which are still the subject of numerous studies because they involve a field of velocity which is that of transport aircraft and numerous devices involving fluids at high velocity such as turbomachines.

25.7. Critical Mach number of an airfoil

We consider an airfoil placed at a fixed incidence α in a uniform subsonic flow with the Mach number M_∞. We start from a situation where M_∞ is small, for example, 0.3. As we know, the flow accelerates on the upper side of the airfoil and locally reaches Mach numbers greater than M_∞. In particular, there is a point A where the Mach number is maximum (see Figure 25.6a). Gradually increasing M_∞, correspondingly, the maximum Mach number M_A on the airfoil increases and we reach a situation where M_A is equal to 1, M_∞ *remaining subsonic*. For any subsequent increase in M_∞, a pocket with supersonic flow forms on the upper side of the airfoil (see Figure 25.6c).

The Mach number M_∞ for which the sonic condition appears in a single point A is called the *critical Mach number* of the airfoil and will be denoted as M^* (see Figure 25.6b). The concept of the critical Mach number is very important because for $M_\infty > M^*$ the aerodynamic performances of the airfoil, particularly its drag, will experience profound changes.

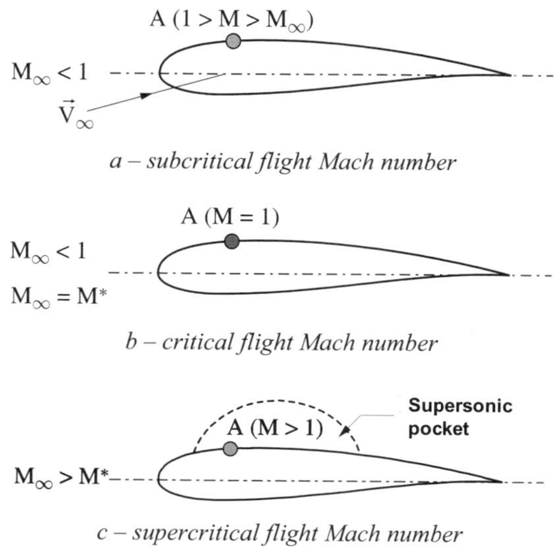

Figure 25.6. *Critical Mach number of an airfoil*

A thick airfoil tends to significantly disturb the flow by inducing a strong expansion on the upper side leading, for a fixed upstream Mach number, to high local Mach number values. Consequently, a local sonic Mach number will be reached for a relatively low value of M_∞. In contrast, for a thin airfoil, less perturbing for the flow, the acceleration on the upper side is smaller and the local Mach numbers lower. Hence, a critical Mach number is greater compared to thick airfoils. Thus, there would be an advantage to use a thin airfoil to induce a significant lift while achieving a critical Mach number close to 1. In fact, for multiple reasons, there are limits to the thinness of an airfoil:

– the integrity of the wing to aerodynamic forces, especially to bending moment, requires an internal structure capable of withstanding these efforts, and hence a design requiring a certain space. The wing of the *Airbus A380*, which has a wingspan of 80 m and a surface of 800 m^2, must bear a lift force of about 6×10^6 N (in common language 600 tons). Its structure must withstand such a force and associated bending moments,

– the commands to steer the flaps during takeoff or landing and to retract them for cruise flight are housed in the wing. Also, many hydraulic and electrical circuits pass in the wing,

– a part of the fuel reserves is housed in the wing.

For these reasons, the relative thickness of the wing profile can hardly be less than 12%. Hence, the use of so-called transonic supercritical airfoil (see below).

25.8. Transonic flow around an airfoil

Figure 25.7 shows a sequence of schlieren visualizations illustrating the passage of the "sound barrier" over a *NACA 64A010* airfoil tested in an Onera wind tunnel. The first picture, made for an upstream Mach number $M_0 = 0.76$, reveals the emergence of a small supersonic region at the upper side, highlighted by rapid local density variations. We also note the expansion area at the leading edge as well as the wake of the airfoil. The Mach number $M_0 = 0.76$ is slightly greater than the critical Mach number of the airfoil.

The compressibility phenomena are accentuated when the upstream Mach number increases, a shock wave becoming clearly visible for $M_0 = 0.84$. At Mach 0.90, the flow is supersonic over most of the upper surface and a shock appears on the lower side of the airfoil. At $M_0 = 0.96$, separation occurs near the trailing edge with the formation of a small lambda structure (see section 15.8). For higher Mach numbers, the flow is supersonic over the whole airfoil, except in the region upstream of the rounded leading edge. Because the flow of the upper and lower surfaces requires taking a common direction downstream of the trailing edge, two oblique confluence shocks form. The lambda structure of the shocks is a consequence of the pressure imposed downstream.

Figure 25.8 shows the change of the pressure coefficient on the *NACA 0012* symmetric airfoil at zero incidence for increasing values of the upstream Mach number. These results were obtained by solving Euler equations using the *elsA* code developed at Onera. When M_0 is less than 0.76, the pressure distribution is characterized by a rapid expansion in the leading edge region followed by a gradual compression on the airfoil aft part. Greater than $M_0 = 0.76$, a local compression develops which is indicative of the formation of a shock wave. When the Mach number continues to increase, the shock intensifies and moves toward the trailing edge which is reached for M_0 a little over 0.9. The transition to transonic is thus characterized by an extension of the expansion zone on the airfoil, which results in a displacement with a large amplitude of the center of pressure. This dramatic change in the structure of the flow is the cause of severe aerodynamic problems, poorly understood until the 1950s. These could result in a loss of control of the plane due to an inversion effect phenomenon caused by the change of the pressure distributions on the control surfaces.

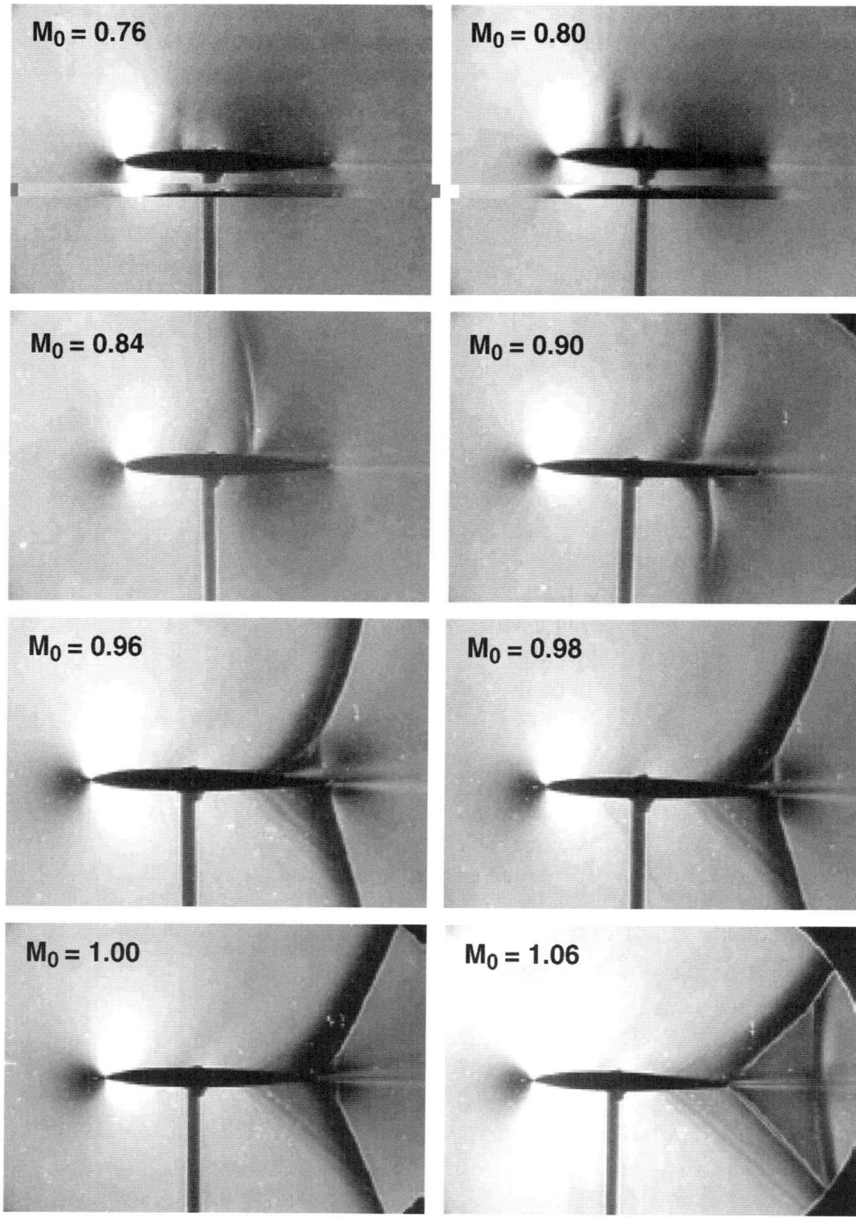

Figure 25.7. *Transonic passage around the airfoil NACA 64A010 at an incidence of 3° (document Onera)*

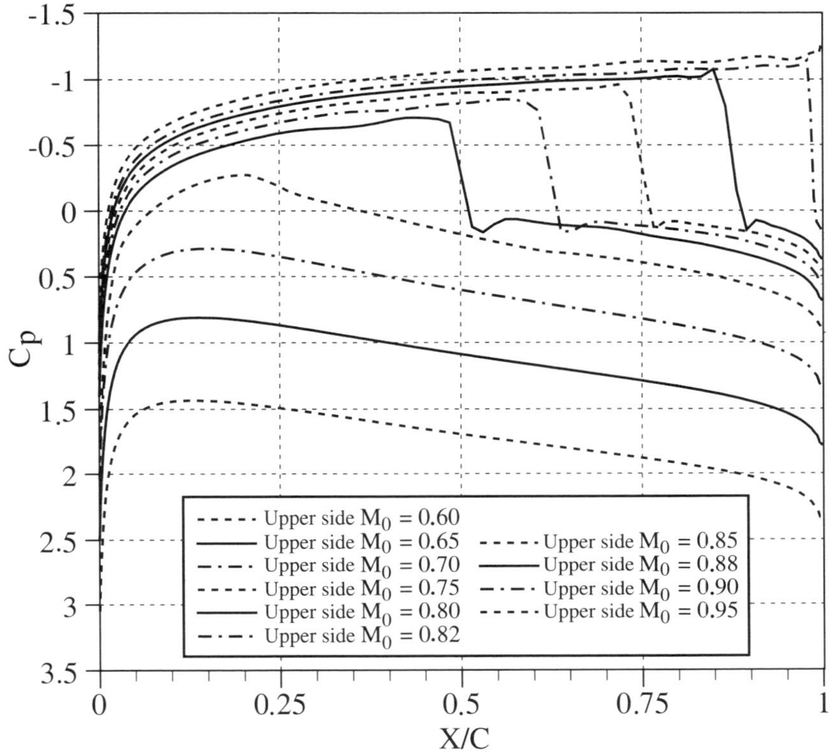

Figure 25.8. *Change of the pressure coefficient distribution over a NACA 0012 airfoil at zero incidence in the transonic domain (Euler calculations with the Onera code elsA)*

As we know, in the absence of viscosity, the drag of a body is zero (paradox of d'Alembert). This is shown in Figure 25.9 where the drag coefficient C_D of the airfoil is plotted as a function of the upstream Mach number M_0. We verify that as M_0 is less than 0.76, then $C_D = 0$. The increase in C_D, occurring for $M_0 > 0.76$, reveals the presence of an increasingly intense shock wave in the flow which, as a source of entropy, is the cause of a drag called *wave drag* (see Chapter 2). Rapid movement of the center of pressure and onset of wave drag are two of the most striking features of transonic flows.

Figure 25.10 shows how the shock wave forms in transonics. The flow in the supersonic pocket expands along left running characteristics (η). These waves are reflected from the sonic line, boundary of the supersonic domain, as compression waves (ξ) as the sonic line is an isobar (isentropic flow). It is the focalization of the right running waves (ξ) which will form the transonic shock (see Chapter 3).

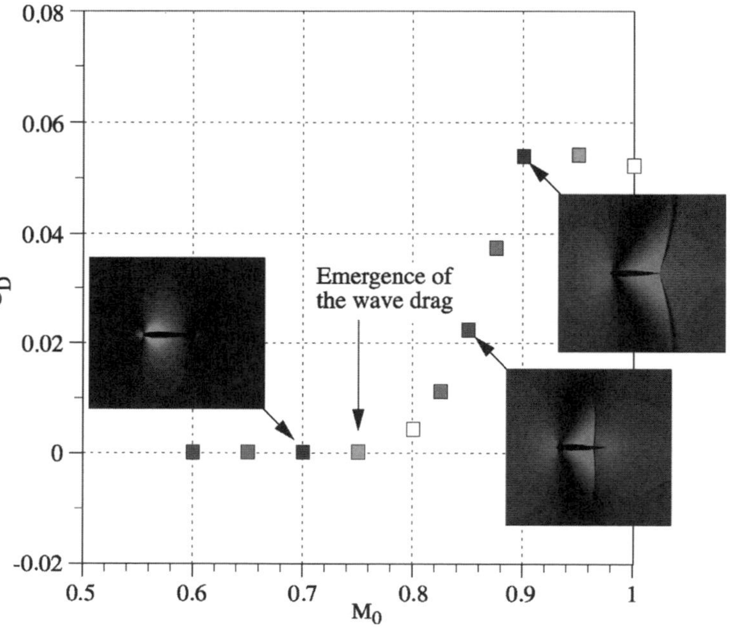

Figure 25.9. *Onset of the wave drag for the airfoil NACA0012 (Euler calculations with the Onera code elsA)*

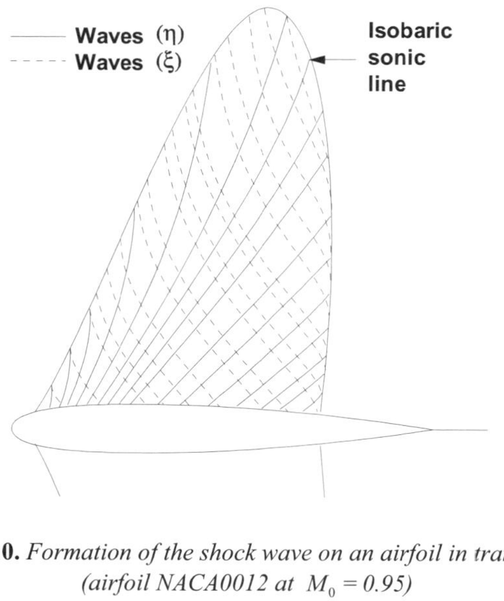

Figure 25.10. *Formation of the shock wave on an airfoil in transonic flow (airfoil NACA0012 at $M_0 = 0.95$)*

25.9. Drag divergence Mach number

We consider a conventional subsonic airfoil tested in a wind tunnel by varying the upstream Mach number M_∞ of the flow while maintaining the incidence fixed and we plot the curve C_D as a function of M_∞ (see Figure 25.11).

For incidences corresponding to cruising conditions at fully subsonic flight, the drag coefficient C_D remains almost constant when the Mach number M_∞ increases. This plateau C_D is maintained until the critical Mach number M^*. Beyond M^*, C_D begins to increase with M_∞, at first slowly, then rapidly. This rise begins from a Mach number value M_{div} which is called the *drag divergence Mach number*. If M_∞ increases beyond M_{div}, the drag coefficient undergoes a steep increase to pass through a maximum $C_{D(\max)}$ when M_∞ is close to 1. Then, in the entirely supersonic domain, C_D decreases. This sudden increase in the drag, with the passage through a maximum, is popularly known as the "sound barrier." It explains the difficulties encountered in the early 1950s to fly aircraft at Mach numbers greater than 1. For a conventional subsonic profile, $C_{D(\max)}$ can reach 10 times the value of C_D in subsonics, and hence the corresponding increase in propulsion power required to accelerate the plane beyond Mach 1.

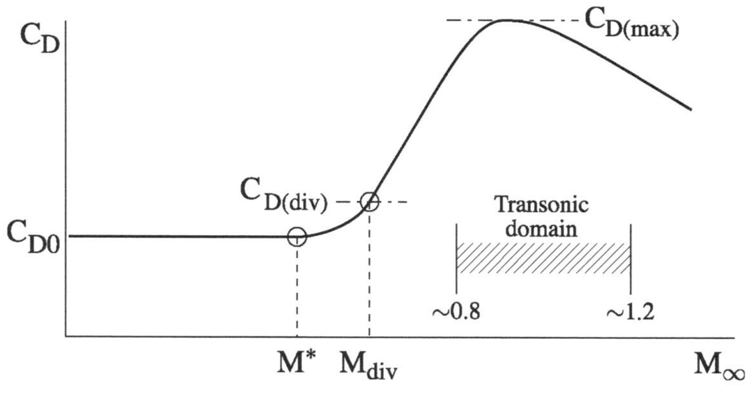

Figure 25.11. *Drag divergence Mach number*

The drag divergence of a conventional subsonic airfoil is due to the following phenomena of which we have already spoken. From $M_\infty = M^*$ and beyond, a supersonic flow pocket is formed on the upper side of the airfoil. In this zone, the Mach number can reach values between 1.2 and 1.4. We have seen that this supersonic region is terminated by a shock wave. It is this shock, which contributes to a dramatic increase in the drag:

– first as an entropy source, which gives rise to what we call the wave drag,

– then by its interaction with the boundary layer of the airfoil which can cause separation, leading to a rapid growth of the so-called friction drag (see section 2.5.2).

Similar phenomena can occur on the lower surface, contributing to a further increase in drag. If the maximum Mach number in the supersonic pocket is not too high (around 1.2), we can consider defining the airfoil contour to cause a return to subsonics through an *isentropic compression*; i.e. by avoiding the focalization of the waves (ξ) (see Figure 25.10). This kind of behavior is difficult to achieve, so that on most airfoils the transition from supersonic to subsonic is accomplished irreversibly through a shock wave.

For a well-designed wing, with properly defined airfoils, the shock forming in normal cruise flight is rather weak and contributes little to the drag of the aircraft. In *off-design* flight conditions, an incursion into the high transonic domain can take place. There results the intensification of the shock wave with a risk of shock-induced separation if the local Mach number is close to 1.4. If the separation is extended, the shock is driven by an oscillatory motion at low frequency, a phenomenon known as *transonic buffeting*. This may cause awkward movements of the aircraft and can compromise its structural integrity.

25.10. Supercritical airfoils

We recall the Breguet formula given in section 2.2 which plays a key role in defining the operational performance of an airplane. If L denotes the range, m_1 the aircraft weight at takeoff, m_2 the landing weight, C_s specific fuel consumption of the engines (fuel mass consumed per time and thrust *units*), g the acceleration of gravity, a_∞ the velocity of sound at the flight altitude, f the aerodynamic lift-over-drag ratio, and M_∞ the flight Mach number, this formula is written as:

$$L = a_\infty \frac{f \ M_\infty}{g \ C_s} \ \ln\left(\frac{m_1}{m_2}\right).$$

The aim being to carry the maximum payload over the longest possible distance, optimization will play on the dry mass (i.e. cell mass plus payload: m_2) and the grouping:

$$\frac{f \, M_\infty}{C_s}.$$

An efficiency gain is obtained by increasing the fuel efficiency of the engines, improving the lift-over-drag ratio, finally increasing the cruise Mach number without diminishing the lift-over-drag ratio, i.e. without increasing the drag. Therefore, we will seek to fly at a high as possible Mach number M_∞ without going too far into the transonic domain where the drag will increase dramatically.

Thus, when designing airliner wings, one of the objectives is to increase the drag divergence Mach number M_{div} in such a way as to increase the cruise velocity, without unduly penalizing drag. For conventional airfoils, the Mach number M_{div} follows closely the critical Mach number M^*. Thus, it is similar to trying to increase M^*. This goal can be achieved by using thinner profiles, but the thickness of a profile cannot be too small, for the reasons we have seen. Hence the idea of using airfoils with standard thickness (12 to 14%) but with a shape such that the drag divergence Mach number M_{div} will be postponed as much as possible without having to increase M^*.

We will therefore develop airfoils where supersonic velocities occur for almost the same value of M_∞ as for conventional airfoils, but which are such that phenomena causing the drag divergence will be reduced or postponed to higher Mach numbers M_∞. To maintain a good lift, we adopt a curvature of the leading edge accelerating the flow to supersonic in order to quickly create a strong depression. Then, we give the upper side an almost flat contour to limit the amplitude of the subsequent expansion of the supersonic flow. The pressure distribution on the upper surface is thus characterized by the existence of a quasi-plateau until the terminal shock which causes the flow to return to subsonics. The expansion of the leading edge is sometimes followed by a slight compression in order to reduce the shock intensity without losing too much lift. The lower side is also modified to increase the overpressures (we remember that the lift of the profile is represented by the area limited by the pressure distributions over the upper and lower surfaces, see section 2.5.3). Such airfoils, commonly used today on airliners, are called *supercritical airfoils*.

Figure 25.12 shows the shape of the supercritical airfoil *Onera OAT15A*. We note the almost flat upper side and the particular shape of the rear part of the lower side in the trailing edge region whose concave shape creates a lift increasing compression. This airfoil is compared to the conventional *NACA 2412* profile with 12% relative thickness (see Figure 25.14b) whose upper side is convex and the lower side nearly flat.

Figure 25.13 compares the distributions of the pressure coefficient on a classic and a supercritical airfoil. On the supercritical airfoil (see Figure 25.13b), we note the very particular form of the pressure distribution on the upper surface where the shock forms at about 76% of the chord. Comparing these distributions with those of a conventional airfoil (see Figure 25.13a), we understand the lift gain obtained by the supercritical airfoil. For these calculation conditions (upstream Mach $M_\infty = 0.75$ and incidence $\alpha = 3°$), the lift coefficient is equal to 1.229 for the supercritical airfoil and 0.720 for the conventional airfoil. Obtaining a similar lift with a conventional airfoil would require a much broader supersonic expansion and, consequently, a more intense shock, resulting in a significant drag and the risk of boundary layer separation.

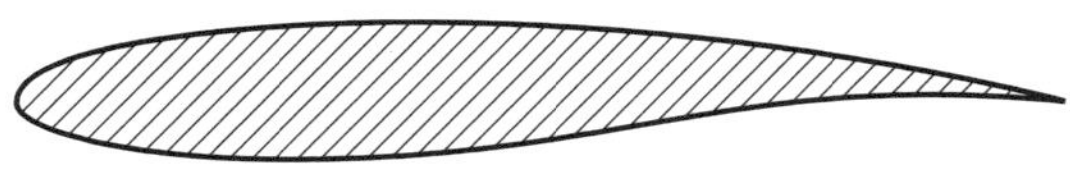

a – supercritical airfoil Onera OAT15A

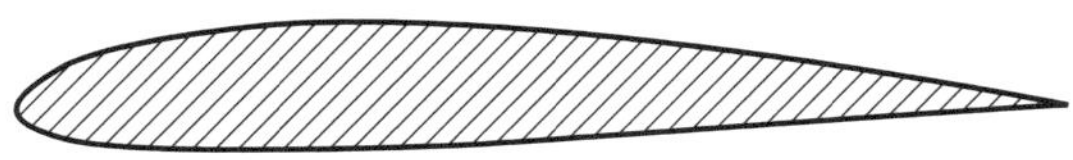

b – conventional airfoil NACA 2412

Figure 25.12. *Comparison of a supercritical and a conventional airfoil*

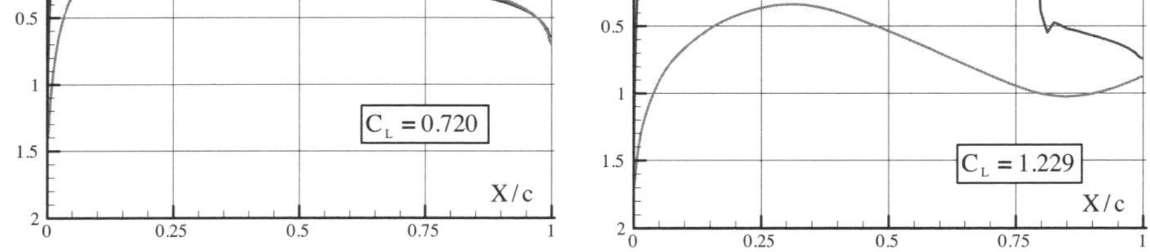

a – conventional airfoil *b – supercritical airfoil Onera OAT15A*

Figure 25.13. *Distribution of the pressure coefficient on transonic airfoils, $M_\infty = 0.75, \alpha = 3°$ (Euler calculation by the Onera code elsA)*

In Figure 25.14, we compare the Mach number distributions around the *ONERA OAT15A* airfoil at a Mach number $M_\infty = 0.75$ for an incidence of $\alpha = 1°$, close to the conditions of cruise flight (see Figure 25.14a), and for an incidence of $\alpha = 3°$, corresponding to *off-design* conditions (see Figure 25.14b). In the first case $C_L = 0.852$, in the second $C_L = 1.229$, this higher lift value being paid for by a much higher drag coefficient and thus lower lift-over-drag ratio.

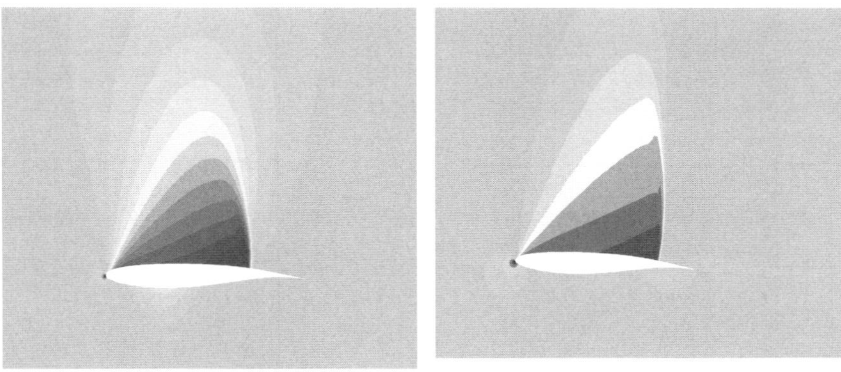

a – incidence: 1° *b – incidence: 3°*

Figure 25.14. *Mach number distribution around the supercritical airfoil Onera OAT15A, upstream Mach number: $M_\infty = 0.75$ (Euler calculation by the Onera code elsA)*

The calculation of the performance of a supercritical airfoil is difficult because of the very slowly changing shape of the upper surface, as a small error in predicting the viscous effects (boundary layer) causes a large difference in the position of the shock, leading to a large lift coefficient difference. The sensitivity of a supercritical airfoil to viscous effects is shown in Figure 25.15 where the pressure distributions on the airfoil *OAT15A* for $M_\infty = 0.75$ and $\alpha = 1°$ are plotted first within the non-viscous fluid approximation (Euler calculation, Figure 25.15a) and then by taking viscous effects into account (Navier–Stokes calculations, Figure 25.15b).

Because of the presence of a boundary layer, the relative position of the shock wave moves from $\frac{x}{c} \approx 0.70$, for the non-viscous model, to $\frac{x}{c} \approx 0.58$ for the viscous case, the lift coefficient dropping from $C_L = 0.852$ to $C_L = 0.771$.

Furthermore, the rear part of the upper surface of a supercritical airfoil must be carefully shaped to avoid separation of the boundary layer, which has been subjected to the destabilizing influence of the shock wave.

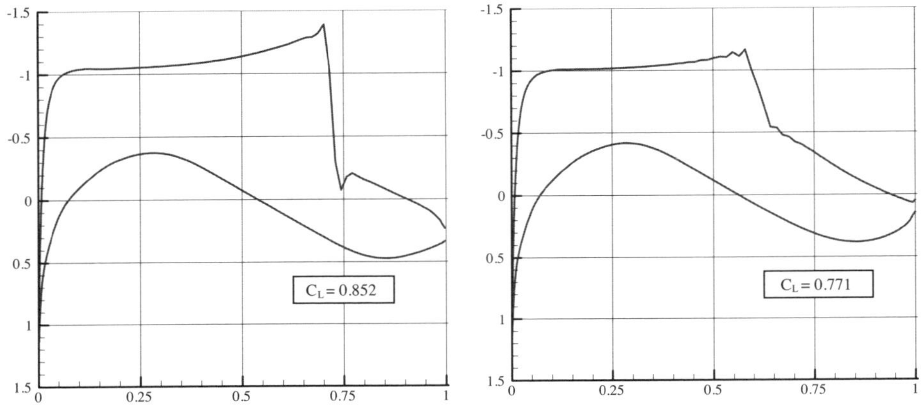

a – non-viscous fluid (Euler equations) *b – viscous fluid (Navier–Stokes equations)*

Figure 25.15. *Distribution of the pressure coefficient on the supercritical airfoil Onera OAT15A,* $M_\infty = 0.75, \alpha = 1°$ *(calculations by the Onera code elsA)*

The picture in Figure 25.16 is an interferential schlieren of the transonic field past a supercritical airfoil studied in one of the transonic wind tunnels of the Onera Meudon-Center.

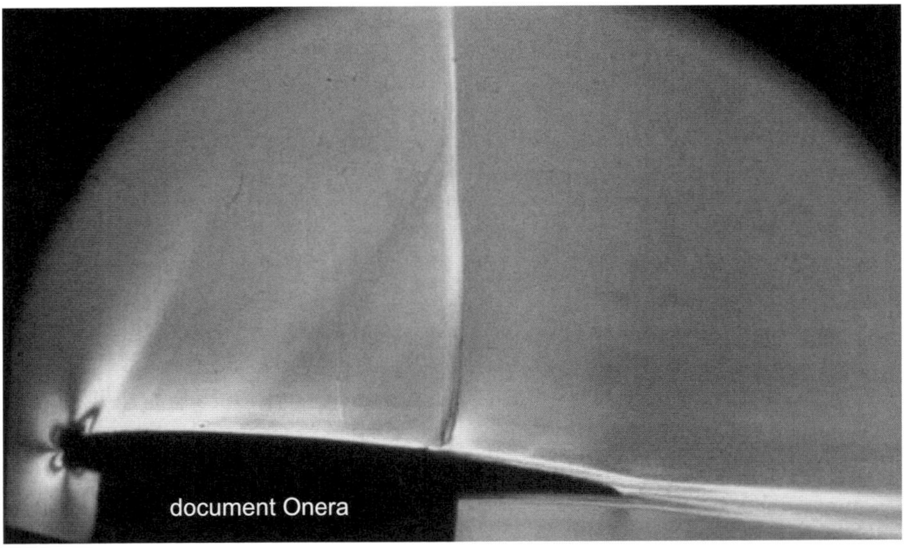

Figure 25.16. *Interferential schlieren of the flow past a supercritical airfoil (document Onera)*

General Conclusion

The field of compressible aerodynamics, or gas dynamics, is not only that of planes, missiles, space launchers, and projectiles flying at supersonic velocities, that of planetary probes entering into the atmosphere at hypersonic Mach numbers, but also that of present transonic transport aircraft. In the domain of propulsion, it concerns air intakes, turbomachines, propulsive nozzle, or supersonic diffusers. Unsteady propagation mechanism and shock wave formation directly pertain to gas dynamics. The application field of unsteady theory is particularly vast. It includes the propagation of waves generated by a sudden energy release (explosion), supersonic boom, or shock wave formation due to the entering into a tunnel of a high-speed train, to cite a few examples. We should also mention auto-induced fluctuations in a flow, like transonic buffeting or air-intake buzz. Such phenomena, which are the source of numerous inconveniences (noise, vibrations, loss of control, etc.) are the subject of special attention from aerodynamicists. Acoustics belongs to these domains; however this discipline by itself is a science justifying separate developments; this is only evoked in this treatise.

The presented notions constitute the theoretical basis for the study of flows reaching speeds high enough to induce substantial variations of the density. This fact requires us to take into account, in the formulation of the problem, the energy equation, and hence a strong coupling between fluid mechanics and thermodynamics leading to remarkable results. Theory allows us to establish basic relations for the flow evolution and shows the drastic changes in the flow behavior which occur when its speed goes beyond the sound velocity. In this treatise, we have established the basic properties of the steady supersonic flows and explained the mechanism by which perturbations are propagated along waves or characteristics. In some circumstances, this propagation mechanism can lead to the formation of shock waves, whose presence is perhaps the most striking feature of supersonic flows. This basic knowledge is essential for an understanding of the physics of supersonic flows in which perturbations induced by a vehicle only affect a limited downstream domain and in which concentrated pressure variations can exist in the form of expansion fans and/or shock waves.

We have shown that there is a close parallelism between steady two-dimensional supersonic flows and one-dimensional unsteady flows. In these two cases, the phenomena are governed by differential systems with similar properties. In addition to its interest for analyzing the propagation process, the unsteady approach enables a deep understanding of the structure of steady flows. Indeed, in the real world, phenomena have a beginning followed by an establishment time (transient period), before reaching (but not always) a steady state. The nature of the steady state frequently depends on the unsteady phase which has led to its establishment. This is the case of a supersonic nozzle starting process, for example.

We have considered flows with one or two spatial dimensions, which is a severe restriction since real flows are most often three-dimensional. In fact, these simplified analyses allowed establishing, with a minimum of mathematical formalism, general fundamental results and understanding basic phenomena on cases conceptually simple. Then, the passage to three dimensions is greatly facilitated.

This treatise is also restricted to the study of compressible aerodynamic flows in the context of the non-viscous fluid approximation. This hypothesis leads to a "truncated" description of the real flows since it ignores the fact that there always exist regions where viscosity cannot be neglected; with the penalty that some basic physical constraints are violated (like the no slip condition on a wall). Fortunately, in usual aerodynamic applications, where the Reynolds number is very high, viscosity effects are confined within a region of very small size (compared to that of the vehicle): boundary layers, wakes, and shock waves. Consequently, the vast majority of the flow behaves like a non-viscous fluid amenable to the presented theories.

The non-viscous flow model provides a powerful tool to interpret and predict the flow in many circumstances. However, it does not allow the prediction of transfer phenomena such as the skin friction and/or wall heat flux. This fact excludes the capacity to predict the friction drag of a vehicle or the heat transfer between the flow and its surface. In addition, as pointed out in the general introduction, there are circumstances where viscous effects play a predominant role, leading to massive separation with formation of intense vortical structures. These circumstances are a major concern for aerodynamicists since their occurrence puts a limit to the achievable performance (maximum incidence of a wing before buffeting and stall, compressor surge, to cite two important examples). The dominant influence of viscous effects is permanent in the aerodynamics of terrestrial vehicles whose shape has poor aerodynamic efficiency, for obvious reasons. Then, a purely non-viscous analysis is meaningless and we have to take into account the true nature of the flow. But this is another story…

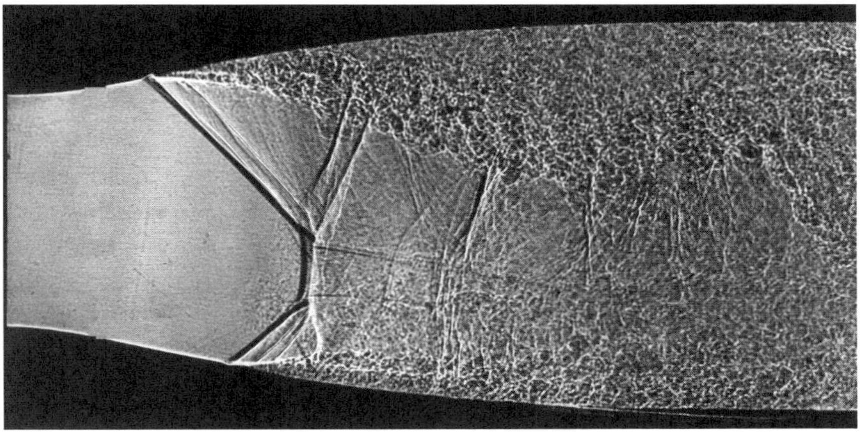

Massive separation in an overexpanded supersonic nozzle; the flow is dominated by viscous effects (document Onera)

Appendix 1

Review of Mathematical Notations and Relations

The purpose of this appendix is to facilitate the understanding of theoretical developments by recalling some mathematical notions and relations used in fluid mechanics. These reminders are restricted to the mathematical tools used in this treatise.

A.1. Scalars, vectors, and tensors

The behavior of a fluid is described by calling upon mathematical concepts with proper properties and submitted to well-defined operation rules. Thus, we introduce:

– One-dimensional scalar quantities represented by a number: this is the case of the pressure p, density ρ, temperature T, etc.

– Vectors which in a three-dimensional space are represented by three numbers, namely, their components in some coordinate system: this is the case of the velocity $\vec{V}$, acceleration $\vec{\gamma}$, force $\vec{F}$, tension $\vec{P}$, etc.

In a Cartesian system of coordinates, with unit vectors $\vec{i}, \vec{j}, \vec{k}$ along directions Ox, Oy, Oz, a vector will be noted:

$$\vec{V} = u\vec{i} + v\vec{j} + w\vec{k}$$

where u, v, w are the vector components in the chosen axes system. One can use the tensorial notation by noting u_i the components of $\vec{V}$ and x_i the coordinates (x, y, z), the suffix i taking the values 1, 2, 3 in a three-dimensional space. Then, a symbol such as u_i designates a vector with components:

$$u_1,\ u_2\ ,u_3.$$

The module or norm of the vector is defined by:

$$\left|\vec{V}\right| = \sqrt{u^2 + v^2 + w^2} \quad or \quad \left|\vec{V}\right| = \sqrt{u_1^2 + u_2^2 + u_3^2}.$$

– Tensors of order 2 which are represented by matrices with nine components in the three-dimensional space (a tensor is not a matrix). This is the case of the strain and stress tensors.

For example:

$$\overline{\overline{T}} = \begin{pmatrix} \tau_{xx} & \tau_{xy} & \tau_{xz} \\ \tau_{yx} & \tau_{yy} & \tau_{yz} \\ \tau_{zx} & \tau_{zy} & \tau_{zz} \end{pmatrix}$$

is the viscous stress tensor. One demonstrates that in fluid mechanics the stress tensor is symmetrical; which means that:

$$\tau_{xy} = \tau_{yx},\ \tau_{xz} = \tau_{zx},\ \tau_{zy} = \tau_{yz}.$$

Also note:

$$\overline{\overline{T}} = \begin{pmatrix} a_{11} & a_{12} & a_{13} \\ a_{21} & a_{22} & a_{23} \\ a_{31} & a_{32} & a_{33} \end{pmatrix} = a_{ij}$$

with i varying from 1 to 3 and j from 1 to 3. Thus, the symbol a_{ij} designates an order 2 tensor with nine components. If the tensor is symmetrical, then: $a_{ij} = a_{ji}$.

The trace of the tensor is the sum of its diagonal elements:

$$trace\left(\overline{\overline{T}}\right) = a_{11} + a_{22} + a_{33}.$$

The unit tensor, noted $\overline{\overline{I}}$, has all its components equal to zero except those of the main diagonal which are equal to 1:

$$\overline{\overline{I}} = \begin{pmatrix} 1 & 0 & 0 \\ 0 & 1 & 0 \\ 0 & 0 & 1 \end{pmatrix}.$$

We sometimes note that $\overline{\overline{I}} = \delta_{ij}$, with $\delta_{ij} = 1$ if $i = j$ and $\delta_{ij} = 0$ if $i \neq j$. The tensor δ_{ij} is called the *Kronecker delta*. Thus, in the expression of the tension vector $\vec{P}$ representing the contact actions within a fluid, one isolates the pressure by writing:

$$\overline{\overline{P}} = -p\overline{\overline{I}} = -p\delta_{ij}.$$

A tensor such as $\overline{\overline{P}}$ is said to be spherical.

These mathematical concepts are defined from invariance properties in a change of coordinate system. Thus, a scalar is independent of the coordinate system and the norm of a vector is invariant. Tensors have more complex invariance properties which we will not give here. In particular, the trace of a tensor is an invariant.

To simplify calculations and formulas, we agree that a repeated suffix means a summation with respect to this suffix. Thus, the notation $u_i u_i$ means $u_1^2 + u_2^2 + u_3^2$: which is the expression of the square of the norm of the vector. If u_i is a velocity vector, this quantity is a scalar (for example, the double of the specific kinetic energy). As a general rule, the repetition of a suffix lowers the order of the mathematical quantity: applied to a vector, it gives a scalar, applied to a tensor, a vector. A double suffix repetition on a term involving tensors and vectors gives a scalar. For example, in the energy equation, the term:

$$\frac{\partial \left(u_j \tau_{ij} \right)}{\partial x_i}$$

where the indices *i* and *j* are repeated has the dimension of a scalar (here the power density or power per unit volume).

A.2. Some common operators

– *Scalar product of two vectors*

This is defined by:

$$\vec{a}.\vec{b} = a_1 b_1 + a_2 b_2 + a_3 b_3$$

or with the tensorial notation: $\vec{a}.\vec{b} = a_i b_i$, the suffix repetition indicating summation with respect to *i*. This repetition has lowered the dimension since the scalar product involving two vectors is a scalar.

– *Vectorial product of two vectors*

It is defined by:

$$\vec{a} \times \vec{b} = \begin{pmatrix} a_1 \\ a_2 \\ a_3 \end{pmatrix} \times \begin{pmatrix} b_1 \\ b_2 \\ b_3 \end{pmatrix} = \begin{pmatrix} a_2 b_3 - a_3 b_2 \\ b_1 a_3 - a_1 b_3 \\ a_1 b_2 - b_1 a_2 \end{pmatrix} = \varepsilon_{ijk} a_j b_k$$

where ε_{ijk} is the permutation symbol: $\varepsilon_{ijk} = 1$ for a direct permutation over the indices, $\varepsilon_{ijk} = -1$ for an indirect permutation. The result of a vectorial product is a vector

– *Vector gradient of a function*

If $f(x, y, z)$ is a scalar function of the variables (x, y, z), the gradient of f is a vector with components:

$$\overrightarrow{grad} \;\; f(x, y, z) \equiv \vec{\nabla} f = \begin{pmatrix} \dfrac{\partial f}{\partial x} \\ \dfrac{\partial f}{\partial y} \\ \dfrac{\partial f}{\partial z} \end{pmatrix} = \frac{\partial f}{\partial x_i}.$$

Symbol $\vec{\nabla}$, or nabla operator, is a symbolic vector with components $\frac{\partial}{\partial x}, \frac{\partial}{\partial y}, \frac{\partial}{\partial z}$ or $\frac{\partial}{\partial x_i}$. Its introduction allows a more compact writing of formulas. The variation of the function f between points $P(x, y, z)$ and $P'(x+dx, y+dy, z+dz)$ is such that:

$$df = \frac{\partial f}{\partial x} dx + \frac{\partial f}{\partial y} dy + \frac{\partial f}{\partial z} dz.$$

Since:

$$\left(PP'\right)^2 = ds^2 = dx^2 + dy^2 + dz^2$$

and designating by $\overrightarrow{ds}$ (components dx, dy, dz) the displacement vector $\overrightarrow{PP'}$, we see that:

$$df = \overrightarrow{grad} f \,.\, \overrightarrow{ds}.$$

The variation of the function f between points P and P' is equal to the scalar product of the vector $\overrightarrow{grad} f$ by the displacement vector between P and P': $\overrightarrow{ds}$. Consequently, if the function f is constant on a curve (C), then $df = 0$ for a displacement $\overrightarrow{ds}$ on (C) and:

$$\overrightarrow{grad} f \;.\; \overrightarrow{ds} = 0.$$

The vector $\overrightarrow{grad} f$ is normal to the constant-value curves of the function *f*.

If $\Phi(x, y, z)$ is a function potential for velocities and $\vec{n}$ the unit normal vector to a surface, then the relation:

$$\overrightarrow{grad}\Phi\left(x, y, z\right).\vec{n} = \vec{V}.\vec{n} = 0$$

expresses the slip condition on the obstacle.

– Divergence of a vector

This operator is defined by:

$$div\ \vec{V} \equiv \vec{\nabla}.\vec{V} = \begin{pmatrix} \dfrac{\partial}{\partial x} \\ \dfrac{\partial}{\partial y} \\ \dfrac{\partial}{\partial z} \end{pmatrix} . \begin{pmatrix} u \\ v \\ w \end{pmatrix} = \frac{\partial u}{\partial x} + \frac{\partial v}{\partial y} + \frac{\partial w}{\partial z} = \frac{\partial u_i}{\partial x_i}.$$

Applied to a vector, the result of the divergence operator is a scalar (suffix i repetition).

– Divergence of a tensor

The divergence of an order 2 tensor is defined by:

$$div\ \overline{\overline{T}} = \begin{vmatrix} \dfrac{\partial a_{11}}{\partial x_1} + \dfrac{\partial a_{12}}{\partial x_2} + \dfrac{\partial a_{13}}{\partial x_3} \\ \dfrac{\partial a_{21}}{\partial x_1} + \dfrac{\partial a_{22}}{\partial x_2} + \dfrac{\partial a_{23}}{\partial x_3} \\ \dfrac{\partial a_{31}}{\partial x_1} + \dfrac{\partial a_{32}}{\partial x_2} + \dfrac{\partial a_{33}}{\partial x_3} \end{vmatrix} = \frac{\partial a_{ij}}{\partial x_j}.$$

The result is a vector since in the above expression there is a repetition of the suffix j, and hence summation over j.

– Tensorial product

The product of two order 2 tensors is a tensor of same order whose expression is:

$$c_{ij} = a_{ik} b_{jk}.$$

The development of this expression does not present any difficulty. We frequently use the contracted product of a tensor and a vector which is defined by:

$$\begin{vmatrix} a_{11} & a_{12} & a_{13} \\ a_{21} & a_{22} & a_{23} \\ a_{31} & a_{32} & a_{33} \end{vmatrix} . \begin{vmatrix} b_1 \\ b_2 \\ b_3 \end{vmatrix} = \begin{vmatrix} a_{11}b_1 + a_{12}b_2 + a_{13}b_3 \\ a_{21}b_1 + a_{22}b_2 + a_{23}b_3 \\ a_{31}b_1 + a_{32}b_2 + a_{33}b_3 \end{vmatrix} = a_{ij} b_j.$$

The result is a vector (suffix j repetition). One finds this operation in the relation expressing the tension $\vec{P}$ as a function of the stress tensor $\overline{\overline{T}}$ and the unit normal vector $\vec{n}$ of the surface on which the tension applies, i.e.,

$$\vec{P} = -\overline{\overline{T}}.\vec{n}.$$

The tensorial product is frequently represented by the symbol $\otimes$.

– *Curl of a vector*

This operation associates to a vector another vector defined by:

$$\overrightarrow{curl}\,\vec{V} = \vec{\nabla} \times \vec{V} = \begin{vmatrix} \dfrac{\partial}{\partial x} \\ \dfrac{\partial}{\partial y} \\ \dfrac{\partial}{\partial z} \end{vmatrix} \times \begin{vmatrix} u \\ v \\ w \end{vmatrix} = \begin{vmatrix} \dfrac{\partial w}{\partial y} - \dfrac{\partial v}{\partial z} \\ \dfrac{\partial u}{\partial z} - \dfrac{\partial w}{\partial x} \\ \dfrac{\partial v}{\partial x} - \dfrac{\partial u}{\partial y} \end{vmatrix}.$$

The curl vector is formally obtained by forming the vectorial product of the nabla operator with the vector $\vec{V}$.

One demonstrates that the curl of a gradient is zero.

$$\overrightarrow{curl}\left(\overrightarrow{grad}\,f\right) = \vec{\nabla} \times \left(\overrightarrow{grad}\,f\right) = \begin{vmatrix} \dfrac{\partial}{\partial x} \\ \dfrac{\partial}{\partial y} \\ \dfrac{\partial}{\partial z} \end{vmatrix} \times \begin{vmatrix} \dfrac{\partial f}{\partial x} \\ \dfrac{\partial f}{\partial y} \\ \dfrac{\partial f}{\partial z} \end{vmatrix} = \begin{vmatrix} \dfrac{\partial^2 f}{\partial z \partial y} - \dfrac{\partial^2 f}{\partial y \partial z} \\ \dfrac{\partial^2 f}{\partial x \partial z} - \dfrac{\partial^2 f}{\partial z \partial x} \\ \dfrac{\partial^2 f}{\partial x \partial y} - \dfrac{\partial^2 f}{\partial y \partial x} \end{vmatrix} = 0.$$

This property is used in the calculation of irrotational flows, the condition of zero rotational (zero curl of the velocity) being automatically satisfied by introduction of a potential function for the velocity.

The divergence of the curl of a vector is zero, since:

$$div\left(\overrightarrow{curl}\;\vec{V}\right) = \frac{\partial}{\partial x}\left(\frac{\partial w}{\partial y} - \frac{\partial v}{\partial z}\right) + \frac{\partial}{\partial y}\left(\frac{\partial u}{\partial z} - \frac{\partial w}{\partial x}\right) + \frac{\partial}{\partial z}\left(\frac{\partial v}{\partial x} - \frac{\partial u}{\partial y}\right).$$

Hence, by virtue of the cross partial derivative properties:

$$div\left(\overrightarrow{curl}\ \vec{V}\right)=\frac{\partial^2 w}{\partial x\partial y}-\frac{\partial^2 v}{\partial x\partial z}+\frac{\partial^2 u}{\partial y\partial z}-\frac{\partial^2 w}{\partial y\partial x}+\frac{\partial^2 v}{\partial z\partial x}-\frac{\partial^2 u}{\partial z\partial y}=0.$$

– *Laplace operator*

The Laplace operator is the result of the divergence of a gradient vector. Hence:

$$\nabla^2 f(x,y,z)=\frac{\partial^2 f}{\partial x^2}+\frac{\partial^2 f}{\partial y^2}+\frac{\partial^2 f}{\partial z^2}.$$

With other notations:

$$\nabla^2 f=\vec{\nabla}.\vec{\nabla}f, \qquad \nabla^2 f=\frac{\partial}{\partial x_i}\left(\frac{\partial f}{\partial x_i}\right)=\frac{\partial^2 f}{\partial x_i\partial x_i}.$$

The introduction of a velocity potential $\Phi(x,y,z)$ enables us to satisfy the irrotationality condition. To represent a velocity field, $\overrightarrow{grad}\,\Phi(x,y,z)$ must satisfy the condition of zero divergence (in incompressible flows), which leads to the Laplace equation:

$$div\left[\overrightarrow{grad}\ \ \Phi(x,y,z)\right]=\frac{\partial^2\Phi}{\partial x^2}+\frac{\partial^2\Phi}{\partial y^2}+\frac{\partial^2\Phi}{\partial z^2}=0.$$

A.3. Some classic theorems

– *Divergence theorem*

Let us consider a closed surface (S) bounding a volume (υ) and a vector field $\vec{V}$ continuous and differentiable in (υ) and on (S). If $\vec{n}$ designates the unit normal vector to the surface element ds, then:

$$\iiint_{(V)} div\vec{V}\,dv=\iiint_{(V)}\vec{\nabla}.\vec{V}\,dv=\iint_{(S)}\vec{V}.\vec{n}\,ds$$

or:

$$\iiint_{(V)} \frac{\partial u_i}{\partial x_i} \; dv = \iint_{(S)} u_i.n_i \; ds.$$

The divergence theorem transforms a volume integral into a surface integral thus resulting in an expression for the flux of the vector $\vec{V}$ through (S).

– *Stokes theorem*

Let us consider a surface (S) supported by a closed curve (C). The Stokes theorem gives a relation between the flux of the curl of $\vec{V}$ through (S) (or the surface integral of the curl of $\vec{V}$ on (S)) and the circulation of the velocity $\vec{V}$ along (C)

$$\iint_{(S)} \overrightarrow{curl} \; \vec{V} \; ds = \oint_{(C)} \vec{V}.\vec{t} \; dl.$$

Here, $\vec{t}$ is a unit vector tangent to (C). Thus, we demonstrate that the integral of a vector along a closed surface (C) is equal to the flux of its curl through any surface (S) supported by (C).

A.4. Kinematic relations

– *Particle derivative or material derivative or total derivative*

In a moving fluid, one considers a quantity f convected with the velocity $\vec{V}$ of the fluid. Let us evaluate the variation df of f during the time interval dt. For a given spatial position $\vec{x}$, this quantity varies not only with time (except if the flow is steady) but also due to the fact that during the interval dt the quantity f has been entrained by the fluid. Hence, its position having changed by a quantity $d\vec{x}$, the total variation df is expressed by:

$$df(\vec{x},t) = \frac{\partial f(\vec{x},t)}{\partial t} dt + \frac{\partial f(\vec{x},t)}{\partial \vec{x}} d\vec{x}$$

or:

$$\frac{df(\vec{x},t)}{dt} = \frac{\partial f(\vec{x},t)}{\partial t} + \frac{\partial f(\vec{x},t)}{\partial \vec{x}} \frac{d\vec{x}}{dt}.$$

However: $\dfrac{d\vec{x}}{dt} = \vec{V}$ is equal to the flow velocity and $\dfrac{\partial f(\vec{x},t)}{\partial \vec{x}} = \overrightarrow{grad} f$. Hence, the expression:

$$\boxed{\frac{df}{dt} = \frac{\partial f}{\partial t} + \vec{V}.\overrightarrow{grad} f = \frac{\partial f}{\partial t} + u_i \frac{\partial f}{\partial x_i}.}$$

– *Expression of the acceleration*

The material derivative applied to the velocity vector $\vec{V}$ gives the acceleration vector $\vec{\gamma}$. Hence:

$$\vec{\gamma} \equiv \frac{d\vec{V}}{dt} = \frac{\partial \vec{V}}{\partial t} + \overrightarrow{grad}\ \vec{V}.\ \vec{V}.$$

The term $\overrightarrow{grad} \vec{V}$ is a tensor with components:

$$\overrightarrow{grad}\ \vec{V} = \begin{vmatrix} \dfrac{\partial u}{\partial x} & \dfrac{\partial u}{\partial y} & \dfrac{\partial u}{\partial z} \\ \dfrac{\partial v}{\partial x} & \dfrac{\partial v}{\partial y} & \dfrac{\partial v}{\partial z} \\ \dfrac{\partial w}{\partial x} & \dfrac{\partial w}{\partial y} & \dfrac{\partial w}{\partial z} \end{vmatrix} = \frac{\partial u_i}{\partial x_j}.$$

The term $\overrightarrow{grad} \vec{V}.\vec{V}$ is the contracted product of the tensor $\overrightarrow{grad} \vec{V}$ by the vector $\vec{V}$.

Now we develop the expression for $\vec{\gamma}$:

$$\gamma_x = \frac{\partial u}{\partial t} + u\frac{\partial u}{\partial x} + v\frac{\partial u}{\partial y} + w\frac{\partial u}{\partial z}$$

$$\gamma_y = \frac{\partial v}{\partial t} + u\frac{\partial v}{\partial x} + v\frac{\partial v}{\partial y} + w\frac{\partial v}{\partial z}$$

$$\gamma_z = \frac{\partial w}{\partial t} + u\frac{\partial w}{\partial x} + v\frac{\partial w}{\partial y} + w\frac{\partial w}{\partial z}.$$

Let us consider the component γ_x which can be written as:

$$\gamma_x = \frac{\partial u}{\partial t} + u\frac{\partial u}{\partial x} + v\frac{\partial u}{\partial y} + w\frac{\partial u}{\partial z} = \frac{\partial u}{\partial t} + u\frac{\partial u}{\partial x} + v\frac{\partial v}{\partial x} + w\frac{\partial w}{\partial x} - v\frac{\partial v}{\partial x} + v\frac{\partial u}{\partial y} - w\frac{\partial w}{\partial x} + w\frac{\partial u}{\partial z}.$$

We introduce the vorticity vector:

$$\vec{\omega} = \frac{1}{2}\overrightarrow{curl}\,\vec{V}$$

whose components are:

$$2\omega_x = \frac{\partial w}{\partial y} - \frac{\partial v}{\partial z}$$

$$2\omega_y = \frac{\partial u}{\partial z} - \frac{\partial w}{\partial x}$$

$$2\omega_z = \frac{\partial v}{\partial x} - \frac{\partial u}{\partial y}.$$

The expression of γ_x is written as:

$$\gamma_x = \frac{\partial u}{\partial t} + u\frac{\partial u}{\partial x} + v\frac{\partial v}{\partial x} + w\frac{\partial w}{\partial x} - 2\omega_z v - 2w\omega_y.$$

By making the same calculation for the other components of $\vec{\gamma}$, we obtain the relation:

$$\boxed{\vec{\gamma} \equiv \frac{d\vec{V}}{dt} = \frac{\partial \vec{V}}{\partial t} + \overrightarrow{grad}\left(\frac{V^2}{2}\right) + 2\left(\vec{\omega}\times\vec{V}\right) = \frac{\partial \vec{V}}{\partial t} + \overrightarrow{grad}\left(\frac{V^2}{2}\right) + \overrightarrow{curl}\;\vec{V}\times\vec{V}.}$$

This relation plays an important role in fluid mechanics since it shows that the hypothesis of a zero curl (or vorticity) for the velocity leads to an important simplification of the motion equations. In reality, this hypothesis can be applied to many flows; occurrence of a vorticity vector being linked to entropy production in the flow, which itself depends on well-specific phenomena, such as the action of viscosity.

Appendix 2

Table of Useful Relations for the Fundamental Concepts of Aerodynamics

Aerodynamic coefficients		
Normal force	$C_N = \dfrac{F_N}{\frac{1}{2}\rho_\infty V_\infty^2 S}$	
Axial force	$C_A = \dfrac{F_A}{\frac{1}{2}\rho_\infty V_\infty^2 S}$	
Lift	$C_L = \dfrac{F_L}{\frac{1}{2}\rho_\infty V_\infty^2 S}$	
Drag	$C_D = \dfrac{F_D}{\frac{1}{2}\rho_\infty V_\infty^2 S}$	
Transfer relations	$F_L = F_N \cos\alpha - F_A \sin\alpha$ $F_D = F_N \sin\alpha + F_A \cos\alpha$	
Moment	$C_m = \dfrac{M}{\frac{1}{2}\rho_\infty V_\infty^2 SL}$	
Lift-over-drag ratio (L/D)	$f = C_L / C_D$	
Bréguet range equation	$L = \dfrac{f V_\infty}{g C_s} \ln\left(\dfrac{m_1}{m_2}\right)$	$L = a_\infty \dfrac{f M_\infty}{g C_s} \ln\left(\dfrac{m_1}{m_2}\right)$

Aerodynamic coefficients (continued)		
Pressure coefficient	$C_p = \dfrac{p - p_\infty}{\frac{1}{2}\rho_\infty V_\infty^2}$	$C_p = \dfrac{2}{\gamma M_\infty^2}\left(\dfrac{p}{p_\infty} - 1\right)$
Skin friction coefficient	$C_f = \dfrac{\tau_w}{\frac{1}{2}\rho_\infty V_\infty^2}$	
Heat transfer coefficient	$C_h = \dfrac{\dot{q}_w}{\rho_e V_e \left(h_f - h_w\right)}$	$C_h = \dfrac{\dot{q}_w}{\rho_e V_e C_p \left(T_f - T_w\right)}$

Aerodynamic forces on a two-dimensional airfoil with unit span
Normal force
$F_N = \int_{LE}^{TE} \left(-p_u \cos\theta + \tau_u \sin\theta\right) ds + \int_{LE}^{TE} \left(p_l \cos\theta + \tau_l \sin\theta\right) ds$
Axial force
$F_A = \int_{LE}^{TE} \left(p_u \sin\theta + \tau_u \cos\theta\right) ds + \int_{LE}^{TE} \left(-p_l \sin\theta + \tau_l \cos\theta\right) ds$
Lift
$F_L = \left[\int_{LE}^{TE} -p_u \cos\theta\, ds + \int_{LE}^{TE} p_l \cos\theta\, ds\right]\cos\alpha - \left[\int_{LE}^{TE} p_u \sin\theta\, ds - \int_{LE}^{TE} p_l \sin\theta\, ds\right]\sin\alpha$ $+\left[\int_{LE}^{TE} \tau_u \sin\theta\, ds + \int_{LE}^{TE} \tau_l \sin\theta\, ds\right]\cos\alpha - \left[\int_{LE}^{TE} \tau_u \cos\theta\, ds + \int_{LE}^{TE} \tau_l \cos\theta\, ds\right]\sin\alpha$
Drag
$F_D = \left[\int_{LE}^{TE} -p_u \cos\theta\, ds + \int_{LE}^{TE} p_l \cos\theta\, ds\right]\sin\alpha + \left[\int_{LE}^{TE} p_u \sin\theta\, ds - \int_{LE}^{TE} p_l \sin\theta ds\right]\cos\alpha$ $+\left[\int_{LE}^{TE} \tau_u \sin\theta\, ds + \int_{LE}^{TE} \tau_l \sin\theta\, ds\right]\sin\alpha + \left[\int_{LE}^{TE} \tau_u \cos\theta\, ds + \int_{LE}^{TE} \tau_l \cos\theta\, ds\right]\cos\alpha$
Pressure drag
$F_D = \left[\int_{LE}^{TE} -p_u \cos\theta\, ds + \int_{LE}^{TE} p_l \cos\theta\, ds\right]\sin\alpha + \left[\int_{LE}^{TE} p_u \sin\theta\, ds - \int_{LE}^{TE} p_l \sin\theta ds\right]\cos\alpha$
Friction drag
$F_D = \left[\int_{LE}^{TE} \tau_u \sin\theta\, ds + \int_{LE}^{TE} \tau_l \sin\theta\, ds\right]\sin\alpha + \left[\int_{LE}^{TE} \tau_u \cos\theta\, ds + \int_{LE}^{TE} \tau_l \cos\theta\, ds\right]\cos\alpha$
Lift of a thin airfoil at small incidence
$F_L \cong F_N = \int_{LE}^{TE} \left(p_l - p_u\right) dx$
Pitching moment with respect to the airfoil leading edge
$M_{LE} = \int_{LE}^{TE} \left[\left(p_u \cos\theta - \tau_u \sin\theta\right) x + \left(p_u \sin\theta + \tau_u \cos\theta\right) y\right] ds$ $+ \int_{LE}^{TE} \left[\left(-p_l \cos\theta - \tau_l \sin\theta\right) x + \left(-p_l \sin\theta + \tau_l \cos\theta\right) y\right] ds$

Thermodynamic Review		
First principle	$\frac{dE_T}{dt} = W' + Q'_C + D'_D + Q'_E + E'_Q + E'_M$	
Internal energy	$\frac{dE}{dt} + W_{\text{int}} = Q'_C + D'_D + Q'_E + E'_Q + E'_M$	$e_T = e + \frac{V^2}{2} \rightarrow e = e_T - \frac{V^2}{2}$
Enthalpy	$H = E + pV$	$h = e + \frac{p}{\rho} = e + p\,v$
Second principle	$T\frac{dS}{dt} \geq Q'_C + Q'_E$	$T\frac{dS}{dt} = Q'_C + Q'_E + Q'_I$
Second principle	*Adiabatic evolution* $s_2 - s_1 \geq 0$	*Isentropic evolution* $s_2 - s_1 = 0$
Expression for entropy	$T\frac{ds}{dt} = \frac{dh}{dt} - \frac{1}{\rho}\frac{dp}{dt}$	
Equation of state for a perfect gas	$\frac{p}{\rho} = \frac{\Re}{M}T$	$\Re = 8.3145\ J/mole/K$ *For air:* $\frac{\Re}{M} = r = 287\ J/kg/K$
Mayer relation	$C_p - C_v = \frac{\Re}{M} = r$	*For air:* $C_p = 1003\ J/kg/K$, $C_v = 716\ J/kg/K$
Isentropic exponent	$\gamma = \frac{C_p}{C_v}$	*For air:* $\gamma = 1.4$
Specific heats	$C_p = \frac{\gamma r}{\gamma - 1}$	$C_v = \frac{r}{\gamma - 1}$
Internal energy, enthalpy for a perfect gas	$e(T) = \int_{T_0}^{T} C_v(T)dT + e_0(T_0)$	$h(T) = \int_{T_0}^{T} C_p(T)dT + h_0(T_0)$

Thermodynamic Review (continued)		
Entropy for a perfect gas	$s(T,p)=\int_{T_0}^{T} C_p \frac{dT}{T} - r\ln\frac{p}{p_0} + s_0(T_0,p_0)$	
Entropy for a perfect gas	$s(T,\rho)=\int_{T_0}^{T} C_v \frac{dT}{T} - r\ln\frac{\rho}{\rho_0} + s_0(T_0,\rho_0)$	
Internal energy, enthalpy for a calorically perfect gas	$e = C_v T + e_0$	$h = C_p T + h_0$
Internal energy and enthalpy	$e = \frac{r}{\gamma-1} T + e_0$	$h = \frac{\gamma r}{\gamma-1} T + h_0$
Entropy of a calorically perfect gas	$s_2 - s_1 = C_p \ln\frac{T_2}{T_1} - r\ln\frac{p_2}{p_1}$	$s_2 - s_1 = C_v \ln\frac{T_2}{T_1} - r\ln\frac{\rho_2}{\rho_1}$
Isentropic relations	$\frac{p_2}{p_1} = \left(\frac{\rho_2}{\rho_1}\right)^{\gamma} = \left(\frac{T_2}{T_1}\right)^{\frac{\gamma}{\gamma-1}}$	
Sound velocity	$a = \left(\frac{\partial p}{\partial \rho}\right)_s^{1/2}$	*Calorically perfect gas* $a^2 = \gamma \frac{p}{\rho}$ *or* $a^2 = \gamma r T$

Fundamental equations in Cartesian coordinates		
Continuity equation	$\frac{\partial \rho}{\partial t} + \frac{\partial}{\partial x_i}(\rho u_i) = 0$	*Incompressible flow:* $\nabla.\vec{V} = 0$ *or* $\frac{\partial u_i}{\partial x_i} = 0$
Equation of motion	$\frac{\partial(\rho u_i)}{\partial t} + \frac{\partial}{\partial x_j}(\rho u_i u_j) = \frac{\partial T_{ij}}{\partial x_j} + \rho f_i$	*Stress tensor:* $T_{ij} = -p\,\delta_{ij} + \tau_{ij}$
Viscosity tensor	$\tau_{ij} = \lambda \frac{\partial u_k}{\partial x_k}\delta_{ij} + \mu\left(\frac{\partial u_i}{\partial x_j} + \frac{\partial u_j}{\partial x_i}\right)$	*Stokes hypothesis:* $3\lambda + 2\mu = 0$
Sutherland formula for the viscosity of air	$\mu = 1.454\times10^{-6}\frac{T^{3/2}}{T+110.4}$	$in: N\times s/m^2$

Fundamental equations in Cartesian coordinates (continued)		
Energy equation with internal energy	$\frac{\partial}{\partial t}\left[\rho\left(e+\frac{V^2}{2}\right)\right]+\frac{\partial}{\partial x_i}\left[\rho\, u_i\left(e+\frac{V^2}{2}\right)\right]=$ $-\frac{\partial(pu_i)}{\partial x_i}+\frac{\partial}{\partial x_i}\left(u_j\tau_{ij}-\dot{q}_i\right)+\rho f_i u_i$	*Fourier law:* $\dot{q}_i=-\kappa\frac{\partial T}{\partial x_i}=-\frac{\mu C_p}{\Pr}\frac{\partial T}{\partial x_i}$ $\Pr=\frac{\mu C_p}{\kappa}$
Energy equation with enthalpy	$\frac{dh_T}{dt}=\frac{1}{\rho}\frac{\partial p}{\partial t}+\frac{1}{\rho}\frac{\partial}{\partial x_i}\left(u_j\,\tau_{ij}-\dot{q}_i\right)+f_i\,u_i$	$h_T\equiv h_{st}=e+\frac{V^2}{2}+\frac{p}{\rho}$ *Stagnation enthalpy*
Energy equation with entropy	$T\frac{ds}{dt}=\frac{1}{\rho}\left(\tau_{ij}\frac{\partial u_j}{\partial x_i}-\frac{\partial \dot{q}_i}{\partial x_i}\right)-\alpha_i\frac{d\xi_i}{dt}$	
Lighthill equation	$\frac{\partial^2\rho}{\partial t^2}-a_0^2\Delta\rho=\frac{\partial^2\Theta_{ij}}{\partial x_j\partial x_i}$	*Lighthill tensor:* $\Theta_{ij}=\rho u_i u_j+\left(p-a_0^2\rho\right)\delta_{ij}-\tau_{ij}$
Fundamental equations. Intrinsic form		
Continuity equation	$\frac{\partial\rho}{\partial t}+\nabla.\left(\rho\,\vec{V}\right)=0$	
Equation of motion	$\frac{\partial\left(\rho\,\vec{V}\right)}{\partial t}+\nabla.\left[\left(\rho\,\vec{V}\right)\otimes\vec{V}\right]-\nabla.\overline{\overline{T}}=0$	
Viscous stress tensor	$\overline{\overline{T}}=-\left(p+\frac{2}{3}\mu\,\nabla.\vec{V}\right)\overline{\overline{I}}+\mu\left[\nabla\,\vec{V}+\left(\nabla\,\vec{V}\right)^T\right]$	
Energy equation	$\frac{\partial\left(\rho\,e_T\right)}{\partial t}+\nabla.\left(\rho\,e_T\,\vec{V}\right)+\left(\nabla.\dot{\vec{q}}-\nabla.\vec{V}.\overline{\overline{T}}\right)=0$	
Fourier law	$\dot{\vec{q}}=-\kappa\nabla T=-\frac{\mu\,C_p}{\Pr}\nabla\,T$	

Aerodynamic forces exerted on a vehicle		
Dynalpy	$\vec{D} = \vec{P} + \left(\rho \vec{V}.\vec{n}\right)\vec{V}$	
Dynalpy theorem	$\vec{R} = -\iint_{(S_1)} \vec{D}\, dS$	*Force exerted by the fluid on a body*
Thrust of a nacelle	$\vec{R} = -\left[\iint_{(S_2)} \vec{D}\, dS + \iint_{(S_1)} \vec{D}\, dS\right]$	*Thrust of the nozzle:* $F = \left(\iint_{(S_2)} \vec{D}\, dS\right).\vec{x}$ *Captation drag:* $T = -\left(\iint_{(S_1)} \vec{D}\, dS\right).\vec{x}$
Drag of a vehicle	$C_D = 2\iint_{(S_2)} \frac{\rho u}{\rho_\infty V_\infty}\left(1 - \frac{u}{V_\infty}\right)\frac{dS}{A}$	*In the plane of Trefftz*
Drag of a vehicle	$C_D = \frac{2}{A}\iint_{(S_2)}\left[\frac{1}{\gamma M_\infty^2}\left(1 - \frac{p}{p_\infty}\right) + \frac{\rho u}{\rho_\infty V_\infty}\left(1 - \frac{u}{V_\infty}\right)\right] dS$	*In a downstream plane close to the model*
Oswatitsch relation	$T = \frac{1}{V_\infty}\iint_{(S)}\left(T_\infty \delta s - \delta h_{st}\right) dq_m$	$h_{st} \equiv h_T$ *Stagnation enthalpy*

Remarkable numbers		
Knudsen number	$Kn = \frac{\lambda}{L}$	$Kn = 1.48\frac{M}{\mathrm{Re}}$
Mach number	$M = \frac{V}{a}$	*Compressibility effects*
Crocco number	$Cr = \frac{V}{V_{\lim}}$	$Cr = \sqrt{\frac{\frac{\gamma-1}{2}M^2}{\left(1+\frac{\gamma-1}{2}M^2\right)}}$
Reynolds number	$\mathrm{Re} = \frac{\rho V L}{\mu}$	*Viscous effects*
Prandtl number	$\Pr = \frac{\mu C_p}{\kappa}$	*Thermal conduction*

Remarkable numbers (continued)		
Péclet number	$Pe = \text{Re} \times \text{Pr} = \dfrac{\rho V L C_p}{\kappa}$	*Dissipation/thermal conduction*
Froude number	$Fr = \dfrac{V}{\sqrt{Lg}}$	*Heavy fluid*
Damköhler number	$Da = \dfrac{KL}{V}$	K: *rate of formation of the considered species (dimension:* $[T]^{-1}$)
Strouhal number	$Su = \dfrac{fL}{V}$	*Unsteady effects*
Lewis number	$Le = \dfrac{\kappa}{\rho D_j C_p}$	D_j *: diffusion velocity of species (j)*
Schmidt number	$Sc = \dfrac{\nu}{D_j}$	$Sc = Le\,\text{Pr}$
Stanton number	$St = \dfrac{\dot{q}_w}{\rho_e U_e (h_f - h_w)}$	*Wall heat transfer coefficient*

Fundamental concepts in fluid mechanics		
Equation of a trajectory	$\dfrac{dx}{u(x,y,z,t)} = \dfrac{dy}{v(x,y,z,t)} = \dfrac{dz}{w(x,y,z,t)} = dt$	
Equation of a streamline	$\dfrac{dx}{u(x,y,z,T)} = \dfrac{dy}{v(x,y,z,T)} = \dfrac{dz}{w(x,y,z,T)}$	
Expression of acceleration	$\dfrac{d\vec{V}}{dt} = \dfrac{\partial \vec{V}}{\partial t} + \overrightarrow{grad}\,\dfrac{V^2}{2} + \overrightarrow{curl}\,\vec{V} \times \vec{V}$	
Gradient in Cartesian coordinates	$\dfrac{\partial f}{\partial x}, \dfrac{\partial f}{\partial y}, \dfrac{\partial f}{\partial z}$	
Gradient in cylindrical coordinates	$\dfrac{\partial f}{\partial r}, \dfrac{1}{r}\dfrac{\partial f}{\partial \theta}, \dfrac{\partial f}{\partial z}$	
Gradient in spherical coordinates	$\dfrac{\partial f}{\partial r}, \dfrac{1}{r}\dfrac{\partial f}{\partial \varphi}, \dfrac{1}{r \sin\varphi}\dfrac{\partial f}{\partial \theta}$	

Fundamental concepts in fluid mechanics (continued)		
Rotational vector in Cartesian coordinates	$\left(\frac{\partial w}{\partial y}-\frac{\partial v}{\partial z}\right)\vec{i}+\left(\frac{\partial u}{\partial z}-\frac{\partial w}{\partial x}\right)\vec{j}+\left(\frac{\partial v}{\partial x}-\frac{\partial u}{\partial y}\right)\vec{k}$	*Planar two-dimensional flow: the rotational is normal to the plan of the flow*
Rotational vector in cylindrical coordinates	$\Omega_r=\frac{1}{r}\frac{\partial V_z}{\partial \theta}-\frac{\partial V_\theta}{\partial z}$ $\Omega_\theta=\frac{\partial V_r}{\partial z}-\frac{\partial V_z}{\partial r}$ $\Omega_z=\frac{1}{r}\frac{\partial}{\partial r}\left(rV_\theta\right)-\frac{1}{r}\frac{\partial V_r}{\partial \theta}$	*Axisymmetric flow around axis Oz: the rotational is along Oz*
Rotational vector in spherical coordinates	$\Omega_r=\frac{1}{r\sin\varphi}\frac{\partial\left(\sin\varphi V_\theta\right)}{\partial \varphi}-\frac{1}{r\sin\varphi}\frac{\partial V_\varphi}{\partial \theta}$ $\Omega_\theta=\frac{1}{r\sin\varphi}\frac{\partial V_r}{\partial \theta}-\frac{1}{r}\frac{\partial\left(rV_\theta\right)}{\partial r}$ $\Omega_z=\frac{1}{r}\frac{\partial}{\partial r}\left(rV_\theta\right)-\frac{1}{r}\frac{\partial V_r}{\partial \theta}$	
Crocco relation	$\overrightarrow{grad}\, h_{st}=T\,\overrightarrow{grad}\, s+\vec{V}\times\overrightarrow{curl}\,\vec{V}$	*If isenthalpic flow:* $T\,\overrightarrow{grad}\, s+\vec{V}\times\overrightarrow{curl}\,\vec{V}=0$
Circulation	$\Gamma=-\oint_{(C)}\vec{V}.\vec{t}\,dl$	
Stokes theorem	$\Gamma=-\oint_{(C)}\vec{V}.\vec{t}\,dl=-\iint_{(S)}\overrightarrow{curl}\,\vec{V}.\vec{n}\,dS$	
Kelvin theorem	$\frac{d\Gamma}{dt}=-\oint_{(C)}T\,ds$	*If isentropic flow:* $\frac{d\Gamma}{dt}=0$
Stream function in Cartesian coordinates	$\frac{\partial\psi}{\partial y}=\rho\,u,\quad \frac{\partial\psi}{\partial x}=-\rho\,v$	
Stream function in cylindrical coordinates	$V_r=\frac{1}{r}\frac{\partial\Psi}{\partial\theta}\quad V_\theta=-\frac{\partial\Psi}{\partial r}$	
Potential function in Cartesian coordinates	$u=\frac{\partial\Phi}{\partial x},\quad v=\frac{\partial\Phi}{\partial y},\quad w=\frac{\partial\Phi}{\partial z}$	
Potential function in cylindrical coordinates	$V_r=\frac{1}{r}\frac{\partial\Phi}{\partial\theta},\quad V_\theta=\frac{\partial\Phi}{\partial r}$	

Appendix 3

Table of Useful Relations for Stationary One-dimensional Flows and Discontinuity Surfaces

One-dimensional non-viscous, adiabatic, steady flows		
Continuity equation	$q_m = \rho VA = constant$	$\frac{d\rho}{\rho} + \frac{dV}{V} + \frac{dA}{A} = 0$
Equation of motion	$VdV + \frac{dp}{\rho} = 0$	*Incompressible (Bernoulli relation):* $p_{st} = p + \rho_0 \frac{V^2}{2} = Cst$
Energy equation	$d\left(e + \frac{V^2}{2} + \frac{p}{\rho}\right) = 0$	*Stagnation enthalpy:* $h_{st} = h + \frac{V^2}{2} = Cst$
Entropy equation	$Tds = 0 \quad hence \quad ds = 0$	
Hugoniot relation	$\frac{dV}{V}\left(1 - \frac{V^2}{a^2}\right) + \frac{dA}{A} = 0$	$\frac{dV}{V}\left(1 - M^2\right) + \frac{dA}{A} = 0$

One-dimensional non-viscous, adiabatic, steady flows (continued)		
Hugoniot theorem	$M<1\rightarrow\frac{dV}{dA}<0 \quad M>1\rightarrow\frac{dV}{dA}>0$	*throat:* $M=1\rightarrow dA=0$
Flow velocity	$V=\sqrt{2\left(h_{st}-h\right)}$	$V_{\lim}=\sqrt{2\,h_{st}}$
Crocco number	$Cr=\frac{V}{V_{\lim}}=\sqrt{\frac{\frac{\gamma-1}{2}M^2}{1+\frac{\gamma-1}{2}M^2}}$	
Energy equation or Saint-Venant equation	$V^2+\frac{2}{\gamma-1}a^2=\frac{2}{\gamma-1}a_{st}^2$	$V^2+\frac{2}{\gamma-1}a^2=\frac{\gamma+1}{\gamma-1}a_c^2$
Isentropic evolution: temperature	$\frac{T}{T_{st}}=\left(1+\frac{\gamma-1}{2}M^2\right)^{-1}$	
Isentropic evolution: pressure	$\frac{p}{p_{st}}\equiv\pi\left(M,\gamma\right)=\left(1+\frac{\gamma-1}{2}M^2\right)^{-\frac{\gamma}{\gamma-1}}$	
Isentropic evolution: density	$\frac{\rho}{\rho_{st}}=\left(1+\frac{\gamma-1}{2}M^2\right)^{-\frac{1}{\gamma-1}}$	
Area–Mach number relation	$\frac{A}{A_c}\equiv\Sigma\left(M,\gamma\right)=\left(\frac{2}{\gamma+1}\right)^{\frac{\gamma+1}{2(\gamma-1)}}\frac{1}{M}\left(1+\frac{\gamma-1}{2}M^2\right)^{\frac{\gamma+1}{2(\gamma-1)}}$	
Expression for the velocity	$\frac{V}{a_{st}}=\frac{M}{\left(1+\frac{\gamma-1}{2}M^2\right)^{\frac{1}{2}}}$	$\frac{V}{a_c}=\frac{M\left(\frac{\gamma+1}{2}\right)^{\frac{1}{2}}}{\left(1+\frac{\gamma-1}{2}M^2\right)^{\frac{1}{2}}}$

Application for the calculation of supersonic propulsive nozzles		
Mass flow of a sonic throat	$q_m = \frac{\delta(\gamma)}{\sqrt{C_p T_{st}}} p_{st} A_c$	$\delta(\gamma) = \frac{\gamma}{\sqrt{\gamma - 1}} \left(\frac{\gamma+1}{2}\right)^{-\frac{\gamma+1}{2(\gamma-1)}}$
Thrust of a nozzle in the vacuum	$F = \left(p_E + \rho_E V_E^2\right) A_E$	$F = p_E \left(1 + \gamma_E M_E^2\right) A_E$
Thrust function	$\Phi(M,\gamma) = \pi(M,\gamma)\Sigma(M,\gamma)\left(1+\gamma M^2\right)$	$\pi(M,\gamma) = p / p_{st}$ $\Sigma(M,\gamma) = A / A_c$
Expression for the vacuum thrust	$F = \Phi(M_E,\gamma_E)\, p_{st}\, A_c$	
Thrust of a conical nozzle	$F = \frac{1+\cos\Theta}{2}\left(p_s + \rho_s V_s^2\right) A_S$	$F = \frac{1+\cos\Theta}{2} \Phi(M_s,\gamma_s)\, p_{st}\, A_S$
Thrust at the pressure p_a	$\overline{F} = \left[\rho_E V_E^2 + (p_E - p_a)\right] A_E$	$\overline{F} = q_m V_E + (p_E - p_a) A_E$
Thrust at adaptation	$\overline{F} = q_m V_E$	
Bernoulli formula for the velocity	$V = \sqrt{\frac{2(p_{st} - p)}{\rho_0}}$	
Venturi formula	$V_1 = \sqrt{\frac{2(p_2 - p_1)}{\rho_0\left[1 - \left(\frac{A_1}{A_2}\right)^2\right]}}$	

One-dimensional flow with friction and heat transfer		
Nackab relation for the pressure loss coefficient	$\lambda = \left\{2Log_{10}\left[0.27\frac{k}{D_h} + \frac{2.51}{\mathrm{Re}\left(0.4\mathrm{Re}^{-0.3} + 0.0053\right)^{\frac{1}{2}}}\right]\right\}^{-2}$	$D_h = \frac{4A}{L_h}$
Generalized Hugoniot relation	$\frac{dV}{V}\left(1 - M^2\right) + \frac{dA}{A} = \frac{1}{\rho a^2}\left[\left(1 + \frac{K^2}{\rho T}\right)df + \frac{K^2}{T}d\dot{q}\right]$	
Generalized Hugoniot relation for a calorically perfect gas	$\frac{dV}{V}\left(1 - M^2\right) + \frac{dA}{A} = \frac{\gamma}{\rho a^2}df + \frac{\gamma - 1}{a^2}d\dot{q}$	

One-dimensional flow with friction and heat transfer (continued)		
Pressure loss due to friction	$\dfrac{p_{i_2}}{p_{i_1}} = \exp\left[-\dfrac{\gamma}{2}\int_{x_1}^{x_2} M^2 \lambda \dfrac{dx}{D}\right]$	$\lambda = 4C_f$
Function $g(M,\gamma)$	$g(M,\gamma) = -\dfrac{1}{2\gamma M^2} + \dfrac{\gamma+1}{4\gamma}\ln\left[\dfrac{\left(1+\dfrac{\gamma-1}{2}M^2\right)}{M^2}\right]$	
Mach number at the exit of a tube of length L	$\dfrac{L}{D} = \dfrac{x_2}{D} - \dfrac{x_1}{D} \cong \dfrac{2}{\lambda}\left[g(M_2) - g(M_1)\right]$	*Allows calculating* M_2 *knowing the entry Mach number* M_1
Length of choking by viscous effects	$\dfrac{L}{D} \cong \dfrac{2}{\lambda}\left[g(1) - g(M_1)\right]$	
General equations	$\dfrac{dV}{V} = \dfrac{\gamma M^2}{\left(1-M^2\right)}dX + \dfrac{1}{\left(1-M^2\right)}\dfrac{d\dot{q}}{C_pT}$	$dX = \dfrac{\lambda}{2}\dfrac{dx}{D_h}$
	$\dfrac{dp}{p} = -\gamma M^2 \dfrac{1+(\gamma-1)M^2}{\left(1-M^2\right)}dX - \dfrac{\gamma M^2}{\left(1-M^2\right)}\dfrac{d\dot{q}}{C_pT}$	
	$\dfrac{dT}{T} = -\gamma\dfrac{(\gamma-1)M^4}{\left(1-M^2\right)}dX + \dfrac{1-\gamma M^2}{\left(1-M^2\right)}\dfrac{d\dot{q}}{C_pT}$	
	$\dfrac{dM}{M} = \gamma M^2 \dfrac{1+\dfrac{(\gamma-1)}{2}M^2}{\left(1-M^2\right)}dX + \dfrac{1+\gamma M^2}{2\left(1-M^2\right)}\dfrac{d\dot{q}}{C_pT}$	

Application to supersonic ejectors		
Relation for the pressure ratio	$\dfrac{p_{st_s}}{p_{st_p}} = \dfrac{\mu\sqrt{\tau}}{\lambda-1}\dfrac{\Sigma\left(M_{0_s},\gamma_s\right)}{\Sigma\left(M_{0_p},\gamma_p\right)}$	
Relation for the pressure ratio	$\dfrac{p_{st_m}}{p_{st_p}} = \dfrac{1+\mu}{\lambda}\sqrt{\tau_m}\dfrac{\Sigma\left(M_m,\gamma_m\right)}{\Sigma\left(M_{0_p},\gamma_p\right)}$	p_{st_m}: *stagnation pressure of the mixed flux*
Geometrical constraint	$\dfrac{\Sigma\left(M_{1_p},\gamma_p\right)}{\Sigma\left(M_{0_p},\gamma_p\right)} = \lambda - (\lambda-1)\dfrac{\Sigma\left(M_{1_s},\gamma_s\right)}{\Sigma\left(M_{0_s},\gamma_s\right)}$	$\lambda = \dfrac{A_m}{A_{0_p}},\quad \dfrac{A_{0_s}}{A_{0_p}} = \lambda - 1$

Application to supersonic ejectors (continued)		
Geometrical constraint	$\frac{\Sigma\left(M_{1_p},\gamma_p\right)}{\Sigma\left(M_{0_p},\gamma_p\right)}=\lambda-\frac{(\lambda-1)}{\Sigma\left(M_{0_s},\gamma_s\right)}$	*Secondary flow choking:* $M_{1_s}=1$
Dynalpy conservation	$\mu\sqrt{\tau}=\frac{\Phi\left(M_{1_p},\gamma_p\right)-\Phi\left(M_{0_p},\gamma_p\right)}{\Phi\left(M_{0_s},\gamma_s\right)-\Phi\left(M_{1_s},\gamma_s\right)}$	
Dynalpy conservation	$\Phi\left(M_m,\gamma_m\right)=\frac{\Phi\left(M_{0_p},\gamma_p\right)+\mu\sqrt{\tau}\Phi\left(M_{0_s},\gamma_s\right)}{(1+\mu)\sqrt{\tau_m}}$	$\tau=\frac{\delta^2\left(\gamma_p\right)}{\delta^2\left(\gamma_s\right)}\frac{h_{st_s}}{h_{st_p}}=\frac{\delta^2\left(\gamma_p\right)}{\delta^2\left(\gamma_s\right)}\frac{C_{p_s}T_{st_s}}{C_{p_p}T_{st_p}}$
Energy conservation	$(1+\mu)\frac{\delta^2\left(\gamma_m\right)}{\delta^2\left(\gamma_p\right)}\tau_m=1+\mu\frac{\delta^2\left(\gamma_s\right)}{\delta^2\left(\gamma_p\right)}\tau$	$\delta(\gamma)=\frac{\gamma}{\sqrt{\gamma-1}}\left(\frac{\gamma+1}{2}\right)^{-\frac{(\gamma+1)}{2(\gamma-1)}}$

Discontinuity surfaces: shock wave and slip line		
Rankine–Hugoniot equations	$\rho\vec{V}.\vec{n}=constant$ $p\vec{n}+\left(\rho\vec{V}.\vec{n}\right)\vec{V}=constant$ $\rho\vec{V}.\vec{n}\left(h+\frac{V^2}{2}\right)=constant$	$\left[\rho\vec{V}.\vec{n}\right]=0$ $[p]\vec{n}+\rho\vec{V}.\vec{n}\left[\vec{V}\right]=0$ $\rho\vec{V}.\vec{n}\left[h+\frac{V^2}{2}\right]=0$
Oblique shock equations	$\rho_1V_{n_1}=\rho_2V_{n_2}$ $V_{t_1}=V_{t_2}$ $p_1+\rho_1V_{n_1}^2=p_2+\rho_2V_{n_2}^2$ $h_1+\frac{V_1^2}{2}=h_2+\frac{V_2^2}{2}\rightarrow h_{st_1}=h_{st_2}$	*or:* $h_1+\frac{V_{n_1}^2}{2}=h_2+\frac{V_{n_2}^2}{2}$
Relations for the slip line	$V_{n_1}=V_{n_2}=0$ $V_{t_1}\neq V_{t_2}$ $p_1=p_2$ $\rho_1\neq\rho_2$, $T_1\neq T_2$, $h_{st_1}\neq h_{st_2}$	
Normal shock equations	$\rho_1V_1=\rho_2V_2$ $p_1+\rho_1V_1^2=p_2+\rho_2V_2^2$ $h_1+\frac{V_1^2}{2}=h_2+\frac{V_2^2}{2}$	

Discontinuity surfaces: shock wave and slip line (continued)		
Prandtl relation	$V_1 V_2 = a_c^2$	$M_1^* M_2^* = 1$
Pressure ratio	$\frac{p_2}{p_1} = 1 + \frac{2\gamma}{\gamma+1}\left(M_1^2 - 1\right)$	
Density ratio	$\frac{\rho_2}{\rho_1} = \frac{(\gamma+1) M_1^2}{(\gamma-1) M_1^2 + 2}$ $\frac{\rho_1}{\rho_2} = \frac{\gamma-1}{\gamma+1}\left[1 + \frac{2}{(\gamma-1) M_1^2}\right]$	*if* $M_1 \to \infty$ $\frac{\rho_2}{\rho_1} \to \frac{\gamma+1}{\gamma-1}$ ($\frac{\rho_2}{\rho_1} = 6$ *if* $\gamma = 1.4$)
Temperature ratio	$\frac{T_2}{T_1} = \left[1 + \frac{2\gamma}{\gamma+1}\left(M_1^2 - 1\right)\right] \times \left[\frac{2 + (\gamma-1) M_1^2}{(\gamma+1) M_1^2}\right]$	
Mach number downstream of the shock	$M_2^2 = \frac{1 + \frac{\gamma-1}{2} M_1^2}{\gamma M_1^2 - \frac{\gamma-1}{2}}$	*if* $M_1 \to \infty$ $M_2 \to \sqrt{\frac{\gamma-1}{2\gamma}}$ ($M_2 = 0.378$ *if* $\gamma = 1.4$)
Stagnation pressure ratio	$\frac{p_{st_2}}{p_{st_1}} = \left[1 + \frac{2\gamma}{\gamma+1}\left(M_1^2 - 1\right)\right]^{-\frac{1}{\gamma-1}} \times \left[1 - \frac{2}{\gamma+1}\left(1 - \frac{1}{M_1^2}\right)\right]^{-\frac{\gamma}{\gamma-1}}$	
Entropy rise	$\frac{s_2 - s_1}{C_v} = \ln\left\{\left[1 + \frac{2\gamma}{\gamma+1}\left(M_1^2 - 1\right)\right] \times \left[1 - \frac{2}{\gamma+1}\left(1 - \frac{1}{M_1^2}\right)\right]^{\gamma}\right\}$	
Rayleigh formula	$\frac{p_{st_2}}{p_1} = \left[1 + \frac{2\gamma}{\gamma+1}\left(M_1^2 - 1\right)\right]^{-\frac{1}{\gamma-1}} \times \left[1 - \frac{2}{\gamma+1}\left(1 - \frac{1}{M_1^2}\right)\right]^{-\frac{\gamma}{\gamma-1}} \times \left(1 + \frac{\gamma-1}{2} M_1^2\right)^{\frac{\gamma}{\gamma-1}}$	
Oblique shock equation	$\frac{\tan(\sigma - \varphi)}{\tan\sigma} = \left(\frac{\gamma-1}{\gamma+1}\right)\left[1 + \frac{2}{(\gamma-1) M_1^2 \sin^2\sigma}\right]$	
Normal Mach numbers	$M_{n_1} = M_1 \sin\sigma$	$M_{n_2} = M_2 \sin(\sigma - \varphi)$

Shock angle for the maximum deflection	$\sin^2\sigma_{\max}=\frac{1}{4\gamma M_1^2}\left\{(\gamma+1)M_1^2-4+\sqrt{(\gamma+1)\left[(\gamma+1)M_1^4+8(\gamma-1)M_1^2+16\right]}\right\}$	
Shock angle for a sonic downstream state	$\sin^2\sigma_{son}=\frac{1}{4\gamma M_1^2}\left\{(\gamma+1)M_1^2-(3-\gamma)+\sqrt{(\gamma+1)\left[(\gamma+1)M_1^4-2(3-\gamma)M_1^2+(\gamma+9)\right]}\right\}$	
Conical flow	$\frac{\partial z}{\partial\theta}=\gamma M^2\frac{\sin\varphi\sin(\theta-\varphi)}{\left[M^2\sin(\theta-\varphi)^2-1\right]}\frac{1}{\sin\theta}$ $\frac{\partial\varphi}{\partial\theta}=\frac{\sin\varphi\cos(\theta-\varphi)}{\left[M^2\sin^2(\theta-\varphi)-1\right]}\frac{1}{\sin\theta}$	$z=\ln p$

Shock structure		
Continuity equation	$\frac{\partial\rho}{\partial t}+\frac{\partial}{\partial x}(\rho u)=0$	
Equation of motion	$\frac{\partial}{\partial t}(\rho u)+\frac{\partial}{\partial x}\left(\rho u^2+p-\frac{4}{3}\mu\frac{\partial u}{\partial x}\right)=0$	
Energy equation	$\frac{\partial}{\partial t}(\rho e_T)+\frac{\partial}{\partial x}\left(\rho u e_T+pu-\frac{4}{3}\mu u\frac{\partial u}{\partial x}-\kappa\frac{\partial T}{\partial x}\right)=0$	
Entropy variation	$\left[\rho u(s_2-s_1)\right]=\int_{-\infty}^{+\infty}\frac{1}{T^2}\left[\frac{4}{3}\mu T\left(\frac{du}{dX}\right)^2+\kappa\left(\frac{dT}{dX}\right)^2\right]dX$	*Stationary shock*

Supersonic air intakes		
Mass flow coefficient	$\varepsilon=\frac{A_0}{A_1}$	
Efficiency	$\eta=\frac{p_{st_2}}{p_{st_0}}$	
Mass flow formula	$q_m=\frac{\delta(\gamma)}{\sqrt{C_p T_{st}}}p_{st}A_c$	$\delta(\gamma)=\frac{\gamma}{\sqrt{\gamma-1}}\left(\frac{\gamma+1}{2}\right)^{-\frac{\gamma+1}{2(\gamma-1)}}$
Invariant	$p_{st_0}A_{c_0}=p_{st_2}A_{c_2}$	$T_{st}=constant$
Fundamental relation	$\varepsilon=\frac{\Sigma(M_0)}{\Sigma(M_2)}\eta\frac{A_2}{A_1}$	

Pitot-type air intake		
Sub-critical regime	$\varepsilon < 1$	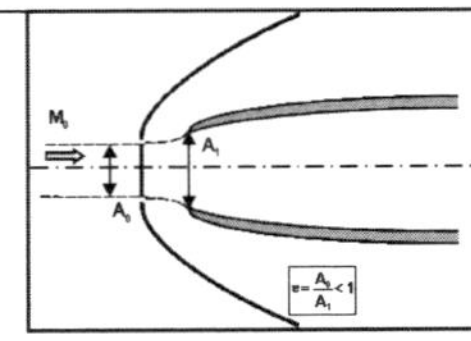
Critical regime	$\varepsilon = 1$	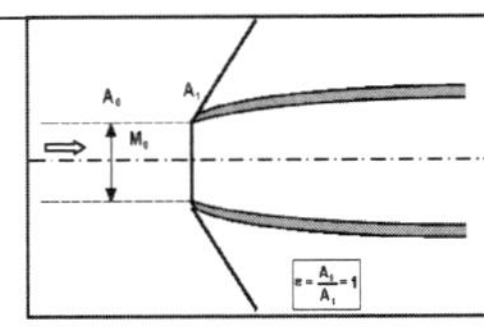
Super-critical regime	$\varepsilon = 1$	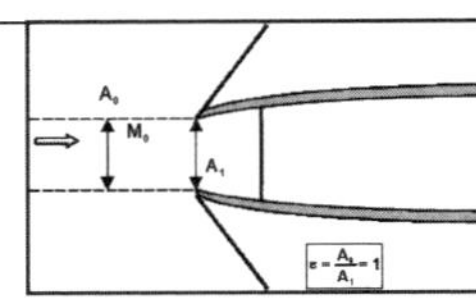
Ramp-type air intake		
Sub-critical regime	$\varepsilon = \dfrac{A_0}{A_1} < 1$	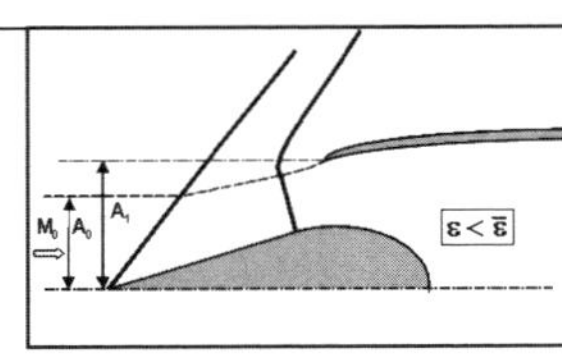
Critical regime	$\varepsilon = \overline{\varepsilon} < 1$	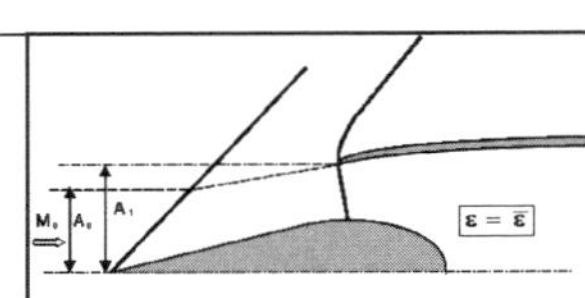
Super-critical regime	$\varepsilon = \overline{\varepsilon} < 1$	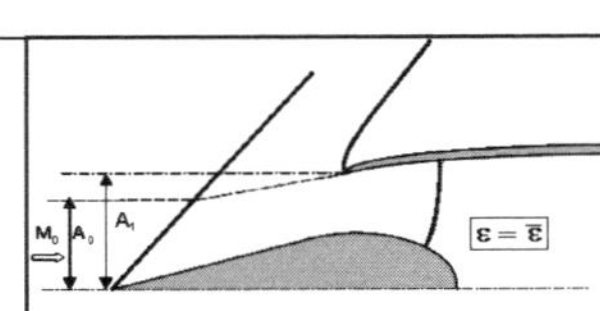
Adapted regime	$\varepsilon = \overline{\varepsilon} = 1$	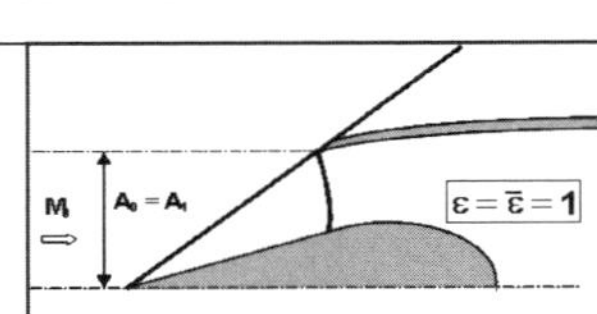

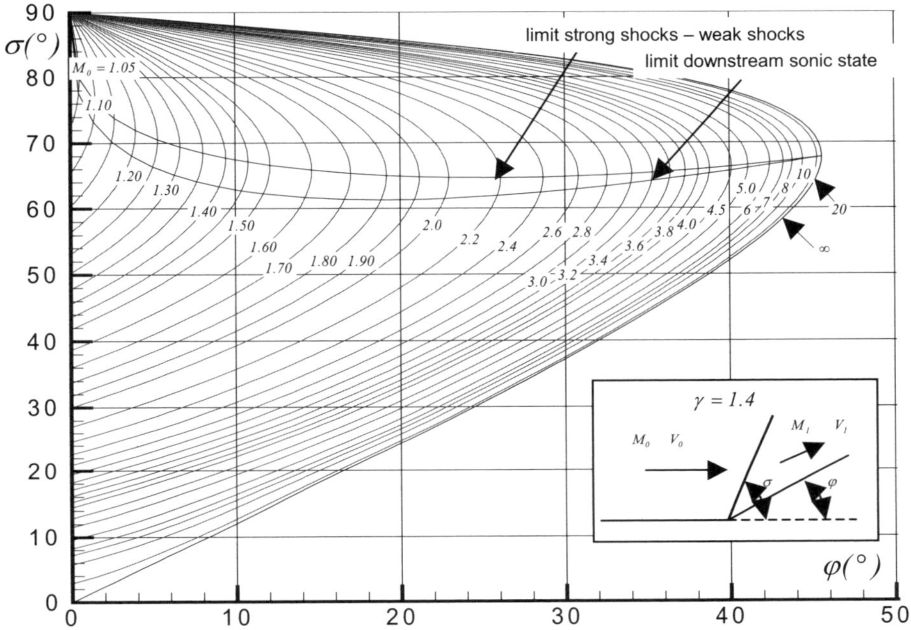

Oblique shock polars in the plane deflection – shock angle

Appendix 4

Table of Useful Relations for Applications of the Theory of Characteristics and Transonic Flows

Two-dimensional Euler equations in intrinsic coordinates		
Continuity equation	$\frac{\partial \rho V}{\partial x_1} + \rho V \frac{\partial \varphi}{\partial y_1} = -\varepsilon \frac{\rho V \sin\varphi}{y}$	*two-dimensional planar:* $\varepsilon = 0$ *axisymmetric:* $\varepsilon = 1$
Equation of motion along the velocity	$\rho V \frac{\partial V}{\partial x_1} + \frac{\partial p}{\partial x_1} = 0$	
Equation of motion along the normal to the velocity	$\rho V^2 \frac{\partial \varphi}{\partial x_1} + \frac{\partial p}{\partial y_1} = 0$	
Intrinsic equation 1	$\left(\frac{V^2}{a^2} - 1\right)\frac{\partial p}{\partial x_1} + \rho V^2 \frac{\partial \varphi}{\partial y_1} = -\varepsilon \rho V^2 \frac{\sin\varphi}{y}$	
Intrinsic equation 2	$\rho V^2 \frac{\partial \varphi}{\partial x_1} + \frac{\partial p}{\partial y_1} = 0$	
Entropy equation	$\frac{\partial s}{\partial x_1} = 0$	

Characteristic directions and equations		
Mach angle	$\sin\alpha = \dfrac{1}{M}$ *or* $\alpha = \sin^{-1}\dfrac{1}{M}$	
Characteristic directions	$\psi = \varphi + \alpha$ *for the left running* (η) $\psi = \varphi - \alpha$ *for the right running* (ξ)	
Characteristic equations	$\dfrac{\sqrt{M^2-1}}{\gamma p M^2}\dfrac{\partial p}{\partial \eta} + \dfrac{\partial \varphi}{\partial \eta} = -\varepsilon\dfrac{\sin\varphi\sin\alpha}{y}$ *on* (η) $\dfrac{\sqrt{M^2-1}}{\gamma p M^2}\dfrac{\partial p}{\partial \xi} - \dfrac{\partial \varphi}{\partial \xi} = -\varepsilon\dfrac{\sin\varphi\sin\alpha}{y}$ *on* (ξ)	*two-dimensional planar:* $\varepsilon = 0$ *axisymmetric:* $\varepsilon = 1$
Characteristic equations: other form	$\dfrac{\sin\alpha\cos\alpha}{\gamma p}\dfrac{\partial p}{\partial \eta} + \dfrac{\partial \varphi}{\partial \eta} = -\varepsilon\dfrac{\sin\varphi\sin\alpha}{y}$ *on* (η) $\dfrac{\sin\alpha\cos\alpha}{\gamma p}\dfrac{\partial p}{\partial \xi} - \dfrac{\partial \varphi}{\partial \xi} = -\varepsilon\dfrac{\sin\varphi\sin\alpha}{y}$ *on* (ξ)	*two-dimensional planar:* $\varepsilon = 0$ *axisymmetric:* $\varepsilon = 1$
Characteristic equations: two-dimensional planar flow, $\gamma = constant$	$\dfrac{\sqrt{M^2-1}}{\gamma M^2}\dfrac{dp}{p} + d\varphi = 0$ *on* (η) $\dfrac{\sqrt{M^2-1}}{\gamma M^2}\dfrac{dp}{p} - d\varphi = 0$ *on* (ξ)	
Busemann pressure number	$\omega(M,\gamma) = \sqrt{\dfrac{\gamma+1}{\gamma-1}}\tan^{-1}\sqrt{\dfrac{\gamma-1}{\gamma+1}\left(M^2-1\right)} - \tan^{-1}\sqrt{M^2-1}$	$\gamma = constant$
Integrated equations	$\omega(M,\gamma) - \varphi = constant$ *on* (η) $\omega(M,\gamma) + \varphi = constant$ *on* (ξ)	*two-dimensional planar flow* $\gamma = constant$

<table>
<tr><th colspan="3">Numerical method of characteristics</th></tr>
<tr><td>Linearized equations</td><td>$$Z_{13}\left(z_3^{(n)}-z_1\right)-\left(\varphi_3^{(n)}-\varphi_1\right)=N_{13}\delta\xi^{(n-1)}$$ $$Z_{23}\left(z_3^{(n)}-z_2\right)+\left(\varphi_w^{(n)}-\varphi_2\right)=N_{23}\delta\eta^{(n-1)}$$</td><td>$$Z=\frac{\sin\alpha\cos\alpha}{\gamma}$$ $$N=-\varepsilon\frac{\sin\alpha\sin\varphi}{y}$$</td></tr>
<tr><td>Interpolation formula for entropy</td><td>$$s_3=\frac{s_1\sin\left(\alpha_{23}\right)\Delta\eta+s_2\sin\left(\alpha_{13}\right)\Delta\xi}{\sin\left(\alpha_{23}\right)\Delta\eta+\sin\left(\alpha_{13}\right)\Delta\xi}$$</td><td>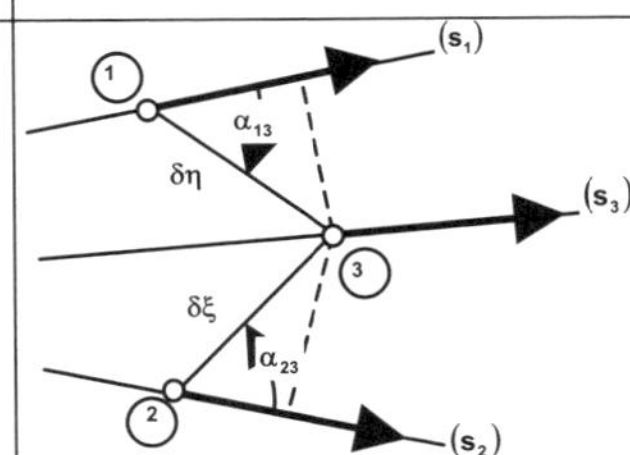
</td></tr>
<tr><td>Limit form on an axisymmetry axis</td><td>$$Z\,\delta z-2\delta\varphi=0$$</td><td></td></tr>
<tr><td>Linearized equation for a point on an axisymmetry axis</td><td>$$z_3=z_1-\frac{1}{2}\left[\varphi_1\left(\frac{1}{Z_1}+\frac{2}{Z_3}\right)-\frac{N_1}{Z_1}\delta\xi\right]$$</td><td></td></tr>
<tr><td>Mass flow through a characteristic</td><td>$$\overline{q}_m=2\pi\int_0^{(\xi)_{p0}}\rho\;a\;y\;d\xi$$</td><td>Axisymmetric flow</td></tr>
<tr><td>Normal point in a flow field</td><td>$$Z_{13}\left(z_3^{(n)}-z_1\right)-\left(\varphi_3^{(n)}-\varphi_1\right)=N_{13}\delta\xi^{(n-1)}$$ $$Z_{23}\left(z_3^{(n)}-z_2\right)+\left(\varphi_w^{(n)}-\varphi_2\right)=N_{23}\delta\eta^{(n-1)}$$</td><td>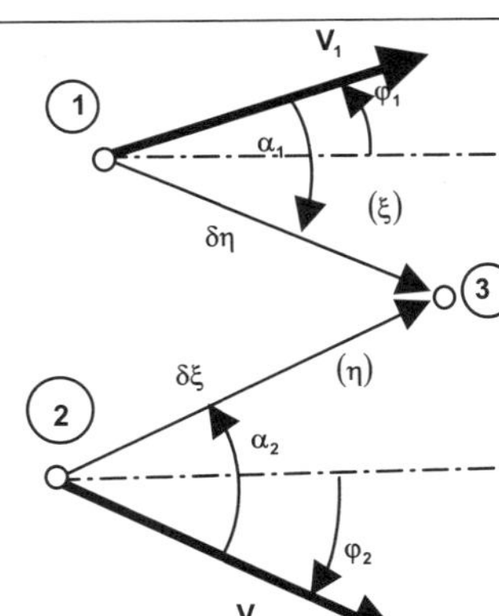
</td></tr>
</table>

Numerical method of characteristics

Point on a wall	$Z_{23}\left(z_3^{(n)} - z_2\right) + \left(\varphi_p^{(n)} - \varphi_2\right) = N_{23}\delta\eta^{(n-1)}$	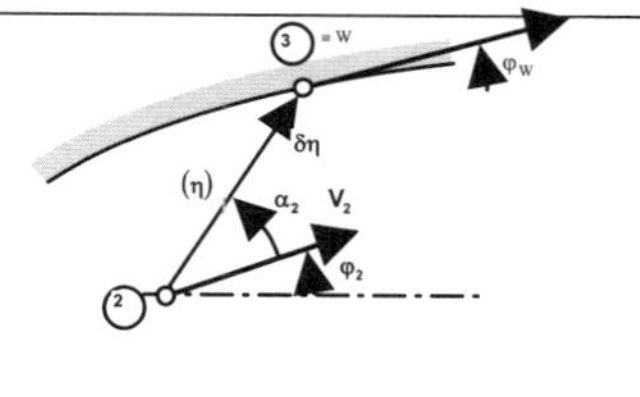
Point on a jet boundary	$Z_{23}\left(z_3^{(n)} - z_J\right) + \left(\varphi_3^{(n)} - \varphi_2\right) = N_{23}\delta\eta^{(n-1)}$	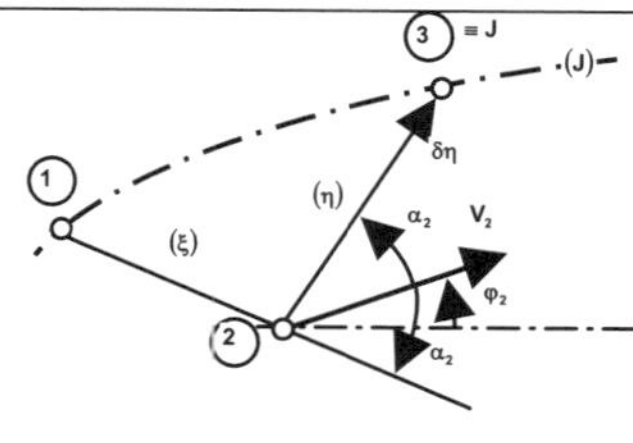
Point on an axisymmetry axis	$z_3 = z_1 - \dfrac{1}{2}\left[\varphi_1\left(\dfrac{1}{Z_1} + \dfrac{2}{Z_3}\right) - \dfrac{N_1}{Z_1}\delta\xi\right]$	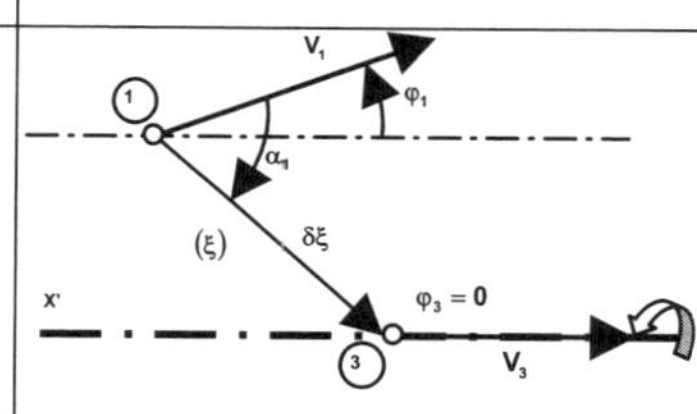

Flow with shock waves: the rotational method of characteristics

Origin of a shock induced by: *– a wall discontinuity,* *– a pressure discontinuity.*	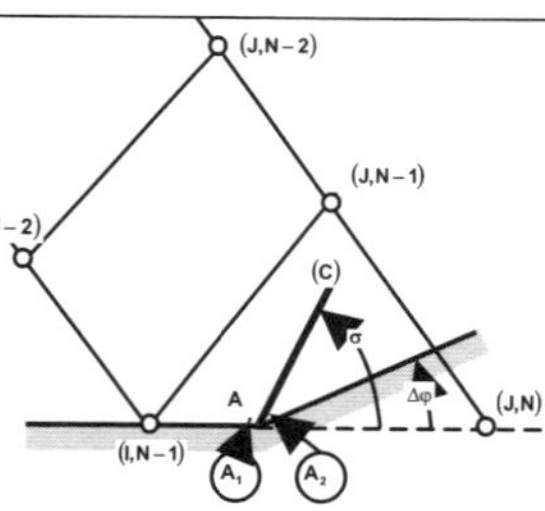	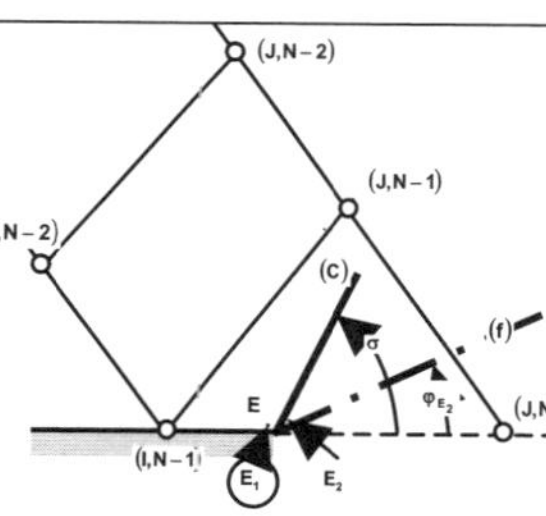
First point of a shock induced by: *– a wall discontinuity,* *– a pressure discontinuity.*	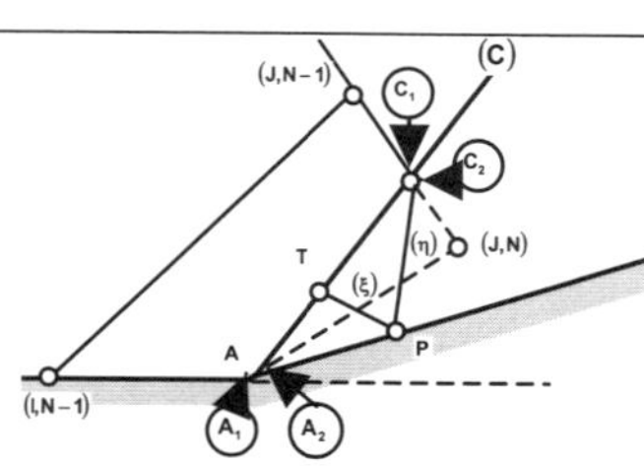	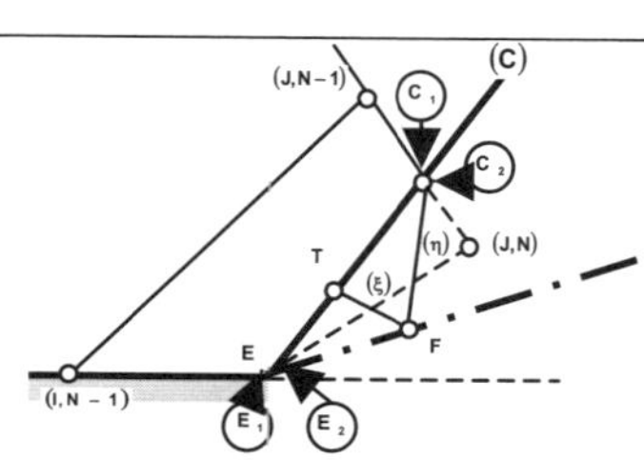

Flow with shock waves: the rotational method of characteristics (continued)		
Focalization shock. *– detection of shock origin,* *– first point on the shock.*	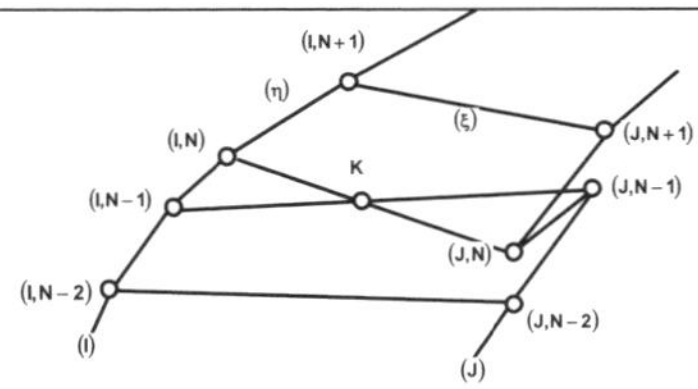	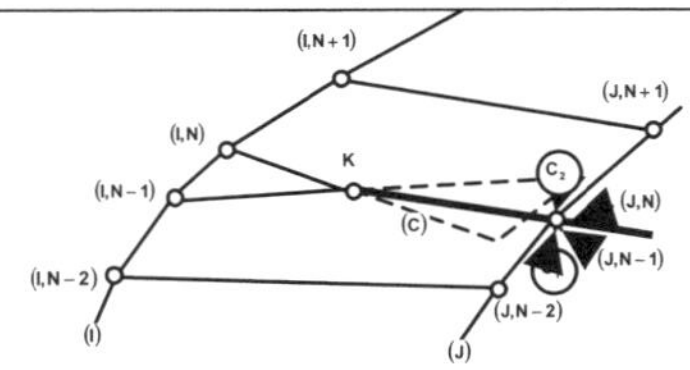
Normal point on a shock wave		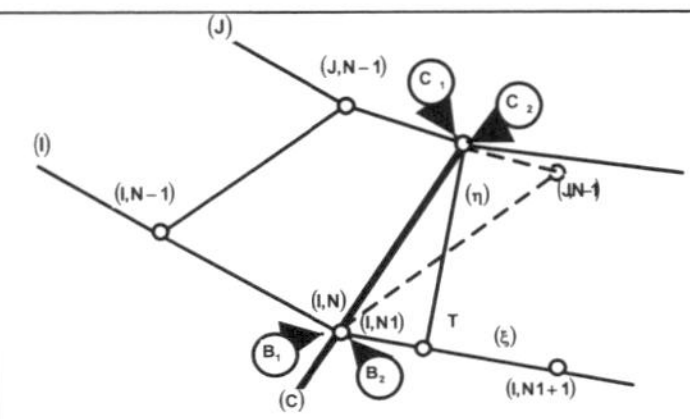

Transonic domain of a nozzle	
Compressible potential equation	$\left[a^2-\left(\frac{\partial\Phi}{\partial x}\right)^2\right]\frac{\partial^2\Phi}{\partial x^2}+\left[a^2-\left(\frac{\partial\Phi}{\partial y}\right)^2\right]\frac{\partial^2\Phi}{\partial y^2}-2\left(\frac{\partial\Phi}{\partial x}\right)\left(\frac{\partial\Phi}{\partial y}\right)\left(\frac{\partial^2\Phi}{\partial x\partial y}\right)=-\varepsilon\frac{a^2}{y}\left(\frac{\partial\Phi}{\partial y}\right)$ $\varepsilon=0$ *: two-dimensional planar,* $\varepsilon=0$ *: axisymmetric*
<u>*Carrrière formulas*</u> *Axisymmetric flow*	$\bar{u}=1+\bar{x}+\frac{\gamma+1}{4}\bar{y}^2\left[1+(2\gamma-1)\bar{x}+3\gamma(\gamma-1)\bar{x}^2+(\gamma-1)(4\gamma^2-2\gamma-1)\bar{x}^3\right]$ $+\frac{(\gamma+1)^2}{64}\bar{y}^4\left[6\gamma-1+(36\gamma^2-19\gamma-4)\bar{x}\right]$ $\bar{v}=\frac{\gamma+1}{2}\bar{y}\left[\bar{x}+\frac{2\gamma-1}{2}\bar{x}^2+\gamma(\gamma-1)\bar{x}^3+\frac{(\gamma-1)(4\gamma^2-2\gamma-1)}{4}\bar{x}^4\right]$ $+\frac{(\gamma+1)^2}{16}\bar{y}^3\left[1+(6\gamma-1)\bar{x}+\frac{36\gamma^2-19\gamma-4}{2}\bar{x}^2\right]$ $+\frac{(\gamma+1)^3(6\gamma-1)}{128}\bar{y}^5$
<u>*Carrière formulas*</u> *two-dimensional planar flow*	$\bar{u}=1+\bar{x}+\bar{y}^2\left[1.2+2.16\bar{x}+2.016\bar{x}^2+1.9392\bar{x}^3\right]$ $+\bar{y}^4\left[2.256+13.622\bar{x}\right]$ $\bar{v}=\bar{y}\left[2.4\bar{x}+2.16\bar{x}^2+1.344\bar{x}^3+0.9696\bar{x}^4\right]$ $+\bar{y}^3\left[0.96+9.24\bar{x}+27.245\bar{x}^2\right]+3.871\bar{y}^5$

Transonic domain of a nozzle (continued)	
Sauer formulas *two-dimensional planar flow:* $\varepsilon = 0$ *Axisymmetric flow:* $\varepsilon = 1$	$\overline{u} = 1 + \overline{x} + \dfrac{(\gamma+1)\overline{y}^2}{2(1+\varepsilon)}$ $\overline{v} = \dfrac{(\gamma+1)\overline{x}\,\overline{y}}{(1+\varepsilon)} + \dfrac{(\gamma+1)^2\overline{y}^3}{2(1+\varepsilon)(3+\varepsilon)}$

One-dimensional non-viscous unsteady flows		
Continuity equation	$\dfrac{\partial}{\partial t}(\rho A) + \dfrac{\partial}{\partial x}(\rho u A) = 0$	
Equation of motion	$\rho \dfrac{du}{dt} = -\dfrac{\partial p}{\partial x}$	
Energy equation	$\dfrac{ds}{dt} = 0$	

Isentropic flow in a cylindrical tube		
Continuity equation	$\dfrac{1}{\rho}\dfrac{\partial \rho}{\partial t} + \dfrac{u}{\rho}\dfrac{\partial \rho}{\partial x} + \dfrac{\partial u}{\partial x} = 0$	
Equation of motion	$\dfrac{\partial u}{\partial t} + u\dfrac{\partial u}{\partial x} + \dfrac{1}{\rho}\dfrac{\partial p}{\partial x} = 0$	
Progressive wave (η) *velocity*	$\dfrac{dx}{dt} = u + a$	
Regressive wave (ξ) *velocity*	$\dfrac{dx}{dt} = u - a$	
Compatibility relations	$\dfrac{1}{\rho a}\dfrac{\delta^+ p}{\delta t} + \dfrac{\delta^+ u}{\delta t} = 0 \text{ on wave } (\eta)$ $\dfrac{1}{\rho a}\dfrac{\delta^- p}{\delta t} - \dfrac{\delta u^-}{\delta t} = 0 \text{ on wave } (\xi)$	
Integrated forms and Riemann invariants	$P + u = 2\mu = \text{constant on } (\eta)$ $P - u = 2\lambda = \text{constant on } (\xi)$	$P(p) = \int^p \dfrac{dp}{\rho a}$

Isentropic flow in a cylindrical tube		
Characteristic equation on (η)	$\frac{2}{\gamma-1}a+u=2\mu=constant\ on\ (\eta)$	*Calorically perfect gas*
Characteristic equation on (ξ)	$\frac{2}{\gamma-1}a-u=2\lambda=constant\ on\ (\xi)$	*Calorically perfect gas*
Unsteady limit velocity	$U_{\lim}=\frac{2}{\gamma-1}a_1$	
Acoustic impedance	$\frac{\delta u_{42}}{\delta u_{31}}=\frac{2Z_1}{Z_2+Z_1}$ *and* $\frac{\delta p_{42}}{\delta p_{31}}=\frac{2Z_2}{Z_2+Z_1}$	$Z=\rho a$. *If* $Z_1=Z_2$: *acoustic adaptation*

Unsteady shock wave		
Unsteady shock equations	*continuity*: $\rho_1 D=\rho_2(D-w)$ *motion*: $p_2-p_1=\rho_1 Dw$ *energy*: $h_1+\frac{D^2}{2}=h_2+\frac{1}{2}(D-w)^2$	*D*: *shock wave velocity*
Shock Mach number	$M_s=\frac{D}{a_1}$	
Pressure ratio	$\frac{p_2}{p_1}=1+\frac{2\gamma}{\gamma+1}(M_s^2-1)=\frac{2}{\gamma+1}\left(\gamma M_s^2-\frac{\gamma-1}{2}\right)$	
Density ratio	$\frac{\rho_2}{\rho_1}=\frac{(\gamma+1)M_s^2}{2+(\gamma-1)M_s^2}$	
Temperature ratio	$\frac{T_2}{T_1}=\left[1+\frac{2\gamma}{\gamma+1}(M_s^2-1)\right]\left[\frac{2+(\gamma-1)M_s^2}{(\gamma+1)M_s^2}\right]$	
Velocity difference	$\frac{w}{a_1}=\frac{2}{(\gamma+1)}\left(M_s-\frac{1}{M_s}\right)$	
Equation for the shock Mach number	$M_s^2-\frac{\gamma+1}{2}\left(\frac{V_P}{a_1}\right)M_s-1=0$	V_P : *piston velocity*
Downstream Mach number	$M_s=\sqrt{\frac{1}{2\gamma}\left[(\gamma+1)\Pi+(\gamma-1)\right]}$	$\Pi=p_2/p_1$

Entropy rise	$\frac{s_2 - s_1}{C_v} = \ln\left\{\Pi\left[\frac{(\gamma-1)\Pi+(\gamma+1)}{(\gamma+1)\Pi+(\gamma-1)}\right]^{\gamma}\right\}$	
Pressure jump in a shock reflection	$\Pi_2 = \frac{(3\gamma-1)\Pi_1-(\gamma-1)}{(\gamma-1)\Pi_1+(\gamma+1)}$	$\Pi_1 = p_2 / p_1$ $\Pi_2 = p_3 / p_1$
Shock tube relation	$\frac{p_4}{p_1} = \frac{\frac{\gamma_1-1}{\gamma_1+1}\left(\frac{2\gamma_1}{\gamma_1-1}M_{s_1}^2-1\right)}{\left[1-\frac{\gamma_4-1}{\gamma_1+1}\frac{a_1}{a_4}\left(M_{s_1}-\frac{1}{M_{s_1}}\right)\right]^{\frac{2\gamma_4}{\gamma_4-1}}}$	
Limit Mach number	$M_{s(\lim)} = \frac{1+\sqrt{1+4\alpha^2}}{2\alpha}$	$\alpha = \frac{\gamma_4-1}{\gamma_1+1}\frac{a_1}{a_4}$

Flow in a duct with variable section		
Continuity equation	$\frac{\partial(\rho A)}{\partial t}+\frac{\partial(\rho u A)}{\partial x}=0$	
Equation of motion	$\frac{\partial(\rho u A)}{\partial t}+\frac{\partial}{\partial x}\left[A\left(p+\rho u^2\right)\right]=p\frac{\partial A}{\partial x}$	
Energy equation	$\frac{\partial(\rho e_T A)}{\partial t}+\frac{\partial}{\partial x}\left[\rho A u\left(e_T+\frac{p}{\rho}\right)\right]=-p\frac{\partial A}{\partial t}$	$e_T = e+\frac{u^2}{2}$
Stability condition	$\frac{\Delta t}{\Delta x}<\frac{1}{\left(\lvert u+a\rvert+\lvert u-a\rvert\right)}$	$(\Delta t)_{\lim}=\frac{\Delta x}{\left(\lvert u+a\rvert+\lvert u-a\rvert\right)}$
Courant–Friedrich–Lewy stability criterion (CFL)	$CFL=\frac{\Delta t}{(\Delta t)_{\lim}}$	

Spherical two-dimensional flow		
Continuity equation	$$\frac{\partial \rho}{\partial t}+\frac{\partial(\rho V)}{\partial r}+\frac{2\rho V}{r}=0$$	
Equation of motion	$$\frac{\partial(\rho V)}{\partial t}+\frac{\partial}{\partial r}\left(p+\rho V^2\right)+\frac{2\rho V^2}{r}=0$$	
Energy equation	$$\frac{\partial(\rho e_T)}{\partial t}+\frac{\partial}{\partial r}\left[\rho V\left(e_T+\frac{p}{\rho}\right)\right]+\frac{2(\rho e_T V+pV)}{r}=0$$	

Some properties of transonic flows		
Equation for the perturbation potential	$$\left(1-M_\infty^2\right)\frac{\partial^2\Phi'}{\partial x^2}+\frac{\partial^2\Phi'}{\partial y^2}=0$$	
The Prandtl–Glauert rule: pressure coefficient	$$C_p=\frac{C_{p(inc)}}{\sqrt{1-M_\infty^2}}$$	
The Prandtl–Glauert rule: lift and moment coefficients	$$C_L=\frac{C_{z(inc)}}{\sqrt{1-M_\infty^2}},\quad C_m=\frac{C_{m(inc)}}{\sqrt{1-M_\infty^2}}$$	
Supersonic small perturbations: pressure coefficient	$$C_p=\frac{2\,\theta}{\sqrt{M_\infty^2-1}}$$	
Supersonic small perturbations: lift and drag coefficients	$$C_L=\frac{4\varepsilon}{\sqrt{M_\infty^2-1}},\quad C_D=\frac{4\varepsilon^2}{\sqrt{M_\infty^2-1}}$$	ε *: flat plate incidence*
Transonic small perturbations: potential equation	$$\left[1-M_\infty^2-(\gamma+1)M_\infty^2\frac{\partial\Phi'}{\partial x}\right]\frac{\partial^2\Phi'}{\partial x^2}+\frac{\partial^2\Phi'}{\partial y^2}=0$$	

Bibliography

This bibliography includes general books on aerodynamics or fluid mechanics as well as handbooks on gas dynamics, in which the reader will find more detailed and specialized information on the subject. Here are also cited publications, whose information helped writing the present book or from where certain points have been mentioned but not developed in further detail. It is the case of the boundary layer, which is mentioned many times with regard to viscous effects, and of the numerical approach, which we have mainly used to illustrate some of the theoretical developments.

General aerodynamics and fluids mechanics

J.D. ANDERSON Jr., *Hypersonic and High Temperature Gas Dynamics*, McGraw-Hill, New York, 1989.

J.D. ANDERSON Jr., *Fundamentals of Aerodynamics*, 3rd edition, McGraw-Hill, Series in Aeronautical and Aerospace Engineering, New York, 2001.

J.J. BERTIN, M.L. SMITH, *Aerodynamics for Engineers*, 3rd edition, Prentice-Hall, 1998.

E.A. BRUN, A. MARTINOT-LAGARDE, J. MATHIEU, *Mécanique des Fluides*, Dunod, Paris, 1968.

S. CANDEL, *Mécanique des Fluides,* 2nd edition, Dunod, Paris, 1995.

P. GERMAIN, P. MULLER, *Introduction à la Mécanique des Milieux Continus*, Masson, Paris, 1980.

E.M. GREITZER, C.S. TAN, M.B. GRAF, *Internal Flow, Concepts and Applications*, Cambridge University Press, 2004.

W.D. HAYES, R.F. PROBSTEIN, *Hypersonic Flow Theory*, vol. 1 (*Inviscid Flows*), Academic Press, New York, 1966.

P. HUERRE, *Mécanique des Fluides*, vol. 1, Département de Mécanique, Ecole Polytechnique, 1994.

A.M. KUETHE, C.-Y. CHOW, *Foundations of Aerodynamics,* 4th edition, John Wiley and Sons, 1986.

L. LANDAU, E. LIFCHITZ, *Mécanique des Fluides*, Editions Mir, Moscow, 1971.

H.W. LIEPMANN, A. ROSHKO, *Elements of Gas Dynamics*, John Wiley and Sons, 1957.

M.J. LIGHTHILL, "On sound generated aerodynamically, I. General theory", *Proceedings of the Royal Society of London.* A 211, pp. 564–587, 1952.

M.J. LIGHTHILL, "On sound generated aerodynamically, II. Turbulence as a source of sound", *Proceedings of the Royal Society of London*, A 222, pp. 1–32, 1954.

R. (VON) MISES, *Mathematical Theory of Compressible Fluid Flow,* vol. 3, Academic Press, New York, 1958.

K. OSWATITSCH, *Der Luftwiderstand als Integral des Entropiestromes* (Drag Expressed as the Integral of Entropy Flux), presented by Ludwieg Prandtl (Kaiser-Wilhelm Institute for Research), 1945.

P. REBUFFET, *Aérodynamique Expérimentale*, vol. 2, Dunod, Paris, 1966.

I. RYHMING, *Dynamique des Fluides,* Presses Polytechniques et Universitaires Romandes, 1985.

R. SAUER, *Méthodes Mathématiques de la Dynamique des Gaz*, Béranger, Paris, 1950.

A.H. SHAPIRO, *The Dynamics and Thermodynamics of Compressible Fluid Flow*, vols. 1 and 2, The Ronald Press Company, New York, 1953.

W.R., SEARS (Ed.), *General Theory of High Speed Aerodynamics*, Princeton University Press, 1954.

A. SELLIER, *Ecoulements Compressibles et Fluides Hétérogènes: Une Introduction,* Département de Mécanique, Ecole Polytechnique, 1998.

D.J. TRITTON, *Physical Fluid Dynamics,* Second edition, Oxford Science Publications, 1988.

Gas dynamics

A. BONNET, J. LUNEAU, *Aérodynamique. Théories de la Dynamique des Gaz*, Cépaduès-Editions, Toulouse, 1989.

P. CARRIÈRE, *Dynamique des Gaz,* vols. 1 et 2, Ecole Nationale Supérieure des Techniques Avancées, Paris, 1972.

G. EMMANUEL, *Gas Dynamics: Theory and Applications*, AIAA Education Series, 1986.

A. FERRI, *Elements of Aerodynamics of Supersonic Flows*, The MacMillan Company, New York, 1949.

W.G. VINCENTI, C.H. KRUGER, *Introduction to Physical Gas Dynamics*, John Wiley and Sons, New York, 1965.

M.J. ZUCROW, J.D. HOFFMAN, *Gas Dynamics*, vols. 1 and 2, John Wiley & Sons, 1977.

Thermodynamics

H.D. BAEHR, K. STEPHAN, *Heat and Mass Transfer*, Springer-Verlag, 1996.

S. FABRÉGA, *Thermodynamique Macroscopique*, Techniques de l'Ingénieur, A 223, 1978.

E.A. GUGGENHEIM, *Thermodynamics*, North Holland, 1957.

H. LUMBROSO, *Thermodynamique. Problèmes Résolus*, Dunod, Paris, 2000.

I. PRIGOGINE, D. KONDEPUDI, *Thermodynamique. Des Moteurs Thermiques aux Structures Dissipatives*, Odile Jacob, Paris, 1999.

M. ROY, *Thermodynamique Macroscopique. Notions Fondamentales*, Dunod, Paris, 1964.

R. SUARDET, *Thermodynamique*, Tec & Doc, Lavoisier, Paris, 1988.

M.W. ZEMANSKY, *Heat and Thermodynamics*, McGraw-Hill, New York, 1943.

Shock waves

G. BEN-DOR, I. OZER, T. ELPERIN (Eds.), *Hand Book of Shock Wave*, Academic Press, 2001.

D. BERSHADER, R. HANSON (Eds.), *Shock Waves and Shock Tubes*, Stanford University Press, Stanford, 1986.

A.L. JAUMOTTE (Ed.), *Chocs et Ondes de Choc. Tome I: Aspects Fondamentaux, Tome II: Applications Techniques*, Masson, Paris, 1971.

J.W. MACCOLL, "The conical shock wave formed by a cone moving at a high speed", *Proceedings of the Royal Society of London*, A 159, pp. 459–472, 1937.

J. (VON) NEUMANN, "The point source solution", *Collected Works of J. von Neumann*, Pergamon, Oxford, 1963.

G.I. TAYLOR, J.W. MACCOLL, "Air pressure on a cone moving at high speeds", *Proceedings of the Royal Society of London*, A 139, pp. 279–311, 1933.

Interferences, shock reflections

G. BEN-DOR, *Shock Wave Reflection Phenomena*, Springer Verlag, 1991.

B. CHANETZ, T. POT, M. LEFEBVRE, P. BOUCHARDY, "Fundamental study of shock/shock interferences in low density flow: flowfield measurements by DLCARS", *Rarefied Gas Dynamics* (R. Brun, R. Campargue, R. Gatignol, J.-C. Lengrand, Eds.), Cépaduès-Editions, Toulouse, 1999.

B. CHANETZ, J. DÉLERY, "Experimental aspects of code validation in hypersonic flows", *24th Rarefied Gas Dynamics Symposium*, Bari, Italy, July 10–16, 2004; and *American Institute of Physics Conference Proceedings*, vol. 762 (M. Capitelli, Ed.), pp. 1309–1318, Melville, New York, 2005.

A. CHPOUN, D. PASSEREL, G. BEN-DOR, "Reconsideration of oblique shock wave reflection in steady flows. Part I: experimental investigation", *Journal of Fluid Mechanics*, vol. 301, pp. 19–35, 1995.

J. DÉLERY, "Shock phenomena in high speed aerodynamics: still a source of major concern", *The Aeronautical Journal of the Royal Aeronautical Society*, pp. 19-34, 1999.

J. DÉLERY, B. CHANETZ, "Experimental aspects of code verification/validation. Application to internal aerodynamics", VKI Lecture Series on *Verification and Validation of Computational Fluid Dynamics*, 2000.

J. DÉLERY, "Shock/shock and shock wave/boundary layer interactions in hypersonic flows", AGARD-FDP-VKI Special Course on *Aerothermodynamics of Hypersonic Vehicles*, 1988.

A. DURAND, B. CHANETZ, R. BENAY, A. CHPOUN, "Investigation of shock waves interference and associated hysteresis effect at variable Mach upstream flow", *Shock Waves*, vol. 12, pp. 469–477, 2003.

B. EDNEY, "Anomalous heat-transfer and pressure distributions on blunt bodies at hypersonic speeds in the presence of an impinging shock", FFA Report no. 115, 1968.

H.G. HORNUNG, H. OERTEL, R.J. SANDEMAN, "Transition to Mach reflection of shock waves in steady and pseudosteady flow with and without relaxation", *Journal of Fluid Mechanics*, vol. 90, part 3, pp. 541–560, 1979.

H. LI, A. CHPOUN, G. BEN-DOR, "Analytical and experimental investigations of the reflection of axisymmetric shock waves in steady flow", *Journal of Fluid Mechanics*, vol. 390, pp. 25–43, 1999.

Method of characteristics

P. CARRIÈRE, *Méthodes Théoriques d'Étude des Écoulements Supersoniques*. Publication Scientifique et Technique du Ministère de l'Air n° 339a, 1964.

A. FERRI, *Application of the Method of Characteristics to Supersonic Rotational Flow,* NACA Technical Report 841, 1946.

A. FERRI, "The method of characteristics: general theory of high speed aerodynamics", *High Speed Aerodynamics and Jet Propulsion*, Section G, vol. 6, pp. 583–669, Princeton University Press, Princeton, 1954.

M. HOLT, "The method of characteristics for steady supersonic rotational flow in three dimensions", *Journal of Fluid Mechanics*, vol. 1, part 4, pp. 409-423, October 1956.

R.E. MEYER, "The method of characteristics", *Modern Development in Fluid Dynamics of High Speed Flow*, Oxford University Press, pp. 71–104, 1953.

Nozzles, ejectors, flow conduct, and air intake

A. BOURGOIN, P REIJASSE, "Experimental analysis of unsteady separated flows in a supersonic planar nozzle", *Shock Waves*, vol. 14, no. 4, pp. 251–258, 2005.

W.L. CHOW, A.L. ADDY, "Interaction between primary and secondary streams of supersonic ejector systems and their performance characteristics", *AIAA Journal*, vol. 2, no. 4, pp. 686-695, 1964.

J. DÉLERY, *Aérodynamique Interne. Tuyères et Arrière-Corps*, Département propulsion énergétique, Ecole Nationale Supérieure de l'Aéronautique et de l'Espace (Supaéro), 1988.

A. LALLEMAND, *Ecoulement des Fluides. Ecoulements en Conduite. Réseaux*, Techniques de l'Ingénieur, BE 8 161.

J. FABRI, J. PAULON (1957). *Théorie et Expérimentation des Éjecteurs Supersoniques*, Note Technique Onera, no. 36, English translation: *Theory and Experiments on Supersonic Air-to-Air Ejectors*, NACA TM 1410 (1958).

P. REIJASSE, F. BOUVIER, P. SERVEL, "Experimental and numerical investigation of the cap-shock structure in over-expanded thrust-optimized nozzles". D. Zeitoun, J. Périaux, J.-A. Désidéri, M. Marini (Eds.), *West-East High Speed Flow Fields Conference: Aerospace Applications from High Subsonic to Hypersonic Regime*, pp. 338–345, CIMNE, Barcelona, 2003.

R. SAUER, *General Characteristics of the Flow Through Nozzles at Near Critical Speeds*, NACA TM No. 1147, 1947.

J. SEDDON, E.L. GOLDSMITH, *Intake Aerodynamics*, Collins, London, 1985.

Numerical approach

L. CAMBIER, J.-P. VEUILLOT, "Status of the *elsA* CFD software for flow simulation and multidisciplinary applications", AIAA Paper 2008-664, *46th AIAA Aerospace Sciences Meeting and Exhibit*, Reno, January 2008.

R. COURANT, K.O. FRIERICHS, H. LEWY, "Uber die Differenzengleichungen der Mathematischen Physiks," *Annals of Mathematics*, vol. 100, pp. 32-74, 1928.

C. HIRSCH, *Numerical Computation of Internal and External Flow,* vols. 1 and 2, John Wiley and Sons, 1988.

R.W. MACCORMACK, "Numerical solution of the interaction of a shock wave with a laminar boundary layer", *Lecture Notes in Physics*, vol. 8, Springer-Verlag, 1970.

M.N. PLOOSTER, "Shock waves from line sources. Numerical solutions and experimental measurements", *The Physics of Fluids*, vol. 13, no. 11, pp. 2665–2675, 1970.

P.J. ROACHE, *Fundamentals of Computational Fluid Dynamics*, Hermosa Publishing, 1998.

H. STEINER, W. GRETLER, T. HIRSCHLER, "Numerical solution for spherical laser-driven shock waves", *Shock Waves*, vol. 8, pp. 337–349, 1998.

Transonic aerodynamics

J.D. COLE, L.P. COOK, *Transonic Aerodynamics*, North Holland, 1986.

I.M. HALL, "Transonic flow in two-dimensional and axially-symmetric nozzles", *Quarterly Journal of Mechanics and Applied Mathematics*, vol. 15, part 4, pp. 487–508, 1962.

K.G. GUDERLEY, *The Theory of Transonic Flows*, Pergamon Press, 1962.

M. MURMAN, J.D. COLE, "Calculation of Plane Steady Transonic Flows", *AIAA Journal*, vol. 9, no. 1, pp. 114-121, January 1971.

H. SOBIECZKY (Ed.), *Symposium Transsonicum IV*, IUTAM Symposium, Göttingen, Germany, Kluwer Academic Publishers, 2003.

R.T. WHITCOMB, L.R. CLARK, "An airfoil shape for efficient flight at supercritical mach numbers", NASA, TMX-1109, 1965.

J. ZIEREP, H. OERTEL (Eds.), *Symposium Transsonicum III*, IUTAM Symposium, Göttingen, Springer-Verlag, 1989.

Boundary layer

J. COUSTEIX, *Couche Limite Laminaire*, Cépaduès-Editions, Toulouse, 1988.

J. COUSTEIX, *Turbulence et Couche Limite*, Cépaduès-Editions, Toulouse, 1989.

H. SCHLICHTING, *Boundary Layer Theory*, 7th edition, McGraw-Hill, New York, 1979.

Miscellaneous

A. BONNET, J. LUNEAU, *Tables de Détente ou Compression Isentropique et Tables de Choc Plan ou Conique de Révolution,* 3rd edition, Cépaduès-Editions, Toulouse, 1996.

Staff Equations, Tables and Charts for Compressible Flows, NACA Report 1135, 1953.

M. VAN DYKE, *An Album of Fluid Motion*, Parabolic Press, Stanford, 1982.

Index

A, B

B

C

G, H

I

K, L

M

S

T

U, V

W,Y